# THE EAST AFRICAN GREAT LAKES: LIMNOLOGY, PALAEOLIMNOLOGY AND BIODIVERSITY

# ADVANCES IN GLOBAL CHANGE RESEARCH

VOLUME 12



*The titles published in this series are listed at the end of this volume.*

# THE EAST AFICAN GREAT LAKES: LIMNOLOGY, PALAEOLIMNOLOGY AND BIODIVERSITY

Edited by

Eric O. Odada

and

Daniel O. Olago

*Pan African START Secretariat,*
*Department of Geology,*
*University of Nairobi, Kenya*

KLUWER ACADEMIC PUBLISHERS
DORDRECHT / BOSTON / LONDON

A C.I.P. Catalogue record for this book is available from the Library of Congress.

ISBN 1-4020-0772-8

Published by Kluwer Academic Publishers,
P.O. Box 17, 3300 AA Dordrecht, The Netherlands.

Sold and distributed in North, Central and South America
by Kluwer Academic Publishers,
101 Philip Drive, Norwell, MA 02061, U.S.A.

In all other countries, sold and distributed
by Kluwer Academic Publishers,
P.O. Box 322, 3300 AH Dordrecht, The Netherlands.

*Printed on acid-free paper*



Printed in the Netherlands.

# CONTENTS

**Hydrology and Physical Limnology**



**Biodiversity, Food Webs and Fisheries**



**Sedimentary Processes, Paleoclimate and Paleoenvironment**

# INTRODUCTION

The Second International Symposium on the East African Lakes was held from 10-15 January 2000 at Club Makokola on the southern shore of Lake Malawi. The symposium was organized by the International Decade for the East African Lakes (IDEAL), a research consortium of African, European and North American scientists interested in promoting the investigations of African Great Lakes as archives of environmental and climatic dynamics. Over one hundred African, European and North American scientists with special expertise in the tropical lakes participated in the symposium which featured compelling presentations on the limnology, climatology, palaeoclimatology and biodiversity of the East African Lakes. It is their papers that comprise this book.

The large lakes of East Africa are important natural resources that are heavily utilized by their bordering countries for transportation, water supply, fisheries, waste disposal, recreation and tourism. The lakes are unique in many ways: they are sensitive to climatic change and their circulation dynamics, water-column chemistry and biological complexity differ significantly from large lakes at higher latitudes; they have long, continuous, high resolution records of past climatic change; and they have rich and diverse populations of endemic organisms. These unique properties and the significance of the palaeolimnological records demand and attract research interest from around the world.

IDEAL research is contributing to our understanding of basic limnological processes in the African Great Lakes and how physical dynamics drive their biogeochemistry and thus rendering them sensitive, compared to temperate great lakes, to climatic and anthropogenic change. Recent studies indicate that Lake Victoria has undergone dramatic shifts in the lake ecosystem caused by the introduction of the Nile Perch in 1950s and of the water hyacinth during the past five years. The lake also dried up completely prior to 12,400 years BP. Thus, the hundreds of species of fish in modern Lake Victoria may have evolved within the last 12,400 years; this is the fastest rate of vertebrate species evolution ever recorded.

Elsewhere in East Africa, high resolution studies of past climate change in Lake Naivasha, Kenya and in Lake Malawi have shown a distinct Little Ice Age in tropical Africa. Evidence for the Younger Dryas even in tropical Africa has also been documented in the sediment record of Lake Albert. More recent studies have demonstrated that Lake Malawi was at a low stand during the LGM like all the African lakes in the Northern Hemisphere. This lake was previously known to have been low in the early Holocene and around 35ka but was believed to have been at a high stand during the LGM.

Lake Malawi and Lake Tanganyika are aquatic island systems of elevated endemic biodiversity providing extraordinary conditions to study evolutionary biology. In these lakes we have the unique opportunity to investigate the dynamics of evolutionary and ecological change. Patterns of speciation, the origin of major morphological evolution, and the origin of major reorganizations in community structures can all be investigated in a comparative setting in these two lakes. The sedimentary record of these lakes offers us an opportunity to resolve both evolutionary and ecological changes in their biota at time scales of decades, centuries, millennia, to millions of years. Despite their long histories and geological similarities, the patterns of diversity and genetic differentiation of the biota differ dramatically between Lakes Malawi and Tanganyika. Both lakes were colonized by cichlid fishes, thiarid gastropods and ostracode crustaceans, but these exemplar taxa currently have contrasting aspects in the two lakes. Approximately 1000 fish species are estimated to have evolved within the cradle of Lake Malawi, which is approximately 10 per cent of all freshwater fish species in the world. Despite their astonishing multitude, these species encompass a rather modest degree of molecular genetic and morphological change. The fishes in Lake Tanganyika are genetically and morphologically much more diverse than those in Lake Malawi, yet total only 300 species. In Lake Tanganyika about 240 out of 250 species of prosobranch gastropods and ostracode crustaceans are unique to that lake, and like the cichlid fish, form numerous distinct, divergent lineages. The living prosobranch gastropod fauna of Lake Malawi has undergone only limited differentiation and few if any endemic ostracodes are reported from this lake.

The papers presented in this book provide a comprehensive coverage of the large lakes of East African Rift Valley, touching on climate, limnology, palaeoclimatology, sedimentation processes, biodiversity and management issues of these lakes. The papers show that high quality, globally relevant research can be, and is being done in Africa. The call from African researchers is for their international colleagues and the science funding agencies to move from a position where they see their interactions in Africa essentially as "capacity building" to one of partnership and "capacity recognition" with capacitating where necessary and effective. African and developed world science administrators must work together to sustain the scientific capacity which has been built in Africa, instead of tacitly allowing it to migrate to Europe and North America. The world needs it to stay home.

Eric O Odada.

# ACKNOWLEDGEMENTS


The Second International Symposium on the East African Lakes on the limnology, climatology and palaeoclimatology of the East African Lakes was held in Malawi at Club Makokola on the palm-fringed southern shore of Lake Malawi. The symposium was sponsored by the John D. and Catherine T. MacArthur Foundation, The International START Secretariat through NORAD funding for global environmental change research and capacity building in sub-Saharan Africa, and the PAGES International Project Office. We are grateful for the generous support that made the symposium and the publication of this volume possible.

The symposium was organized by IDEAL in collaboration with the University of Malawi, the Large Lakes Observatory, University of Minnesota, USA and the Pan-African START Secretariat, University of Nairobi, Kenya. We are particularly grateful to the Department of Geology at Bunda College University of Malawi for providing all the logistical support for the symposium. We are also very grateful to the management of the Club Makokola for providing excellent conference facilities at this picturesque resort on the southern shore of Lake Malawi.

Tom Johnson of the Large Lakes observatory, University of Minnesota spent much of him time providing valuable advice to the organizing committee for the symposium. We also benefited considerably from the advice provided by the IDEAL Steering Committee. The authors of the articles in this volume are to be commended for their commitment to research on the African Great Lakes, despite the logistic challenges that accompany such efforts. We greatly appreciate the efforts of all the following reviewers, who substantially improved the quality of the papers published in this book.

Eddie Allison, P. Anadon, Daniel Ariztegui, Mamboudou Ba, Richard Back, Michael Biryabarema, Gary Bowen, Don Branstrator, Arthur S. Brooks, John Bullister, Peter Casper, Tesfaye Cherinet, Steven Dadzie, Jean-Pierre Descy, Cynthia Ebinger, Hubert Gallee, A.T. Grove, Paul Hamblin, Robert Hecky, Karin Holmgren, Andrew Hudson, Ken Irvine, Emi Ito, Thomas C. Johnson, G. Khroda, Seth Kisia, Koen Martens, Henry Lamb, Suzanne Levine, Stephen McCord, Sarah Metcalf, Joseph Mworia-Maitima, Isaac Nyambok, S. Odingo, R. Okoola, Anne-Marie Oldewage, A.O. Opere, Adrian G. Parker, Ingemar Renberg, Louis Scott, F. Semazzi, William Smethie Jr., J. Sutcliffe, David Swayne, Michael Talbot, Jean-Jacques Tiercelin, Dirk Verschuren, Martin Vollmer.

# Geological and Structural Setting of the East African Lakes

# A 300 MILLION YEARS HISTORY OF RIFT LAKES IN CENTRAL AND EAST AFRICA: AN UPDATED BROAD REVIEW

J.-J. TIERCELIN[1] AND K.-E. LEZZAR[2]

[1]*UMR CNRS 6538 "Domaines Océaniques", Institut Universitaire Européen de la Mer, Place Nicolas Copernic, 29280 Plouzané, France*

[2]*Department of Earth Sciences, Syracuse University, 204 Heroy Geology Laboratory, Syracuse, New York 13244-1070, USA.*

**ABSTRACT**

Over the past 300 Ma, the complex geological history of the African continental plate has been marked by an important development of lacustrine domains of various size. We present in this paper an updated broad review of major Central and East African lakes initiation, development or extinction during this 300 Ma-long period of time before present. Today, more than 35 wide/deep and narrow/shallow lakes occupy a wide surface area in Eastern and Central Africa between longitudes 30° E and 40° E. They belong to the 3000-km long East African Rift System (EARS) which goes from the Afar Depression to the north down to Lake Malawi in the south. These lakes have been formed as the result of a combination of interacting forcing factors such as major rifting processes (faulting and volcanic activities), that started to develop from about 40 Ma ago in Southwest Ethiopia and Northern and Central Kenya, and regional or global climate changes. The largest and deepest lakes, not necessarily the oldest, have been formed during the last 10 Ma along the Western Branch of EARS, while the smallest lakes of the Eastern Branch and Main Ethiopian Rift are the remnants of lakes developed during Pliocene-Pleistocene times. The rifting phase that is affecting today the Central and Eastern regions of Africa has been preceded by two other major rifting phases that started during Upper Palaeozoic times at about 280



*E.O. Odada and D.O. Olago (eds.),*
*The East African Great Lakes: Limnology, Palaeolimnology and Biodiversity,* 3–60.

Ma, and during Mesozoic times at about 160 Ma, respectively, as well as several more local events. Their duration have been at least two times longer than the duration of the present East African Rift System formation. Tens of lakes developed during these two oldest phases of rifting in Central and Southeastern Africa, displaying limnogeological characteristics similar to some of the modern lakes of the EARS.

# 1. INTRODUCTION

"It is often forgotten that the key prerequisite for a lake is a hole in the ground" (I. Price, in Summerhayes, 1988).

Among all tectonic processes that control the evolution of the Earth, rifting mechanism is the one that will always create deep faulted holes in the crust. Consequently, lakes of various size and depth easily develop in this type of structural environment. For this reason, the eastern and central regions of the African continent are today characterized by several tens of lakes that display a nearly indescribable variety of morphological, geological, physical, chemical and biological characters. These lakes are all linked to the continental-scale tectonic structure of the East African Rift System (EARS) that crosscuts Eastern and Central Africa from the triple junction of the Red Sea and Gulf of Aden oceanic ridges with the Afar Rift at about 10° latitude North to more than 30° latitude South, ending on land with the Lake Malawi Basin to the east and the Okavango Rift to the southwest (Figure 1).

The present-day general architecture of the East African Rift System is made of a series of relatively narrow (40-70 km-wide) faulted troughs that are distributed in two distinct branches, eastern and western, all over a length of more than 4000 km. These troughs are generally half-grabens bounded by major high-angle boundary faults on one side and a faulted flexural margin on the other, or less frequently grabens bounded by two faults of similar importance. More than 35 "tectonic lakes" lie today in these faulted depressions (Figure 1B). Some lakes form within several of those basins and are generally connected by structural highs defined as "accommodation zones" or "transfer faults" (e.g., Rosendahl, 1987; Chorowicz, 1989). On the other hand, a few "volcanic (caldera or crater) lakes" are associated with volcanic features that are also related to rift structuration (Figure 1B).

Within the Eastern Branch, lakes are generally small (maximum 30 x 20 km), shallow (from 5- to 50 m water depth), and have more or less saline characteristics, mainly because they lie in areas under semi-arid climatic conditions and/or have no outlet. Lake Turkana, at the northern end of the Kenya Rift, is the exception in terms of dimensions, with a length of 250 km and a maximum water depth of 125 m (Figure 1B). Consequently to the intensive volcanism (from Oligocene time) that characterizes the Eastern Branch (Baker *et al.*, 1972; Mohr, 1982) (Figure 2), there are several crater lakes of various size and water depth, the largest and deepest being Lake Shala in the Main Ethiopian Rift (Figure 1B). The Western Branch has little volcanism, that started in the Late Miocene and was restricted to four main areas, from south to north: Rungwe, South Kivu, Virunga, and Toro-Ankole (Ebinger, 1989a, b) (Figure 2). It contains several large and deep lakes along its length, mainly with freshwater characteristics as a result of more humid regional climatic conditions

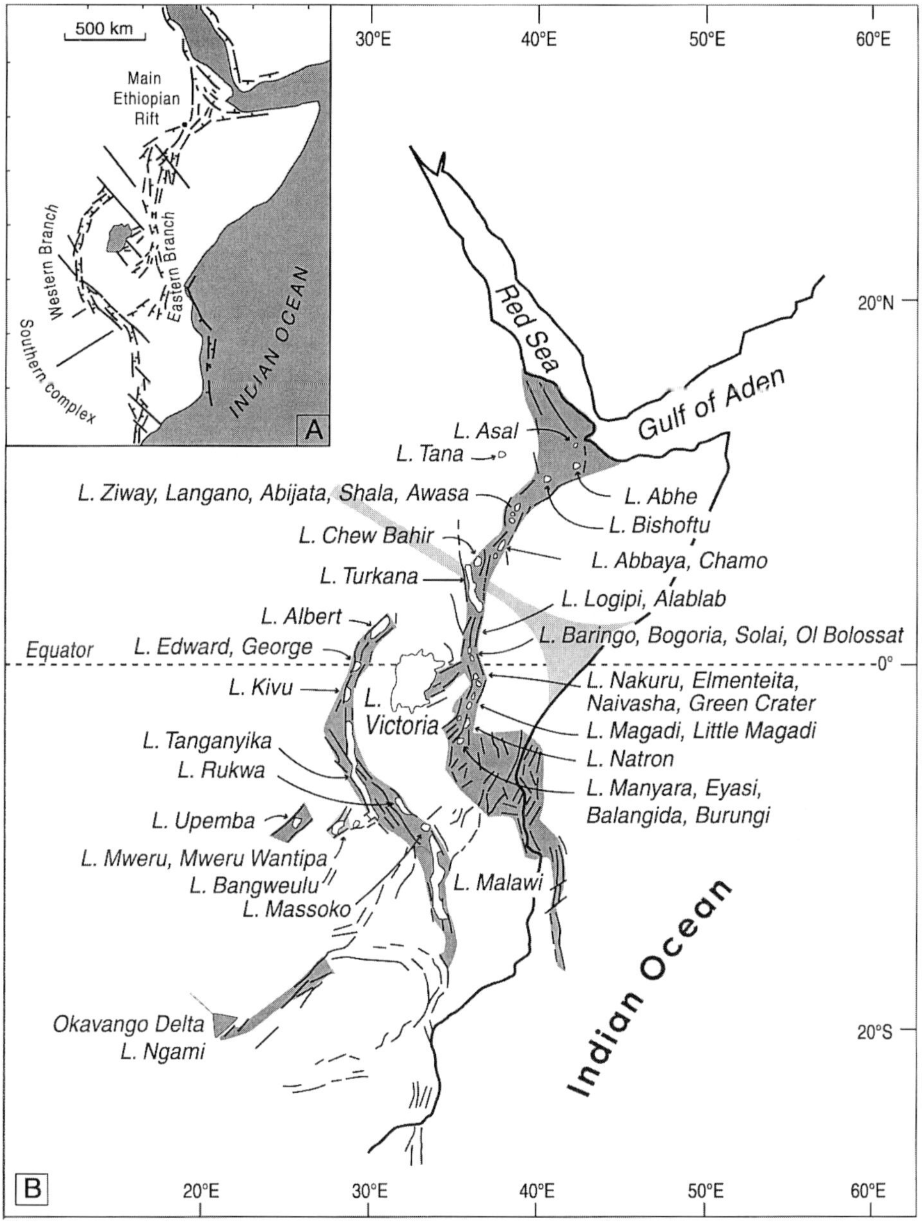


*Figure 1*.A. General tectonic architecture of the East African Rift System (EARS). B. Location of the present-day tectonic and volcanic lakes within the EARS.

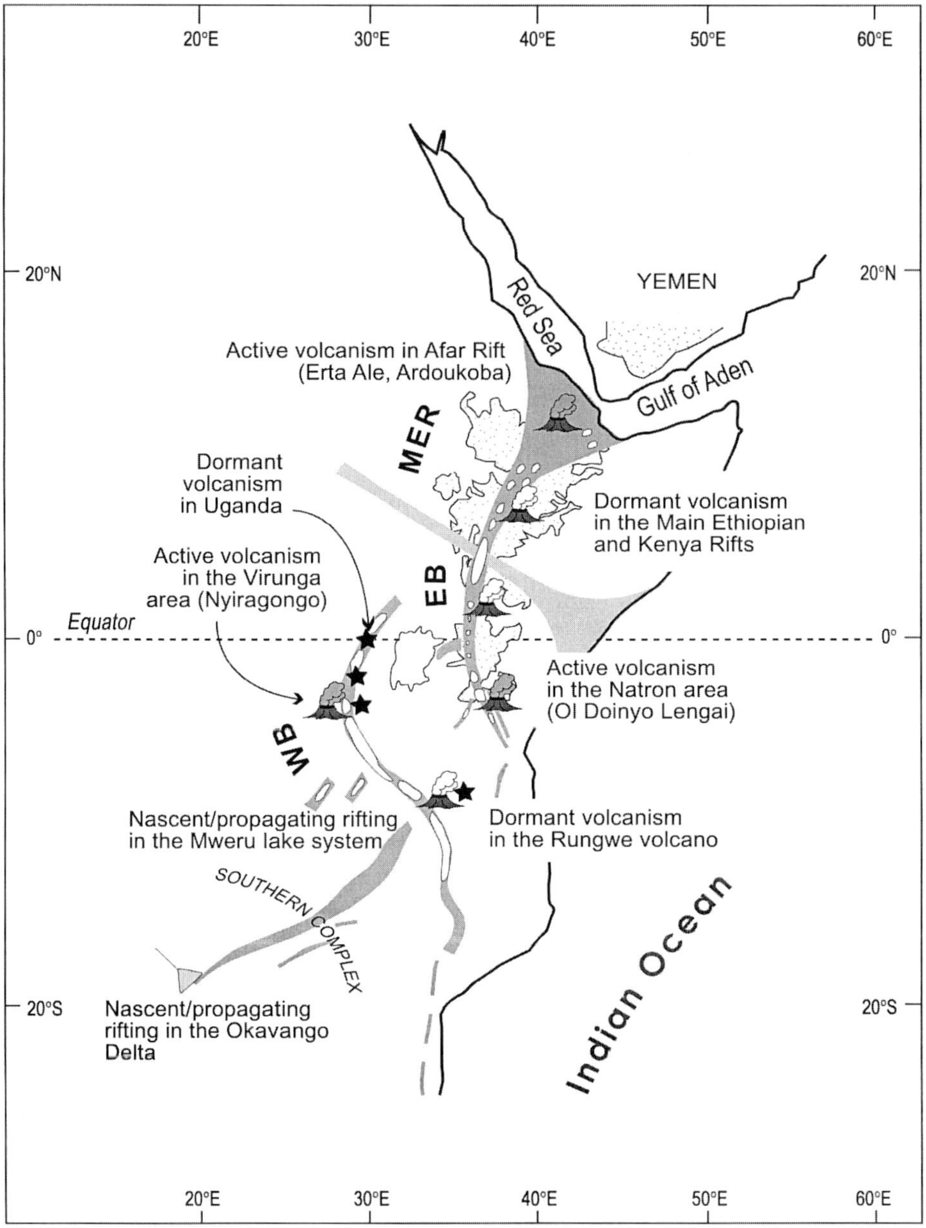


*Figure 2*. Spatial distribution of Cenozoic volcanism in East Africa, Main Ethiopian Rift (MER) and Eastern Branch (EB), and Yemen (dotted area), and major Cenozoic volcanic centres in the Western Branch (WB) (black stars), from north to south: Toro-Ankole, Virunga, South Kivu, and Rungwe (modified after Baker *et al.*, 1972). Present-day active or dormant volcanic centres in the East African Rift System and areas of nascent/propagating rifting forming the "Southern Complex" are indicated (after Mondeguer *et al.*, 1989; Modisi *et al.*, 2000).

and the existence of major, generally permanent outlets (Figure 1B). Lake Tanganyika is the deepest, with a maximum water depth of 1470 m (Capart, 1949), followed by the 770-m-deep Lake Malawi. The exception in the Western Branch is Lake Rukwa, only a few metres deep (Delvaux *et al.* 1998; Thévenon *et al.*, in press) (Figure 1B). Crater lakes are not dominant in the Western Branch, and are mainly located in the Rungwe volcanic range north of Lake Malawi (Lake Massoko) (Winckel, 1998; Gibert *et al.*, in press) (Figures 1B and 2).

Occupying a particular position between the uplifted border plateaus of the Eastern and Western Branches of EARS close to the equator is Lake Victoria, the largest lake in the tropics and the third largest in the world with a surface area of 68800 $km^2$ (Beadle, 1981). Lake Victoria resulted from river reversal and ponding as a consequence of rift margin uplift (Bishop and Posnansky, 1960; Ebinger, 1989b; Scholz *et al.*, 1990, 1998), and thus can also be described as a "tectonically induced lacustrine system" (Figure 1B).

The geological history of the present-day lakes in Eastern and Central Africa is quite complex. Many of the small lakes (~100-200 $km^2$ surface) of the Eastern Branch have been formed during relatively recent times (mostly since the Middle Pleistocene), while the largest and deepest lakes of the Western Branch are known to have formed since Middle Miocene times (Cohen *et al.*, 1993; Ebinger *et al.*, 1993c; Lezzar *et al.*, 1996; Lezzar *et al.*, in press). As a result of the increasing interest of the petroleum industry for continental rift basins and lake series during the last 15 years, a considerable amount of data, especially geophysical and drilling data, added significantly to the understanding of extensional provinces of the East African Rift System. Gravity, seismic reflection and magneto-telluric surveys revealed thick (5- to 7-km) sedimentary sequences below the water body of several large lakes (Rosendahl *et al.*, 1986; Rosendahl *et al.*, 1988) or, in some areas of the rift, hidden deep half-grabens containing several km-thick lacustrine and fluvial series, often interbedded with volcanic products (Morley *et al.*, 1992; Ebinger and Ibrahim, 1994; Upcott *et al.* 1996; Mugisha *et al.*, 1997; Morley, 1999, Wescott *et al.*, 1999; Hautot *et al.*, 2000; Karner *et al.*, 2000). Chronological data deduced from few petroleum wells and more or less continuous field sections indicate that some lake basins resulted from extensional episodes during Palaeogene time (and interpreted as the early beginnings of the East African Rift System).

However, much older lake basins of various size and depth are known to have resulted from at least two older major rifting phases that have acted through the African continent, mainly during Upper Palaeozoic (from Permian to Lower Jurassic - also known as Karoo period-) and Mesozoic times (from Late Jurassic to Cretaceous). These phases are closely related to the progressive breakup of Gondwanaland (Fairhead, 1988). These Karoo or Mesozoic basins are generally only imaged through reflection seismics or petroleum wells. In Central and Eastern Africa, the Karoo basins are generally oriented NE-SW, NW-SE, or N-S. Their present outlines, even reduced in size by uplift and erosion, largely exceed the size of the East African Rift System. The second major phase of rifting that affected Central-Eastern Africa is closely linked to the breakup of South America from west and southwest Africa. It resulted in the development during Late Jurassic-Cretaceous times of continental-scale rift systems known as the West and Central Africa Rift Systems (WCARS) (Fairhead, 1988; Bosworth, 1992; Guiraud and Maurin, 1992). In Central and Eastern Africa, rift structures belonging to the Central Africa Rift System (CARS) extend along the NW-SE tectonic trend in Sudan (South Sudan and Nile rift basins) and in Northern Kenya with the Anza Rift (Bosworth, 1992).

A broad review of the structural and sedimentary history of these present to past rift basins and associated lake environments that characterize the eastern and central regions of Africa over 300 millions of years is presented in this paper.

## 2. AFRICAN RIFT LAKES ASSOCIATED WITH TERTIARY RIFTING

### 2.1 The Palaeogene to Early Miocene Rift Lakes

The early beginnings of the East African Rift System and its associated lacustrine sedimentary basins are known to be located in the Palaeocene-Eocene times in an area extending between Southern Sudan, Northern-Central Kenya and Southern Ethiopia, that was partially structured by early rifting between Late Jurassic and Cretaceous (Bosworth, 1992). At about 65 Ma, active extension mechanisms affected this region (Schull, 1988; Morley *et al.*, 1992; Mugisha *et al.*, 1997; Morley, 1999; Wescott *et al.*, 1999; Hautot *et al.*, 2000). This event has been identified to have lasted over 30-40 Ma (up to Miocene time) but with a strong diachronism between the different basins of the system (Bosworth, 1992). During this period, huge volumes of volcanic material was erupted in Northern Kenya, Southern/Northern Ethiopia, and Yemen (between 45 and 29 Ma) (Figure 3) but the timing relationship between volcanism and rifting is not well known in many areas (e.g., Zanettin *et al.*, 1983; Berhe *et al.*, 1987; Ebinger *et al.*, 1993; Baker *et al.*, 1996; Hofmann *et al.*, 1997; Ebinger and Sleep, 1998). In Southern Ethiopia, such volcanic activity resulted between ~ 45 and 37 Ma in the extrusion of a 1-km-thick sequence of flood basalts, accompanied with little or no extension (Ebinger *et al.*, 1993). In Central and Northern Ethiopia, close to the Red Sea rift margin, more than a 2 km thick sequence of fissure basalts erupted over a period of 1 Ma at about 30 Ma (e. g., Hofmann *et al.*, 1997) (Figure 3). In Yemen, basaltic continental flood volcanism began between 30.9 and 29.2 Ma in northwestern and southwestern regions, while rhyolitic volcanism started at 29.3-29.0 Ma throughout Yemen. Volcanic activity extended up to 26.5 Ma throughout Yemen (Baker *et al.*, 1996). At abouth the same period (Oligo-Miocene boundary), major rifting of the continental lithosphere began between Africa and Arabia, forming the initial Red Sea Rift (Camp and Roobol, 1992), and extended to the northwest, thus forming the Gulf of Suez (Garfunkel and Bartov, 1977) (Figure 4).

From Palaeocene/Late Eocene to Miocene, renewed extension and subsidence and at some places wrench and compressional tectonics affected large areas of the

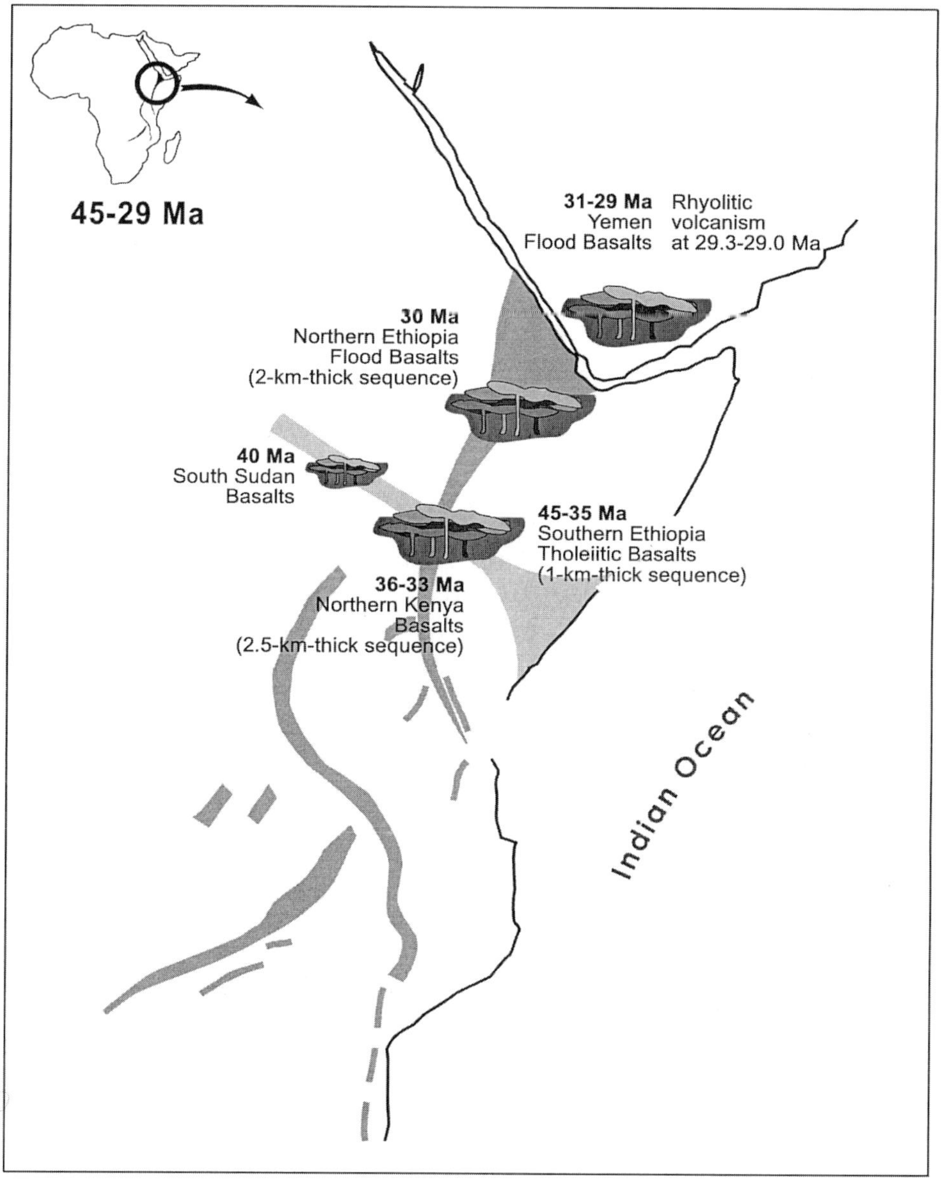


*Figure 3*. Distribution map of Early Tertiary (45-29 Ma) magmatism in Eastern Africa and Arabia.

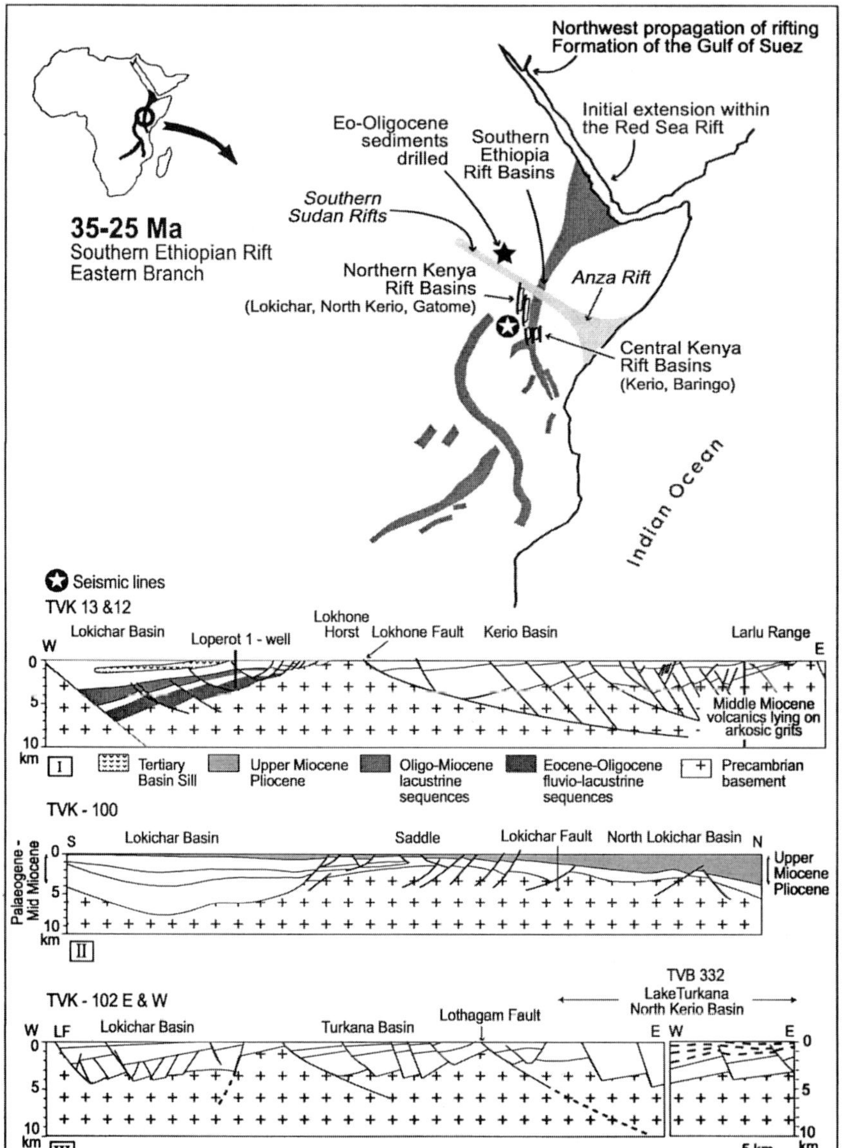


*Figure 4*. Distribution map of the Early-Middle Tertiary rift basins in Northern-Central Kenya within the frame of the Cretaceous rifts of Southern Sudan and Anza (Northern Kenya) (after Morley *et al.*, 1999a; Ebinger *et al.*, 2000; Hautot *et al.*, 2000), and regional geological cross-sections of the Early-Middle Tertiary rift basins of the Turkana area, Northern Kenya, based on seismic lines. I: The Lokichar and North Kerio rift basins; II: The Lokichar and North Lokichar rift basins; III: The North Lokichar, Turkana and North Kerio rift basins (from Morley *et al.*, 1999c).

Late Jurassic/Cretaceous Southern Sudan and Anza rift basins (Bosworth and Morley, 1994; Morley *et al.*, 1999a). Palaeocene/Late Eocene to Middle Miocene lacustrine and fluviatile sequences are known from the Sudan and Anza Rifts (Guiraud *et al.*, 1992; McHargue *et al.*, 1992). In Southern Sudan, a large number of exploration wells penetrated a 13-km-thick sedimentary series of Cretaceous-Tertiary age, including 5.5 km of Tertiary rocks. In this series, Palaeocene sandstones relate to an extensive alluvial-plain environment, while thick Late Eocene-Oligocene sediments illustrate fluvial-floodplain and lacustrine environments that developed during a final phase of rifting, in association with discrete evidence of basaltic volcanism during Late Eocene (Vail, 1978; Schull, 1988) (Figures 3 and 4).

Several sedimentary (half-graben) basins initiated as early as Palaeogene time in Southern Ethiopia (Ebinger and Ibrahim, 1994; Ebinger *et al.*, 2000) and Northern and Central Kenya (Baker and Wohlenberg, 1971; Morley *et al.*, 1992; Mugisha *et al.*, 1997; Hautot *et al.*, 2000) (Figure 4). In Northern Kenya, immediately west of the present-day Lake Turkana, the oldest rift structures initiated during this Early Tertiary rift phase are represented by the N-S oriented Lokichar, North Kerio and (possibly) Gatome rift basins (Morley, 1999; Morley *et al.*, 1999c; Wescott *et al.*, 1999) (Figure 4). Among them, the Lokichar sedimentary basin is a 60 x 30 km-sized and 7 km-deep east-facing half-graben bounded to the west by a prominent east-dipping listric fault named the Lokichar Fault (Morley *et al.*, 1992; Morley, 1999; Morley *et al.*, 1999c) (Figure 5). At the same period, tectonic movements along faults adjacent to the Lokichar Fault (the Lokhone and Lothagam Faults) resulted in the formation of the North Kerio Basin, expected to be filled by Oligo-Miocene sediments that are only expressed on seismic data (Morley *et al.*, 1992; Wescott el al., 1993; Morley *et al.*, 1999c) (Figure 4). Further north, in the Gatome Basin (Figure 4), a major half-graben basin has been imaged by reflection seismics below a ~3.5 km-thick series of basalt and rhyolite flows, the age of which ranges between Oligocene and Early Miocene (Walsh and Dodson, 1969; Dodson, 1971; Zanettin *et al.*, 1983; Brotzu *et al.*, 1984). A Palaeogene (more likely) or Cretaceous (less likely) age is suggested by Wescott *et al.* (1999) for this lower Gatome rift basin. Sediments are suspected to fill this half-graben but the lack of seismic reflections due to thick volcanic cover prevents any seismic/environmental facies interpretation. Further south, between the equator and 1° latitude North, extension mechanisms also occurred during Palaeogene time (Mugisha *et al.*, 1997; Hautot *et al.*, 2000), contemporaneous with rifting in Northern Kenya, Southern Sudan and Ethiopia (Morley *et al.*, 1992; Hendrie *et al.*, 1994). This resulted in the initiation of the two oldest deep rift basins of the Central Kenya Rift, the Kerio Basin to the west, and the Baringo Basin to the east (Figure 6).

The Lokichar Basin contains a 7-km-thick sedimentary series that mainly includes interbedded lacustrine and fluvio-lacustrine sediments Palaeogene to Middle Miocene in age (Morley *et al.*, 1992), a part of which has been calibrated by the ~3-km-deep Loperot-I exploration well (Morley *et al.*, 1999c). Alternating packages of sandstones and claystone/shale intervals, known as "Turkana Grits" (Arambourg, 1935, 1943; Arambourg and Wolff, 1969) (Figure 5), indicate successive fluvial,

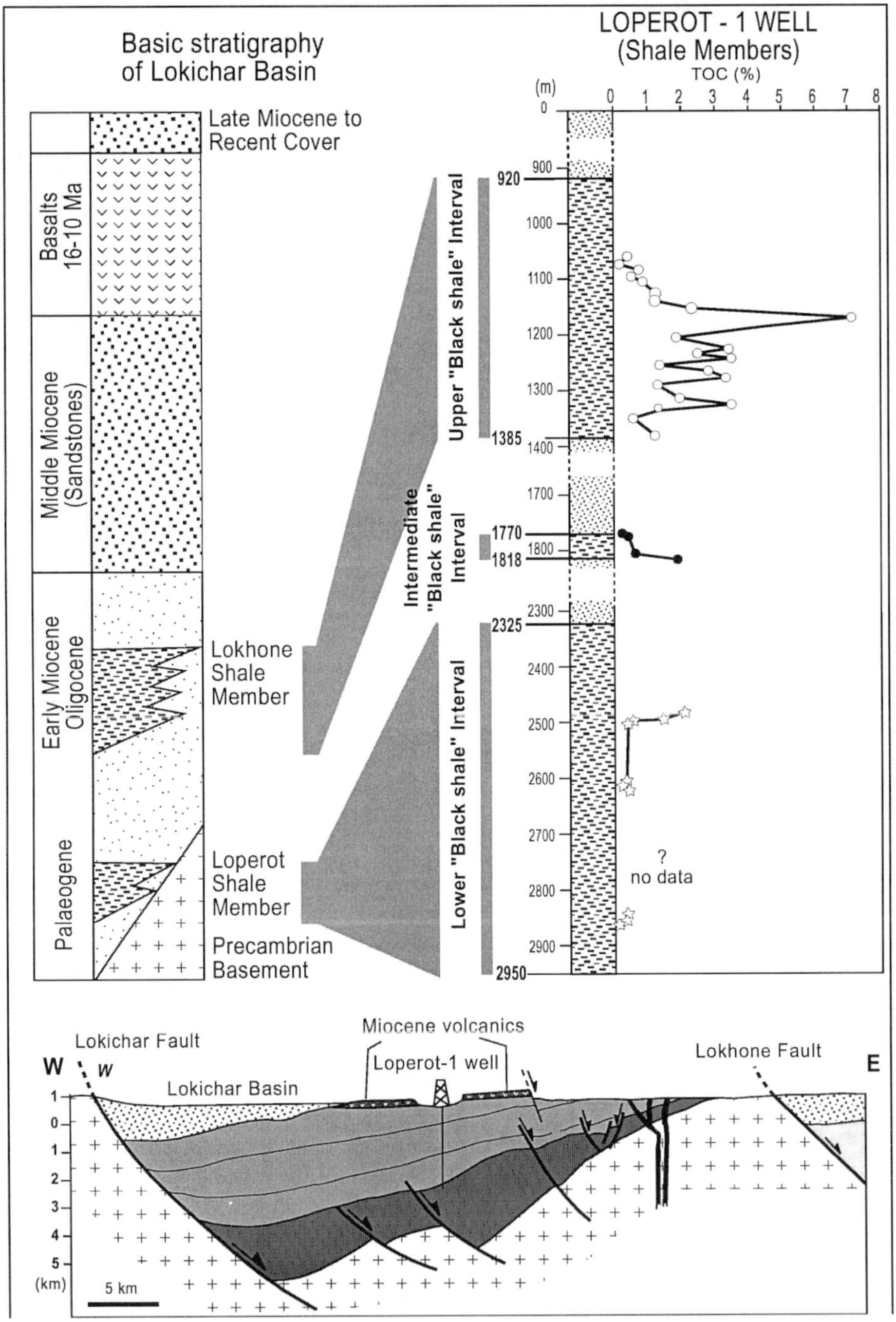


*Figure 5.* General tectonic structure of the Lokichar Basin, and stratigraphic and geochemical data from the Loperot-1 well (after Morley *et al.*, 1999c; Tiercelin *et al.*, submitted).

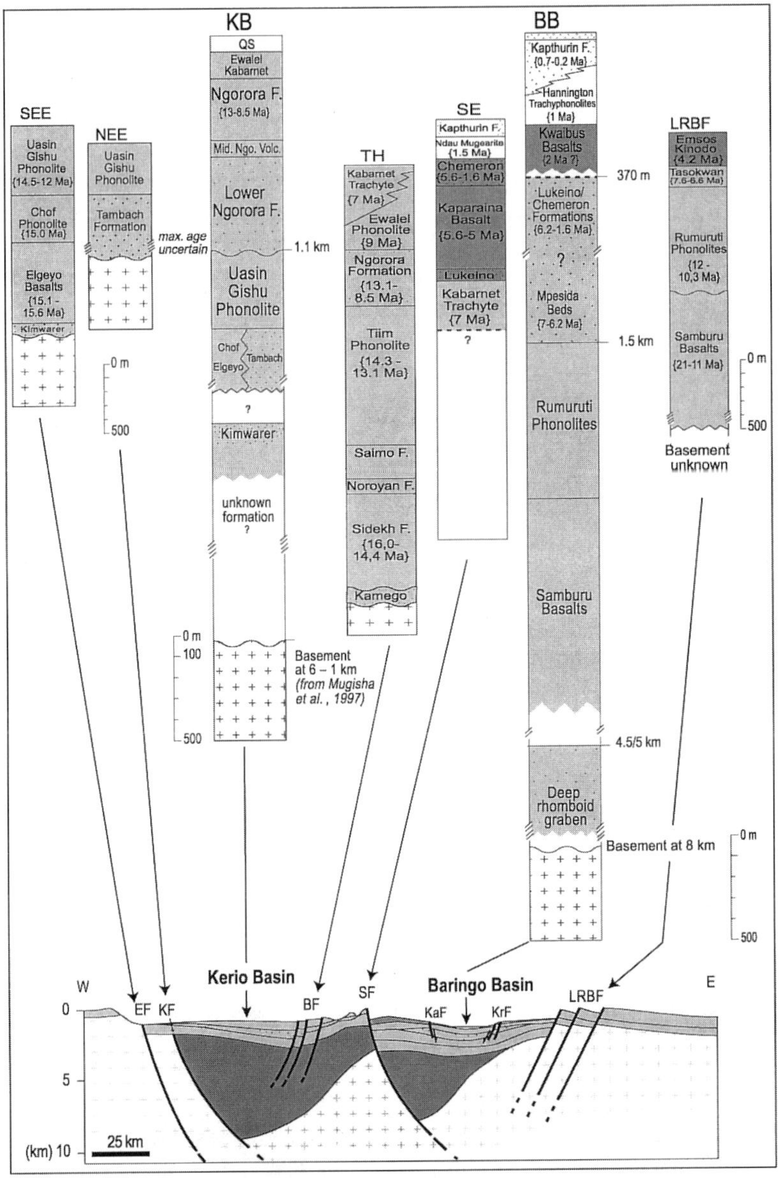


*Figure 6.* Structural cross-section through the Central Kenya Rift showing the deep structure of the Palaeogene Kerio and Baringo Basins, and general stratigraphy of the volcanic and sedimentary series filling these basins from Palaeogene to Recent (SEE, NEE: South and North Elgeyo Escarpment; KB: Kerio Basin; TH: Tugen Hills; SE: Saimo Escarpment; BB: Baringo Basin; EF: Elgeyo Fault; KF: Kerio Fault; BF: Barwesa Fault; SF: Saimo Fault; KaF: Kapthurin Fault; KrF: Karau Fault; LRBF: Laikipia Rift Border Fault (from Mugisha *et al.*, 1997; Hautot *et al.*, 2000).

fluvio-deltaic and open lacustrine conditions, possibly as a consequence of climatic changes and/or fault movements along the main Lokichar Fault (Morley *et al.*, 1992; Morley, 1999). The Lokichar lakes extended from Late Oligocene to end of Middle Miocene. Fresh-water depositional environments as well as anoxic lake bottom conditions are indicated by well-preserved remains of the green algae *Pediastrum* and *Botryococcus* (Reynolds, 1980) that have been found in the organic-rich (1-17 % TOC, average 4.5 %) shale intervals (maximum thickness 1.3 km) (Morley *et al.*, 1992; Morley *et al.*, 1999c; Tiercelin *et al.*, submitted) (Figure 5). Modern equivalents of the Lokichar lakes could be found among the present-day lakes of the East African Rift System, especially with the 75-km-long, 35-km-wide and 120-m-deep, permanently stratified Lake Edward (Western Branch) (Hecky and Degens, 1973; Lehman, this volume) (Figure 1B), or the central (East-Kigoma) sub-basin of Lake Tanganyika (Figure 1B) where similar black organic-rich shales were cored (Tiercelin and Mondeguer, 1991; Tiercelin *et al.*, 1994). Further south, in the Central Kenya Rift, exposed lake- and fluvial-type sediments of possible Palaeogene age are green laminated shales and sandstones forming the Kimwarer and Kamego formations that outcrop in the Kerio and Baringo Basins, respectively (Renaut *et al.*, 1999; Hautot *et al.*, 2000) (Figure 6). These two formations represent the upper part of a several-km-thick sediment pile of possible Palaeogene age, that is only illustrated by geophysical methods (Mugisha *et al.*, 1997; Hautot *et al.*, 2000) (Figure 6). They lie at the base of a nearly entire rift-fill sequence (known as the Tugen Hills sequence) that is considered as one of the best exposed successions of Neogene sediments in the East African Rift System (e.g., Andrews and Banham (eds.), 1999). Similarly, to the north, in the Lokichar, North Kerio and Gatome Basins, the oldest parts of the rift-fill sections are only displayed on seismic lines (Figure 4). All these "geophysically-identified" sediments probably represent the five or six foremost lakes created during the Early Cenozoic phase of rifting in Central and Eastern Africa.

## 2.2 The Early Miocene to Late Pliocene Rift Lakes

### 2.2.1 From Early to Upper Miocene

From the beginning of the Miocene period (at about 24 Ma), rifting processes in Central and Eastern Africa were dominated by an intense basaltic volcanic activity that affected the northern and central parts of the Kenya Rift as well as the Southern Main Ethiopian Rift (Ebinger *et al.*, 1993b; Ebinger and Ibrahim, 1994). In Central Kenya, thick basaltic and phonolitic series more or less extensively capped the Kerio and Baringo Basins (Samburu, Sidekh and Elgeyo Formations) between 23 and 10 Ma (Chapman and Brook, 1978; Chapman *et al.*, 1978; Hill *et al.*, 1986) (Figures 6 and 7A). During this period, fluvio-lacustrine sedimentation continued to develop in the Kerio Basin of Central Kenya, with the deposition of the 400-m-thick Tambach Formation dated between 16 and 14 Ma (Renaut *et al.*, 1999) (Figure 7B).

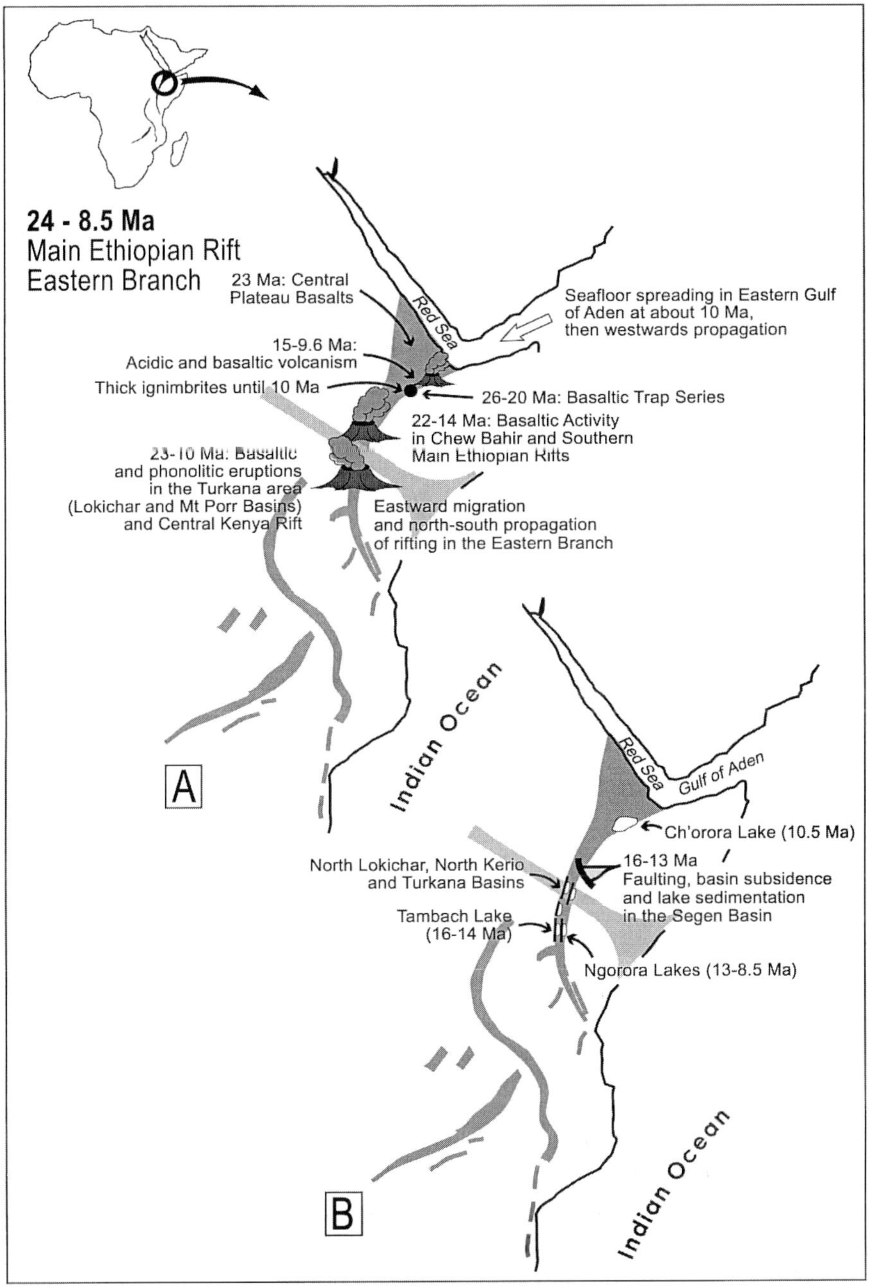


*Figure 7.* A, B: Distribution map of volcanism, faulting and lake basin initiation in the Main Ethiopian Rift and Eastern Branch of EARS between 24 and 8.5 Ma.

Sedimentological and mineralogical studies by Ego (1994) indicate that the Tambach palaeolake fluctuated in surface level and in chemistry, being at least for part of its history a hydrologically closed lake basin in a region with high net evaporation. Eruption of the 500-m-thick Uasin Gishu Phonolites at 14.5-12 Ma in the Kerio Basin ended the life of the Tambach Lake (Lippard, 1973; Chapman and Brook, 1978). At the same period, basaltic flows also erupted in the Turkana area, at 18.5 Ma in the Mount Porr/Kajong Basin (Savage and Williamson, 1978; Williamson and Savage, 1986; Wescott *et al.*, 1993), and between 12.5 and 10.7 Ma in the Lokichar Basin (Morley *et al.*, 1992) (Figure 7A), thus marking the end of fluvio-deltaic and lake environments in these areas.

At the north end of the Kenya Rift (Chew Bahir Rift) as well as in the Southern Main Ethiopian Rift, considerable basaltic volcanic activity associated with basin deepening and sedimentation started at 22 Ma and lasted up to 14 Ma, leading to accumulation of an unknown thickness of sedimentary and volcanic rocks (WoldeGabriel *et al.*, 1990; Ebinger *et al.*, 1993b) (Figure 7A). In the Segen Basin of Ethiopia (Figure 7B), lacustrine mudstones onlapping phonolite lavas dated at about 16-13 Ma indicate that lakes developed in subsiding rift basins by Middle Miocene times (WoldeGabriel *et al.*, 1991; Ebinger *et al.*, 2000). Further North, in the eastern part of the proto-Afar Depression, rifting was restricted to initial volcanic activity that resulted in the deposition of the thick basaltic Trap Series at about 26-20 Ma (Black *et al.*, 1975; Hofmann *et al.*, 1997), and the Central Plateau Basalts at 23 Ma (George *et al.*, 1998). Silicic outpourings of very thick ignimbrites persisted in this region until 10 Ma (Ukstins *et al.*, submitted) (Figure 7A).

Between 15 and 9.6 Ma, major acidic and basaltic volcanic activity associated with faulting resulted in the development of restricted sedimentary basins along the nascent south-east Ethiopian Escarpment (Figure 7A). There, sedimentation linked to this combined fault/volcanic activity was mainly characterized by diatomites and clays, and a few sands and conglomerates, alternating with numerous tuff and pumice deposited during episods of acidic volcanism. This series, known as the Ch'orora Formation (Sickenberg and Schönfeld, 1975), is dated at about 10.5 Ma and has yielded a rich vertebrate fauna (Tiercelin *et al.*, 1979) (Figure 7B). It relates to the development of small and temporary lake environments, essentially characterized by high diatom productivity, possibly as a consequence of a high silica input in the lake waters consecutive to the fall of vitric tuffs and pumices in the lake basin. The Ch'orora Formation is the only lacustrine series of Middle Miocene age known in the southern part of the Afar Depression. Further north, sea floor spreading started about 10 Ma ago in the Eastern Gulf of Aden, and propagated westwards (Cochran, 1981) (Figure 7A).

By Middle Miocene, rift activity migrated northward then eastward in the Northern Kenya Rift. While tectonic activity and lake sedimentation continued in the North Kerio Basin, deformation along the major Lokichar Fault propagated northwards, resulting in the Late Miocene time in the development of a new string of half-grabens basins comprising: the North Lokichar Basin, about 65-km long and superimposed on the previous Lothidok Basin, the northern end of the North Kerio Basin, and the Turkana Basin (Boschetto, 1988; Boschetto *et al.*, 1992; Morley,

1999; Morley *et al.*, 1999c) (Figure 7B). Lacustrine and fluvial environments represented by alternating shales and sandstones developed in these basins above Middle Miocene volcanics. Further south, in Central Kenya, following the period of magmatism leading to the deposition of the Uasin Gishu Phonolites (14.5-12 Ma), a second phase of rift tectonics affected the Kerio Basin, initiating the deposition of the Ngorora fluvio-lacustrine formation in an almost 100-km long and 40-km wide faulted basin (Figures 7 and 8). A recent re-interpretation by Hautot *et al.* (2000) of the seismic data from the Kerio Basin (Mugisha *et al.*, 1997) proposed a much thicker (>800-m-thick) Ngorora Formation than previously advocated (400-m-thick) by Bishop and Pickford (1975) (Figure 8). The lower half of this thicker Ngorora Formation is only "geophysically" represented while the upper 400 m of sediments outcrop largely in the Tugen Hills region between the Kerio and Baringo Basins. This upper sediment pile is essentially formed of volcaniclastic and fluviatile sediments with two predominantly lacustrine members (known as Members C and E; Pickford, 1978a; Renaut *et al.*, 1999) (Figure 8A), and is dated between 13.1 Ma and 8.5 Ma (Hill, 1999). This indicates a period of deposition of about 4.6 Ma for this quite unique, in terms of fossil faunas assemblages, fluvio-lacustrine formation (Hill, 1995; Hill, 1999). Figure 8B, C, D illustrates the Kerio-Baringo Basins palaeogeography during Ngorora Member C times (Pickford, 1978a), with evidence of three distinct, quite small palaeolakes: Lake Waril to the west, lying along the foot of the major Elgeyo rift escarpment, Lake Kapkiamu that occupies a small, elongated graben on the eastern flank of the Kerio half-graben, and Lake Kabarsero lying along the foot of the Saimo Fault in the Baringo Basin. Lake Kapkiamu is described as a strongly saline and alkaline lake throughout much of its history, certainly hydrologically closed with a small catchment area (Pickford, 1978a). Lake Waril and Lake Kabarsero were fresh during Member C times. At times Lake Kapkiamu was relatively fresh, or reduced to a shallow playa-lake, thus resembling the modern lakes Nakuru or Elmenteita in the Central Kenya Rift (Renaut *et al.*, 1999) (Figure 1B). Organic-rich laminated shales (up to 4 % TOC) found in Member C (Figure 8A) suggest that Lake Kapkiamu was perennial and stratified with anoxic bottom waters (Ego, 1994), a possible analogue being the modern Lake Bogoria Basin in the Central Kenya Rift (Tiercelin and Vincens, 1987; Renaut and Tiercelin, 1994) (Figure 1B). Palaeobotanical assemblages from this region at the time of Member C suggest an arid open woodland (Jacobs and Kabuye, 1987; Jacobs and Deino, 1996; Kingston, 1999), that is comparable with the modern Bogoria Lake environment. The maximum of lake extension was reached at the upper part of the Ngorora Formation, with a quite large freshwater lake that overflowed the Saimo/Tugen Hills faulted block towards the Baringo Basin (Figure 8C). The life of this Ngorora Lake ended at around 9-8.5 Ma when extensive phonolite flows of the Ewalel Formation flooded from the south of the Central Kenya Rift over the whole Kerio half-graben as well as a large part of the Baringo Basin (Figure 8D).

At about 800 km west of the Kenya Rift, in an area prefiguring the nascent Western Branch of the East African Rift System, rifting processes also occurred during the Middle Miocene period, almost 20 Ma later than initial volcanism and

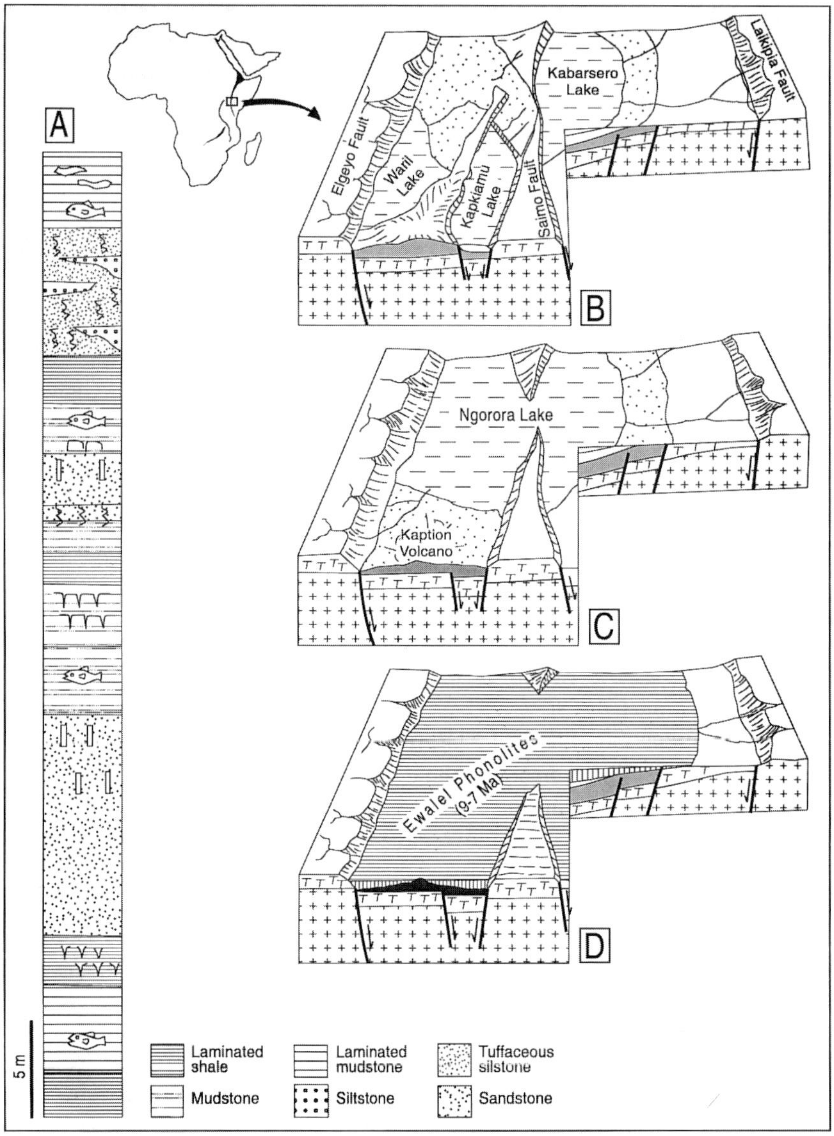


*Figure 8.* The Miocene fluvio-lacustrine Ngorora Formation, Tugen Hills, Central Kenya Rift. A: Lithostratigraphy of the lacustrine Member C of the Ngorora Formation (after Renaut *et al.*, 1999). B, C, D: Three stages of the tectono-sedimentary history of the Ngorora lake basin. Stage B corresponds to the period represented by Member C, with the co-existence of three tectonically-controlled small lakes: Lake Kapkiamu, a saline/alkaline lake, and Lakes Waril and Kabarsero, both freshwater lakes. Stage C is characterized by the filling up of the Lakes Kabarsero and Kapkiamu basins, and a major westward and eastward transgression of the Ngorora lake domain over the tectonic boundary of the Saimo Fault. Stage D shows the complete flooding of the Ngorora Basin with the extensive Ewalel phonolite floods at ~9 Ma (modified from Pickford, 1978a).

faulting in the Eastern Branch. In the Virunga region, immediately north of the present-day Lake Kivu, and in the Southern Kivu-Rusizi region, initial volcanism preceded or was concurrent with faulting at about 12-10 Ma (Ebinger, 1989a; Pasteels *et al.*, 1989; Kampunzu *et al.*, 1998) (Figures 2 and 9). Sediments indicating the development of the foremost rift lakes in the proto-Western Branch of EARS have been identified north of the Virunga volcanic mountains, in the Albertine Rift (Figure 9). There, rifting processes are estimated to start at about 8 Ma (Foster *et al.*, 1997) and generated an asymmetric rift basin filled by a maximum of 4.6 km of sediment located within the central region of the present-day Lake Albert (Karner *et al.*, 2000) (Figure 1B). On the eastern shore of the lake, borehole data have indicated at least 2.4 km of fluvio-lacustrine sedimentary sequences (Davies, 1951; Brown, 1955; Harris *et al.*, 1956; McConnell, 1972), with basal fluvial sequences dated of Early Miocene age. The earliest lacustrine environments to develop in this area are dated at ~8 Ma, and extend up to 4 Ma (Hopwood and Lepersonne, 1953; Bishop, 1965; Pickford *et al.*, 1989, 1990). Immediately south, rifting initiated during the same period in the Semliki Valley, Lake George and Lake Edward areas, with the development of several lake basins complicated by uplift related to the wide Rwenzori granitic horst block, which forms today the 5,200-m-high Rwenzori Mountains (McConnell, 1972; Ebinger, 1989a; Laerdal and Talbot, in press) (Figure 9). Geophysical studies indicate the presence of as much as 4 km of sediments in the Edward half-graben (Upcott *et al.*, 1996). To the south of the Kivu volcanic province (Figure 2), an orthogonal-to-the-rift-axis extension contemporary to the volcanic activity in the Virunga and Southern Kivu-Rusizi region, resulted in the opening of the central segment of the proto-Lake Tanganyika between 12 and 9 Ma, as indicated by the deposition of the oldest rift lake sediments over a very broad and flat prerift surface identified as the Nyanja Event (Rosendahl *et al.*, 1986; Morley, 1988; Rosendahl *et al.*, 1988; Cohen *et al.*, 1993; Lezzar *et al.*, 1996; Lezzar *et al.*, in press) (Figure 9). Such development of the central part of the proto-Tanganyika Basin marks the propagation of the rifting processes along the NW-SE tectonic trend. At about 8.6 Ma, faulting associated with initial volcanic activity in the proto-Rungwe volcanic province resulted in a broad asymmetric lake basin prefiguring the northern part of Lake Malawi (Ebinger *et al.*, 1989; 1993). Faulting also concentrated along the NW-SE tectonic trend of the TRM (Tanganyika-Rukwa-Malawi) Fault Zone (Chorowicz *et al.*, 1983; Tiercelin *et al.*, 1988). It resulted in the reactivation of the major N130°-trending Lupa Fault initiated during Karoo times and delineating the half-graben structure of the Rukwa rift basin (Pierce and Lipkov, 1988; Morley *et al.*, 1999b) (Figures 2 and 9).

### 2.2.2 From Upper Miocene to Late Pliocene

From Upper Miocene, rifting processes affecting Eastern Africa were strongly marked by propagation and migration mechanisms (e.g., Tiercelin, 1981; Morley *et al.* , 1992; Ebinger *et al.*, 2000). In the Turkana region of the Northern Kenya Rift,

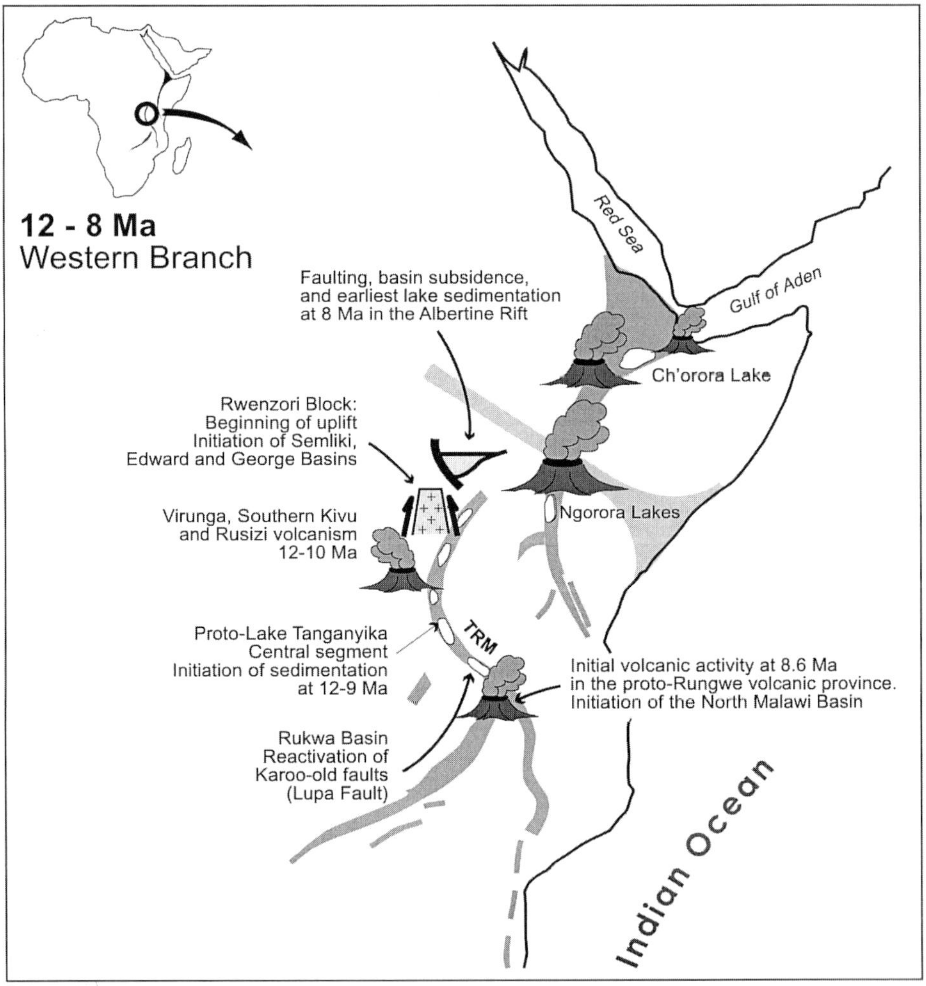


*Figure 9.* Distribution map of volcanism, faulting and lake basin initiation in the proto-Western Branch of the East African Rift System between 12 and 8 Ma.

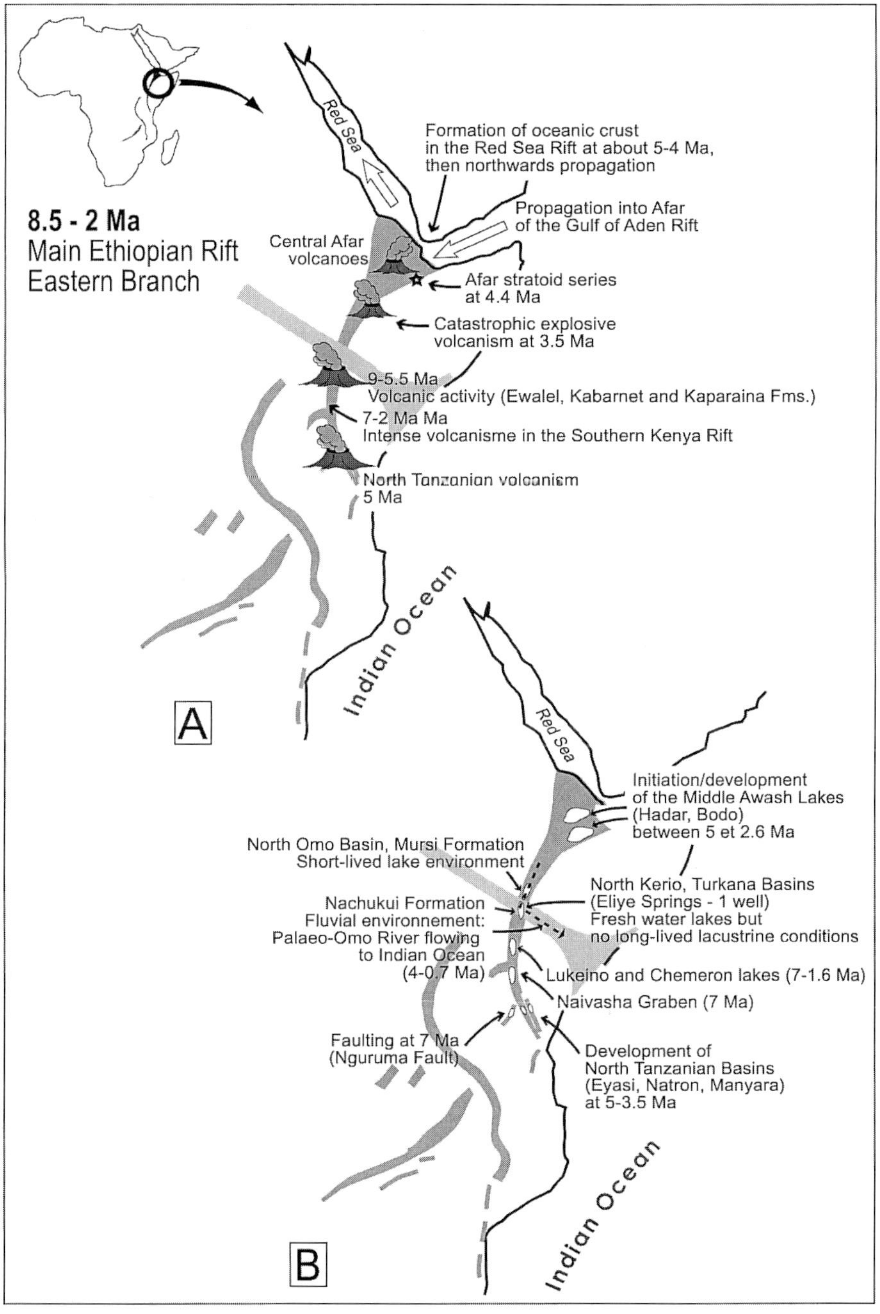


*Figure 10.* A: Distribution map of volcanism, and B: Lake basin initiation in the Main Ethiopian Rift and Eastern Branch of EARS between 8.5 and 2 Ma.

extensional strength migrated north and east, with a Late Miocene-Pliocene extension phase affecting the North Lokichar, North Kerio and Turkana sedimentary basins (Morley *et al.*, 1999c) (Figure 10). The upper part of the North Kerio basin fill is represented by a Late Miocene-Pliocene sedimentary sequence, mainly fluvio-deltaic deposits overlying Middle-Late Miocene volcanics in the area of Lothagam Hill (Patterson *et al.*, 1970; McDougall and Feibel, 1999). Almost a 3-km-thick series od sands and shales of Late Miocene-Pliocene age has been drilled in the North Kerio basin infill by the Eliye Springs-1 exploration well (Morley *et al.*, 1999c) (Figure 10). It indicates that fluvial environment predominated in this area during Late Miocene-Pliocene, and was temporarily associated with ephemeral lacustrine systems as suggested by offshore reflection seismic data (Dunkelman *et al.*, 1988; 1989). The presence of *Pediastrum* and *Botryococcus* algae in the shale intervals indicate the existence of freshwater lacustrine conditions (Morley *et al.*, 1999c). However, there is no indication that long-lived lacustrine conditions (representing a proto-lake Turkana) were established in this area during the Late Miocene-Pliocene. The upper part of the section sampled by the Eliye Springs-1 well might partially correspond to the 730-m-thick Nachukui Formation that oucrops on the northwestern shore of the present-day Lake Turkana (Brown and Feibel, 1988; Feibel *et al.*, 1991). This formation covers the period between >4 Ma and <0.7 Ma and demonstrates prior to 2 Ma the existence of wide alternating fluvial/floodplain environments linked to a palaeo-Omo River flowing from the north through the Turkana Basin toward the Indian Ocean (Brown and Feibel, 1988) (Figure 10A). These floodplain sediments have recently yielded the oldest hominid stone tools (dated at 2.34 Ma) found in the Kenya Rift (Roche *et al.*, 1999). From about 2 Ma, permanent lacustrine environments represent the first pulses of the modern Lake Turkana.

Immediately north, at the hinge between the northern end of the Turkana Basin and the Main Ethiopian Rift, thin fluviolacustrine strata of the Mursi Formation indicate that a rift lake had also formed briefly just before 4.2 Ma in the North Omo Basin (Brown and Nash, 1976; Ebinger *et al.*, 2000) (Figure 10B). In the Main Ethiopian Rift, the Mid-Late Miocene period was marked by the full development of the eastern and western faulted margins of the rift, with the evolution from alternating half-grabens to a symmetrical graben that was fully defined by 3.5 Ma (Davidson and Rex, 1980; WoldeGabriel *et al.*, 1990). A paroxysmic ignimbritic eruption occurred on the rift floor at this time, creating a giant (more than 30 km in diameter) and now buried caldera (WoldeGabriel, 1987) (Figure 10A). Vertical displacement of up to 2 km occurred in the central sector of the Main Ethiopian Rift at this time (WoldeGabriel *et al.*, 1990, 1992).

Further north, in the Middle Awash Depression of the Afar Rift, a few tens-m thick sedimentary series was deposited from Late Miocene to Early Pliocene. Brown clays, tuffaceous sands, diatomaceous lake clays and diatomites (the Bodo Beds; Kalb *et al.*, 1980) indicate alternating swamp, lake margin and open (shallow or deep) lake environments (Williams *et al.*, 1986) (Figure 10B). The top of this series is calibrated by the deposition of the Cindery Tuff at about 3.85 Ma (Hall *et al.*, 1984). A few tens of km north, in the Hadar region, lake environments alternating with fluvial systems developed between 4 and 2.6 Ma (Tiercelin, 1986) (Figure 10B).

The Bodo and Hadar lake shorelines are well-known to have hosted Early Pliocene hominids, and also have revealed the oldest (2.6-2.8 Ma) stone tools found in the East African Rift System (Roche and Tiercelin, 1977, 1980; Johanson *et al.*, 1982; White *et al.*, 1994; WoldeGabriel *et al.*, 1994). By the same period, intense fissural volcanism started at about 4.4 Ma and resulted in the deposition of the 1.5-km-thick Afar Stratoid Series (Aronson *et al.*, 1977; Manighetti *et al.*, 1998) (Figure 10B). In the Red Sea Rift, oceanization began at about 4-5 Ma, then essentially propagated northwards (Camp and Roobol, 1992) (Figure 10A). Rift tectonics was certainly responsible for the development of these Middle Awash lakes. However, regional climatic variations evidenced in Eastern Afar and Southeastern Ethiopian Plateau during this period (Williams *et al.*, 1979; Gasse *et al.*, 1980) are possibly linked to the Middle Awash lake level changes that are clearly expressed in the stratigraphic record. Evidence for major down-faulting of at least 1 km after 2.9 Ma in the Hadar area is provided by sedimentological, micropalaeontological and palynological data (Tiercelin, 1986; Bonnefille *et al.*, 1987; WoldeGabriel *et al.*, 1994).

To the south, in the Central Kenya Rift, the period from about 9 Ma to 6.5 Ma coincides with a substantial gap in the Tugen Hills sedimentary sequence (Hill, 1999). This is probably related to the renewal of volcanic activity with the deposition of the Ewalel Phonolites between 9 and 7 Ma (Chapman and Brook, 1978) (Figure 6D), and the Kabarnet Trachytes at 7-6.7 Ma (Chapman *et al.*, 1978) (Figure 10A). From 6.5 Ma, lake environments developed in the eastern flank of the Tugen Hills range (Baringo region), as indicated by the 140-m-thick Lukeino Formation, interbedded between the Kabarnet (7 Ma) and the Kaparaina (5.5 Ma) lava flows (Figures 6 and 10). Sediments, mainly diatomaceous tuffs and sometimes laminated shales, and some algal limestones demonstrate the existence of a quite extensive freshwater lake lying along the eastern foot of the active Tugen Hills faults (Pickford, 1978b). Faulting at the foot of the Tugen Hills associated with the eruption of Kaparaina Basalts at about 5.5-5.3 Ma ended the life of the Lukeino Lake.

Immediately above the Lukeino Formation and Kaparaina Basalts, recurrent lake environments are represented within the Chemeron Formation (Figures 6 and 10B) that, as the Ngorora Formation, extends for a considerable range of time, from about 5.6 to 1.6 Ma (Martyn and Tobias, 1967; Hill, 1999). Two lakes were lying at the same time on both sides of the Kaparaina anticline. These are Lake Kipcherere to the west, along the faulted foot of the Tugen Hills, and Lake Chemeron to the east, facing the foot of the Laikipia Border Fault escarpment. The Chemeron Lake area probably hosted the earliest hominid known so far (Hill, 1985). At about the same period (~7-5 Ma), rifting started to migrate further south, forming the southern segment of the Central Kenya Rift (Crossley, 1980; Baker, 1986; Foster *et al.*, 1997). Rifting propagated down to the extreme end of the Eastern Branch of EARS, where the Kenya Rift diverges from a single, 50-km-wide graben to a 200-km-wide complex zone in Northern Tanzania (Figure 1A). In the Lake Naivasha area, which represents today one of the most important drainage environment of the Central Kenya Rift, rift tectonics started at about 7 Ma (Figure 10B). This rifting phase occurred with the initiation of an east-facing half-graben, rapidly filled by Upper Miocene-Pliocene phonolites and trachytes flows (Baker and Mitchell, 1976; Roessner and Strecker, 1997). Volcanic activity persisted in this area up to 3.4 Ma. During the same time period, major faulting occurred along the Sattima Fault forming the eastern margin of the rift, converting the early half-graben to a graben (Baker, 1986; Baker *et al.*, 1988). Late Pliocene faulting (between 3.4 and 2 Ma) has internally structured the graben, producing a deep (at least 400 m) inner basin in the Naivasha region. Trachytic flood lavas filled this graben almost completely over the period 2.0-1.7 Ma (Baker *et al.*, 1988). Volcanism was so intense in this southern segment of Central Kenya Rift between 7 and 2 Ma that no long-lived lake environment developed and no thick sedimentary infill accumulated in the active graben despite a denudation rate of 20 m per one million years (Roessner and Strecker, 1997).

Further south, in Southern Kenya/Northern Tanzania, rift tectonics resulted by 7 Ma in the initiation of the main Nguruma Fault, prefiguring the western border of the rift (Baker *et al.*, 1978; Baker, 1986). To the South, basaltic volcanism and sedimentation started in isolated basins at about 5 Ma (Figure 10A), in the Eyasi and Natron-Manyara proto-rifts, as indicated by fluvio-lacustrine sediments and volcaniclastic products dated at about 4.9 Ma. But no major fault-bounded depressions apparently existed at this time (Dawson, 1992; Foster *et al.*, 1997) (Figure 10B). Lacustrine sedimentation also occurred in the region of Olduvai Gorge and west of Lake Natron, where the Laetolil Beds started to accumulate at 4.32 Ma in a mainly fluvial environment alternating with some lake environments, and continued until 2 Ma associated with volcanic activity (Hay, 1987).

In the Western Branch of EARS, rifting mechanisms continued by Upper Miocene time, with the initiation of sedimentation in the South Rusizi half-graben, the predecessor of the North Lake Tanganyika rift basin at about 7.4 Ma (Lezzar *et al.*, 1996). The nearby Rusizi Basin and the Lake Kivu Basin initiated and formed between 7.5 and 4 Ma, contemporary with the culmination of volcanic activity centred between 6 and 5.5 Ma in the Virunga-Kivu province (Degens *et al.*, 1971, 1973; Ebinger, 1989a; Pasteels *et al.*, 1989) (Figure 11A). From ~4.9 Ma up to 3.6 Ma, extension migrated northward and eastward in the proto-Lake Tanganyika Basin, and lake sedimentation commenced considerably later within the Burundian lake basin side of the North Kigoma half-graben (Lezzar *et al.*, 1996; Lezzar, 1997) (Figure 11B). At this time, the proto-Lake Tanganyika corresponded to an ~400-km-long lake resembling the present-day lake between 3° and 7° latitude S. Changes in lake level during this 7.5 to 5 Ma period resulted in an increase of downwarping but also as a consequence of major climatic changes. Low lake levels in the proto-North-Tanganyika Basin between 7 and 5 Ma have to be related to the decline in global temperature that affected Africa during this period (Van Zinderen Bakker and Mercer, 1986; Cohen *et al.*, 1997). Expansion of proto-Lake Tanganyika occurred then at about 3.6 Ma, correlative with a major expansion of lacustrine conditions in the Lake Malawi Basin recorded from the fossiliferous Chiwondo Beds (Bromage, 1995), as a consequence of a wetter climate in Central and Eastern Africa during the Mid-Pliocene. The southern end of Lake Tanganyika, today represented by the N160°-trending Mpulungu sub-basin (Mondeguer *et al.*, 1989; Tiercelin and Mondeguer, 1991), started to form between 4 and 2 Ma (Cohen *et al.*, 1993) (Figure

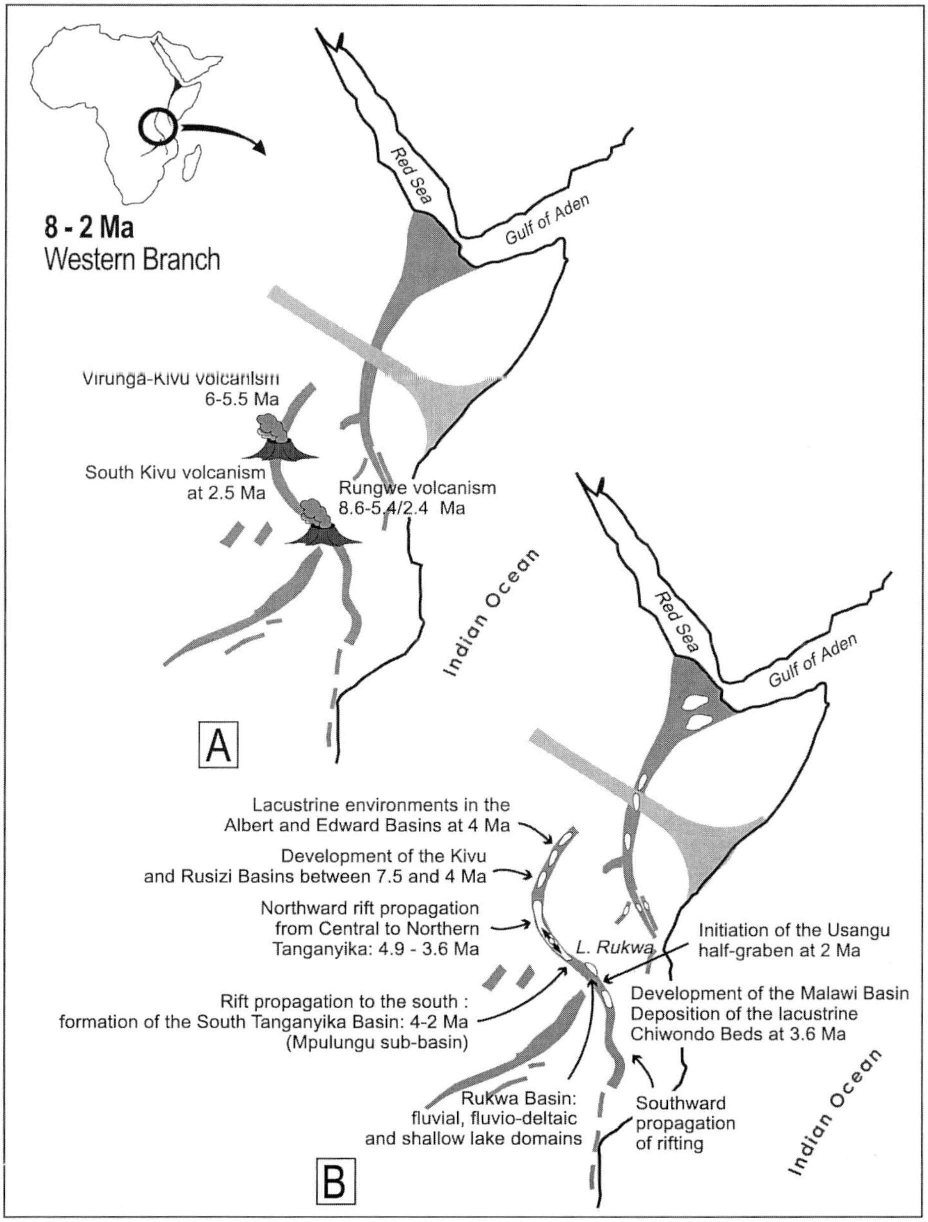


*Figure 11.* Distribution map of volcanism, faulting and lake basin initiation in the Western Branch of EARS between 8 and 2 Ma.

11B). Immediately East, tectonic activity is also demonstrated in the nearby N140°-trending Rukwa Rift by the well-marked angular unconformity existing between the lower part of the basin infill, formed by Karoo series (Dypvik *et al.*, 1990; Kilembe and Rosendahl, 1992; Morley *et al.* 1999b), and the overlying Red Sandstone/Lake beds formations (Morley *et al.*, 1989; Wescott *et al.*, 1991) (Figure11B). Controversy about the age of the Red Sandstone Formation, possibly Cretaceous (Dixey, 1928; Spence, 1954) or Miocene in age, is now clarified by the discovery of fossil wood that indicates a Tertiary age for this formation (Damblon *et al.*, 1998), and also by micropalaeontology and palynology results from exploration wells that yielded an Upper Miocene age (Wescott *et al.*, 1991). With a total thickness ranging from 1300 to 2000 m, the Red Sandstone/Lake Beds formations, mainly formed by lens-shaped sand bodies with thinner shale intervals, indicate that initial sedimentation associated to Tertiary rifting in the Rukwa Basin mainly related to a wide fluvial/fluvio-deltaic environment, sometimes associated with fluctuating shallow lacustrine domains. This type of environment persisted in the Rukwa region from Upper Miocene to Holocene-Present, with sediment transfer controlled by tectonic activity and mainly issued from the active Rungwe volcanic massif to the southeast (Delvaux *et al.*, 1998) (Figure 11B).

## 2.3 The Pleistocene to Present Rift Lakes

The Late Pliocene-Pleistocene to Recent period in Eastern and Central Africa is certainly by far the best and extensively studied along the two main branches of the East African Rift System. Proposed models integrate lake environment changes and their relationships with regional and global climatic variations, as well as recent tectonics and volcanic activity. Major questions regarding the past and future of humankind are closely linked to the knowledge of this geological or pre-historical period. The first question concerns the evolution of Hominids in Eastern-Central Africa during the last 2 Ma in connection with major changes in their habitat, as a result of interplay between forcing factors as climate variations, volcanic activity and recent rift tectonics. The second question directly concerns the present-day condition and future evolution of tens of small and large lake basins that have developed during the Pleistocene in the East African Rift System. Major tectonic and/or volcanic events are known to have occurred in the EARS between 2 and 0.5 Ma (Figure 12), creating new lake basins, or bringing basins formed during Mio-Pliocene times to their final morphology.

In the Eastern Branch (Kenya Rift), flood volcanism persisted from the whole rift from 2-1.5 Ma, and ended at about 0.8-0.3 Ma (Baker *et al.*, 1988), followed by several close phases of rift deformation and migration (Chapman *et al.*, 1978; Le Gall *et al.*, 2000). Such tectonic phases were associated with the formation of a suite of large, axial trachytic caldera volcanoes (Suswa, Longonot, Menengai, etc....) (Figure 12) and several other smaller volcanic complexes. It resulted in the development on the rift floor of inner troughs of widths ranging from 15 to 35 km (Chapman *et al.*, 1978; Dunkley *et al.*, 1993). Today, lakes with an area of less than 200 km$^2$ lie

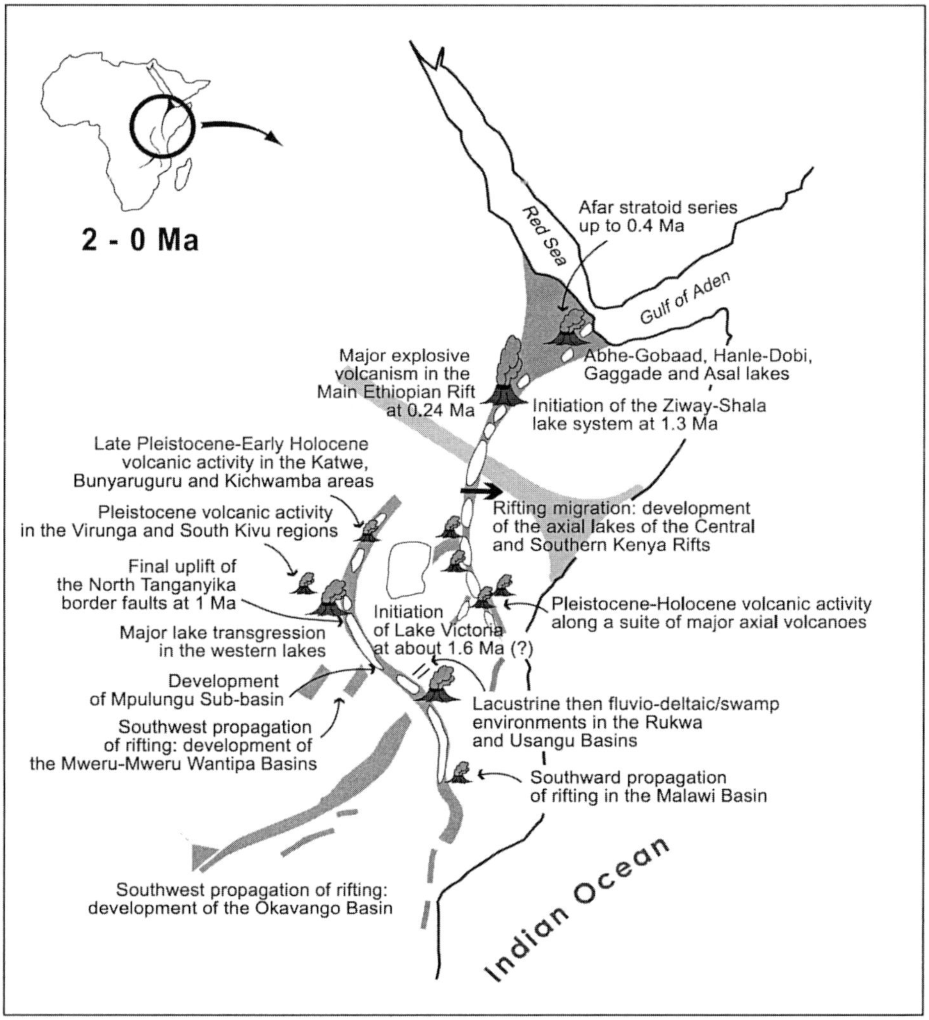


*Figure 12.* Distribution map of volcanism and lake basin initiation in the East African Rift System between 2 and 0 Ma, with indication of areas of present-day nascent/propagating rifting.

within 30- to 60-km-long tectonically-controlled segments of the rift floor, while alluvial deposits accumulate between the major axial volcanoes and alluvial fans developed along the faulted margins of the inner troughs (Tiercelin and Vincens, 1987; Dunkley *et al.*, 1993). Present-day lakes such as Lake Baringo or Lakes Nakuru, Elmenteita and Naivasha (Figure 1B) are the remnants of largest lake domains developed during Lower-Middle Pleistocene times and represented by rare sedimentary deposits sometimes uplifted or down-faulted along the inner trough margins (Richardson, 1966; Richardson and Richardson, 1972; Kamau, 1974; Renaut, 1982). Within the Baringo Basin in the Central Kenya Rift, the Kapthurin sedimentary formation illustrates a 'Central Rift Lake' that existed from about 700 ka to less than 200 ka (Tallon, 1976, 1978; Cornelissen *et al.*, 1990) (Figure 12). It has been formed in a structural setting roughly similar to that of the modern Lake Baringo. Kapthurin sediments that crop out along the western faulted margin of the Baringo Basin are predominantly of fluvial origin, possibly accumulated by rift-lateral rivers flowing into the inner trough. The lacustrine sediments associated to this 'Central Rift Lake' are supposed to be down-faulted along the axis of the inner trough. Vertebrate faunal evidence (Tallon, 1978) and sediment mineralogy from rare outcrops (Renaut, 1982; Renaut *et al.*, 1999) suggest that the Kapthurin palaeolake was alternatively fresh or saline/alkaline. This 'Central Rift Lake' started to retreat by ~200 ka, resulting in the development of a wide alluvial plain and the permanence at the northern end of the basin of a small lake prefiguring the present-day Lake Baringo. Other large and quite deep lakes developed in the Kenya Rift during Middle-Upper Pleistocene. Those lake systems were found in the Nakuru-Naivasha Basin (Central Kenya) (Thompson and Dodson, 1963; Richardson and Richardson, 1972; Kamau, 1974), and in the Suguta Rift Valley (Northern Kenya) where fossil beaches and stromatolites have been dated at 121 ka (Casanova, 1986; Casanova *et al.*, 1988; Bosworth and Maurin, 1993). They were also investigated in the Magadi-Natron Basin at the southern end of the Kenya Rift, where the large Lake Oloronga developed at >780 ka (Eugster, 1980, 1986; Casanova, 1987; Casanova and Hillaire-Marcel, 1987) (Figure 12).

In the Main Ethiopian Rift, volcanic activity persisted from 1.6-0.83 Ma to 0.20 Ma (Morton *et al.*, 1979; Mohr *et al.* 1980; WoldeGabriel *et al.*, 1990; Boccaletti *et al.*, 1999). Together with this volcanic activity, structural deformation was confined at this time to a NNE-trending structure formed by a line of hundreds of young faults defined as the Wonji Fault Belt (WFB) (Mohr, 1960; 1987) (Figures 12 and 13). Regional rift volcano-tectonic paroxysms occurred at 1.3-1.05 Ma and at 0.30-0.20 Ma along this structure, resulting in the development of several large and deep calderas on the rift floor. These are the Awasa and Corbetti Calderas in the south, and the Gademotta, O'a and Alutu Calderas in the north (Laury and Albritton, 1975; Mohr *et al.*, 1980) (Figure 13). Around ~0.60 to 0.30 Ma, basin subsidence and associated fault development resulted in the initiation of the Lake Abijata Main Basin, and the smaller Ziway basin to the north. This basin development occurred over a pre-existing depression formed by the superimposed remants of the previously formed Munesa and Gademotta Calderas (Le Turdu *et al.*, 1999) (Figure 13). This tectono-volcanic depression flooded very rapidly, forming the proto-lake Abijata. By

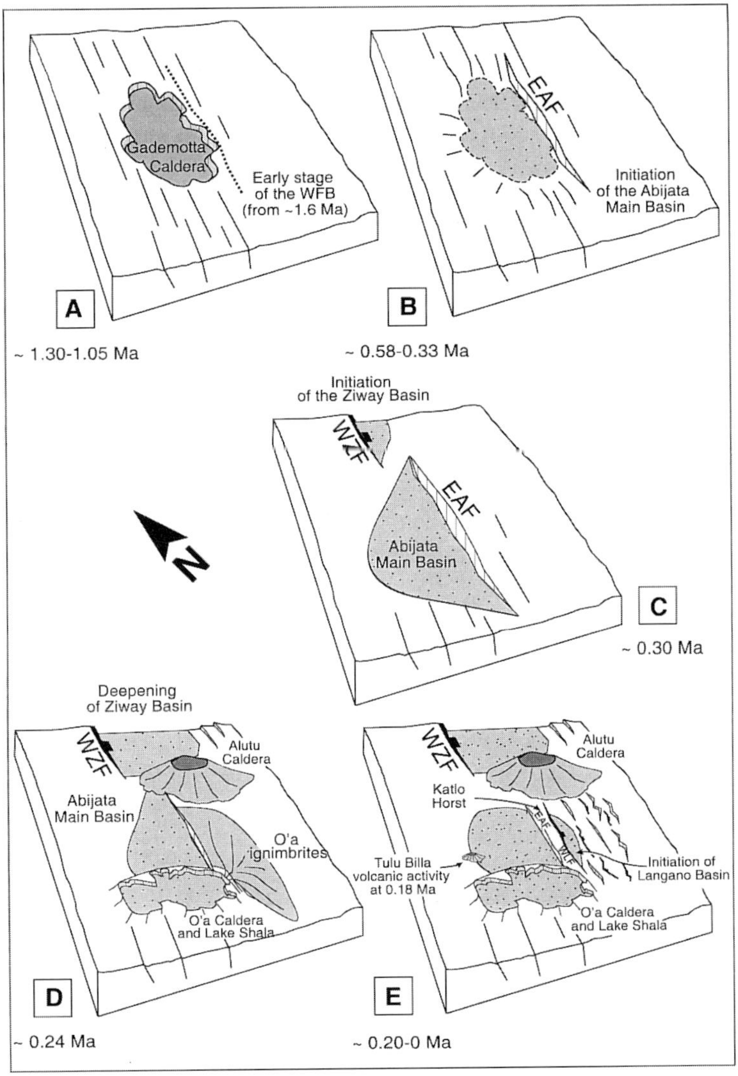


*Figure 13.* Schematic block-diagrams of the evolution of the Ziway-Shala lake basin from Early Pleistocene to Present. A: 1.30-1.05 Ma: formation of the Gademotta Caldera and early stage of faulting of the Wonji Fault Belt (WFB). B: 0.58-0.33 Ma: initiation of the Abijata Main Basin, bounded to the east by the nascent East Abijata Fault (EAF). C: 0.30 Ma: active subsidence of the Abijata Main Basin and initiation of the Ziway half-graben, bounded by the West Ziway Fault (WFZ) to the north. D: 0.24 Ma: regional volcano-tectonic activity, resulting in the formation of the Alutu and O'a Calderas, accompanied by eruption of wide ignimbritic bodies. Initiation of the Shala caldera lake. E: 0.20-0 Ma: eastward migration of rifting processes, development of the West Langano fault (WLF) and formation of the Lake Langano Basin (after Le Turdu *et al.*, 1999).

0.25 Ma, ignimbrite activity affected the O'a and Corbetti/Awasa volcanoes, resulting in large caldera-type collapse that led to the instantaneous formation of the large Awasa and Shala caldera lakes (Figures 1B and 13). From ~0.20 Ma, fault movement along the NNE the Wonji Fault Belt trend resulted in the shaping of the present Lake Ziway Basin, and in the initiation of the east-dipping Langano half-graben and the proto-lake Langano (Figure 13). At this time, the suite of lakes lying along the major axis of the Main Ethiopian Rift was formed, and started to fluctuate during the Upper Pleistocene and Holocene periods (Gasse and Street, 1978; Street, 1979; Le Turdu *et al.*, 1999).

Further north, rifting processes became very active in the Afar Depression, with the development of an intense grid faulting pattern aligned along the NW and NNW trends and affecting the Afar Stratoid Series lavas deposited between 4.4 and 0.4 Ma (Manighetti *et al.*, 1998) (Figure 12). This faulting activity migrated eastward from the center of the Afar Depression, occupied during Upper Pliocene by the Middle Awash lakes, and resulted between Lower Pleistocene and Holocene in the formation of the Lower Awash (Abhe-Gobaad, Hanle-Dobi, Gaggade) Basins (Gasse, 1977; Tiercelin *et al.*, 1980) (Figure 12). A progression in the age of the tectonic activity which affected the different basins is observed. Most of the basins remained stable since the Middle-Upper Pleistocene while Holocene to recent volcano-tectonic activity concentrated in the NW-trending Ghoubbet-Asal Rift, resulting in the formation of the saline Lake Asal (Gasse and Fontes, 1989) (Figure 12).

Westward, in the Kivu-Tanganyika segments of the Western Branch of the EARS (between 2° S and 4° S) (Figure 12), it has been demonstrated that at ~2 Ma, a major episode of rifting associated with intense volcanism in the Kivu Province resulted in the final development of the Kivu and North Tanganyika lake basins (Ebinger *et al.*, 1989a, b; Lezzar *et al.*, 1996). At ~1 Ma, a last focused rifting phase resulted in the uplift of rift shoulder escarpments along the major border faults structuring the North Tanganyika Basin (the Uvira Border Fault System, and the East and West Ubwari Faults) (Lezzar *et al.*, 1996; Cohen *et al.*, 1997) (Figure 12). On the other hand, southward rift propagation along the Lake Tanganyika Rift resulted in the final morphology of the Mpulungu sub-basin at the southern end of Lake Tanganyika from ~2 Ma up to Present. In the Rukwa Rift, faulting continues to develop, controlling axial drainage systems that prevent the development of deep lacustrine environments. As a result of rotation of the regional extension direction from E-W to NW-SE (Delvaux *et al.*, 1992; Ring *et al.*, 1992), the Usangu Flats Basin started subsiding at about 2 Ma along the NE-SW trend (Ebinger *et al.*, 1989), perpendicular to the axis of the Rukwa and Northern Malawi Basins (Figure 12). From reflection seismic data, an idealized stratigraphy of the 1-km-thick basin infill suggest that lake environments existed during the first phase of basin development, followed by thickest sediments indicating fluvio-deltaic and swamp environments (Harper *et al.*, 1999).

By the same period, young rift lake basins started to develop along two newly formed rift branches that started to open along the NE-SW tectonic trend, thus forming the "Southern Complex" (Mondeguer et al., 1989) (Figures 1 and 12): 1) at the southern end of the Tanganyika Rift, where the Mweru-Mweru Wantipa fault zone developed perpendicular to the main axis of the South Tanganyika Basin. This faulted zone corresponds to a system of parallel horsts and grabens and "en échelon" grabens representing pull-apart troughs with little subsidence, resulting of extension and strike-slip motions, and presently occupied by the shallow Mweru and Mweru Wantipa lakes (Mondeguer *et al.*, 1989; Mondeguer, 1991); 2) at the northern end of the Lake Malawi Rift, where the NE-SW tectonic trend extends through the Permo-Triassic Luangwa and Kariba grabens (Banks *et al.*, 1995) towards the Okavango Rift in Botswana. There, recent active faulting created a 100-km-wide and 300-m-deep half-graben occupied today by the wide Okavango intracontinental delta and the intermittent Lake Ngami (Modisi *et al.*, 2000) (Figures 1B and 12). The "Southern Complex" is at the present time an area of high seismicity (Shudofsky, 1985), and gravimetric and high heat flow anomalies (Ballard and Pollack, 1988) demonstrate the migration of extensional deformation southwesterly towards Southern Africa (Vail, 1968; Reeves, 1972; Girdler *et al.*, 1969).

Basin subsidence and rift tectonics continued up to 450 ka. A major lake level rise resulted at this time in the flooding of essentially the entire modern Lake Tanganyika, in correlation with other high lake stands evidenced in the Eastern Branch of EARS, contemporary with an increase in intensification of the Asian monsoon (DeMenocal *et al.*, 1993; Cohen *et al.*, 1997). The tectonic and limnological history of the other large lakes of the Western Branch for the last million years is presently less documented than Lake Tanganyika or Lake Malawi (Lezzar *et al.*, 1996; Soreghan *et al.*, 1999). Major transgressive and regressive phases affected the different lake domains of both branches of EARS, in reaction to a series of humid and dry phases that characterized Eastern and Central Africa from Middle-Late Pleistocene up to the present-day, correlating with fluctuations of the Asian monsoon (Gasse and Street, 1978; Lézine and Casanova, 1991). Tectonic deformations continued during the Late Pleistocene-Holocene time interval, but with minor effects compared to the previous periods. A contemporary and important volcanic activity persisted in the Eastern and Western Branches, mainly resulting in the building of small volcanic craters, associated to the large axial volcanoes in the Kenya Rift or Main Ethiopian Rift (Figure 12), or located on high relief accommodation zones as the ones separating the Lakes Kivu, Edward and Albert Basins in the Western Branch (Reynolds, 1984; Rosendahl, 1987; Boven et al, 1998; Demant *et al.*, 1994; Brooks *et al.*, 1995; Yamba and Boven, 1998) (Figure 12). Some of these craters were rapidly occupied by small lakes such as the Bischoftu Lakes in the Main Ethiopian Rift, Lake Massoko in the Rungwe volcanic massif (Winckel, 1998; Gibert *et al.*, in press), or the Green Crater Lake in the Naivasha Basin, Central Kenya Rift (Figure 1B).

Among all these rift lakes formed during successive episodes of rifting, one exception is Lake Victoria, that is known to be relatively young as compared with the other large lakes occupying the East African Rift System (Figure 12). Lake Victoria straddles an ancient drainage system that flowed from the east to the west and was modified by uplift on the shoulders of the Lake Albert rift basin. Although this tectonic activity may have begun in the Miocene, river downcutting was maintained westward until the Pleistocene, when flow reversal created a lake substantially larger than the present one. Precisely when this occurred in the Pleistocene is not known: 1) possibly during Early to Middle Pleistocene age (1.6 to 0.8 Ma), as suggested by lacustrine sequences 100 m above the present lake near the Kavirondo Gulf on the Kenyan side of the lake; or 2) during Middle to Late Pleistocene (younger than 0.8 Ma), on the basis of lake series exposed 130 m above the present lake, in the Kagera River Valley. During the Late Pleistocene-Holocene period, Lake Victoria fluctuated in response to palaeoclimatic events of regional importance, with two periods of more-or-less complete desication during the Late Pleistocene, also identified in the nearby Lake Albert (Beuning *et al.*, 1997; Talbot and Laerdal, 2000). Today, like the other lakes of the East African Rift system, Lake Victoria is facing a strong rise in sediment accumulation rates, probably as a result of man's influence in the lake and its catchment (Alin, 1999; Talbot and Laerdal, 2000).

At the present time, rifting processes are known to be active along many segments of the East African Rift System. As indicated by the distribution of seismicity recorded in East Africa during the last 100 years (Nusbaum *et al.*, 1993), the Eastern Branch and Main Ethiopian Rift can be described as "seismically quiet", while the Western Branch and the Southern Complex appear considerably more active. There, present-day tectonic activity and basin subsidence contribute to maintain deep basins and associated lakes, or to create and control new lake basins. Among the other expressions of rifting, active or dormant volcanism characterizes several segments of the EARS (Fig. 2), often associated with hydrothermal activity that plays a major role in controlling the geochemical characteristics of the small or large lakes of the East African Rift System (Tiercelin *et al.*, 1993; Coussement *et al.*, 1994).

## 3. AFRICAN LAKES ASSOCIATED WITH UPPER PALAEOZOIC RIFT TECTONICS

At about 280 Ma, at the beginning of the Permo-Carboniferous (Kreuser, 1995), major rifting systems concentrated on the eastern margin of Africa, along the axis of the future Central Atlantic Ocean, and in the Tethys domain in relation with the breakup of Gondwana supercontinent (Guiraud *et al.*, 1992). In Southeastern Africa, the onset of this Karoo rifting occurred during Late Carboniferous and Permian times, with the initiation of wide and deep rift basins such as the Central Kalahari Basin in Southern Botswana, or the Tuli half-graben in Northern Botswana (e.g., Lambiase, 1989; Banks *et al.*, 1995; Le Gall *et al.*, submitted) (Figure 14). A large volume magmatic event occurred in the Early Jurassic, resulting in the formation of the Karoo Large Igneous Province of Southern Africa and the associated giant dolerite dyke swarms of Northern Botswana and Zimbabwe, dated between 190 and 170 Ma (Burke and Dewey, 1973; Dingle *et al.*, 1983; Reeves, 1978; Duncan *et al.*, 1997; Le Gall *et al.*, submitted) (Figure 14). This Early Jurassic magmatism postdates the onset of extension in both flexural and rift basins of Karoo age.

In Central and Eastern Africa, the Karoo rifting phase was marked by the development of three large groups of rift systems oriented N-S, NE-SW and NW-SE,

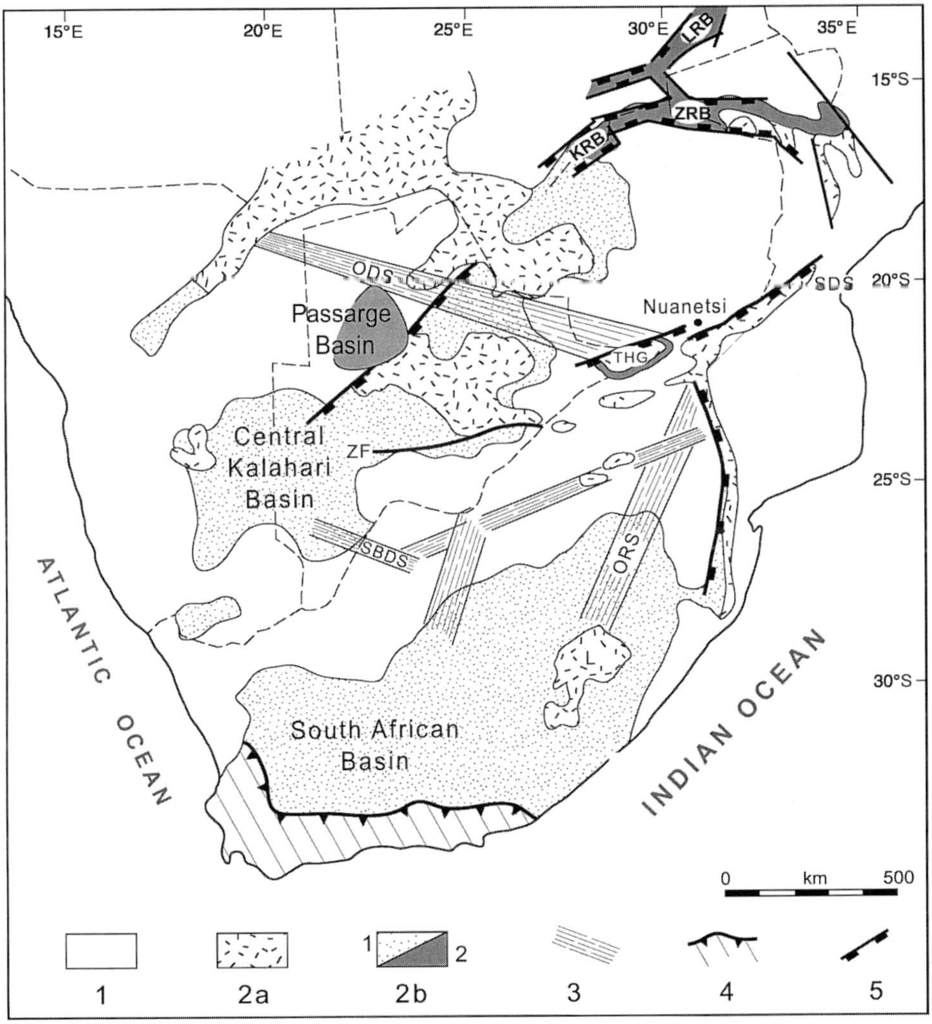


*Figure 14.* Major Karoo structures (volcanics, dyke swarms and sedimentary basins) in Central-Eastern and Southern Africa. 1. Basement and post-Karoo cover; 2. Karoo units, a. Volcanic complexes, b1. Flexural basins, b2. Rifted basins; 3. Major dyke swarms. ODS: Okavango Dyke Swarm; ORS: Olifants River Dyke Swarm; SBDS: South Botswana Dyke Swarm; SDS: Sabi Dyke Swarm; 4. Cape fold belt; 5. Major faults; ZF: Zoetfontain Fault; THG: Tuli half-graben (from Le Gall *et al.*, submitted).

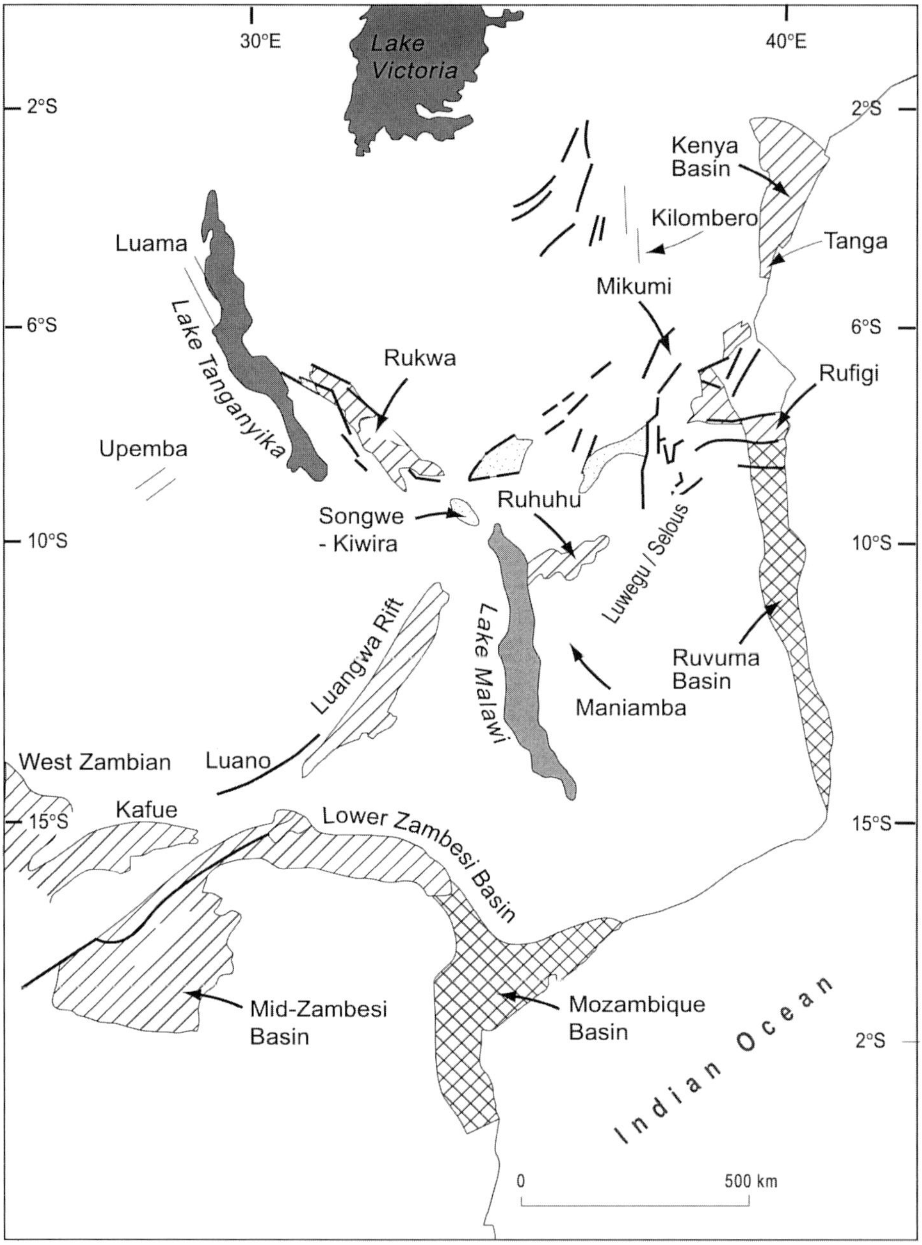


*Figure 15.* Location map of Central-Eastern and Southeastern African Karoo rift basins and pericratonic basins along the East African coast line (Ruvuma and Mozambique Basin) (modified from Kreuser, 1995).

respectively, as the result of the influence of continent-scale basement features initiated during the successive Proterozoic orogenes (McConnell, 1950; Daly, 1988; Daly *et al.*, 1989) (Figure 15). The NE-SW rift group comprises successively the Rufiji, Ruhuhu, Luwegu (synonymous for Selous) and Maniamba Basins east of 34° longitude E (Kreuser, 1995), and the Upemba, Luangwa, Kafue and Kariba Basins west of 34° longitude E (Figure 15). The NW-SE-trending rift group includes the Luama Trough, located at 6° latitude south on the west side of Lake Tanganyika (Fourmarier, 1935), and the Rukwa and Songwe-Kiwira Basins in Southern Tanzania (Dypvik *et al.*, 1990; Morley *et al.*, 1990; Kilembe and Rosendahl, 1992; Rosendahl *et al.*, 1992; Morley *et al.*, 1999b). Further north, in Northeastern Kenya, considerable thickness of Carboniferous(?)/Triassic rocks underlies parts of the Mesozoic Anza Rift in the Mandera Lugh region (Morley *et al.*, 1999a). The N-S oriented rift system is mainly represented by the Kilombero Rift in Central Tanzania (Nilsen *et al.*, 1999), that extends further north into the Tanga and Kenya Basins (Kreuser, 1995) (Figure 15). All these Karoo rift basins are geologically not as well known as the other African rift basins of Cenozoic age, with the exception of some major structures that have been intensely studied for oil exploration. Field investigations, seismic reflection and exploration wells data have shown that shallow or deep water rift lake environments developed within almost these 20 rift basins at various periods between 280 and 245 Ma, this in alternation with fluvial and fluvio-deltaic environments (Kreuser, 1995).

One of the best documented of these rift basins is the Luangwa Valley Basin, that extends from 10° 30' S up to 14° 00' S in Southern Zambia (Figures 15 and 16). Structurally, this basin comprises two opposing half-grabens/sub-basins about 200 km long and 60 km wide, respectively (Figure 16A, B), and containing up to 8 km of Permo-Triassic fluvial, deltaic and lacustrine sediments (Banks *et al.*, 1995) (Figures 16B and 17). As suggested by seismic facies distribution, these two half-grabens were separate features occupied by at least two lakes - forming the Madumabisa lake system - for much of Permian time. These lakes became connected during high lake stands, possibly forming between Lower and Upper Permian ( at ~ 260-250 Ma) a single, >400-km-long lake (Banks *et al.*, 1995) (Figure 17). Other lakes developed in this region during the same period, for example in the Ruhuhu Basin where black shales were deposited at about 280 Ma in a deep anoxic lake environment (Kreuser, 1990), while Upper Permian time (~250 Ma) was characterized by a rapid infill of these lakes by conglomerates and playa deposits, indicating a major change in the regional hydrology and climate (Kaaya, 1992). Other rift basins such as the Luwegu/Selous Rift (Figure 15) initiated in the Upper Permian with fluvio-deltaic clastics including a lacustrine black shale member (Kreuser, 1983). This basin extended up to the Triassic/Jurassic boundary with >4-km-thick fluvial (braided stream) and shallow lacustrine deposits. These deposits relate the presence of extensive channels over wide alluvial plains where shallow lakes with restricted size formed occasionally (Hankel, 1987).

To the north-west of the Luangwa Rift, the present-day Rukwa Rift represents a NW-SE-trending segment of the Western Branch of the EARS (Figure 15), bounded

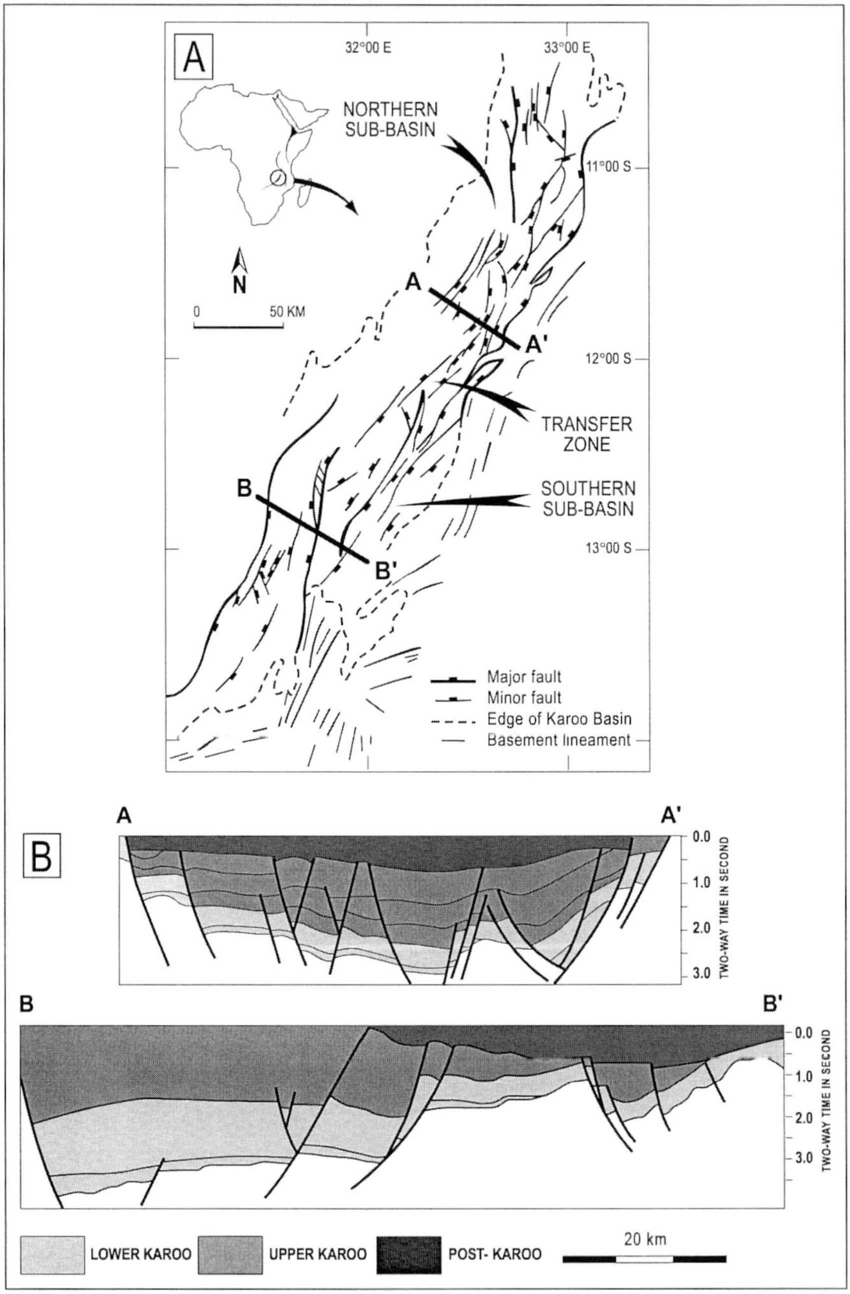


*Figure 16.* A. Simplified structural map of the Karoo Luangwa rift basin, Zambia, showing the two opposed half-grabens separated by a major transfer zone. B. Cross-sections through the northern and southern sub-basins of the Luangwa Basin, showing the major structural and stratigraphic units deduced from reflection seismic data.

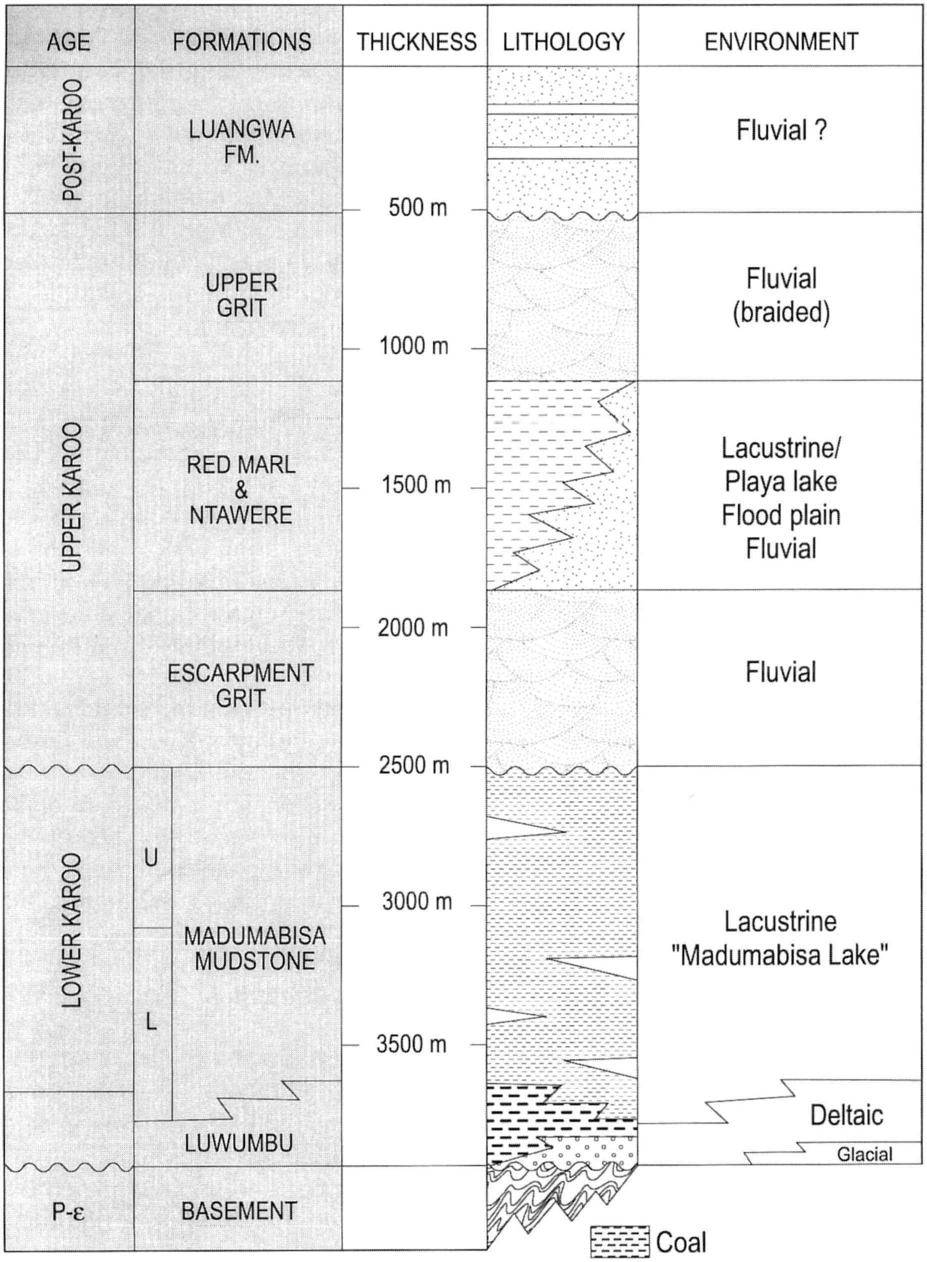


*Figure 17.* Lithostratigraphic column of the syn-rift and post-rift infill of the Luangwa Basin, indicating the major Madumabisa lacustrine phase (modified after Banks *et al.*, 1995).

to the north-east by the 200-km-long Lupa border fault. Seismic reflection and well data indicate that the Rukwa Rift started to form during Karoo times, with indication of Permian age from palynological well studies (Wescott *et al.*, 1991; Morley *et al.*, 1992). Tectonic movements were mostly concentrated along the Lupa Fault and extended up to recent times, resulting in the development of a 11-km-deep basin (Morley *et al.*, 1992; Delvaux *et al.*, 1998; Morley *et al.*, 2000). Sediments that accumulated in the nascent Rukwa Rift were mainly of continental origin, sandstones, shales and coal, forming a sequence greater than 1800 m in the Galula coal field (Spence, 1954). Seismic reflection data indicate that the Karoo sediments thicken into the Lupa Fault, attaining maximum thickness of 3-3.5 km (Morley *et al.*, 1992). Such deposits indicate a predominance of fluvial systems in this nascent rift basin, with only an occasional development of minor marshy or lacustrine environments as suggested by the deposition of coal and shale intervals in the Karoo sequence.

Thus, more than 20 shallow and/or deep lakes alternated with giant fluvial and fluvio-deltaic systems within Southeastern and Southwestern Africa from 280 to 205 Ma, accompanying the initial break up of Gondwanaland (Norton and Sclater, 1979; Simpson *et al.*, 1979) (Figures 14 and 15). Among them, the largest and deepest lakes, some being quite similar in size with the present-day large lakes of the East African Rift System, were probably long-lived (at least a 5- to 10-Ma-long life for the Madumabisa Lake in the Luangwa Basin), and characterized by anoxic conditions as shown by the accumulation of thick series of black, organic-rich shales (as in the Ruhuhu, Luama or Kilombero lake basins) (Figure 15) or oxic conditions as illustrated by the 2- to 5-km-thick, grey or red/brown Madumabisa Mudstone that accumulated in the Madumabisa Lake (Banks *et al.*, 1995) (Figure 17). Lake Tanganyika or Lake Malawi in the Western Branch of EARS can be considered as possible modern analogs of the Permo-Triassic Ruhuhu, Luama or Kilombero Lakes. Oxic bottom conditions existing in the present-day Lake Turkana (Eastern Branch of EARS) (Figure 1B) can be compared to the conditions prevailing in Lake Madumabisa during the deposition of the Madumabisa Mudstone. Lakes with restricted size and shallower depth formed as a consequence of regional tectonic uplift reactivating the accommodation zones cross-cutting the basins, or resulting from strong changes in climate.

In terms of spatial scale, the distribution of Karoo rift basins in Central-Eastern Africa is comparable in length to the Main Ethiopian Rift/Eastern Branch of EARS (Figure 1A) but shows a wider, fan-shaped distribution of the different basins. In terms of time scale, Karoo rifting appears to have extended over a period of ~80 My, about two times the present-day life of the East African Rift System which has yet to enter thermal subsidence phase. Nevertheless, important sedimentary gaps in such continental series and resolution in geochronology data have to be taken into account for this large difference in longevity between the Karoo and Tertiary East African Rift Systems. Lake environments that characterize the Karoo rifts and the Main Ethiopian Rift/Eastern Branch of EARS strongly differ in size but also in terms of limnological characteristics, probably as a result of strong differences in climate conditions and volcanic activity prevailing in these two rift systems during Permo-

Triassic and Tertiary times, respectively. Better analogs for several of the Karoo rift lakes can be provided by the large lakes of the Western Branch of EARS.

## 4. AFRICAN LAKES ASSOCIATED WITH MESOZOIC RIFT TECTONICS

In the Late Jurassic (160 Ma), a second major and complex episode of crustal extension began again in Central-Eastern Africa, marked by the development of other large systems of continental rifts, collectively named the West and Central African Rift System (WCARS) (Bermingham *et al.*, 1983; Browne and Fairhead, 1983; Fairhead, 1986). This system extended from Southern Kenya Central Sudan (Jorgensen and Bosworth, 1989; Bosworth, 1992), where it connected with the West African Rift System (Benkhelil and Robineau, 1983; Allix and Popoff, 1983) (Figure 18A). Such rift tectonic events accompanied the Africa-South America plates breakup, and spanned a period of >100 My, much longer than rift basin development in the East African Rift System (roughly 35 My; Morley *et al.*, 1992). Events were divided in several individual extensional phases, each occupying time periods of several tens of millions years (Bosworth, 1992). The general distribution of Mesozoic rift basins in Central-Eastern Africa is shown on Figure 18A, B (Browne and Fairhead, 1983; Genik, 1992). The South and Northeastern Sudan rift systems comprise a minimum of five NW-SE trending rift basins formed as a consequence of NE-SW extension processes (Schull, 1988; Bosworth, 1992) (Figure 18B, C). At the extreme south of Sudan, the rift basins show E-W trends and are interpreted as a structural link (known as the South Sudan Shear) with the large Anza rift basin that extends over 500 km along a NW-SE trend from Northern Kenya to the Lamu Embayment (Reeves *et al.*, 1987; Greene *et al.*, 1991; Bosworth, 1992) (Figure 18B).

The Sudan and Anza Rifts have been quite intensively studied for oil exploration, with major oil discoveries in the Sudan Rifts (Vail, 1978; Schull, 1984, 1988). The Northern Sudan basins (Blue Nile, White Nile and Melut Rifts) (Figure 18B) are characterized by relatively simple, large-scale rift geometries, generally asymmetric half-grabens 20-to 50-km-wide (Schull, 1988). Syn-rift sequences of sandstone, siltstone and shale have been dated quite precisely from Jurassic(?)-Early Cretaceous to Early-Middle Tertiary using several 3.5-km-deep exploration wells (Bosworth, 1992). The oldest (Jurassic?) penetrated rocks are nonmarine siltstones and claystones, overlain by more than 6 km of Cretaceous sediments that accumulated in the deepest troughs. Two cycles of deposition, directly relating to rifting phases, have been identified. The first cycle extended from Neocomian to Cenomanian (100-140 Ma), and was characterized by alternating alluvial, fluvial-floodplain and lacustrine environments. The Aptian-Early Albian period (about 10 My-long period) corresponded to the greatest lake development, with the deposition of more than 1800 m of organic-rich lacustrine claystones and shales (Abu Gabra Formation; Muglad Rift) (Figure 18B) that form the primary source rock of the Sudan basins. The second depositional cycle, related to the second rifting phase, extended from

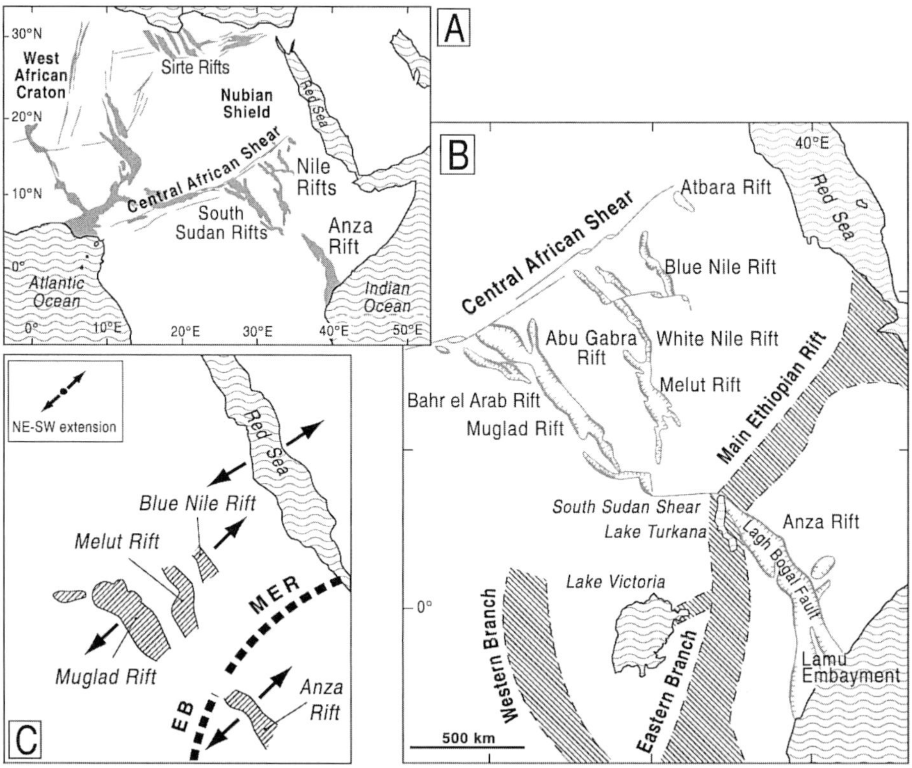


*Figure 18.* A: Location of the Mesozoic South Sudan, Nile and Anza Rifts within the West and Central African Rift Systems (after Bosworth, 1992; Guiraud and Maurin, 1992). B: Distribution of the Late Jurassic-Cretaceous rift basins in Sudan and Kenya (modified after Bosworth, 1992), compared to the general trend of the Eastern Branch (EB) and Main Ethiopian Rift (MER), East African Rift System. C: Regional tectonic map of Central and Eastern Africa, showing the major zones of extension during Late Jurassic-Cretaceous, indicated by black arrows). Stippled areas indicate the Sudan and Anza rift basins (modified after Schull, 1988).

Turonian to Late Senonian, and was accompanied by minor volcanism at about 90 Ma. Floodplain and lacustrine environments developed, with sediments accumulating in shallow and well-oxygenated waters as suggested by their low organic carbon content. The Cretaceous ended with the wide development of fluvial and alluvial fan environments (Schull, 1988).

To the southwest, the Anza Rift has a length of 500 km and a width of 130 km, bounded to the northeast by the major, NW-SE-trending Lagh Bogal Fault (Figures 18 and 19A). A minimum thickness of 7 km of syn-rift fill is indicated by seismic data, while gravity and magnetic data suggest more than 10 km (Reeves *et al.*, 1987; Bosworth, 1992; Bosworth and Morley, 1994; Morley *et al.*, 1999a). Such dimensions are similar to those of the Tanganyika Rift that is 600 km long, but only 80 km at its widest extent. Several exploration wells >4-km-deep have been drilled (Morley *et al.*, 1999a) (Figure 19A, B), and indicate active rifting have commenced by the Neocomian (145 Ma) and continued more or less continuously into the Early Tertiary (50-35 Ma) (Winn *et al.*, 1993; Bosworth and Morley, 1994). The sandstone and shale series drilled at different places in the Anza Rift are mostly Campanian to Maastrichtian in age (70 Ma) (Figure 19B) and are interpreted as deposited in lacustrine, lacustrine-deltaic, fluvial/flood plain and eolian environments. Palaeogeographic reconstructions based on well and seismic data have been proposed by Bosworth and Morley (1994) (Figure 20), showing that between Neocomian and Campanian, lacustrine environments were restricted to small or medium-size lakes (max. 200 km in length; quite similar in size with the present-day Lake Turkana), often distributed within complex axial fluvial systems (Figure 20). Dark shale intervals have been identified and interpreted as indicating low-oxygen conditions similar to those existing in the present-day medium to large lakes of the East African Rift System (e.g., Lakes Edward, Tanganyika, Malawi). Gray shales are indicative of open, oxygenized lacustrine conditions that can be illustrated by shallow depth areas in present-day Lake Tanganyika, or by the present-day Lake Turkana or even smaller lakes such as Lake Baringo (Tiercelin, 1981; Tiercelin *et al.*, 1994) (Figure 1B). During Cenomanian (at about 100 Ma), marine environments succeeded to lakes, suggesting episodic opening to the Indian Ocean via the Lamu Embayment (Figure 20), as shown by the several hundreds meters of deep water, marine shales drilled in the N'dovu-1 well (Figure 19B). A brief marine incursion occurred again during Lower Maastrichtian at ~70 Ma (Morley *et al.*, 1999a). Such a link with the Indian Ocean possibly persisted during Miocene times as suggested by the fossil marine whale found in the western Turkana region near Loperot (Maglio, 1969; Mead, 1975), and by marine microfossils found in Miocene sediments in the Hot Hori-1 well (Morley *et al.*, 1999a) (Figure 19). Such opening of the Anza Graben to the Indian Ocean can be explained by the complex structural evolution and thermal subsidence that continued to affect the graben during the Tertiary.

Accompanying the early opening of the South and Equatorial Atlantic oceans, the development of West and Central African Rift Systems spanned a period of >100 My, much longer than rift basin development in the East African Rift System (roughly 40 My). The W and CAR Systems extend over the whole African continent,

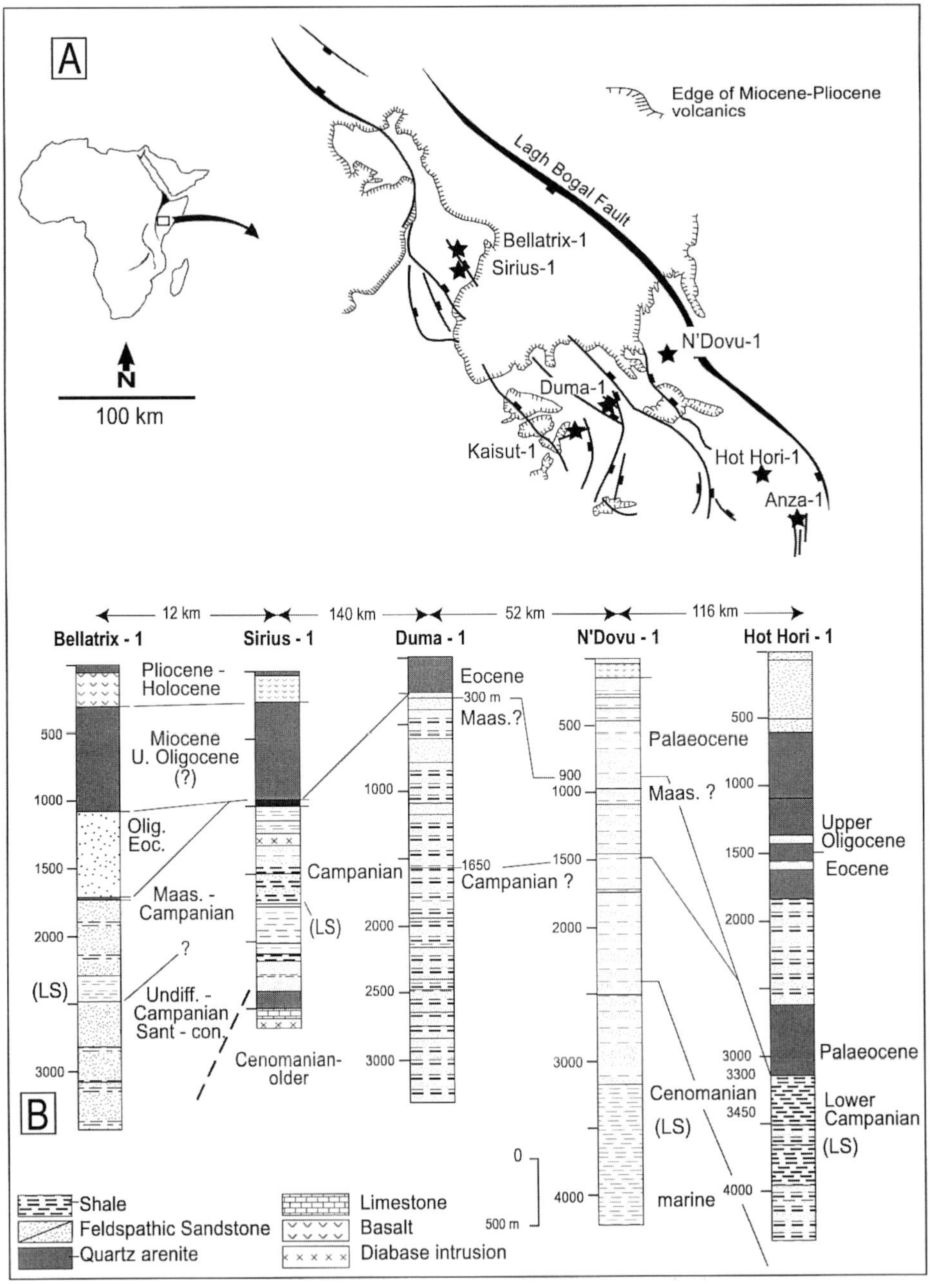


*Figure 19.* A: Schematic structural map of the Anza Rift, with indication of the exploration wells (from Bosworth and Morley, 1994). B. Stratigraphic cross-sections through the Central Anza exploration wells, indicating the location and inferred ages for the shale intervals (LS) that indicate the development of wide lacustrine domains (modified from Winn *et al.*, 1993; Morley *et al.*, 1999a).

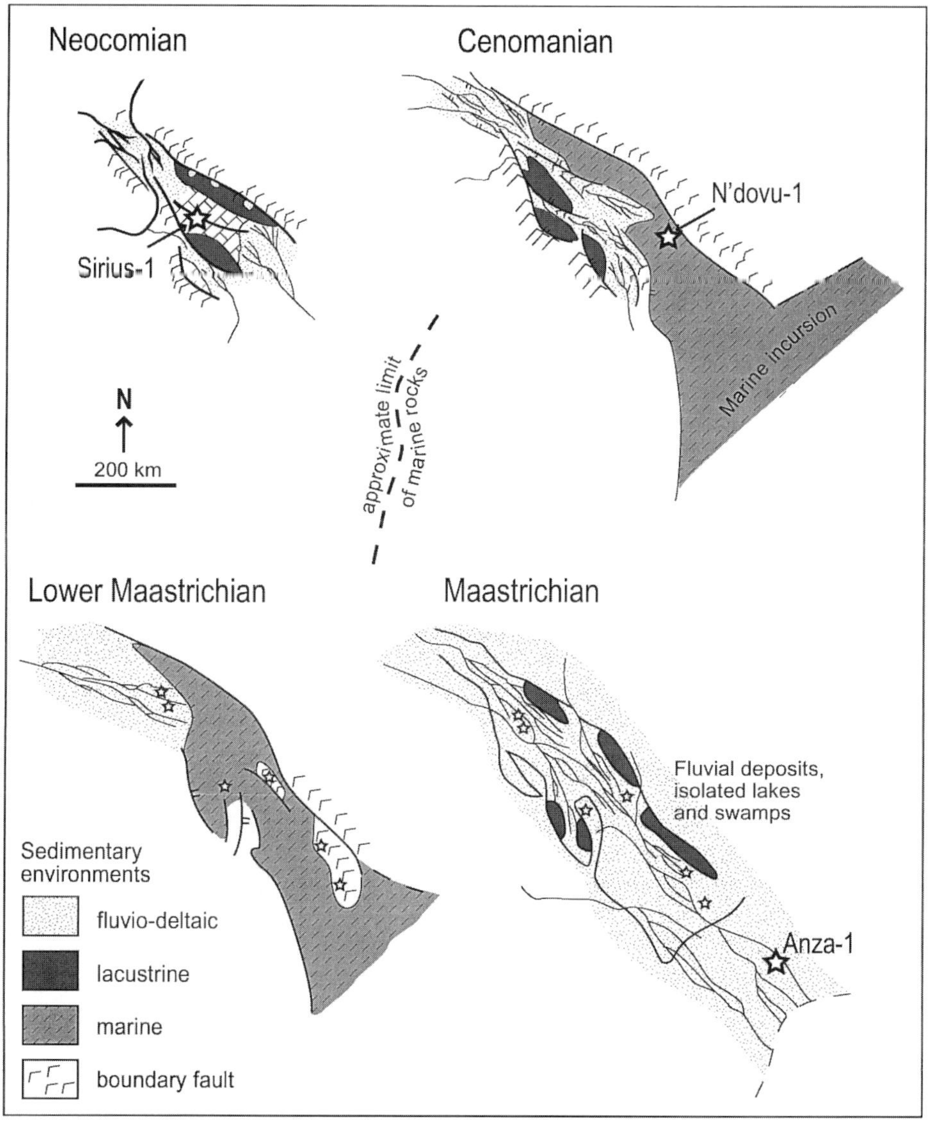


*Figure 20.* Palaeogeographic reconstructions of the Anza Rift environments between Neocomian and Maastrichtian (modified from Bosworth and Morley, 1994; Morley *et al.*, 1999a).

from the Benoue Trough of Nigeria to the Indian Ocean. Shear tectonics combined with extensional tectonics in the development of the Benoue Trough (Fairhead, 1988) while extensional mechanisms resulted in the series of NW-SE-trending basins of Southern Sudan and Northern Kenya. In terms of sedimentary environments, the Sudan and Anza rift basins display a large variety of lakes in terms of size and spatial distribution, demonstrating similarities with the Northern and Central Kenya Rift during Oligocene and Miocene times.

## 5. CONCLUSIONS

When looking through this 300 millions years long geological history in the Central-Eastern part of the African continent, it appears that tectonic deformation has been almost continuous. Only minor periods of tectonic quiescence, when deformation was replaced by intense volcanic activity, occur. Broad areas in Central-Eastern Africa were occupied by Late Palaeozoic to Mesozoic rift basins, largely exceeding the size of the present East African Rift System. More than ten wide half-graben basins developed in Eastern-Southern Africa during Permo-Trias time. Giant fluvial systems alternating with shallow/playa or deep lakes occupied these basins from ~280 Ma up to ~200 Ma, accumulating several km of sediments, mainly fluvial and deltaic sandstones, lacustrine mudstones and floodplain and playa lake shales over a 80 Ma-long period, two-times longer than the duration of the Cenozoic East African Rift System. From Jurassic time, tectonic inversion, rift abortion or major flood volcanism marked the end of lakes and fluvial systems in this area.

From the Late Jurassic at ~160 Ma, a second major phase of crustal extension resulted further North, in Central-Eastern Africa, in the development of other large continental rift systems forming the Central African Rift System. It has occurred over a period of 100 Ma, nearly three times longer than the East African Rift System. Several lakes up to 200 km long occupied more or less synchronously a series of parallel half-grabens in South Sudan, or developed along the 500-km-long Anza Rift, inter-fingering with complex fluvial systems. More than 2 km of lacustrine shales accumulated in these basins during a 10 Ma-long major phase of lacustrine development. These processes and period of time are comparable to the duration of initation and development of the Lake Tanganyika rift basin.

From Early Tertiary (~40 Ma), rift tectonics continued to affect parts of the Sudan and Anza Rifts, and migrated southward and westward to form the earliest half-grabens/lake basins of the Cenozoic East African Rift System. Today, only a part of the present-day Lake Turkana and the small Baringo Lake can be considered as remnants of these earliest basins. Within a time period about 40 Ma-long, more than 30 lakes formed and fluctuated according to the volcano-tectonic evolution of the rift and climatic changes at a regional and global scale. The oldest and largest lakes of the East African Rift System belong to the Western Branch, while in the Eastern Branch and the Main Ethiopian Rift, recent volcanic products in the form of lava flows or wide axial volcanoes occupy large areas of the rift floor, allowing very small accommodation space for lake environments to develop. The small lakes of the

Eastern Branch/Main Ethiopian Rift fluctuated many times during Pleistocene-Holocene times, often reaching the desiccation term, or overflowing beyond high relief accommodation zones separating smaller lake subbasins and thus forming quite large and deep lakes. Today those large or small, freshwater or saline lakes of Eastern and Central Africa are strongly suffering from anthropogenic activities such as watershed deforestation or water pumping for intensive irrigation, that will irreparably force them to fill completely or to dry out.

At the present day, rifting processes are continuing all along the rift system, maintaining a strong tectonic control on sediment transfer toward the lacustrine basins. Two new NE-SW-trending rift branches are presently forming perpendicular to the Western Branch. Young sedimentary basins, occupied by the Mweru and Mweru Wantipa Lakes as well as by the large intracontinental Okavango Delta, develop in this tectonically active context that represents a perfect example of an intracontinental rift in its early stage of evolution. Beyond the tectonically induced changes to the lakes and their catchments, that occur at a time scale of more than $10^4$ years, more redoubtable are the anthropogenic effects that have already modifed the life of certain lakes like Lake Tanganyika or Lake Victoria during the last 100 years.

## ACKNOWLEDGEMENTS

The authors wish to thank E. Odada, D. Olago and T.C. Johnson for their kind invitation to contribute to this volume on East African Great Lakes. Special thanks are due to C. Ebinger and one anonymous reviewer for very stimulating scientific suggestions, and aid in improving the text. Contribution of B. Coléno for drafting and editing was greatly appreciated.

## REFERENCES

Alin, S.R., Cohen, A.S., Bills, R., Gashagaza Masta Mukwaya, Michel, E., Tiercelin, J.-J., Martens, K., Coveliers, P., Mboko Sima Keita, West, K., Soreghan, M., Kimbadi Sona and Ntakimazi, G. (1999) Effects of landscape disturbance on animal communities in Lake Tanganyika, East Africa, *Conservation Biology* **13**, 1017-1033.

Allix, P. and Popoff, M. (1983) Le Crétacé inférieur de la partie nord-orientale du fossé de la Bénoué (Nigeria) : un exemple de relation étroite entre tectonique et sédimentation, in M. Popoff and J.-J. Tiercelin (eds.) Rifts et Fossés anciens, *Bull. Centres Rech. Explor.-Prod. Elf Aquitaine* **7**, 349-359.

Andrews, P. and Banham, P. (eds.) (1999) Late Cenozoic Environments and Hominid Evolution: a tribute to Bill Bishop, Geological Society, London.

Arambourg, C. (1935) Esquisse géologique de la bordure occidentale du Lac Rodolphe. Mission Scient. Omo 1932-3, *Mus. Natn. Hist. Nat. Paris* **I (1)**, 59 pp.

Arambourg, C. (1943) Contribution à l'étude géologique et paléontologique du Bassin du lac Rodolphe et de la Basse Vallée de l'Omo. Mission Scientifique de l'Omo, *Mus. Natn. Hist. Nat. Paris* **I 1(2)**, 157-230.

Arambourg, C. and Wolff, R.G. (1969) Nouvelles données paléontologiques sur l'âge des grès du Lubur (Turkana grits) à l'Ouest du lac Rodolphe, *C. R. Soc. géol. France* **6**, 190-202.

Aronson, J.L. Schmitt, T.J., Walter, R.C., Taieb, M., Tiercelin, J.-J., Johanson, D.C., Naeser, C.W. and Nairn, A.E.M. (1977) New geochronologic and paleomagnetic data for the hominid-bearing Hadar Formation of Ethiopia, *Nature* **267**, 323-327.

Baker, B.H. (1986) Tectonics and volcanism of the Southern Kenya Rift Valley and its influence on rift sedimentation, in L.E. Frostick, R.W. Renaut, I. Reid and J.-J. Tiercelin (eds.), Sedimentation in the African Rifts, *Geol. Soc. London Spec. Publ.* **25**, 45-57.

Baker, B.H. and Wohlenberg, J. (1971) Structure and evolution of the Kenya Rift Valley, *Nature* **229**, 538-542.

Baker, B.H. and Mitchell, J.G. (1976) Volcanic stratigraphy and geochronolgy of the Kedong-Olorgesailie area and the evolution of the south Kenya rift valley, *J. Geological Society, London* **132**, 467-484.

Baker, B.H., Williams, L. A. J., Miller, J.A. and Fitch, F.J. (1971) Sequence and geochronology of the Kenya Rift volcanics, *Tectonophysics* **11**, 191-215.

Baker, B.H., Mohr, P.A. and Williams, L.A.J. (1972) Geology of the Eastern Rift System of Africa, *Geol. Soc. Am. Spec. Paper* **136**, 67 p.

Baker, B.H., Crossley, R. and Goles, G.G. (1978) Tectonic and magmatic evolution of the southern part of the Kenya rift valley, in E.R. Neumann and I.B. Ramberg (eds.), *Petrology and Geochemistry of Continental Rifts*, D. Reidel Publishing Co., Holland, pp. 29-50.

Baker, B.H., Mitchell, J.G. and Williams, L.A.J. (1988) Stratigraphy, geochronology and volcano-tectonic evolution of the Kedong-Naivasha-Kinangop region, Gregory Rift Valley, Kenya, *J. Geological Society, London* **145**, 107-116.

Baker, J., Snee, L. and Menzies, M. (1996) A brief Oligocene period of flood volcanism in Yemen: implications for the duration and rate of continental flood volcanism at the Afro-Arabian triple junction, *Earth Planet. Sci. Lett.* **138**, 39-55.

Ballard, S. and Pollack, H.N. (1988) Modern and ancient geotherms beneath Southern Africa, *Earth Planet. Sci. Lett.* **88**, 132-142.

Banks, N.L., Bardwell, K.A. and Musiwa, S. (1995) Karoo Rift basins of the Luangwa Valley, Zambia, in J.J. Lambiase (ed.), Hydrocarbon Habitat in Rift Basins, *Geological Society Special Publication* **80**, 285-295.

Beadle, L.C. (1981) *The Inland Waters of Tropical Africa*, London, Longman, 433 pp.

Benkhelil, J. and Robineau, B. (1983) Le fossé de la Bénoué est-il un rift? in M. Popoff and J.-J. Tiercelin (eds.) Rifts et Fossés anciens, *Bull. Centres Rech. Explor.-Prod. Elf Aquitaine* **7**, 315-321.

Berhe, S.M., Desta, B., Nicoletti, M. and Teferra, M. (1987) Geology, geochronology and geodynamic implications of the Cenozoic magmatic province in W and SE Ethiopia, *J. Geological Society, London* **144**, 213-226.

Bermingham, P.M., Fairhead, J.D. and Stuart, G.W. (1983) Gravity study of the Central African rift system: a model of continental disruption. 2. the Darfur domal uplift and associated Cenozoic volcanism, in P. Morgan (ed.), Processes of continental rifting, *Tectonophysics* **94**, 205-222.

Beuning, K.R.M., Talbot, M.R. and Kelts, K. (1997) A revised 30,000 year paleoclimatic and paleohydrologic history of Lake Albert, East Africa, *Palaeogeography, Palaeoclimatology, Palaeoecology* 136, 259-279.

Bishop, W.W. (1965) Quaternary Geology and Geomorphology in the Albertine Rift Valley, Uganda, *Geol. Soc. Am. Special Paper* **84**, 293-321.

Bishop, W.W. and Pickford, M. (1975) Geology, fauna and palaeoenvironments of the Ngorora Formation, Kenya Rift Valley, *Nature* **254**, 185-192.

Bishop, W.W. and Posnansky, M. (1960) Pleistocene environments and early man in Uganda, *Uganda Journal* **24**, 41-61.

Boccaletti, M., Mazzuoli, R., Bonini, M., Truad T. and Abebe, B. (1999) Plio-Quaternary volcanotectonic activity in the northern sector of the Main Ethiopian Rift: relationships with oblique rifting, *J. African Earth Sciences* **29 (4)**, 679-698.

Bonnefille, R., Vincens, A. and Buchet, G. (1987) Palynology, stratigraphy and palaeoenvironment of a Pliocene hominid site (2.9-3.3 m.y.) at Hadar, Ethiopia, *Palaeogeography, Palaeoclimatology, Palaeoecology* **60**, 249-281.

Boschetto, H.B. (1988) Geology of the Lothidok range, Northern Kenya, MS thesis, Dept. Geol. and Geophys., University of Utah.

Boschetto, H.B., Brown, F.H. and McDougall, I.M. (1992) Stratigraphy of the Lothidok Range, northern Kenya, and K/Ar ages of its Miocene primates, *J. Human Evolution* **22**, 47-71.

Bosworth, W. (1992) Mesozoic and early Tertiary rift tectonics in East Africa, *Tectonophysics* **209**, 115-137.

Bosworth, W. and Morley, C.K. (1994) Structural and stratigraphic evolution of the Anza rift, Kenya, in C. Prodehl, G.R. Keller and M.A. Khan (eds.), Crustal and Upper Mantle Structure of the Kenya Rift, *Tectonophysics* **236**, 93-115.

Bosworth, W. and Maurin, A. (1993) Structure, geochronology and tectonic significance of the northern Suguta Valley (Gregory Rift), Kenya, *J. Geol. Soc., London* **150**, 751-762.

Boven, A., Pasteels, P., Punzalan, L. E., Yamba, T. K. and Musisi., J. H. (1998) Quaternary perpotassic magmatism in Uganda (Toro-Ankole Volcanic Province): age assessment and significance for magmatic evolution along the East African Rift, *J. African Earth Sciences* **26**, 463-476.

Bromage, T. (1995) Paleobiogeography of the Malawi Rift: Vertebrate and invertebrate faunal contexts of the Chiwondo Beds, northern Malawi, *J. Human Evolution* **28**, 37-58.

Brooks, A. S., Helgren, D. M., Cramer, J. S., Franklin, A., Hornyak, W., Keating, J. M., Klein, R. G., Rink, W. J., Schwarcz, H., Smith, J. N. L., Stewart, K., Todd, N. E., Verniers, J. and Yellen, J. E. (1995) Dating and Context of Three Middle Stone Age Sites with Bone Points in the Upper Semliki Valley, Zaire, *Science* **268**, 548-553.

Brotzu, P., Morbidelli, L., Nicoletti, M., Piccirillo, E.M. and Traversa, G. (1984). Miocene to Quaternary volcanism in eastern Kenya: sequence and geochronology, *Tectonophysics* **101**, 75-86.

Brown, J.M. (1955) *A brief note on the thickness of the Albert rift sediments*. Report no.JMB/27, Geological Survey of Uganda.

Brown, F.H. and Feibel, C.S. (1988) "Robust" hominids and Plio-Pleistocene paleogeography of the Turkana Basin, Kenya and Ethiopia, in F.E. Grine (ed.), *Evolutionary history of the "Robust" Australopithecines*, Aldine de Gruyter, New York, pp. 325-341.

Brown, F. and Nash, W. (1976) Radiometric dating and tuff mineralogy of Omo Group deposits, in Y. Coppens *et al.* (eds.), *Stratigraphy, Paleoecology and Evolution in the Lake Rudolf Basin*, Univ. of Chicago Press, 50-63.

Browne, S.E. and Fairhead, J.D. (1983) Gravity study of the Central African Rift System: a model of continental disruption. 1. The Ngaoundere and Abu Gabra Rifts, in P. Morgan (ed.), Processes of continental rifting, *Tectonophysics* **94**, 187-203.

Burke, K. and Dewey, J.F. (1973) Two plates in Africa during the Cretaceous?, *Nature* 249, 313-316.

Camp, V.E. and Roobol, M.J. (1992) Upwelling asthenosphere beneath western Arabia and its regional implications. *J. Geophysical Research* **97**, 15255-15271.

Capart, A. (1949) Sondages et carte bathymétrique du Lac Tanganyika. Résultats scientifiques de l'exploration hydrobiologique du Lac Tanganyika (1946-1947), *Institut Royal des Sciences Naturelles de Belgique* **2**, 1-16.

Casanova, J. (1986) Les stromatolites continentaux : Paléoécologie, Paléohydrologie, Paléoclimatologie. Application au Rift Gregory, Thèse Etat Sciences, Univ. Aix-Marseille II, 256 pp.

Casanova, J. (1987) Stromatolites et hauts niveaux lacustres pléistocènes du bassin Natron-Magadi (Tanzanie-Kenya), *Sci. Géol. Bull.* **40**, 135-153.

Casanova, J. and Hillaire-Marcel, C. (1987) Chronologie et paléohydrologie des hauts niveaux quaternaires du bassin Natron-Magadi (Tanzanie-Kenya) d'après la composition isotopique ($^{18}O$, $^{13}C$, $^{14}C$, U/Th) des stromatolites littoraux, *Sci. Géol. Bull.* **40**, 121-134.

Casanova, J., Hillaire-Marcel, C., Page, N., Taieb, M. and Vincens, A. (1988) Stratigraphie et paléohydrologie des épisodes lacustres du Quaternaire récent du rift Suguta (Kenya), *C. R. Acad. Sci. Paris* **307**, 1251-1258.

Chapman, G.R. and Brook, M. (1978) Chronostratigraphy of the Baringo Basin, Kenya Rift Valley, in W.W. Bishop (ed.), *Geological Background to Fossil Man*, Geological Society, London, Scottish Academic, 207-223.

Chapman, G.R., Lippard, S.J. and Martyn, J.E. (1978) The stratigraphy and structure of the Kamasia Range, Kenya Rift Valley, *J. Geological Society, London* **135**, 265-281.

Chorowicz, J. (1989) Transfer and transform fault zones in continental rifts: examples in the Afro-Arabian Rift System. Implications of crust breaking, *J. African Earth Sciences* **8**, 203-214.

Chorowicz, J., Le Fournier, J., Le Mut, C., Richert, J.P., Spy-Anderson, F.L. and Tiercelin, J.-J. (1983) Observation par télédétection et au sol de mouvements décrochants NW-SE dextres dans le secteur transformant Tanagnyika-Rukwa-Malawi du rift Est-africain, *C. R. Acad. Sci. Paris* **296**, 997-1002.

Cochran, J. (1981) The Gulf of Aden: structure and evolution of a young ocean basin and continental margin, *J. Geophysical Research* **86**, 263-287.

Cohen, A. S., Soreghan, M. J. and Scholz, C. A. (1993) Estimating the age of formation of lakes: An example from Lake Tanganyika, East African Rift system, *Geology* **21**, 511-514.

Cohen, A. S., Lezzar, K. E., Tiercelin, J.-J. and Soreghan, M. (1997) New palaeogeographic and lake level reconstructions of Northern Lake Tanganyika: Implications for tectonic, climatic and biologic evolution in a rift lake, *Basin Research* **9**, 107-132.

Cornelissen, E., Boven, A., Dabi, A., Hus, J., Ju Yong, K., Keppens, E., Langohr, R., Moeyersons, J., Pasteels, P., Pieters, M., Uytterschaut, H., Van Noten, F., Workineh, H. (1990) The Kapthurin Formation revisited, *Afr. Arch. Review* **8**, 23-75.

Coussement, C., Gente, P., Rolet, J., Tiercelin, J.-J., Wafula, M. and Buku, S. (1994) The North Tanganyika hydrothermals fields, East African Rift system. Their tectonic control and relationship to volcanism and rift segmentation, *Tectonophysics* **237**, 155-173.

Crossley, R. (1980) Structure and volcanism in the S. Kenya rift, in *Geodynamic evolution of the Afro-Arabic Rift System*. Roma: Accademia Nazionale dei Lincei, Atti dei convegni Lincei **47**, 89-98.

Daly, M.C. (1988) Crustal Shear Zones in Central Africa: a Kinematic Approach to Proterozoic Tectonics, *Episodes* **11**, 5-11.

Daly, M.C., Chorowicz, J. and Fairhead, J.D. (1989) Rift basin evolution in Africa: the influence of reactivated steep basement shear zones, in M.A. Cooper and G.D. Williams (eds.), Inversion Tectonics, *Geol. Soc. London Spec. Publ.* **44**, 309-334.

Damblon, F., Gerrienne, Ph., D'Outrelepont, H., Delvaux, D., Beeckman, H. and Back, S. (1998) Identification of a fossil wood specimen in the Red Sandstone Group of southwestern Tanzania: stratigraphical and tectonic implications, *J. African Earth Sciences* **26**, 387-396.

Davidson, A. and Rex, D.C. (1980) Age of volcanism and rifting in southern Ethiopia, *Nature* **283**, 657-658.

Davies, K.A. (1951) The Uganda section of the western rift, *Geological Magazine* **88**, 377-385.

Dawson, J.B. (1992) Neogene tectonics and volcanicity in the northern Tanzania sector of the Gregory rift valley: Contrasts with the Kenya rift sector, *Tectonophysics* **204**, 81-92.

Degens, E.T., Deuser, W.G., Von Herzen, R.P., Wong, H.K., Wooding, F.B., Jannasch, H.W. and Kanwisher, J.W. (1971) Lake Kivu Expedition: Geophysics, hydrography, sedimentology, *Techn. Rep. Woods Hole Oceanogr. Inst. Ref.* **71-52**, 1-20.

Degens, E.T., Von Herzen, R.P. and Wong, H.K. (1973) Lake Tanganyika. Water chemistry, sediments, geological structure, *Naturwissenschaften* **58**, 229-241.

Delvaux, D., Levi, K., Kajara, R. and Sarota, J. (1992) Cenozoic paleostress and kinematic evolution of the Rukwa-North Malawi rift valley (East African Rift System), *Bull. Centres Rech. Explor.-Prod. Elf Aquitaine* **16**, 383-406.

Delvaux, D., Kervyn, F., Vittori, E., Kajara, R.S.A. and Kilembe, E. (1998) Late Quaternary tectonic activity and lake level change in the Rukwa Basin, *J. African Earth Sciences* **26**, 397-421.

Demant, A., Lestrade, P., Lubala, R. T, Kampunzu, A. L. and Durieux, J. (1994) Volcanological and petrological evolution of Nyiragongo volcano, Virunga volcanic field, Zaire, *Volcanology* **56**, 47-61.

DeMenocal, P.B., Ruddiman, W.F. and Pokras, E.M. (1993) Influences of high- and low latitude processes on African terrestrial climate: Pleistocene eolian records from equatorial Atlantic Ocean Drilling Program Site 663, *Paleoceanography* **8**, 209-242.

Dingle, R.V., Siesser, W.G. and Newton, A.R. (1983) *Mesozoic and Tertiary Geology of Southern Africa*, Balkeme, Rotterdam, 375 pp.

Dixey, F. (1928) The Dinosaur Beds of Lake Nyasa, *Transactions of the Geological Society of South Africa* **16**, 55-66.

Dodson, R.G. (1971) Geology of the area south of Lodwar. *Rep. Geol. Surv. Kenya* **87**, 36 pp.

Duncan, R.A., Hooper, P.R., Rehacek, J., Marsh, J.S. and Duncan, A.R. (1997) The timing and duration of the Karoo igneous event, southern Gondwana, *J. Geophysical Research* **102**, 18,127-18,138.

Dunkelman, T.J., Karson, J.A. and Rosendahl, B.R. (1988) Structural style of the Turkana Rift, Kenya, *Geology* **16**, 258-261.

Dunkelman, T.J., Rosendahl, B.R. and Karson, J.A. (1989) Structure and stratigraphy of the Turkana rift from seismic reflection data, *J. African Earth Sciences* **8**, 489-510.

Dunkley, P.N., Smith, M., Allen, D.J. and Darling, W.G. (1993) The geothermal activity and geology of the northern sector of the Kenya Rift Valley, Bristish Geological Survey Research Report SC/93/1, 185 pp.

Dypvik, H., Nesteby, H., Ruden, F., Aagaard, P., Johansson, T., Msindai, J. and Massay, C. (1990) Upper Paleozoic and Mesozoic sedimentation in the Rukwa-Tukuyu Region, Tanzania, *J. African Earth Sciences* **11**, 437-456.

Ebinger, C.J. (1989a) Geometric and kinematic development of border faults and accommodation zones, Kivu-Rusizi rift, Africa, *Tectonics* **8**, 117-133.

Ebinger, C. J. (1989b) Tectonic development of the western branch of the East African rift system, *Bull. Geol. Soc. Amer.* **101**, 885-903.

Ebinger, C.J. and Ibrahim, A. (1994) Mutiple episodes of rifting in Central and East Africa: A re-evaluation of gravity data, *Geol. Rundsch.* **83**, 689-702.

Ebinger, C.J. and Sleep, N. (1998) Cenozoic magmatism throughout East Africa resulting from impact of a single plume, *Nature* **395**, 788-791.

Ebinger, C.J., Crow, M.J., Rosendahl, B.R., Livingstone, D.A. and Le Fournier, J. (1984) Structural evolution of Lake Malawi, Africa, *Nature* **308**, 627-629.

Ebinger, C.J., Rosendahl, B.R. and Reynolds, D.J. (1987) Tectonic model of the Malawi rift, Africa, *Tectonophysics* **141**, 215-235.

Ebinger, C. J., Deino, A. L., Drake, R. E. and Tesha, A. L. (1989) Chronology of Volcanism and Rift Basin Propagation: Rungwe Volcanic Province, East Africa, *J. Geophysical Research* **94**, 15785-15803.

Ebinger, C.J., Yemane, T., Woldegabriel, G., Aronson, J.L. and Walter, R.C. (1993b) Late Eocene-Recent volcanism and faulting in the southern main Ethiopian rift, *J. Geological Society, London* **150**, 99-108.

Ebinger, C.J., Deino, A.L., Tesha, A.L., Becker, T. and Ring, U. (1993c) Tectonic controls on rift basin morphology: evolution of the Northern Malawi (Nyasa) Rift, *J. Geophysical Research* **98**, 821-836.

Ebinger, C.J., Yemane, T., Harding, D.J., Tesfaye, S., Kelley, S. and Rex, D.C. (2000) Rift deflection, migration, and propagation: Linkage of the Ethiopian and eastern rifts, Africa, *Geol. Soc. Am. Bull.* **112**, 163-176.

Ego, J.K. (1994) Sedimentology and diagenesis of Neogene sediments in the central Kenya Rift Valley, MSc thesis, University of Saskatchewan.

Eugster, H.P. (1980) Lake Magadi, Kenya and its precursors, in A. Nissenbaum (ed.), *Hypersaline Brines and Evaporites*, Developments in Sedimentology No. 28, Elsevier, Amsterdam, pp. 195-232.

Eugster, H.P. (1986) Lake Magadi, Kenya: a model for rift valley hydrochemistry and sedimentation, in L.E. Frostick, R.W. Renaut, I. Reid and J.-J. Tiercelin (eds.), Sedimentation in the African Rifts, *Geol. Soc. London Spec. Publ.* **25**, 177-189.

Fairhead, J.D. (1986) Geophysical controls on sedimentation within the African Rift Systems, in L.E. Frostick, R.W. Renaut, I. Reid and J.-J. Tiercelin (eds.), Sedimentation in the African Rifts, *Geol. Soc. London Spec. Publ.* **25**, 19-27.

Fairhead, J.D. (1988) Late Mesozoic rifting in Africa, in W. Manspeizer (ed.) Triassic-Jurassic Rifting, Continental Breakup and the Origin of the Atlantic Ocean and Passive Margins, Part B, *Developments in Geotectonics 22*, Elsevier, 821-831.

Feibel, C.S. (1988) Paleoenvironments of the Koobi Fora Formation, Turkana Basin, Northern Kenya, PhD Thesis, Utah University.

Feibel, C.S., Harris, J.M. and Brown, F.H. (1991) Neogene paleoenvironments of the Turkana Basin, in J.M. Harris (ed.), Koobi Fora Research Project, Volume 3, Stratigraphy, Artiodactyls and Paleoenvironments, Oxford, Clarendon Press, pp. 321-346.

Foster, A., Ebinger, C., Mbede, E. and Rex, D. (1997) Tectonic development of the northern Tanzanian sector of the East African Rift System, *J. Geological Society, London* **154**, 689-700.

Fourmarier, P. (1935) Note au sujet du bassin charbonnier du Tanganyika, *Inst. Roy. Colonial Belge* BVI, 486-493.

Garfunkel, Z. and Bartov, Y. (1977) The tectonics of the Suez rift, *Geol. Surv. Isr. Bull.* **71**, 44 pp.

Gasse, F. (1977) Evolution of Lake Abhé (Ethiopia and TFAI), from 70,000 b.p., *Nature* **265**, 42-45.

Gasse, F. and Fontes, J.C. (1989) Palaeoenvironments and palaeohydrology of a tropical closed lake (Lake Asal, Djibouti) since 10,000 yr B.P., *Palaeogeography, Palaeoclimatology, Palaeoecology* **69**, 67-102.

Gasse, F. and Street, A. (1978) Late Quaternary lake-level fluctuations and environments of the northern Rift Valley and Afar region (Ethiopia and Djibouti). *Palaeogeography, Palaeoclimatology, Palaeoecology* **24**, 279-325.

Gasse, F., Richard, O., Robbe, D., Rognon, P. and Williams, M.A.J. (1980). Evolution tectonique et climatique de l'Afar central d'après les sédiments plio-pléistocènes, *Bull. Soc. géol. France* **6**, 987-1001.

Genik, G.J. (1992) Regional framework and structural aspects of rift basins in Niger, Chad and the Central African Republic (C.A.R.), in P.A. Ziegler (ed.), Geodynamics of Rifting, Volume II. Case History Studies on Rifts: North and South America and Africa, *Tectonophysics* **213**, 169-185.

George, R., Rogers, N. and Kelley, S. (1998) Earliest magmatism in Ethiopia: Evidence for two mantle plumes in one flood basalt province, *Geology* **26**, 923-926.

Gibert, E., Bergonzini, L., Massault, M. and Williamson, D. (in press) AMS-$^{14}$C chronology of 40.0 cal. kyr BP continuous deposits from a crater lake (Lake Massoko, Tanzania): Modern water balance and environmental implications, *Palaeogeography, Palaeoclimatology, Palaeoecology Special Issue*, 2nd International Congress of Limnogeology, Plouzané, May 1999.

Girdler, R.W., Fairhead, J.D., Searle, R.C. and Sowerbutts, W.T.C. (1969) Evolution of rifting in Africa, *Nature* **224**, 1178-1182.

Greene, L.C., Richards, D.R. and Johnson, R.A. (1991) Crustal structure and tectonic evolution of the Anza rift, northern Kenya, *Tectonophysics* **197**, 203-211.

Guiraud, R. and Maurin, J.C. (1992) Early Cretaceous rifts of Western and Central Africa: an overview, in P.A. Ziegler (ed.), Geodynamics of Rifting, Volume II. Case History Studies on Rifts: North and South America and Africa, *Tectonophysics* **213**, 153-158.

Guiraud, R., Binks, R.M., Fairhead, J.D. and Wilson, M. (1992) Early Cretaceous Rifts of Western and Central Africa, in P.A. Ziegler (ed.), Geodynamics of Rifting, Volume II. Case History Studies on Rifts: North and South America and Africa, *Tectonophysics* **213**, 227-234.

Hall, C.M., Walter, R.C., Westgate, J.A. and York, D. (1984) Geochronology, stratigraphy and geochemistry of Cindery Tuff in Pliocene hominid-bearing sediments of the Middle Awash, Ethiopia, *Nature* **308**, 26-31.

Hankel, O. (1987) Lithostratigraphic subdivision of Karoo rocks of the Luwegu Basin, Tanzania and their biostratigraphic classification based on micro-macroflora, fossil wood and vertebrates, *Geologische Rundschau* **76**, 539-565.

Harper, R.M., Stone, D.M. and Morley, C.K. (1999) Geophysics of the Usangu Flats, Tanzania, in C.K. Morley (ed.), *Geoscience of Rift Systems-Evolution of East Africa*, AAPG Studies in Geology **44**, 111-114.

Harris, N., Pallister, J.W. and Brown, J.M. (1956) Oil in Uganda, *Geol. Surv. Uganda* **9**, 33 pp.

Hautot, S., Tarits, P., Whaler, K., Le Gall, B., Tiercelin, J.-J. and Le Turdu, C. (2000) Deep structure of the Baringo Rift Basin (Central Kenya) from three-dimensional magnetotelluric imaging: Implications for rift evolution, *J. Geophys. Res.* **105**, 493-518.

Hay, R.L. (1987) Geology of the Laetoli area, in M.D. Leakey and J.M. Harris (eds.), *Laetoli, a Pliocene site in northern Tanzania*, Clarendon Press, p. 23-47.

Hecky, R.E. and Degens, E.T. (1973) Late Pleistocene-Holocene chemical stratigraphy and paleolimnology in the Rift Valley Lakes of Central Africa, *Technical Report, Woods Hole Oceanographic Institution* **1-73-28**, 1-144 p.

Hendrie, D.B., Kusznir, N.J., Morley, C.K. and Ebinger, C.J. (1994) Cenozoic extension in northern Kenya: a quantitative model of rift basin development in the Turkana region, in C. Prodehl, G.R. Keller and M.A. Khan (eds.), Crustal and Upper Mantle Structure of the Kenya Rift, *Tectonophysics* **236**, 409-438.

Hill, A. (1985) Early hominid from Baringo, Kenya, *Nature* **315**, 222-224.

Hill, A. (1995) Faunal and environmental change in the Neogene of East Africa: evidence from the Tugen Hills sequence, Baringo District, Kenya, in E.S. Vrba, G.H. Denton, T.C. Partridge and L.H. Burckle (eds.), *Paleoclimate and Evolution, with Emphasis on Human Origins*, Yale University, New Haven, 178-193.

Hill, A. (1999) The Baringo Basin, Kenya: from Bill Bishop to BPRP, in P. Andrews and P. Banham (eds.), *Late Cenozoic Environments and Hominid Evolution: a tribute to Bill Bishop*, Geological Society, London, 85-97.

Hill, A., Curtis, G. and Drake, R. (1986) Sedimentary stratigraphy of the Tugen Hills, Baringo, Kenya, in L.E. Frostick, R.W. Renaut, I. Reid and J.-J. Tiercelin (eds.), Sedimentation in the African Rifts, *Geol. Soc. London Spec. Publ.* **25**, 285-295.

Hofmann, C., Courtillot, V., Féraud, G., Rochette, P., Yirgu, G., Ketefo, E. and Pik, R. (1997) Timing of the Ethiopian flood basalts event and implications for plume birth and global change, *Nature* **389**, 838-841.

Hopwood, A.T. and Lepersonne, J. (1953) Présence de formations d'âge Miocène inférieur dans le fossé tectonique du lac Albert et de la Basse Semliki (Congo Belge), *Ann. Soc. Géol. Belgique* **77**, 83-113.

Jacobs, B.F. and Kabuye, C.H.S. (1987) A middle Miocene (12.2 Ma) forest in the East African Rift Valley, Kenya, *J. Human Evolution* **6**, 147-155.

Jacobs, B.F. and Deino, A.L. (1996) Test of climate-leaf physionomy regression models, their applications to two Miocene floras from Kenya, and $^{40}Ar/^{39}Ar$ dating of the late Miocene Kapturo site, *Palaeogeography, Palaeoclimatology, Palaeoecology* **123**, 259-271.

Johanson, D.C., Taieb, M. and Coppens, Y. (1982) Pliocene hominids from the Hadar Formation, Ethiopia (1973-1977): stratigraphic, chronologic, and paleoenvironmental contexts, with notes on Hominid morphology and systematics, *Am. J. Phys. Anthropol.* **57**, 373-402.

Jorgensen, G.J. and Bosworth, W. (1989) Gravity modelling in the Central African Rift System, Sudan: rift geometries and tectonic significance, *J. African Earth Sciences* **8**, 283-306.

Kaaya, C.Z. (1992) Depositional environment of Late Permian Karoo beds in the Ruhuhu Basin and Mikumi area of Tanzania, *Geologisches Institut der Universität zu Köln, Sonderveröffentlichungen* **83**, 156 pp.

Kalb, J.E., Wood, C.B., Smart, C., Oswald, E.B., Assefa Mabrete, Sleshi Tebedge and Whitehead, P. (1980) Preliminary geology and palaeontology of the Bodo D'Ar hominid site, Afar, Ethiopia, *Palaeogeography, Palaeoclimatology, Palaeoecology* **30**, 107-120.

Kamau, C. (1974) The Lake Naivasha Basin, Report, Geography Department, Nairobi University.

Kampunzu, A.B., Bonhomme, M.G. and Kanika, M. (1998) Geochronology of volcanic rocks and evolution of the Cenozoic Western Branch of the East African Rift System, *J. African Earth Sciences* **26**, 441-461.

Karner, G.D., Byamungu, B.R., Ebinger, C.J., Kampunzu, A.B., Mukasa, R.K., Nyakaana, J., Rubondo, E.N.T. and Upcott, N.M. (2000) Distribution of crustal extension and regional basin architecture of the Albertine rift system, East Africa, *Marine and Petroleum Geology* **17**, 1131-1150.

Kilembe, E.A. and Rosendahl, B.R. (1992) Structure and stratigraphy of the Rukwa rift, in C.J. Ebinger, H.K. Gupta and I.O. Nyambok (eds.), Seismology and related Sciences in Africa, *Tectonophysics* **209**, 143-158.

Kingston, J.D. (1999) Environmental determinants in early hominid evolution: issues and evidence from the Tugen Hills, Kenya, in P. Andrews and P. Banham (eds.), *Late Cenozoic Environments and Hominid Evolution: a tribute to Bill Bishop*, Geological Society, London, 69-84.

Kreuser, T. (1983) Stratigraphie der Karoo-Becken in Ost-Tansania, *Geologisches Institut der Universität zu Köln, Sonderveröffentlichungen* **45**, 212 pp.

Kreuser, T. (1990) Permo-Trias im Ruhuhu-Becken und anderen Karoo-Becken von SE-Africa, *Geologisches Institut der Universität zu Köln, Sonderveröffentlichungen* **75**, 132 pp.

Kreuser, T. (1995) Rift to drift evolution in Permian-Jurassic basins of East Africa, in J.J. Lambiase (ed.), Hydrocarbon Habitat in Rift Basins, *Geol. Soc. London Spec. Publ.* **80**, 297-315.

Laerdal, T. and Talbot, M.R. (in press) Structure and neotectonics in the Edward and George Basins, Uganda-Congo, East Africa, *Palaeogeography, Palaeoclimatology, Palaeoecology Special Issue*, 2nd International Congress of Limnogeology, Plouzané, May 1999.

Lambiase, J.J. (1989) The framework of African rifting during the Phanerozoic, *J. African Earth Sciences* **8**, 183-190.

Laury, R.L. and Albritton, C.C. Jr. (1975) Geology of Middle Stone Age archaeological sites in the Main Ethiopian Rift Valley, *Geol. Soc. Am. Bull.* **86**, 999-1011.

Le Gall, B., Tiercelin, J.-J., Richert, J.P., Gente, P., Sturchio, N.C., Stead, D., Le Turdu, C. (2000). A morphotectonic study of an extensional fault zone in a magma-rich rift: the Baringo Trachyte Fault System, central Kenya Rift, *Tectonophysics* **320**, 87-106.

Le Gall, B, Féraud, G., Tshoso, G., Kampunzu, A.B., Tiercelin, J.-J., Bertrand, H., Modisi, M., Dyment, J. and Maia M. The Okavango giant dike swarm of northern Botswana: New $^{39}Ar$-$^{40}Ar$ radiometric dating and structural implications for Gondwana breakup, submitted to *Tectonophysics*.

Le Turdu, C., Tiercelin, J.-J., Gibert, E., Travi, Y., Lezzar, K.-E., Richert, J.P., Massault, M., Gasse, F., Bonnefille, R., Decobert, M., Gensous, B., Jeudy, V., Tamrat, E., Umer Mohammed, M., Martens, K., Atnafu, B., Chernet, T., Williamson, D. and Taieb, M. (1999) The Ziwai-Shala lake basin system, Main Ethiopian Rift: Influence of volcanism, tectonics, and climatic forcing on basin formation and sedimentation, *Palaeogeography, Palaeoclimatology, Palaeoecology* **150**, 135-177.

Lézine, A.M. and Casanova, J. (1991) Correlated oceanic and continental records demonstrate past climate and hydrology of North Africa (0-140 ka), *Geology* **19**, 307-310.

Lezzar, K.E. (1997) Evolution du Bassin Nord du Lac Tanganyika, Rift Est-Africain, depuis le Miocène Supérieur. Analyse de la cinématique du rifting et des processus sédimentaires à partir de données de sismique réflexion, imagerie satellitaire et carottages, Thèse de Doctorat, Université de Bretagne Occidentale.

Lezzar, K.-E., Tiercelin, J.-J., De Batist, M., Cohen, A.S., Bandora, T., Van Rensbergen, P., Le Turdu, C., Mifundu, W. and Klerkx, J. (1996) New seismic stratigraphy and Late Tertiary history of the North Tanganyika Basin, East African Rift System, deduced from multichannel and high-resolution reflection seismic data and piston core evidence, *Basin Research* **8**, 1-28.

Lezzar, K.-E., Tiercelin, J.-J., Le Turdu, C., Cohen, A.S., Reynolds D. J., Le Gall, B. and Scholz, C. (in press). Control of normal fault interaction on the distribution of major Neogene sedeimentary depocenters, Lake Tanganyika, East African Rift, in J. Underhill (ed.), The Structure and Sequence Stratigraphy of Rift Systems, *AAPG Memoir*.

Lippard, S.J. (1973) Plateau phonolite lava flows, Kenya, *Geological Magazine* **110**, 543-549.

Maglio, V.J. (1969) A shovel-tusked Gomphotere from the Miocene of Kenya, *Mus. Comp. Zool. Breviora* **310**, 1-10.

Manighetti, I., Tapponier, P., Courtillot, V., Gruszow, S. and Gillot, P.-Y. (1998) Propagation of rifting along the Arabia-Somalia plate boundary, *J. Geophys. Res.* **102**, 2681-2710.

Martyn, J.E. and Tobias, P.V. (1967) Pleistocene deposits and new fossil localities in Kenya, *Nature* **215**, 476-480.

McConnell, R.B. (1950) Outline of the geology of Ufipa and Ubende, *Geological Surv. Tanganyika Bull.* **19**, 62 p.

McConnell, R.B. (1972) Geological development of the Rift System of Eastern Africa, *Bull. geol. Soc. Am.* **83**, 2549-2572.

McDougall, I. and Feibel, C.S. (1999) Numerical age control for the Miocene-Pliocene succession at Lothagam, a hominoid-bearing sequence in the northern Kenya Rift, *J. Geological Society, London* **156**, 731-745.

McHargue, T.R., Heidrick, T.L. and Livingston, J.E. (1992) Tectonostratigraphic development of the Interior Sudan rifts, Central Africa, in P.A. Ziegler (ed.), Geodynamics of Rifting, Volume II. Case History Studies on Rifts: North and South America and Africa, *Tectonophysics* **213**, 187-202.

Mead, J.G. (1975) A fossil beaked whale (Cetacea: Ziphiidae) from the Miocene of Kenya, *Jl. Palaeo.* **49**, 745-751.

Modisi, M.P., Atekwana, E.A., Kampunzu, A.B. and Ngwisanyi, T.H. (2000) Rift kinematics during the incipient stages of continental extension: evidence from the nascent Okavango rift basin, northwest Botswana, *Geology* **28**, 939-942.

Mohr, P.A. (1960) Report on a geological excursion through Southern Ethiopia, *Bull. Geophys. Observ., Addis Ababa University* **2**, 9-19.

Mohr, P.A. (1982) Musing on continental rifts, in G. Palmason (ed.), *Continental and Oceanic Rifts*, vol. 8, Geodynamics Series, pp. 293-309. American Geophysicists' Union, Washington, DC.

Mohr, P. (1987) Patterns of faulting in the Ethiopian rift valley, *Tectonophysics* **143**, 169-179.

Mohr, P., Mitchell, J.G. and Raynolds, R.G.H. (1980). Quaternay volcanism and faulting at O' A caldera, Central Ethiopia rift, *Bulletin Volcanologique* **43-1**, 173-189.

Mondeguer, A. (1991) Bassins sédimentaires en contexte extensif et décrochant: l'exemple du Complexe des Fossés Sud-Tanganyika, Rift Est-africain, Thèse Université de Bretagne Occidentale, Brest, 258 pp.

Mondeguer, A., Ravenne, C., Masse, P. and Tiercelin, J.-J. (1989) Sedimentary basins in an extension and strike-slip background: the "South Tanganyika troughs complex", East African Rift, *Bull. Soc. géol. France* **8**, 501-522.

Morley, C.K. (1988) Variable extension in Lake Tanganyika, *Tectonics* **7**, 785-801.

Morley, C.K. (1999) Marked along-strike variations in dip of normal faults - the Lokichar fault, N. Kenya rift: a possible cause for metamorphic core complexes, *J. Structural Geology* **21**, 479-492.

Morley, C.K., Wescott, W.A., Cunningham, S.M. and Harper, R.M. (1989) Recent exploration in the Lake Rukwa Area, East African Rift, 28th International Geological Congress, Abstracts, 2-462-463, Washington, D.C..

Morley, C.K., Nelson, R.A., Patton, T.L. and Munn, S.G. (1990). Transfer zones in the East African rift sytem and their relevance to hydrocarbon exploration in rifts, *Am. Assoc. Pet. Geol. Bull.* **74**, 1234-1253.

Morley, C.K., Wescott, W.A., Stone, D.M., Harper, R.M., Wigger, S.T. and Karanja, F.M. (1992) Tectonic evolution of the northern Kenyan Rift, *J. Geological Society, London* **149**, 333-348.

Morley, C.K., Bosworth, W., Day, R.A., Lauck, R., Bosher, R., Stone, D.M., Wigger, S.T., Wescott, W.A., Haun, D. and Bassett, N. (1999a) Geology and Geophysics of the Anza Graben, in C.K. Morley (ed.), *Geoscience of Rift Systems-Evolution of East Africa*, AAPG Studies in Geology **44**, 67-90.

Morley, C.K., Wescott, W.A., Harper, R.M. and Cunningam, S.M. (1999b) Geology and Geophysics of the Rukwa Rift, in C.K. Morley (ed.), *Geoscience of Rift Systems-Evolution of East Africa*, AAPG Studies in Geology **44**, 91-110.

Morley, C.K., Wescott, W.A., Stone, D.M., Harper, R.M., Wigger, S.T., Day, R.A. and Karanja, F.M. (1999c) Geology and Geophysics of the western Turkana basins, in C.K. Morley (ed.), *Geoscience of Rift Systems-Evolution of East Africa*, AAPG Studies in Geology **44**, 19-54.

Morley, C.K., Vanhauwaert, P. and De Batist, M. (2000) Evidence for high frequency cyclic fault activity from high resolution reflection survey, Rukwa Rift, Tanzania, *J. Geol. Soc., London* **157**, 983-994.

Morton, W.H., Rex, D.C., Mitchell, J.G. and Mohr, P. (1979) Riftward younging of volcanic units in the Addis Ababa region, Ethiopian Rift Valley, *Nature* **280**, 284-288.

Mugisha, F., Ebinger, C.J., Strecker, M. and Pope, D. (1997) Two-stage rifting in the Kenya Rift: implications for half-graben models, *Tectonophysics* **278**, 61-81.

Nilsen, O., Dypvik, H., Kaaya, C. and Kilembe, E. (1999) Tectono-sedimentary development of the (Permian) Karoo sediments in the Kilombero Rift Valley, Tanzania, *J. African Earth Sciences* **29**, 393-409.

Norton, I.O. and Sclater, J.G. (1979) A model for the evolution of the Indian Ocean and the breakup of Gonwanaland, *J. Geophys. Res.* **84**, 6803-6830.

Nusbaum, R.L., Girdler, R.W., Heirtzler, J.R., Hutt, D.J., Green, D., Millings, V.E., Schmoll, B.S. and Shapiro, J. (1993) The distribution of earthquakes and volcanoes along the East African Rift System, *Episodes* **16**, 427-432.

Pasteels, P., Villeneuve, M., De Paepe, P. and Klerkx, J. (1989) Timing of the volcanism of the southern Kivu province: implications for the evolution of the western branch of the East African Rift system, *Earth Planet. Sci. Lett.* **94**, 353-363.

Patterson, B., Behrensmeyer, A.K. and Sill, W.D. (1970) Geology and fauna of a new Pliocene locality in north-western Kenya, *Nature* **226**, 918-921.

Pickford, M. (1978a) Geology, palaeoenvironments and vertebrate faunas of the mid-Miocene Ngorora Formation, Kenya, in W.W. Bishop (ed.), *Geological Background to Fossil Man*, Geological Society, London, Scottish Academic, 237-262.

Pickford, M. (1978b) Stratigraphy and mammalian palaeontology of the late-Miocene Lukeino Formation, Kenya, in W.W. Bishop (ed.), *Geological Background to Fossil Man*, Geological Society, London, Scottish Academic, 263-278.

Pickford, M., Senut, B., Roche, H., Mein, P., Ndaati, G., Obwona, P. and Tuhumwire, J. (1989) Uganda Palaeontology Expedition : résultats de la deuxième mission (1987) dans la région de Kisegi-Nyabusosi (Bassin du Lac Albert, Ouganda), *C. R. Acad. Sci. Paris* **308**, 1751-1758.

Pickford, M., Senut, B., Tiercelin, J.-J., Kasande, R. and Obwona, P. (1990) Résultats de la troisième mission (1988) de l'Uganda Palaeontology Expedition: Régions de Nkondo-Sebugoro et de Hohwa (Bassin du Lac Albert, Ouganda), *C. R. Acad. Sci. Paris* **311**, 737-744.

Pierce, J. and Lipkov, L. (1988) Structural interpretation of the Rukwa Rift, Tanzania, *Geophysics* **53**, 824-836.

Reeves, C. (1972) Rifting in Kalahari?, *Nature* **237**, 95-96.

Reeves, C. (1978) A failed Gondwana spreading axis in southern Africa, *Nature* **273**, 222-223.

Reeves, C.V., Karanja, F.M. and Macleod, I.N. (1987) Geophysical evidence for a failed Jurassic Rift and triple junction in Kenya, *Earth Planet. Sci. Lett.* **81**, 229-311.

Renaut, R.W. (1982) Late Quaternary geology of the Lake Bogoria fault-trough, Kenya Rift Valley, Unpublished PhD thesis, University of London, 498 pp.

Renaut, R.W. and Tiercelin, J.-J. (1994) Lake Bogoria, Kenya Rift Valley-A sedimentological overview, *Sedimentology and Geochemistry of Modern and Ancient Saline Lakes*, SEPM Special Publication **50**, 101-123.

Renaut, R.W., Ego. J., Tiercelin, J.-J., Le Turdu, C. and Owen, R.B. (1999) Saline, alkaline palaeolakes of the Tugen Hills-Kerio Valley region, Kenya Rift Valley, in P. Andrews and P. Banham (eds.), *Late Cenozoic Environments and Hominid Evolution: a tribute to Bill Bishop*, Geological Society, London, 41-58.

Reynolds, C.S. (1980) Phytoplankton assemblages and their periodicity in stratifying lake systems, *Holartic Ecology* **3**, 141-159.

Reynolds, D. J. (1984) Structural and Dimensional Repetition in Continental Rifts, Master Thesis, Department of Geology, Duke University, USA.

Richardson, J.L. (1966) Changes in level of Lake Naivasha, Kenya, during postglacial times, *Nature* **209**, 290-291.

Richardson, J.L. and Richardson, A.E. (1972). History of an African rift lake and its climatic implications, *Ecological Monographs* **42**, 499-534.

Ring, U., Betzler, C. and Delvaux, D. (1992) Normal vs. strike slip faulting during rift development in East Africa: The Malawi Rift, *Geology* **20**, 1015-1018.

Roche, H. and Tiercelin, J.-J. (1977). Découverte d'une industrie lithique ancienne in-situ dans la formation d'Hadar, Afar Central, Ethiopie, *C. R. Acad. Sci. Paris* **3284**, 1871-1874.

Roche, H. and Tiercelin, J.-J. (1980) Industries lithiques de la formation Plio-Pléistocene d'Hadar : campagne 1976, in R.E. Leakey and B.A. Ogot (eds.), *Proc., VIIIth Panafrican Congress of Prehistory and Quaternary studies* (1977), pp. 194-199. Nairobi: The International Louis Leakey Memorial Institute of African Prehistory.

Roche, H., Delagnes, A., Brugal, J.-P., Feibel, C., Kibunjia, M., Mourre, V. and Texier, P.-J. (1999) Early hominid stone tool production and technical skill 2.34 Myr ago in West Turkana, Kenya, *Nature* **399**, 57-60.

Roessner, S. and Strecker, M.R. (1997) Late Cenozoic tectonics and denudation in the Central Kenya Rift: quantification of long-term denudation rates, in K. Fuchs, R. Altherr, B. Müller and C. Prodehl (eds.), Structure and dynamic processes in the lithosphere of the Afro-Arabian rift system, *Tectonophysics* **278**, 83-94.

Rosendahl, B.R. (1987) Architecture of continental rifts with special reference to East Africa, *Ann. Rev. Earth Planet. Sci.* **15**, 445-503.

Rosendahl, B.R., Reynolds, D.J., Lorber, P.M., Burgess, C.F., McGill, J., Scott, D., Lambiase, J.-J. and Derksen, S.J. (1986) Structural expressions of rifting: lessons from Lake Tanganyika, Africa, in L.E.

Frostick, R.W. Renaut, I. Reid and J.-J. Tiercelin (eds.), Sedimentation in the African Rifts, *Geol. Soc. London Spec. Publ.* **25**, 29-43.

Rosendahl, B. R., Versfelt, J. W., Scholz, C. A., Buck J. E. and Woods L. D. (1988) Seismic atlas of Lake Tanganyika, East Africa, Project PROBE Geophysical Atlas Series, folio 1, Durham, North Carolina, Duke University.

Rosendahl, B.R., Kilembe, E. and Kaczmarick, K. (1992) Comparison of the Tanganyika, Malawi, Rukwa and Turkana Rift zones from analyses of seismic reflection data, in P.A. Ziegler (ed.), Geodynamics of Rifting, Volume II. Case History Studies on Rifts: North and South America and Africa, *Tectonophysics* **213**, 235-256.

Savage, R.J.G. and Williamson, P.G. (1978) The early history of the Turkana depression, in W.W. Bishop (ed.), *Geological Background to Fossil Man*, Scottish Academic, Edinburgh, 375-394.

Scholz, C.A., Rosendahl, B.R., Versfelt, J.W. and Rach, N. (1990) Results of high-resolution echo-sounding of Lake Victoria, *J. African Earth Sciences* **11**, 25-32.

Scholz, C.A., Johnson, T.C., Cattaneo, P., Malinga, H. and Shana, S. (1998) Initial results of 1995 IDEAL seismic reflection survey of Lake Victoria, Uganda and Tanzania, in J.T. Lehman (ed.), *Environmental Change and Response in East African Lakes*, Kluwer, Dordrecht, 47-58.

Schull, T.J. (1984) Oil exploration in the non-marine rift basins of interior Sudan, *Can. Soc. Petrol. Geol. Reservoir* **12(8)**, 1-2.

Schull, T.J. (1988) Rift basins of interior Sudan: petroleum exploration and discovery, *American Assoc. Petrol. Geol. Bull.* **72**, 1128-1142.

Shudofsky, G. N. (1985) Source mechanisms and focal depths of East African earthquakes using Rayleigh-wave inversion and body-wave modelling, *Geophys. J. Roy. Astr. Society* **83**, 563-614.

Sickenberg, A. and Schönfeld, M. (1975) The Chorora Formation. Lower Pliocene limnical sediments in the southern Afar (Ethiopia), in A. Pilger and A. Rösler (eds.), *Afar Depression of Ethiopia*, Stuttgart (Schweizerbart), pp. 277-284.

Simpson, E. S. W., Sclater, J. G., Parsons, B., Norton, I., Meinke, L. (1979) Mesozoic magnetic lineations in the Mozambique Basin, *Earth Planet. Sci. Lett.* **43**, 260-264.

Soreghan, M.J., Scholz, C.A. and Wells, J.T. (1999) Coarse-grained, deep-water sedimentation along a border fault margin of Lake Malawi, Africa: Seismic stratigraphic analysis, *J. Sedimentary Research* **69**, 832-846.

Spence, J. (1954) The geology of the Galula Coalfield, Mbeya District, *Tanganyika Geol. Surv. Dep. Bull.* **25**, 34 pp.

Street, F.A. (1979) Late Quaternary lakes in the Ziway-Shala Basin, Southern Ethiopia, Unpublished PhD thesis, University of Cambridge, 475 pp.

Summerhayes, C.P. (1988) Prediicting palaeoclimates, in A.J. Fleet, K. Kelts and M.R. Talbot (eds.), Lacustrine Petroleum Source Rocks, *Geol. Soc. London Spec. Publ.* **40**, 77-78.

Talbot, M.R. and Laerdal, T. (2000) The Late Pleistocene-Holocene palaeolimnology of Lake Victoria, east Africa, based upon elemental and isotopic analyses of sedimentary organic matter, *J. Paleolimnology* **23**, 141-164.

Tallon, P.W.J. (1976) *The stratigraphy, palaeoenvironments and geomorphology of the Pleistocene Kapthurin Formation, Kenya*, PhD thesis, University of London.

Tallon, P.W.J. (1978) Geological setting of hominid fossils and Acheulian artefacts from the Kapthurin Formation, Baringo District, Kenya, in W.W. Bishop (ed.), *Geological Background to Fossil Man*, Geological Society, London, Scottish Academic, 361-378.

Thevenon F., Williamson, D. and Taieb, M. (in press) A 22 kyr BP sedimentological record of Lake Rukwa (8° S, SW Tanzania): environmental, chronostratigraphic and climatic implications, *Palaeogeography, Palaeoclimatology, Palaeoecology Special Issue*, 2nd International Congress of Limnogeology, Plouzané, May 1999.

Thompson, A.O. and Dodson, R.G. (1963) Geology of the Naivasha area, *Geological Survey of Kenya Report* **55**.

Tiercelin, J.-J. (1981) Rifts continentaux. Tectonique, climats, sédiments. Exemples: la sédimentation dans le nord du Rift Gregory (Kenya) et dans le Rift de l'Afar (Ethiopie) depuis le Miocène, Thèse Doct. Etat, Univ. Aix-Marseille II, 260 pp.

Tiercelin, J.-J. (1986) The Pliocene Hadar Formation, Afar Depression of Ethiopia, in L.E. Frostick, R.W. Renaut, I. Reid and J.-J. Tiercelin (eds.), Sedimentation in the African Rifts, *Geol. Soc. London Spec. Publ.* **25**, 253-265.

Tiercelin, J.-J. and Mondeguer, A. (1991) The geology of the Tanganyika Trough, East African Rift, in G. W. Coulter (ed.), *Lake Tanganyika and its Life*, British Museum Natural History Publications and Oxford University Press, p. 7-48.

Tiercelin, J.-J. and Vincens, A. (coordonateurs) (1987) Le demi-graben de Baringo-Bogoria, Rift Gregory, Kenya. 30.000 ans d'histoire hydrologique et sédimentaire, *Bull. Centres Rech. Explor.-Prod. Elf-Aquitaine* **11**, 249-540.

Tiercelin, J.-J., Michaux, J. and Bandet, Y. (1979) Le Miocène supérieur du Sud de la dépression de l'Afar, Ethiopie : sédiments, faunes, âges isotopiques, *Bull. Soc. géol. France* **(7)**, 255-258.

Tiercelin, J.-J., Chorowicz, J., Bellon, H., Richert, J.P., Mwanbene, J.T. and Walgenwitz, F. (1988) East african rift system: offset, age and tectonic significance of the Tanganyika-Rukwa-Malawi intracontinental transcurrent fault zone, *Tectonophysics* **148**, 241-252.

Tiercelin, J.-J., Cohen, A.S., Soreghan, M.J. and Lezzar, K.E. (1994) Pleistocene-Modern deposits of the Lake Tanganyika Rift Basin, East Africa: A modern analog for lacustrine source rocks and reservoirs, in A.J. Lomando, B.C. Schreiber and P.M. Harris (eds.), *Lacustrine Reservoirs and Depositional Systems*, SEPM Core Workshop **19**, 37-59.

Tiercelin, J.-J., Le Fournier, J., Herbin, J.P. and Richert, J.P. (1980) Continental rifts: modern sedimentation, tectonic and volcanic control. Example from the Bogoria-Baringo graben, Gregory Rift, Kenya, in *Geodynamic evolution of the Afro-Arabic Rift System*, Roma: Accademia Nazionale dei Lincei, Atti dei convegni Lincei **47**, 143-164.

Tiercelin, J.-J., Pflumio, C., Castrec, M., Boulègue, J., Gente, P., Rolet, J., Coussement, C., Stetter, K. O., Huber, R., Buku, S. and Mifundu, W. (1993) Hydrothermal vents in Lake Tanganyika, East African Rift system, *Geology* **21**, 499-502.

Tiercelin, J.-J., Potdevin, J.L., Le Gall, B., Talbot, M.R., Rio, A., Le Hérissé, A. and Brosse, E. Lacustrine source rocks and fluvio-lacustrine reservoirs in the Cretaceous to Tertiary Turkana Depression, northern Kenya Rift: Petrology and diagenesis, submitted to *Am. Assoc. Pet. Geol. Bull.*.

Ukstins, I., Renne, P., Wolfenden, E., Baker, J., Ayalew, D. and Menzies, M. Matching conjugate volcanic rifted margins: Ar/Ar chronostratigraphy of pre- and syn-rift bimodal flood volcanism in Ethiopia and Yemen, submitted to *Earth Planet. Sci. Lett.*.

Upcott, N.M., Mukasa, R.K., Ebinger, C.J. and Karner, G.D. (1996) Along-axis segmentation and isostasy in the Western Rift, East Africa, *J. Geophys. Res.* **101**, 3247-3268.

Vail, J.R. (1968) The southern extension of the East African Rift system, and related igneous activity, *Geol. Rundsch.* **57**, 601-614.

Vail, J.R. (1978) Outline of the geology and mineral deposits of the Democratic Republic of the Sudan and adjacent areas, *Institute of Geological Sciences Overseas Geology and Mineral Resources* **49**, 68 p.

Van Zinderen Bakker, E.M. and Mercer, J.H. (1986) Major Late Cainozoic events and palaeoenvironmental changes in Africa viewed in a world wide context, *Palaeogeography, Palaeoclimatology, Palaeoecology* **56**, 217-235.

Wescott, W.A., Krebs, W.N., Engelhardt, D.W. and Cunningham, S.M. (1991) New biostratigraphic age dates from the Lake Rukwa rift basin in western Tanzania, *Am. Assoc. Pet. Geol. Bull.* **75**, 1255-1263.

Wescott, W.A., Morley, C.K. and Karanja, F.M. (1993) Geology of the "Turkana Grits" in the Lariu Range and Mt. Porr areas, southern Lake Turkana, Northwestern Kenya, *J. African Earth Sciences* **16**, 425-435.

Wescott, W.A., Wigger, S.T., Stone, D.M. and Morley C.K. (1999) Geology and Geophysics of the Lotikipi Plain, in C.K. Morley (ed.), *Geoscience of Rift Systems-Evolution of East Africa*, AAPG Studies in Geology **44**, 55-65.

White, T.D., Suwa, G. and Asfaw, B. (1994) *Australopithecus ramidus*, a new species of early hominid from Aramis, Ethiopia, *Nature* **371**, 306-312.

Williams, M.A.J., Williams, F.M., Gasse, F., Curtis, G.H. and Adamson, D.A. (1979) Plio-Pleistocene environments at Gadeb prehistoric site, Ethiopia, *Nature* **282**, 29-33.

Williams, M.A.J., Getaneh Assefa and Adamson, D.A. (1986) Depositional context of Plio-Pleistocene hominid-bearing formations in the Middle Awash valley, southern Afar Rift, Ethiopia, in L.E. Frostick, R.W. Renaut, I. Reid and J.-J. Tiercelin (eds.), Sedimentation in the African Rifts, *Geol. Soc. London Spec. Publ.* **25**, 241-251.

Williamson, P.G. and Savage, R.J.G. (1986) Early rift sedimentation in the Turkana basin, northern Kenya, in L.E. Frostick, R.W. Renaut, I. Reid and J.-J. Tiercelin (eds.), Sedimentation in the African Rifts, *Geol. Soc. London Spec. Publ.* **25**, 267-283.

Winckel, A. (1998) Essais de bilans hydrologiques de deux lacs tanzaniens (Lacs Massoko et Rukwa), Mémoire de DEA, Université Paris-Sud, 46 pp.

Winn, R.D., Steinmetz, J.C. and Kerekgyarto, W.L. (1993) Stratigraphy and rifting history of the Mesozoic-Cenozoic Anza Rift, Kenya, *Am. Assoc. Pet. Geol. Bull.* **77**, 1989-2005.

WoldeGabriel, G. (1987) Volcanotectonic history of the central sector of the Main Ethiopian Rift: A geochronological, geochemical, and petrological approach, Ph.D. Thesis, Cleveland, Ohio, Case Western Reserve University, 410 pp.

WoldeGabriel, G., Aronson, J.L. and Walter, R.C. (1990) Geology, geochronology, and rift basin development in the central sector of the Main Ethiopia Rift, *Geol. Soc. Am. Bull.* **102**, 439-458.

WoldeGabriel, G., Yemane, T., Suwa, G., White, T. and Asfaw, B. (1991) Age of volcanism and rifting in the Burji-Soyoma area, southern Main Ethiopian Rift: Geo- and biochronologic data, *J. African Earth Sciences* **13**, 437-447.

WoldeGabriel, G., Walter, R.C., Aronson, J.L. and Hart W.K. (1992) Geochronology and distribution of silicic volcanic rocks of Plio-Pleistocene age from the central sector of the Main Ethiopian Rift, *Quat. Int.* **13/14**, 69-76.

WoldeGabriel, G., White, T.D., Suwa, G., Renne, P., De Heinzelin, P., Hart, W.K., and Heiken, G. (1994) Ecological and temporal placement of early Pliocene hominids at Aramis, Ethiopia, *Nature* **371**, 330-333.

Yamba, T.K. and Boven, A. (1998) Evolution Pliocène et Quaternaire du remplissage sédimentaire dans le sud du bassin du lac Edouard, branche occidentale du Rift Est-Africain, *J. African Earth Sciences* **26**, 423-439.

Zanettin, B., Justin Visentin, E., Bellieni, G., Piccirillo, E.M. and Rita, F. (1983) Le volcanisme du Bassin du Nord-Turkana (Kenya) : Age, succession et évolution structurale, in M. Popoff and J.-J. Tiercelin (eds.), Rifts et Fossés anciens, *Bull. Centres Rech. Explor.-Prod.Elf-Aquitaine* **7**, 249-255.

# Climate Dynamics and Climate Variability in the East African Lakes Region

# EXTREME RAINFALL EVENTS AND LAKE LEVEL CHANGES IN EAST AFRICA: RECENT EVENTS AND HISTORICAL PRECEDENTS

DECLAN CONWAY

*Lecturer in Natural Resources, School of Development Studies, University of East Anglia, Norwich NR4 7TJ, United Kingdom*

## 1. INTRODUCTION

Towards the end of 1961 an extreme rainfall event occurred that extended over much of East Africa stretching across the Indian Ocean to India. This event caused widespread flooding, rapid and prolonged increases in the levels of many lakes in East Africa and significant economic disruption (Odingo, 1962; Mörth, 1967). During the last few months of 1997, in a similar fashion to 1961, heavy rainfall caused flooding across East Africa (FAO/GIEW, 1998; Birkett *et al.*, 1999). The principal driving mechanism of these extreme events has recently been established, a dipole reversal in atmospheric circulation and Indian Ocean sea surface temperatures (SSTs, Webster *et al.*, 1999; Saji *et al.*, 1999). Rapid progress is being made in understanding the dynamics of the event, for instance Latif *et al.* (1999) have directly related rainfall anomalies with Indian Ocean SSTs in ensemble general circulation model (GCM) experiments. Little is known, however, about the spatial and temporal characteristics of the hydrometeorological anomalies associated with these and other dipole events in the Indian Ocean. For instance, over India, rainfall in 1961 was the highest on record (Grove, 1996; Saji *et al.*, 1999) but in 1997 totals were close to normal (Janowiak and Xie, 1999). Other major SST gradient reversals



*E.O. Odada and D.O. Olago (eds.),*
*The East African Great Lakes: Limnology, Palaeolimnology and Biodiversity,* 63–92.

apart from 1961 and 1997 have occurred (for instance 1967 and 1994) but it is not clear to what extent other SST gradient reversals have been associated with the type of anomalies that occurred in 1961 and 1997.

This chapter provides a comprehensive assessment of the 1961 and 1997 hydrometeorological anomalies, their spatial and temporal nature, and surveys the historical records for possible precedents to the 1961 and 1997 events. The chapter is organised as follows: section 2 reviews the literature on the causes of the 1961 and 1997 events and section 3 describes the datasets used for the analysis. In section 4 the spatial and temporal characteristics of rainfall during both events are analysed and historical rainfall records are surveyed for other events with similar spatio-temporal characteristics. The impacts of the 1961 event on East African lake levels are well documented (Grove, 1996; Nicholson, 1997a, 1997b) and are reviewed briefly in section 5 along with a more detailed analysis of impacts on river flows, particularly in the Nile and Congo river basins. Section 5 also examines the causes of sustained high levels in Lake Victoria following the 1961 event, to compare with the likely duration of the impacts of the 1997 event. The implications of these events for water resource management and some conclusions are presented in section 6.

## 2. TWO EXTREME HYDROMETEOROLOGICAL ANOMALIES IN EAST AFRICA: 1961 AND 1997

### 2.1 1961

The 1961 event over East Africa was widely documented at the time (Odingo, 1962) and in subsequent analyses (Lamb, 1966; Mörth, 1967; Kite, 1981; Flohn, 1987; Grove, 1996). Odingo (1962) described the record rainfalls recorded at many sites in Kenya and the social and economic disruption caused by the event, which followed three consecutive years of low rainfall in parts of Kenya. Extensive flooding occurred in the region, loss of homes and lives, damage to crops, and emergency food had to be flown in to marooned villages. Odingo (1962) estimated the total flood damage costs at the time for Kenya to have been around five million pounds. Most studies of this event have only dealt with the characteristics of the rainfall and hydrological anomalies, although Reverdin *et al.* (1986) and Flohn (1987) did touch upon the climatological origins of the event. Reverdin *et al.* (1986) investigated the interannual variability of surface observations in the equatorial Indian Ocean between 1954 and 1976. They identified 1961, 1967 and 1963 as the three years with the largest cloudiness anomalies, synchronous with positive SST anomalies in the central Indian Ocean and negative anomalies further to the east. They describe the 1961 anomaly in detail and note that the Pacific Ocean winds were not particularly anomalous suggesting that the event was triggered by air-sea interactions in the Indian Ocean. Flohn (1987) associated the episode with a large anomaly of SST, surface winds and convective cloudiness over the western equatorial Indian Ocean.

## 2.2 1997

During the last few months of 1997, widespread heavy rainfall caused flooding across the East and the Horn of Africa. For example, the October to February dry season in 1997-98 was the wettest on record over much of Ethiopia (Conway, 2000) and the levels of many lakes in East Africa rose significantly in response to the heavy rainfall event. The event produced wide-ranging agricultural, hydrological, ecological and economic impacts across the region summarised in Table 1.

*Table 1.* Summary of some of the major impacts during 1997/98 associated with the heavy rainfall event over East Africa adapted from FAO/GIEW (1998).

| Country | Agricultural impacts | Non-agricultural impacts |
|---|---|---|
| Somalia | Crop damage in secondary 'Deyr' season, losses up to 80% in some regions. Replanting after flood recession. | Estimated 2,000 deaths, 250,000 displaced persons, 35,500 animals lost, serious damage to housing and infrastructure. Outbreak of Rift Valley Fever after December 1997. |
| Kenya | 1997/98 maize crop affected during harvest. Worst effects on second-season crops (mid-October to February) - yields reduced by up to 33%. | Serious floods caused loss of life, damage to housing and infrastructure. Outbreak of Rift Valley Fever after December 1997. Beneficial effect on pastures. |
| Tanzania | Localised crop losses and damage to 'Vuli' crop in lowland areas, some gains in yield in highland areas due to higher rainfall. | Disruption of road and rail systems. Beneficial effect on pastures. |
| Uganda | Despite localised crop losses the heavy rains were beneficial for crop development. | Floods and mudslides, mainly in eastern areas, caused loss of life, damage to housing and infrastructure. |
| Ethiopia | Disrupted crop harvesting processes | Extensive flooding in southeast areas bordering Somalia and Kenya, caused loss of life, damage to housing and infrastructure and displacement of people. Over 12,000 animals lost, 30,000 ha inundated. |
| Eritrea | Spoilage of harvested crops, grain output expected to be reduced. | High levels of local locust outbreaks reported. |

The event in 1997 prompted studies into its nature and causes (Birkett *et al.*, 1999; Chambers *et al.*, 1999; Saji *et al.*, 1999 and Webster *et al.*, 1999) and has renewed interest in the earlier 1961 event. Both Webster *et al.* (1999) and Saji *et al.* (1999) show the importance of the Indian Ocean circulation itself for generating these anomalies, independent of more distant phenomena such as the El Niño

Southern Oscillation (ENSO). Webster *et al.* (1999) propose a mechanism to account for the 1997-98 event (and hence also earlier events) based on a characteristic internal mode of the Indian Ocean climate system. Beginning in June (July) 1997, strong, cool (warm) SST anomalies developed in the eastern (western) Indian Ocean reaching a maximum of below –2°C in November (over +2°C, February 1998). These anomalies produced a reversal of the usual SST gradient (in absolute temperatures) and the normal westerly winds were replaced by easterly winds. The sea surface height was depressed in the east and generally higher in the west. Saji *et al.* (1999) define the reversal in SST gradient using the difference in SST anomaly between the tropical western Indian Ocean and the tropical south-eastern Indian Ocean and refer to it as a dipole mode index (DMI). Rainfall anomalies during the event (gauge-measured and out-going longwave radiation estimates) were above average in the western and northwestern Indian Ocean and below average in the east (Sumatra and Malaysian Archipelago), associated with ascendance in the west and subsidence in the east (Webster *et al.* 1999).

## 2.3 Disentangling the Role of ENSO

Webster *et al.* (1999) suggest the changes were not an exaggerated response to the strong 1997-98 El Niño. Many of the SST and sea surface height changes can be explained by changes occurring in the surface winds. Many of the features of the 1997 event were different to those usually associated with a Pacific warm event. Correlation between mean equatorial SST in the Indian Ocean and ENSO is +0.52 but only +0.19 between ENSO and the equatorial SST gradient. The correlation between East Africa 'short rains' (October-November) and the equator Indian Ocean SST gradient is +0.62. The circulation and SST patterns in 1997 were similar to those of other events, particularly 1961. Webster *et al.* (1999) identify 16 years between 1950 and 1998 in which the equatorial SST gradient (or DMI) reversed for at least one month. Only three of these were El Niño years and none La Niña.

Latif *et al.* (1999) use ensemble GCM experiments with prescribed SST anomalies to show that the strong rainfall anomalies in December and January 1997-98 (note that this does not include October or November 1997) were driven by Indian Ocean anomalies and not directly by El-Niño related SST anomalies in the equatorial Pacific. Although positive rainfall anomalies over East Africa are associated with warm ENSO extremes, Latif *et al.* (1999) found SST anomalies, as measured by the Niño 3 index, only accounted for 20 per cent of the December to January rainfall variability. Indian Ocean SST anomalies, particularly in the western and south western tropical Indian Ocean, account for up to 60 per cent of the rainfall variance (Latif *et al.* 1999). They discuss the potential for seasonal climate forecasting given that there is some predictability in Indian Ocean SSTs which respond to tropical Pacific SST fluctuations with a lag of several months. However, a significant component of Indian Ocean SST fluctuations are not related to ENSO and the 1997-98 SST anomaly pattern was not typical of the more usual SST anomalies observed during ENSO events (Venzke *et al.*, 2000). Latif *et al.* (1999) cite work from

Grötzner *et al.* (in press) which does suggest, however, that the December-January 1997-98 anomaly was forced remotely by SST anomalies in the Pacific.

During El Niño events, Reason *et al.* (2000) found more zonal easterly flow anomalies occurred over the central Indian Ocean, across generally warm SST anomalies leading to enhanced rainfall production. Using satellite altimeter and wind observations Chambers *et al.* (1999) related Indian Ocean warming in 1994 and 1997 with westerly wind anomalies in the central Indian Ocean connected with the ENSO. These anomalies were suggested to excite downwelling Rossby waves which move westward and depress the southwestern Indian Ocean thermocline.

Using an Ocean GCM, Murtugudde *et al.* (2000) studied the basic processes that caused warm and cold SST anomalies during the 1997 event. They found similarity with the evolution of other weaker Indian Ocean events suggesting that they represent a natural mode of oscillation in the Indian Ocean. They suggest this oscillation is externally forced by ENSO but also speculate that it is generated by ocean-atmosphere interactions internal to the Indian Ocean. Time series of SST anomalies for the west (5°S-5°N 45°-55°E) and east (5°S-5°N 90°-100°E) Indian Ocean show significant warm western anomalies tend to occur together with strong cool eastern events (r = -0.44 for 1958-97, significant at the 10 per cent level). The eastern cooling was greater than -1°C in 1961, 1994 and 1997 and greater than -0.5° in many other years. Many of the stronger cooling episodes coincide with ENSO years but the correlation between zonal and meridional wind anomalies with the Southern Oscillation Index is not very large (0.57 and -0.51, respectively) and many events do not occur during major ENSO events.

In conclusion, the role ENSO plays in either initiating or influencing these events in the Indian Ocean is at present unclear, particularly given that 1961 was not an El Niño year, and this is an important question for further research.

## 2.4 Longer Time Scales

There are very few high-resolution long term climate series for the region pre-dating 1900 to indicate whether similar extreme events have occurred in the past. Nicholson's (1997a and b) reconstructions of East African lake levels for the 18$^{th}$ to 20$^{th}$ centuries and the Nile series of annual flood maxima at Cairo back to the 7$^{th}$ century are the most suitable records, although new high-resolution reconstructions are beginning to appear. Recently, Verschuren *et al.* (2000) have produced a 1,100 year decadal resolution rainfall and drought reconstruction inferred from Lake Naivasha in Kenya. A 194-year annual record of coral growing at Malindi, Kenya shows changes in SSTs coherent with instrumental and proxy records of tropical Pacific climate variability (Cole *et al.*, 2000). An overall warming trend of 1.3°C has occurred since 1840, much of it in the recent period. The record shows strong association with a 147 year record from the Seychelles (Charles, *et al.* 1997). Both annual records, from November to the following October, show strong peaks around 1878, a year with an extremely high Niño 3.4 SST anomaly and high all-India rainfall. Smaller peaks occur around 1915-20, but in the early 1960s the Malindi record reaches the minimum of a 20-year cooling trend with a large cold SST anomaly around 1963-64, whereas the Seychelles record shows no particular patterns. Neither record extends to 1997-98. Both records show stronger coherence with Pacific Ocean SSTs than with all-India and East African rainfall indices.

## 3. DATA AND METHODS

Rainfall series from Hulme (1994, updated) and the merged satellite and station dataset from Xie and Arkin (1998) are used for the analysis. For spatial analysis of the 1961 event and others pre-1995, all stations with at least 15 years complete record between 1961 and 1990 are used to define percentage rainfall anomalies from the 1961-90 baseline for the period of interest. This extensive subset (sampling during the period with greatest station coverage in East Africa) is used to give the best possible spatial and temporal coverage. Only a few stations have temporal coverage from 1960 up to 1998 which precludes extensive direct comparison of the 1961 and 1997 events using station data alone. Therefore, the Xie and Arkin dataset (1998, for the period 1979-1999) is also used for analysis of the 1997 event. To identify other events with similar temporal characteristics, zonally averaged (calculated over 5° zones from 25° East to the coast) rainfall series are derived from Hulme (1994, updated) for the period 1900 to 1998. The series are calculated as an unweighted average of all gauges with at least 15 years data between 1961 and 1990 using the mean of the percentage departures from each station's 1961-1990 mean to take the series back to 1900 as described in Jones and Conway (1997). Many of the series have very few gauges contributing during the first and last decades of this period and should be interpreted with caution. The first decade in these series is shown with dashed lines in Figures 1 and 2, the final decade is not shown but is discussed in section 4.2.

Riverflow and lake level data for the Nile are primarily from Hurst and Phillips (1933 plus supplements). The Congo, Oubangui and Tana riverflow series are from UNESCO (1995, Congo updated from Laraque *et al.*, 1998) and the Zambezi series from Grove (1996). Table 2 lists the river and lake gauge locations and basin areas.

## 4. RAINFALL: THE 1961 AND 1997 EVENTS AND HISTORICAL PRECEDENTS

### 4.1 Rainfall in 1961

Annual and October-November (ON) rainfall series for nine 5° zonally averaged areas (from 20-15°N to 20-25°S) are shown in Figures 1 and 2, respectively. In the annual series 1961 stands out as extremely wet between 5°N and 15°S (the wettest year on record in three series, and second wettest in the 10-15°S series). Rainfall is close to average in the other zones, except 20-15°N where it was amongst the ten wettest years on record. A similar pattern is exhibited during ON for most regions, excepting 10-5°N (slightly more extreme) and 5-10°S (less extreme, third wettest on record). The ON 1961 anomaly is also very marked when September and December are included (not shown) with anomalies showing similar patterns, most pronounced at 10-5°N and between 5-10°S and 10-15°S. Rainfall in the other seasons during 1961 (not shown) was unremarkable except in June to September (JJAS) between 5°N and 15°S where it was well above average, but not exceptional. Rainfall during January and February and the March to May (MAM) long rains in 1962 was fairly unexceptional, except at 10-15°S, where rainfall was amongst the wettest ten years in both seasons.

*Table 2.* Characteristics of lake level and river flow series. 1 = Main Nile, 1871-1910 Aswan (dashed line), 1890-1939 Wadi Halfa, 1940-64 Kajnarty, 1965-90 Dongola, 2 = Blue Nile, 1900-11 Khartoum (dashed line), 1912-63 Roseires, 1964-97 El Deim, 3 = Runoff is calculated as the difference between river flow at the outlet to Lake Victoria and downstream at Mongalla, just upstream of the Sudd swamps on the White Nile. This stretch includes gains and losses through Lakes Kyoga, Edward, George and Albert and tributary inflows. 4 = 1961-90 or earlier, depending upon record length. 5 = 1962-64. 6 = Per cent difference between 1961-90 and 1961-64. MAR = mean annual river flow ($m^3s^{-1}$).

| River / Lake | Gauge location | Lat. (N°) | Lon. (°E) | Period of record | Basin area ($km^2$) | MAR 1961-1990[4] | MAR 1961-1994 | Per cent change[6] |
|---|---|---|---|---|---|---|---|---|
| Main Nile | Dongola[1] | 19.10 | 30.32 | 1890-1990 | - | 2745 | 3227 | +18 |
| Atbara | 3 kilo | 7.41 | 34.00 | 1903-1981 | 137000 | 296 | 434 | +47 |
| Blue Nile | El Deim[2] | 11.23 | 34.98 | 1912-1997 | 195000 | 1458 | 1823 | +25 |
| Sobat | Hillet Doleib | 9.20 | 31.38 | 1905-1979 | 231000 | 437 | 527 | +21 |
| Equatorial | Owen Falls | 0.43 | 33.23 | 1905-1982 | 293000 | 314 | 429 | +37 |
| Lakes[3] | Mongalla | 5.12 | 31.46 | | | | | |
| Lake Victoria | Owen Falls | 0.43 | 33.23 | 1896-1997 | 258000 | 1174 | 1394[5] | +19 |
| Tana | Garissa | 0.45°S | 39.70 | 1934-1975 | 42220 | 203 | 282 | +39 |
| Congo | Kinshasa | 4.30°S | 15.30 | 1903-1994 | 3475000 | 42,091 | 50,600 | +20 |
| Oubangui | Bangui | 4.37 | 18.58 | 1911-1990 | 500000 | 3716 | 5110 | +38 |
| Zambezi | Livingstone / Victoria Falls | 17.95 | 25.90 | 1907/8-1993/4 | 360683 | 1172 | 1586 | +35 |

Maps of percentage rainfall anomalies in 1961 for MAM, JJAS, ON and annual periods are shown in Figure 3. The 1961 event covered an area stretching from southern Sudan and the Eritrean coast in the North to northern Zambia and Zimbabwe in the South (no data for Mozambique). The event was centred over eastern Uganda, southern Kenya and northern Tanzania, peaked in November 1961 and ranged in magnitude from slightly dry conditions over southeast Sudan to 200 per cent increases in annual rainfall over Kenya, Eritrea and Somalia in 1961.

Seasonal anomalies were greatest in ON 1961, ranging up to 400 per cent over Lake Turkana (southwest Ethiopia) and the East African coast. The ON anomaly covered Uganda, Kenya, and northern Tanzania up to central Ethiopia and across Somalia. Dry conditions occurred over Sudan and near normal conditions south of 15°S. High annual rainfall totals in 1961 extend to 20°S due to high rainfall during JJAS over large parts of the region and particularly over Malawi and Zimbabwe. Whether this earlier heavy rainfall was a southerly precursor of the ON event or unconnected with it is not clear. Figure 4 shows the time series of mean (1961-1990) and monthly rainfall 1960-1965 averaged over the 5° zones. High rainfall was experienced during the JJAS season over 20-15°N and 15-10°N with near normal conditions after September. It is only the regions which generally experience substantial rainfall during the short rains period (ON) between 10°N and 10°S that record large monthly anomalies. There is some indication of a southwards transgression in high rainfall following the movement of the ITCZ; from August at 5-0°N, peaking in November over 5°N to 5°S, in December over 5°S to 15°S where it also continued into January (contributing to high annual rainfall for 1962 over 10°-15°S) but dissipating south of 15°S.

## 4.2 Rainfall in 1997

For 1997 the zonal average series have only between two and 11 stations with data and the 20-15°N and 0-5°S series have no data, which restricts analysis of the event and extensive comparison with 1961. Most of the series become rather erratic after 1990 due to the low numbers of stations with data and are therefore not shown in Figures 1 and 2 after this date. The series, however, do show the following for 1997. The anomaly is most pronounced between 10°N and 10°S, particularly at 0-5°S (wettest year on record, although note that only two stations contribute to the series in this year), and 10-5°N (third wettest on record and higher than 1961). Rainfall during 1997 was also well above normal from 10°S to 25°S, more so than in 1961 (except for 10-15°S). The anomalies during ON are more apparent than in the annual series, particularly at 15-10°N to 10-5°N and 20-25°S. Rainfall was close to normal between 5-10°S and 15-20°S.

For the period September to December the anomalies are less extreme at 15-10°N and 0-5°S and 20-25°S, more extreme at 5-10°S and similar across the other zones with data. Figure 5 shows percentage rainfall anomalies for stations with data in 1997 (mainly in Ethiopia, Zambia and Zimbabwe). The highest anomalies occurred during JJAS over Zambia and during ON over Ethiopia and at Dar es Salaam in Tanzania. Figure 1 shows there is reasonable agreement between the station only and the merged satellite and station zonal time series in terms of mean rainfall amounts and temporal variability at 20-15°N, 15-10°N, 5-10°S, 10-15°S and 20-25°S. For the four other zones there are marked differences between the station series and the merged series, particularly in rainfall amounts. In the merged annual data 1997 is only exceptional over 5-0°N and 0-5°S and, to a lesser extent, 5-10°S (6th wettest year in 21 years).

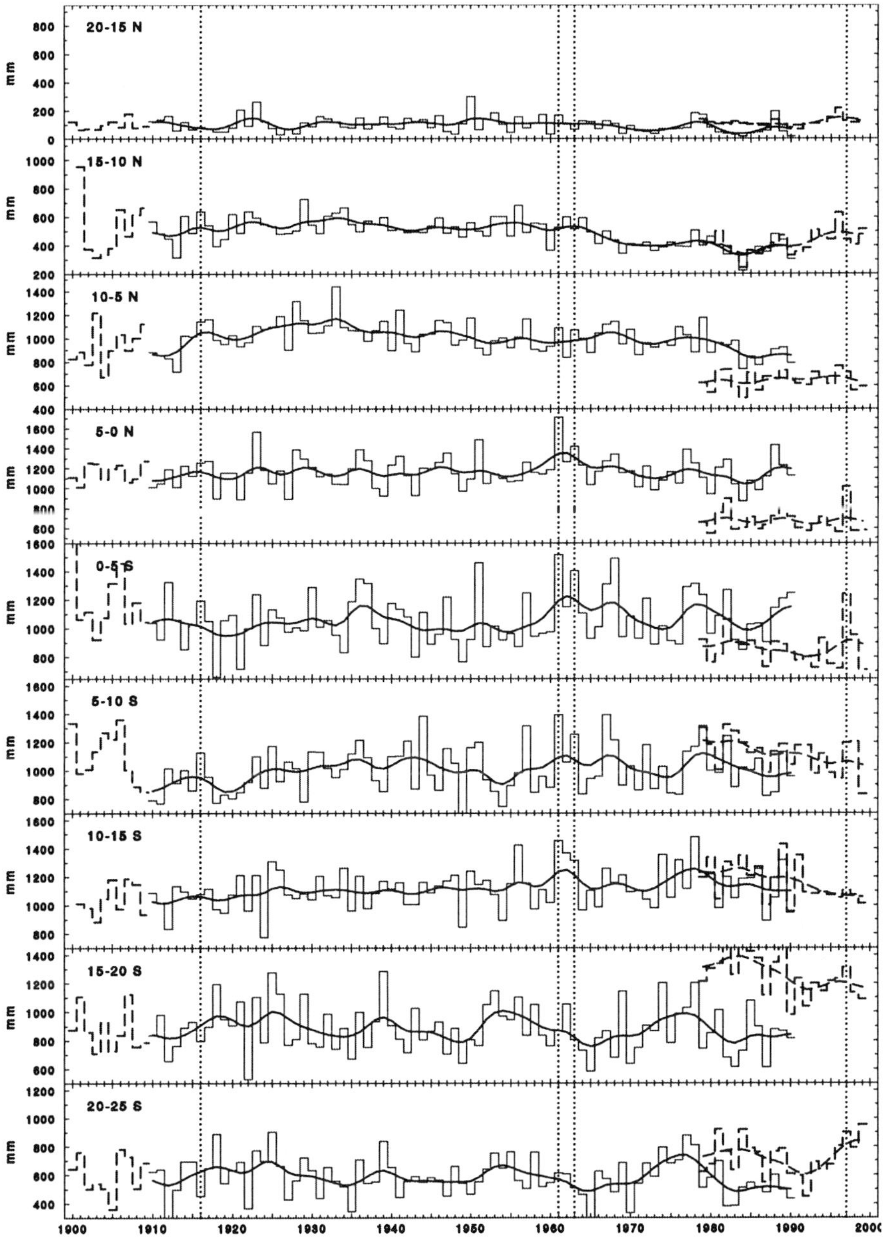


*Figure 1.* Time series of annual rainfall averaged over 5° zones, east of 25° East, 1901 to 1999. Dashed line from 1901 to 1909 represents period with very limited data availability. Data from 1901 to 1990 are from Hulme (1994) with a 10-year gausian fitter. Dashed line from 1979 to 1999 is from Xie and Arkin (1998), with 10 year gausian fitter. Dotted vertical lines highlight four years, 1916, 1961, 1964 and 1997.

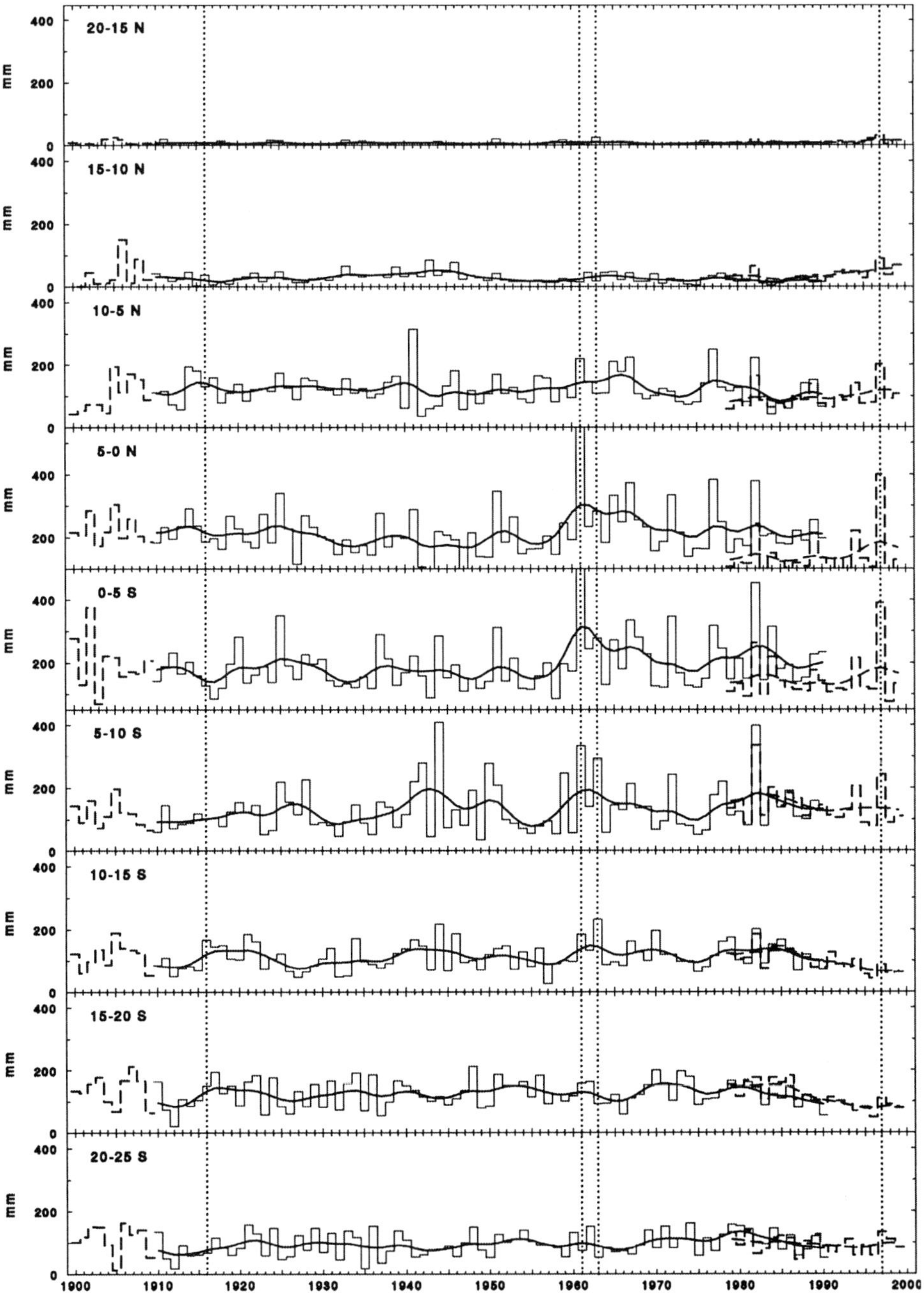


*Figure 2.* Time series of October – November rainfall averaged over 5° zones, east of 25° East, 1901 to 1999. See caption to Figure 1.

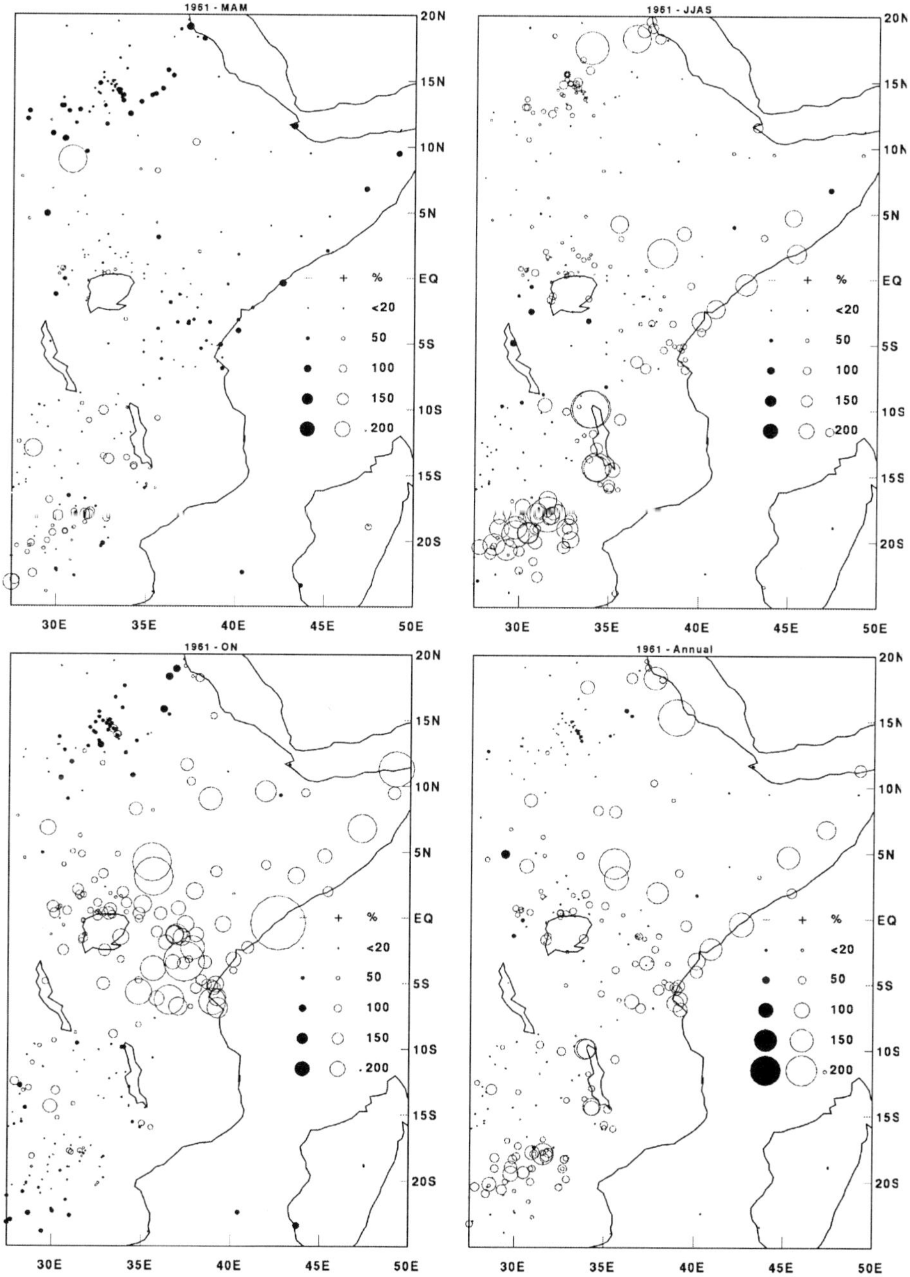


*Figure 3.* Rainfall anomalies in 1961 expressed as percentage departure from 1961-1990 mean for stations with available data. MAM = March to May, JJAS = June to September, ON = October to November.

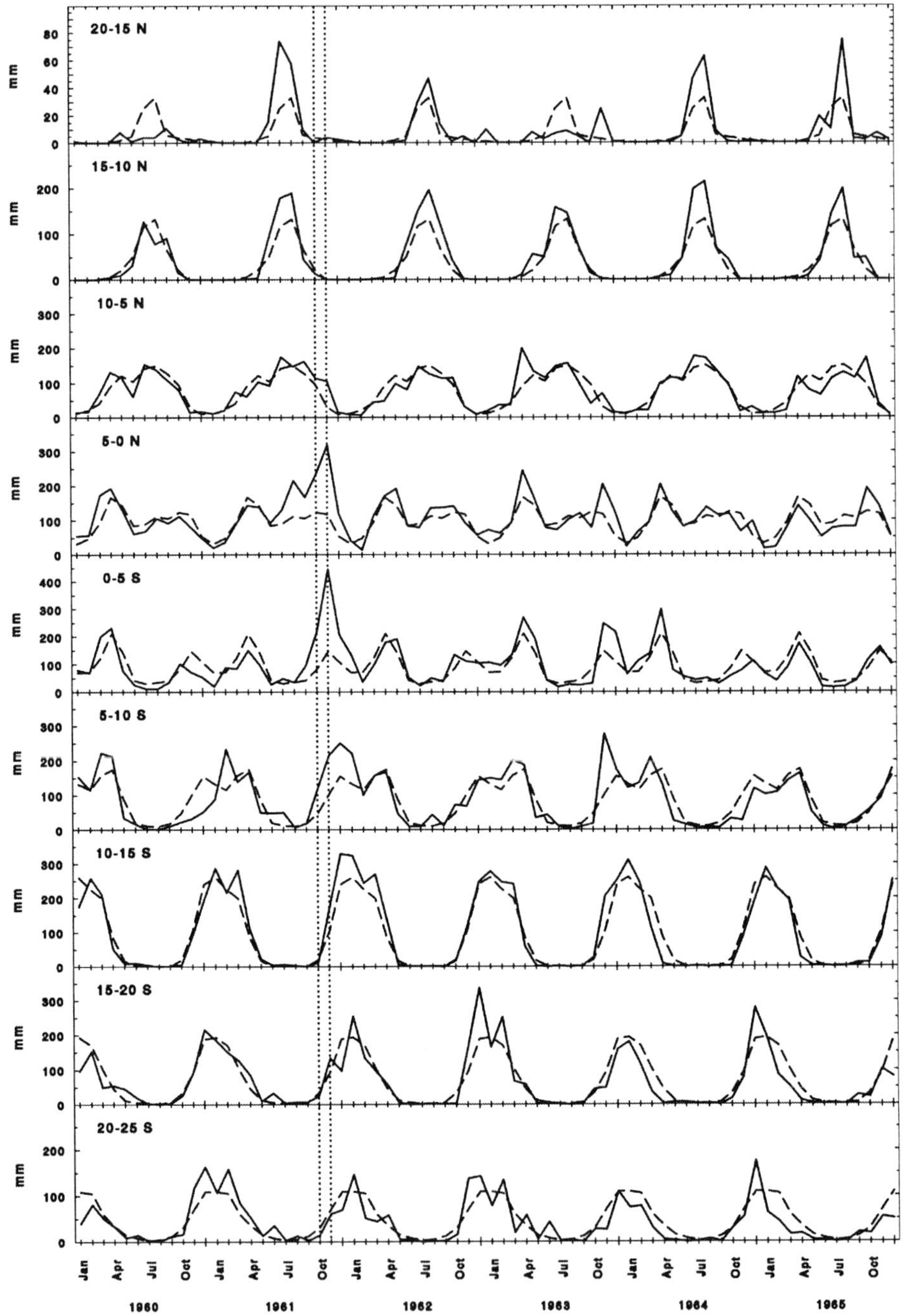


*Figure 4.* Time series of mean (1961-1990) and monthly (1960-1965) rainfall averaged over 5° zones. Note different scales. Dotted vertical lines highlight October and November, 1961.

Agreement between the two data sets is better for ON (Figure 2) in terms of rainfall amount and variability with large errors occurring in magnitude but not variability, over 5-0°N and 0-5°S. The event is more apparent over 5-15°N and 5-10°S. Per cent monthly anomalies from the 21-year mean in Figure 6 show the anomaly actually extends further north but is obscured in the annual totals because of the low rainfall normally recorded during this period north of 15°N. The event is also apparent further south over 5-10°S. The event is contemporaneous across the region beginning in October and lasting through until February 1998, except 20-15°N where the event disappears after November 1997.

## 4.3 HISTORICAL PRECEDENTS

On rainfall totals shown in Figure 2 highlight other years with extreme anomalies contemporaneous across much of the region. Notable anomalies occur in 1925 (15°N-10°S), 1944 (restricted to 0°S-15°S), 1951 (5°N-10°S), 1963 (10°N-15°S), 1967 (10°N-10°S), 1977 (10°N-10°S), 1982 (10°N-15°S and 15°N-15°S in the Xie and Arkin data), and 1994 (restricted to 10°N-5°S, 10°N-10°S S in the Xie and Arkin data). Only ON in 1963 and 1982 come close to matching the spatial extent and magnitude of the 1961 event. Time series for other seasons (January to February, MAM and JJAS, not shown here) do not show particularly unusual rainfall during any of these years.

Seasonal and annual anomalies are shown for 1982, which is the closest year to ON 1961, in Figure 7. Rainfall during ON was very high over Kenya and Tanzania and the Horn (over 400 per cent above average in places) and generally above average across the whole region. In the other seasons rainfall was less extreme, and generally below normal. The ON spatial pattern is similar to 1961 between 10°N-15°S but with wet conditions extending to 20°N (1961 restricted to 10°N-15°S). Figure 8 shows the percentage rainfall anomalies in other years with very wet ON; 1925, 1963, 1967, and 1977. All four years show reduced rainfall south of about 15°S and wet to extremely wet conditions north of this point. Some particularly large anomalies occurred over southern Sudan in 1925, where normal ON rainfall is extremely low. ON 1967 was wet across the region but with most stations recording anomalies of around 100-150 per cent, except for very wet conditions for two gauges near Lake Malawi and most gauges in Ethiopia.

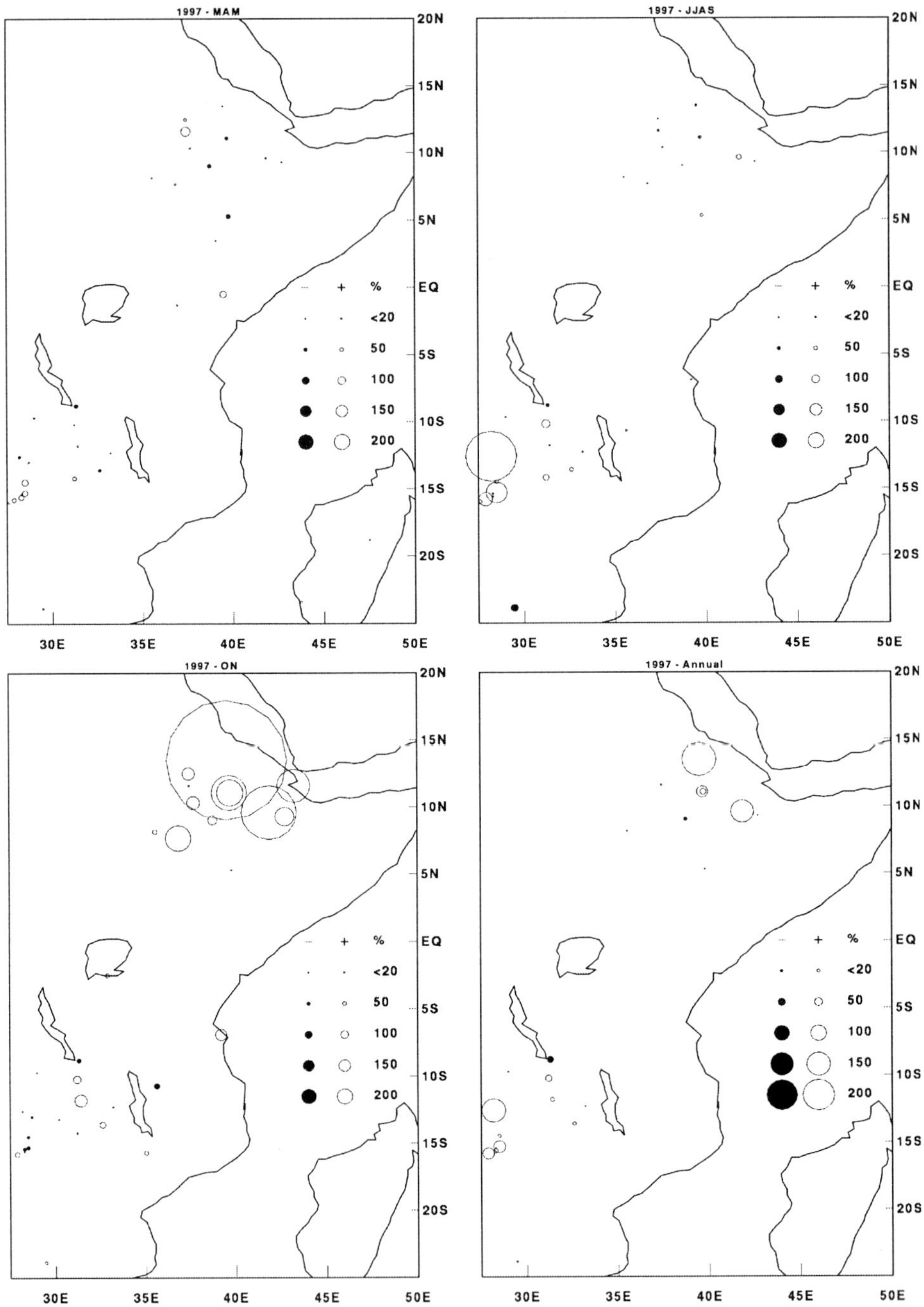


*Figure 5.* Rainfall anomalies in 1997 expressed as percentage departure from 1961-1990 mean for stations with available data. MAM = March to May, JJAS = June to September, ON = October to November.

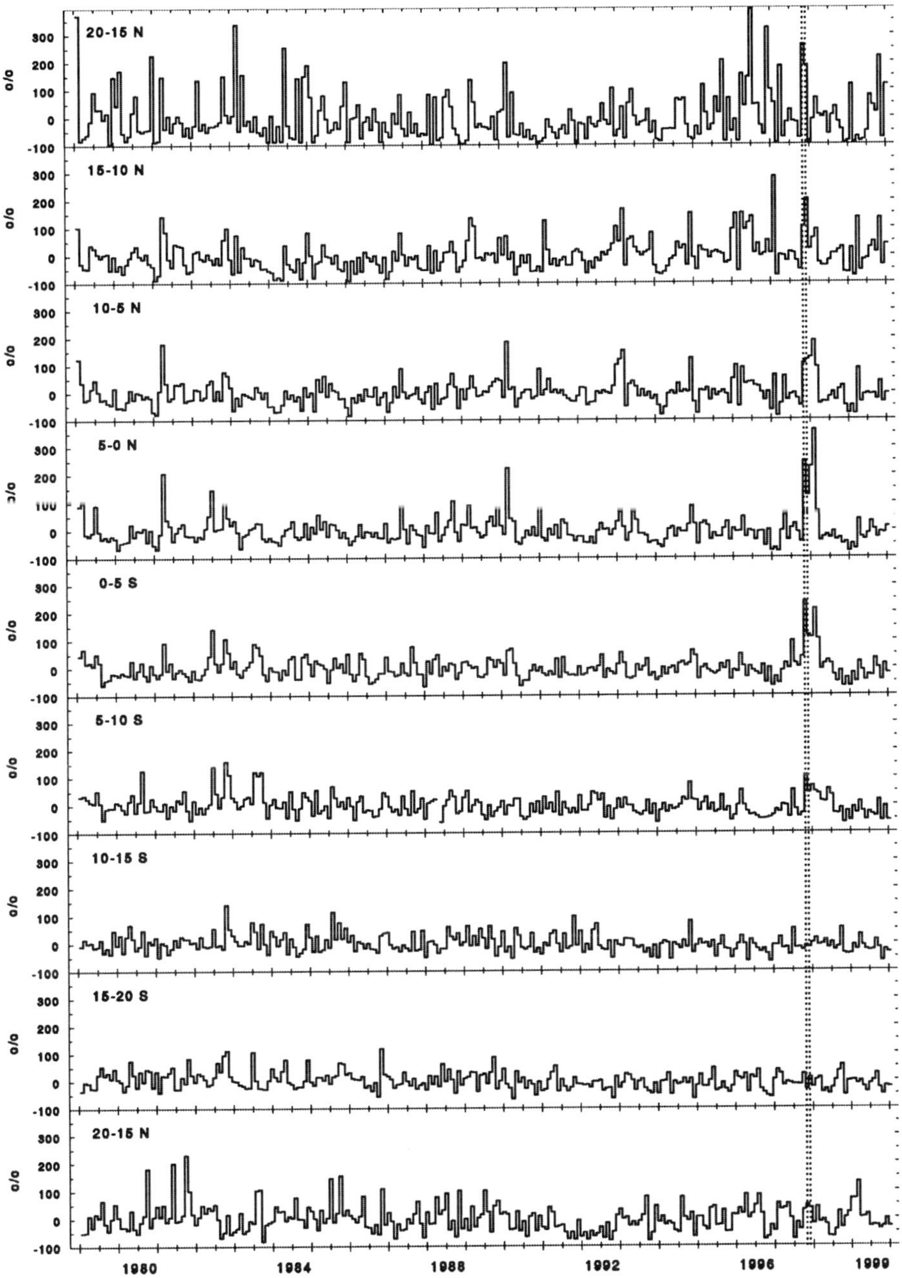


*Figure 6.* Time series of monthly rainfall percentage anomalies (from 1979-1999, 21 year mean) averaged over 5° zones, data from Xie and Arkin (1998). Dotted vertical lines highlight October and November, 1997.

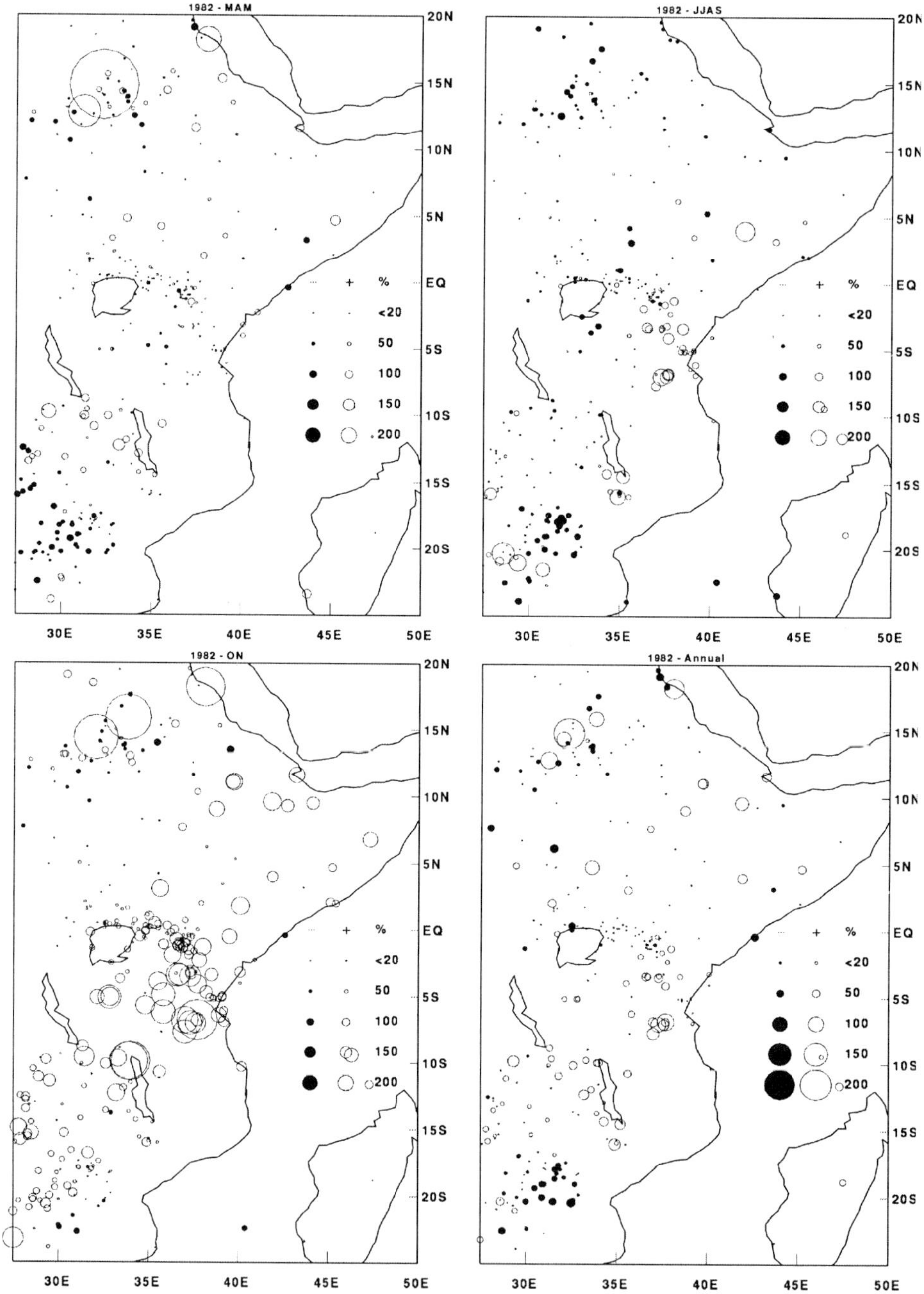


*Figure 7.* Rainfall anomalies in 1982 expressed as percentage departure from 1961-1990 mean for stations with available data. MAM = March to May, JJAS = June to September, ON = October to November.

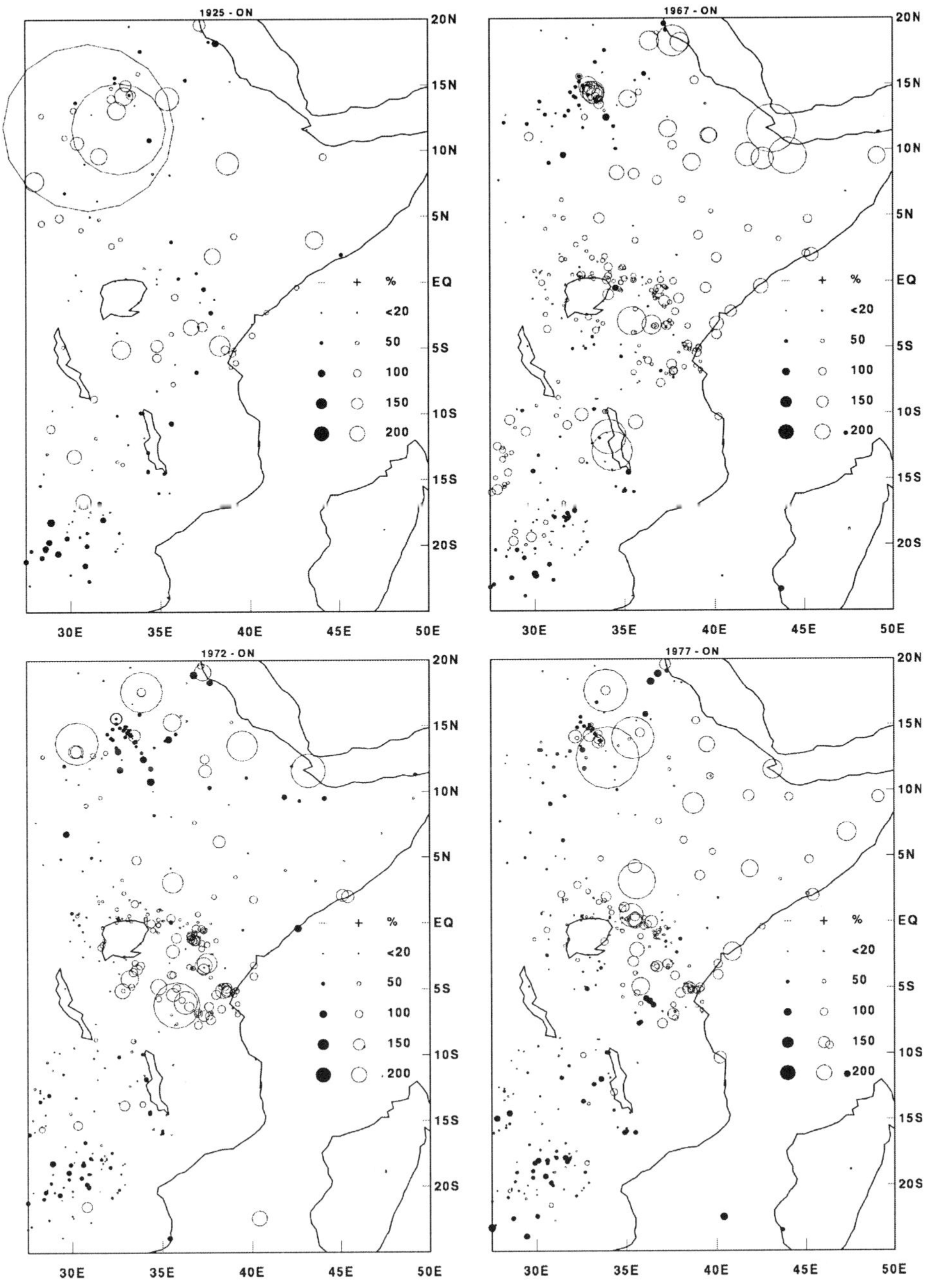


*Figure 8.* ON rainfall anomalies in 1941, 1973, 1967 and 1977, expressed as percentage departure from 1961-1990 mean for stations with available data. MAM = March to May, JJAS = June to September, ON = October to November.

# 5. HYDROLOGY: THE 1961 AND 1997 EVENTS AND HISTORICAL PRECEDENTS

This section analyses river flow records for areas to the north, west and south of the East African lakes for which the impacts of the 1961 event are well documented (Grove, 1996; Nicholson, 1997a and b for 1961, see also Birkett *et al.*, 1999 for 1997 impacts, and Sutcliffe and Parks, 1999 for a review of river flows in the Nile Basin). Annual river flow series are shown in Figure 9 and monthly mean (1960-90) and monthly flows (1960-65) are shown in Figure 10.

## 5.1 River Flow and Lake Level Series

### 5.1.1 Main Nile

The period of extremely high flows prior to 1899 has been the subject of a number of studies especially concerning the accuracy of the early gauge data. Of note during this high period are the floods of 1878 and 1879. Flohn and Burkhardt (1985) used correlation between dry season flows at Aswan and end of year Lake Victoria lake levels to reconstruct lake levels back to the beginning of the Aswan record, 1870, which suggested very high levels during 1878.

Sutcliffe and Parks (1999) identified periods of high flows at Aswan (January to May, indicative of high Lake Victoria levels) during 1879, 1895-97, 1917-1918 and similar increases upstream at Dongola after 1961.

### 5.1.2 Atbara and Blue Nile

Both records show high flows during 1903-05 and 1916-17, unexceptional flows from 1961-64 in the Atbara and high Blue Nile flows in 1961 due to monsoon rainfall during JAS. Neither river records unusual flows during ON 1961. Some stage readings do exist for the Blue Nile at Khartoum prior to 1900, for the years between 1869 and 1883 (Walsh *et al.*, 1994). These indicate floods were more frequent and severe in the period 1869-1883 than at any time during the 20th Century, with 1878 producing the highest flood level on record.

### 5.1.3 Sobat

The Sobat catchment lies between roughly 5°N and 10°N where annual rainfall was not particularly high during 1961 but ON rainfall was amongst the five highest on record. The Sobat displays similar flood years to the Blue Nile, in 1904-05 and 1917-18 and 1962 (one year later than Blue Nile floods). Riverflows were above average during 1962 due to very high flows between December 1961 and March 1962 (Figure 10). Monthly flows were high throughout 1962 and October 1964 to

February 1965. During years of heavy rainfall flooding may occur from the Sobat's tributaries draining Ethiopia feeding into the Machar marshes and other wetlands in southeast Sudan. Delayed response at the Sobat mouth may result from contributions from the upper Pibor which lies near Lake Kyoga (Sutcliffe and Parks, 1999) and small headwater wetlands in the upper Baro in the Ethiopian highlands (Dixon, 2000).

### 5.1.4 Equatorial Lakes

Runoff from the Equatorial Lakes catchment along the Nile region north of Lake Victoria and south of the Sudd swamps was extremely high in 1916-18. The monthly series shows a massive increase in flows occurred from August 1961, three months earlier than Lake Victoria (see below), and produced high annual runoff in 1961, whereas the increase in annual runoff from Lake Victoria only becomes notable by 1962 (Figure 9). The return in runoff rates to magnitudes typical of the pre 1961 event was faster than in Lake Victoria. Downstream impacts of high river flows along the White Nile were significant. For instance, the Sudd swamps, where from 1950-52 to 1980 the area of permanent swamp increased from 2800 $km^2$ to 16600 $km^2$ and the area of seasonal swamp increased from 11200 $km^2$ to 14000 $km^2$ (Sutcliffe and Parks, 1987).

### 5.1.5 Lake Victoria

Lake Victoria levels had two short-lived high periods during 1903-07 and 1916-18 but the annual series is dominated by the sudden change from 1962 onwards (Figure 9).

A significant and sustained increase in Lake Victoria levels and outflows occurred in late 1961 and has been fully documented in Kite (1981) and Piper *et al.* (1986). Lake Victoria levels peaked in 1964 and decreased steadily except for increases in 1978-79 and 1990-91 but even by 1997 they remained well above their pre-1961 levels.

### 5.1.6 Tana

The Tana river has the smallest gauged catchment area of the rivers examined here and provides the clearest and most short-lived example of riverflow response to the 1961 event. The anomaly lasts from October to December, peaking in November and generating the second highest annual flow on record. River flows in 1963 and 1964 were also high.

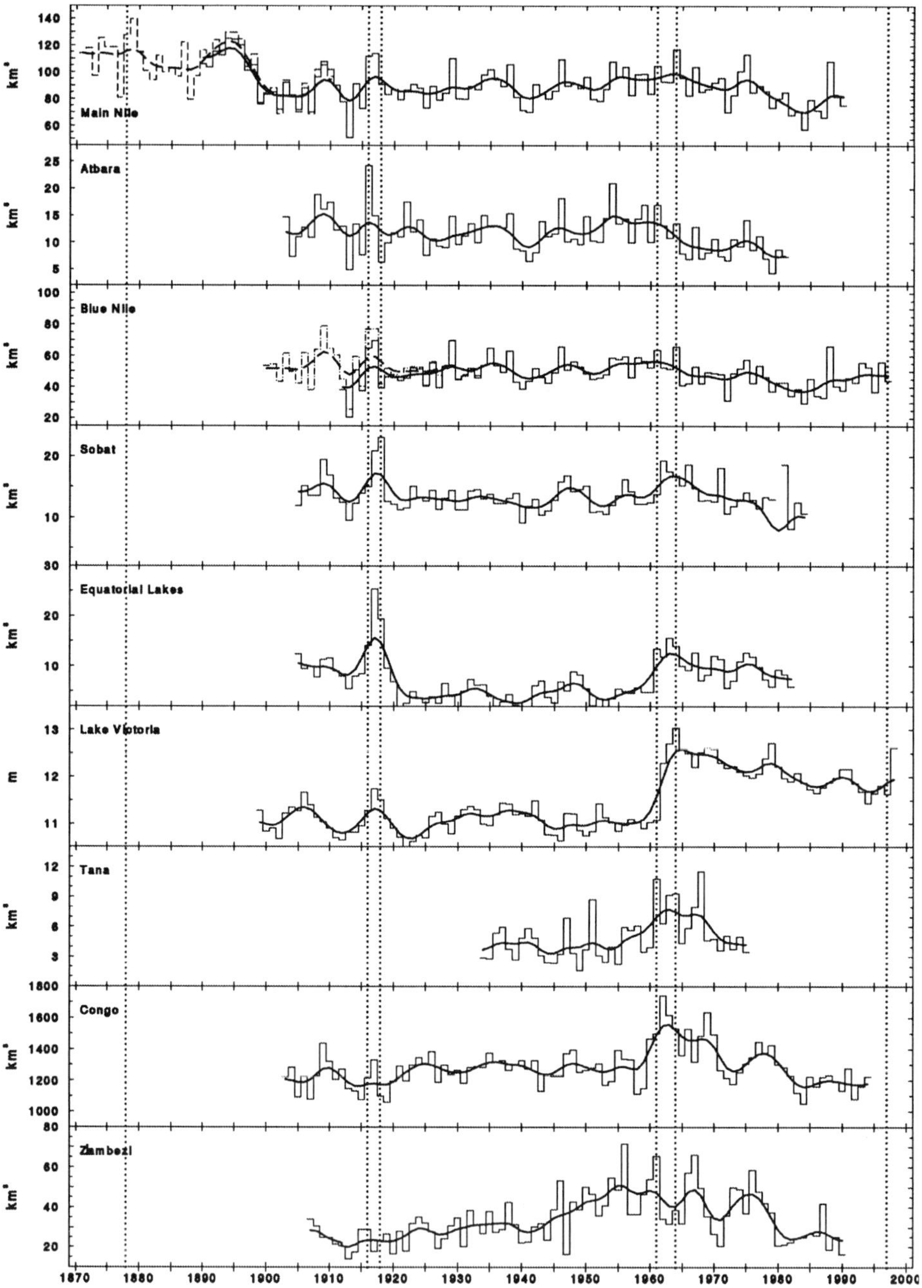


*Figure 9.* Time series of annual riverflows for various periods. Note different scales. Dotted vertical lines highlight five years, 1878, 1916, 1961, 1964 and 1997. See Table 2 for description of series.

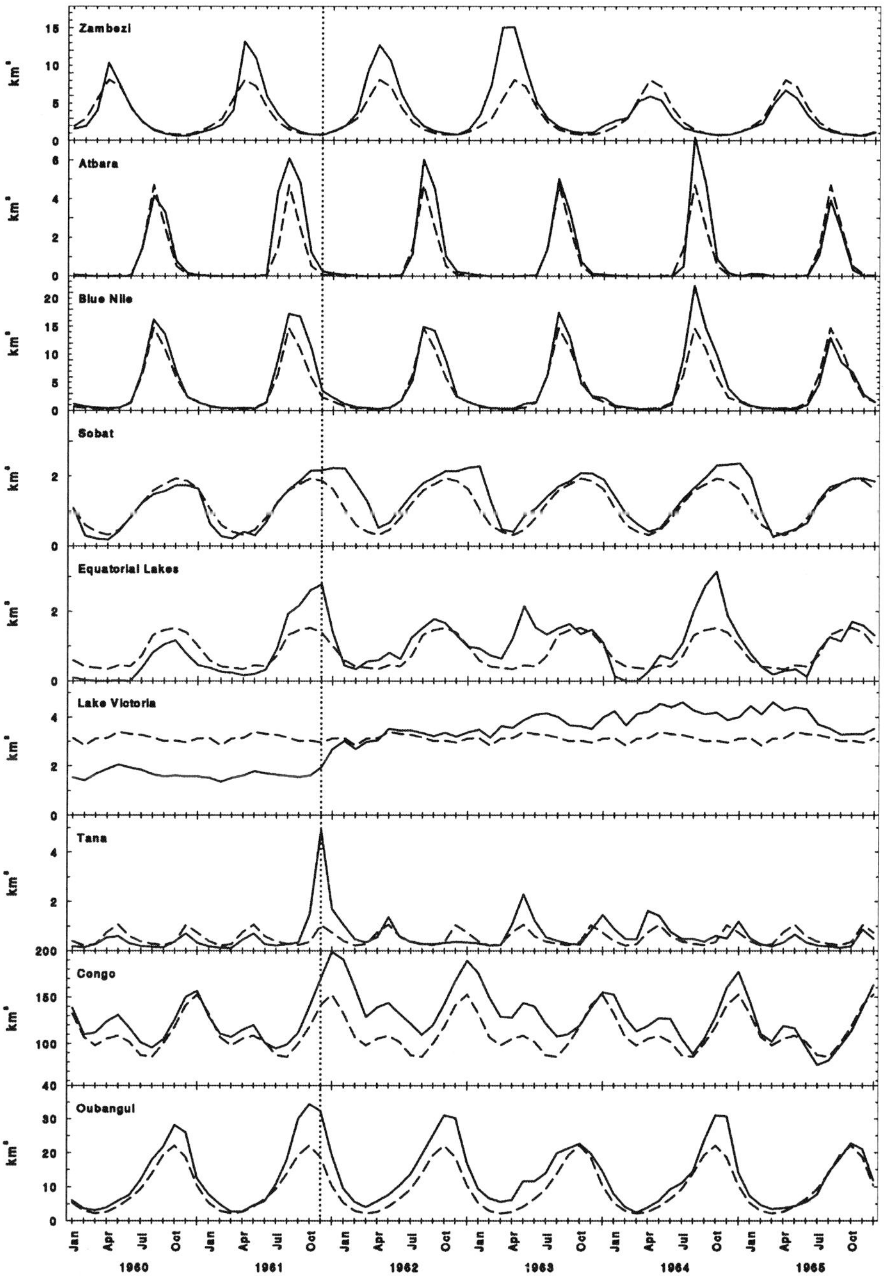


*Figure 10.* Time series of mean (1961-1990) and monthly (1960-1965) riverflow. Note different scales and different rivers compared to Figure 9. Dotted vertical line highlights November 1961. See Table 2 for description of series.

### 5.1.7 Congo and Oubangui

Small floods occurred in 1909 and, as in the White and Blue Niles, 1917. October, November and December 1961 produced the highest monthly river flows on record in the Congo basin making 1961 the sixth highest annual flow on record. 1962 recorded record monthly flows from January to October and produced an even higher annual flow (over 30 per cent higher than the long-term mean). River flows remained exceptionally high until the end of 1964, probably maintained by a combination of the following: high rainfall in 1962-64; storage within this vast complex river system with areas of seasonal and permanent wetland; and possibly additional inputs from Lake Tanganyika via the river Lukuga outlet as occurred in 1878 (section 5.3). Annual river flow in the Oubangui, a major tributary of the Congo river, is very similar (annual flows not shown) except that the significant anomalies begin one month earlier (September 1961, Figure 10). Oubangui river flows remained exceptionally high until November 1964.

### 5.1.8 Zambezi

Further to the south in the Zambezi basin river flows show quite different temporal variability to the other series. 1916-18 are unexceptional years and the prolonged increase in flows from 1940 to 1956 does not feature in any of the other series. 1961 was the third highest flood on record but was followed by three very dry years. Nicholson (1999) also notes differences in behaviour of the southerly lakes Malawi and Rukwa compared to lake level changes in the more northerly East African lakes.

## 5.2 Impacts of the 1997 Event

Hydrological data are only available for the Blue Nile and Lake Victoria in 1997. Blue Nile flows were unexceptional in 1997, much lower than in 1961, but rainfall over the catchment during the following dry season (October to February) was the highest on record (Conway, 2000). A significant rise in Lake Victoria level occurred in 1998, a rise of approximately 1.0m, slightly less than the 1.2m rise from 1961 to 1962 (mean annual levels) which was the largest annual increase on record. Whether levels remain high as after 1961, or will fall more rapidly as after the rises in 1903-07 and 1916-18 will largely depend upon subsequent rainfall over the lake basin. Figure 1 suggests that unlike after 1961 which was followed by a sequence of wet years, rainfall during 1998 and 1999 has not been particularly high over the main contributing zone 5°N-5°S. Satellite altimetry data also show similar Lake Victoria level increases to the observations during 1997 and 1998 (a rise of ~ 1.7 m by 1998, Birkett *et al.*, 1999) but were unavailable after November 1998. Birkett *et al.* (1999) also identified large increases in other lake levels across East Africa as a result of the heavy rainfall, for instance Lake Tanganyika rose by ~ 2.1 m and Lake Malawi by ~

1.8 m and a rise in the levels of the Sudd swamps. IRD (1999) also showed very high river flows for the Congo at Kinshasa during December 1997-March 1998, and during January-March 1998 when they were almost as high as during the same months in 1961-62.

## 5.3 Historical Precedents

Apart from the extreme floods in 1916-18 there are no other twentieth century events similar to the wide-ranging and extreme characteristics of the 1961-64 period. There is not enough evidence here to compare in detail the extent of the 1997 event with 1961 in terms of extreme floods and prolonged impacts on lake levels. The 1916-18 episode was smaller in magnitude and extent, mainly limited to the area of Lake Victoria, southern Sudan and southwestern Ethiopia and marked by 10 per cent to 40 per cent increases in rainfall between October 1916 and September 1917 (Conway, 1997). The largest seasonal increases (up to 100 per cent) occurred from December 1916 to February 1917. The most likely contender is 1878 which was a year of record floods on the Blue Nile, very high Lake Victoria levels and is generally noted as an extremely wet year across the region. Nicholson (1999), for example, cites the explorer Thompson as reporting Lake Tanganyika's maximum level was in 1878. Lake Tanganyika has an intermittent outlet dependent on lake level into the Lukuga river which discharges into the Congo. From documentary evidence Nicholson (1999) notes that 1878 was a year of very heavy rains which opened the Lukuga outlet which reportedly led to flooding in the Congo river. Lyons (1906) reports travellers' observations of high rainfall and lake levels in both MAM and ON seasons in 1877 and that the *'lake rose three feet* [0.91m] *above its usual June maximum in August and September, 1878, in consequence of the heavy rains'* (p.35). Lyons makes no mention of the ON rains in 1878. It is noteworthy that 1878 was also the year of a major ENSO event.

## 5.4 The Persistence of High Lake Victoria Levels After 1961

Lake Victoria is unusual amongst the lakes in East Africa for the prolonged duration of high lake levels following 1961. Water balance studies of Lake Victoria have shown that the increase in lake levels was largely due to extremely high rainfall over the lake and catchment in ON 1961 followed by three further wet years in 1962, 1963 and 1964. Piper *et al.* (1986) found it was not possible, however, to account fully for the rise in lake levels using a water balance approach because their estimate of lake rainfall, which was calculated from an average of eight lakeside gauges, tended to underestimate over-lake rainfall (Datta, 1981). This underestimate has recently been confirmed by Nicholson *et al.* (2000) using satellite cold-cloud data to estimate rainfall over the lake and land area. Lake rainfall is found to be a relatively constant 321mm higher than rainfall over the land area which allows for an improved simulation of observed Lake Victoria levels. Conway and Hulme (1993) calculated that the 2.25m increase in lake level between 1961-64 would account for roughly eight years of the excess outflows after levels peaked in 1964 and together with generally higher rainfall after 1964 than before 1961, this could account for the sustained high levels. Lake Victoria is likely to have a long response time to perturbations to its water balance given its vast area and wetlands around the lake perimeter and in upper tributaries such as the Kagera. Indeed, using a theoretical model of Lake Victoria and the downstream lakes and Sudd swamps, Sene (2000) obtains a period of 10-20 years for Lake Victoria levels to decline towards their initial equilibrium values following a perturbation.

The increase in rainfall subsequent to 1961 has been noted in other studies, although Rodhe and Virji (1976) found no evidence of long-term trends in annual rainfall at six rain gauges around Lake Victoria. Farmer (1981) found evidence of a step-like increase in mean ON rainfall over the Kenyan area of Lake Victoria in the early 1960s. Farmer postulated that a positive feedback, created by increased lake surface area generating increased local moisture and rainfall, could have triggered an increase in the Lake basin rainfall. Hulme *et al.* (in press) found a positive trend in annual rainfall over East Africa of +10 per cent per century (based on 1900-1998). Table 3 shows statistically significant increases in mean ON rainfall over 5°N to 5°S from 1931-60 to 1961-90 (ON 1931-90 trend is 1.2mm per year). Large increases in JJAS and ON rainfall occurred over 5-10°S, and annual increases over 10-15°S. In contrast, northwards of 5°N the zonal average time series show substantial decreases in rainfall between both periods in nearly all seasons.

# 6. DISCUSSION AND CONCLUSIONS

## 6.1 Discussion

These results go some way towards understanding the 1961 and 1997 extreme events particularly in terms of their spatio-temporal characteristics, hydrological impacts and possible precedents. However, questions remain about their causes, frequency, and association with ENSO and the Indian Ocean DMI.

Both events wcrc associated with a dipole-like periodic reversal of Indian Ocean SSTs that is independent of the ENSO. Webster *et al.* (1999) identified a total of 16 years between 1950 and 1998 in which the equatorial SST gradient (or DMI) reversed for at least a month. Saji *et al.* (1999) identified six extreme dipole events since 1958 in 1961, 1967, 1972, 1982, 1994 and 1997 with smaller events in 1977 and 1992. These years all coincide with wet ON seasons (Figure 2) but none have produced hydrological impacts on the scale of those that occurred in 1961 and 1997.

It is not entirely clear to what extent these Indian Ocean events have been influenced by the effects of ENSO and Pacific Ocean SSTs. 1961 was a non-El Niño year but 1997 and 1878 (the most likely historical precedent, in terms of hydrological impacts) were both strong El Niño years.

*Table 3.* Seasonal and annual 30-year mean rainfall for zonal average time series. Percentage difference between 1931-60 and 1961-90 is shown, bold figures represent significant differences between the means at the 5% level, using a one-tailed t-test. Trend is shown using correlation with time over 1931-1990, bold figures represent significant correlation at the 5% level.

| | | Mean 31-60 | Mean 61-90 | % diff. | Correlation with time |
|---|---|---|---|---|---|
| 15-20°N | MAM | 5 | 3 | **-44** | -0.23 |
| | JJAS | 98 | 68 | **-31** | **-0.30** |
| | ON | 7 | 6 | -9 | -0.07 |
| | Annual | 116 | 82 | **-29** | **-0.34** |
| 10-15°N | MAM | 56 | 28 | **-50** | **-0.47** |
| | JJAS | 442 | 365 | **-17** | **-0.57** |
| | ON | 34 | 24 | **-30** | **-0.45** |
| | Annual | 536 | 421 | **-21** | **-0.71** |
| 5-10°N | MAM | 283 | 254 | **-10** | **-0.31** |
| | JJAS | 575 | 526 | **-9** | **-0.35** |
| | ON | 118 | 126 | 7 | -0.04 |
| | Annual | 1035 | 954 | **-8** | **-0.50** |
| 0-5°N | MAM | 414 | 405 | -2 | -0.05 |
| | JJAS | 403 | 392 | -3 | -0.10 |
| | ON | 187 | 241 | **29** | 0.20 |
| | Annual | 1153 | 1176 | 2 | -0.03 |
| 0-5°S | MAM | 473 | 475 | 0 | -0.03 |
| | JJAS | 141 | 151 | 6 | -0.03 |
| | ON | 164 | 229 | **40** | 0.21 |
| | Annual | 1027 | 1109 | **8** | 0.06 |
| 5-10°S | MAM | 454 | 421 | -7 | **-0.26** |
| | JJAS | 48 | 62 | **30** | 0.00 |
| | ON | 126 | 149 | 18 | 0.10 |
| | Annual | 1016 | 1040 | 2 | -0.07 |
| 10-15°S | MAM | 284 | 301 | 6 | 0.07 |
| | JJAS | 5 | 6 | 6 | -0.06 |
| | ON | 110 | 125 | 13 | 0.08 |
| | Annual | 1108 | 1163 | **5** | 0.13 |
| 15-20°S | MAM | 164 | 157 | -5 | -0.06 |
| | JJAS | 21 | 18 | -13 | -0.05 |
| | ON | 130 | 127 | -2 | -0.04 |
| | Annual | 886 | 857 | -3 | -0.02 |
| 20-25°S | MAM | 113 | 108 | -4 | -0.09 |
| | JJAS | 24 | 28 | 16 | 0.14 |
| | ON | 93 | 103 | 10 | 0.18 |
| | Annual | 582 | 566 | -3 | -0.04 |

ENSO has varying length lagged effects on Indian Ocean SSTs and a known influence on rainfall in the region; it accounts for around 20 per cent of variability in October-December East African rainfall. Of the 16 years between 1950 and 1998 that Webster *et al.* (1999) identified with SST reversals only three of these were El Niño years and none La Niña. Already there are many examples where impacts of the 1997 event in East Africa and the Indian Ocean have been, possibly wrongly, attributed directly to the ENSO, e.g. the extreme rainfall and floods (FAO/GIEW, 1998). Further work is therefore necessary to disentangle the effects of the ENSO and its influence on Indian Ocean activity and the dipole mode.

The events of 1961 and 1997 and the possibility of similar events occurring in the future raises the question of their predictability and has implications for longer-term resource management strategies in the region. Many studies have investigated the predictability of the East African short rains (e.g. Mutai *et al.*, 1998 and Kabanda and Jury, 1999) without focusing on the 1961 or 1997 events. Saji *et al.* (1999) identify a SST anomaly pattern build up in the Indian Ocean as a precursor to the 1997 event and Latiff *et al.* (1999) discuss the possibility of prediction given the couple of months time lag between tropical Pacific anomalies and Indian Ocean anomalies. Given that Indian Ocean dipole reversals have occurred many times since 1958, but have not always been associated with extreme rainfall anomalies, the ability to forecast impacts may be limited but warrants further investigation.

There is also the possibility that climate change may alter the nature and frequency of Indian Ocean dipole events and their relationship with rainfall over East Africa. Whether such events in the past may have left a signal in environmental proxy records (either directly through prolonged lake level changes or indirectly through changes in lake sediment delivery and primary productivity) also needs to be considered when such records are interpreted (Cole *et al.*, 2000; Verschuren *et al.*, 2000).

The immediate hydrological impacts of such events include disruption and damage due to temporary inundation of lakeside and wetland areas and river flooding. Longer term management implications revolve around the dynamic nature of water resources over time and the need for flexible management systems that consider the inherent uncertainty in the resource base. This undermines traditional assumptions of reliable yields for planning water supply projects. For example, the dynamic nature of wetlands in response to these events is quite dramatic. In the Sudd swamps in southern Sudan the area of permanent swamp increased from 2800 $km^2$ to 16600 $km^2$ and the area of seasonal swamp increased from 11200 $km^2$ to 14000 $km^2$ after the early 1960s (Sutcliffe and Parks, 1987). Fluctuating lake levels also present management challenges and opportunities for fishing and lakeside activities (Sarch and Allison, in press). Further study of the impacts and response strategies associated with these events is necessary to provide an indication of the region's vulnerability and adaptive capacity in relation to present day climate variability and also future climate change.

## 6.2 Conclusions

The hydrometeorological events of 1961 and 1997 show similar spatial and temporal characteristics insofar as they are directly comparable given the limited number of observations for 1997-98. Both events occurred mainly during ON and primarily affected the area between 10°N and 10°S and from at least 25°E to the Indian Ocean. The events are primarily located over regions where ON rainfall forms part of the normal rainfall season. However, some areas, north of 10°N, recorded heavy rainfall during ON 1997, months that are normally relatively dry.

The spatio-temporal characteristics of the 1961 and 1997 events are quite exceptional in terms of magnitude and extent during the 20$^{th}$ century. Zonal average (5°) rainfall series for the 20$^{th}$ century highlight other years with very wet ON seasons spread over large regions but none as extreme in magnitude nor area as 1961 and 1997.

Riverflow records dating from around 1900 for East, Central and the Horn of Africa show no other events similar to the wide-ranging and extreme floods in 1961, continuing into 1962. The series show other common features (except for the Zambezi), such as high floods in 1916-18 in the Nile basin, particularly the Equatorial Lakes stretch of the White Nile, but they were shorter-lived and smaller in extent than the floods in 1961-62. The 1961 event affected riverflows as far north as the Sobat (up to 10°N) but it is not clear whether high riverflows further north in the Blue Nile and Atbara during 1961 were connected with the event as they occurred earlier, during June-August. The event spread west over the Congo basin and as far south as Lake Malawi, but not southwest over the Zambezi headwaters.

Over the four years 1961-64 the cumulative excess river flow (above the 1961-90 mean) in the White Nile upstream of the Sudd (at Mongalla), Blue Nile, Atbara, Congo, Tana and Zambezi rivers was 11,324$m^3s^{-1}$ (357$km^3$). This amounts to roughly four times the mean annual flow of the Nile and almost one third of the Congo.

There is not enough data for 1997 to fully assess the extent of this event in terms of impacts on riverflows and the possibility of prolonged impacts on lake levels. However, by 1998 the effects upon Lake Victoria and other East African lakes levels had been similar in magnitude to those of 1961-62, and very high flows were recorded in the Congo basin, along with widespread flooding and crop damage in the region, as in 1961.

The closest historical precedent in terms of hydrological effects appears to be 1878, which was a year of record floods on the Blue Nile, and very high Lake Victoria levels. From documentary evidence Nicholson (1999) notes that 1878 was a year of very heavy rains which opened Lake Tanganyika's outlet to the Congo system via the Lukuga leading to flooding downstream.

Many of the rivers had record flows in 1961 followed by high (sometimes higher) flows in 1962. Lake Victoria levels did not peak until 1964, remained above their pre-1961 level until 1997 and remain high. The persistence of high Lake Victoria levels is most likely related to the combined effects of large catchment size and potential for storage, very high rainfall in 1963 and, between 5°N and 5°S, a statistically significant increase in ON rainfall from 1931-60 to 1961-90 of about 35 per cent.

## REFERENCES


Birkett, C.M., Murtugudde, R. and Allan, J.A. (1999) Indian Ocean climate event brings floods to East Africa's lakes and Sudd Marsh. *Geophys. Res. Lett.* **26**, 1031-1034.

Chambers, D.P., Tapley, B.D. and Stewart, R.H. (1999) Anomalous warming in the Indian Ocean coincident with El Niño. *J. Geophys. Res.* **104**, 3035-3047.

Charles, C.D., Hunter, D.E. and Fairbanks, R.G. (1997) Interaction Between the ENSO and the Asian Monsoon in a Coral Record of Tropical Climate. *Science* **277**, 925-928.

Cole, J.E., Dunbar, R.B., McClanahan, T.R. and Muthiga, N.A. (2000) Tropical pacific forcing of decadal SST variability in the Western Indian Ocean over the past two centuries. *Science* **287**, 617-619.

Conway, D. (1997) A spatial and temporal analysis of two extreme rainfall episodes in East Africa: 1916-1917 and 1961-1964. Fifth Int. Conf. on Southern Hemisphere Meteorology and Oceanography, Pretoria 1997. AMS Pre-print volume, pp. 158-159.

Conway, D. (2000) The climate and hydrology of the Upper Blue Nile, Ethiopia. *Geog. J.* **166**, 49-62.

Conway, D. and Hulme, M. (1993) Recent fluctuations in precipitation and runoff over the Nile subbasins and their impact on Main Nile discharge. *Climatic Change* **25**, 127-151.

Datta, R.R. (1981) Certain Aspects of Monsoonal Precipitation Dynamics over Lake Victoria, in Sir James Lighthill and R.P. Pearce (eds.), *Monsoon Dynamics*. Cambridge University Press, Cambridge, pp. 333-349.

Dixon, A. (2000) *Indigenous hydrological knowledge and the management of wetlands in Illubabor, Ethiopia*. Unpublished Ph.D. thesis, University of Huddersfield.

FAO/GIEW (1998) Heavy rains attributed to El Niño cause extensive crop damage in parts of Eastern Africa. www.fao.org/WAICENT/faoinf...english/alertes/1998/sreaf981.htm

Farmer, G. (1981) *Regionalisation and Study of An Alleged Change in the Regional Climatology of East Africa*. Unpublished Ph.D. Thesis, University of Sheffield, U.K.

Flohn, H. (1987) East African Rains of 1961/62 and the Abrupt Change of the White Nile Discharge. *Palaeoecology of Africa* **18**, 3-18.

Flohn, H. and Burkhardt, T.H. (1985) Nile Runoff at Aswan and Lake Victoria: A Case of Discontinuous Climate Time Series. *Zeitschr. Gletscherkde. Glaziolgeol.* **21**, 125-130.

Grötzner, A., Latif, M. and Dommenget, D. (1999): Atmospheric response to sea surface temperature anomalies during El Niño 1997/1998. *Q. J. Roy. Met. Soc.* **126**, 2175-2198.

Grove, A.T. (1996) African river discharges and lake levels in the Twentieth Century, in T. C Johnson and E. O. Odada (eds.), *The Limnology, Climatology and Paleoclimatology of the East African Lakes*, Gordon and Breach, The Netherlands, pp. 95-100.

Hulme, M. (1994) Validation of large-scale precipitation fields in General Circulation Models, in M. Desbois and F. Desalmand (eds.), *Global Precipitation and Climate Change*, NATO ASI Series, Vol. 126, Springer-Verlag, Berlin, Heidelberg, pp. 387-405.

Hulme, M., Doherty, R., Ngara, T., New, M. and Lister, D. (in press) African Climate Change: 1900-2100. *Clim. Res.*, in press.

Hurst, H.E. and Phillips, P. (1933) *The Nile Basin, Volume IV. Ten-day Mean and Monthly Mean Discharges of the Nile and its Tributaries.* Ministry of Public Works, Physical Department, Cairo.

IRD (1999) Observatoire hydrologique regional de l'Afrique de l'Ouest et Centrale. Accessed online at http://ohraoc.orstom.bf/HTML/ETUDES/HYDRO/BRAZZA.HTM. Accessed on 19/05/99.

Janowiak, JE and Xie, P. (1999) CAMS-OPI: A global satellite-rain gauge merged product for real-time precipitation monitoring applications. *J. Climate* **12**, 3335-3342.

Jones, P.D. and Conway, D. (1997) Precipitation in the British Isles: an analysis of area-average data updated to 1995. *Int. J. Climatol.* **17**, 427-438.

Kabanda, T.A. and Jury, M.R. (1999) Inter-annual variability of short rains over northern Tanzania. *Clim. Res.* **13**, 231-241.

Kite, G.W. (1981) Recent Changes in Level of Lake Victoria. *Hydrol. Sci. Bull.* **26**, 233-243.

Lamb, H.H. (1966) Climate in the 1960s. *Geog. J.* **132**, 183-212.

Laraque A., Maziezoula, B., Orange, D., Olivry, J. C. (1998) Origine des variations de débits du Congo à Brazzaville durant le XXième siècle, in E. Servat, D. Hughes, Jean-Marie Fritsch and M. Hulme (eds.), *Water Resources Variability in Africa during the XXth Century*, IAHS Publ. no. 252, pp. 171-180.

Latif M., Dommenget D., Dima M., Grotzner A. (1999) The role of Indian Ocean sea surface temperature in forcing east African rainfall anomalies during December-January 1997/98. *J. Climate* **12**, 3497-3504.

Lyons, H.G. (1906) *The Physiography of the River Nile and its Basin.* Royal Met. Soc. Cairo.

Mörth, H.T. (1967) *Investigations into the Meteorological Aspects of the Variations in the Level of Lake Victoria.* East African Met. Dept. Memoirs 4, 1-22.

Murtugudde, R., McCreary Jr., J.P. and Busalacchi, A.J. (2000) Oceanic processes associated with anomalous events in the Indian Ocean with relevance to 1997-1998. *J. Geophys. Res.* **105**, 3295-3306.

Mutai, C.C., Ward, M.N. and Colman, A.W. (1998) Towards the prediction of the East Africa short rains based on sea-surface temperature-atmosphere coupling. *Int. J. Climatol.* **18**, 975-997.

Nicholson, S.E. (1997a) Historical fluctuations of Lake Victoria and other lakes in the northern Rift Valley of East Africa, in Lehman, J.T. (ed.), *Environmental change and response in East African lakes.* Kluwer, Dordrecht, pp. 7-35.

Nicholson, S.E. (1997b) Fluctuations of Rift Valley lakes Malawi and Chilwa during historical times: a synthesis of geological, archaeological and historical information, in Lehman, J.T. (ed.), *Environmental change and response in East African lakes.* Kluwer, Dordrecht, pp. 207-231.

Nicholson, S.E. (1999) Historical and modern fluctuations of lakes Tanganyika and Rukwa and their relationship to rainfall variability. *Climatic Change* **41**, 53-71.

Nicholson, S.E., Yin, X. and Ba, M.B. (2000). On the feasibility of using a lake water balance model to infer rainfall: an example from Lake Victoria. *Hyd. Sci. J.* **45**, 75-95.

Odingo, R.S. (1962) The abnormal and unseasonal rains in East Africa. *The Geographical Review* **52**, 440-442.

Piper, B.S., Plinston, D.T. and Sutcliffe, J.V. (1986) The Water Balance of Lake Victoria. *Hydrol. Sci. J.* **31**, 25-37.

Reason, C.J.C., Allan, R.J., Lindesay, J.A. and Ansell, T.J. (2000) ENSO and climatic signals across the Indian Ocean Basin in the global context: Part I, interannual composite patterns. *Int. J. Climatol.* **20**, 1285-1327.

Reverdin, G., Cadet, D.L. and Gutzler, D. (1986) Interannual displacements of convection and surface circulation over the equatorial Indian Ocean. *Quart. J. R. Met. Soc.* **112**, 43-67.

Rodhe, H. and Virji, H. (1976) Trends and Periodicities in East African Rainfall Data. *Monthly Weather Review* **104**, 307-315.

Saji, N.H., Boswami, B.N., Vinayachanran, P.N. and Yamagata, T. (1999) A Dipole Mode in the Tropical Indian Ocean. *Nature* **401**, 360-363.

Sarch, M-T. and Allison, E. (in press) Fluctuating fisheries in Africa's inland waters: well adapted livelihoods, maladapted management. Proceedings of International Institute of Fisheries Economics and Trade (IIFET), 10th International Conference. Eugene, Oregon, July 2000.

Sene, K.J. (2000). Theoretical estimates for the influence of Lake Victoria on flows in the Upper White Nile. *Hyd. Sci. J.* **45**, 125-145.

Sutcliffe, J.V. and Parks, Y.P. (1987) Hydrological Modelling of the Sudd and Jonglei Canal. *Hydrol. Sci. J.* **32**, 143-159.

Sutcliffe, J.V. and Parks, Y.P. (1999) *The Hydrology of the Nile.* IAHS Special Publication No. 5. IAHS Press, Wallingford, Oxfordshire.

UNESCO (1995) *Discharge of selected rivers of Africa.* UNESCO Studies in hydrology No 52.

Venzke, S., Latif, M. and Villwock, A. (in press): The coupled GCM ECHO-2. Part II: Indian Ocean response to ENSO. *J. Climate*, in press

Verschuren D. Laird, K.R. and Cumming, B.F. (2000) Rainfall and drought in equatorial east Africa during the last 1,100 years. *Nature* **403**, 410-413.

Walsh, R.P.D., Davies, H.R.J. and Musa, S.B. (1994) Flood frequency and impacts at Khartoum since the early nineteenth century. *Geog. J.* **160**, 266-279.

Webster, P.J., Moore, A.M., Loschnigg, J.P. and Lebden, R.R. (1999) Coupled Ocean-Atmosphere Dynamics in the Indian Ocean During 1997-98. *Nature* **401**, 356-360.

Xie, P.P. and Arkin, P.A. (1998) Global monthly precipitation estimates from satellite-observed outgoing longwave radiation. *Journal of Climate* **11**, 137-164.

# MESOSCALE PATTERNS OF RAINFALL, CLOUDINESS AND EVAPORATION OVER THE GREAT LAKES OF EAST AFRICA

SHARON E. NICHOLSON and XUNGANG YIN
*Department of Meteorology, Florida State University, Tallahassee, Florida 32306, USA*

## ABSTRACT

By employing remote sensing techniques, the characteristics of rainfall, cloudiness and evaporation over Lakes Victoria, Tanganyika and Malawi are studied. There exist diurnal cycles induced by the interaction between lake/land breeze and the lower level southeasterlies over each lake, particularly Lake Victoria. Generally, maximum convection/rainfall is in the night to early morning on the west side of the lake while on the other side of the lake it is in the afternoon to early evening. The mean annual over-lake rainfall is greater than mean annual catchment rainfall by 30% and 20% for Lakes Victoria and Tanganyika, respectively. Lake Victoria is divided into four quadrants based on the fact that convection is stronger in the northern quadrants and that the diurnal cycles in the east and west quadrants are quite different. Convective activity is compared among the four quadrants and the catchment. Temporal and spatial distribution of cloud cover resembles that of convection. Annual mean cloudiness is 0.5, 0.45 and 0.38 for the above three lakes, respectively. For Lake Victoria both the monthly and annual mean cloudiness maps demonstrated a daytime cloudiness minimum and a nighttime cloudiness maximum over the lake. Evaporation is calculated for each lake by energy-budget and combined-Penman methods. Both a fixed cloudiness scenario and a varied cloudiness scenario are used as cloud cover input. Calculated values range from 1537 to 1669 mm/year for Lake Victoria, 1559 to 1721 mm/year for Lake Tanganyika and 1698 to 1901 mm/year for Lake Malawi. In order to investigate the influence of



*E.O. Odada and D.O. Olago (eds.),*
*The East African Great Lakes: Limnology, Palaeolimnology and Biodiversity,* 93–119.

cloudiness, evaporation calculation is also carried out in each quadrant and in each pixel of Lake Victoria. The results show that evaporation rate is highest in the middle of the lake and decreases in both the east and the west directions.

# 1. INTRODUCTION

Straddling the equator from the northern tip of Lake Turkana to the southern tip of Lake Malawi, the East African lakes (Figure 1) provide a key to understanding climate variability in the tropics (Nicholson and Flohn, 1980; Kite, 1981; Piper *et al.*, 1986; Nicholson, 1998; Nicholson *et al.*, 2000). Their collective records can yield a spatially and temporally detailed picture of the region's long-term environmental

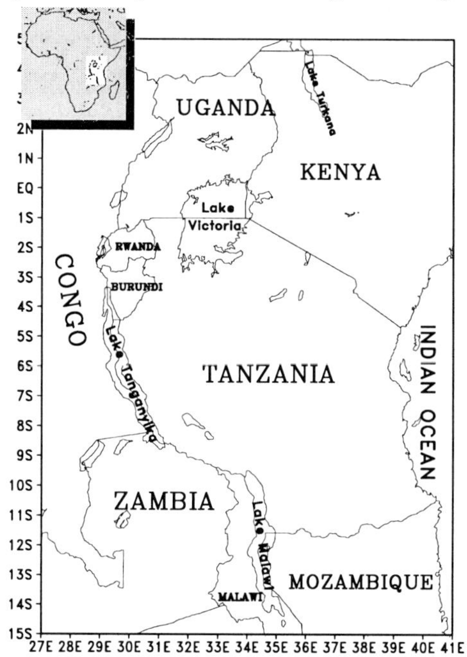


*Figure 1.* Map of the East African Lakes.

histories. In the absence of tectonic activity and human influence, the water balance process of a lake is directly driven by climatic parameters through their impacts on each term of the water budget. Therefore, an in-depth study of the water balance terms, such as rainfall and evaporation, is essential to a successful water balance model.

The East African lakes, particularly those large ones, alter the regional climates, as shown in several water balance studies. In an earlier work on Lake Victoria, Flohn and Fraedrich (1966) described a diurnal circulation system and demonstrated that rainfall over the lake surface is essentially controlled by the convergence associated with the nocturnal land breeze component of this system. Because the lake is warmer than the air, strong thunderstorms develop over the center of the lake. A recent study of Lake Victoria by Ba and Nicholson (1998) concluded that the over-lake rainfall is more than the catchment rainfall by approximately 26%. The spatial and temporal inhomogeneities in the cloudiness distribution have a significant implication on the evaporation. Yin and Nicholson (1998) pointed out that cloudiness is the most influential factor in determining evaporation. In addition, the lake/land breeze and cloudiness over these lakes can influence the local ecosystem and geochemical processes through the water circulation and net radiation.

In practice, the most important water balance terms, over-lake rainfall and evaporation, are also the most difficult ones to estimate. For some historical reasons, the climatic and hydrologic data from these lakes are scarce and fragmentary, even in the recent decades. Since the early 20th century, there have been many efforts to estimate the water balance of the East African lakes. Some of the published works for Lake Victoria (e.g., Hurst, 1952; EAMD, 1961, 1968; de Baulny and Baker, 1970; WMO, 1974; Hastenrath and Kutzbach, 1983; Spigel and Coulter, 1996; Howell *et al.*, 1988; Flohn, 1983; Flohn and Burkhardt, 1985; Kite, 1982; Piper *et al.*, 1986; Balek, 1977; Sene and Plinston, 1994) are reviewed in Yin and Nicholson (1998). Valuable studies on the other two large lakes, Lakes Tanganyika and Malawi, have also been published in the past few decades (e.g., Livingstone, 1965; Scholz and Rosendahl, 1988; Calder *et al.*, 1995; Savijärvi, 1997; Bergonzini *et al.*, 1997). In those results, the over-lake rainfall and evaporation values are very uncertain. Even for basically the same period, estimates by different authors may differ considerably.

Reasons for the uncertainties of rainfall estimates are varied. However, one pervasive problem is that most of the previous estimates are based on data from either the shoreline stations or the few island rainfall records. This is inadequate, particularly for a lake with a vast surface area over which rainfall varies nonlinearly. The problem is illustrated by the work of de Baulny and Baker (1970), who assessed rainfall over Lake Victoria from gauges at eight lakeshore stations. The calculations suggested a rainfall minimum over the lake, whereas other studies (e.g., EAMD, 1968; Datta, 1981) showed a maximum over the lake. The latter studies also showed the complexity of the problem by evaluating the diurnal cycle of rainfall. They found a broad early morning to midday maximum in rainfall frequency over the western shore, an afternoon maximum on the eastern shore, and an early morning maximum frequency over the center of the lake.

Likewise, calculating evaporation with cloudiness at shoreline stations, as many authors have done, can introduce error due to the different cloud amounts and diurnal cycles over the lake and over the catchment. As with rainfall, Lake Victoria experiences a daytime cloudiness minimum and a nighttime cloudiness maximum (NOAA-USAF, 1971; Kayiranga, 1991; Ba and Nicholson, 1998). This means more net radiation and thus higher evaporation is expected from the lake. Yin and Nicholson (1998) showed that a 10% cloud cover increase could result in a decrease of evaporation by 200-400 mm/year. Another source of error is the use of a mean cloud cover in evaporation calculations. Because the effect of cloudiness on the surface radiation balance is non-linear, failure to account for the strong diurnal and seasonal cycles in the over-lake cloudiness can introduce substantial error in the estimates. The empirical formulae governing the radiation budget, which will be introduced in the following section, demonstrate this fact.

The above problems cannot be solved with the current surface observational network. To clearly derive a 3-dimensional (spatial plus temporal) picture of the over-lake convection/cloudiness or rainfall/evaporation, much higher resolution data are desirable. At present, this can only be done by using satellite data.

In this work, firstly, the temporal and spatial distribution of convection/rainfall is analyzed for Lakes Victoria, Tanganyika and Malawi. Secondly, we derive the over-lake and catchment rainfall relationship for each of the lakes so that over-lake rainfall can be calculated from the catchment rainfall through the established regression equations. Thirdly, cloudiness is calculated and analyzed for the lakes. Finally, evaporation is estimated for each lake using two independent methods and two cloudiness scenarios. For Lake Victoria the analysis is carried out in greater detail. First of all, cloudiness amount and diurnal cycles are evaluated for and compared among the four quadrants. Evaporation over the lake is calculated at the pixel scale ($0.25° \times 0.25$, or ca. $30 \times 30$ km$^2$), in order to produce monthly maps. For each of the four quadrants of Lake Victoria, evaporation is estimated as to study the geographical difference caused by the cloudiness distribution.

## 2. DATA AND METHODOLOGY

### 2.1 Estimates of Convection and Cloudiness

Both convective activity and cloudiness are estimated from ISCCP (International Satellite Cloud Climatology) B2 format infrared (IR) data derived from Meteosat (ESOC, 1986). The data have a spatial resolution of $0.25° \times 0.25°$, equivalent to about 30 km, and a temporal resolution of 3-hour commencing with approximately 01:30 LST (Local Solar Time). The raw data appear as IR counts between 0 and 255. Using the Meteosat calibration formula, the counts are converted to brightness temperatures. The result is the mean of a five-year period consisting of 1984, 1985, 1986, 1993 and 1994. These years were selected in order to obtain a sampling of data that spans more than a decade and includes both wet and dry years.

Convection is studied by the concept of cold cloud frequency, which assumes that a pixel has convection if and only if the observed infrared (IR) brightness temperature is lower than a predefined threshold temperature. Then, for a certain period, the frequency of convection can be calculated and is used as an indicator of the strength of convective activity (Ba *et al.*, 1995; Ba and Nicholson, 1998). In the study of Yin (2000), the threshold temperature is determined by regression of station rainfall and corresponding cold cloud frequency. For Lake Victoria it is -39°C and for Lakes Tanganyika and Malawi it is -36°C. With the help of the relationship between rainfall and cold cloud frequency, mean over-lake rainfall is derived from mean catchment rainfall through a regression technique (Nicholson *et al.*, 2000).

Since the above relationship is linear, the spatial distribution of rainfall resembles that of convection. Therefore, the characteristics of rainfall and convection are analyzed together.

Cloudiness is estimated by the infrared threshold method (Rossow *et al.*, 1985; Kidder and Vonder Harr, 1995). Based on the ISCCP cloud algorithm, the estimation proceeds in three steps. First, a clear sky map is constructed by way of the 15-day maximum brightness temperature for each pixel. This is based on the fact that clear sky is much warmer than cloudy sky and also the variation of clear sky is much less. A pixel's status is indicated as either cloudy (100% cover) or clear (0% cover). Second, a threshold temperature is determined. For the IR brightness temperature over the continent, a temperature difference of 12 K is selected to distinguish cloudy or clear. The threshold is designed to take into account both instrument noise and uncertainty in the clear-sky values. If a pixel's brightness temperature is lower than its clear sky temperature constructed in the first step by 12 K, it is labeled as cloudy. This is different from the threshold value adopted in our earlier work (Yin *et al.*, 2000) since our experiment shows that by using the 12 K threshold, the calculated annual mean cloudiness for Lakes Victoria and Tanganyika is closer to the available observations. This is demonstrated in the results section of this paper. Finally, mean cloudiness is calculated, both for each pixel and for a specified area. For each month, after the half-month mean cloudiness is estimated for each observation time of a day, the first and the second half-month cloudiness are averaged to represent the monthly mean cloudiness for this observation time. The overall monthly mean cloud cover is an average of all the eight observation times. In this work, mean cloudiness is averaged over the five-year period to represent the long-term mean.

## 2.2 Estimates of Evaporation

To cross-validate the evaporation estimates, both energy-budget and combined-Penman formulae are used in the study. Details of the two methods were given in Yin and Nicholson (1998) and Yin *et al.* (2000). Hence only the most basic equations and a summary of input data are presented here.

### 2.2.1 Radiation Calculation

Radiation balance is the basis for evaporation calculation. Net short wave radiation is estimated by a formula given by Black (1956):

$$SW = I_0 \times (0.803 - 0.34c - 0.458c^2)(1 - \alpha) \tag{1}$$

where $I_0$ is the solar insolation at the top of the atmosphere, $c$ is the cloudiness and $\alpha$ is the water surface albedo calculated as (Anderson, 1954):

$$\text{low cloudiness } (\leq 0.55) \qquad \alpha = 2.17 S_a^{-0.96}$$

$$\text{high cloudiness } (>0.55) \qquad \alpha = 0.78 S_a^{-0.68}$$

where $S_a$ is the Sun's altitude. For the three large lakes, the calculated albedo is roughly between 0.06 and 0.07. This is higher than the commonly used value of 0.06 for this region. However, evaporation is insensitive to albedo variation in the above range (Yin, 2000).

Net long wave radiation is estimated by Budyko's equation (1974):

$$LW = \varepsilon\sigma(t_w + 273.15)^4(0.39 - 0.05\sqrt{e_a})(1 - 0.53c^2) \qquad (2)$$

where $\varepsilon$ is the water surface emissivity, $\sigma$ is the Stefan-Boltzman constant, $t_w$ is the water temperature and $e_a$ is the vapor pressure of the air at 2 m height. Then, net radiation gained by the water surface is calculated as:

$$R_{net} = SW - LW \qquad (3)$$

### 2.2.2 Energy-Budget Method

The essence of the energy-budget approach is that, under steady-state conditions, the net radiation at the water surface is balanced by the latent heat used to evaporate water from the lake and by the sensible heat transfer between the lake and the atmosphere. The basic energy balance equation is rewritten through the use of the Bowen ratio, $B$, defined as the ratio of sensible heat transfer to latent heat transfer, to yield the following equation for evaporation, $E$:

$$E = \frac{R_{net}}{\rho_w(1+B)L_v} \qquad (4)$$

wherein $\rho_w$ is the mass density of water and $L_v$ is the latent heat of vaporization calculated from the water temperature $t_w$:

$$L_v = 597.3 - 0.564t_w \qquad (5)$$

The Bowen ratio is evaluated using "bulk formulae" through the following equation:

$$B = \frac{C_a P(t_w - t_a)}{0.622 L_v (e_{sat} - e_a)} \tag{6}$$

where $C_a$ is the heat capacity of air, $t_a$ is the air temperature and $P$ is the atmospheric pressure at the elevation of the lake. The saturated vapor pressure $e_{sat}$ is calculated as:

$$e_{sat}(t_w) = 6.11 e^{\frac{17.3 t_w}{t_w + 237.3}} \tag{7}$$

### 2.2.3 Combined-Penman Method

The combined-Penman approach estimates evaporation by the following equation (Penman, 1948):

$$E = \frac{R_d \Delta + \gamma E_a}{\Delta + \gamma} \tag{8}$$

which is a combination of the energy-balance method with the concept of mass-transfer. The parameters in the equation are expressed as follows:

$$R_d = \frac{R_{net}}{\rho_w L_v} \tag{9}$$

$$\Delta = \frac{25083}{t_a + 237.3} 6.11 e^{\frac{17.3 t_w}{ta + 237.3}} \tag{10}$$

$$\gamma = \frac{C_a P}{0.622 L_v} \tag{11}$$

$$E_a = (0.013 + u_2)(e_{sa} - e_a) \tag{12}$$

where $u_2$ is the wind speed at 2 m height expressed in km/day, $e_{sa}$ is the saturated vapor pressure of a water surface at the air temperature $t_a$.

### 2.2.4 Input Data

Input data of evaporation calculation for the three lakes are summarized in Tables 1, 2 and 3. Water temperature $t_a$ is taken from Talling's publication (1969). Solar insolation $I_0$ is taken from a table in Black (1956) and has been interpolated for the latitudes of the lakes. The rest of the data, air temperature $t_w$, air vapor pressure $e_a$ and wind speed $u_2$, are the corresponding average values from lakeshore stations found in FAO (1984). There are six stations (Jinja, Entebbe, Mwanza, Musoma, Bukoba, Kisumu) for Lake Victoria; two stations (Kalemie and Kigoma) for Lake Tanganyika and three stations (Cobue, Nkhata Bay and Karonga) for Lake Malawi. An analysis of Lake Victoria by Yin and Nicholson (1998) and Yin *et al.* (2000) shows that data averaged from the shoreline stations are generally in close agreement with the available measurements from the open lake or on the islands.

*Table 1.* Lake Victoria input data for evaporation calculation. ($t_w$: water temperature; ta: air temperature; $e_a$: air vapor pressure; $I_0$: solar insolation at the top of the atmosphere; $u_2$: wind speed at 2 m height over the water surface).

| Month | $t_w$ (°C) | $t_a$ (°C) | $e_a$ (mb) | $I_0$ ($Wm^{-2}$) | $u_2$ ($ms^{-1}$) |
|---|---|---|---|---|---|
| January | 25.5 | 22.6 | 19.7 | 418 | 1.90 |
| February | 26.3 | 22.8 | 19.9 | 431 | 1.95 |
| March | 26.3 | 22.9 | 20.5 | 433 | 1.97 |
| April | 25.8 | 22.5 | 21.3 | 419 | 1.87 |
| May | 25.4 | 22.2 | 21.0 | 393 | 1.83 |
| June | 25.3 | 21.8 | 19.4 | 378 | 2.08 |
| July | 24.7 | 21.3 | 18.3 | 381 | 2.08 |
| August | 24.8 | 21.6 | 18.7 | 404 | 2.00 |
| September | 24.7 | 22.1 | 19.0 | 425 | 2.05 |
| October | 24.9 | 22.7 | 19.5 | 430 | 1.98 |
| November | 25.4 | 22.7 | 19.7 | 420 | 1.90 |
| December | 25.7 | 22.4 | 19.9 | 411 | 1.78 |

*Table 2.* Lake Tanganyika input data for evaporation calculation. (symbols meanings are the same as given in Table 1)

| *Month* | $t_w$ (°C) | $t_a$ (°C) | $e_a$ (mb) | $I_0$ ($Wm^{-2}$) | $u_2$ ($ms^{-1}$) |
|---|---|---|---|---|---|
| January | 26.0 | 23.6 | 23.2 | 436 | 1.65 |
| February | 26.1 | 23.4 | 23.1 | 442 | 1.65 |
| March | 26.8 | 23.6 | 23.5 | 431 | 1.65 |
| April | 26.8 | 23.5 | 23.5 | 404 | 1.65 |
| May | 26.5 | 23.5 | 22.0 | 371 | 1.85 |
| June | 26.2 | 22.4 | 18.9 | 352 | 2.10 |
| July | 25.5 | 22.0 | 17.3 | 259 | 2.40 |
| August | 25.2 | 23.1 | 17.6 | 287 | 2.35 |
| September | 25.7 | 24.3 | 19.2 | 416 | 2.20 |
| October | 25.7 | 24.6 | 21.1 | 433 | 2.00 |
| November | 26.0 | 23.3 | 22.5 | 434 | 1.80 |
| December | 26.1 | 23.0 | 23.0 | 433 | 1.65 |

*Table 3.* Lake Tanganyika input data for evaporation calculation. (symbols meanings are the same as given in Table 1).

| *Month* | $t_w$ (°C) | $t_a$ (°C) | $e_a$ (mb) | $I_0$ ($Wm^{-2}$) | $u_2$ ($ms^{-1}$) |
|---|---|---|---|---|---|
| January | 28.5 | 24.5 | 25.0 | 456 | 1.10 |
| February | 28.7 | 24.6 | 25.5 | 452 | 1.03 |
| March | 28.2 | 24.4 | 24.9 | 426 | 1.06 |
| April | 28.2 | 24.1 | 23.0 | 386 | 1.30 |
| May | 27.0 | 22.5 | 19.3 | 344 | 1.46 |
| June | 26.0 | 21.0 | 16.2 | 321 | 1.50 |
| July | 24.1 | 20.7 | 16.9 | 331 | 1.66 |
| August | 23.6 | 21.4 | 16.5 | 464 | 1.73 |
| September | 24.6 | 23.3 | 17.3 | 404 | 1.80 |
| October | 26.3 | 25.5 | 19.3 | 436 | 1.90 |
| November | 27.7 | 26.2 | 22.3 | 450 | 1.73 |
| December | 28.4 | 25.2 | 24.6 | 457 | 1.33 |

# 3. RESULTS

## 3.1 Convective Activity over the Lakes

Over Lake Victoria, there are two rainy seasons each year. One is in March-May and is traditionally named as "long-rains". The other is in October-November and is called as "short-rains". Over the other two basins, a single rainy season appears in November-April for Lake Tanganyika and in December-March for Lake Malawi. To elucidate the characteristics of convection and rainfall, two maximum-rainfall months for each lake are selected to show the temporal and spatial distribution of the cold cloud frequency, which is an equivalent measure of convection and rainfall.

### 3.1.1 Lake Victoria

Figure 2 presents the April and November convection/rainfall for Lake Victoria. The contour is a representative of strength and the shading shows the time when the maximum frequency is observed. There are two prominent features, one is the enhancement of convection/rainfall over the lake compared to the surrounding catchment, and the other is the difference of the diurnal cycles over different parts of the lake. Stronger convection/rainfall over the lake is due to the wind convergence and to the warmer water surface. In the nighttime, land breezes from around the catchment converge over the center of the lake, inducing upward air motion, which is at the same time intensified by the warm water body. Low-level southeasterlies play an important role in the process of convection/rainfall formation: on the one hand, it joins the over-lake convergence; on the other hand it displaces the convergent center to the northwest of the lake.

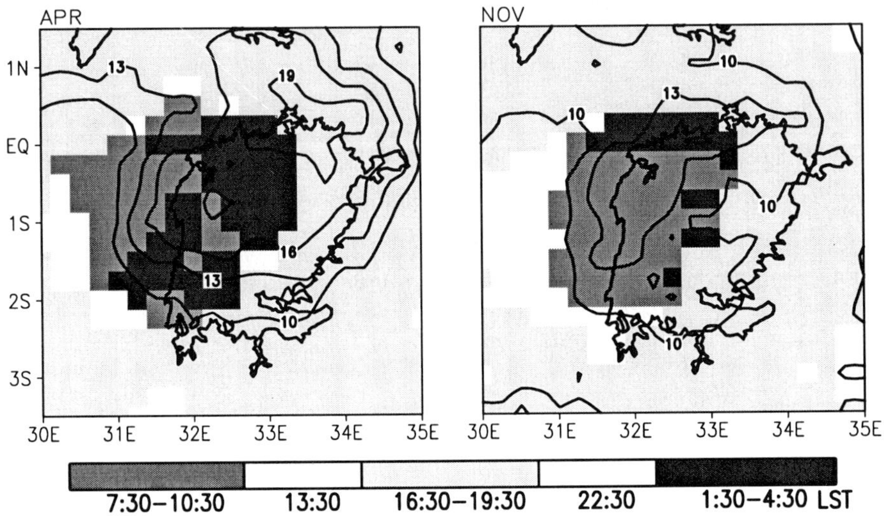


*Figure. 2.* Lake Victoria convection in April (left panel) and November (right panel). Contour: frequency of convection (×100); shading: time of maximum frequency of convection.

In the long-rains represented by April, maximum frequency of convection/rainfall exists in the late night between 01:30 to 04:30 LST over the western lake and in the morning between 07:30 to 10:30 LST over the western catchment. In the short-rains represented by November, maximum frequency is generally observed in the morning over the western part of the lake, extending to the western catchment. This pattern is best developed in November. In December (not shown), this nocturnal maximum moves to the middle of the lake and occurs some hours earlier. In both rainy seasons over the eastern lake and much of the catchment area, maximum frequency of convection/rainfall is in the afternoon to early evening between 16:30 to 19:30 LST.

The transient zone of convection/rainfall pattern from the east to the west in Lake Victoria is roughly along the 33°E, and there is also a difference between the north and the south. Accordingly, Lake Victoria is divided into four quadrants along the 33.125°E longitude line and the 1.125°S latitude line, as shown in Figure 3. The Lake Victoria basin is represented by the unshaded area on this figure. The areal mean frequency by month and by hour in each of the four quadrants and the catchment is shown in Figure 4. The contrast between quadrants 1 and 2 is sharp. In quadrant 1, the highest frequency is in the long-rains (March-April) around 05:00 LST and the second highest one is in the short-rains (October-November) between 08:00 and 11:00 LST. In quadrant 2, the March-April maximum is in the evening around 20:00 LST, while the second maximum is in November around 17:00 LST. Convection in quadrants 3 and 4 resembles that of quadrants 1 and 2, respectively.

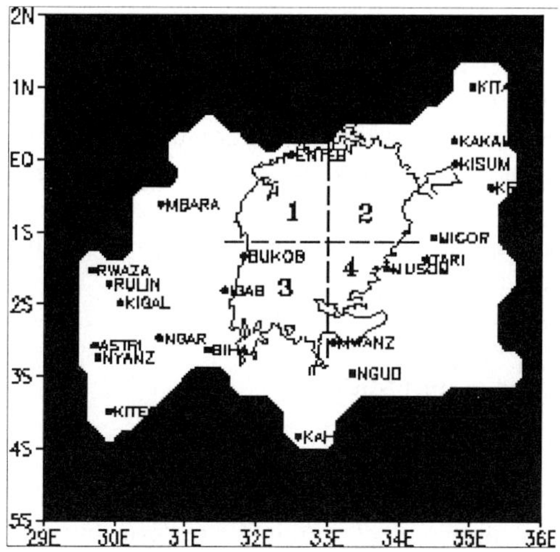


*Figure 3.* Map of the basin of Lake Victoria. The whole basin is the unshaded area. The four quadrants are indicated by the numbers over the lake. Catchment rainfall stations used in this study are also shown.

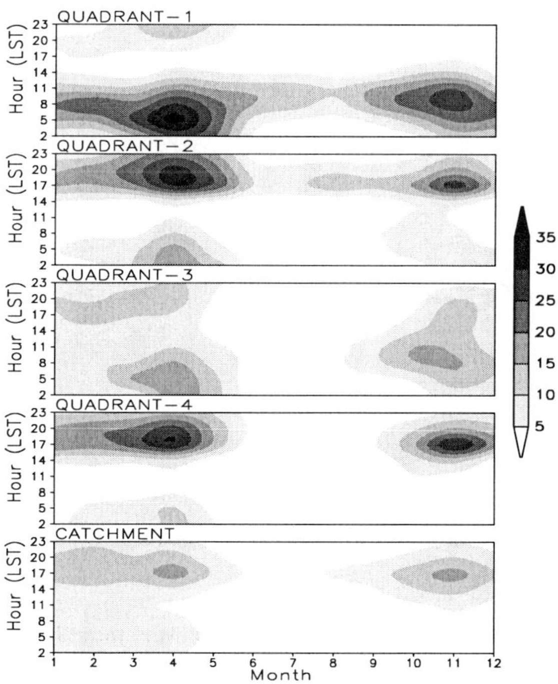


*Figure 4.* Frequency of convection (×100) by hour and month in each quadrant and the catchment of Lake Victoria.

However, the frequency in quadrant 3 is very low (weak convection) while in quadrant 4 is quite high (strong convection), particularly the November maximum which is stronger than its counterpart in quadrant 2. Over the catchment, convection frequency pattern is very similar to that of quadrants 2 and 4, but much weaker. The two weak maximum frequency centers are in April and November and are all found in the late afternoon around 17:00 LST. During the dry season in June-August, convection is weak in the first two quadrants and is almost disappeared in the last two quadrants as well as in the catchment.

### 3.1.2 Lakes Tanganyika and Malawi

Figures 5 and 6 are the January and February monthly mean frequency of convection/rainfall for Lakes Tanganyika and Malawi, respectively. Both lakes have an observable effect on the convection/rainfall pattern, but the features are not as obvious as those of Lake Victoria.

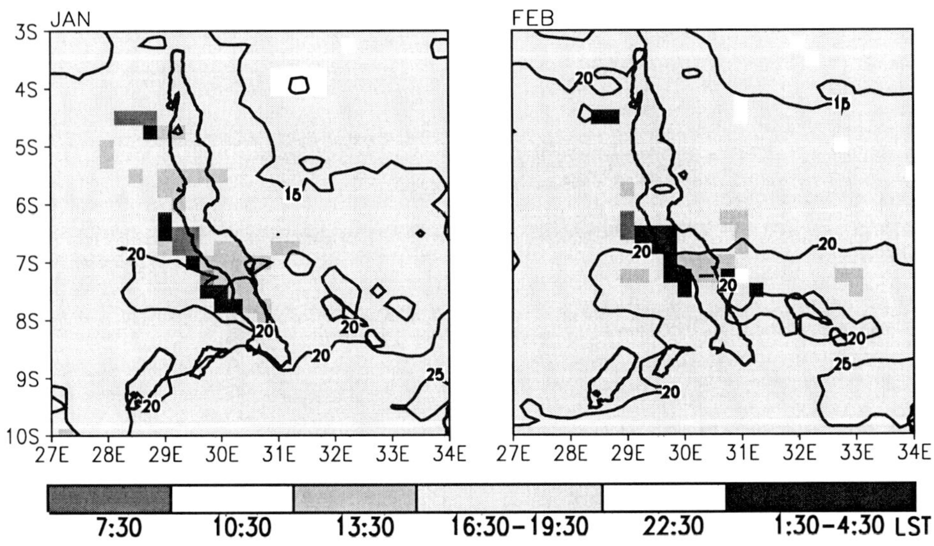


*Figure 5.* Same as Fig. 2 except for Lake Tanganyika in January and February.

Over the catchment of each lake, maximum frequency is mainly in the afternoon between 16:30 to 19:30 LST. This is common for all the three lakes. Over the south of Lake Tanganyika, the time of maximum frequency of convection/rainfall is often in the early afternoon around 13:30 LST, which is earlier than most of the surrounding catchment. In Figure 5, between 8°S through 6°S, along the western lakeshore, convection/rainfall is frequent in the late night to early morning. This feature is more significant for Lake Malawi between 13°S through 10°S, as shown in Figure 6. Compared to the catchment, over-lake enhancement of convection/rainfall is not observed for Lake Malawi but can be detected for Lake Tanganyika throughout the rainy season from November to April. We will further discuss this issue in the next section using the lake-land rainfall regression equations.

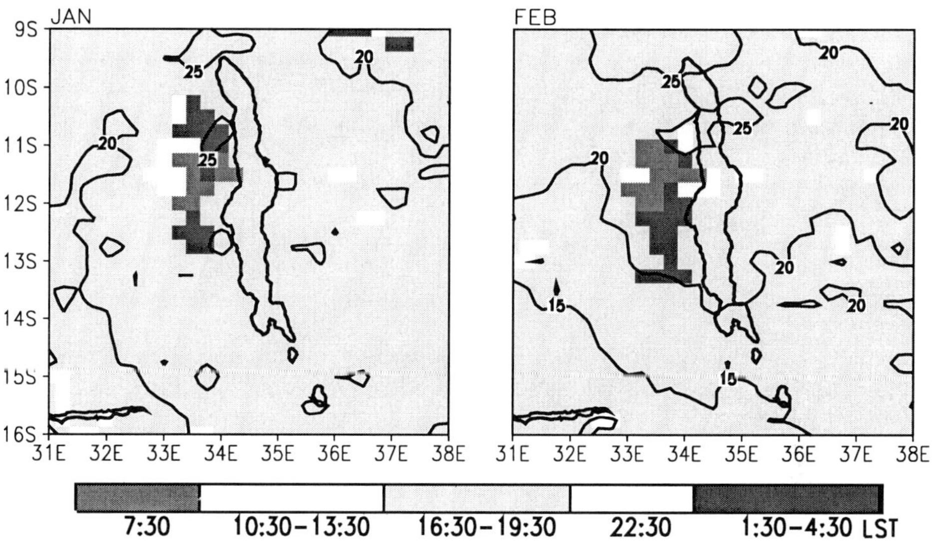


*Figure 6.* Same as Fig. 2 except for Lake Malawi in January and February.

## 3.2 Rainfall over the Lakes and the Catchments

By using the equations (Yin, 2000) relating rainfall to cold cloud frequency, the five-year mean rainfall is calculated for the three basins and shown in Figure 7. The rainfall pattern is simple and clear in Lake Victoria basin. The maximum occurs in the northwestern part of the lakeshore with an annual mean over 1800 mm. In the south and east, the isohyets are nearly parallel to the coastline. These features indicate that the regional climate pattern is modified by the large lake. In the basin of Lake Tanganyika, although the rainfall pattern is not that well organized, the lake's influence on rainfall distribution is visible via the isohyets, which basically lie parallel to the lakeshore. Rainfall decreases almost linearly from the west to the east with maxima around 1300-1500 mm/year over the western catchment of the lake. Lake Malawi does not influence the regional rainfall significantly. The rainfall maximum is over the northwestern catchment and the amount is comparable to that of Lake Tanganyika.

The patterns shown in Figure 7 are based on the five years 1984, 1985, 1986, 1993 and 1994 and solely on cold cloud frequency, as described in section 3.1 and in

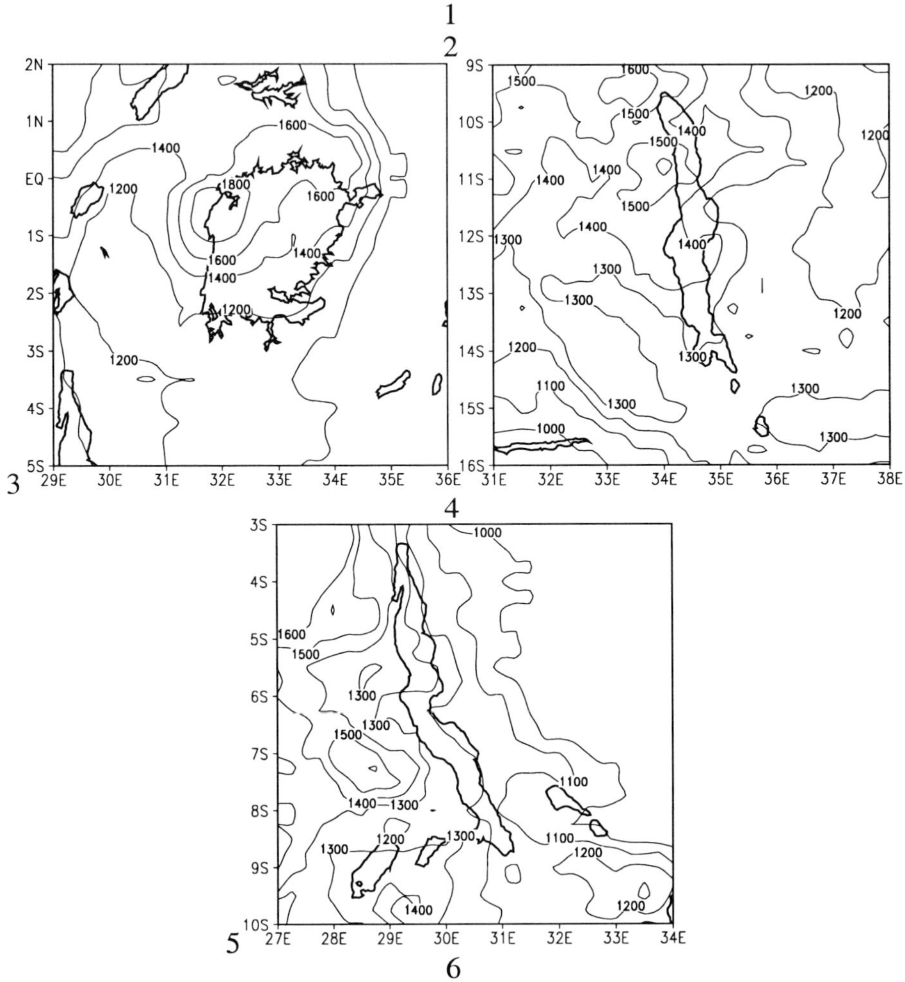


*Figure 7.* Five-year mean rainfall calculated from cold cloud frequency. Upper left panel: Lake Victoria; upper right panel: Lake Tanganyika; lower panel: Lake Malawi.

greater detail in Nicholson *et al.* (2000). In order to determine the long-term mean over-lake rainfall and the relationship between catchment rainfall $P_c$ and over-lake rainfall P, regression equations relating these two variables were derived for the period 1984 to 1994 (the period for which we had both satellite and gauge data). This allowed the calculation of a long-term mean of rainfall over the lakes and to compare the over-lake and catchment rainfall for all three lakes.

The three regression equations are listed in Table 4. These suggest that the over-lake rainfall exceeds catchment rainfall by 30% for Lake Victoria and 20% for Lake Tanganyika. For Lake Malawi, over-lake and catchment rainfall are roughly equal.

To obtain longer-term means, the catchment means for the period 1956 to 1978 were calculated from gauge data and substituted for catchment $P_c$ in these equations. For Lake Victoria, the calculation was carried out by applying a Polygon algorithm to the mean annual rainfall at 22 stations in the catchment (see Nicholson *et al.* 2000). For the other lakes, the calculation was carried out by arithmetically averaging the mean annual rainfall at 15 stations and 14 stations in the catchments of Lakes Tanganyika and Malawi, respectively.

*Table 4.* Regression equations expressing over-lake rainfall by catchment rainfall for the three great lakes.

| Lake | Equation |
|---|---|
| Lake Victoria | $P = 1.353P_c - 87$ |
| Lake Tanganyika | $P = 1.108P_c + 96$ |
| Lake Malawi | $P = 1.014P_c - 17$ |

The above yields an over-lake average for the period 1956 to 1978 of 1780 mm, 1302 mm, and 1350 mm, respectively, for Lakes Victoria, Tanganyika and Malawi, respectively. For these same lakes, catchment rainfall in this period averaged 1380 mm, 1088, and 1348 mm, respectively. Thus, the input to the lake by rainfall is considerably higher over Lake Victoria than over Lake Malawi, although their surrounding regions receive about the same amount of rainfall. On the other hand, the rainfall input into Lake Tanganyika is about the same as that over Lake Malawi, but the catchment of Lake Tanganyika is much drier. These results have important implications for water balance studies.

For Lake Victoria, a final calculation is presented that might be further useful in such studies. The over-lake rainfall, calculated as described above, is given in Table 5 for each year from 1931 to 1994. The results indicate a minimum annual value of 1356 mm in 1943 and a maximum of 2448 mm in 1961.

*Table 5.* Annual mean over-lake rainfall calculated from catchment rainfall for Lake Victoria.

| *year* | *P (mm)* | *year* | *P (mm)* | *year* | *P (mm)* | *year* | *P (mm)* |
|---|---|---|---|---|---|---|---|
| 1931 | 1696 | 1947 | 1813 | 1963 | 2136 | 1979 | 1646 |
| 1932 | 1633 | 1948 | 1467 | 1964 | 1886 | 1980 | 1538 |
| 1933 | 1505 | 1949 | 1445 | 1965 | 1606 | 1981 | 1691 |
| 1934 | 1572 | 1950 | 1582 | 1966 | 1668 | 1982 | 1862 |
| 1935 | 1610 | 1951 | 2132 | 1967 | 1752 | 1983 | 1613 |
| 1936 | 1762 | 1952 | 1483 | 1968 | 1971 | 1984 | 1410 |
| 1937 | 1878 | 1953 | 1474 | 1969 | 1574 | 1985 | 1742 |
| 1938 | 1508 | 1954 | 1513 | 1970 | 1846 | 1986 | 1702 |
| 1939 | 1391 | 1955 | 1615 | 1971 | 1612 | 1987 | 1704 |
| 1940 | 1702 | 1956 | 1549 | 1972 | 1839 | 1988 | 1943 |
| 1941 | 1882 | 1957 | 1662 | 1973 | 1765 | 1989 | 1761 |
| 1942 | 1606 | 1958 | 1539 | 1974 | 1525 | 1990 | 1753 |

| *year* | *P (mm)* | *year* | *P (mm)* | *year* | *P (mm)* | *year* | *P (mm)* |
|---|---|---|---|---|---|---|---|
| 1943 | 1356 | 1959 | 1479 | 1975 | 1799 | 1991 | 1803 |
| 1944 | 1647 | 1960 | 1607 | 1976 | 1637 | 1992 | 1482 |
| 1945 | 1508 | 1961 | 2448 | 1977 | 2064 | 1993 | 1440 |
| 1946 | 1577 | 1962 | 2046 | 1978 | 1917 | 1994 | 1810 |

## 3.3 Cloud Cover over the Lakes

The diurnal and seasonal cycles of cloudiness are very similar to those of convection. The occurrence times of the maximum for both cloudiness and convection are nearly identical. This is conceivable since convection is part of cloud cover. Generally, strong convection is accompanied by larger scale cloud cover. In this study, the diurnal cycle of cloudiness is more concerned with its application in evaporation estimation. For convenience, we define daytime as the hours between 06:15-18:15 LST, which includes four satellite observations at 07:30, 10:30, 13:30 and 16:30 LST, and the remaining hours are defined as night.

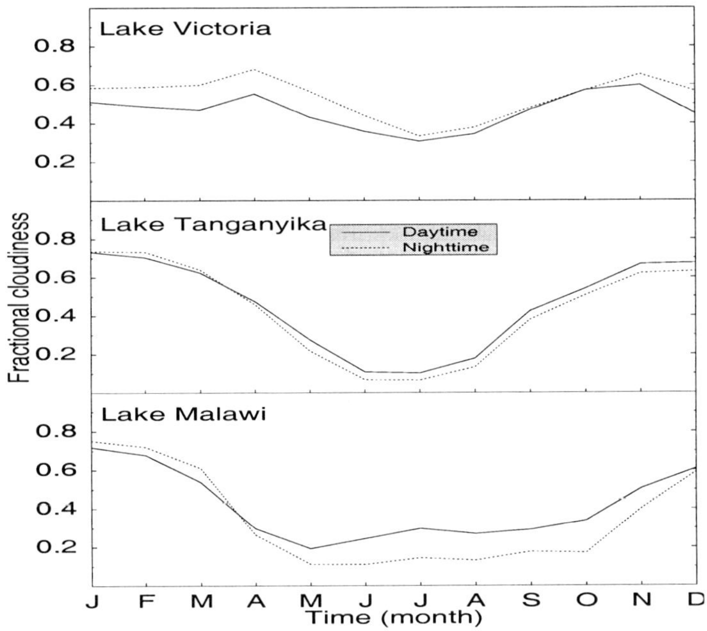


*Figure 8.* Monthly mean diurnal cloudiness over the three lakes.

Figure 8 is the monthly mean daytime and nighttime fractional cloud cover over the three lakes. The diurnal cycle is different between Lake Victoria and the other two lakes. For Lake Victoria, the nighttime cloudiness is more than the daytime cloudiness almost throughout the year, particularly in the rainy seasons. For Lakes

Tanganyika and Malawi, except during the second half of the rainy season between January to March, nighttime cloudiness is always less than daytime cloudiness. This diurnal cycle difference has great effect on the local evaporation because of the fact that while the long wave radiation from the water surface is a function of 24-hour mean cloudiness, the short wave radiation received by the water surface is affected by the daytime cloudiness only.

For the year as a whole, the daytime and nighttime cloudiness in percent is summarized in Table 6. The daily mean estimates of Lakes Victoria and Tanganyika are identical to the previous published values (Atkinson and Sadler, 1970; Hastenrath and Kutzbach 1983; Bergonzini *et al.*, 1997).

*Table 6.* Annual mean diurnal cloudiness for the three lakes.

| | Lake Victoria | Lake Tanganyika | Lake Malawi |
|---|---|---|---|
| Daytime | 46% | 46% | 42% |
| Nighttime | 54% | 43% | 35% |
| Daily mean | 50% | 45% | 38% |

Figure 9 presents the geographical distribution of the monthly mean and the annual mean diurnal cloudiness over Lake Victoria. The two months are selected since April has the strongest and October has the weakest diurnal cloudiness contrast over the lake. In both months and the annual mean, daytime cloudiness minimum is obvious. The minimum cloud cover zone is in the middle of the lake in the southwest-northeast direction, which is parallel to the lake's eastern and western shorelines. The nighttime cloudiness maximum is always in the northern part of the lake, particularly in April and the annual mean. This is consistent with the convection pattern shown in Figure 2.

Monthly diurnal cloudiness in the four quadrants and the catchment of Lake Victoria is presented in Figure 10. In quadrants 1 and 2, nighttime cloudiness is greater than daytime cloudiness in all months. In quadrant 3 between June and October and in quadrant 4 between July and November, daytime cloudiness is marginally higher. For the annual mean, which is plotted the figure and also listed in Table 7 together with the daily mean, the nighttime cloudiness is more than the daytime cloudiness by 9 and 12% in the first two quadrants but only 3 and 5% in the last two quadrants. As a comparison, the catchment always has a daytime cloud cover maximum. The northern quadrants (1 and 2) have the same nighttime cloudiness of 57% but the times of maximum occurrence are different (Yin *et al.*, 2000). In quadrant 1, nighttime cloud cover is most prevalent in the later nocturnal hours, while in quadrant 2 it favors the earlier hours. Similarly, the southern quadrants (3 and 4) have the same nighttime cloudiness of 50% but the maximum is in the early night in quadrant 3 and in the late night in quadrant 4. Although the daytime cloudiness difference among the four quadrants is less than 3%, the analysis in the next section show that its impact on evaporation is significant.

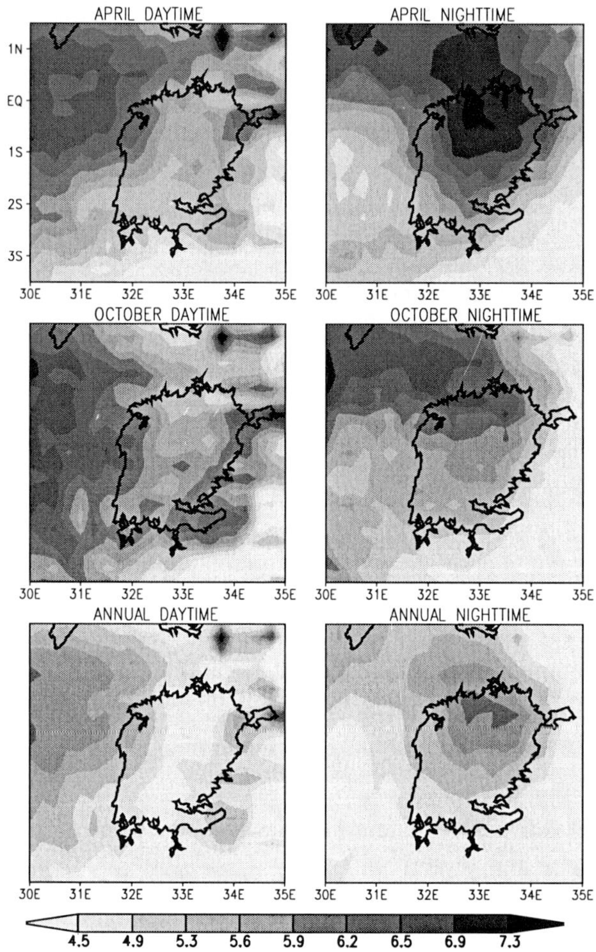


*Figure 9.* Daytime and nighttime geographical cloudiness patterns over Lake Victoria basin. Upper panel: April; middle panel: October; lower panel: annual mean

*Table 7.* Annual mean daytime, nighttime and daily cloudiness for each quadrant of Lake Victoria.

| | Quadrant 1 | Quadrant 2 | Quadrant 3 | Quadrant 4 |
|---|---|---|---|---|
| Day time | 48% | 45% | 45% | 47% |
| Night time | 57% | 57% | 50% | 50% |
| Daily | 53% | 51% | 47% | 49% |

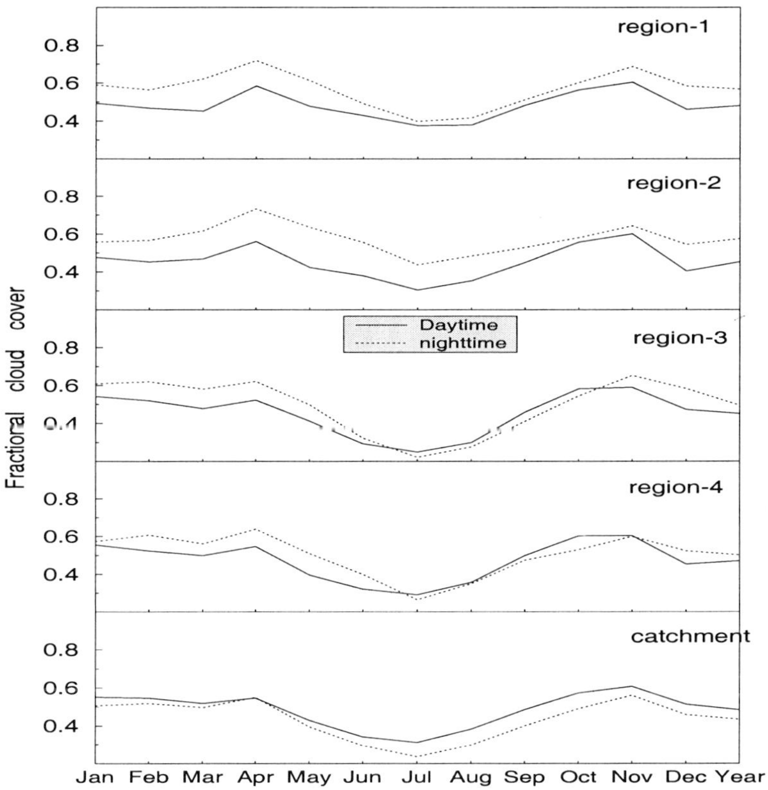


*Figure 10.* Monthly mean diurnal cloudiness in the four quadrants and the catchment of Lake Victoria.

## 3.4 Estimates of Evaporation

Because of the uncertainty in the cloud cover and the high sensitivity of the evaporation to cloud cover (Yin, *et al.*, 2000), evaporation is evaluated in two scenarios for comparison. The first scenario is called `fixed cloudiness', in which a constant cloudiness equal to the annual mean is assumed for each lake, i.e., cloudiness is invariant in a year. The second scenario is called `varied cloudiness', in which the monthly mean cloudiness calculated in this work is adopted and the diurnal cycle is taken into account in the radiation estimation. In order to study the effects of geographical differences in cloudiness, evaporation in each pixel and also in each quadrant of Lake Victoria is calculated separately.

### 3.4.1 Fixed Cloudiness

In this scenario, cloudiness is set as the calculated annual mean value of 0.5, 0,45 and 0.38 for Lakes Victoria, Tanganyika and Malawi, respectively. Albedo is also fixed at 0.07 for the three lakes. This value is very close to the result calculated by Anderson's equation, but is higher than the value of 0.06, which is commonly used by other authors. The calculated monthly and annual evaporation is listed in Table 8. For each of the lakes, minimum evaporation is in the southern winter around June. Since cloudiness is constant, the most determinant parameter is the solar insolation at the top of the atmosphere ($I_0$), which has a winter minimum.

*Table 8.* Evaporation estimated by energy-budget and combined-Penman methods with fixed cloudiness.

| Month | Lake Victoria | | Lake Tanganyika | | Lake Malawi | |
|---|---|---|---|---|---|---|
| | $E_{energy}$ | $E_{penman}$ | $E_{energy}$ | $E_{penman}$ | $E_{energy}$ | $E_{penman}$ |
| Jan | 135 | 137 | 162 | 154 | 186 | 177 |
| Feb | 126 | 129 | 147 | 142 | 166 | 157 |
| Mar | 141 | 144 | 156 | 151 | 170 | 162 |
| Apr | 130 | 130 | 138 | 135 | 142 | 140 |
| May | 122 | 123 | 126 | 129 | 117 | 122 |
| June | 109 | 115 | 106 | 118 | 95 | 105 |
| July | 112 | 119 | 112 | 128 | 108 | 114 |
| Aug | 124 | 129 | 132 | 146 | 187 | 178 |
| Sep | 133 | 135 | 150 | 156 | 157 | 156 |
| Oct | 143 | 144 | 169 | 165 | 189 | 187 |
| Nov | 132 | 134 | 155 | 149 | 191 | 167 |
| Dec | 130 | 132 | 156 | 150 | 192 | 183 |
| Annual | 1537 | 1572 | 1709 | 1721 | 1901 | 1868 |

### 3.4.2 Varied Cloudiness

Yin *et al.* (2000), using two artificial cloudiness scenarios, demonstrated that the diurnal cycle of cloudiness has a significant impact on evaporation. With daytime cloudiness but clear skies at night, evaporation is greatly suppressed. When the diurnal cycle is reversed, evaporation can be as high as 2300 mm/year.

In this scenario, the diurnal variation of the cloudiness is considered in the monthly evaporation calculation in such a way that for short wave radiation, only the daytime mean cloudiness is used while for the long wave radiation, the 24-hour mean cloudiness is applied. Evaporation resulting from this scenario is shown in Table 9. Different from fixed cloudiness, the varied cloudiness scenario finds its maximum evaporation in the winter season around August. This is because the solar insolation and cloudiness both function as determinant parameters.

*Table 9*. Evaporation estimated by energy-budget and combined-Penman methods with varied cloudiness.

| Month | Lake Victoria | | Lake Tanganyika | | Lake Malawi | |
|---|---|---|---|---|---|---|
| | $E_{energy}$ | $E_{penman}$ | $E_{energy}$ | $E_{penman}$ | $E_{energy}$ | $E_{penman}$ |
| Jan | 133 | 136 | 77 | 82 | 87 | 90 |
| Feb | 132 | 135 | 80 | 83 | 90 | 90 |
| Mar | 154 | 155 | 109 | 111 | 133 | 129 |
| Apr | 120 | 121 | 133 | 131 | 157 | 153 |
| May | 140 | 138 | 159 | 157 | 144 | 145 |
| June | 138 | 140 | 158 | 161 | 110 | 117 |
| July | 153 | 154 | 169 | 176 | 118 | 122 |
| Aug | 163 | 161 | 188 | 190 | 213 | 198 |
| Sep | 144 | 144 | 158 | 162 | 176 | 172 |
| Oct | 124 | 128 | 144 | 145 | 201 | 198 |
| Nov | 107 | 113 | 93 | 96 | 157 | 158 |
| Dec | 144 | 144 | 91 | 94 | 126 | 127 |
| Annual | 1652 | 1669 | 1559 | 1589 | 1712 | 1698 |

### 3.4.3 Geographical Variation of Evaporation over Lake Victoria

Monthly evaporation at each pixel of Lake Victoria is calculated based on the pixel's monthly mean cloudiness so that a geographical evaporation map can be constructed for the lake in each month. Among all the parameters, in their possible variation ranges, cloudiness is the most sensitive one for the evaporation estimation (Yin and Nicholson, 1998). Therefore, to simplify the process, monthly mean cloudiness at each pixel is from the result of this work, while for all the other parameters except albedo the monthly mean values of the whole lake are adopted. Albedo is fixed as 0.07. In this way, the influence of the diurnal cycle on evaporation is analyzed in a geographical view. The result in Figure 11 is from the energy-budget method, but the combined-Penman approach indicates a similar pattern. Not surprisingly, the evaporation pattern resembles the daytime cloud cover pattern. Two features are immediately noticeable. First, in the east-west direction of the map, there is an enhancement of evaporation in the middle of the lake in each month. The intensity of the maximum is essentially determined by cloudiness not only because it is the most influential factor for evaporation but also it is the most variable parameter for Lake Victoria, which is on the Equator. As a consequence, evaporation is high in the December-March season and around August, when in the dry seasons the cloud cover is very low. Evaporation is lowest in the two rainiest months, April and November, which have the highest cloudiness. Second, the central maximum migrates from south to north between the May through August period and the December through March period. This shift is caused by the migration of the ITCZ. When the ITCZ is north of equator in the May-August season, the southern lake sky is clearer than the northern lake sky. Thus the highest evaporation is in the south. This is reversed in the December-March season.

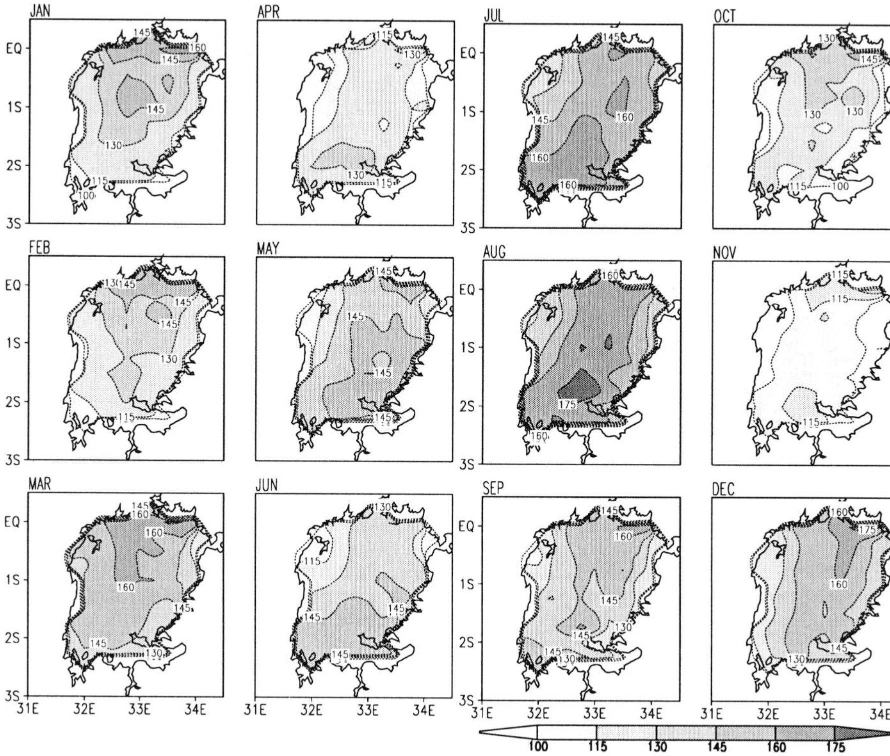


*Figure 11.* Geographical distribution of monthly mean evaporation over Lake Victoria.

In order to see the influence of the diurnal cycle on evaporation, calculation is also carried out for each quadrant of Lake Victoria in two cases based on the quadrant's monthly mean cloudiness. In the first case, a 24-hour mean cloudiness is applied to the estimation of both long wave and short wave radiation. In the second case, long wave radiation is still estimated by the 24-hour mean cloudiness but short wave radiation is calculated by the daytime cloudiness only. The second case is more realistic. For convenience albedo is fixed at 0.07 in both cases. As pointed out earlier, albedo is not a sensitive parameter in its variant range 0.06-0.07. Table 10 shows the calculated annual mean evaporation in the four quadrants of Lake Victoria via both the energy-budget method and combined-Penman method.

First, we notice that for each quadrant and by each method, evaporation calculated from case 2 is higher. The reason is that there is a nighttime cloudiness maximum over the whole of Lake Victoria. Compared to case 1, case 2 uses lower cloudiness in calculating short wave radiation and thus results in higher net radiation, which increases evaporation. Because it has the strongest diurnal cycle, quadrant 2 has the largest difference between the two cases, 199 mm/year by the energy-budget method and 169 mm/year by the combined-Penman method. The second largest difference, 129 and 153 mm/year, is found in quadrant 1, which has the second strongest diurnal cycle. Differences in the last two quadrants are small, generally between 40 to 80 mm/year. Under the same circumstance, combined-Penman evaporation is higher than energy-budget evaporation, but no more than 50 mm/year. In case 1, the two southern quadrants have higher evaporation, while in case 2 the two east quadrants have higher evaporation. The reason is clear; case 1 depends on the 24-hour mean cloudiness, which is lower in the southern quadrants, while case 2 mainly depends on the daytime cloudiness, which is lower in the eastern quadrants.

*Table 10.* Evaporation of the four quadrants of Lake Victoria as estimated by the energy-budget method and the combined-Penman method. In case 1, 24-hour mean cloudiness is used to calculate both long wave and short wave radiation; in case 2, only day time cloudiness is used to calculate short wave radiation.

| Method | Quadrant | 1 | 2 | 3 | 4 |
|---|---|---|---|---|---|
| Energy-budget | Case 1 | 1446 | 1481 | 1576 | 1546 |
| | Case 2 | 1599 | 1680 | 1654 | 1597 |
| Combined-Penman | Case 1 | 1495 | 1524 | 1605 | 1580 |
| | Case 2 | 1624 | 1693 | 1671 | 1624 |

# 4. SUMMARY AND CONCLUSIONS

The East African Lakes alter the regional climate. The effects are particularly strong for Lake Victoria, the largest of these lakes. However, some degree of influence is also apparent over Lakes Tanganyika and Malawi. The lakes modify the diurnal cycles of cloudiness and convective activity, and thereby modify the rainfall regime over the lakes. For each lake, the over-lake rainfall and catchment rainfall are highly correlated and the result shows that Lake Victoria and Lake Tanganyika actually enhance rainfall, compared to that in the surrounding catchment. The enhancement is roughly 30% for Lake Victoria and 20% for Lake Tanganyika.

Lake Victoria has a very obvious effect on the diurnal cycle of convective activity. A complex geographical pattern results from the interaction between the lake/land breeze and the prevailing easterlies. Over the eastern part of the lake and the catchment, these wind systems converge during the daytime, producing maximum frequency of convection in the afternoon to evening hours. Over the western part of the lake and the catchment, the convergence and convective maximum appears in the night to morning hours. Because the easterlies in this area are actually south-easterlies, convective activity over the lake is found to be distinctive in not only the east-west direction but also the north-south direction. While the diurnal timing of convection seems to vary mostly in the east-west direction, the amount of rainfall varies in the north-south direction. In general, there is more convection and rainfall in the north and the highest rainfall occurs in the northwest. Mean annual rainfall averages 1780 mm over the lake.

The diurnal cycle in the basins of Lakes Tanganyika and Malawi is not as well organized, but it exists, particularly over the western catchment of each lake during the night to morning hours, as expected. Mean annual rainfall averages 1302 mm over Lake Tanganyika and there is little spatial variation in rainfall amount over the lake; however, there is a small area in the west-central part of the lake where the rainfall occurs predominantly at night. Over Lake Malawi, mean annual rainfall averages 1350 mm. There is also a large area west of the lake with nocturnal rains.

The influence of the diurnal cycle is further evidenced in the evaporation regime over the lakes. Because of the different influences of daytime cloudiness, which influence short wave and long wave radiation, and nocturnal cloudiness, which only influences long wave, we tabulated these separately and utilized both in calculating evaporation. Over Lake Victoria, nighttime cloudiness is greater than daytime cloudiness in almost all months, while over the other two lakes this is reversed. The annual mean cloudiness is calculated as 0.50, 0.45 and 0.38 for Lakes Victoria, Tanganyika and Malawi, respectively. Over Lake Victoria, there is a daytime cloudiness minimum over the lake in all months and the nighttime cloudiness is enhanced over the lake, particularly in the northern part.

Evaporation over the three lakes was calculated from both the energy-budget and combined-Penman methods. The results of the two methods are very close. The influence of the diurnal cycle was shown by testing two cloudiness scenarios: a fixed cloudiness scenario, in which the annual mean is used for all the months, and a varied cloudiness scenario, in which a seasonal and diurnal cycle is assumed. In the latter, only daytime cloudiness is used to assess short wave radiation, but daily mean cloudiness is used to assess long wave radiation. The calculated evaporation differed greatly between the two cloudiness scenarios. For example, for Lakes Victoria, Tanganyika and Malawi, respectively, annual evaporation calculated by the energy-budget approach is 1537, 1709 and 1901, respectively, using the fixed cloudiness scenario, compared to 1652, 1559 and 1712 mm in the varied cloudiness scenario. These values are comparable with some often-referenced published results (WMO, 1981; Bultot, 1965; Bergonzini et al., 1997; Drayton, 1979; Owen et al., 1990). The geographical variations in cloudiness also produce a pronounced spatial variation in evaporation, particularly for Lake Victoria. There, the daytime minimum in cloud cover over the lake tends to enhance evaporation over the center of the lake.

The spatial and seasonal variations in cloudiness, rainfall and evaporation must be taken into account in numerous environmental problems related to the lakes. They are important in water balance assessment, as well as in assessment of biological productivity and biochemical processes. This is particularly true for Lake Victoria. Clearly, more direct monitoring of the climatic environment over the lakes are desirable, as meteorological data from the lakeshores are very unrepresentative of the lake environment.

## ACKNOWLEDGMENTS

This study was supported by a grant from the National Science Foundation (No. ATM9417063). A NASA fellowship (NGT5-30186) provided partial support for Dr. Yin's contribution. We would also like to acknowledge the support of IDEAL in attending the Malawi symposium where some of these results were presented.

## REFERENCES

Anderson, E.R., (1954) Energy-budget studies, in: *Water Loss Investigations: Lake Hefner Studies,* Tech Report, Prof. Paper 269, Geol. Survey, U.S. Dept. Interior, pp. 71-119.

Atkinson, G. and Sadler, J.C. (1970) *Mean Cloudiness and Gradient-Level Wind Charts Over Tropics.* Technical Report No. 215, USAF, Air Weather Service.

Ba, M.B. and Nicholson, S.E. (1998) Analysis of convective activity and its relationship to the rainfall over the Rift Valley lakes of East Africa during 1983 1990 using the METEOSAT infrared channel. *J. Climate Appl. Meteor.* **10**, 1250-1264.

Ba, M.B., Frouin, R., and Nicholson, S.E. (1995) Satellite-derived interannual variability of West African rainfall during 1983-88. *J. Appl. Meteor.* **34**, 411-431.

Balek, J. (1977) *Hydrology and Water Resources of Tropical Africa,* Elsevier, Amsterdam.

Bergonzini, L., Chalié, F., and Gasse, F. (1997) Paleoevaporation and paleoprecipitation in the Tanganyika basin at 18,000 years B.P. inferred from hydrologic and vegetation proxies. *Quat. Res.* **47**, 295-305.

Black, J.N. (1956) The distribution of solar radiation over the Earth's surface, in: *Wind and solar energy: Proceedings of the New Delhi. Arid Zone Research* **7**, UNESCO, Paris, pp. 238.

Budyko, M.I. (1974) *International Geophysics Series, Vol 18: Climate and life*, Academic Press, New York.

Bultot, F. (1965) A propos de l'évaporation du lac Tanganika. *Acad. R. Sci. Qutre-Mer (Brussels) Bull Séances*, **1965**(4), 1226-1241.

Calder, I.R., Hall, R.L., Bastable, H.G., Gunston, H.M., Shela, O. , Chirwa, A., and Kafundu, R. (1995) The impact of land use change on water resources in sub-Saharan Africa: a modelling study of Lake Malawi. *J. Hydrol.* **170**, 123-135.

Datta, R.K. (1981) Certain aspects of monsoonal precipitation dynamics over Lake Victoria, in J. Lighthill and R.P. Pearce (eds), *Monsoon Dynamics*, Cambridge University Press, pp. 333-349.

de Baulny, H.L., and Baker, D. (1970) *The Water Balance of Lake Victoria*, Technical Note, Water Development Department, Republic of Uganda.

Drayton, R.S. (1979) *A study of the causes of the abnormally high levels of Lake Malawi in 1979*, Department of Lands, Valuation and Water, Lilongwe, Malawi.

EAMD (1961) *Frequency of surface wind speeds and directions,* EAMD, Nairobi.

EAMD (1968) *Tables Showing the Diurnal Variation of Precipitation in East Africa and Seychelles.* Tech. Memo. 10, East African Meteorological Department, Nairobi, Kenya, 49pp. [Available from Kenya Meteorological Department, P.O. Box 30259, Nairobi, Kenya.].

ESOC (1986) METEOSAT contribution to ISCCP. ESA-ESOC Publ. [Available from METEOSAT Exploitation Project Office, European Space Operations Center, Robert-Bosch-Strase 5, 6100 Darmstadt, Germany.]

FAO (1984) *Agroclimatological data*, Rome.

Flohn, H. (1983) Das Katastrophenregen 1961/2 und die Wasserbilanz des Viktoria-See-Gebietes. *Wissenschaftliche Berichte des Meteorologischen Instituts der Universität Karlsruhe* **4**, 17-34

Flohn, H., and Burkhardt, T. (1985) Nile runoff at Aswan and Lake Victoria: a case of a discontinuous climate time series. *Z. Gletscherkunde und Glazialgeologie* **21,** 125-130.

Flohn, H. and Fraedrich, K. (1966) Tagesperiodische Zirkulation und Niederschlagsverteilung am Victoria--See (Ostafrika). *Meteorologische Rundschau* **19**, 157-165.

Hastenrath, S. L. and Kutzbach, J.E. (1983) Paleoclimatic estimates from water and energy budgets of East African lakes. *Quat. Res.* **19**, 157-165.

Howell, P.P., Lock, M., and Cobb, S. (1988) *The Jonglei Canal: Impact and Opportunity*. Cambridge University Press, Cambridge.

Hurst, H.E. (1952) *The Nile*. Constable, London.

Kayiranga, T. (1991) Observation of convective activity from satellite data over the Lake Victoria region in April 1985. *Veille Climatique Satellitaire*. **37**, 44-55.

Kidder, S.Q. and Vonder Harr, T.H. (1995) *Satellite Meteorology: An Introduction*. Academic Press, Inc, San Diego.

Kite, G.W. (1981) Recent changes in level of Lake Victoria. *Hydrol. Sci. Bull.* **26**, 233-243.

Kite, G.W. (1982) Analysis of Lake Victoria levels. *Hydrol. Sci. Bull.* **27**, 99-110.

Livingstone, D.A. (1965) Sedimentation and the history of water level change in Lake Tanganyika. *Limnology and Oceanography* **10**, 607-610.

Nicholson, S.E. (1998) Historical fluctuations of Lake Victoria and other lakes in the northern Rift Valley of East Africa, in J.T. Lehman (ed.), *Environmental Change and Response in East Africa Lakes*, Kluwer, Dordrecht, pp. 7-35.

Nicholson, S.E. and Flohn, H. (1980) African environmental and climatic changes and the general atmospheric circulation in late Pleistocene and Holocene. *Climatic Change* **2**, 313-348.

Nicholson, S.E., Yin, X., and Ba, M. (2000) Use of a water balance model to interpret historical fluctuations of Lake Victoria. *Hydrol. Sci. J* **45**(1), 75--96.

NOAA-USAF (1971) *Global Atlas of Relative Cloud Cover 1967-1970*. Washington DC.

Owen, R.B., Crossley, R., Johnson, T.C., Tweddle, D., Kornfield, I., Davison, S., Eccles, D.H., and Engstrom, D.E., (1990) Major low levels of Lake Malawi and implications for speciation rates in cichlid fishes. *Proceedings of the Royal Society of London* **B 240**, 519-553.

Penman, H.L. (1948) Natural evaporation from open water, soil and grass. *Proc. Roy. Soc.* London Ser. A, **76**, 372-383.

Piper, B.S., Plinston, D.T., and Sutcliffe, J.V. (1986) The water balance of Lake Victoria. *Hydrol. Sci. J.* **31**(1), 25-37.

Rossow, W.B., Mosher, F., Kinsella, E., Arking, A., Desbois, M., Harrison, E., Minnis, P., Ruprecht, E., Seze, G., Simmer, C., and Smith, E. (1985) ISCCP cloud algorithm intercomparison. *J. Clim. Appl. Met.* **24**, 877-903.

Savijärvi, H. (1997) Diurnal winds around Lake Tanganyika. *Q.J.R. Meteorol. Soc.* **123**, 901-918.

Scholz, C.Z. and Rosendahl, B.R. (1988) Low level stand in lakes Malawi and Tanganyika, East Africa, delineated with multifold seismic data. *Science* **240**, 1645-1648.

Sene, K.J. and Plinston, D.T. (1994) A review and update of the hydrology of Lake Victoria in East Africa. *Hydrol. Sci. J* **39**(1), 47-63.

Spigel, R.H. and Coulter, G.W. (1996) Comparison of hydrology and physical limnology of the East African Great Lakes: Tanganyika, Malawi, Victoria, Kivu and Turkana (with reference to some North American Great Lakes), in T.C. Johnson and E. Odada (eds), *The Limnology, Climatology and Paleoclimatology of the East African Lakes*, Gordon and Breach, Amsterdam, 103-139.

Talling, J.F. (1969) The incidence of vertical mixing, and some biological and chemical consequences, in tropical African lakes. *Verh. Int. Ver. Limnol.* **17**, 998-1012.

WMO (1974) *Hydrometeorological survey of the catchments of Lakes Victoria, Kyoga and Albert.* RAF 66-025, Technical Report 1, WMO, Geneva.

WMO (1981) *Hydrometeorological survey of the catchments of Lakes Victoria, Kyoga and Mobutu Sese Seko.* WMO, Geneva.

Yin, X. (2000) *The Water Balance of the East African Great Lakes*, Florida State University Ph.D. Dissertation, Tallahassee, Florida.

Yin, X. and Nicholson, S.E. (1998) The water balance of Lake Victoria. *Hydrol. Sci. J.* **43**(5), 789-812.

Yin, X., Nicholson, S.E., and Ba, M. (2000) On the diurnal cycle of cloudiness over Lake Victoria and its influence on evaporation from the lake. *Hydrol. Sci. J.* **45**(3), 407-424.

# OBSERVATIONS, EVAPORATION AND PRELIMINARY MODELLING OF OVER-LAKE METEOROLOGY ON LARGE AFRICAN LAKES

P. F. HAMBLIN[1], P. VERBURG[2], P. ROEBBER[3], H.A. BOOTSMA[4] and R.E. HECKY[1]

[1]*Environment Canada, National Water Research Institute, Burlington, Ontario, L7R 4A6 Canada*

[2]*Department of Biology, University of Waterloo, Waterloo, Ontario, Canada*

[3]*Department of Mathematical Sciences, University of Wisconsin – Milwaukee, USA*

[4]*Great Lakes WATER Institute, University of Wisconsin – Milwaukee, USA*

**ABSTRACT**

Water quality models of lakes require accurate specification of the advective and turbulent transport fields. These are usually obtained from lake hydrodynamic models. In turn, hydrodynamic models require accurate specification of meteorological forcing. Uncertain specification of meteorological forcing over large lakes is one of the main reasons for the lack of correspondence between three-dimensional hydrodynamic models and observations of lake currents, temperatures and water levels. This is especially the case for intermontane lakes where sheltering effects of the surrounding topography disturb the air flow and generate such other mesoscale meteorological features as slope winds which can reinforce lake breezes.

Direct observations of meteorological variables on lakes are sparse in the tropics. We present here the results of such observations for Lakes Malawi/Nyasa and Tanganyika. During 1998-1999 a roving meteorological station was mounted aboard the research vessel, R/V *Usipa*, on Lake Malawi/Nyasa. Ship velocity and position were recorded, thus allowing winds to be measured aboard the moving platform. On Lake Tanganyika similar data were recorded at two moored meteorological buoys for substantial



*E.O. Odada and D.O. Olago (eds.),*
*The East African Great Lakes: Limnology, Palaeolimnology and Biodiversity,* 121–151.

periods over a period of four years. An examination of the longest running series of winds and air temperatures over Lake Malawi/Nyasa showed no obvious interannual differences in wind speed although air temperatures in the second half of 1999 were cooler than in the same period in 1998. On Lake Tanganyika wind speeds decreased between 1993 and 1996 but air temperatures were highest in 1995. Based on spectral analysis, both lakes illustrate a strong diurnal signal of wind components and air temperatures. Calculations of an average evaporation rate for Lake Malawi/Nyasa based on observed meteorological data from all temporal scales and three different calculation methods resulted in a mean of 6.4 ± mm/d. Diurnally fluctuating meteorological conditions accounted for 36% of the total evaporation. Wet and dry season evaporation rates were compared for the two extremities of Lake Tanganyika and found to be higher in south and during the dry season. Preliminary results of an application of a three-dimensional mesoscale meteorological model to Lake Malawi/Nyasa are compared to direct over-lake observations of a number of forcing parameters required by hydrodynamic models. Comparisons of over-lake winds show that modeled winds are superior by three statistical measures to those interpolated from a limited number of shore-based stations.

Key Words: Large lake meteorology, evaporation, climatology, modeling

# 1. INTRODUCTION

With the increase in computational power, three-dimensional mathematical modeling of water quality parameters is becoming feasible in lakes. Water quality models are coupled with hydrodynamic models through water temperatures and the advection and turbulent transport terms all of which must be calculated by hydrodynamic models. In the case of a large lake, such as Lake Michigan, Belitsky *et al.* (2000) clearly showed the sensitivity of the agreement between an unstratified three-dimensional hydrodynamic model and observed velocity profiles to a more accurate specification of the wind field. The surprising fact about their comparison was that a hindcast by a mesoscale meteorological model improved wind field specification over the open lake when compared to the observed wind field based on interpolation from a dozen shore-based wind stations. African lakes of interest generally have far fewer than a dozen meteorological stations on the shoreline. A more effective method of evaluating the hindcast meteorological parameters would be to compare mesoscale model output directly with observations taken over the lake on either a meteorological buoy or a ship.

Water quality models are presently under development in all the Great Lakes of Africa to study various environmental problems. These models require accurate hydrological and meteorological input as driving forces. As model time steps are as short as several minutes, model forcing must be highly resolved in time. This is especially the case for wind. Mortimer (1979) stated that wind stress and its horizontal distribution over the whole water surface is the critical variable and usually the least well defined. Ideally, hydrodynamic models would be driven by accurate meteorological models of winds and other relevant parameters but such applications are still rare. The only known application of a three-dimensional mesoscale meteorological model to Africa is that of Mukabana and Pielke (1996) who applied the Regional Atmospheric Modeling System (RAMS) model to Kenya and Lake Victoria. The grid resolution was 20 km, a spacing likely too coarse for 50 to 60 km wide Rift Valley lakes. Moreover, their model hindcasts were not evaluated over Lake Victoria. One of the goals of this study is to improve the understanding of the over-lake meteorology of Lakes Tanganyika and Malawi/Nyasa in order to more accurately specify the driving terms of water quality models. In this study we present a summary of the recent meteorological observations recorded on Lakes Malawi/Nyasa and Tanganyika. The focus, where possible, will be on measurements over the water distant from shoreline influences. Similarly, meteorological conditions over lakes define their heat and water budgets. With estimated evaporation rates of up to two metres per year (Spigel and Coulter, 1996) evaporation is the dominant term in the water balance of the African Great Lakes. Accurate estimates of evaporation are also required for the estimation of heat budgets. Calculating evaporation is one of the subjects of the present paper.

There has been little prior study of the meteorology of Lakes Malawi/Nyasa and Tanganyika. Bootsma (1993) and Patterson and Kachinjika (1995) briefly summarized the known climatology of Lake Malawi/Nyasa and presented monthly means of air temperatures and winds at several meteorological stations inland from the shoreline over a 14-yr period. They stated that a hot rainy season extends from October to April characterized by the generally light northerly *Mpoto* winds followed by a cool dry season accompanied by the frequent strong south/south-easterly winds known locally as *Mwera*. Hamblin *et al.* (2002a) plotted 10-min wind readings at one station nearby the lake and daily averaged longitudinal wind component at another for selected periods. Hamblin *et al.* (2002b) provided a detailed analysis of the over-lake wind, temperature and vapor pressure fields. Verburg (1997) and Verburg *et al.* (1998) discussed in detail the over-lake wind and air temperature fields for Lake Tanganyika with particular attention to their daily variations.

Outlines of Lakes Tanganyika and Malawi/Nyasa and their drainage basins are shown in Figures 1 and 2. Evident from the major topographic features is that these lakes are surrounded by steep sides and mountain ranges typical of Rift Valley lakes. Another goal of this study is to examine the influence of basin topography on the atmospheric boundary layer over the lake surface.

## 2. METHODS AND PRIMARY OBSERVATIONS

### 2.1 Field Measurements

At four stations located within several hundred meters of the high-water mark on the shoreline of Lake Malawi/Nyasa (see Figure 3) complete sets of surface meteorological observations were taken at 10-min (Chilumba) or 30-min (Senga Bay, Itungi and Likoma Island) intervals starting in early 1997 and continuing until mid-1999.

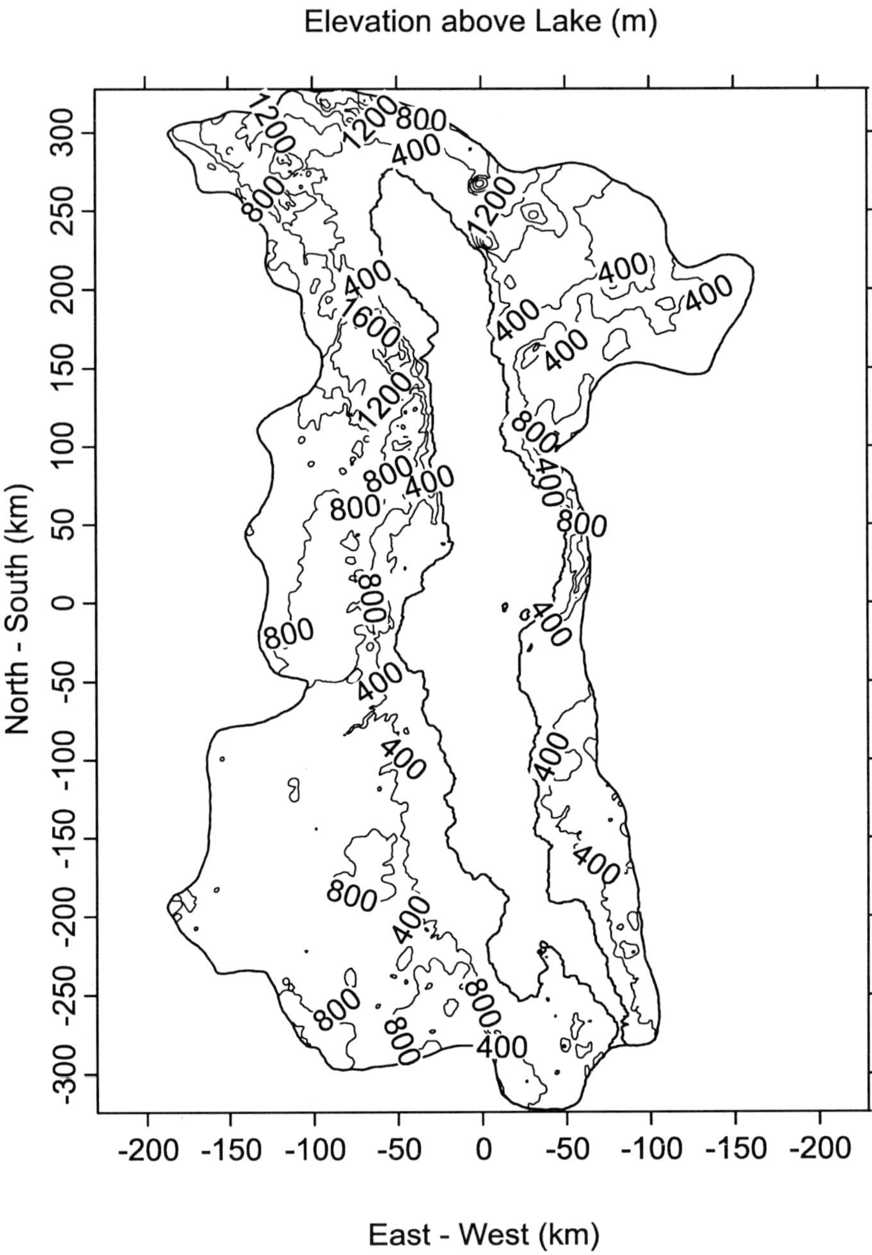


*Figure 1.* Lake Malawi/Nyasa and its basin topography in metres above the lake surface (471m above sea level). Origin is at 12.5° S, 35.° E. Cartographic data are taken from TOPO30 (National Geophysical Data Center).

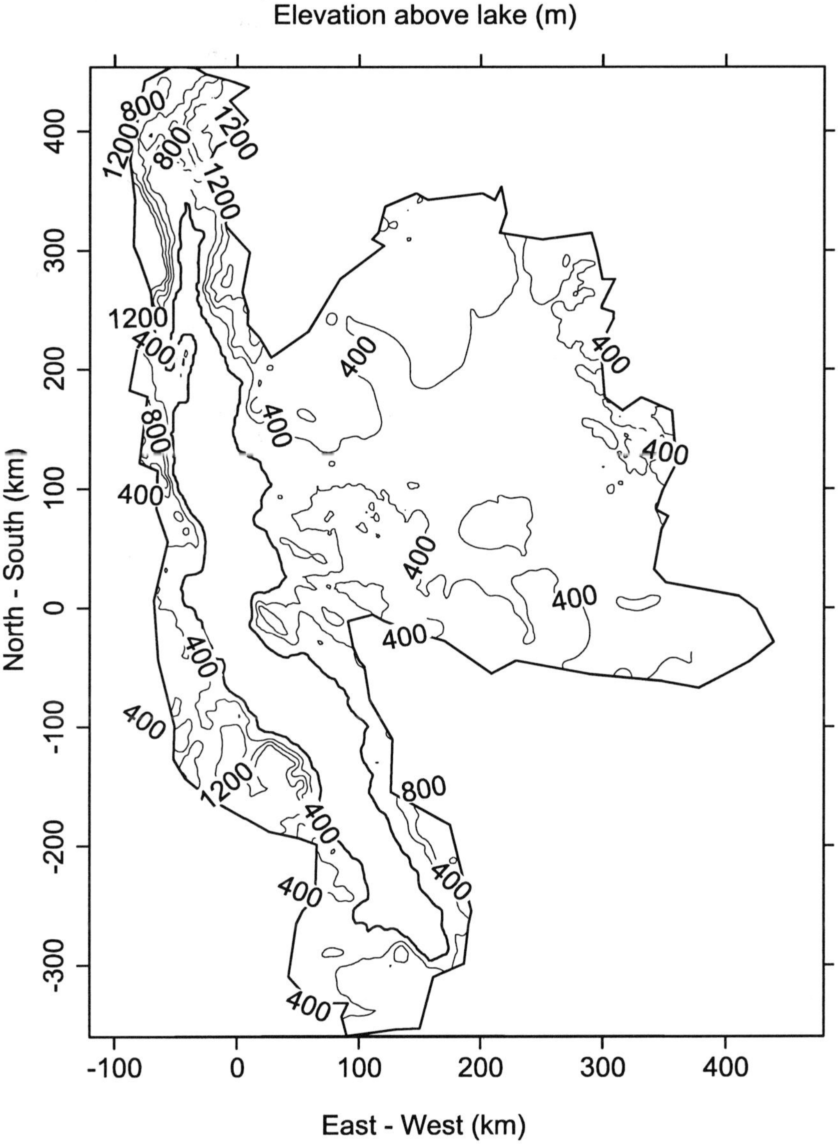


*Figure 2.* Lake Tanganyika and its basin topography in metres above the lake surface (774m above sea level). Origin is at 6° S, 29.5° E. Cartographic data are taken from TOPO30 (National Geophysical Data Center).

During this period a roving meteorological station was operated, concurrently with eight full-lake research surveys by the R/V *Usipa,* at 10-min sampling intervals. A track of one of the cruises is shown in Figure 3. At typical ship speeds meteorological samples were recorded approximately 2 km apart. Between surveys while the R/V *Usipa* was berthed at Monkey Bay, data were also recorded. These data were eliminated from the analysis due to the concern that the local meteorology may be disturbed by the steep shorelines of the bay. A fairly representative coverage of the lake is evident (Figure 3). Beside the ship track, selected times and dates are indicated on the figure. Successful cruises took place from January to early May, which is normally the wet season. The ship velocities (speed and direction), as determined by an automatically recording Global Positioning System, were subtracted from the raw wind data to correct for platform motion. Next, the 10-min data were averaged to 30-min periods for comparison with the land-based stations. A total of 3316 30-min intervals were processed. Unlike the shore-based stations water temperatures were recorded aboard the R/V *Usipa* while underway. For details of the evaluation and correction of the novel water temperature data and the classification of the data into nine zones along and across the lake, the reader is referred to Hamblin *et al.* (2002b).

Open-lake winds and air temperatures were measured at two recording meteorological buoys located well offshore in the southern and northern portions of Lake Tanganyika as seen in Figure 4. They were operated from March 1993 to October 1996 but sporadic breakdowns interrupted the continuity of the records. The recording interval was 60-min at Mpulungu and 30-min at Kigoma. In order to calculate evaporation the over-lake wind and air temperatures were supplemented by the extrapolation of relative humidity from the nearest land stations at Mpulungu and Bujumbura to the buoys. The details of the extrapolation procedure are given by Verburg (1997) and Verburg *et al.* (1998). Both methods of field measurement are illustrated schematically in Figure 5. It is noted that the shipboard measurement heights are at the standard height of 10m above the lake surface (Lake Malawi/Nyasa) whereas the buoy heights are at a height of 2.6 m (Lake Tanganyika).

## 2.2 Methods used for Estimating Evaporation

Evaporation from the surface to the air over Lakes Tanganyika and Malawi/Nyasa was calculated by three methods currently used by limnologists. All methods are variations on the bulk transfer method. First, in the simplest method evaporation is assumed to be proportional to the wind speed at the measurement height of the specific humidity and the difference between the specific humidity at that height and the water surface (Fischer *et al.,* 1979). The specific humidity of the air is calculated from the measured air temperature and the relative humidity while that at the water surface is assumed to be completely saturated at the water surface temperature. The constant of proportionality has been slightly adjusted from the value suggested by Fischer *et al.* (1979) based on the applications to numerous thermodynamic calculations in lakes by the model DYRESM (e.g. Hamblin *et al.*, 1999).

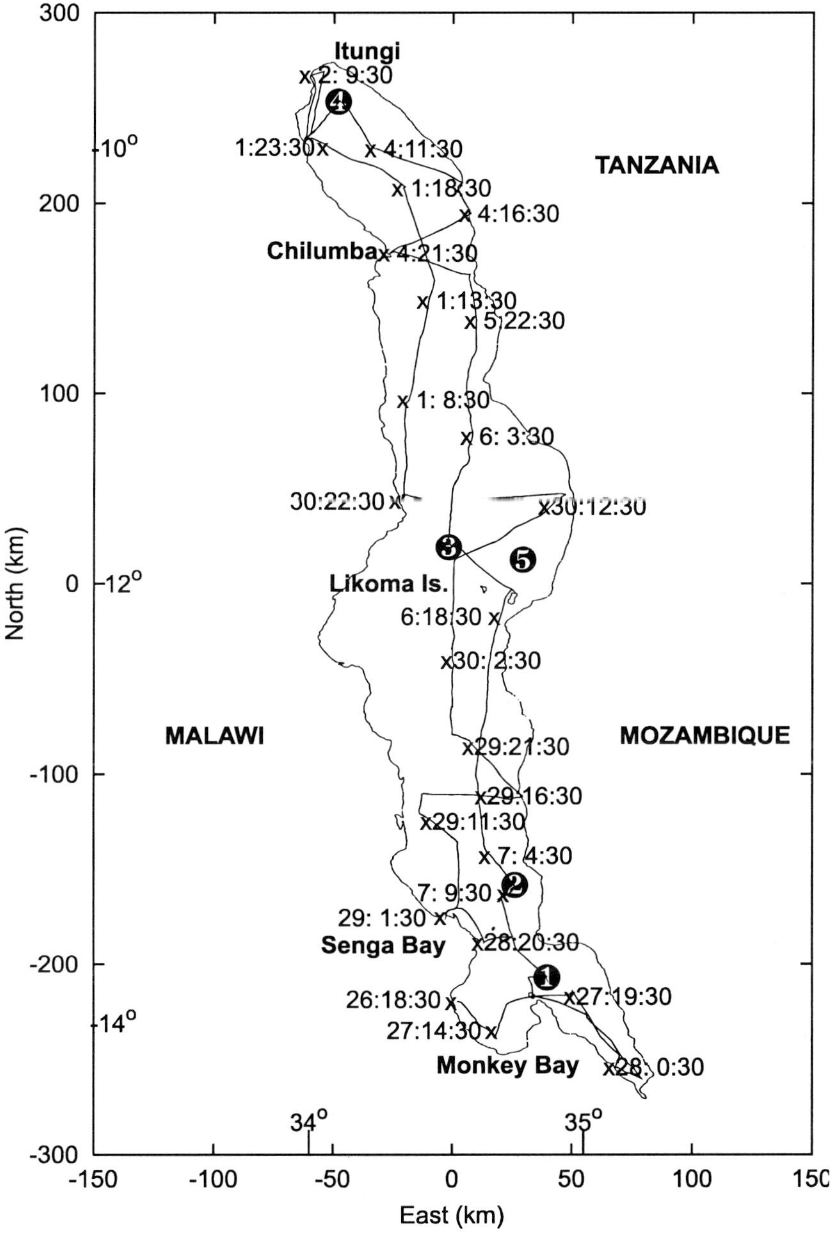


*Figure 3*. Meteorological and thermistor chain (solid circle) station locations and a R/V *Usipa* Track for April 26 to May 7, 1998. Date: hour: minute are indicated.

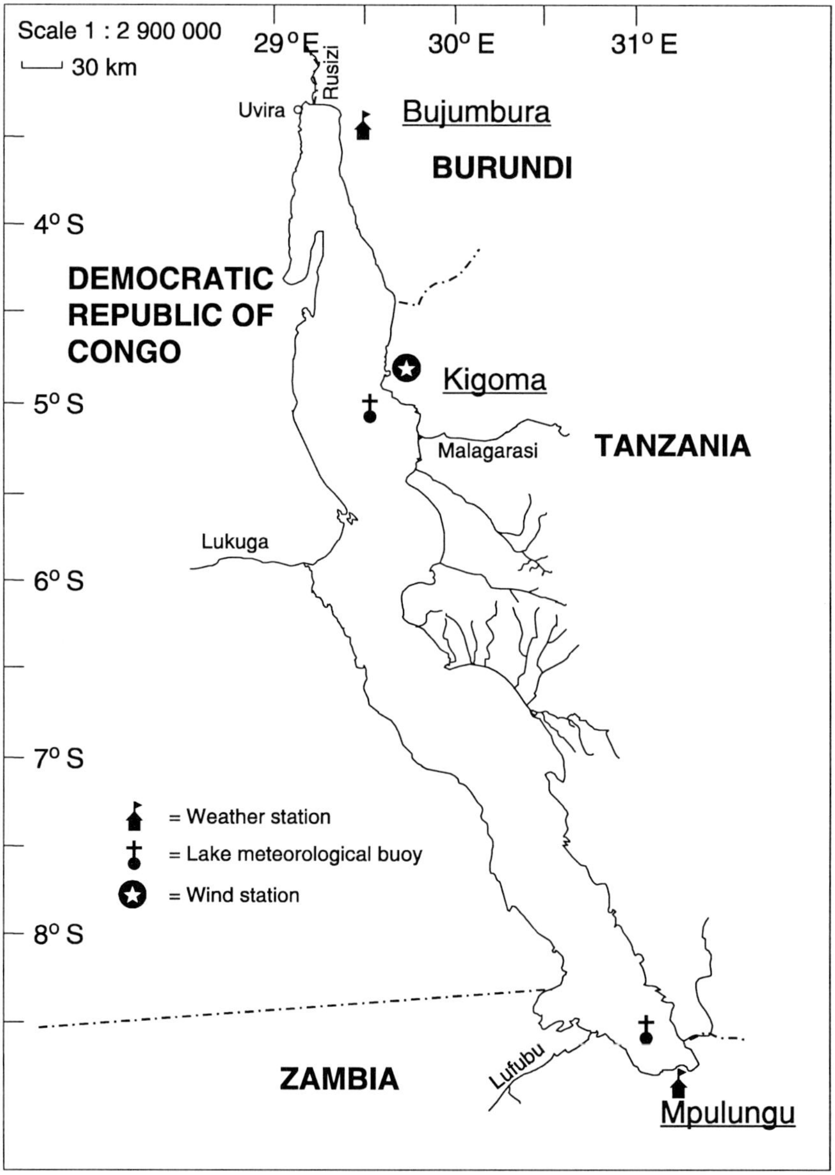


*Figure 4*. Meteorological station locations on Lake Tanganyika (Verburg and Hecky, 2002).

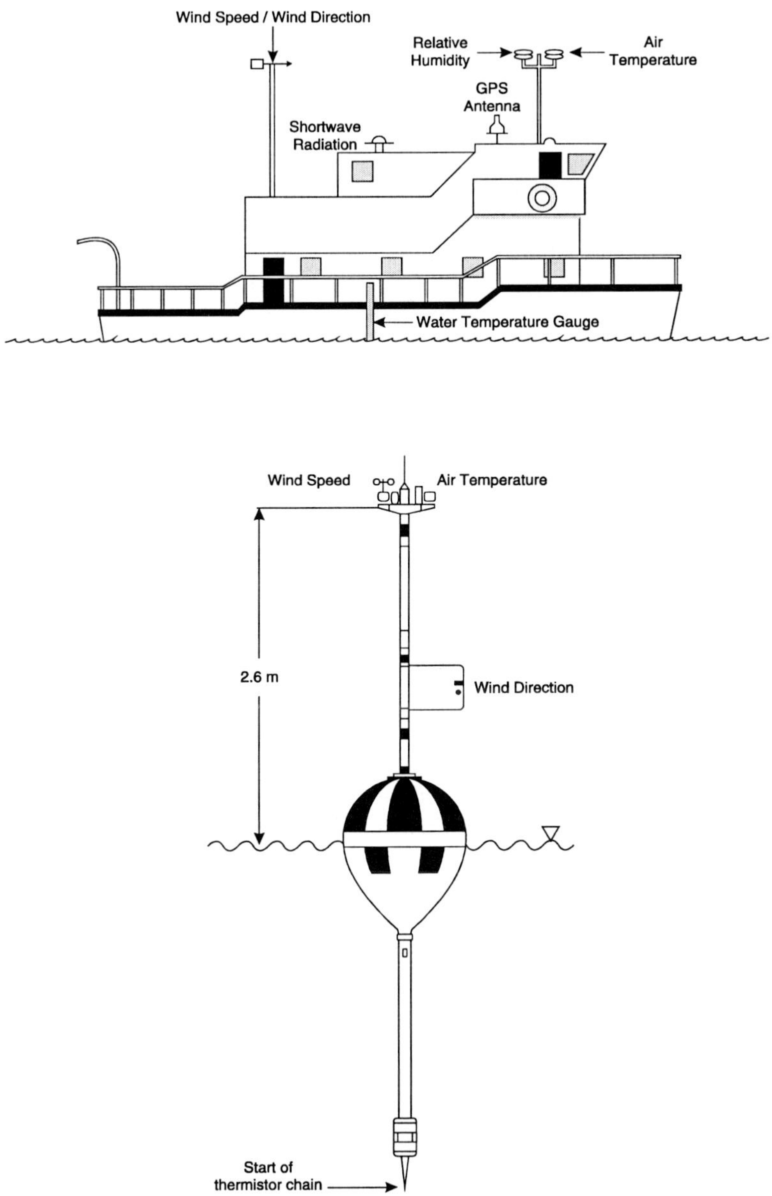


*Figure 5*. Schematic of over-lake meteorological measurement methods; upper panel, R/V *Usipa* Lake Malawi/Nyasa, lower, a Lake Tanganyika meteorological buoy.

It is noted that the latent heat flux given by Fischer *et al.* is converted to an evaporation rate for comparison with the other methods reported in the literature by dividing the flux by the product of the latent heat of evaporation and the water density.

The second method summarized by Chow *et al.* (1988) takes into account the additional factors of the roughness of the surface boundary layer and the measurement height provided that wind, relative humidity and air temperature are all measured at the same level. In the case of the R/V *Usipa* these heights were nearly identical at 10 m above the water surface, the standard height for surface boundary layer measurements. The roughness factor, $z_0$, must be specified by the user. As the authors suggest a wide range of possible roughness lengths over water it was decided to use the $z_0$ calculated as a byproduct of the third method by Liu *et al.* (1979) (order $10^{-4}$m) for Lake Malawi/Nyasa. For Lake Tanganyika the evaporation was calculated by the Chow *et al.* (1988) method using a roughness factor of $1X10^{-4}$m. Liu *et al.'s* (1979) method is the most sophisticated taking into account boundary layer stability based on the relative wind strength and air-water temperature difference, variable boundary roughness over water as a function of wind speed and the individual heights of all three input variables.

## 2.3 Mesoscale Modeling

A two-day pilot experiment was conducted on the three-dimensional mesoscale meteorological model, MM5, (Grell *et al.,* 1994) for Lake Malawi/Nyasa and its basin. Simulations over the January 12-14, 1999 period used this Pennsylvania State University /National Center for Atmospheric Research (PSU/NCAR) 5th generation model. It was run from a cold start with initial and lateral boundary conditions provided by the National Centers for Environmental Prediction global mandatory-level analyses (2.5° latitude by 2.5° longitude) maintained at NCAR. The simulations were conducted in a doubly nested, two-way interactive mode (Zhang *et al.,* 1986), such that conditions in the inner domain feed back to the "mother" domain, and vice versa, with matching at the nest boundary. The outermost domain (D1), with 12 km grid spacing and time step of 30s, was designed to capture the synoptic scale environment and physiography within the Lake Malawi basin and surrounding region, while the innermost 4 km domain (D2) was established to resolve mesoscale details of the lake meteorology. D2 had a time step of 10s. The Kain-Fritsch convective parameterization was used in D1, while an explicit moisture scheme that includes prognostic equations for cloud water, ice, rainwater and snow was employed on both domains (Reisner *et al.,* 1998). No convective parameterization was used in D2, where the resolution is sufficient to begin to explicitly resolve convective details. Radiative processes were handled using a cloud-radiation scheme in which diurnally varying shortwave and longwave radiative fluxes interact with explicit cloud and clear air, while the surface fluxes are used in the ground energy budget calculations (Dudhia, 1989). The planetary boundary layer (PBL) was modeled using the high-resolution Blackadar scheme (Zhang and Anthes, 1982) coupled with a 5-layer soil model (Dudhia, 1996). Physiographic and land use patterns were back interpolated from high-resolution data sets (approximately 1 km topographic, see Figure 1, and landuse) to the model grids. Vertical sigma levels were arranged such that the model output is available on a total of 23 levels with a relative concentration at the lowest levels in order to resolve planetary boundary layer structure. Only surface model outputs were saved for later comparison with field observations.

# 3. RESULTS

## 3.1 Annual and Interannual Variability

Due to frequent disruptions in the Lake Tanganyika buoy data it was difficult to compare one year with another. Also, no large ENSO events occurred during the 4-yr study period. Consequently, an annual cycle was assembled for each station as completely as possible. Meteorological conditions at the north end of Lake Tanganyika are summarized in Figure 6, which presents monthly total rainfall (means for 1973 to 1993) and 21-d running averages of solar radiation (shortwave), air pressure, air temperature, wind speed and relative humidity at Bujumbura in 1995. For the Mpulungu land station (1996) and south-end buoy wind and air temperature (1995), similar variables are compared in Figure 7.

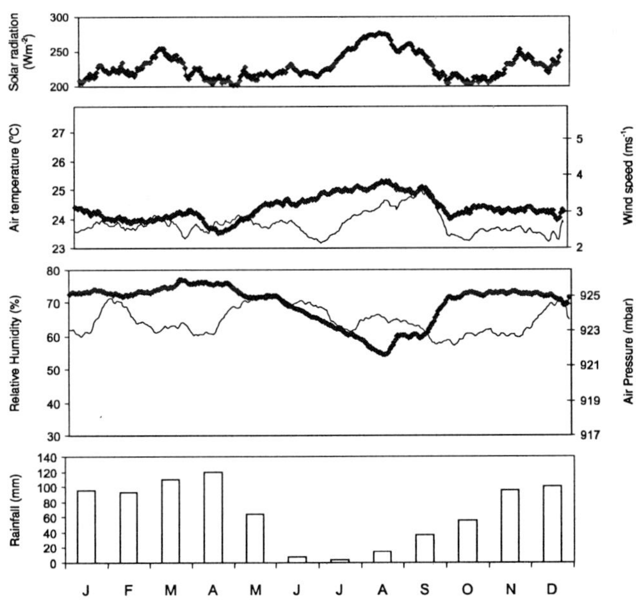


*Figure 6.* Meteorological conditions at the north end of Lake Tanganyika. Monthly total rainfall (means for 1973 to 1993) and 21-d running averages of solar radiation (shortwave), air pressure, air temperature, wind speed and relative humidity at Bujumbura (1995). Heavy lines are wind speed or relative humidity.

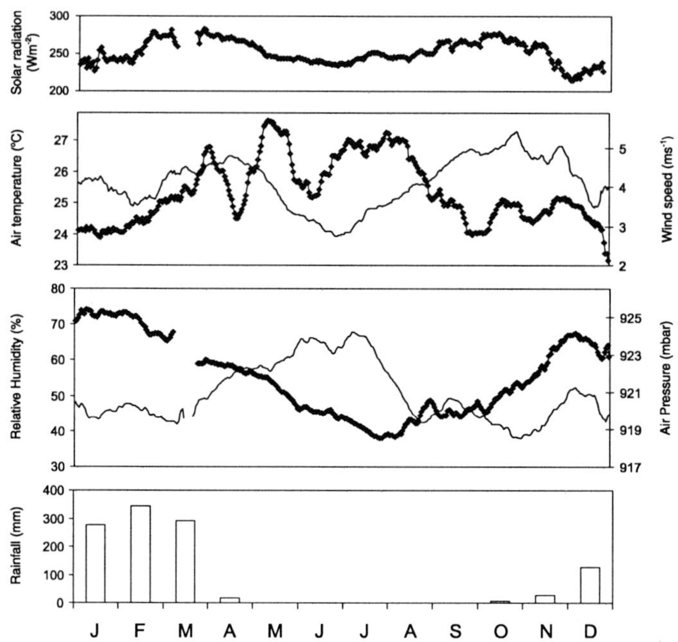


*Figure 7.* Same as Figure 6 but for Mpulungu land station (1996) and south end buoy wind and air temperature (1995). Heavy lines are wind speed or relative humidity.

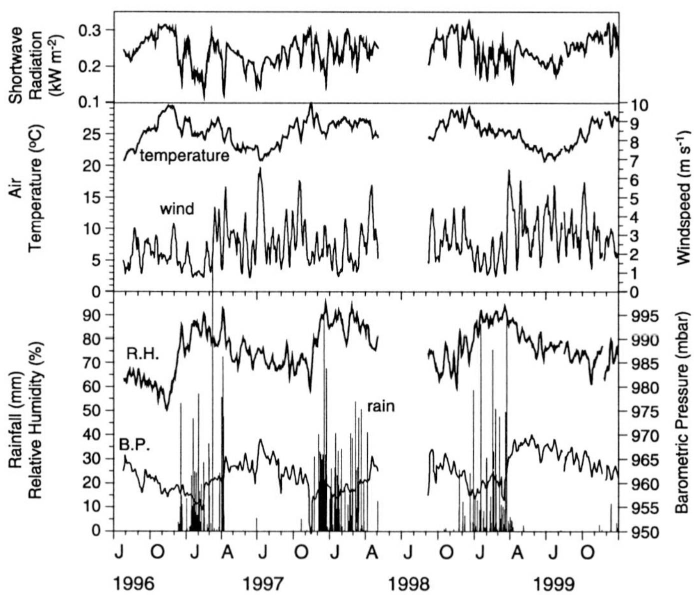


*Figure 8.* Meteorological conditions at Senga Bay smoothed by a 7-d moving average except for rainfall. Note that rainfall for 1998/99 is substituted by the station at Salima (5 km distant) due to a faulty gauge.

Due to the long intervals between full-lake cruises, the data collected aboard the research vessel on Lake Malawi/Nyasa were unsuitable for long-term analysis. Rather, the most continuous series of meteorological data was collected at the land-based station at Senga Bay. While it was anticipated that this data set might capture two opposite climatological extremes, the warmer-than-usual El Niño considered to peak in late1997 followed by the cooler La Niña event thought to peak in late 1998, the comparison of daily averaged winds in Figure 8 does not suggest an obvious trend. On the other hand, the air temperatures are somewhat higher in 1998 and lower in the latter half of 1999 suggesting the expected trend in temperature although delayed. No obvious trends in precipitation nor other meteorological variables were observed for the years sampled.

## 3.2 Diurnal Analysis

To illustrate the dominance of diurnal variability in meteorological data it is convenient to use a method of time series analysis known as spectral analysis. Unfortunately, again the shipboard data are unsuitable for this method so instead, the three concurrent land stations were chosen for analysis for Lake Malawi/Nyasa.
For ease of computation over the approximately 120-day period of analysis the north and east components of wind were first averaged from 30-min to 3-hr intervals. Figure 9 compares the autospectra at each location and for east and north wind components. All three land stations demonstrate remarkably high energies at the diurnal period. Only for the Senga Bay alongshore (north) wind component does the energy at very long periods exceed the diurnal period energy. The Likoma Island easterly diurnal component is the most energetic of all corresponding to an average amplitude of 2 m $s^{-1}$. The partition of energy into two bands, namely diurnal and very low frequency suggests that the subsequent analysis can be focused on each of the two highly energetic bands while other regions of the spectrum can be safely ignored. In the case of Lake Tanganyika, sufficiently long continuous records were available at the two buoys spanning each of the rainy and dry seasons. Spectra of winds and air temperatures at the southern buoy, Mpulungu, demonstrate dominant diurnal variability in Figure 10. According to Savijarvi (1997) dry season conditions are most favorable to diurnal winds. Remarkably, the diurnal winds are nearly as strong during the wet season, being only about 40% weaker. A similar plot at the buoy offshore of Kigoma (not shown) does not reveal any appreciable spectral differences from Figure 10.

The R/V *Usipa* wind data are displayed as sets of wind vectors every three hours over a daily period in Figure 11. The shipboard data are classified into nine geographical zones with the average position in each zone indicated by the origin of the wind vector in Figure 11. When there are fewer than ten samples in a class vectors are not plotted. While only at midnight were all zones represented, there is an indication of near surface diurnal wind divergence (winds off the lake) during the daytime from 9 to 15 hr and nightly wind convergence (winds off the land) from 21 to 6 hr. Within the limitations of the data set there is no indication that the diurnal wind has a latitudinal variation in strength in Figure 11.

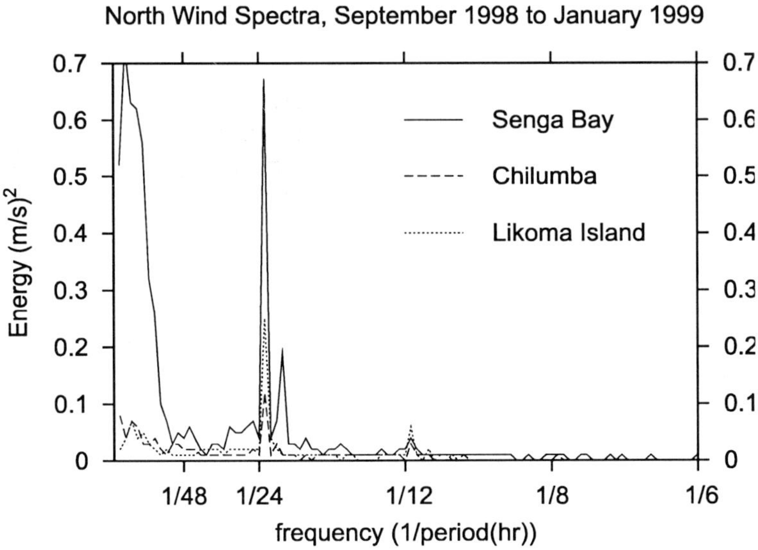


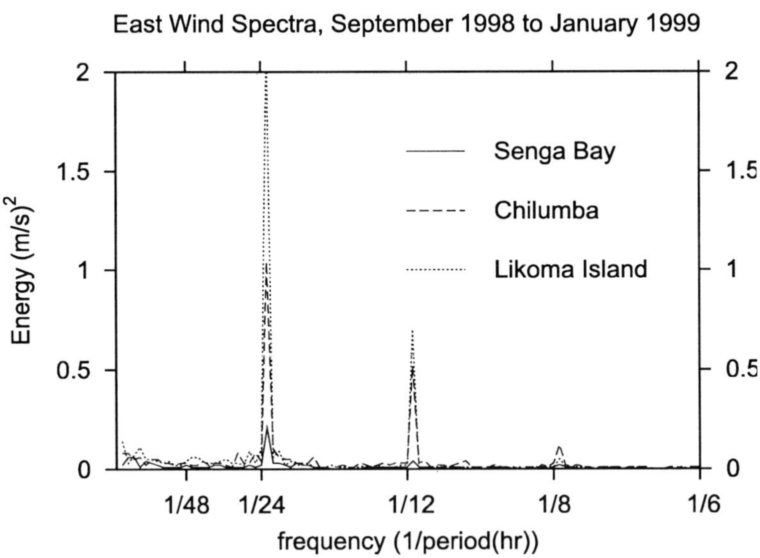


*Figure 9.* Spectra of wind components at three shore-based stations, Lake Malawi/Nyasa. The period analysed is from mid September 1998 to mid January, 1999.

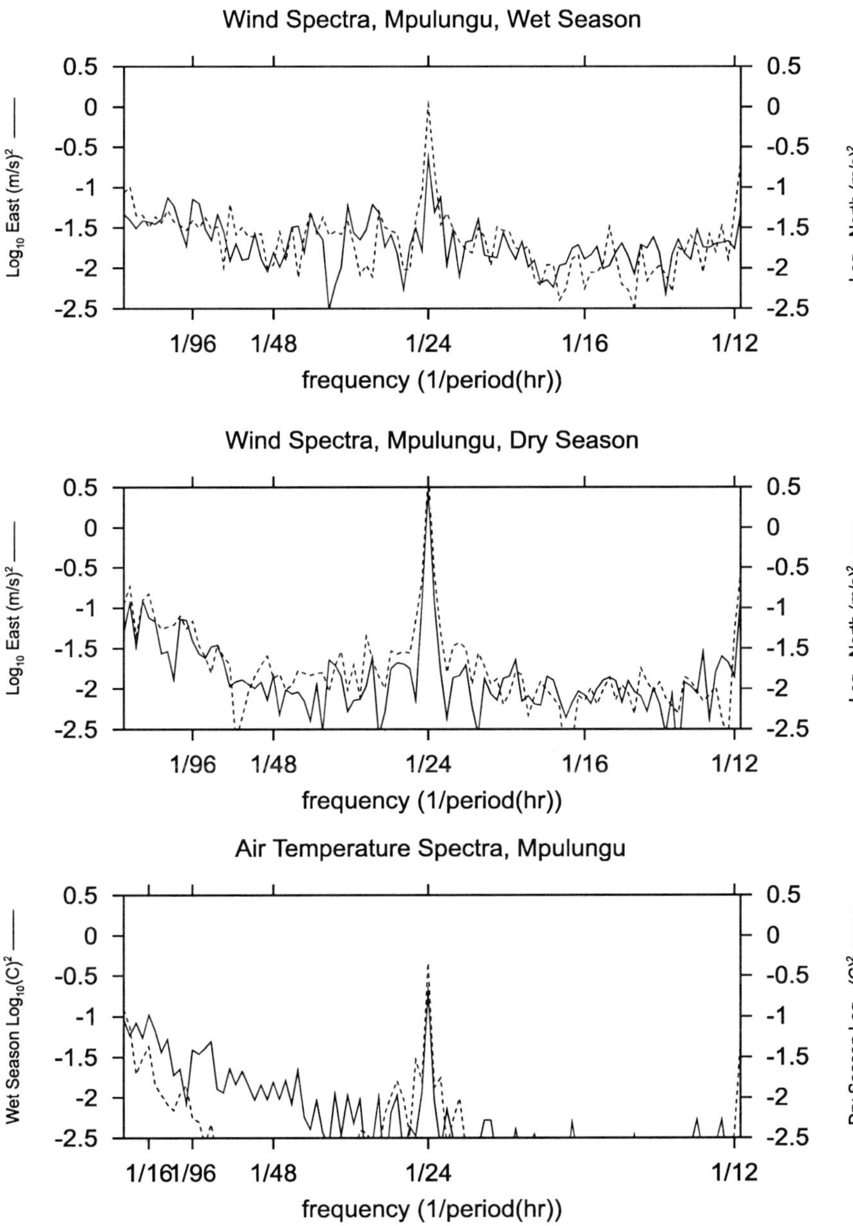


*Figure 10.* A comparison of spectra of wind components and air temperatures at the Mpulungu buoy station, Lake Tanganyika for the rainy and dry seasons. The spectral density is logarithmic.

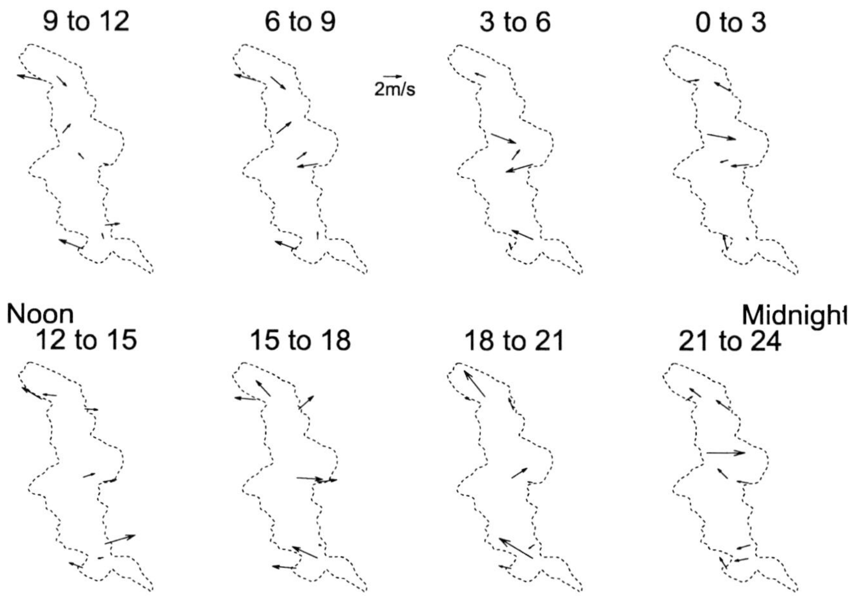


*Figure11.* Average observed diurnal winds over 3-hr interval in each zone for Lake Malawi/Nyasa.

Air temperature is a key variable for both evaporation and specification of the driving terms for water quality models. Its diurnal behaviour in both lakes is given in detail by Hamblin *et al.* (2002b) and Verburg (1997). The mid-lake diurnal temperature variation in Lake Malawi/Nyasa based on the roving shipboard data is less than the nearshore, indicating some moderation of the air mass over the lake. This is true also for Lake Tanganyika (Verburg, 1997).

## 3.3 Evaporation

First, the variation of evaporation rates are compared for two selected months in Lake Tanganyika at the north (Bujumbura) and south (Mpulungu buoy) in Figure 12. The consistently higher evaporation in the south suggests that there is a latitudinal gradient in evaporation. This is likely due to air mass modification by the prevailing southerly winds.

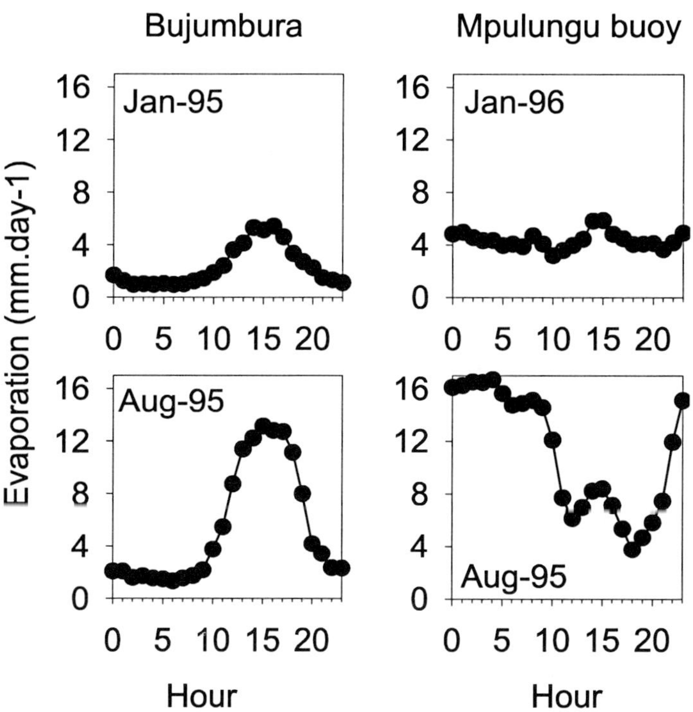


*Figure 12.* Comparisons of evaporation over the day at the north (Bujambura) and south (Mpulungu buoy), Lake Tanganyika, during the dry seasons, August 1995 and wet season January 1996.

Daily rates of evaporation are compared by zone for Lake Malawi/Nyasa in Figure 13. There appear to be three diurnal modes, one fairly constant throughout the day tending to occur on the eastern portion, another with peak evaporation around 15 hr (the most prevalent and occurring at all mid-lake zones) and finally, one peaking at night found only in the southwest zone. Figure 14 demonstrated a similar behaviour for Lake Tanganyika but in this case, by season and spatial position. The constant response occurred only during the wet season, January 1996 at Mpulungu in the south whereas the nightly peaking mode occurred during the dry season, August 1995. The evaporation in the north always peaks in the afternoon but is, not surprisingly, higher in the dry season. By reference to the monthly averaged plots of wind speed, air and water temperature and relative humidity of Verburg (1997) it is apparent that the daily cycle of evaporation is mainly determined by wind speed variations.

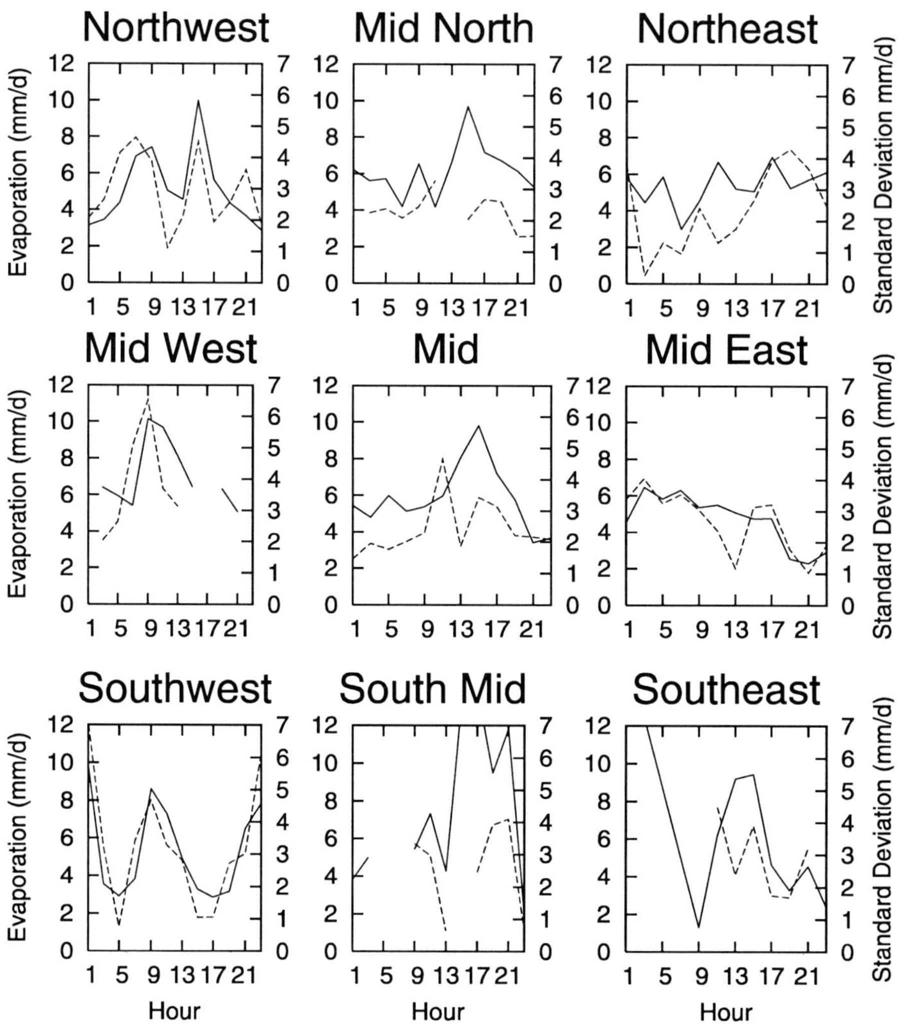


*Figure 13.* Daily evaporation fluctuations by zones for all cruises for Lake Malawi/Nyasa. Standard deviations (dashed lines) are not defined if there are fewer than four values in a class. Some classes are not present.

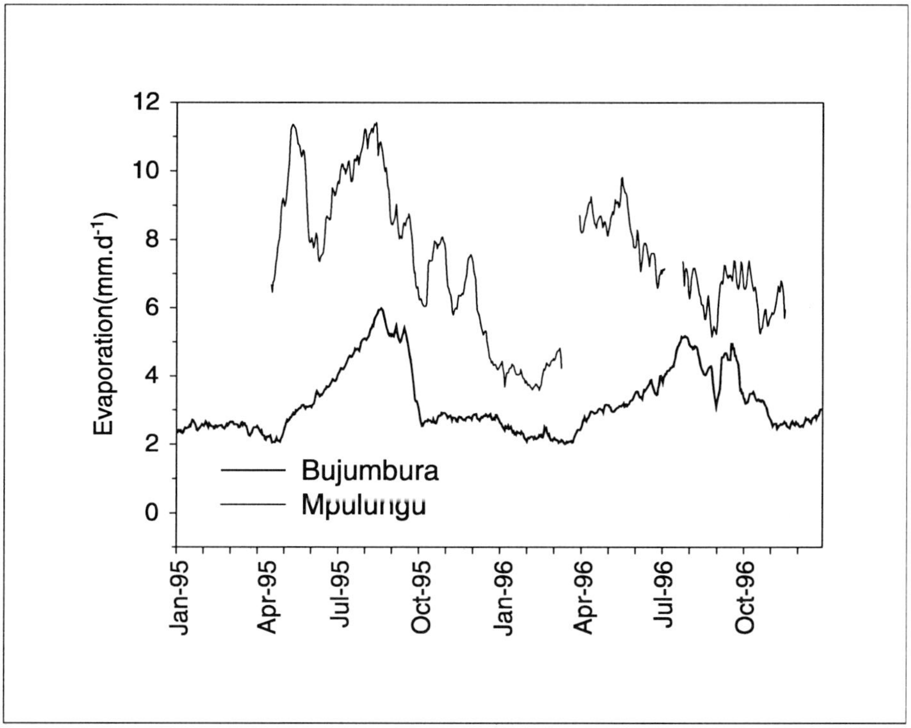


*Figure14.* Twenty-one-day running averages of evaporation calculated for Bujumbura and Mpulungu buoys.

The lakewide spatial distribution of evaporation in Lake Malawi/Nyasa plotted in Figure 15 is based on horizontally smoothed estimates of evaporation in contrast to the above diurnal variations. Shipboard evaporation estimates are positive as might be expected during and just after the wet season as air temperatures (not shown) were from 2 to 4°C cooler than the lake surface water temperatures (see Hamblin *et al.*, 2002b). Hamblin *et al.* (2002b) compared estimates of lakewide evaporation rates based on temporally smoothed wind, air temperature and vapor pressures with those based on raw data to study the role played by diurnal processes in evaporation. They found that evaporation based on daily or longer term averaged quantities could underestimate evaporation by as much as 36%. The peak evaporation rate to the north of Senga Bay is probably due to the high wind speeds occurring there. Similarly, sensible heat flux which is also proportional to the wind would be underestimated based on smoothed or daily winds. In Lake Tanganyika evaporation was underestimated by 10 to 15% if monthly means were used.

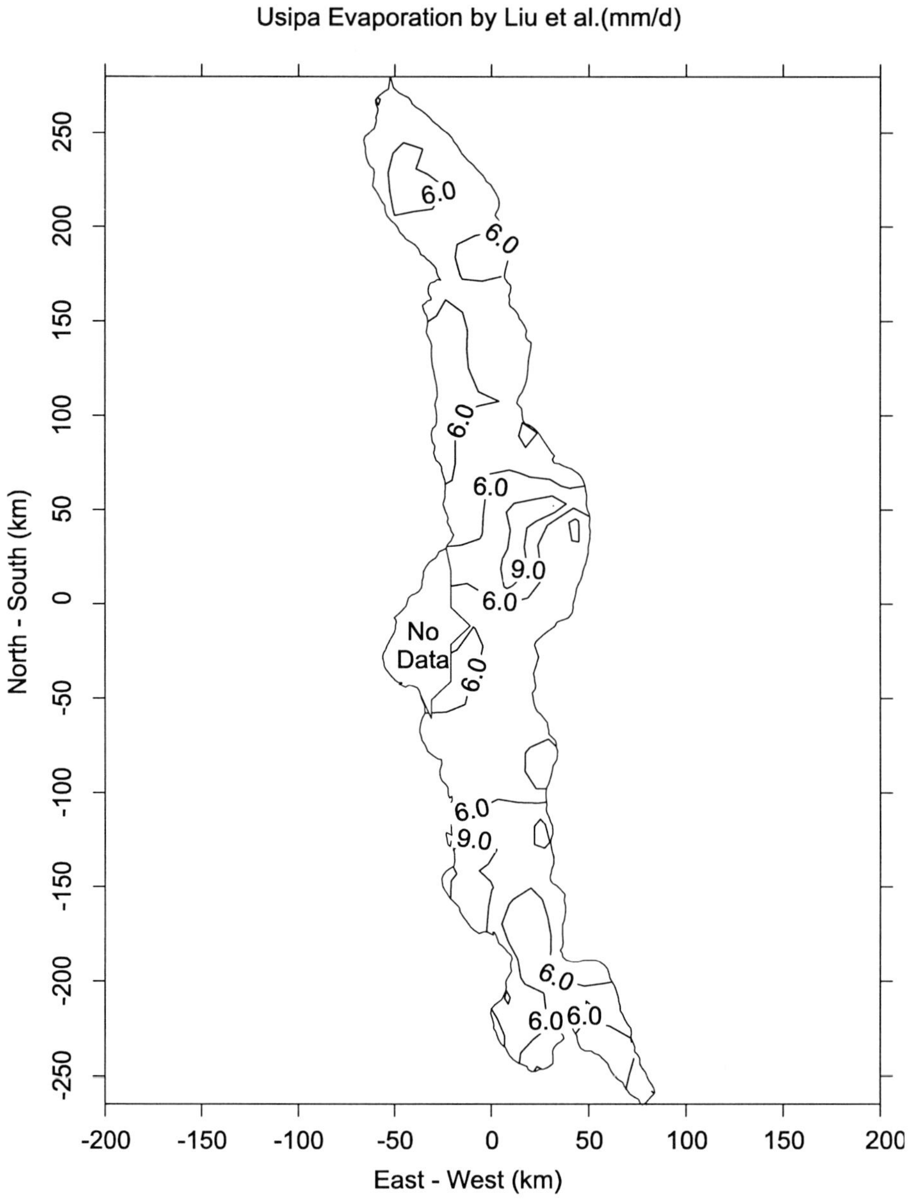


*Figure 15.* Contours of evaporation rate (mm/d) for Lake Malawi/Nyasa based on Liu *et al.* (1979) and unsmoothed data for the R/V *Usipa* cruises covering generally the wet season.

The lakewide rates of evaporation for Lake Malawi/Nyasa calculated by three variations on the bulk transfer method are summarized in Table 1. The Fischer *et al.* (1979) result is slightly higher than that of the Liu *et al.* (1979) method and is apt to be higher due to the fact that in most prior applications of the Fischer *et al.* (1979) expression the winds were measured at a height lower than 10 m. The Chow *et al.* (1988) expression may be less accurate as it is generalized for use over land as well as water. The most accurate, therefore, is apt to be the Liu *et al.* (1979) method. Averaging the three methods but excluding the temporally smoothed value for the reasons discussed above gives an overall estimate of the rate of evaporation of $6.4 \pm 1$ mm $d^{-1}$ for the periods sampled, generally encompassing the rainy season.

*Table 1.* A comparison of lakewide evaporation methods for Lake Malawi/Nyasa.

| Method | Evaporation (mm/d) | Standard Deviation (mm/d) |
|---|---|---|
| Fischer *et al.* (1979) | 7.6 | 3.3 |
| Fischer smoothed data | 4.1 | 2.4 |
| Liu *et al.* (1979) | 6.5 | 2.9 |
| Chow *et al.* (1988) with Liu $z_0$ | 5.2 | 2.3 |

## 3.4 Mesoscale Modeling

Savijarvi (1997) was one of the first to apply a mesoscale meteorological model to a large African Lake. Since the model was two-dimensional in an east-west plane across the middle of Lake Tanganyika, it would be difficult to compare its results with meteorological observations at a point. Such a model provides little practical information on the lake surface specification of wind stress and energy fluxes needed by the hydrodynamic modeler. Nonetheless, the model demonstrated an organized mesoscale wind circulation in the cross-lake vertical plane, especially during the dry season. As well, the interaction of the diurnal circulation and the southeast trade winds was captured and was found to enhance the eastern shore's lake breeze system. Savijarvi (1997) was able, with the aid of the mesoscale model to estimate the portion of the diurnal circulation attributable to land-water temperature differences, to tradewind interactions and to slope winds on the steep sided Lake Tanganyika.

For the comparison of modeled variables with over-lake observations on Lake Malawi/Nyasa, meteorological conditions were observed along the track shown in Figure 16 which is, for the most part, beyond the influence of shoreline disturbances. Modeled and observed winds, air temperatures, relative humidities, solar radiation and barometric pressures are compared in Figure 17 at the grid point closest to the research vessel's location at a given time. The agreement is good for wind speed and

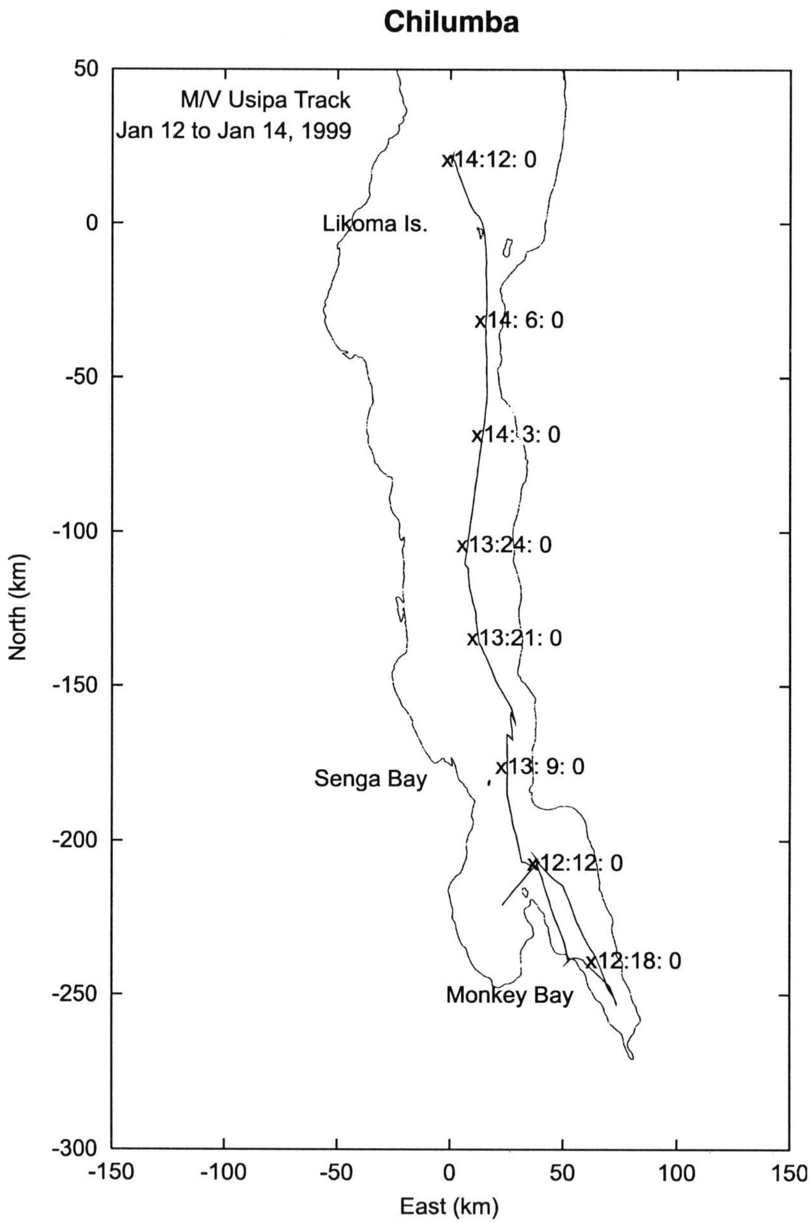


*Figure 16.* R/V *Usipa* track, Lake Malawi/Nyasa, for 2-d modeling period.

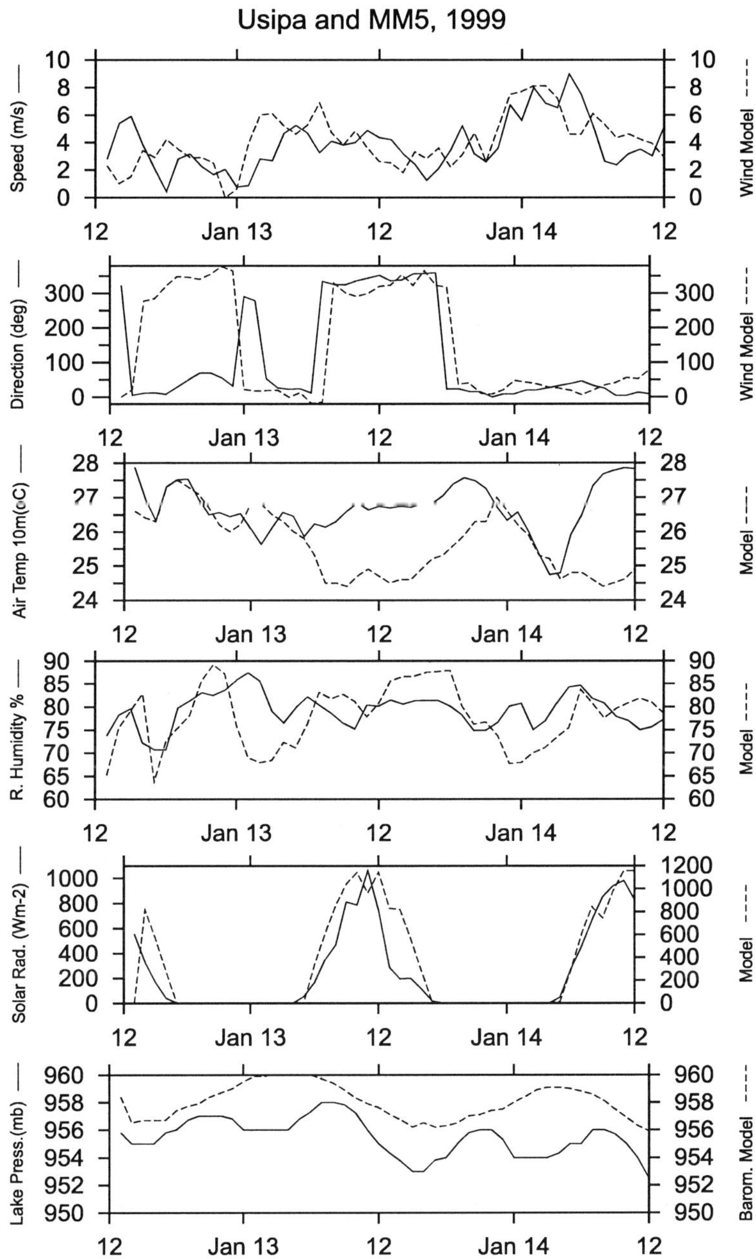


*Figure 17.* Comparison of observed (solid line) and modeled (dashed line) wind speed, direction, air temperature, and barometric pressure along track of Figure 16.

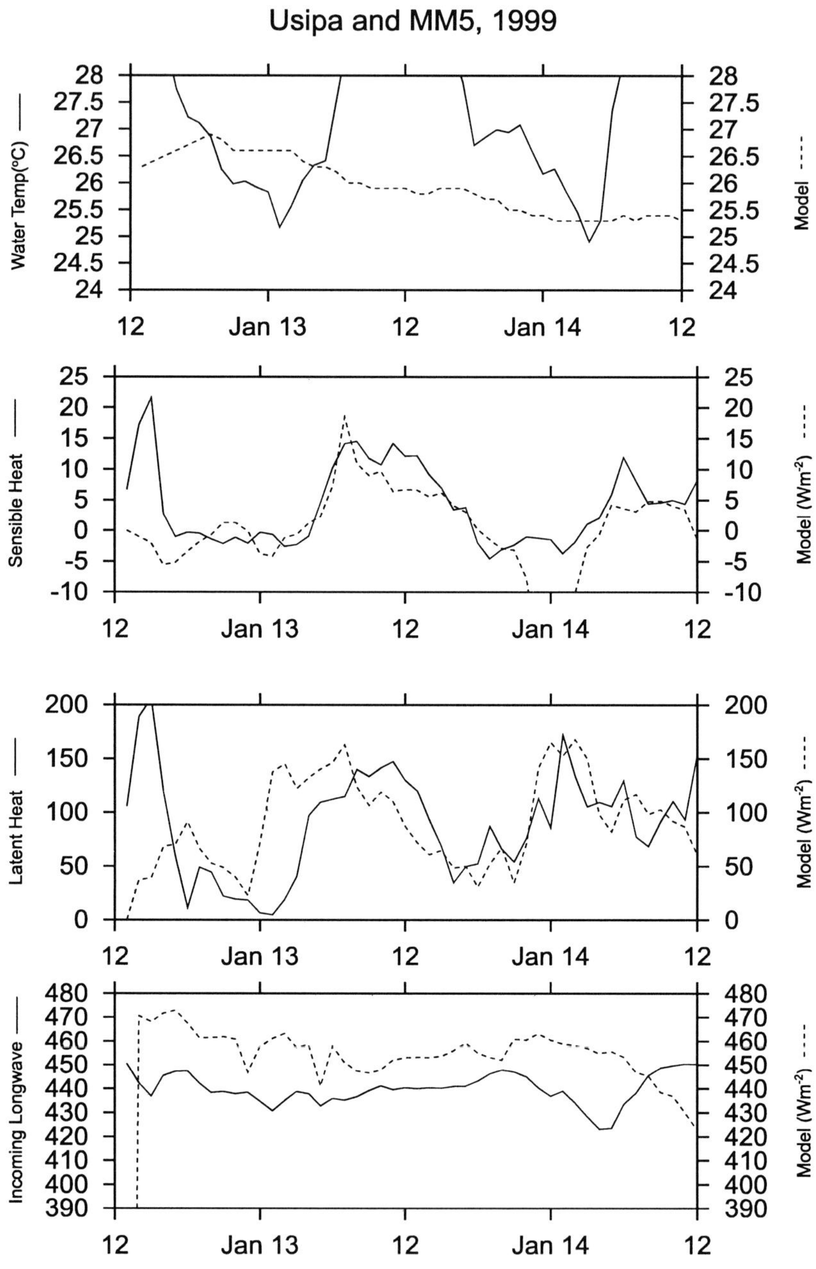


*Figure 18.* Comparison of observed (solid line) and modeled (dashed line) modeled energy fluxes along track of Figure 16.

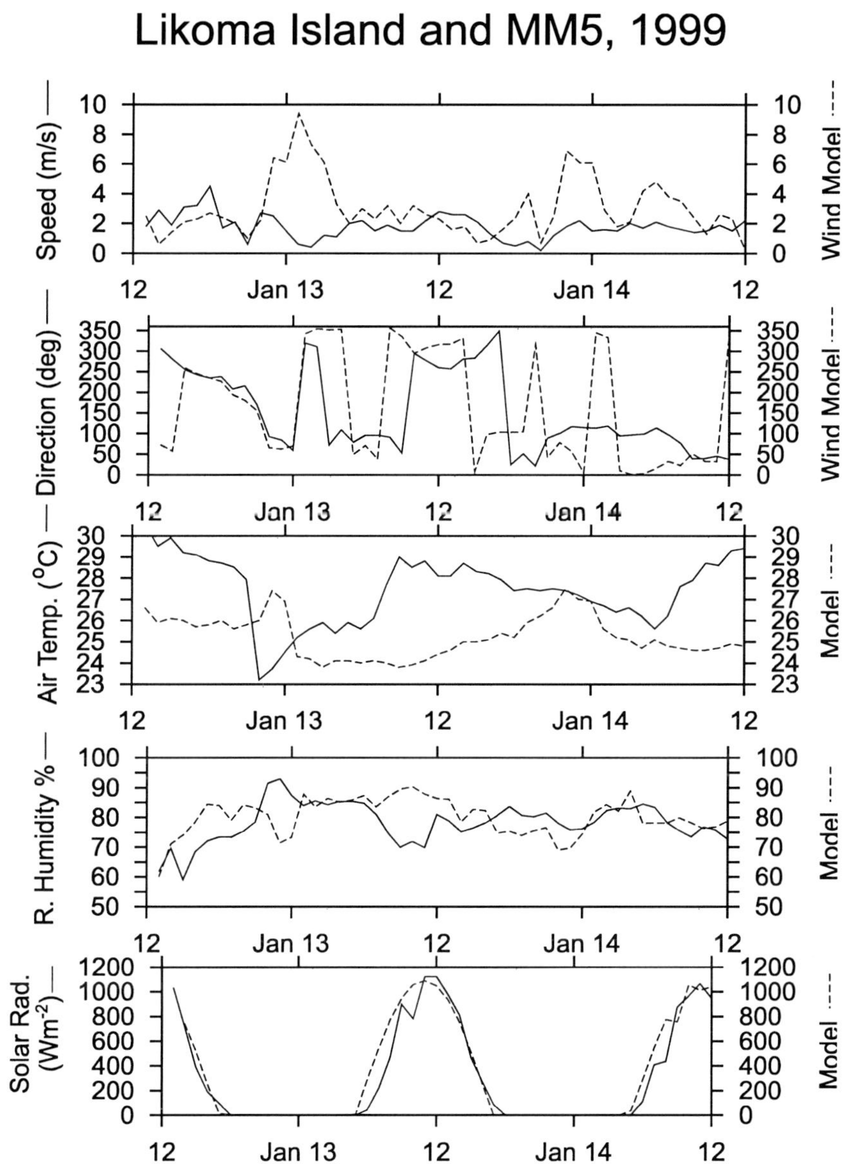


*Figure 19.* Same as Figure 17 but for the Likoma Island station.

direction and relative humidity but poorer for air temperature. As well, shortwave radiation agreed well with that observed aboard the ship. Air temperature is important for the lake hydrodynamic modeler as it enters into expressions for the sensible and latent heat fluxes. These fluxes are also output by the model and compared favorably with those calculated from observed water and air temperatures and relative humidity as is evident in Figure 18. On a daily basis the incoming longwave energy flux is the most energetic of all the downward energy fluxes. As it was not measured aboard the R/V *Usipa* it had to be estimated from an empirical expression involving the unmeasured cloud cover. The observed longwave curve in Figure 18 is based on a value of 100% cloud cover which is supported by examination of the visible bands of the AVHRR local area coverage satellite images obtained from the National Ocean and Atmospheric Administration's Satellite Active Archive System. During the two-day comparison clouds completely covered the southern zone of the lake. In contrast to the open lake meteorological data comparisons, modeled and observed data were found to correspond poorly with one another at the three shoreline stations. As an example of a well exposed station, the Likoma Island comparison is given in Figure 19.

Quantitative comparisons of shipboard, modeled and shore-based winds are provided in Table 2. The nearest neighbor statistics refer to estimates of over-lake wind components interpolated from the shore-based observations. In most hourly samples interpolated winds at the ship were based on weighting factors proportional to the inverse distance squared from the ship to the land station up to a distance of 150 km. This eliminated the Chilumba station and occasionally either the Likoma Island or the Senga Bay station from the interpolation procedure depending on the location of the ship. It is useful to compare the results by an number of statistics. The percentage of explained variability from rest was used by Beletsky *et al.* (2000) to compare modeled and observed current components and corresponds to the ability of either the model or shoreline observations to account for the variability of the observed wind field. The correlation coefficient is defined in the standard way. All three statistics are consistent with the agreement between the curves on Figure 17. Further, they indicate that the modeled wind field is superior to extrapolations from a limited number of shore-based observations. A negative explained variability means that the result is worse than no prediction.

*Table 2*. Comparison of observed and modeled winds. U is the east component of wind and V, the north.

| Model Type | Error (m/s) | Percentage of Explained RMS Variability from rest | Correlation Coefficient U,V |
|---|---|---|---|
| Nearest Neighbor (150 km) | 3.7 | -198 | 0.32, 0.68 |
| Mesoscale Met Model, MM5 original constant LST | 3.6 | 33 | 0.74, 0.6 |

On a qualitative basis the modeled winds may be compared to the observed winds in each of the nine zones and 3-hr intervals over a day shown in Figure 11. Similarly to the shipboard observations the model shows in Figure 20 divergent or onshore winds during the day and convergent or offshore winds during the night.

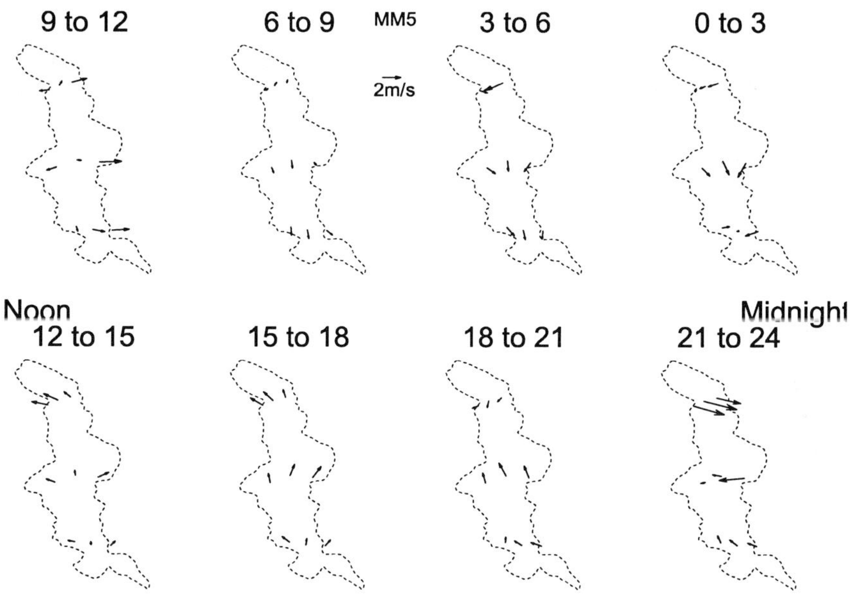


*Figure 20.* Same as Figure 11 but for modeled winds over a 2-d period.

## 4. DISCUSSION

The lack of evidence for signatures in the precipitation, air temperature and wind strength of the 1997-1998 ENSO event is surprising. It is possible that Lake Malawi/Nyasa is situated at node between two types of climatic response to the El Niño/ La Niña cycle, one to the north and the other south of the lake.

Unlike the case for Lake Tanganyika, moored meteorological buoys were not used on Lake Malawi/Nyasa. However, with the use of an automated system of recording position and platform motion, it has been demonstrated that a reasonably comprehensive data set can be collected aboard a roving vessel. However, this approach is far from ideal when compared to a network of meteorological buoys. It suffers from a lack of synopticity and may be biased to daylight hours and fair weather. At this point, equipment failures have made it impossible to establish the seasonal differences in open lake meteorology for Lake Malawi/Nyasa that have been demonstrated for Lake Tanganyika. It is likely that dry season evaporation rates would be higher than those observed herein. It is possible that some uncorrected water temperature errors of unknown origin remained in the shipboard observations which could have affected evaporation estimates.

It is possible to compare the evaporation rates estimated in the present study by variants of the aerodynamic method with those based on the water balance method. Owen *et al.* (1990) estimated an evaporation rate of 5.2mm $d^{-1}$ for Lake Malawi/Nyasa based on water level recession over the dry season, assuming no groundwater inflow but allowing for precipitation and river inflows and outflows. Spigel and Coulter (1996) estimated 4.2mm $d^{-1}$ from the mean annual water balance. As the variation between two applications of the water balance method of 1mm/d is about the same as between the aerodynamic method the agreement is reasonable; that is, the two independent methods agree to within the uncertainty of the estimates. The average rate of evaporation of this study ($6.4 \pm$ 1mm $d^{-1}$) is comparable (perhaps fortuitously) to the wet season evaporation rate quoted by Patterson and Kachinjika (1995) as measured at Salima by pan evaporation, a station situated 5 km inland from Senga Bay.

The over-lake meteorological data sets collected by meteorological buoys and the R/V *Usipa* demonstrate the importance of the diurnal weather system in these two Rift Valley lakes. This characteristic is in contrast to weather systems over the Laurentian Great Lakes, where diurnal winds are barely detectable in offshore wind spectra (Hamblin, 1987, Hamblin and Elder 1973). However, there are certain similarities to the shoreline winds on the temperate zone intermontane Lake Geneva. Lemmin and D'Adamo (1996) found some stations where the cross-shore diurnal energy exceeded the long-term energy but not in the longshore component unlike Lakes Tanganyika and Malawi/Nyasa. Diurnal winds have typical amplitudes of 2 to 3 m $s^{-1}$ compared to the strength of the long-term wind of 2 to 4 m $s^{-1}$. However, at the extremities of the lake where smoothed winds are 2m $s^{-1}$ or less, diurnal winds predominate. Savijarvi (1997) has pointed out that for Rift Valley lakes the usual land/lake air temperature contrast driving the diurnal wind pattern is augmented by a valley or slope wind component. This slope component is evidently solely responsible for the nocturnal wind blowing from the land in equatorial lakes (Savijarvi, 1997). The addition of slope winds to those driven by land-lake thermal contrasts likely accounts for the dominance of the diurnal wind systems on the two lakes of interest. Evaporation estimates may be seriously in error if the diurnal system is not taken into account. Interestingly, the east-west asymmetry in the diurnal wind system observed in Lake Malawi/Nyasa is in accordance with the mesoscale modeling results of Savijarvi (1997) which took into account the interaction of the diurnal wind system with the south-east trade winds.

The lack of low frequency fluctuations in the along-shore wind spectra, the reduction of wind speed at the extremities and the tendency of wind to blow along the major axis are indications of strong topographic sheltering by the lakes' basins. For a smaller lake Hamblin *et al.* (1999) demonstrated directly the sheltering effect of surrounding topography by a comparison of winds on the shoreline with winds in the middle and how the reduction of over-lake wind fields conformed to boundary layer theory.

Hamblin (1987) found that the results of an application of a wind-driven hydrodynamic model to Lake Erie were degraded when a dozen shoreline wind stations were included along with six moored stations in the interpolation of over-lake winds when compared to those for the six meteorological buoys alone. Likely, local sheltering at the shoreline sites renders these stations less useful for hydrodynamic modeling purposes than well situated meteorological buoys. The 4-km grid spacing and 1-hr output employed in this application of a mesoscale meteorological model appear to be adequate for capturing the complex atmospheric circulation over the surface of a 60-km wide lake but the MM5 simulation may be less realistic for nearshore locations.

# 5. CONCLUSIONS AND RECOMMENDATIONS FOR FURTHER STUDY

The pronounced diurnal wind and temperature systems suggest a strong mesoscale organization of the wind and temperature fields. A three-dimensional extension of the type of two-dimensional mesoscale meteorological model outlined by Savijarvi (1997) has been shown to be in reasonable agreement with over-lake field data and thus could potentially yield valuable additional information to the sparse meteorological coverage that we measured in our field experiments. In particular, it could be used on a practical basis to specify the meteorological forcing to lake hydrodynamic models. An evaluation of a similar pilot application to Lake Tanganyika would be most instructive. In other African Great Lakes where there are no regular networks of meteorological buoys and few shore-based stations, the preliminary analysis reported herein suggests that three-dimensional mesoscale meteorological models can provide useful information on the forcing needed by the lake hydrodynamic and water quality modeler.

Comparison of modeled and observed cloud cover over the lake as might be inferred from short wave radiation or satellite imagery needs to be undertaken. As well, the coupling of either observed or modeled lake surface temperature into the lower boundary condition of the mesoscale model should be examined.

The dominance of the diurnal variation of meteorological variables not only at the shoreline but in the offshore areas of the open lake has been demonstrated. This factor has implications for the estimation of the lake's heat budget and hydrodynamics. Neglect of shorter term winds would lead to an underestimation of the latent and sensible heat fluxes and bias the results of hydrodynamic models.

Savijarvi (1997) has shown from a mesoscale model that the diurnal atmospheric circulation over Lake Tanganyika is dominant and is composed of a slope component and a land-lake component, both of which reinforce one another. The observed over-lake wind fields of two African Great Lakes reported in the present study support this conclusion.

In the short-term while funding is being established for lakewide networks of meteorological buoys in the African Great Lakes, it is recommended that both research vessels and other ships of opportunity such as lake ferries be instrumented with recording meteorological systems similar to the one reported in the present study. Before undertaking such measurements it is strongly recommended that the source of error in underway temperature measurement be identified and corrected in applications where moored data are unavailable.

## ACKNOWLEDGEMENTS

We are grateful for the assistance of Mr. S. Smith, Captain M. Day and the crew of R/V *Usipa* in the field program. Environment Canada, the U.S. National Science Foundation, the Canadian International Development Agency, the Global Environmental Fund, the Department for International Development, U.K. and the Research Project for the Management of the Fisheries of Lake Tanganyika, Food and Agriculture Organization, GCP/RAF/271/FIN provided funding for this study. A. Piggott assisted in the extraction of the TOPO30 data. D.H.Eccles is thanked for insightful discussions on lake meteorology and S. McCord and an anonymous reviewer for their improvements to the manuscript.

## REFERENCES

Beletsky, D., Schwab, D.J., McCormick, M.L., Miller, G.S., Saylor, J.H. and Roebber, P.J. (2000) Hydrodynamic modelling for the 1998 Lake Michigan coastal turbidity plume event, *Proc. Sixth Estuarine and Coastal Modelling Symposium*, New Orleans. Am. Soc. Civil Engrg., pp. 597-613.

Bootsma, H.A. (1993) Algal dynamics in an African great lake and their relation to hydrographic and meteorological conditions. Doctoral Dissertation. University of Manitoba, Winnipeg, Canada. p. 311.

Chow, V.T., Maidment, D.R. and Mays, L.W. (1988). *Applied hydrology*, McGraw-Hill.

Dudhia, J. (1989) Numerical study of convection observed during the Winter Monsoon Experiment using a mesoscale two-dimensional model. *J. Atmos. Sci.* **46**, 3077-3107.

Dudhia, J. (1996) A multi-layer soil temperature model for MM5. Preprints, Sixth Annual PSU/NCAR Mesoscale Model Users' Workshop, Boulder CO, National Center for Atmospheric Research, 49-50.

Fischer, H.B., List, E.J, Koh, T.Y.C., Imberger, J and Brooks, N.H. (1979) *Mixing in inland and coastal waters*. Academic Press New York, 483p.

Grell, G.A., Dudhia, J. and Stauffer D.R. (1994) A description of the fifth generation Penn State/NCAR mesoscale model (MM5). NCAR Tech. Note NCAR/TN-398 STR.

Hamblin, P.F., (1987) Meteorological forcing and water level fluctuations on Lake Erie. *J. Great Lakes Res.* **13**, 436-454.

*Hamblin, P.F., Bootsma, H.A. and Hecky, R.E.(2002a) Modeling nutrient upwelling in Lake Malawi/Nyasa* Submitted to the J. Great Lakes Research.

Hamblin, P.F., Bootsma,H.A. and Hecky, R.E. (2002b) Surface meteorological observations over Lake Malawi/Nyasa. *Submitted to the J. Great Lakes Research.*

Hamblin, P.F., Stevens, C. L and Lawrence, G. A. (1999) Simulation of vertical transport in a mining pit lake. *J. Hydraulic Res.* **125** (10), 1029-1038.

Hamblin, P.F. and Elder, F.C. (1973) A preliminary investigation of the wind stress field over Lake Ontario. Proc. Of the 16th Conf. Great Lakes Res. pp 723-734. Internat. Assoc. Great Lakes Res.

Lemmin, U. and D'Adamo, N. (1996) Summertime winds and direct cyclonic circulation: observations from Lake Geneva. *Ann. Geophysicae* **14**, 1207-1220.

Liu, T.W., Katsaros, K.B. and Businger, J.A. (1979) Bulk parameterization of air-sea exchanges of heat and water vapour including the molecular constraints at the interface. *J. Atmos. Sci.* **36** (9), 1722-1735.

Mortimer, C.H. (1979) Strategies for coupling of data collection and analysis with dynamic modelling of lake motion, in W.H. Graf and C.H. Mortimer (eds.), *Hydrodynamics of lakes*. Elsevier Amsterdam.

Mukabana, J.R. and Pielke, R.A. (1996) Investigating the influence of synoptic-scale monsoonal winds and mesoscale numerical model. *Mon. Weather Rev.* **124** (2), 224-243.

Owen, R.B., Crossley, R., Johnson, T.C., Tweddle, D., Kornfield, I., Davison, D.H., Eccles and Engstrom, D.E. (1990) Major low levels of Lake Malawi and their implications for speciation rates in cichlid fishes. *Proc. R. Soc. Lond.* ***B*** **240**, 519-553.

Patterson, G, and Kachinjika, O. (1995) Limnology and phytoplankton ecology, in A. Menz (ed.), *The Fishery Potential and Productivity of the Pelagic Zone of Lake Malawi/Niassa*. Chatham, UK: Natural Resources Institute, pp. 1-67.

Reisner, J. , Rasmussen, R.M. and Bruintjes, R.T. (1998) Explicit forecasting of supercooled liquid water in winter storms using the MM5 mesoscale model. *Quart.J. Roy. Meteor. Soc.* **125B**, 1071-1108.

Savijarvi, H. (1997) Diurnal winds around Lake Tanganyika. *Quart. J. Royal Met. Soc.* **123**, 901-918.

Spigel, R.H. and Coulter, G. W. (1996) Comparison of hydrology and physical limnology of the East Africa great lakes: Tanganyika, Malawi, Victoria, Kivu and Turkana (with reference to some North American great lakes, in T.C. Johnson and E.O. Odada (eds.), *Limnology, Climatology and Paleoclimatology of the East African Lakes.* Gordon and Breach, Toronto, pp. 103-139.

Verburg, P. (1997) Lake Tanganyika hydrodynamics and meteorology: the diel cycle. Food and Agricultural Organization of the United Nations. GCP/RAF/271/FIN-TD/73.

Verburg, P. and Hecky, R.E. (2002) Wind patterns, evaporation and related physical variables in Lake Tanganyika. Submitted to *J. Great Lakes Research.*

Verburg, P., Kakogozo, B., Makasa, L., Muhoza, S. and Tomba, J.M. (1998) Hydrodynamics of Lake Tanganyika 1993-1996, Synopsis and Interannual Comparisons. Food and Agricultural Organization of the United Nations. GCP/RAF/271/FIN-TD/87.

Zhang, D.L. and Anthes, R.A. (1982) A high-resolution model of the planetary boundary layer sensitivity tests and comparisons with SESAME-79 data. *J. Appl. Meteor.* **21**, 1594-1609.

Zhang, D.L., Chang, H.R., Seaman, N.L.,Warner, T.T. and Fritsch, J.M. (1986) A two-way interactive nesting procedure with variable terrain resolution. *Mon. Wea. Rev.* **114**, 1330-1339.

# DEVELOPMENT OF A COUPLED REGIONAL CLIMATE SIMULATION MODEL FOR THE LAKE VICTORIA BASIN

YI SONG, FREDRICK H. M. SEMAZZI and LIAN XIE

*Department of Marine, Earth and Atmospheric Sciences, North Carolina State University, Raleigh, NC 27695-8208, USA*

## ABSTRACT

A three dimensional dynamical model for Lake Victoria based on the Princeton Ocean Model (POM) model has been developed. In this formulation the standard version of the POM model is modified by replacing the open boundaries with a closed coastline and adopting the bathymetry of Lake Victoria. The lake model has 9 equal vertical sigma levels and the horizontal resolution is 20 km. The model's upper boundary conditions are based on momentum, sensible heat and radiative energy fluxes derived from double-nested simulations of the NCAR regional climate model (RegCM2) or idealized surface boundary forcing. The idealized simulations are 360-day model simulations to investigate, (i) the sensitivity of the lake's circulation on variations in the shear and direction of the idealized upper boundary wind stress forcing, (ii) the response to abrupt change in the direction of surface wind stress forcing to assess the memory of the lake, and (iii) the sensitivity of the lake's circulation on its vertical temperature stratification. The RegCM2 model output is used for constructing the upper boundary conditions of the lake model for the October-November-December seasonal "short-rains" of eastern Africa. In these experiments we apply selective and systematic suppression of radiation, heat and momentum fluxes contributions to the upper boundary forcing to determine their relative importance. The results show that the bathymetry and geometry of the lake play a fundamental role in determining the climatology of Lake Victoria. The regions of upwelling or downwelling are determined by wind stress and nearshore



*E.O. Odada and D.O. Olago (eds.),*
*The East African Great Lakes: Limnology, Palaeolimnology and Biodiversity,* 153–186.

bathymetry. The momentum memory of Lake Victoria is about two weeks long. There exists Kelvin wave like disturbances in the thermocline that are trapped along the coast and propagate clockwise around Lake Victoria with periodicity of about 30 days. The oscillations entirely disappear in the case of the isothermal conditions. The lake surface circulation is characterized by anti-clockwise circular motion in response to the predominantly easterly surface winds. The three-dimensional model produces a surface temperature pattern indicative of horizontal lake water mixing characterized by a horizontal spiral pattern in the temperature field. This is associated with the spreading of a pool of warm water across the northern section of the lake. This pattern is not present in the one-dimensional model. Comparison of the POM model simulation results with the one dimensional thermal diffusion lake model, used in the NCAR RegCM2 regional climate model, indicates that the former produces more realistic results.

This work represents an important phase towards the development of a fully coupled regional climate simulation model for the Lake Victoria basin.

# 1. INTRODUCTION

Lake Victoria (Figure 1) is the largest freshwater lake in the tropics and it is the second in the world after the Great Lakes in the USA. It straddles the Equator and extends across the borders of Kenya, Tanzania and Uganda in East Africa. The lake is about 400 km long and 240 km wide and its only major outlet is the river Nile which is the longest river in the World.

The seasonal climate over the Lake Victoria basin is primarily governed by the passage of the intertropical convergence zone (ITCZ). The ITCZ separates the NE and SE monsoons (Nicholson *et al.*, 1996). The ITCZ crosses East Africa twice every year, once during April-May and again during October-November. This migration of the ITCZ is responsible for the two main rainfall peaks, each year in most regions of eastern Africa, which occur in March to May, and Mid-October to early December. These seasons are commonly known as the long rains and short rains of Eastern Africa, respectively. Lake Victoria also experiences the two-rainfall maxima (Faris, 1997). In this investigation we focus on the short rain season (October to December) of 1988. The short rain season has been chosen because it has greater spatial coherence than the long rains of March through May. In this preliminary inquiry we focus on 1988 because it exhibited near normal climate conditions (Ininda, 1994). Furthermore, the short rains are closely linked to the ENSO global climate anomaly conditions than the long rains (Semazzi and Indeje, 1999; Indeje *et al.,* 2000). Since the ENSO climate signal is strong over Eastern Africa, we envisage that Lake Victoria has strong response to it. This paper discusses some of our work toward the goal of adopting the modeling approach in investigating the predictability of the ENSO signal and other teleconnection climate signals within the lake and the surrounding regions of Eastern Africa.

There are three types of wave motion associated with lakes, namely, the Poincare free internal waves, the Kelvin internal waves, and the topographic waves (Csanady, 1967, 1968). The Poincare waves extend across the entire lake basin within the thermocline. The largest scale of this type of waveform has maximum wave amplitude on opposite sides of the lake with a node at the center. The internal Kelvin waves are coastally trapped in the thermocline and progress cyclonically in the Northern Hemisphere and in the opposite direction in the Southern Hemisphere. The topographic waves have a barotropic response. These are vorticity waves occurring only in the presence of significant earth's rotation and sloping bathymetry. In this paper, we also discuss model evidence of wave motion in Lake Victoria which we believe is important in the development comprehensive understanding of the variability and predictability of the regional climate of Eastern Africa.

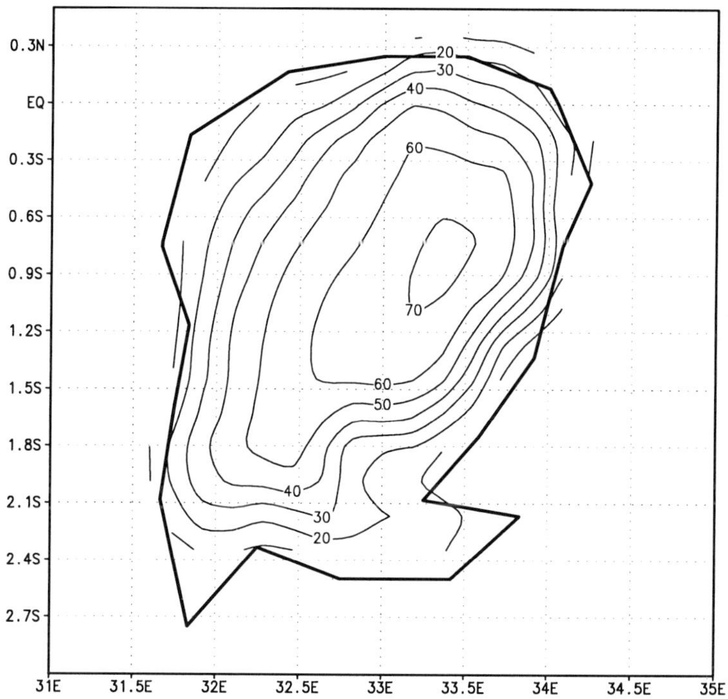


*Figure 1.* Lake Victoria Bathymetry with horizontal resolution of 20 km (redrawn from Johnson, 1997). The thick line corresponds to the perimeter of Lake Victoria. Units in meters.

On longer timescales, several studies (Stager *et al.*, 1986; Kendall, 1969) have shown that Lake Victoria has experienced large surface water level fluctuations during the past 25,000 years. Stager *et al.* (1986) have estimated that the water level was only 26m between 14,750 and 13,700 BP. Such paleoclimatic changes could be prescribed in regional models to infer the corresponding changes in catchment rainfall and lake circulation, and thereby provide improved interpretation of the paleoclimate theories which have been proposed over the years. Potential applications of regional climate modeling of the Lake Victoria basin include, (i) prediction of fish environments and population dynamics, (ii) prediction of lake transportation of potentially highly toxic chemical affluent from the food processing, textile production, and cement industries around the lake shores of Lake Victoria, (iii) prediction of the migration patterns of the Water Hyacinth (Eichornia Crossipes), which begun invading the lake during the late 1980s, (iv) provision of marine and meteorological advisories for ship navigation considering a series of previous lake circulation and weather-related catastrophic accidents, (v) provision of guidance in the design and management of hydroelectric power plants, and (vi) climate change and biodiversity studies. It is therefore a matter of high priority to develop a comprehensive numerical simulation model for Lake Victoria and its coupling with the regional atmospheric circulation and hydrology.

# 2. DESCRIPTION OF THE ATMOSPHERIC AND LAKE MODELS

## 2.1 Regional Climate Model

In this section we briefly describe the main attributes of the North Carolina State University (NCSU) version of the National Center for Atmospheric Research (NCAR) regional climate model (RegCM2; Semazzi, 1999; Sun *et al.*, 1999ab). RegCM2 has been adapted for Eastern Africa using the 60-km horizontal resolution. In the vertical, the atmosphere is stratified into 15 layers. The sensitivity of the model's performance on the vertical and horizontal resolution was extensively examined in Sun *et al.* (1999a,b). It was optimized to take into account the existing computing limitations and the ability of the model to reproduce observed features of the climate variability of Eastern Africa. The model's initial and lateral boundary conditions are taken from European Center for Medium Range Weather Forecasting (ECMWF), 6-hourly, analyzed atmospheric observational data. To evaluate the regional climate model's performance in reproducing the observed precipitation over eastern Africa, regional averages of simulated and observed precipitation were compared over nine homogeneous climate sub-regions constructed by Indeje *et al.* (2000), based on cluster analysis techniques.

A striking feature of the comparison between the model and the observed rainfall (not shown) is that the model reproduces the month-to-month changes (increase or decrease) of the rainfall for all the 9 regions. This demonstrates the ability of the model to resolve the complex migration patterns of the Inter-Tropical Convergence Zone (ITCZ) over such complicated terrain, vegetation, and land-water contrasts over eastern Africa. Contemporary global climate models cannot produce such geographical details because of their coarse resolution. We consider this performance of the regional climate model as an important step toward the application of RegCM2 in the prediction of the climate of the region. The model also reproduces the day-to-day and year-to-year changes (not shown) in rainfall over the catchment region of Lake Victoria (Sun *et al.*, 1999a,b). Furthermore, we have also compared the diurnal variations of the rainfall at several locations around the lake (Figure 2). The model accurately reproduces the observed diurnal asymmetries in the rainfall distribution over and around Lake Victoria. Figure 2 also shows the simulated hourly precipitation in November 1988 interpolated from the simulation in which the RegCM2 model in its standard configuration is coupled to the one-dimensional lake model. Over the west (Kishanda, Masaka and Entebbe) and central region (Kahunda) of the lake, the precipitation occurs mostly during early morning and little or no precipitation during the afternoon and evening. Over the eastern part of the lake (Muhuru Bay and Kadenge), the maximum precipitation occurs around midnight, and thereafter it decreases sharply with minimum precipitation around noon.

To assess the performance of the model in reproducing the diurnal cycle we compare the numerical results in Figure 2 with the observations in Figure 3. In qualitative terms the diurnal cycle of the simulated rainfall (Figure 2) along the lakeshore and Island locations compares well with the observed rainfall (Figure 3). The model rainfall is expressed in mm/month. However, the observations which are based on Datta (1981) are stratified in terms of frequency of rainfall occurrences per hour. This comparison suggests that the RegCM2 model coupled to the one-dimensional lake model produces realistic circulation over the catchment region. This indirect evaluation of the RegCM2 model performance in reproducing the circulation over and around the perimeter of the lake is significant because the momentum, radiation, and heat fluxes from RegCM2 have been used, as proxy for atmospheric forcing in the construction of the upper boundary conditions for the dynamical model of lake Victoria.

In this study we exploit the availability of a few recent studies (Ba and Nicholson, 1998; and Yin and Nicholson, 1998) based on remote sensing to characterize the climate over the lake basin, and therefore provide means to assess the performance of our numerical model. Since observational data for validating model simulations is very limited, we also make use of mechanistic sensitivity model simulations to supplement our analysis. In future, we hope comprehensive meteorological data and marine observations for Lake Victoria will become available for evaluating the high-resolution models used in this study for simulating the climate variability over eastern Africa and the circulation in Lake Victoria. We believe that this study will contribute to the design of future observational platforms for monitoring the variability of climatic conditions in Lake Victoria and the surrounding region.

The representation of Lake Victoria in the standard version of the RegCM2 regional climate model is based on the one-dimensional lake formulation (Hostetler *et al.*, 1993). In their treatment, heat from the Sun's radiation, which is incident at the lake surface, penetrates and exits the lake along vertical columns of water. Although this formulation is satisfactory in reproducing the primary energy exchange between the lake and the atmosphere, it suffers from lack of dynamical mixing. As demonstrated later in this paper, we find that incorporation of the three-dimensional dynamical effects in the representation of the circulation in Lake Victoria results in superior model performance.

Throughout the study, double nested RegCM2 model simulations were used to generate the upper boundary conditions (momentum and heat flux) for the 3-D lake model described in section 2.2. This involves taking output from a coarse resolution (60 km) model run to generate the initial and lateral boundary conditions for the

20km horizontal resolution double nested simulation. In the double nested simulations, the inner domain is centered at (33°E, 1°S), and covers a region of 1660 km by 1480 km compared to 5580 km by 5040 km of the coarse mesh. This high resolution domain covers the lake Victoria catchment. All the other considerations are the same as those of the 60-km resolution model runs.

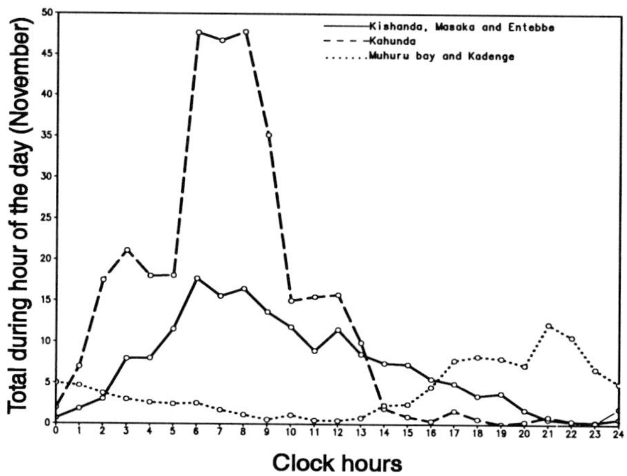


*Figure 2.* Simulated precipitation distribution and diurnal change over lake catchment in November 1988.

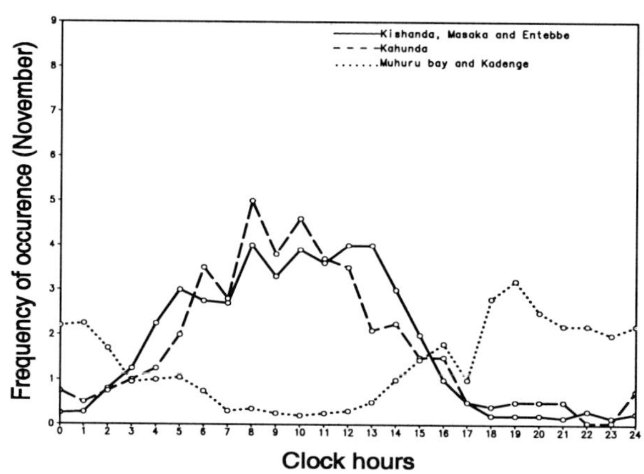


*Figure 3.* Observed mean precipitation distribution and diurnal change over Lake catchment in November (adapted from Datta, 1981).

## 2.2 Lake Model

The Lake model in this study is based on the Princeton Ocean Model (POM) 1997 version (Mellor, 1998). It is a three dimensional, nonlinear, primitive equation finite difference hydrodynamic ocean model. The model employs a mode splitting technique that solves for the barotropic mode for the free surface and vertically averaged horizontal currents, and the baroclinic mode for the full three-dimensional temperature, turbulence, and current structure. The governing equations (1-4) are written in a terrain following $\sigma = (\eta - h)/(H + h)$ vertical coordinate system, where H (x,y) is the bottom topography and η(x,y,t) is the surface elevation and includes a turbulence closure submodel with an implicit time scheme for vertical mixing (Mellor and Yamada, 1974). The equation of state (Mellor, 1991) is used to calculate the density as a function of temperature, salinity, and pressure. A three-time level leapfrog scheme is used for the time differencing. The basic equations of the σ coordinate system may be expressed as follows:

$$\frac{\partial UD}{\partial t}+\frac{\partial U^2D}{\partial x}+\frac{\partial UVD}{\partial y}+\frac{\partial U\omega}{\partial \sigma}-fVD+gD\frac{\partial \eta}{\partial x}+\frac{gD^2}{\rho_0}\int_\sigma^0\left[\frac{\partial \rho'}{\partial x}-\frac{\sigma'}{D}\frac{\partial D}{\partial x}\frac{\partial \rho'}{\partial \sigma'}\right]d\sigma' = \frac{\partial}{\partial \sigma}\left[\frac{K_M}{D}\frac{\partial U}{\partial \sigma}\right]+F_x \tag{1}$$

$$\frac{\partial VD}{\partial t}+\frac{\partial UVD}{\partial x}+\frac{\partial V^2D}{\partial y}+\frac{\partial V\omega}{\partial \sigma}+fUD+gD\frac{\partial \eta}{\partial y}+\frac{gD^2}{\rho_0}\int_\sigma^0\left[\frac{\partial \rho'}{\partial y}-\frac{\sigma'}{D}\frac{\partial D}{\partial y}\frac{\partial \rho'}{\partial \sigma'}\right]d\sigma' = \frac{\partial}{\partial \sigma}\left[\frac{K_M}{D}\frac{\partial V}{\partial \sigma}\right]+F_y \tag{2}$$

$$\frac{\partial TD}{\partial t}+\frac{\partial TUD}{\partial x}+\frac{\partial TVD}{\partial y}+\frac{\partial T\omega}{\partial \sigma}=\frac{\partial}{\partial \sigma}\left[\frac{K_H}{D}\frac{\partial T}{\partial \sigma}\right]+F_T-\frac{\partial R}{\partial \sigma} \tag{3}$$

$$\frac{\partial UD}{\partial x}+\frac{\partial DV}{\partial y}+\frac{\partial \omega}{\partial \sigma}+\frac{\partial \eta}{\partial t}=0 \tag{4}$$

where u, v are, respectively, the velocity components along the x and y axes in the lake, and $D = H + \eta$. The vertical coordinate σ ranges from σ = 0 at $z = \eta$ to σ = -1 at z= -H; ω is the velocity component normal to σ surfaces; T is potential temperature; f is the Coriolis parameter; $\rho_o$ is reference density; $\rho'$ is density at sigma level $\sigma'$; g is gravitational acceleration, t is time; $K_M$ and $K_H$ are vertical eddy viscosity and eddy thermal diffusivity, respectively; $F_x$, $F_y$ and $F_T$ are horizontal viscosity terms for u, v and T, respectively. The surface boundary conditions are given by,

$$\omega = 0 \quad at \quad \sigma = 0; \quad \frac{K_M}{D}\left(\frac{\partial U}{\partial \sigma}, \frac{\partial V}{\partial \sigma}\right) = -(< wu(0), < wv(0) >) \quad at \quad \sigma = 0 \tag{5}$$

where $wu(0)$ and $wv(0)$ are the x and y components of surface momentum fluxes. The first equation specifies no flow normal to the lake surface and the second equation expresses the boundary condition of the shear stresses at the lake surface. The upper boundary condition for temperature is given by,

$$\frac{K_H}{D}\frac{\partial T}{\partial \rho} = -(< w\theta >) \quad at \quad \sigma = 0 \tag{6}$$

$$\omega = 0 \quad at \quad \sigma = -1; \quad \frac{K_M}{D}\left(\frac{\partial U}{\partial \sigma}, \frac{\partial V}{\partial \sigma}\right) = C_Z[U^2 + V^2]^{1/2}(U, V) \quad at \quad \sigma = -1 \tag{7}$$

where

$$C_Z = MAX\left[\frac{k^2}{[\ln\{(1+\sigma_{KB-1})H / z_0\}]^2}, \quad 0.0025\right] \tag{8}$$

The coefficient k=0.4 is the Von Karman constant, $z_0$ is the roughness parameter, and $\sigma_{KB-1}$ represents the σ value at the level just above the bottom boundary. Equations 7 and 8 specify no flow through the bottom and the boundary condition of shear stresses at the bottom. For the temperature field, the following no-flux bottom condition is used.

$$\frac{K_H}{D}\frac{\partial T}{\partial \sigma} = 0 \quad at \quad \sigma = -1 \tag{9}$$

The POM model has been widely used to study major lakes, such as Lake Michigan (O'Connor *et al.*, 1995) and Lake Erie (Kelly *et al.*, 1998). Water temperature and circulation of Lake Michigan and Lake Erie are generally predicted well by POM. In this study, we configure the POM model for the simulation of Lake Victoria, the largest lake in the tropics. A significant difference of freshwater lakes compared to coastal ocean is that they are not influenced by salinity effects and tides. In applying POM model for Lake Victoria, several modifications have been made. First, the model domain is assumed to be completely enclosed by land so that open boundary conditions are not required. Secondly, we ignore the effects of river runoff, evaporation and precipitation on the elevation of the lake. Finally, salinity is set to a constant value of 0.2ppt, which corresponds to freshwater lakes (O'Connor *et al.*, 1994). Our experience indicates that the results based on such low values of salinity are virtually the same as those when the salinity is set to zero.

As seen from the lake topography (Figure 1), the deepest region is located around (33.4°E, 0.7°S). The slope is steeper on the eastern side than on the western side of the lake. In this study we have adopted a rectangular grid system with horizontal grid spacing of 20 km in both the x and y direction which gives 191 grid points covering the entire lake. The model employs 9 sigma levels in the vertical and layer thickness of approximately 8m over the deepest region of the lake and 0.5m over the shallowest locations. The internal and external time steps are 1800s and 60s, respectively. The fluxes of the surface momentum were calculated from the 10-m height winds over lake Victoria, and heat fluxes were calculated by using the 2-m height heat flux parameters based on the output of the double nested RegCM2 runs. To calculate the surface wind stress, a constant drag coefficient $C_d = 0.0015$ was adopted.

*Table 1*. Summary of Numerical Experiments

| | Wind Stress | Heat Flux and Radiation | Initial Temperature Structure |
|---|---|---|---|
| Experiment 1a | Uniform easterly wind | None | Warmer water over the upper layers and colder water with increasing depth |
| Experiment 1b | Uniform northeasterly wind | None | Same as Exp.1a |
| Experiment 1c | Uniform southeasterly wind | None | Same as Exp. 1a |
| Experiment 2a | The direction of wind stress-shear was reversed during experiment | None | Isothermal |
| Experiment 2b | Same as Exp. 2a | None | Same as Exp. 1a |
| Experiment 3 | Simulated wind based on RegCM2 | Simulated heat flux and radiation based on RegCM2 | Simulated temperature structure based on 1-D lake model in RegCM2 |
| Experiment 4a | Same as Exp. 3 | None | Same as Exp. 3 |
| Experiment 4b | None | Same as Exp. 3 | Same as Exp. 3 |

## 3. DESIGN OF EXPERIMENTS

A total of eight experiments have been designed and performed to investigate the circulation and thermodynamics of Lake Victoria based on a 'stand alone' Lake model. Table 1 is a summary of numerical experiments and details of the primary attributes of the eight experiments are presented in section 4.

# 4. LAKE VICTORIA MODEL SIMULATION RESULTS

## 4.1 Idealized Surface Wind Stress Forcing without Horizontal Shear

The POM model has been successfully employed in the Great Lakes Forecast System for short-term integration, however it has not been extensively applied over tropical lakes and in the climate mode. First, we investigate the simple solutions that arise from applying idealized forcing at the top of the lake to understand the basic response to atmospheric wind forcing.

Since the basic wind structure in the tropics is easterly, in Exp. 1a we prescribed simple idealized upper boundary forcing comprised of easterly flow without horizontal shear. The surface momentum flux was set to $W_u=0.0001m^2s^{-2}$ (approximately corresponding to 10 m/s wind). The explicit variable used in the model in the formulation of surface momentum exchange is surface momentum flux. We use the terms "surface momentum flux" and "surface wind stress" interchangeably for the present discussion since they differ only by a simple normalization factor. Figure 4a-c shows the simulated circulation of Lake Victoria at the surface, 10-m depth and 20-m depth, respectively. The lake circulation within the upper 20m is basically barotropic. The mass continuity dynamical constraint, of zero mass flux at boundary, implies that upwelling is initiated along the eastern coastline and downwelling along the western coastline of the lake (Figure 4d). At 20 m depth, the eastward undercurrent splits into two cells, with the anti-clockwise cell to the north and the clockwise cell to the south (Figure 4c).

As noted earlier (section 2) the lake Victoria region is primarily associated with northeasterly monsoon flow during Northern Hemispheric winter, and southeasterly monsoon flow during Northern Hemispheric summer. Thus, two further experiments (Exp. 1b and 1c) forced by northeasterly and southeasterly surface wind stress were carried out with zero heat fluxes at the top. In Exp. 1b, the horizontal orientation of northeasterly surface momentum flux was set to 30° relative to the x-axis and its components are, $W_u=0.0000866m^2s^{-2}$ and $W_v=0.00005m^2s^{-2}$, respectively. The total surface momentum flux is approximately $0.0001m^2s^{-2}$. In Exp. 1c, the southeasterly surface momentum flux components were set to $W_u=0.0000866m^2s^{-2}$ and $W_v=-0.00005m^2s^{-2}$, thus, only the direction of the wind stress on the y-axis was reversed relative to the northeasterly case. Figure 5a-d is the simulated circulation of Lake Victoria driven by the northeasterly wind (Exp. 1b). The main features of the lake circulation are similar to those of the easterly wind case, however the circulation is asymmetric with a broader counter clockwise circulation over the northern part of the lake, and a stronger eastward undercurrent (Figure 5c). On the other hand, with southeasterly surface wind stress forcing (Exp. 1c) the simulated circulation is stronger over the southern sector of the lake and the eastward undercurrent is weaker (Figure6a-d).

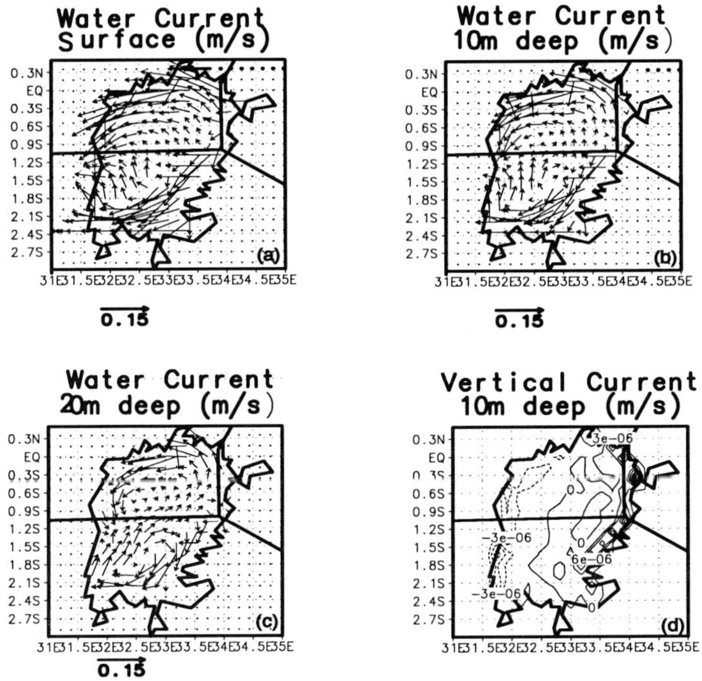


*Figure 4*. Simulated circulation of Lake Victoria driven by easterly surface wind forcing. a). surface current. b) current at 10m depth. c). current at 20m depth. d). vertical velocity at 10m depth (positive values correspond to upward velocity). Units in m/s. Political national boundaries and the perimeter of Lake Victoria are also highlighted in the figure.

In this set of simulations the geometry of the lake plays a significant role. It is wider in the meridional direction than the zonal direction (Figure 1) and it is oriented in the northeast-southwest direction thus somewhat parallel to the northeasterly wind stress in Exp. 1b. Therefore, the northeasterly wind stress exerts greater forcing on the lake flow than the southeasterly wind stress. The resulting lake circulation in Exp. 1b is therefore stronger than that in Exp. 1c. The asymmetric shape of the lake, which is characterized by greater depth and width over the northern sector, may, in part, also account for the differences in the simulated circulation in the three experiments. Since Lake Victoria is situated in the vicinity of the equator (0.3°N-2.4°S) the Coriolis effect is small. Therefore, the circulation changes noted in this section are mainly due to surface wind stress forcing and bottom topography. The accuracy of the bathymetry data therefore plays an important role in customizing the POM model for Lake Victoria.

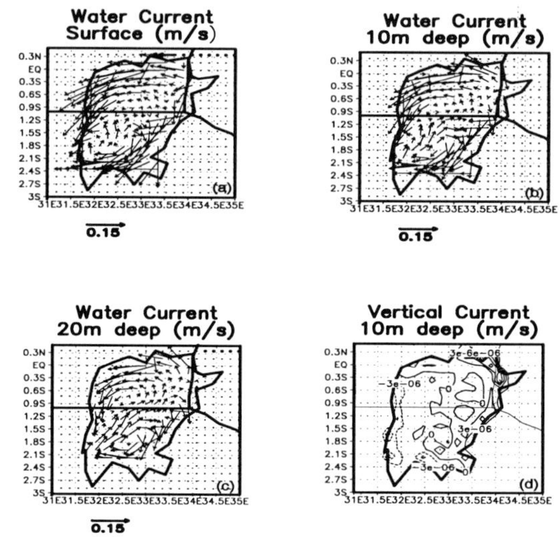


*Figure 5.* Same as Figure 4 except for northeast surface wind stress forcing.

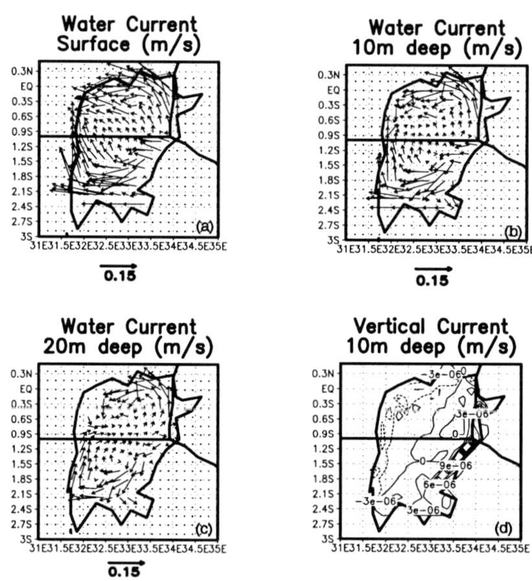


*Figure 6.* Same as Figure 4 except for southeast surface wind stress forcing.

## 4.2 Idealized Surface Wind Stress Forcing with Horizontal Shear

The large heat capacity and density of water tends to cause the temperature of oceans to vary relatively slower and retain larger memory of past conditions than the land and therefore possibly affect the climate on seasonal and longer time-scales. The oceans store large amounts of heat during the summers and release it during the winter. Therefore, the oceans have longer thermodynamic memory than the atmosphere. Lake Victoria is the largest fresh lake in the tropics and considering its spatial scale it would be valuable for prediction purposes to determine if it also exhibits significant ability to retain heat and momentum anomalies which can influence the local and regional climate anomalies during subsequent seasons. In this preliminary inquiry we investigate the ability of Lake Victoria to retain past memory about its circulation. A similar study concerning the role of heat content of the lake is in progress and will be the subject of a separate paper.

Two 360-day numerical simulation experiments starting from motionless flow have been designed and conducted. In each of the two "memory" experiments (Exp. 2a and 2b) the surface wind stress varies only in the north-south direction. In Exp. 2a the basic forcing is easterly along the southern perimeter of the lake, linearly decreases northward to zero near the center and attains maximum westerly intensity along the northern boundary of the lake (Figure 7). The value of the surface momentum flux varied from $0.0001 m^2 s^{-2}$ to 0. The lake model was run for 180 days with this upper boundary forcing. For the rest of the numerical integration the direction of the surface wind stress forcing was reversed to investigate the response of the water circulation to abrupt changes in surface momentum forcing.

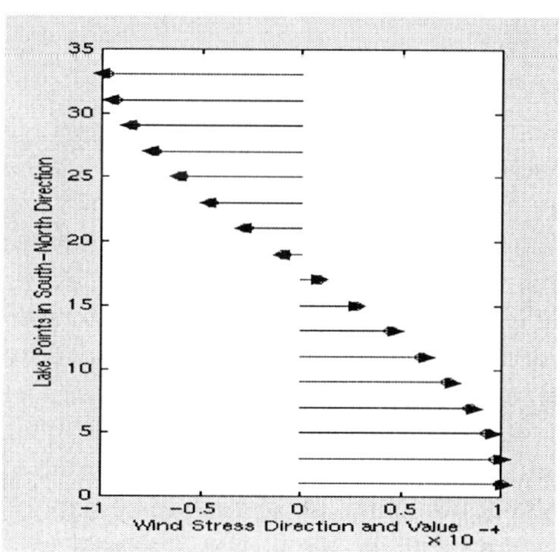


*Figure 7.* Distribution of surface momentum flux in Exp. 2. Units in $m^2 s^{-2}$.

In Exp. 2a the initial temperature over the entire volume of the lake was set to 24°C (Figure 8a) estimated from the equilibrium temperature of the lake in our recent numerical simulation studies (Semazzi, 1999; Sun *et al.,* 1999ab) based on the RegCM2 regional climate model coupled to a one-dimensional lake model.

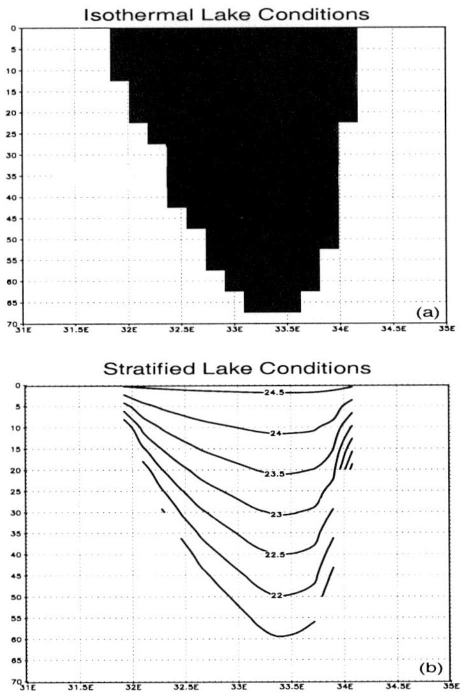


*Figure 8.* Initial vertical temperature distribution along 0.5°S. a) isothermal conditions. b) non-isothermal conditions. Units in °C. Lake's outline suppressed in Figure 8b.

The isothermal conditions are imposed to contrast with the results with the simulation when thermal stratification is introduced in a separate simulation discussed below. The aim is to isolate the effects of vertical temperature variations on the lake's dynamics. Motionless flow was prescribed for the initial conditions of the lake circulation. We can see in Figure 9 (a and b) that the simulation rapidly attains equilibrium in about 7 days. When the direction of the upper boundary surface wind stress forcing is reversed at day-180, the lake's circulation also reverses direction from cyclonic to anticyclonic. Although this is expected because of the reversal in the forcing, it is rather surprising that the circulation readjusts to a new steady state within only two weeks of simulation, around day-194. The isothermal conditions of the lake are maintained throughout the entire 360-day simulation and exhibit virtually no response to the reversal in the surface wind forcing (Figure 9c). Therefore, it is apparent that under isothermal conditions, the memory of the lake is short, on the order of two weeks.

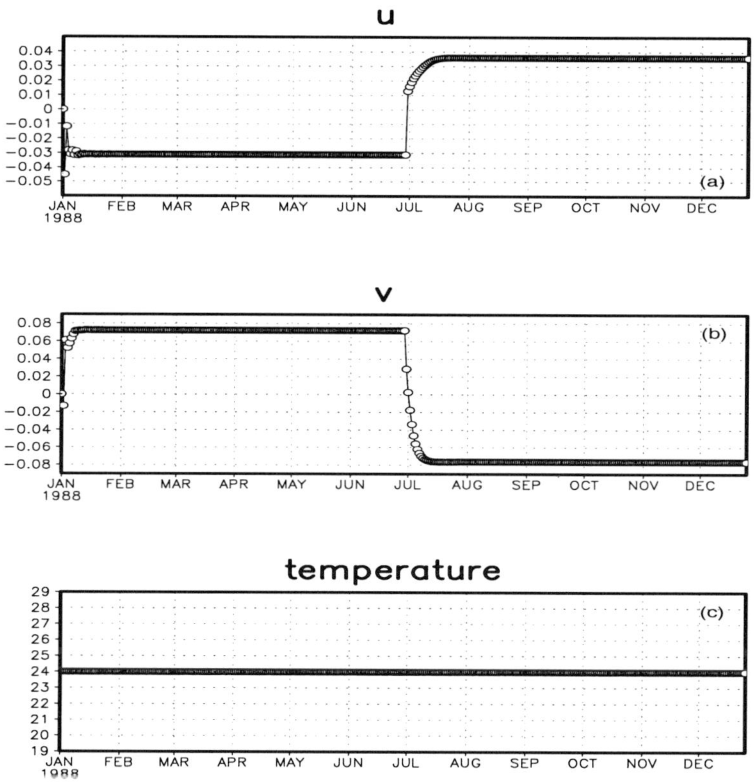


*Figure 9.* Results for experiment 2a in which the Lake Victoria model is driven by the wind stress shown in Figure 7 under isothermal conditions: (a) u component of water current at 5m depth at (33.5°E, 0.5°S). Units in m/s. (b). v component of water current at 5m depth at (33.5°E, 0.5°S). Units in m/s. (c). water temperature at 5m depth at (33.5°E, 0.5°S). Units in °C.

In Exp. 2b the initial temperature of the lake was changed from isothermal conditions by introducing vertical temperature stratification. In the modified initial conditions, the temperature varies from 21.4°C at the bottom to 24.6°C at the surface (Figure 8b), which corresponds to a thermodynamically stable profile. The difference between top and bottom is on the same order as the observed interannual variability (Spigel and Coulter, 1996). The other considerations in the design of this experiment are identical to those in Exp. 2a. The simulated temperature, u component and v component of the flow at 5m depth of the lake around (33.5°E, 0.5°S) are shown in Figure 10a, 10b and 10c, respectively. Starting from the motionless initial conditions the simulation takes about two months to reach steady state, thus much longer than the one week in Exp. 2a where isothermal initial conditions were employed. Although, the physical explanation for this difference is not entirely clear at this time, we envisage that the more stable temperature stratification associated with the non-isothermal conditions and the larger vertical temperature gradient in the shallow water regions in Exp. 2b tend to inhibit initiation of the upwelling and downwelling. Once warm water descends through downwelling and the cold water is raised by upwelling, the stratification of the vertical lake temperature gradient becomes strongly unstable. These unstable conditions promote the developments of wave motion in the lake. The oscillation amplitudes are very large during the stage of strong mixing. Since the initial prescribed vertical temperature gradient is large it prolonged the duration of adjustment to reach equilibrium. Following the reversal in the direction of the upper boundary surface wind stress forcing, a new steady state is achieved in about 14 days and the equilibrium temperature does not change as expected, thus in this respect similar to the results in Exp. 2a where isothermal conditions were assumed. These results indicate that once thermal equilibrium has been established adjustment to changes in the upper boundary wind forcing occurs rapidly, and the memory in the momentum field of the lake is short on the order of 2 weeks.

An important feature of the results that is clearly evident in Figs. 10a-10c is the existence of a 30-45 day oscillation in the u, v, and temperature at 5m depth. The oscillation is synchronous among all the three variables. In the deeper layers, such as at 40m (not displayed), the sign of the oscillation has the opposite sign to the 5m depth signal. This suggests that the oscillations have a coherent three-dimensional return flow structure extending over the entire volume of the lake. To the contrary, we do not observe the oscillation in Exp. 2a where the temperature stratification is isothermal.

Wave motion has been extensively investigated in lakes. For example, Csanady (1967 and 1968) developed a simplified linearized analytical model of the Great Lakes based on a 2-layer circular basin to model the internal baroclinic waves and the surface modes. In the resulting solutions the phase velocity for the surface wave modes is given by,

$$c_1^2 = g(h + h') + O(\varepsilon) \tag{10}$$

The parameter $\varepsilon$ is the baroclinic factor which becomes zero under isothermal conditions. The variables, $h$ and $h'$ correspond to the height perturbation for the top and bottom layers, respectively. The corresponding phase velocity for the internal modes under flat bottom bathymetry conditions is given by,

$$c_1^2 = \varepsilon g \frac{hh'}{h + h'} + O(\varepsilon^2) \tag{11}$$

The latter possibly corresponds to the waves observed in our simulations in Exp. 2b which disappear when $\varepsilon = 0$ under isothermal conditions (Exp. 2a).

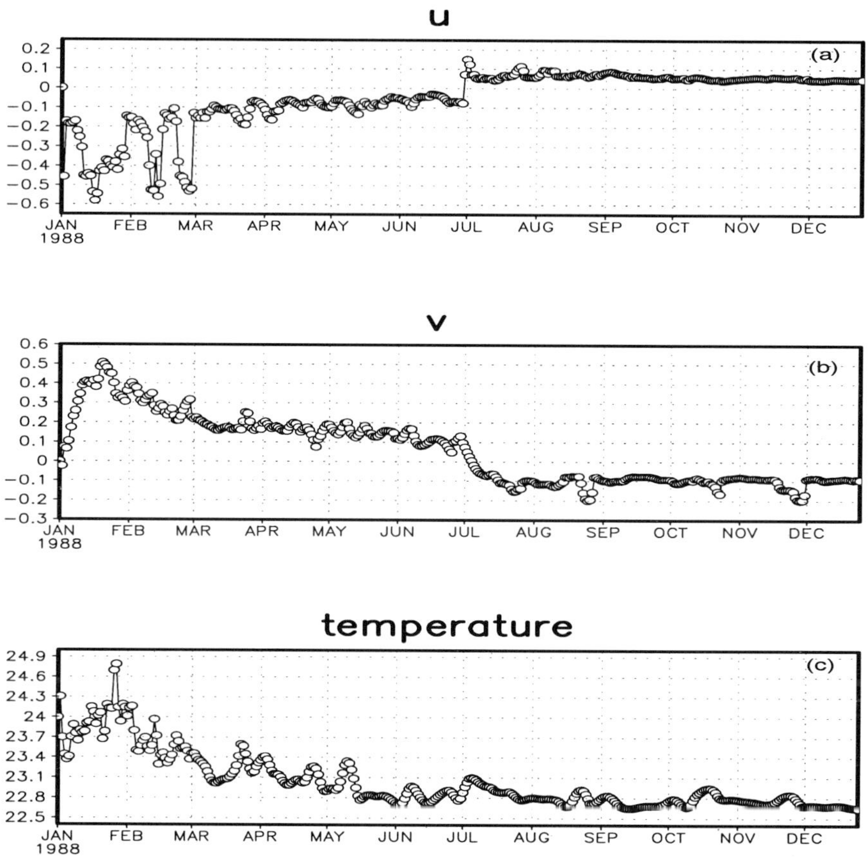


*Figure 10.* Same as Figure 9 except for the non-isothermal initial condition in Figure 8b.

Figures 11a and b are the vertical cross section of the water temperature along 0.5°S from day 181 (wind direction reversed) to day 360, respectively. The vertical temperature settles down well to a quasi-stationary state by day 181 (Figure 11a). With the continuous mixing due to upwelling and downwelling, the lake stratification becomes nearly isothermal except at the deeper layers of the lake (below 30 m) at day 360 (Figure 11b). Meanwhile, the corresponding oscillations progressively become weaker (Figure 10), which indicates that the lake stratification plays a crucial role in the development and maintenance of the wave oscillations. Csanady's theoretical solutions predict that in the southern hemisphere, the wave patterns travel around the circular basin in a clockwise direction as we observe in Exp. 2b (Figure 12), since most of Lake Victoria is situated in the southern hemisphere. Taking the geometry of the lake into account, we plotted the anomalous temperature latitude–time cross sections at 5 m depth from day 181 to day 360, along the 32°E (Figure 12a) and 33.9°E meridians (Figure 12b). The opposite direction in the wave propagation along the east and west coastal regions of the Lake Victoria is clearly evident. Along the western border near 32°E, the oscillation with periodicity of about 30 days travels northward. On the contrary, along the eastern border near 33.9°E, a wave with the approximately the same periodicity travels southward.

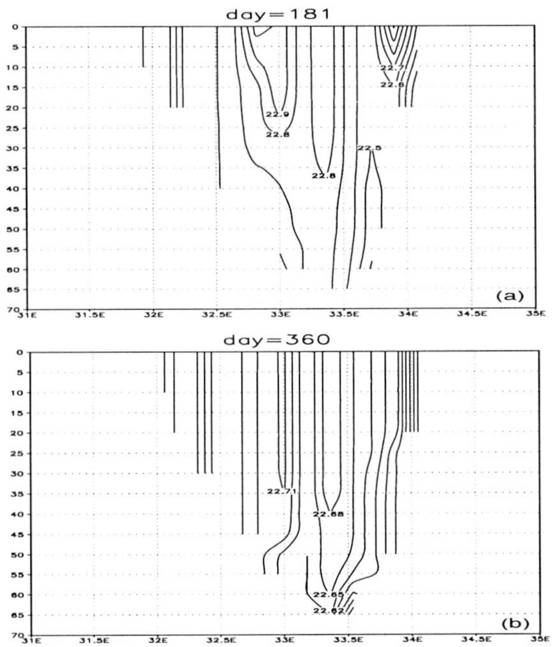


*Figure 11.* Simulated vertical temperature distribution along 0.5°S in Exp. 2b. a) day 181. b) day 360. Units in °C.

Beletsky *et al.* (1997) performed a series of numerical experiments to investigate the numerical simulation of internal Kelvin waves and coastal upwelling fronts. They employed the POM and DIECAST ocean models to study the response of an idealized middle-latitude large circular lake and Lake Michigan to an impulsive wind stress imitating the passage of a weather event. Under steady state conditions resulting from uniform wind forcing, they found that the balance of forces in the region of upwelling is between the wind stress, Coriolis force, and the internal pressure gradient. However, when this balance is disturbed, new balance was established giving rise to the Poincare waves, Kelvin waves and topographic waves. In the northern middle latitudes, Poincare waves are characterized by anticyclonic phase progression with periodicity slightly less than the inertial period, which corresponds to 17.5 h for central Lake Michigan.

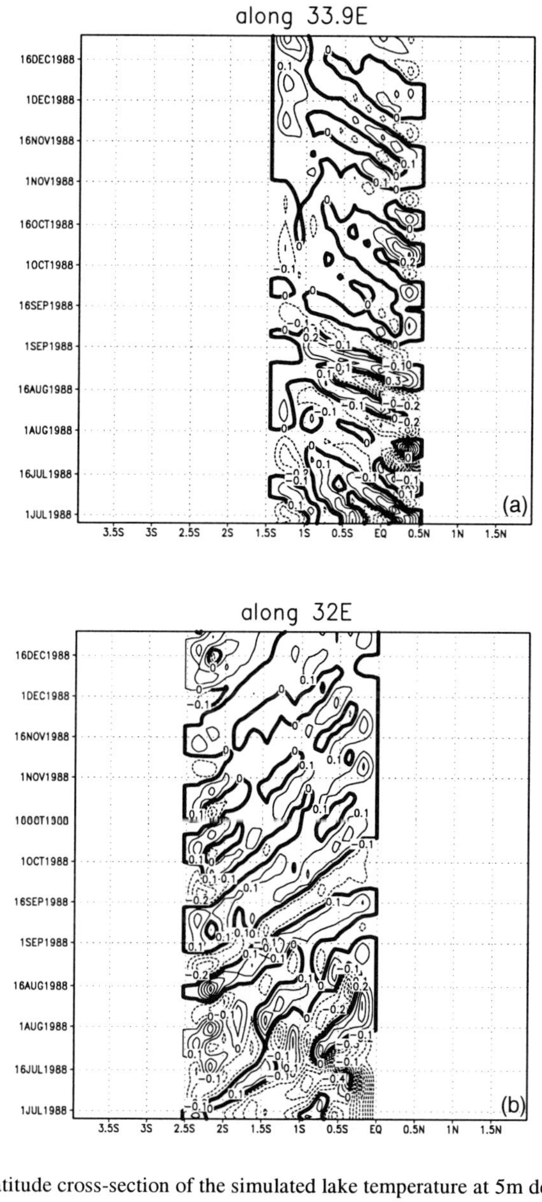


*Figure 12*. Time-latitude cross-section of the simulated lake temperature at 5m depth from day 180 to day 360. Bottom panel: along 32°E, and upper panel: along 33.9°E. Units in °C. The orientation of the contours depicts the direction of wave propagation. In the bottom panel wave propagation is from the south toward the north. The propagation is in the opposite direction in the upper panel.

Poincare waves exhibit a basinwide response with oscillations in the thermocline across the entire lake. The lowest order Poincare wave has maximum amplitude on opposite sides of the lake, with a node at the center. To the contrary, Kelvin wave disturbances in the thermocline are trapped in the coastal region. They progress cyclonically around the lake. As an example, the period of Kelvin wave in Lake Ontario is 25 days (Schwab, 1977). In the Exp. 2b, we note the existence of wave motion with similar features to the numerical solutions obtained by Beletsky *et al.* (1997).

## 4.3 Upper Boundary Forcing based on REGCM2 Fluxes

In (Exp. 3) we adopted the fluxes generated by the RegCM2 regional climate model to construct the upper boundary forcing for the lake Victoria model. The fluxes are the hourly surface wind stress, short wave radiation, longwave radiation, sensible and latent heat fluxes from the 20km resolution inner-nest of the RegCM2 simulation. Figure 13 shows the RegCM2 simulated monthly mean 10-m height wind for December 1988, which is employed as the surface wind stress forcing. As noted earlier the RegCM2 model is coupled to a 1-D lake model. The simulated diurnal surface wind stress is characterized by the divergent motion during the latter part of the day 18UTC (Figure 13a) and convergent motion during the night 24UTC (Figure 13b). Starting from motionless initial state with uniform water temperature of 24°C the lake model was integrated from 01UTC October 1, 1988 to 24UTC December 31, 1988.

### 4.3.1 Control experiment (Upper boundary forcing based on RegCM2 heat and wind fluxes)

Figure 14 shows the monthly mean diurnal variation over the lake surface in December 1988. The contours represent the lake surface temperature pattern and the arrows represent the surface water flow. The surface circulation is stronger during daytime than nighttime. The lake surface temperature is warmer at 12UTC and cooler at 24UTC. The results show that the warmer region is located over the southwestern part of Lake Victoria and the cooler region is located in the northwest region of Lake Victoria. The lake surface circulation characterized by anti-clockwise circular motion in response to the predominantly easterly mean surface wind stress is shown in Figure 13. The close agreement between the surface water circulation pattern and the topography of the lake (Figure 1), indicates that the depth of the lake may play a significant role in determining the climatology of Lake Victoria.

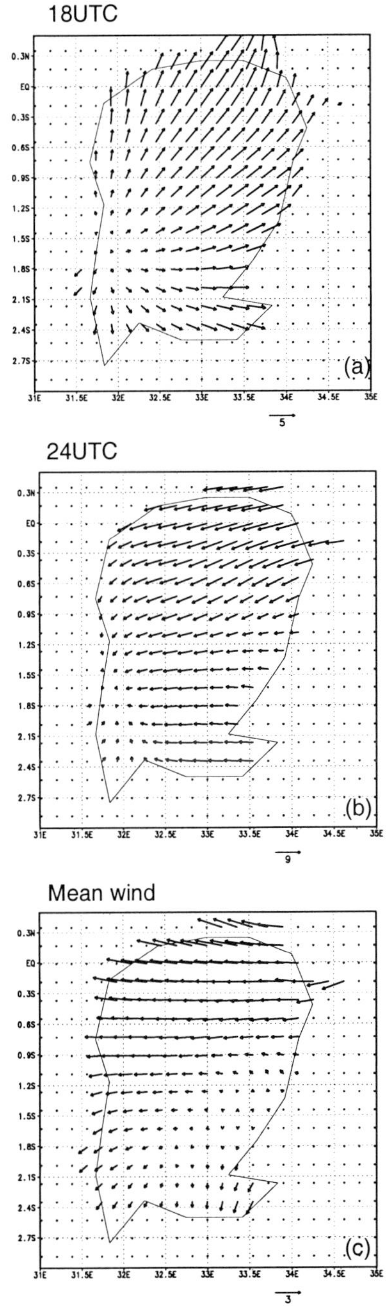


*Figure 13*. RegCM2 simulated wind at 10m height above Lake Victoria in December 1988. a). at 18UTC. b). at 24UTC. c) monthly mean wind. Units in m/s.

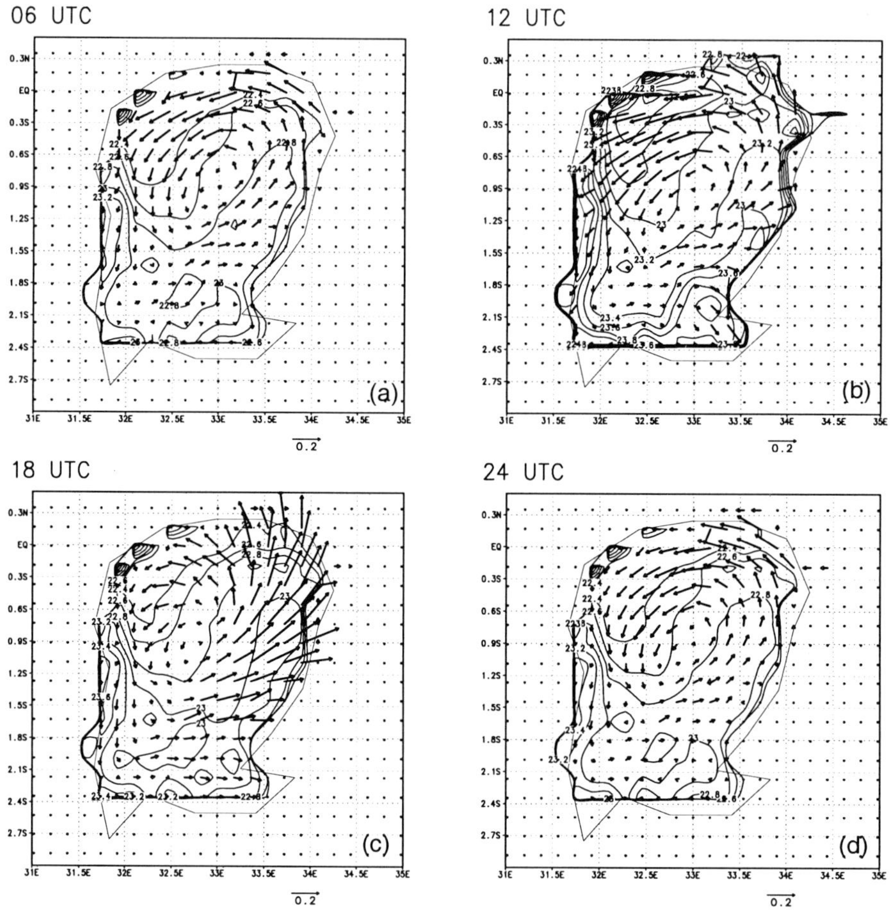


*Figure 14*. Simulated surface circulation and temperature of Lake Victoria driven by the wind stress and heat flux for December 1988. a). at 06UTC. b). at 12UTC. c). at 18UTC. d). at 24UTC. Units in m/s (current) and °C (temperature).

Comparison of the diurnal surface winds and surface water circulation reveals an important phenomena. Although the surface winds reverse direction during the course of the day (Figure 13 a and b) in association with the land/lake breeze, and also dominate the total wind field, the water circulation maintains the same anti-clockwise circulation throughout the day, similar to the mean flow displayed in Figure 15. This observation re-enforces the proposition that the circulation of the lake is primarily controlled by the prevailing wind pattern rather than the component associated with the land/lake breeze. We postulate that the large inertia associated with the water is responsible for the relatively weaker response to diurnal cycle, which dominates the near-surface wind regime. The surface flow transports warmer water northward over the eastern part of the lake and colder water southward in the western part of the lake. This advection helps to mix the warm and cold water horizontally and thus significantly modifies the spatial distribution of the surface water temperature.

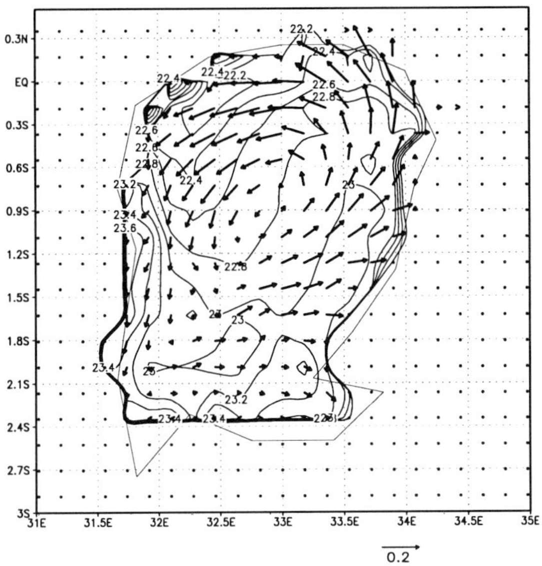


*Figure 15.* Simulated monthly mean surface circulation and temperature of Lake Victoria driven by the wind stress and heat flux in December 1988. Units in m/s (current) and °C (temperature).

It is instructive to note that the largest warming of the lake takes place over the shallowest regions of the lake along the southern and eastern shoreline. These are the regions likely to be responsible for the formation of warm water which then funnels out to influence the temperature in the rest of the lake. Over these regions we envisage that any significant increase in the elevation of the lake could result in extensive increase in the surface area of the lake and induce positive feedback that could, in part, also help to explain the dramatic maintenance of high lake levels observed in connection with the aftermath of abnormally wet rain episodes. The sequence of events that we hypothesize in this mechanism would operate as follows. The initial trigger is an intense rain season e.g., 1898, 1961/62, or 1997/98; this is followed by large increase in the elevation of Lake Victoria; occupation of new coastal territory by the lake occurs likely over the flat regions in the south and east of the lake; this is accompanied by increase in the area of the lake occupied by shallow coastal waters. The water over the swampy coastal region which is originally stagnant could rise high enough to become part of the rest of the lake circulation; this would be followed by formation of warm coastal water due to insolation and the low thermal capacity of the shallow column of water; the new coastal warm pools of water would then be transported toward the interior of the lake and northward where the water is generally colder, by advection driven by surface wind forcing; this would be accompanied by increase in Lake Surface Temperature (LST) and subsequent increase of precipitation over the lake; this would result in the maintenance of the expanded lake area and close the positive feedback loop until a new level of quasi-equilibrium is attained by the lake. Preliminary evidence based on this and other studies that we have undertaken seem to support this hypothesis. However, much work remains to be done to verify it.

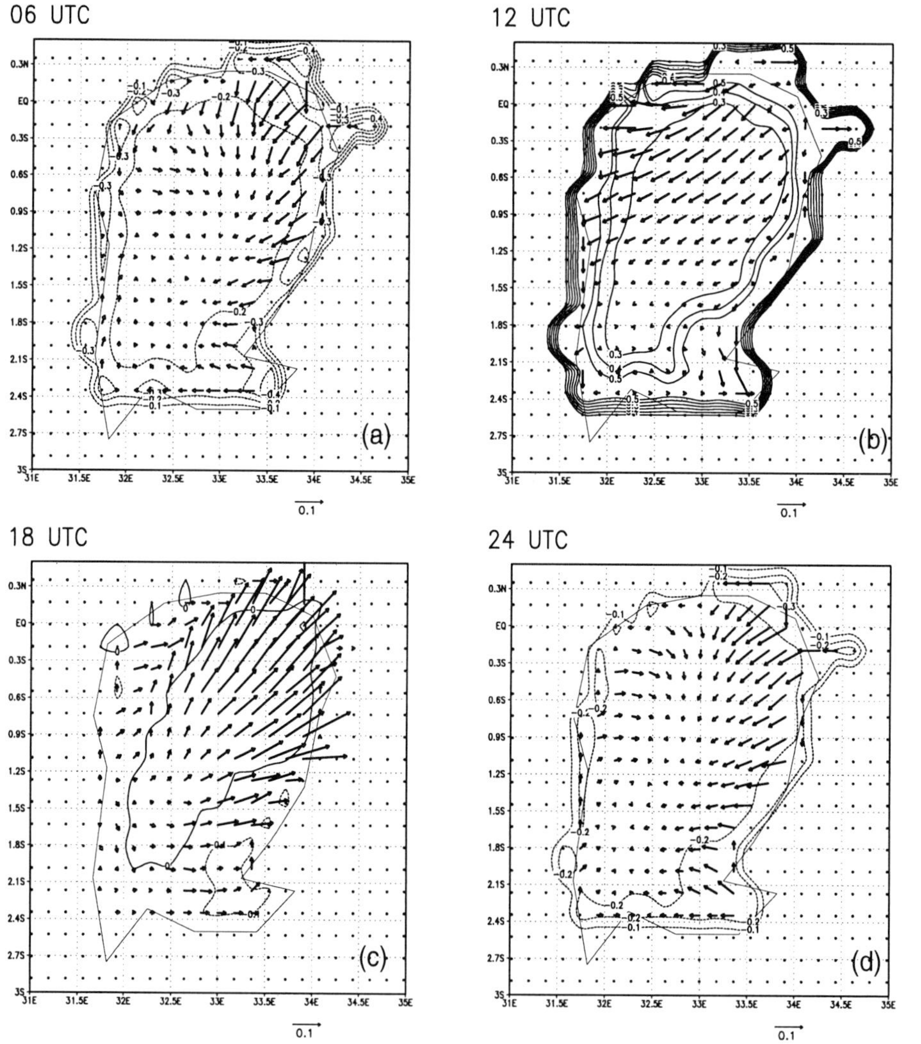


*Figure 16.* Same as Figure 14 except for the anomalous field. a). at 06UTC. b). at 12UTC. c). at 18UTC. d). at 24UTC. Units in m/s (current) and °C (temperature).

Figure 16 shows departures from the monthly mean diurnal cycle at the surface in December 1988. The results are obtained by substracting the entire monthly mean surface temperature and circulation from the monthly mean 06UTC, 12UTC, 18UTC and 24UTC surface temperature and circulation. We can see that the surface water responds strongly to the variations in surface atmospheric forcing, whereby the divergent circulation departures occur during daytime (12UTC and 18UTC) and convergent anomalous circulation during nighttime (24UTC and 06UTC). The surface temperature along the lake lateral boundary increases dramatically at 12 UTC, and decreases at 24 UTC. From daytime to nighttime, the variation in the water temperature at the center of Lake Victoria is relatively insignificant.

Figure 17 is the monthly mean diurnal cycle at the 20m depth in December 1988. The circulation patterns and thermal structures are similar to those at the surface. An anti-clockwise circulation occupies the entire lake and the water circulation transports the warm water northward over the eastern parts and the cold water southward over the western parts of the lake. However, the water is colder and the water currents are weaker than those at the surface (Figure 14). Also, the diurnal cycle is weaker at the 20-m depth than at the lake surface. The experiments described above show that the three-dimensional lake model produces an active diurnal cycle of Lake Victoria.

Next we explored the ability of the model to simulate the seasonal variations of Lake Victoria. Figure 18 (left panels) shows the simulated monthly mean surface water circulation and temperature distribution over Lake Victoria (October-November-December 1988). The contours depict the lake surface temperature and the arrows represent the surface water currents. Figure 18 (right panels) display the corresponding results based on the 1-D thermal lake model in the standard RegCM2 version of the model. Comparison between the results from the 1-D and the 3-D lake model simulations clearly shows that the latter produces much more realistic results. In particular, the 3-D model shows a surface temperature pattern indicative of dynamic mixing characterized by a horizontal spiral pattern, in the temperature field, which is associated with a spreading pool of warm water across the northern section of the lake. This pattern is not present in the 1-D model. These results underscore the need to replace the 1-D formulation of the lake in RegCM2 by the new 3-D formulation. Since the 3-D lake model generates more realistic lake water surface temperature patterns than the 1-D model, it is reasonable to speculate that the coupled RegCM2/3-D lake model would produce more realistic rainfall patterns over and around the lake than the simulations based on the coupled RegCM2-1D lake model. This question will be examined in a separate paper.

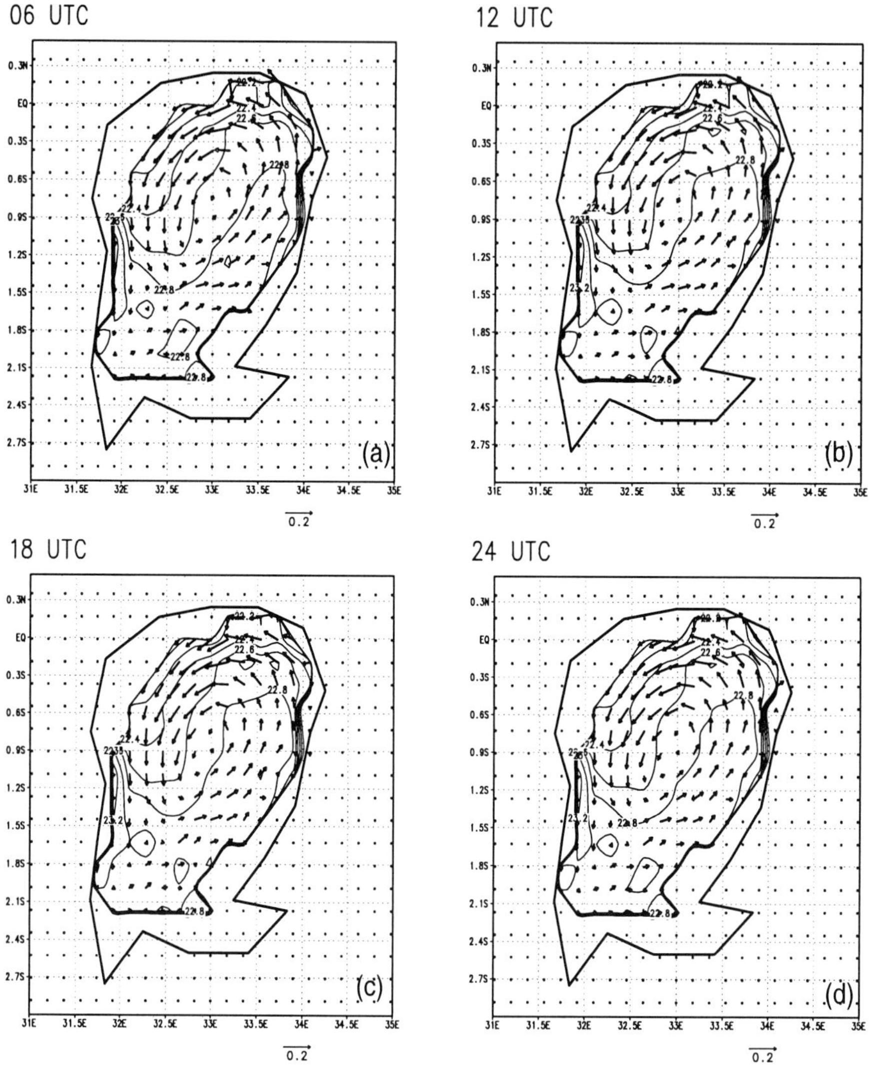


*Figure 17.* Same as Figure 14 except for field at 20m depth.

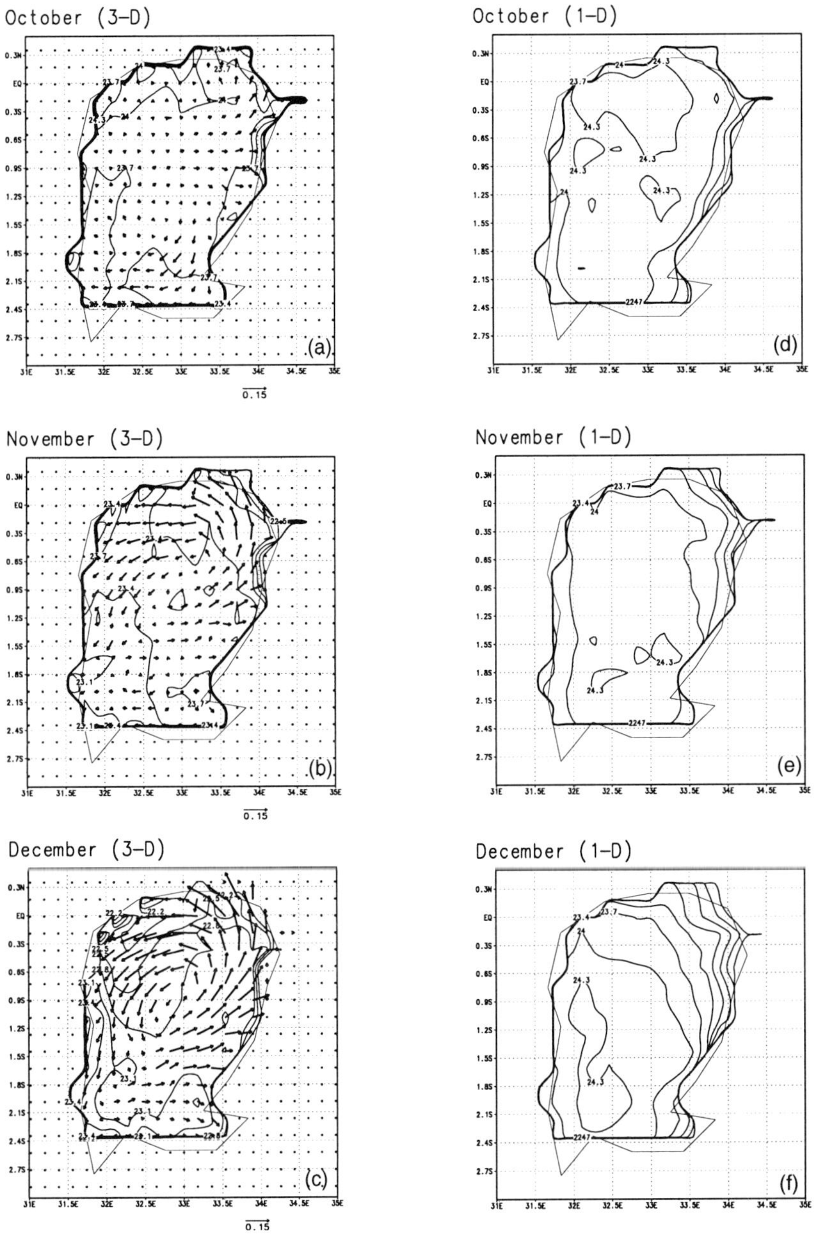


*Figure 18.* The monthly mean surface water circulation and temperature distribution over Lake Victoria (October-December 1988), simulated by the 3-D and 1-D lake model. a). October, b). November, c). and December, based on the 3-D model; d). October, e). November, and, f). December, based on the 1-D model. Units in m/s (current) and °C (temperature).

### 4.3.2 Upper boundary forcing based on RegCM2 wind only (no heat flux forcing)

In order to investigate the influence of surface wind stress forcing, a sensitivity experiment (Exp. 4a) was carried out by using hourly wind stress forcing and suppressing any interfacial heat exchange. The wind stress forcing is same as that in the control experiment (Exp. 3; section 4.3.1). Figure 19 shows the simulated monthly mean diurnal variation of the lake surface currents in December 1988. The arrows represent the surface currents. Similar to that of the control experiment, an anti-clockwise circulation dominates the lake surface flow field. The diurnal cycle (19 a and b) still exists in the lake surface water circulation, however it is weaker than the case when heat flux forcing is included in the boundary forcing (14 c&d). Hence, the mean surface water circulation is primarily determined by wind stress, although heat flux forcing plays the primary role in modulating the temperature diurnal cycle over the lake surface.

### 4.3.3 Upper boundary forcing based on RegCM2 sensible heat, short wave radiation, and long wave radiation fluxes (no surface wind stress forcing)

To investigate the contribution of the heat fluxes in the lake simulation, the hourly short wave radiation, long wave radiation, sensible and latent heat fluxes, which are the same as those in the control experiment (Exp. 3) were employed to force the lake model (Exp. 4b). The wind stress was turned off to isolate the effect of heat fluxes. The results of the fourth month (December) of the simulation are shown in Figure 20. The contours indicate the lake surface temperature and the arrows represent the surface water circulation. The lake surface temperature is relatively warmer at 12UTC and cooler at 24UTC which is a manifestation of the diurnal cycle. In December, the warmer region is located over the southwestern part of the lake and the cooler region is located in the northeast region of the lake, which is consistent with the results based on the 1-D lake model (Figure 18f). However, the corresponding transport of warmer water northward over the eastern region of the lake and colder water southward over the western part of the lake in the control experiment does not exist here. Therefore, the results show that although surface lake temperature is strongly influenced by radiation and heat fluxes at the surface, it is evident that advection by water circulation plays the dominant role in determining its spatial structure.

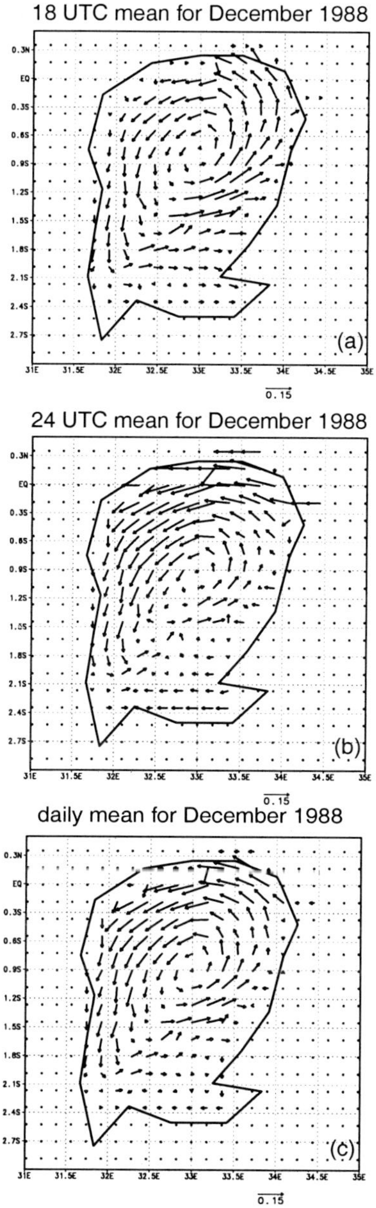


*Figure 19.* Simulated surface circulation of Lake Victoria driven by surface wind stress forcing. a). 12UTC. b). 18UTC. c). mean for December. Units in m/s.

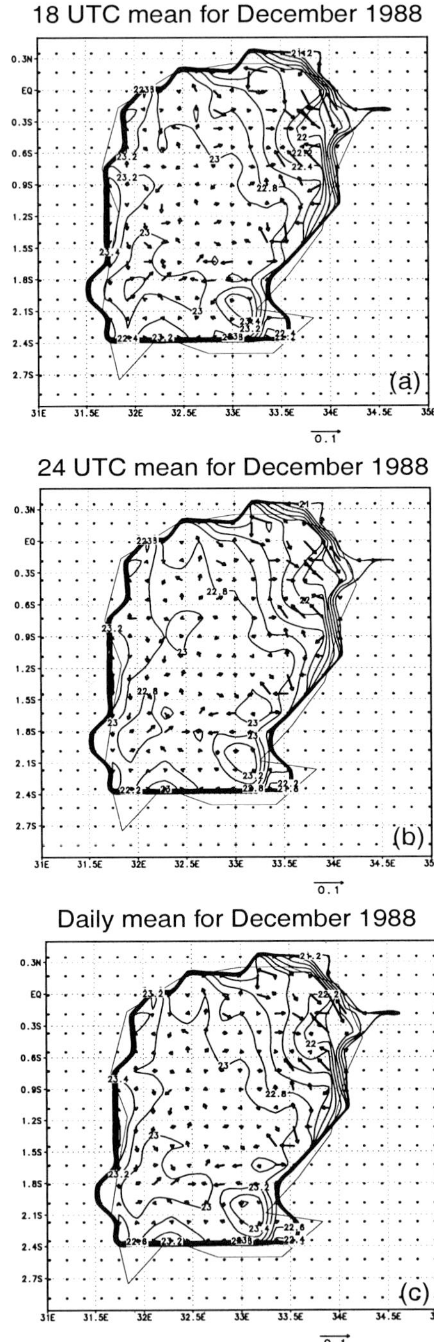


*Figure20.* Same as Figure 12 except driven by heat flux forcing.

# 5. CONCLUSIONS

In this study, we have configured the POM model for Lake Victoria. A number of simulations have been conducted to investigate the dynamics and thermodynamics of Lake Victoria, and its sensitivity to different forcing specifications. The results show that (i) the memory in the momentum of lake Victoria is about two weeks, (ii) there exists a 30-day oscillation in Lake Victoria and the stratified water temperature plays a critical role in the development of the oscillations, (iii) these oscillations are trapped in the thermocline along the coast and progress clockwise, (iv) the depth of the lake plays a significant role in determining the climatology in the water circulation of lake Victoria, (v) the lake surface circulation is characterized by counter clockwise motion in response to the predominantly easterly surface winds, (vi) with time-dependent wind forcing based on the RegCM2 model, although the diurnal component of the surface winds reverse direction in association with the land/lake breeze, the water circulation maintains the same anti-clockwise circulation throughout the day. This observation re-enforces the proposition that the circulation of the lake is primarily controlled by the large-scale wind pattern rather than the component associated with the land/lake breeze. We postulate that the large inertia associated with water is responsible for the weak response to diurnal cycle that dominates the near-surface wind regime. We find that the mean surface water circulation is primarily determined by surface wind stress forcing, however heat flux forcing plays the primary role in modulating its diurnal cycle. Furthermore, the results show that although surface lake temperature is strongly influenced by radiation and heat fluxes, it is evident that advection by water circulation plays the dominant role in determining its spatial structure.

The three-dimensional model produces a surface temperature pattern indicative of horizontal lake water mixing characterized by a horizontal spiral pattern in the temperature field. This is associated with the spreading of the pool of warm water across the northern section of the lake. This pattern is not present in the one-dimensional model. The simulations based on the three dimensional POM model produce more realistic lake temperature conditions than the one-dimensional lake model used in the NCAR regional climate model (RegCM2).

# ACKNOWLEDGEMENTS

This research was supported by the NSF/Climate Dynamics Program, project ATM-9904112. We extend our gratitude to P.-T. Shaw, M. Indeje, G. Pouliot and Chen Zhang who made many useful contributions in the work reported in this paper. The model integrations were performed at the North Carolina Supercomputing Center on the T90 supercomputer. The pre-processing for the RegCM2 model was performed on the NCAR supercomputer. NCAR is supported by the National Science Foundation (NSF). The post-processing of the model output was carried out at the FOAMV Computing Facility at North Carolina State University.

# REFERENCES

Anthes, R. A., Hsie, E. Y. and Kuo, Y. H. (1987) Description of the Penn State/NCAR Mesoscale Model Version 4 (MM4). *NCAR Tech. Note,* NCAR/TN-282+STR.

Beletsky, D., O'Connor, W. P., Schwab, D. J., and Dietrich, D. E. (1997) Numerical Simulation of Internal Kelvin Waves and Coastal Upwelling Fronts. *Journal of Physical Oceanography* **27**, 1197-1215.

Ba, M. B., and Nicholson, S. E. (1998) Analysis of convective activity and its relationship to the rainfall over the rift valley lakes of east Africa during 1983-90 using the meteosat infrared channel. *J. Appl. Meteor.* **37**, 1250-1264.

Bates, G. T., Giorgi, F. and Hostetler, S. W. (1993) Toward the simulation of the effects of the Great Lakes on regional climate. *Mon. Wea. Rev.* **121**, 1373-1387.

Bugenyi, F.W.B., and Magumba, K.M. (1996) The present physicochemical ecology of Lake Victoria, Uganda, in T.C. Johnson and E. Odada (Eds.), *Limnology, Climatology and Paleoclimatology of the East African Lakes*, Gordon and Breach, Amsterdam, pp. 141-154.

Csanady, G. T. (1984) *Circulation in the Coastal Ocean.* D. Reidel, 279 pp.

Datta, R. R. (1981) Certain aspects of monsoonal precipitation dynamics over Lake Victoria, in J. Lighthill and R. P. Pearce (eds.), *Monsoon Dynamics,* Cambridge University Press, pp. 333-349.

Dickinson, R. E., Henderson-Sellers, A. and Kennedy, P. J. (1993) Biosphere-Atmosphere Transfer Scheme (BATS) version 1E as coupled to the NCAR Community Climate Model. *NCAR Tech. Note.*, NCAR/TN-387+STR, 72pp.

Giorgi, F., Marinnucci, R. M. and Bates, G. T. (1993a) Development of a second-generation regional climate model (RegCM2), Part I: Boundary-layer and radiative transfer processes. *Mon. Wea. Rev.* **121**, 2794-2813.

Giorgi, F., Marinnucci, R. M. and Bates, G. T. (1993b) Development of a second-generation regional climate model (RegCM2), Part II: Convective processes and assimilation of lateral boundary conditions. *Mon. Wea. Rev.* **121**, 2814-2832.

Holtslag, A. A. M., de Bruijin, E. I. E. and Pan, H. L. (1990) A high resolution air mass transformation model for short-range weather forecasting. *Mon. Wea. Rev.* **118**, 1561-1575.

Hostetler, S. W., Bates, G. T. and Giorgi, F. (1993) Interactive nesting of a lake thermal model within a regional climate model for climate change studies. *J. Geophys. Res.* **98**, 5045-5057.

Flohn, H. (1987) East African rains of 1961/1962 and the abrupt change of the White Nile discharge. *Palaeoecology of Africa* **18**, 3-18.

Indeje, M., Semazzi, F. H. M. and Ogallo, L. J. (2000) ENSO Signals in East African Rainfall Seasons. *Int. J. Climatol.* **20**, 19-46.

Indeje, M., and Semazzi, F. H. M. (2000) Relationships between QBO in the lower equatorial stratospheric zonal winds and East African seasonal rainfall. *Meteorol. Atmos. Phy.* **73**, 227-244.

Johnson, T. C. (1997) Personal Communication.

Kelly, J. G. W., J. S. Hobgood, J. S., Bedford, K. W. and Schwab, D. J. (1998) Generation of three-dimensional lake model forecasts for Lake Erie. *Weather and Forecasting* **13**, 659-687.

Kendall R. L. (1969) The ecological history of the lake Victoria basin. *Ecological Monographs* **39**, 121-176.

Lehman J. T. (1997) How climate change is shaping Lake Victoria. *IDEAL Bulletin* **Spring 1997**, 1-2.

Livingstone, D. A. (1994) Evolution of African climate, in P. Goldblatt (eds.), *Biological Relationships between Africa and South America*, Yale University Press, New Haven.

Mellor, G. L. (1991) An equation of state for numerical models of oceans and estuaries. *J. Atmos. Ocean Tech.* **8**, 609-611.

Mellor, G. L., and Yamada, T. (1974) A heirachy of turbulence closure models for planetary boundary layers. *J. Atmos. Sci.* **31**, 1791-1806.

Moura, A. D., Bengtsson, L., Buizer, J., Busalacchi, A., Cane, M. A., Lagos, P., Leetma, A., Matsuno, T., Mooney, K., Morel, P., Sarachick, E. S., Shukla, J., Sumi, A. and Petterson, M. (1992) 'International Research Institute for Climate Prediction'. A proposal.

Muwonge M., (1994) Water Hyacinth in lake Victoria and other environmental problems. *The Uganda American Focus* **4**, Number 2.

Nicholson, E. N. (1986) The nature of rainfall variability in Africa south of the equator. *J. Climatol.* **6**, 515-530.

Nicholson, S. E. and Entekhabi, D. (1986) The quasi-periodic behavior of rainfall variability in Africa and its relationship to the Southern Oscillation. *Arch., Met. Geoph. Biocl., Seri. A* **34**, 311.

Nicholson, S. E. and Nyenzi, B. S. (1990) Temporal and spatial variability of SSTs in the tropical Atlantic and Indian Oceans. *Archives for Meteorology, Geophysics, and Bioclimatology, Seri. A*, p. 138-146.

Nicholson, S. E. (1996) A review of Climate Dynamics and Climate Variability in Eastern Africa, in T.C. Johnson and E. Odada (Eds.), *Limnology, climatology and paleoclimatology of the East African lake*, Gordon and Breach, Amsterdam, pp. 57-78.

Nicholson, S. E. (1996) Victoria lake-level modeling aims to predict basin rainfall. *IDEAL Bulletin* **Spring 1996**, pp. 5.

Ochumba, P. B. O. (1996) Measurement of water currents, temperature, dissolved oxygen and winds on the Kenyan Lake Victoria, in T.C. Johnson and E. Odada (eds.), *Limnology, Climatology and Paleoclimatology of the East African Lakes*, Gordon and Breach, Amsterdam, pp. 155-167.

O'Connor, W. P., and Schwab, D. J. (1994) Sensitivity of Great Lakes Forecasting System Nowcasts to Meteorological Fields and Model Parameters, in M. L. Spaulding, K. Bedford, A. Blumberg, R. Cheng, and C. Swanson (eds.), *Estaurine and Coastal Modeling III, Proceedings of the 3rd International Conference, Sept. 8-10, 1993,* Oak Brook, IL, American Society of Civil Engineers, New York, NY, pp. 149-157.

Okeyo, A. E (1987) The influence of Lake Victoria on the convective activities over the Kenya highlands. *J. Meteor. Soc. Japan* **Special Volume** , 689-695.

Schwab, D. J. (1977) Internal free oscillations in Lake Ontario. *Limnol. Oceanogr.* **22**, 700-708.

Schwab, D. J., O'Connor, W. P., and Mellor, G. L. (1995) On the Net Cyclonic Circulation in Large Stratified Lakes. *Journal of Physical Oceanography* **25**, Part II, 1516-1520.

Schwab, D. J., Beletsky, D., O'Connor, W. P., and Dietrich, D. E. (1996) Numerical Simulation of Internal Kelvin Waves with z-level and Sigma Level Models, in M.L. Spaulding and R.T. Cheng (eds.), *Estaurine and Coastal Modeling, Proceedings of the 4th International Conference, Oct. 26-28, 1995, San Diego, CA*, American Society of Civil Engineers, New York, NY, pp. 298-312.

Semazzi, H. F. M., and Indeje, M. (1999) Inter-seasonal variability of ENSO rainfall signal over Africa. *J. Afri. Meteorol. Soc.* **4**, 81-94.

Semazzi, H. F. M. (1999) Development of a regional climate prediction model for the Lake Victoria basin. *IDEAL Bulletin*, **Summer 1999**, pp. 10-11.

Spigel, R. H., and Coulter, G. W. (1996) Comparison of hydrology and physical limnology of the East African Great Lakes: Tanganyika, Malawi, Victoria, Kivu and Turkana (with reference to some

North American Great Lakes), in T.C. Johnson and E. Odada (Eds.), *Limnology, Climatology and Paleoclimatology of the East African Lakes*, Gordon and Breach, Amsterdam, pp. 141-154.

Stager J. C., Reinthal, P. N. and Livingstone, D. A. (1986) A 25,000-year history for lake Victoria, East Africa, and some comments on its significance for the evolution of cichlid dishes. *Freshwater Biology* **16**, 15-19.

Sun, L., Semazzi, F. H. M., Giorgi, F. and Ogallo, L. J. (1999a) Application of NCAR Regional Climate Model to eastern Africa. Part I: Simulation of the Autumn rains of 1988. *J. Geoph. Res.* **104**, 6529-6548.

Sun, L., Semazzi, F. H. M., Giorgi, F. and Ogallo, L. J. (1999b) Modeling study of the interannual variability of the Autumn rains over eastern Africa. *J. Geoph. Res.* **104**, 6549-6562.

Xie, L., Pietrafesa, L., Bohm, J., Zhang, C. and Li, X. (1998) Evidence and mechanism of hurricane Fran-induced ocean cooling in the Charleston trough. *Geophysical Research Letters* **25**, 769-772.

Yin, X., and Nicholson, S. E. (1998) The water balance of Lake Victoria. *Hydrological Sciences Journal* **43**, 789-811.

# Hydrology and Physical Limnology

# A MODELLING APPROACH FOR LAKE MALAWI/NYASA/NIASSA: INTEGRATING HYDROLOGICAL AND LIMNOLOGICAL DATA

D.C.L. LAM[1], L. LEON[2], R. HECKY[2], H. BOOTSMA[3] and R.C. McCRIMMON[1]

[1]*National Water Research Institute, Environment Canada, Burlington, ON, CANADA L7R 4A6*

[2]*University of Waterloo, Waterloo, ON, CANADA N2L 3G1*

[3]*University of Wisconsin-Milwaukee, Milwaukee, WI, U.S.A. 53201*

**ABSTRACT**

This study presents a toolkit approach for linking land- and lake-based data and models to determine the impact of human activities on the water quality of rivers and lakes. The integrated modelling framework was adapted to address specific issues and scenarios. Based on the preliminary results, a hypothetical 50% re-forestation of the Linthipe Watershed in the southern part of Lake Malawi/Nyasa/Niassa may lead to a decrease in the spring peak value of total phosphorus concentration from about 15 μg/L to 10 μg/L in the top layer (0 – 40 m) of the lake's Outlet Basin. Discussions on improvement to future modelling and monitoring programs are also presented.

## 1. INTRODUCTION

Lake Malawi/Nyasa/Niassa (hereafter called Lake Malawi) is one of Africa's critical ecological resources and its aquatic biodiversity has drawn international attention. To protect this valuable resource and to maintain sustainable development, collaborative and cooperative efforts by the three riparian states of Malawi, Tanzania



*E.O. Odada and D.O. Olago (eds.),*
*The East African Great Lakes: Limnology, Palaeolimnology and Biodiversity,* 189–208.

and Mozambique are required, with support from the international community. Catchment issues such as agriculture, deforestation, reforestation, transportation and regional economic development have been identified as key concerns in the area. To study the possible impact of these issues on the environment, a modelling project was conducted by the United Nations University/International Network for Water, Environment and Health, the University of Waterloo and WL|Delft Hydraulics. Supported by the World Bank, the study integrated physical and biogeochemical processes that affect water quality in Lake Malawi and its tributary rivers. The detailed results of the modelling project can be found in the final report (University of Waterloo *et al.*, 2000).

In this paper, we focus on the discussion of one specific aspect of the preliminary model results, namely the investigation of the dependence of lake water quality on human activities in the catchment. Our main purpose is to determine how human activities such as agricultural practices and deforestation may affect the water quality of rivers and streams and subsequently lead to changes in nutrient loading to the lake. Our hypothesis is that the impact from such catchment runoff to the lake will be significant, particularly in the lake basin that receives the increased loading. To confirm this hypothesis, we first make use of all relevant meteorological, land-use, soil type, topographical and other hydrological data and knowledge for the development and calibration of a catchment model to simulate the hydrological transport of non-point and point sources of nutrients from the catchment to the river outlet. We then use meteorological and limnological data, including nutrient loading data, as well as simulated results from a lake hydrodynamic model to develop and calibrate a lake water quality model; this model simulates the transport and dispersion of nutrients within the lake, in order to predict the water quality conditions. The third step involves combining the catchment model and the lake water quality model to simulate the changes in lake water quality due to hypothetical changes in land use practices. Due to the lack of observed data to calibrate and verify the models, this study also aims to identify data and knowledge gaps to improve monitoring and modelling work. The study is therefore not designed to develop a final product for catchment management, but rather to investigate the possible outcomes and to identify the uncertainties in predicting lake water quality by varying land use practices. The emphasis is on the modelling approach and how it can be used to improve the design of future monitoring and modelling studies, with more coordinated efforts to measure and integrate meteorological, hydrological and limnological information to support better management of the lake and its environment.

## 2. MODELLING APPROACH

To select the appropriate modelling approach for this study, we consider the water quality condition in the lake as affected by the nutrient input. As shown in Figure 1, there are three major sources of nutrient inputs to Lake Malawi: land-based discharge, atmospheric deposition and upwelling return. Nutrients from the catchments discharge into the lake and are subject to various physical and biochemical processes. To estimate the impact of these land-based sources of nutrients, we simulate the transport of the nutrients from the catchment into the streams using a hydrological catchment model, the Agricultural Non-point Source (AGNPS) model (Young *et al.*, 1986; Leon *et al.*, 2000). Other inputs to the lake must also be included, such as atmospheric deposition (Bootsma *et al.*, 1996), upwelling return from deep layers of the lake, vertical nutrient distribution and other information about water quality (Hecky *et al.,* 1999). As the nutrients enter the lake, they are carried by lake currents and their distribution is affected by both horizontal and vertical transport and dispersion, as well as the lake's thermal regime. A hydrodynamic model, the DELFT3D model (Delft Hydraulics *et al.*, 1999; Kernkamp *et al.,* 1994), is used to predict these hydrodynamic and thermal regimes. Specifically, it uses advanced turbulence closure models (Delft Hydraulics *et al.*, 1999) for the vertical mixing induced by wind, currents and stable and unstable stratification. The mass and heat fluxes from rivers are included in the model and the sources and sinks can be distributed vertically, e.g. to represent the inflow of cold river water. For the Malawi modelling study, it was modified to become a 2-dimensional (longitudinal and vertical) model, mainly for practical reasons. While the physical processes are taking place, the nutrients also undergo algal uptake, regeneration, sedimentation, nitrification, de-nitrification and other biochemical processes. These processes are simulated by a water quality model, the NWRI (National Water Research Institute) Water Quality Box model (Lam *et al.*, 1987).

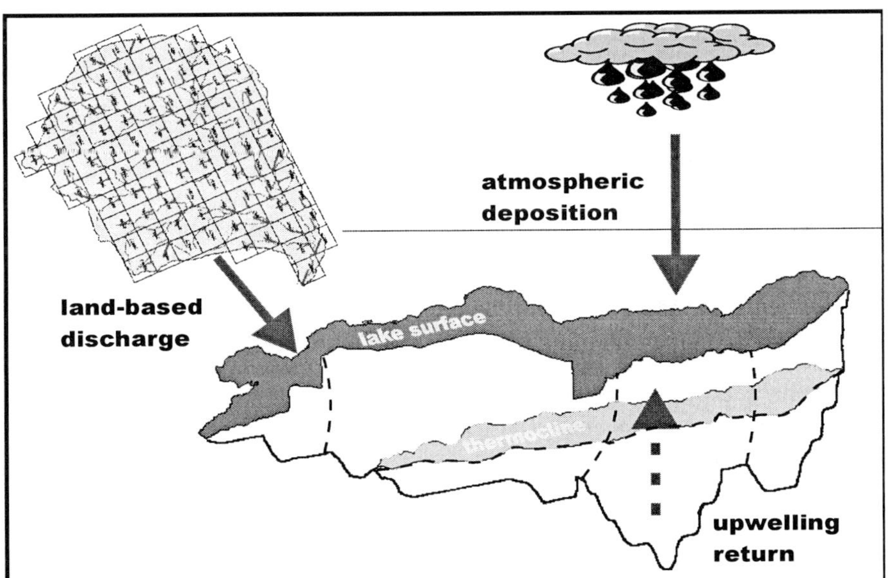


*Figure 1.* Schematic showing the three sources of nutrient input to the productive zone of the lake: land-based discharge, atmospheric deposition and upwelling return.

As shown in Figure 2, the modelling approach consists of incorporating the loading and input data and integrating the catchment and hydrodynamic model results with the lake water quality model to produce the scenario prediction. The integration step is implemented using the RAISON (Regional Analysis by Intelligent Systems ON microcomputers) Decision Support System (Lam *et al.*, 1994). The RAISON system is a generic system that offers input facilities to accept various types of data and maps, which are then stored in its own internal and fully-linked database, map subsystem and graphic components. Often, there are two phases in the construction of a decision support system (Lam, 1997): the technical user interface (TUI) and the public user interface (PUI). The technical user uses the TUI to connect databases, rule-bases and models and control decision processes, such as scenario building and testing. When completed, the TUI may form the basis of the PUI. The PUI is for managers and stakeholders and may even be used in public consultation meetings (Young *et al.*, 2000).

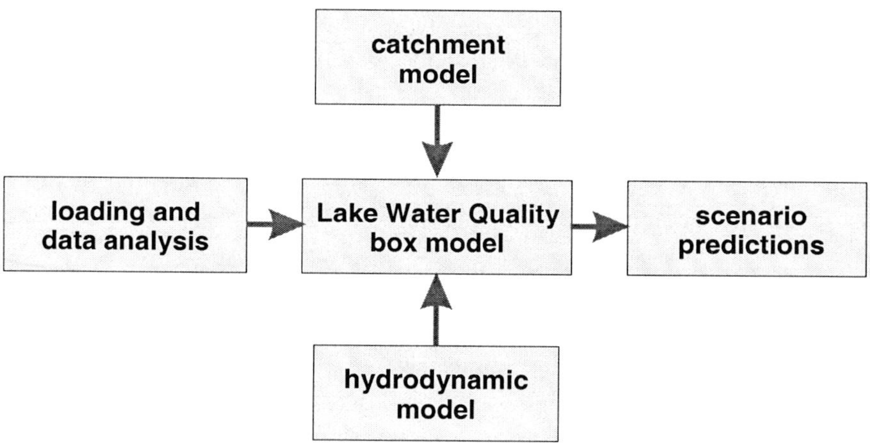


*Figure 2.* Linking input data and model results in RAISON.

In this paper, because we discuss only the linkage between the land-based hydrological and lake water quality models, the emphasis is on the use of TUIs. For example, data files with information on the atmosphere, river, lake and upwelling return from deep layers in the lake are input through TUIs. Also, there are three ways to incorporate models into RAISON. The first way is to run the model separately and feed the model results in as input. This is the method used to incorporate the results from the hydrodynamic (DELFT3D) model in this project. The computed results such as transport and temperature are only required by the lake water quality models. Thus, the hydrodynamic model can be left off-line from the other models, with a weak link to RAISON. The second way is to have the model linked to RAISON, with the model making use of the RAISON databases, maps and graphical facilities. An example of this method is the catchment model (AGNPS), which has its own code but its input and output files are linked through TUIs to the database, map and graphical components of RAISON. This linkage enables the catchment model to communicate directly with the water quality box model in order to generate iterative results for testing the effects of land use scenarios on lake water quality. The third way is to rewrite the code for the model in a programming language (e.g. Visual Basic) which makes it possible to link the model directly. The water quality box model (NWRI Box Model) is such a case, as the original code for this model required modification to accept input from the AGNPS and DELFT3D models, as well as to generate appropriate results for specific scenario predictions, including a new nitrogen process component. The implementation of the modelling approach is shown in Figure 2. Summer *et al.* (1990) used a similar approach linking the AGNPS model to a lake process model, but they did not have a technical interface as implemented here.

## 3. CATCHMENT MODEL

The AGNPS model (Young *et al.*, 1986) is a distributed event-based catchment model that simulates surface runoff, sediment, and nutrient transport, primarily from agricultural catchments. The nutrients include nitrogen (N) and phosphorous (P); both essential plant nutrients and possible major contributors to surface water pollution. Runoff volume and flows are calculated using the Soil Conservation Service (SCS) curve number method (Young *et al.*, 1986). Upland erosion and sediment transport is estimated using a modified form of the Universal Soil Loss Equation, USLE (Wishmeier and Smith, 1978). The sediment transport and depositional relationship, which is based on a steady-state continuity equation, is described by Foster *et al.* (1980). Chemical transport is calculated based on the relationships adapted from the CREAMS (Chemicals, Runoff and Erosion from Agricultural Management Systems) model (Frere *et al.*, 1980 and Knisel, 1980). As a preliminary modelling effort for this project, this model was selected for application to a pilot catchment where sufficient data, in Geographical Information System (GIS) format, were available for model input and calibration. In this study, we chose the Linthipe Watershed as the pilot site, not only because there are sufficient data but also because the catchment provides land-based nutrient input, including phosphorus and nitrogen discharges, to the smallest lake basin on which we can test our hypothesis on the connection between land use and lake water quality.

### 3.1 Input to Catchment Model

Catchments modelled by AGNPS are divided into homogenous square working areas called cells. Subdivision of main cells into smaller sub-cells provides flexibility to account for heterogeneity in the catchment. Due to this spatial segregation, all catchment characteristics are expressed at the grid-cell level, thus requiring the input of spatially distributed data that is handled through the use of a GIS component, such as the one available in RAISON. Cell data include information on values based on topography, soil type, land use and management practices within each cell. Through the AGNPS/RAISON interface (Leon *et al.*, 2000), we can automatically extract the required information for each of these grid cells from GIS map layers as part of the model input processes. In the model, some variables such as land slope and flow direction can be derived from a topography map, while others such as the SCS curve number are functions of soil type and land use. Thus, the automatic extraction of map data, included in the AGNPS Interface, requires a Digital Elevation Model (DEM) file as well as GIS map files for soil type and land use. Figure 3 shows the soil type map, overlain by the grid cells, and Figure 4 shows the land use map for the Linthipe Watershed. While minor adjustments were necessary to convert some local land use types to similar ones specified in the AGNPS model inputs, no further adjustments in the model coefficients defined by the map extraction process was made. The calibration results using these values were found to be adequate for the available data.

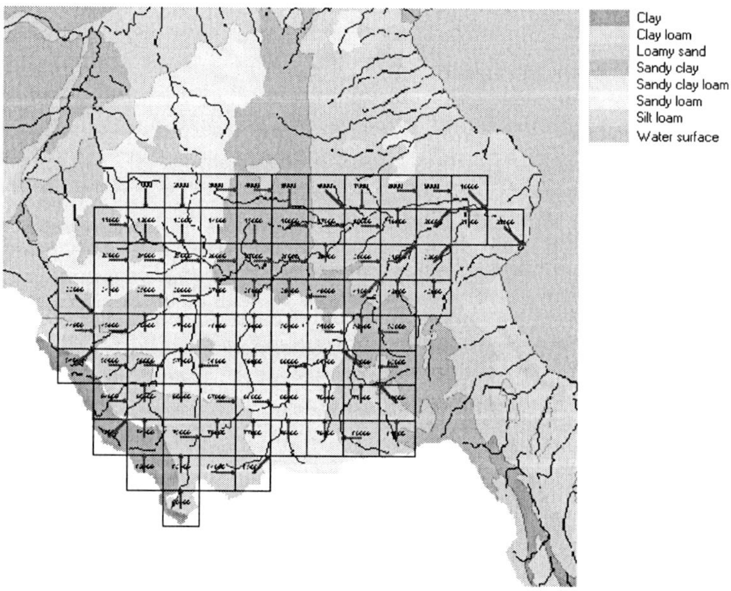


*Figure 3.* Soil type layer map for the Linthipe Watershed area, showing the grid cells and average flow directions used in the AGNPS model.

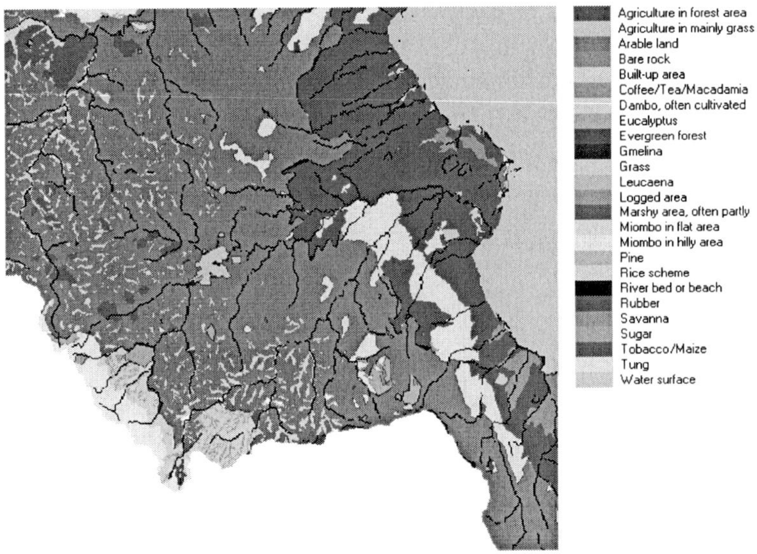


*Figure 4.* Landuse layer map for the Linthipe Watershed area.

## 3.2 Calibrated Results of Catchment Model

Using the input data extracted from GIS maps and other data, such as rainfall events, we can run the AGNPS model from a TUI in the RAISON system. The model output can be displayed as tables or as thematically coloured maps in RAISON, based on the values of the computed results for the grid cells. To demonstrate how the model results compare to actual observations, however, we grouped the results by rainfall events. Figure 5 shows the computed and observed results for the stream runoff flow rate, sediment yield, nitrogen load and phosphorus load. These results are divided into three groups, the low, average and high rainfall events during the year 1997. For example, the median runoff flow is observed at about 250 $m^3/s$ for low rainfall events, at about 475 $m^3/s$ for average rainfall events and at about 890 $m^3/s$ for high rainfall events. As shown in Figure 5, the computed flow rate for low rainfall events agreed well with the observed, but it was greater than the observed values for both average and high rainfall events. Similarly, for the sediment yield the computed values agreed well with the observed for low rainfall events but were lower than the observed for average and high rainfall events. However, as only a few sets of observed data were available for the construction of event statistics, the values of the observed median and range shown are subject to a high level of uncertainty. Due to the data limitations, we did not adjust the model coefficients in the USLE to achieve a better fit between computed and observed results. In general, we are satisfied that both computed and observed runoff and sediment yields increase from low to high rainfall events as they should. On the other hand, the computed and observed results for nitrogen and phosphorus loads are in agreement for all rainfall events, as shown in Figure 5. The results are encouraging in that the information from the land use and soil type maps produced reasonable values for the model coefficients.

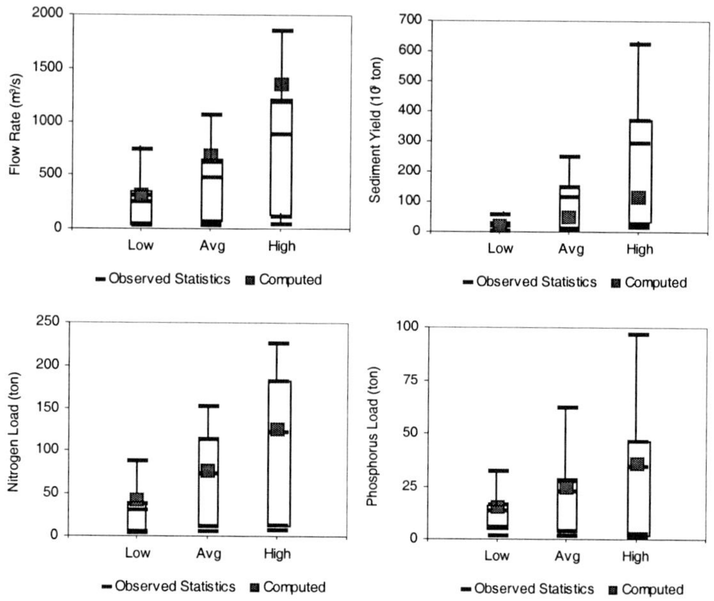


*Figure 5.* Comparison between observed data and AGNPS computed values at the outlet of the Linthipe River for low, average (Avg) and high rainfall events. (For the observed statistics, the bar in the middle of the box indicates the median, the top and bottom bars indicate the 75 and 25 percentiles, and the upper and lower whiskers indicate the 90 and 10 percentiles.).

## 4. LAKE WATER QUALITY MODEL

While the Outlet Basin (Figure 6) in Lake Malawi has sufficient data coverage, the other basins have less data. These other basins were sampled only along the central axis of the lake and only four times each. Some of the stations in the basins were measured as vertical profiles, providing important information on the vertical exchanges. Given the data availability, we have adapted a simple box model (Lam *et al.*, 1987) for the study in this paper to investigate the essential physical and biochemical processes in the four main lake basins (Figure 6) along the longitudinal axis, including the effects of thermal stratification and land-based discharge of nutrients.

## 4.1 The NWRI Water Quality Box Model

The box model adapted for the study in this paper is the NWRI water quality model developed for Lake Erie (Lam *et al.*, 1987). Its development was based on two previous versions (Lam and Halfon, 1978; Simons and Lam, 1980). Its result showed that for long term simulations (i.e. over ten years), long term data are required to calibrate model coefficients such as the settling rate of particulate phosphorus. These versions of the model evolved to become the Lake Erie model, with three horizontal boxes, three vertical layers and three water quality variables: soluble reactive phosphorus (SRP), total phosphorus (TP, as the sum of SRP and OP, where OP is the organic phosphorus) and dissolved oxygen (DO). It predicted the phosphorus dynamics due to nutrient loads and the anoxic conditions under various weather conditions affecting thermal stratification. It was calibrated, validated and post-audited with a total of 16 years of data for Lake Erie (Lam *et al.*, 1987). Thus, this model has been applied to lakes with environmental issues and physical and biochemical processes comparable to those in Lake Malawi.

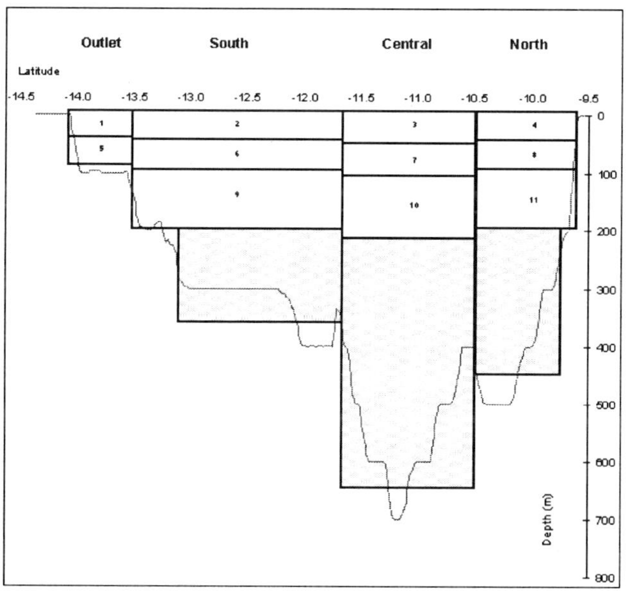


*Figure 6.* Schematic of the boxes used in the water quality box model; the box lengths and layer depths are averaged values; the shaded areas are deep water layers not directly incorporated in this box structure but may serve as internal sources of nutrients.

### 4.1.1 Model Structure

The original box model was configured as a nine box model, with three basins and three layers, for Lake Erie. To adapt it to Lake Malawi, we used an 11-box model for several reasons. As shown in Figure 6, we define four basins in Lake Malawi: Outlet, South, Central and North. There are layers of water deeper than 200m in all basins except the Outlet Basin. In the DELFT3D hydrodynamic model (Delft Hydraulics *et al.*, 1999; Kernkamp *et al.*, 1994), these deep layers are assumed to be isolated from the overlying waters. Thus, for practical purposes they can also be treated as such in the water quality box model, with the exception of the upwelling return of nutrients (Figure 1). Gonfianti *et al.* (1979) and Vollmer *et al.* (2001) showed that the vertical fluxes of nutrients from the deep layer to the overlying layers were possible and significant. To include this input in the water quality box model, we treated it as a source entering from the deep layer to the overlying layer, based on preliminary flux estimates (Vollmer, pers. comm.). For the overlying waters, the seasonal thermocline usually occurs at a depth of 40 to 50m. From previous measurements of pelagic photosynthesis (Degnbol and Mapila, 1982; Bootsma, 1993a), we considered the top layer above the thermocline to be the productive zone. We then defined the middle layer as being between the thermocline and 100m in depth, and the bottom layer as being between 100m and 200m in depth. The next consideration is the output file of lake currents computed by the hydrodynamic model. This file is structured according to the so-called "sigma coordinate", that is, the vertical layers are such that they can expand or contract slightly according to the water movements (Delft Hydraulics *et al.*, 1999; Kernkamp *et al.*, 1994). To define the water transport in the water quality model, the currents computed from the hydrodynamic model are averaged for a number of layers and boxes according to the partitions given in Figure 6. Note that the boxes are all aligned vertically. Some horizontal alignments are, however, slightly off (e.g. between boxes 9 and 10). This slight offset will not cause problems for conserving water or mass, because the transports among the boxes have been directly computed from the hydrodynamic model and satisfy continuity. Adjustments are needed for the transport between the Outlet and the South Basins, because due to its shallowness there are only two layers in the Outlet Basin . While all interfaces are fixed, the lake surface itself is allowed to move up or down to simulate water elevation changes due to wind setup. The water transport is mass conservative, with the inflows balancing the outflows in all boxes in the water quality model.

### 4.1.2 Model Formulation

The model formulation is based on the water quality model ( Lam *et al.*, 1987) for Lake Erie. Basically, the physical transports involve both horizontal advection as used in the hydrodynamic model and vertical transport based on the computed vertical flow. The vertical dispersion coefficient is the value computed in the hydrodynamic model, multiplied by a mixing length scale equal to the vertical distance between adjacent boxes. Thus, the physical transport equations follow those given in Lam *et al.* (1987), except that the internal layer depths are now fixed. Similarly the model formulation for the water quality variables are based on the Lake Erie model , using the algal uptake and regeneration between organic phosphorus (OP) and soluble reactive phosphorus (SRP), with effects on dissolved oxygen (DO) (Figure 7). This figure represents a typical basin (the North, South or

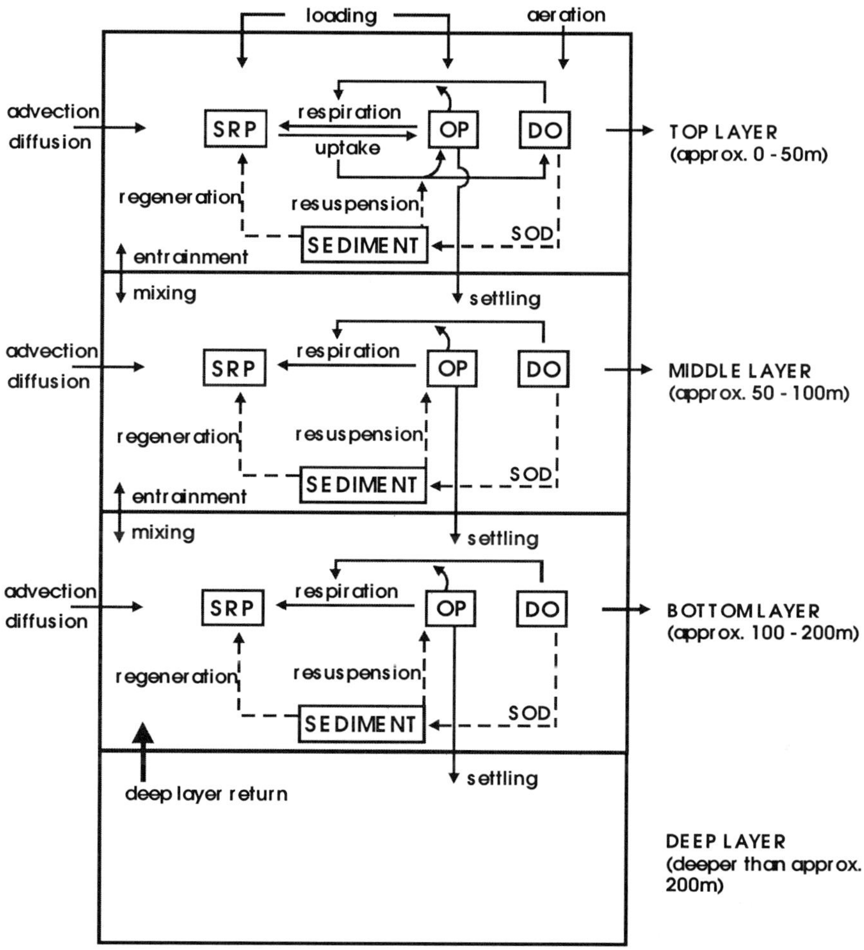


*Figure 7.* Schematic of the physical and biochemical processes for the phosphorus and oxygen dynamics (nitrogen processes not shown) used in the box model for Lake Malawi: SRP (soluble reactive phosphorus), OP (organic phosphorus), with TP (total phosphorus) = SRP + OP; DO (dissolved oxygen), SOD (sediment oxygen demand).

Central). For the Outlet Basin, there are only two layers and the figure would be reduced accordingly. In the other basins the deep water layer forms a fourth layer beneath the hypolimnion which has a sediment layer along the sloping bottom. We have also added a component on nitrate and ammonia uptake and regeneration to the nutrient processes. The detailed model equations for the biochemical processes are provided in the project report (University of Waterloo *et al.*, 2000).

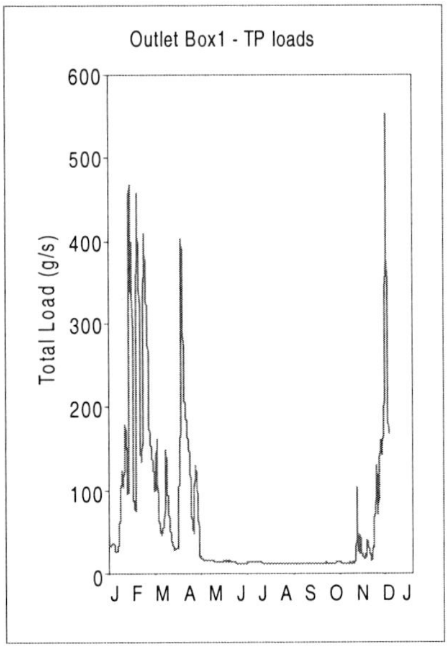


*Figure 8.* Estimated riverine phosphorus loads to outlet box 1 in Lake Malawi for 1997.

## 4.2 Water Quality Model Results

The model was calibrated using the in-lake concentration data for 1997. As the observed data are rather limited, with only two or three observed points per box , we did not calculate the statistical errors between computed and observed results. Instead, we relied on visual comparison. The emphasis here is to show how the water quality model behaves under the current nutrient loading condition and, later, how to link it to the catchment model. For example, Figure 8 shows the riverine phosphorus load to the Outlet Basin for 1997. Figure 9 shows the computed and observed TP concentrations for all basins and layers. The best agreement between observed and computed TP concentrations is in the top layer of the Outlet Basin. The spring peaks (January to April) of the riverine total phosphorus load (Figure 8) to the lake were reflected in the increase in total phosphorus concentration in the upper layer of the Outlet Basin (Figure 9).

Previously, emphasis has been placed on the role of winds and upwelling in controlling nutrient and plankton seasonality in the southern part of the lake (Eccles, 1974; Bootsma, 1993b; Patterson and Kachinjika, 1993). The observed and modelled results presented here suggest that river discharge is also an important factor, at least in the southern end of the lake. The low loading values for May to October lead to a relatively lower level of total phosphorus concentration in the Outlet Basin (Figure 9). Thus, the seasonal trends for river discharge and lake concentration are in good agreement. The agreement between computed and observed results for this basin is due to the availability of more data for model

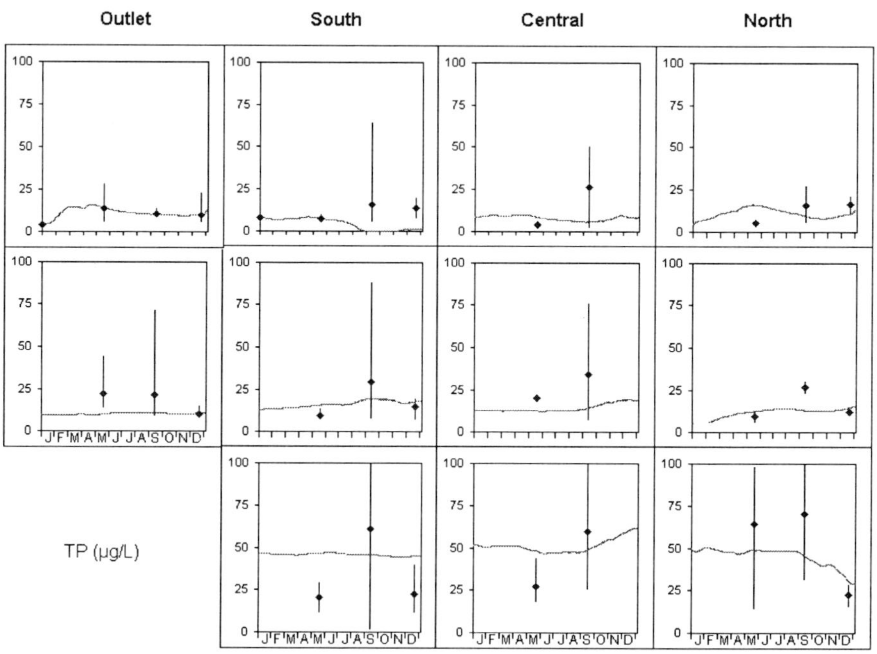


*Figure 9.* Computed total phosphorus concentration (dotted lines as TP in µg/L), vs. observed data (diamonds) and ranges (vertical bars) averaged for all boxes and layers, for 1997.

calibration and a more reliable set of loading estimates. This adds to our confidence in using the box model, linked with the catchment (AGNPS) model, to investigate the effect of land use management strategies for nutrients on this basin. For the other basins, the comparison of computed and observed data is reasonably good. The return of phosphorus from deeper layers (depths > 200m, Figure 1) was indicated in both the simulated and observed data in the bottom layers (depths of 100 to 200m) of the North, Central and South Basins (Figure 9). However, more data are needed to define this deep layer return accurately, and we also require a better understanding of the physical mechanisms affecting the nutrient return. The results for the other variables such as SRP and nitrogen produced similar agreement between computed and observed values (University of Waterloo *et al.*, 2000).

# 5. A LAND USE SCENARIO

Due to the data limitations, the calibrated results obtained to date for both the catchment model (Section 3) and the lake water quality model (Section 4) are preliminary. The model results were within the uncertainties of the observed data most of the time. In particular, good agreement is obtained between the computed and observed results for the total phosphorus concentration in the upper layer of the Outlet Basin, where the input loading and in-lake data were the most adequate. This provides some degree of confidence in using the model to further investigate the effects of land use on input loadings and the subsequent impact on the nutrient levels in this particular lake basin.

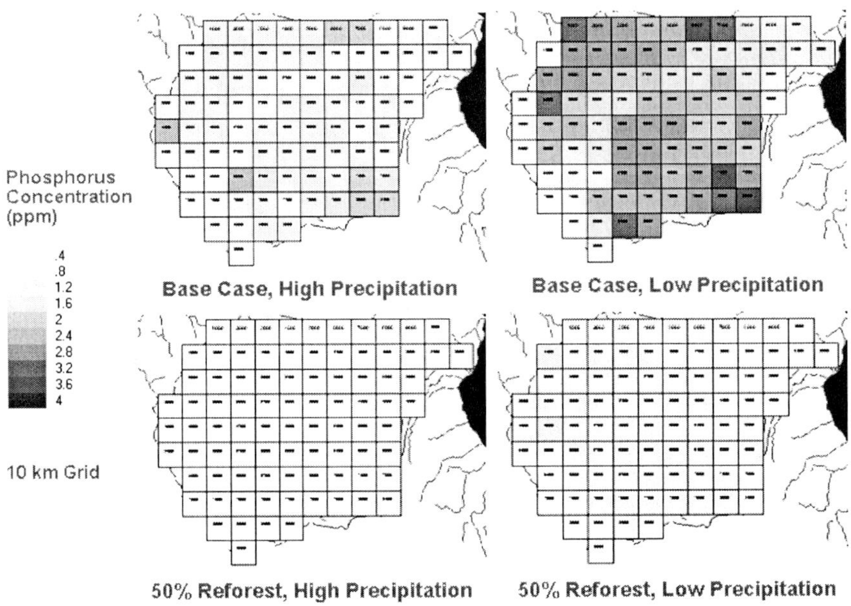


*Figure 10.* Computed stream total phosphorus (TP) concentration (in ppm = mg/L) averaged over each grid cell in the Linthipe Watershed for different scenarios: base case (upper diagrams) and 50% reforestation (lower diagrams) and for high (left side) and low (right side) precipitation events.

Input to the AGNPS catchment model can be altered to generate new results for a number of scenarios. For example, one of the scenarios considered is the hypothetical case in which 50% of the Linthipe Watershed is reforested. This hypothetical change amounts to modifying the land-use map by increasing the forested area in each grid cell by 50%. The changes to the land use pattern in turn affect the computation of the river flow, sediment yield and the nutrient concentrations. Figure 10 shows the predicted total phosphorus concentration for both high and low precipitation events before (base case) and after (50% reforestation) the hypothetical land use change, based on the AGNPS catchment model runs. The reduction in non-point sources of TP is significant for both wet and dry conditions. The reforestation has resulted in lower phosphorus concentrations at the outlet of the catchment.

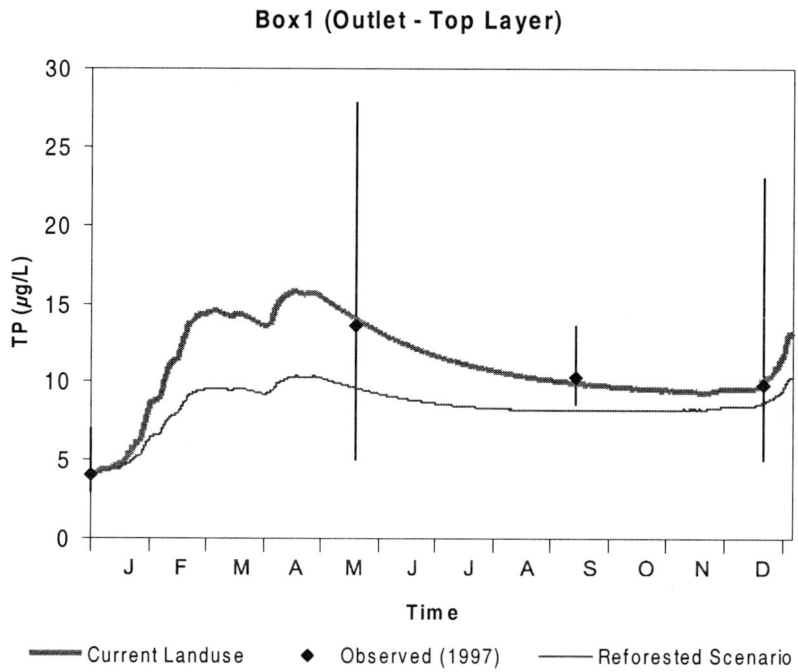


*Figure 11.* Computed total phosphorus (TP) concentration for the top layer in the Outlet Basin using current land use pattern (top line) in the Linthipe Watershed vs. that using a hypothetical 50% reforestation scenario (bottom line), compared to observed data (diamonds) and ranges (vertical bar) in 1997.

This reduction in non-point source input to the rivers in the Linthipe Watershed causes a decrease in the TP load to the Outlet Basin of Lake Malawi. We can link this catchment model result to the box model (Figure 2). We can modify TP load for 1997 accordingly for the 50% reforestation case and save it as a new loading file. To discover the effects of this loading reduction on the lake, we pass this new loading information to the box model via a TUI in RAISON. Figure 11 shows the computed concentration of TP for the top layer in the Outlet Basin, for both the actual TP load and the new load as modified by the output from the catchment model for the 50% reforestation scenario. The reduction in the TP concentration in the lake is significant (Figure 11), from a spring peak value of about 15 μg/L with the actual load, to a value of about 10 μg/L with the modified load. This result demonstrates that land use activities can affect the lake water quality in this area. The changes in phosphorus concentration for the other basins (not shown in Figure 11) are not as great as that in the Outlet Basin, confirming the hypothesis that non-point source input of nutrients has the greatest effect on the immediate receiving impoundment. However, some care is required in interpreting these preliminary results. For example, 1997 was a wetter year than normal and therefore the nutrient loading was probably greater than average (Kingdon *et al.*, 1999). Further scenario simulation and comparison with long term data are required to confirm the predicted results.

## 6. DISCUSSION

The integration of data and models is a complex task. Designing a system that can integrate independent databases and models is only the beginning. The main difficulty is in applying them to practical problems, as required by both public and technical users. The results and examples shown here are preliminary and are mainly intended to invite feedback from water resource managers and potential users. As demonstrated in the examples provided, the technologies to link databases and models are now readily available; the challenge is to make them serve the needs of users. The process of data and knowledge integration is therefore an iterative one: only with feedback from the users can the systems be improved.

As for future needs, the integrated modelling framework (Figure 2) presented here is a viable and growing one. In time, we expect to incorporate additional data and models. For data, additional catchment information is required to complete the entire drainage basin for the lake. For models, we have one catchment model, one hydrodynamic model, and one water quality model. We welcome additional models of these types, particularly those developed and used by regional agencies in the riparian countries. There is a lack of both biological resource and socio-economic models, and water quality models must be extended to include plankton and ultimately fish. Though it is difficult to place a dollar value on environmental quality, the cost of development projects that could affect the environment can be estimated. The decision maker's job is to weigh the benefits of development with the benefits of environmental quality, which may include the preservation of valuable natural resources.

A decision support system for effective management requires all these components to function properly. There are a variety of environmental issues facing Lake Malawi and its catchment; managers and researchers may want to address these issues at different complexities and on different spatial and temporal scales, so the integrated modelling framework must take a "toolkit" approach like as the one we have outlined. While the models may be linked and used together to address complex, ecosystem-scale issues, they may also be used individually to address specific, local issues. This modular approach also simplifies the future modification of models as more data are collected and increases the likelihood that managers will see the models as practical tools relevant to local issues.

The main results achieved in this project are the use of an integrated modelling framework to model non-point source nutrient in a catchment and the linkage of the computed river discharges to a lake water quality model. This work has demonstrated the feasibility of applying advanced technologies in order to integrate models in a uniform platform with GIS capabilities. The use of the TUI interface for the AGNPS model, including the pre- and post-processing tools, allows the user to set up model grids, automatically extract input data and begin analyzing the output. This integration supports the idea that better modelling capabilities need to be combined with the application of new technologies, such GIS capabilities, to resolve problems associated with the ease of model use. One of the major benefits of this system is the automatic data extraction from digital maps, for which two basic procedures are used: one uses DEM files to extract topography-related data such as flow direction and slope, while the other uses land cover and soil type maps with polygon attributes to extract the data relevant to the model. However, there are limitations to such computer automation; the scientific basis of the land cover and soil information must be obtained from research into local land use features and practices to produce the correct values for the model coefficients. For example, local farming practices may affect nutrient discharges, and fertilizer application rates should be estimated with field surveys in the local area to provide the model with adequate applied nutrient rates.

As for the simple water quality box model, it was designed for a preliminary investigation into water quality in Lake Malawi. The existing data on loading and in-lake concentrations are useful to facilitate the modelling study. The model results obtained were reasonable and revealed several interesting interactions between physical and biochemical processes. The thermal stratification of the lake at the 40 to 50m depth plays a significant role in the vertical distribution of nutrients. Likewise, the model demonstrated the importance of significant nutrient fluxes through some form of mixing mechanism across the weak pycnocline (Wüest *et al.*, 1996) at 200m. Some preliminary attempts to include these fluxes were made in the box model, but they require further research. The need for adequate, reliable data is evident in the attempt to link the catchment model for the Linthipe Watershed with the lake water quality model for the Outlet Basin. They generated the best agreement between computed and observed data because there were sufficient data to calibrate both models for this basin.

## 7. CONCLUSIONS

This study shows that the concept of linking different types of models can be implemented, with proper assumptions and integrated modelling tools. The connection between the catchment, hydrodynamic and water quality box models can be facilitated by technical user interfaces that help transfer key data files among the models. Additional models, particularly those used by regional agencies, should be linked to these models to provide relevant answers to water resources and fisheries management questions. Also, application of these or other models to other catchments and lake basins in Lake Malawi is possible, provided there is sufficient data and knowledge to calibrate the models. Additional model calibration for other catchments and basins is required to increase confidence in the use of such an integrated modelling approach for water resource management in Lake Malawi/Nyasa/Niassa.

## ACKNOWLEDGEMENTS

This study is supported as part of a consultancy contract with the World Bank. The authors wish to thank the reviewers of this study and the many colleagues from agencies and universities in Malawi, Tanzania and Mozambique who provided valuable comments during workshop sessions held in Canada and Malawi. They thank J. Smits, H. Kernkamp, A. Markus and others at WL| Delft Hydraulics for their support of hydrodynamic model work required by this study. They also thank Dr. R. Daley, United Nations University, Dr. D. Swayne, University of Guelph, and Dr. C. Mayfield, University of Waterloo, for discussion and support. J. Neysmith, Environment Canada, helped prepare figures and review the manuscript. I. Wong, Environment Canada, prepared the databases. P. Cooley, Centre for Earth Observation Science, University of Manitoba, provided the map files. Dr. M. Vollmer, Scripps Institution of Oceanography, University of California, San Diego, kindly provided the preliminary information for nutrient fluxes from the deep layer.

## REFERENCES

Bootsma, H.A. (1993a) Algal dynamics in an African Great Lake, and their relation to hydrographic and meteorological conditions. Ph.D. thesis, Univ. of Manitoba. 311 p.

Bootsma, H.A. (1993b) Spatio-temporal variation of phytoplankton biomass in Lake Malawi, Central Africa. *Verh. Internat. Verein. Limnol.* **25**, 882-886.

Bootsma, H.A., Bootsma, M.J. and Hecky, R.E. (1996) The chemical composition of precipitation and its significance to the nutrient budget of Lake Malawi/ Nyasa/Niassa, in T.C. Johnson and E.O. Odada (eds.), *The Limnology, Climatology and Paleoclimatology of the East Africa Lakes*, Gordon and Breach, Toronto, pp. 251-265.

Delft Hydraulics, HydroQual and IHE Delft (1999) Preparation of a Preliminary Lake Victoria Physical Processes and Water Quality Model. Final Report. Delft Hydraulics, Report T2274, Delft, The Netherlands.

Degnbol, P. and Mapila, S. (1982) Limnological observations on the pelagic zone of Lake Malawi from 1978 to 1981, in *Fishery Expansion Project, Malawi. Biological Studies on the Pelagic Ecosystem of Lake Malawi.* FI:DP/MLW/75/019 Tech. Rep. 1. FAO, Rome, 182 pp.

Eccles, D.H. (1974) An outline of the physical limnology of Lake Malawi (Lake Nyasa). *Limnol. Oceanogr.* **19**, 730-742.

Gonfiantini, R., Zuppi, G.M., Eccles, D.H. and Ferro, W. (1979) Isotope investigations of Lake Malawi. In: *Isotope in Lake Studies.* International Atomic Energy Agency Panel Proceedings Series STI/PUB/511. pp. 192-205.

Hecky, R.E., Kling, H.J., Johnson, T.C., Bootsma, H.A. and Wilkinson, P. (1999) Algal and sedimentary evidence for recent changes in the water quality and limnology of Lake Malawi/Nyasa/Niassa, *in Water Quality Report, Lake Malawi/Nyasa/Niassa Biodiversity Conservation Project*. SADC/GEF. pp. 191-214.

Foster, G.R., Lane, L.J., Nowlin, J.D., Laflen, J.M., and Young, R.A. (1980) A model to estimate sediment yield from field-sized areas: development of the model, in Knisel, W.G. (ed.) *CREAMS: A Field Scale Model for Chemicals, Runoff, and Erosion from Agricultural Management Systems*, USDA, Cons. Research Report No. 26, pp. 36-63.

Frere, M.H., Ross, J.D., and Lane, L.J. (1980) The nutrient submodel, in *Knisel, W.G. (ed.) 1980. CREAMS: A Field Scale Model for Chemicals, Runoff, and Erosion from Agricultural Management Systems*, USDA, Conservation Research Report No. 26, pp. 65-86.

Kernkamp, H.W.J., Otta, A., Uittenbogaard, R.E. and Veldman, J.J. (1994) Analysis and reliability of simulating seiches with PHAROS and TRISULA (DELFT3D-Flow). Delft Hydraulics, three reports Z568/H1866 (Phases 1 to 3), July (in Dutch).

Kingdon, M.J., H.A. Bootsma, J. Mwita, B. Mwichande, and Hecky, R.E. (1999) River discharge and water quality, in *Water Quality Report, Lake Malawi/Nyasa Biodiversity Conservation Project*. SADC/GEF, pp. 29-69.

Knisel, W.G., (1980) CREAMS: A Field Scale Model for Chemicals, Runoff, and Erosion from Agricultural Management Systems, Cons. Research Report No. 26, USDA, Washington, D.C.

Lam, D.C.L. and Halfon, E. (1978) Model of primary production, including circulation influences, in Lake Superior. *J. App. Math. Modelling* **2**, 30-40.

Lam, D.C.L., Schertzer, W.M. and Fraser, A.S. (1987) A post-audit analysis of the NWRI nine-box water quality model for Lake Erie. *J. Great Lakes Res*. **13**, 782-800.

Lam, D.C.L., Mayfield, C.I., Swayne, D.A. and Hopkins, K. (1994) A Prototype Information System for Watershed Management and Planning, *J. Biol. Sys*. **2**, 499-517.

Lam, D.C.L. (1997) Decision support systems for water resources problems. [Keynote Paper] in D. McDonald and M. McAleer (eds.), *Proc. Int. Congress on Modelling and Simulation,* Modelling and Simulation Society of Australia, A.N.U., Canberra, Australia, pp. 1827-1834.

Leon, L.F., Lam, D.C.L., Swayne, D.A., Farquhar, G.J. and Soulis, E.D. (2000) Integration of a Nonpoint Source Pollution Model with a decision support system. *J. Environmental Modelling & Software* **15**, 249-255.

Patterson, G., and Kachinjika, O. (1993) Effect of wind-induced mixing on the vertical distribution of nutrients and phytoplankton in Lake Malawi. *Verh. Internat. Verein. Limnol*. **25**, 872-876.

Simons, T.J. and Lam, D.C.L. (1980) Some limitations of water quality models for large lakes: a case of Lake Ontario. *Water Resources Res*. **16**, 105-116.

Summer, R.M., Alonso, C.V. and Young, R.A. (1990) Modelling linked watershed and lake processes for water quality management decisions. *J. Environmental Quality* **19**, 421-427.

University of Waterloo, WL| delft hydraulics and The United Nations University (UNU/INWEH) (2000) Preparation of a preliminary physical processes and water quality model for Lake Malawi/Nyasa/Niassa. Final Report, Biology Department, University of Waterloo, Waterloo, Ont. Canada.

Vollmer, M.K., R.F. Weiss, and Bootsma, H.A. (2001) Ventilation of Lake Malawi/Nyasa. (in this IDEAL volume).

Wishmeier, W.H. and Smith, D.D. (1978). Predicting Rainfall Erosion Losses. U.S. Dep. of Agriculture, Agriculture Handbook No. 537, 58pp.

Wüest, A., Piepke, G. and Halfman, J.D. (1996) Combined effects of dissolved solids and temperature on the density stratification of Lake Malawi, in Johnson, T.C. and E.O. Odada (eds.), *The limnology, climatology and paleoclimatology of the East African lakes*. Gordon and Breach, Toronto, pp. 183-202

Young R., Onstad, C., Bosch, D., and Anderson, W. (1986). *Agricultural Nonpoint Source Pollution Model: A Watershed Analysis Tool*, Model Documentation, Agricultural Research Service, U.S. Department of Agriculture, Morris, MN.

Young, W.J., Lam, D.C.L., Ressel, V. and Wong, I.W. (2000) Development of an environmental flows decision support system. *J. Environmental Modelling & Software* **15**, 257-265.

# VENTILATION OF LAKE MALAWI / NYASA

M.K. VOLLMER[1], R.F. WEISS[1] and H.A. BOOTSMA[2]

[1]*Scripps Institution of Oceanography, University of California at San Diego La Jolla, California, 92093–0244, USA*
[2]*Great Lakes WATER Institute, University of Wisconsin - Milwaukee, Milwaukee, Wisconsin, 53204, USA*

**ABSTRACT**

A tracer study was conducted on Lake Malawi/Nyasa, one of the deepest and largest lakes in the world, in order to quantify the renewal rates of the deep water. For this purpose, concentrations of the anthropogenic trace gas chlorofluorocarbon-12 (CFC-12) were measured in water samples which were collected in glass ampoules and analyzed by a new gas chromatographic separation method. Based on measurements of stored duplicate samples, we conclude that the first-order degradation rate for CFC-12 in anoxic water of Lake Malawi/Nyasa lies in the range 0 to 0.012 $yr^{-1}$. The tracer measurements are used in a 3-box mixing model from which average exchange times between the hypolimnion and the metalimnion of 18.5 years and 15.9 years are calculated for the cases of no degradation and maximum degradation in anoxic water, respectively. These exchange times are 2.7 to 2.2 times higher than have been estimated previously based on tritium measurements in 1976 by Gonfiantini and coworkers. The exchange times between the metalimnion and the epilimnion are calculated to be 3.7 years and 3.4 years, again for no degradation and maximum degradation, respectively. These exchange times are comparable to those estimated previously. Volumetrically averaged apparent CFC-12 ages of 8.9 and 21.1 years were calculated for the metalimnion and the hypolimnion, respectively, under the assumption of no degradation. Latitudinal gradients in the CFC-12 and dissolved oxygen concentrations on isopycnal surfaces suggests that the deep water originates predominantly in the southern part of the lake.



*E.O. Odada and D.O. Olago (eds.),*
*The East African Great Lakes: Limnology, Palaeolimnology and Biodiversity,* 209–233.

## 1. INTRODUCTION

Lake Malawi/Nyasa, located in the East African Rift, is meromictic owing to a permanent but periodically weak stratification maintained by small gradients in temperature, salts, and dissolved uncharged species (Wüest *et al.*, 1996). As a consequence of this meromixis and the lake's internal biological cycle, the surface water is depleted in nutrients and biogenic material accumulates and decomposes in the partly anoxic deep water. As a result, the deep water is nutrient-rich, and total dissolved nutrient flux to the surface mixed layer is controlled to a large degree by vertical exchange within the lake. In order to study the cycling of these nutrients, the distribution of pollutants, and the lake's responses to external long-term perturbations, it is essential to know the deep-water renewal rates. In 1976, Gonfiantini *et al.* (1979) conducted a study of Lake Malawi/Nyasa using tritium produced by the atmospheric testing of thermonuclear weapons as a time-dependent tracer. Using a 3-box model, they calculated a 5-year average exchange time of the anoxic hypolimnion with respect to the metalimnion. The renewal time of the metalimnion with respect to the epilimnion was calculated to be 4 years.

The purpose of the current study is to give an independent estimate of deep-water renewal rates using measurements of the gaseous transient tracer $CCl_2F_2$ (chlorofluorocarbon-12 or CFC-12). Anthropogenically produced CFC-12 has been increasing in the atmosphere since its production began in the 1930s. This trace gas dissolves into the lake's surface water from the atmosphere, labeling the water with a time-dependent concentration signal as it is mixed into the interior of the lake. CFC-12 along with other chlorofluorocarbons (CFCs) have been used extensively in oceanographic studies (e.g. Gammon *et al.*, 1982; Weiss *et al.*, 1985; Pickart *et al.*, 1989; Doney and Bullister, 1992) to help understand the circulation and dynamics of the oceans. The successful application of CFC-12 as a limnological tracer in a study of Lake Baikal (Weiss *et al.*, 1991) has led to the application of this technique to Lake Malawi/Nyasa.

While CFC-12 is believed to be stable in oxygenated water, its stability in anoxic water is currently debated (Cook *et al.*, 1995; Oster *et al.*, 1996; Shapiro *et al.*, 1997; Plummer *et al.*, 1998). Laboratory experiments have shown CFC-12 degradation in anoxic soils and sediments (Lovley and Woodward, 1992; Oster *et al.*, 1996) but the applications of these results to an aqueous environment such as Lake Malawi/Nyasa have to be treated with caution as chemical and microbial conditions might differ significantly. Some of the studies in anoxic layers of groundwater (Cook *et al.*, 1995; Oster *et al.*, 1996), of a fjord (Shapiro *et al.*, 1997) and of lake waters (Aeschbach-Hertig *et al.*, 2001) have also suggested CFC-12 degradation under anoxia. However most of these studies are based on mixing models calibrated through inverse methods using other tracers, in particular tritium and helium-3. In these cases, temporal mixing variability together with different temporal evolutions of the atmospheric source functions for tritium and CFC-12 may lead to discrepancies in tracer loading to the water surface which, under the models' steady state assumptions, may show apparent CFC-12 degradation. In order to assess CFC-12 degradation in a manner more applicable to the anoxic water of Lake Malawi/Nyasa, we have studied CFC-12 behavior in duplicate samples stored under anoxic conditions for more than 2 years.

## 2. METHODS

A hydrographic and geochemical survey was conducted on Lake Malawi/Nyasa (Figure 1) in September 1997 aboard R/V *Usipa*. CTD (conductivity-temperature-depth) data were collected by a Seabird SBE-19 profiler, fitted with an oxygen electrode, at 23 routinely visited stations along a north-south transect. At selected stations, the CTD casts were followed by a series of bottle casts, using 5-liter Hydro-Bios bottles, for the analysis of: nutrients, major ions, inorganic carbon species, and the isotopes $\delta^{13}C$ and $\Delta^{14}C$ of dissolved inorganic carbon; the transient tracers CFCs, helium-3 and tritium; and the gases $N_2O$ and $N_2/Ar$. In addition to these selected stations, additional sampling for all the above-mentioned properties was performed at the location with the lake's maximum depth of 703 m (Station 940; 11°8.6'S, 34°19.2'E), northeast of Nkhata Bay. With the exception of helium-3 and tritium, all the measurements have been completed. Here we report the first interpretations of the hydrographic data and the CFC-12 results. CFC samples were collected at 3 stations (Figure 1), taking special precautions to avoid contamination from the sampling bottles and from ambient air (Bullister and Weiss, 1988). O-rings were vacuum-baked at 60°C for several days before installation at each end of the sampling bottles in order to lessen diffusive contamination into the sampled water. For the same purpose the internal rubber band in each bottle was replaced by an epoxy-coated stainless steel spring. Water sample aliquots were sealed into 100 ml custom-made Pyrex glass ampoules, following the technique developed by Busenberg and Plummer (1992) and modified by the Institute of Environmental Physics in Bremen (Klatt, 1997; Bulsiewicz *et al.*, 1998). For this purpose the glass ampoules were connected to the Hydro-Bios bottles via a stainless steel tubing fixture which allowed the flushing and filling of the ampoules. Immediately after filling, the ampoules were taken to the ship's laboratory for flame sealing with a propane-oxygen torch. A headspace was created in the ampoule's stem using ultra-high purity and CFC-free nitrogen gas. This headspace was needed to prevent cracking of the glass during the sealing process and to allow for potential thermal expansion of the water during transportation. The ampoules were kept dark and at ambient temperature while transported to the Scripps Institution of Oceanography for analysis.

Dissolved CFCs were analyzed by electron capture detection (ECD) gas chromatography in a modified purge-and-trap system originally designed by Bullister and Weiss (1988). The instrument was adapted for the ampoule technique, and the NaOH-impregnated absorbent Ascarite was used to remove traces of $H_2S$ (Bullister and Lee, 1995) in order to avoid potential interference with the gases of interest. The water samples were kept in a water bath at 25.0°C before the glass

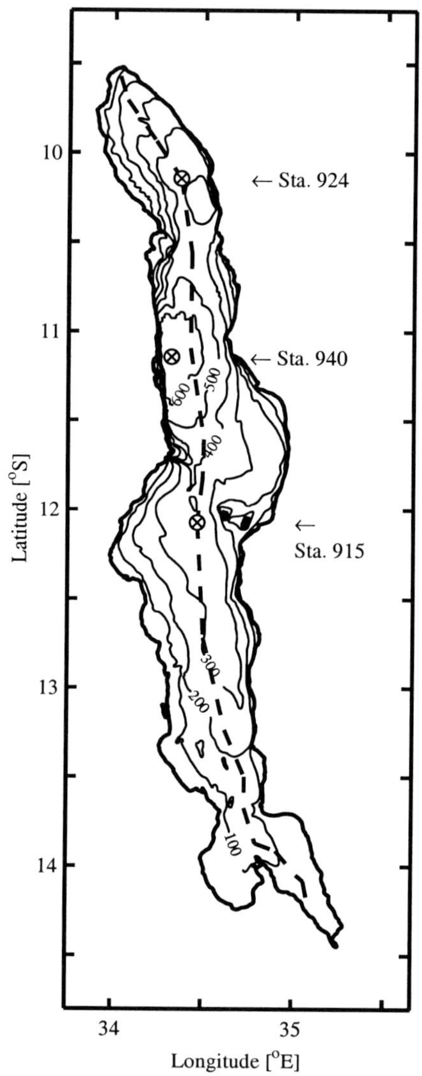


*Figure 1*. Bathymetric map for Lake Malawi/Nyasa including the section of routinely-visited stations (dashed line) and the locations of Stations 915, 924 and 940, where water samples for CFC-12 measurements were taken in September 1997. the 700 m isobath covers a small area in the vicinity of the Station 940 symbol and is not plotted for clarity.

ampoules were opened and aliquots were transferred to the purging chamber. Our first attempt to measure Lake Malawi/Nyasa samples using packed Porasil C columns (Bullister and Weiss, 1988) for chromatographic separation, was hindered by the presence of volatile hydrocarbons at elevated concentrations which interfered with the CFC peaks. The ECD responses to these hydrocarbons were negative and irregular when using P5 (5 % methane in argon) carrier gas, and these interferences affected mainly the quantification of the CFCs by peak area. Three samples were analyzed under these conditions and the results are included in this study, using peak height rather than peak area to determine their CFC concentrations. The chromatographic method was then altered to achieve a better separation of CFC-12 from the hydrocarbons. The sample gases were trapped in a short packed (Porasil C and Porapak T) column held at about -110°C (liquid nitrogen/methanol-ethanol slurry) and released after the trap was immersed into a hot water bath. Separation was achieved using a Porapak Q precolumn (40 cm) and main column (270 cm) at 150°C, followed by a molecular sieve 5A column (85 cm) at 160°C, all packed (80-100 mesh) in 1/8" OD stainless steel tubing. P5 carrier gas was used with a flow rate of 24 ml $min^{-1}$, and the precolumn was back-flushed 88 sec after injection. The gases were detected by a Shimadzu Mini-2 gas chromatograph with an ECD detector at 290°C. This revised chromatographic method did not allow for the measurement of $CCl_3F$ (CFC-11) or $CCl_2FCClF_2$ (CFC-113), a compromise which was acceptable because these compounds are strongly degraded in anoxic waters (Bullister and Lee, 1995; Tanhua, 1997). Instead, the analytical method was optimized to measure CFC-12 and nitrous oxide ($N_2O$), with retention times of 350 sec and 480 sec, respectively. Corrections were made to account for the partitioning of these gases into the headspace of the ampoules. A first set of these samples was analyzed in February 1998. All other samples were stored dark and at room temperature until April 2000, when 33 of the remaining samples were analyzed under similar instrumental conditions in order to explore the possibility of CFC-12 degradation. Results for CFC-12 are reported on the SIO-1998 calibration scale (Prinn *et al.*, 2000). The analytical precisions ($1\sigma$, after applying a $3\sigma$ filter) for the CFC-12 standard gas were 0.3 % and 0.4 % for the Feb. 1998 and April 2000 analysis, respectively. The precisions for CFC-12 in the water samples reported as the mean of the relative standard deviations of duplicate samples during each analytical period were 2.3 % for Feb. 1998 (n = 4) and 0.4 % for April 2000 (n = 3).

## 3. RESULTS

The results for temperature, conductivity, dissolved oxygen and CFC-12 concentration profiles at the stations 915, 940 and 924 are listed in Table 1, and are plotted against depth together with the potential density anomaly in Figure 2. Depth was calculated from pressure using site-specific gravitational accelerations. In general we used the downcast hydrographic properties which we averaged in 5-m depth bins. These results are based on the calibration of the CTD sensors in 1996 and are not corrected for possible sensor drift. We use distinct inflections in the oxygen and the CFC-12 concentration profiles at the deep Station 940 to distinguish three different layers in the lake, following a terminology suggested previously (Gonfiantini *et al.*, 1979). We thus designate the 105 m thick surface layer as the epilimnion, the middle layer as the metalimnion and the deep water below the oxic-anoxic boundary at approximately 220 m as the hypolimnion, realizing that these definitions do not match their more traditional meanings used to describe holomictic lakes.

Water temperature measured by the Seabird SBE-3 sensor is reported on the international temperature scale ITS-90. Potential temperature (Figure 2) was calculated from the CTD in-situ temperature according to Wüest *et al.* (1996) and includes the freshwater-specific adiabatic temperature gradient (Chen and Millero, 1986) but we integrated with respect to pressure instead of depth. The potential temperature decreases gradually throughout the epilimnion from about 25°C to 23.5°C indicating that the winter mixing has ended and spring stratification is well established. Throughout the metalimnion the potential temperature decreases downward with decreasing gradients. The lake reaches nearly isothermal conditions at about 22.7°C in the hypolimnion with a lowest value of 22.671°C found at 680 m at deep Station 940, which corresponds to an in-situ temperature of 22.787°C. In all our observed profiles temperature has an overall stabilizing effect on the water column, and in contrast to other authors (Wüest *et al.*, 1996; Patterson *et al.*, 1997) we found only inconclusive evidence of geothermal heating from inversions in the potential temperature for some of the profiles near the sediment-water interface. We also observed horizontal gradients in potential temperature along the transect of the routinely-visited stations (Figure 3a) resulting in isotherm outcropping in the southern part of the lake. Even though such a contour plot is a snapshot and internal seiching undoubtedly contributes to oscillations of the isotherms, there is support for this to be the prevalent condition in the lake, at least during the months of May to September (Hamblin *et al.*, 1999). The temperature profile for Station 924 differs significantly from the other stations. At this station a strong upwelling of epilimnetic and metalimnetic water is observed. Gradients in potential temperature and other CTD properties are large below a 20 m thick surface cap. Similar upwelling features have been observed in this region by Eccles (1974), who speculated that this may be associated with separate surface circulation cells driven by the local wind pattern, and by Patterson and Kachinjika (1995).

Although on a much finer scale than the upper waters, the abyss of Lake Malawi/Nyasa below a depth of about 300 m shows some horizontal temperature structure also (Figure 3a). Potential temperatures are higher at Stations 940 and 920 and decrease towards the north and south with typical gradients of a few hundredths of a degree per 100 km. The data for Station 940, which are not included in Figure 3, are similar to those at Station 920 for the lower part of the deep water and therefore support these observations. Such a feature could not have been observed during previous studies on the lake due to the lack of sufficient deep-water measurements, and it is therefore unclear whether it is permanent. Effects of internal waves or deep-water cooling from north and south may be possible explanations.

*Table 1.* Lake Malawi/Nyasa profiles for Stations 915, 940 and 924. In-situ temperature, in-situ conductivity and dissolved oxygen concentration from CTD downcast measurements. Potential temperature and reference conductivity $\kappa_{20}$ are given in parentheses. The oxygen concentration measurements in the anoxic layer yielded slightly negative numbers and are not listed here. The CFC-12 concentrations from bottle samples are listed for those analyzed in February 1998 and in parentheses for those analyzed in April 2000. For both sets a correction is included for the loss of CFC-12 into the ampoules' headspaces. The CFC-12 results for the depths 10, 365 and 420 m at Station 924 were measured in Oct. and Nov. 1997 by a different chromatographic separation method (see text).

| **Depth** [m] | ***In-situ* (potential) Temperature** [°C] | ***In-situ* (reference) Conductivity** [μS $cm^{-1}$] | | **Dissolved Oxygen** [μmol $kg^{-1}$] | **CFC-12 Concentration** [pmol $kg^{-1}$] |
|---|---|---|---|---|---|
| | **Station 915** | **[12°18.5' S, 34°29.9' E]** | | **Sep. 14, 1997** | |
| 10 | 24.800 (24.798) | 255.2 | (230.0) | 226.8 | 1.423 |
| 100 | 23.458 (23.441) | 249.7 | (231.4) | 165.9 | 1.364 (1.343) |
| 140 | 23.182 (23.158) | 249.1 | (232.0) | 120.3 | 1.248 (1.259) |
| 180 | 23.069 (23.038) | 248.6 | (232.1) | 96.2 | 1.190 (1.191) |
| 230 | 22.914 (22.875) | 248.2 | (232.4) | 30.6 | 1.072 (1.037) |
| 280 | 22.803 (22.756) | 251.3 | (235.7) | — | 0.602 |
| 360 | 22.744 (22.684) | 252.3 | (236.8) | — | 0.367 (0.354) |
| | **Station 940** | **[11°8.6' S, 34°19.2' E]** | | **Sep. 13, 1997** | |
| 10 | 25.124 (25.123) | 257.3 | (230.3) | 225.9 | 1.422 (1.405) |
| 50 | 24.251 (24.243) | 252.8 | (230.4) | 187.9 | 1.404 |
| 105 | 23.563 (23.544) | 250.6 | (231.7) | 167.6 | 1.369 |
| 120 | 23.365 (23.345) | 250.0 | (232.1) | 121.3 | 1.273 (1.310) |
| 140 | 23.171 (23.146) | 249.4 | (232.4) | 90.7 | 1.198 |
| 170 | 23.068 (23.039) | 249.1 | (232.5) | 70.9 | 1.145 (1.142) |
| 190 | 23.011 (22.979) | 248.8 | (232.5) | 43.2 | 1.067 |
| 200 | 22.987 (22.953) | 248.7 | (232.5) | 30.2 | 1.041 (1.045) |
| 215 | 22.972 (22.935) | 248.6 | (232.5) | 23.0 | 0.998 (0.988) |
| 230 | 22.922 (22.883) | 248.7 | (232.8) | — | 0.908 (0.885) |
| 245 | 22.882 (22.840) | 249.8 | (234.0) | — | 0.821 (0.780) |
| 300 | 22.817 (22.767) | 251.1 | (235.4) | — | 0.573 (0.530) |
| 360 | 22.780 (22.719) | 251.9 | (236.2) | — | 0.474 (0.461) |
| 420 | 22.774 (22.703) | 252.1 | (236.3) | — | 0.425 (0.420) |
| 480 | 22.770 (22.689) | 252.3 | (236.4) | — | 0.401 (0.404) |
| 530 | 22.772 (22.683) | 252.5 | (236.5) | — | 0.389 (0.383) |
| 570 | 22.775 (22.679) | 252.6 | (236.5) | — | 0.382 (0.369) |
| 610 | 22.779 (22.676) | 252.7 | (236.4) | — | 0.378 (0.360) |
| 640 | 22.782 (22.674) | 252.7 | (236.4) | — | 0.373 (0.367) |
| 670 | 22.786 (22.673) | 252.8 | (236.4) | — | 0.360 (0.349) |
| 685 | 22.786 (22.672) | 252.8 | (236.3) | — | 0.359 (0.342) |

| | **Station 924** | **[10°7.3' S, 34°18.4' E]** | | **Sep. 10, 1997** | |
|---|---|---|---|---|---|
| 10 | 24.745 (24.744) | 255.0 | (230.2) | 215.8 | 1.413 |
| 50 | 23.476 (23.467) | 250.0 | (231.6) | 98.5 | 1.256 (1.264) |
| 90 | 23.228 (23.213) | 248.9 | (231.8) | 66.6 | 1.142 (1.154) |
| 120 | 23.125 (23.105) | 248.5 | (231.8) | 36.5 | 1.096 (1.083) |
| 140 | 23.079 (23.055) | 248.4 | (231.9) | 20.7 | 0.995 (0.917) |
| 170 | 23.003 (22.975) | 248.1 | (232.0) | 0.1 | 0.957 (0.933) |
| 190 | 22.952 (22.920) | 248.9 | (233.0) | — | 0.870 (0.854) |
| 215 | 22.919 (22.883) | 249.3 | (233.4) | — | 0.776 |
| 245 | 22.850 (22.809) | 250.3 | (234.6) | — | 0.678 (0.667) |
| 305 | 22.794 (22.743) | 251.0 | (235.4) | — | 0.490 (0.495) |
| 365 | 22.761 (22.701) | 251.7 | (236.2) | — | 0.425 |
| 420 | 22.746 (22.676) | 252.5 | (236.8) | — | 0.352 |
| 460 | 22.753 (22.676) | 252.6 | (236.8) | — | 0.373 |

Unless geothermal warming has been drastically higher in this deep basin in recent years compared to the observations of Von Herzen and Vacquier (1967), this process seems an unlikely explanation for our observed horizontal temperature gradient as heat fluxes are generally low, and near Stations 940 and 920 were among the lowest found in the lake by these authors.

Based on in-situ electrical conductivity measured by the Seabird SBE-4 sensor, we have calculated the reference conductivity, $\kappa_{20}$, and the conductivity-based salinity according to the procedure described by Wüest *et al.* (1996). From the surface to maximum depth, $\kappa_{20}$ and this salinity increased from about 230 to 236 µS cm (Figure 2) and from about 0.210 to 0.215 g kg$^{-1}$, respectively. This salinity has a stabilizing effect on the density throughout the upper part of the water column, with the strongest gradient in the metalimnion and the upper hypolimnion. In the lower hypolimnion near Stations 920 and 940, $\kappa_{20}$ and this salinity are practically uniform. Horizontal gradients of $\kappa_{20}$ are small in general with the exception at Station 924 (Figure 3b). In the epilimnion a slight freshening of the water toward the north can be observed which could be an effect of higher rainfall and larger river inflow in the north, higher rates of evaporation in the south caused by strong southerlies and the resulting upwelling of water with higher conductivities.

We have used the silica concentration profiles (Bootsma and Hecky, 1999a) from this survey to calculate the non-ionic salinity (Wüest *et al.*, 1996). Assuming that the main silica component is silicic acid ($Si(OH)_4$), we find the non-ionic salinity to increase from 1.3 mg kg$^{-1}$ in the epilimnion to 21.4 mg kg$^{-1}$ at maximum depth which, in contrast to the results of Wüest *et al.* (1996), strengthens the stability of the entire water column.

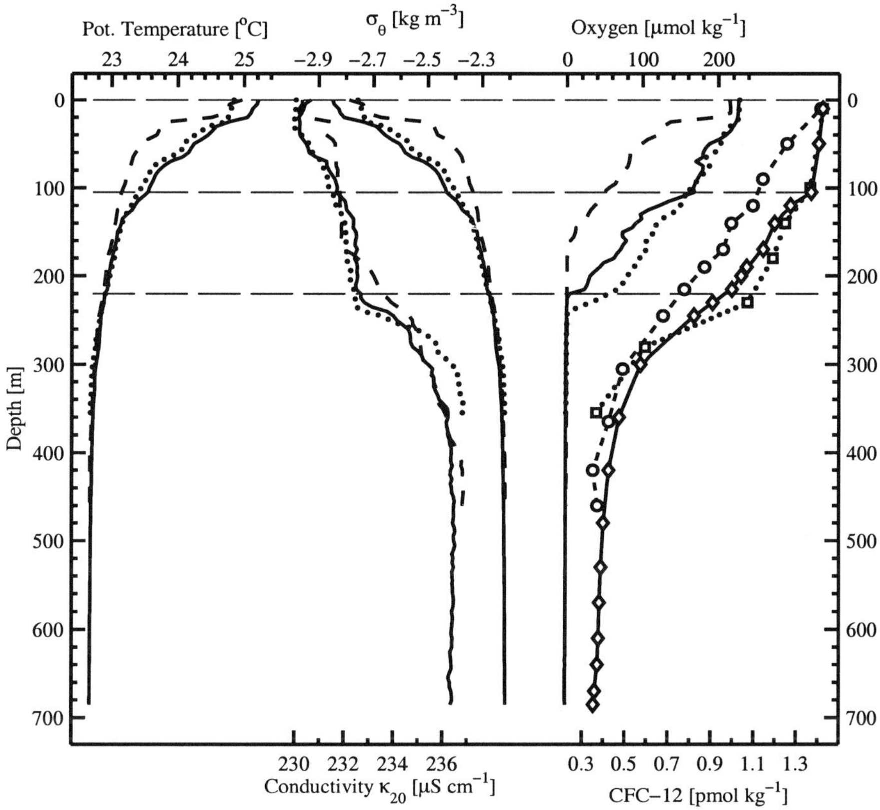


*Figure 2.* Vertical profiles at Stations 915 (dotted lines), 940 (solid lines) and 924 (dashed lines) of potential temperature ($\Theta$), reference conductivity ($\kappa_{20}$), potential density anomaly ($\sigma_\Theta$) and dissolved oxygen concentration from CTD measurements; and of CFC-12 concentration from water sample measurements. The CFC-12 results in this plot are based on the analytical measurements of February 1998. Horizontal dashed lines denote the lake's surface, and the boundaries between the epilimnion and metalimnion at 105 m and between the metalimnion and hypolimnion at 220 m.

Potential temperature and total salinity were used to calculate potential density (Figure 2) profiles (Wüest *et al.*, 1996; Chen and Millero, 1986). The potential density anomaly with respect to the lake surface ($\sigma_\Theta$) increases from about -2.75 kg m$^{-3}$ at the surface to about -2.21 kg m$^{-3}$ at greatest depth. The temperature gradient has a stronger effect on the density stratification than the salinity gradient.

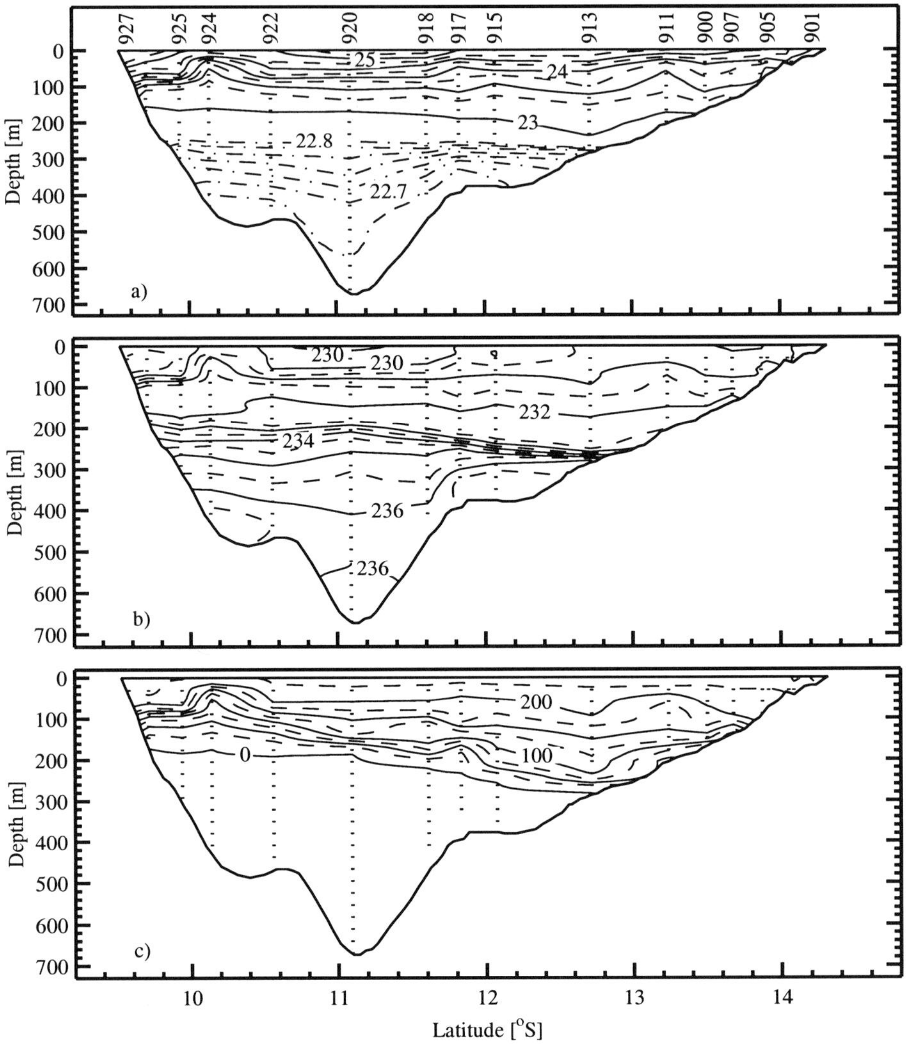


*Figure 3.* Objectively-mapped vertical contour section for CTD properties plotted against latitude south for the Sept. 1997 study on Lake Malawi/Nyasa, including stations along the routinely-visited north-south transect (see Figure 1). Deep Station 940 and stations in the southwest arm are excluded. Vertical dotted lines denote CTD depths at the latitudes of the stations occupied along this cruise track where as selected stations are labeled on the top. a) Potential temperature (°C). Note the 0.25°C isotherm spacing above 23°C, and the 0.02°C spacing below 22.8°C. b) Reference conductivity $\kappa_{20}$ ($\mu S\ cm^{-1}$). The dashed lines at greatest depth near Stations 915 and 924 indicate a conductivity of 236.5 $\mu S\ cm^{-1}$. c) Dissolved oxygen ($\mu mol\ kg^{-1}$).

Between the epilimnetic-metalimnetic boundary and the maximum depth at deep Station 940, temperature, conductivity-based salinity and silica-based salinity contribute to the potential density increase by 95 %, 2 % and 3 %, respectively. The effect of the salinities on the density stratification is insignificant in the epilimnion and largest below the oxic-anoxic boundary where the two salinities contribute equally to the calculated density ratio (Wüest *et al.*, 1996) of about -0.15. Below a depth of about 400 m at Station 940, the conductivity-based density ratio fluctuates around a mean of essentially 0 and the mean silica-based density ratio is about -0.04. In this range the potential density is almost uniform but the stratification is stronger compared to the results of Wüest *et al.* (1996), as is indicated by a higher static stability, $N^2$ (Wüest *et al.*, 1996) which reaches a mean value of 6 x $10^{-7}$ $sec^{-2}$ in the lowest observed 100 m of the lake.

The sound velocity profile (Chen and Millero, 1986) and a preliminary maximum depth for the water column at Station 940 were calculated for the depth of the CTD measurements using the acoustic travel time measured by the ship's echo sounder in this relatively large area of constant and maximum depth. An average sound velocity of 1497.4 m $s^{-1}$ was calculated which includes an extrapolated sound velocity below the maximum depth of the CTD measurement. From this we calculate a maximum depth of 703 m taking into account the vertical displacement of 1.7 m of the echo sounder's transducer from the lake surface which was at the altitude of 473.3 m above sea level (Kingdon *et al.*, 1999) during the expedition. As for potential density, it is likely that non-ionic compounds and dissolved gases alter the sound velocity, thus altering the results for the maximum depth of the lake. Our calculated maximum depth is 3 m less than the maximum depth published by Hutchinson (1957) for an unreported lake stand at the time of these observations. In the rank of the world's deepest lakes, this newly-calculated Lake Malawi/Nyasa maximum depth still exceeds the maximum depth of Lake Issyk Kul (Hutchinson, 1957) by 1 m, or by much more if a recent depth measurement of about 660 m (T. Johnson, pers. comm.) for this brackish lake is used. Therefore Lake Malawi/Nyasa remains the world's 3rd deepest freshwater lake, or the 4th deepest lake if the brackish Caspian Sea is included.

The dissolved oxygen concentration was measured by the Seabird SBE-23B sensor. Dissolved oxygen concentration decreases with depth, with an anoxic layer below 170 m in the north and below 270 m in the south (Figure 3c). Dissolved oxygen is undersaturated throughout the water column. The volume-weighted dissolved oxygen concentration for the top 10 m of the water column is 230 μmol $kg^{-1}$ which corresponds to a saturation of 94 % as calculated from the solubility function of Weiss (1970) at this layer's mean water temperature and salinity and for an annual mean atmospheric pressure (measured at Senga Bay) of 961 mbar (H. Bootsma, unpublished data).

The CFC-12 profiles for Stations 915 and 940 (Figure 2) show concentration decreases from the surface to the maximum depth, with distinct inflections at the interface between the top two layers. These transitions can also be seen in profiles of other properties, such as oxygen. At these 2 stations the epilimnetic CFC-12 concentrations are rather uniform at about 1.4 pmol $kg^{-1}$ ($10^{-12}$ mol $kg^{-1}$), an indication that mixing during the preceding winter extended to a depth of about 100 m. At 10 m depth CFC-12 is saturated to 101 % with respect to a mean 1997 Southern Hemisphere atmospheric CFC-12 background concentration (data from Walker *et al.* (2000) converted to the SIO-1998 scale) of 531 ppt (dry air mole fraction in parts-per-trillion, or parts in $10^{12}$), and a previously-mentioned reduced atmospheric pressure, as calculated from the solubility of CFC-12 (Warner and Weiss, 1985) at the local water temperatures and salinities. In the metalimnion, CFC-12 concentrations decrease to about 1 pmol $kg^{-1}$ at the oxic-anoxic boundary, below which the CFC-12 concentrations decrease steeply to about 0.4 pmol $kg^{-1}$ in the lower part of the hypolimnion. The CFC-12 concentration profile for Station 924 reflects the upwelling observed in the CTD properties. CFC-12 concentrations are lower than at the other two stations at similar depths in the lower epilimnion and in the metalimnion.

In order to estimate CFC-12 degradation in the anoxic water of Lake Malawi/Nyasa, we have measured duplicate samples more than 2 years after the first set of samples was analyzed. With a few exceptions from identical casts, the samples originated from duplicate casts with bottles at the same wire depths. The samples were stored dark and at a room temperature of about 23°C. Analyses were performed under similar conditions as for the first set, although some changes led to improvements in the overall precision of the measurements. For this second set of analyses, 14 out of 33 samples were from originally-oxygenated water in the lake, and the rest were from the anoxic zone. Some of the samples with low initial concentrations of dissolved oxygen may have become anoxic during storage. In particular we assume that this was the case for one sample (Station 924, 140 m) in which all the $N_2O$ had disappeared during storage and this sample was therefore added to the group of anoxic samples. Interestingly this sample showed the largest decrease in CFC-12 concentration.

Most samples showed lower concentrations than their duplicates two years earlier (Figure 4). On average the initially-oxygenated samples yielded CFC-12 concentrations which were lower by 0.002 pmol $kg^{-1}$ (0.22 %), and the concentrations in the anoxic samples were lower by 0.018 pmol $kg^{-1}$ (3.1 %) than the previously analyzed samples. For the anoxic samples at Station 940, the absolute and relative differences between the two sample sets were found to be largest where the vertical concentration gradients are highest. This could be an artifact of slight changes in sampling depth for the two sets of samples taken from different casts during the expedition. For the oxygenated samples, no correlation was found between the observed concentration difference and the original oxygen or CFC-12 concentration. We assume no CFC-12 degradation in these samples, and we therefore conclude that the measured differences in the oxygenated samples are instrumental in nature. If we increase all the values in the second set by 0.002 pmol $kg^{-1}$ in order to remove this offset for the oxygenated samples, then the mean relative difference for the anoxic samples is -2.6 % which corresponds to a degradation rate of 0.012 $yr^{-1}$ assuming this process to be first-order. We interpret this degradation rate with caution and as an upper limit. Analytical uncertainties may well have contributed to the measured differences, as could a possible shift in bottle depth for the separate casts employed for these two sets of samples. In summary, from this study we can neither prove the existence nor the absence of CFC-12 degradation in Lake Malawi/Nyasa and we therefore pursue our calculations for both cases, no degradation and an upper limit of degradation as given above. The profiles in Figure 2 represent the first analytical set of samples listed in Table 1 and do not include a correction for potential degradation during the 5 month storage between dates of sampling and analysis.

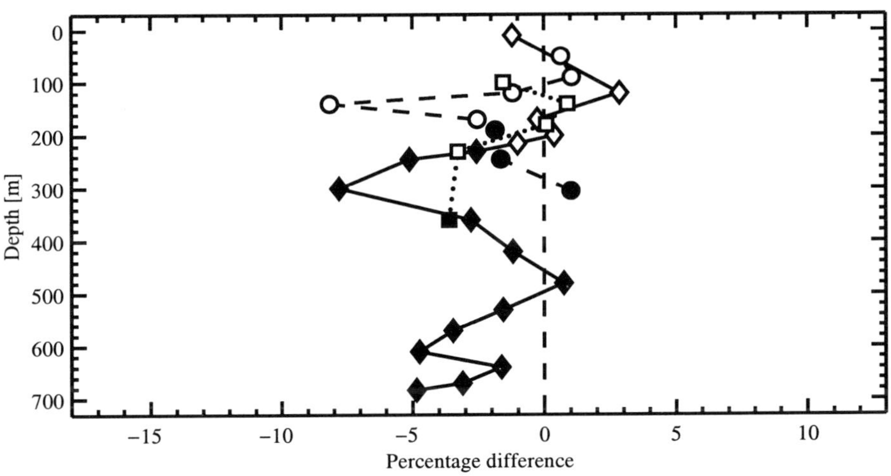


*Figure 4.* Results for the CFC-12 degradation study on Lake Malawi/Nyasa water samples stored in glass ampoules. Percentage differences are expressed as the April 2000 result minus the February 1998 result, divided by the mean of the two results and multiplied by 100. These are shown for Station 915 (squares), Station 924 (circles) and Station 940 (diamonds). Open and filled symbols represent initially-oxygenated and anoxic samples, respectively. A vertical dashed lines is drawn for zero concentration difference.

Our range of CFC-12 degradation rates from zero to 0.012 $yr^{-1}$ for the anoxic water of Lake Malawi/Nyasa is less than most of those found in other anoxic waters. Our results are in agreement with the study of Shapiro *et al.* (1997) who have found "a maximum degradation rate of 0.01 to 0.03 $yr^{-1}$" for an anoxic Norwegian fjord. Their results are based on a comparison between observed CFC-12 concentrations and those modeled using tritium and helium-3. Our results are an order of magnitude smaller compared to Aeschbach-Hertig *et al.* (2001) who have derived a CFC-12 degradation rate of 0.11 $yr^{-1}$ for the anoxic layers of Lake Pavin using a similar method to that of Shapiro *et al.* (1997). If we assumed zero-order CFC-12 degradation kinetics for our samples this would result in a maximum CFC-12 degradation of 0.011 pmol $kg^{-1}$ $yr^{-1}$, again a value which is significantly smaller than that deduced by Aeschbach-Hertig *et al.* (2001) of 0.07 pmol $kg^{-1}$ $yr^{-1}$ for Lake Pavin or than that found by Oster *et al.* (1996) of 0.08 pmol $kg^{-1}$ $yr^{-1}$ for groundwaters. Plummer *et al.* (1998) have found apparent stability of CFC-12 in ampoule samples filled with river water and stored for 2.5 yr under various conditions even though the samples were believed to have turned anoxic. However they also found complete loss of CFC-12 within 1 month of storage in glass ampoules in 2 samples of lake water. We believe that in those studies conducted so far there are significant inconsistencies in the determination of CFC-12 degradation, and that the observed differences may, at least in part, depend on the different chemical and biological conditions in different waters. For the case of Lake Malawi/Nyasa, we realize that only an improved degradation study using stored duplicate poisoned and unpoisoned samples from the same sampling bottles will lead to a clarification of this question.

## 4. MODEL CALCULATIONS

To calculate the exchange rates between the 3 layers we use a one-dimensional 3-box model similar to that used by Gonfiantini *et al.* (1979). The lake is divided vertically into the 3 layers separated at depths of 105 m and 220 m, and the exchange of water between adjacent layers is modeled. In contrast to Gonfiantini *et al.* (1979), who used a reconstructed epilimnetic tritium concentration as a forcing function, our model is forced by the atmospheric CFC-12 concentration evolution and by air-water gas exchange, and therefore the degree of CFC-12 saturation in the epilimnion is allowed to vary in our model. This replaces the approach in similar lake and most oceanic CFC studies, in which the epilimnetic CFC concentrations are assumed to be at a constant degree of saturation with respect to the increasing atmospheric CFC concentration. Such an assumption is justifiable for time periods when the rate of atmospheric increase remained essentially constant. However since the rate of increase of atmospheric CFC-12 has slowed considerably in recent years (Walker *et al.*, 2000) we have chosen to explicitly include the exchange of this tracer between the atmosphere and the epilimnion in our model calculations.

Accordingly, the exchange of water in the model is described by the following three coupled first-order differential equations,

$$\frac{dc_1}{dt} = \frac{V_2}{V_1} k_2 (c_2 - c_1) + \frac{A_s}{V_1} k_p (c_{sat} - c_1) \tag{1}$$

$$\frac{dc_2}{dt} = k_2 (c_1 - c_2) + k_3 \frac{V_3}{V_2} (c_3 - c_2) \tag{2}$$

$$\frac{dc_3}{dt} = k_3(c_2 - c_3) - j \cdot c_3 \tag{3}$$

and the water exchange fluxes between the epilimnion and the metalimnion, $\phi_2$, and between the metalimnion and the hypolimnion, $\phi_3$, are given by $\phi_2 = k_2 \times V_2$ and $\phi_3 = k_3 \times V_3$. The symbols $c_i$ and $V_i$ denote the CFC-12 concentrations and volumes of the epilimnion ($i$ = 1), the metalimnion ($i$ = 2) and the hypolimnion (i = 3), $A_s$ is the lake's surface area, $k_p$ is the air-water transfer velocity in units of m $yr^{-1}$, $k_2$ and $k_3$ are the epilimnetic-metalimnetic and metalimnetic-hypolimnetic exchange coefficients in units $yr^{-1}$, respectively, and $j$ is the first-order degradation rate for CFC-12 in units of $yr^{-1}$. CFC-12 enters the epilimnion through air-water gas exchange following the atmospheric CFC-12 concentration history (Walker *et al.*, 2000), which is used here to calculate the saturation concentration, $c_{sat}$ (Figure 5), for the salinity and pressure conditions described above, and an estimated epilimnetic winter temperature of 23.5°C as derived from a small inflection in the temperature profile at the lower boundary of this layer. We use only the tracer results of the deep Station 940, which we interpolated vertically by a cubic spline function to calculate the CFC-12 inventories in the 3 layers. We exclude Station 924 from this calculation because it reflects a region of strong upwelling which is not representative for this part of the lake (Figure 3). By excluding Station 924 we have also to exclude Station 915 to avoid an inventory bias due to an observed north-south gradient in CFC-12 as shown later. Station 940 appears to be a good average of the lake-wide CFC-12 distribution and due to this station's central position with respect to the north-south extensions, its water column is likely to be least effected by internal waves.

Using the hypsographic curve based on the unpublished bathymetric map of B. Halfman and T. Johnson (Large Lakes Observatory, University of Minnesota, Duluth), we calculate the layer volumes and the volumetrically averaged CFC 12 concentrations in the epilimnion, metalimnion and hypolimnion as: 2.76 x $10^{12}$ $m^3$ and 1.40 pmol $kg^{-1}$, 2.31 x $10^{12}$ $m^3$ and 1.16 pmol $kg^{-1}$, and 2.72 x $10^{12}$ $m^3$ and 0.59 pmol $kg^{-1}$, respectively. These CFC-12 concentrations are based on the Feb. 1998 analytical results. For the calculations where $j$ = 0.012 $yr^{-1}$ these concentrations have been corrected for the 5-month storage of the samples.

CFC-12 concentrations for all 3 layers were calculated according to equations 1 to 3, for 0.1 yr time increments in the atmospheric concentration from 1931 to 1997 (Figure 5). Optimal $k_i$ and $k_p$ values were determined by least squares fitting to the observed mean CFC-12 concentrations. This resulted in exchange times (Table 2) between the layers of 3.7 yr for $k_2^{-1}$ and 18.5 yr for $k_3^{-1}$ assuming no CFC-12 degradation and 3.4 yr for $k_2^{-1}$ and 15.9 yr for $k_3^{-1}$ assuming $j$ = 0.012 $yr^{-1}$. In comparison, using their tritium profile, Gonfiantini *et al.* (1979) estimated $k_2^{-1}$ and $k_3^{-1}$ values of 4 yr and 5 yr, respectively.

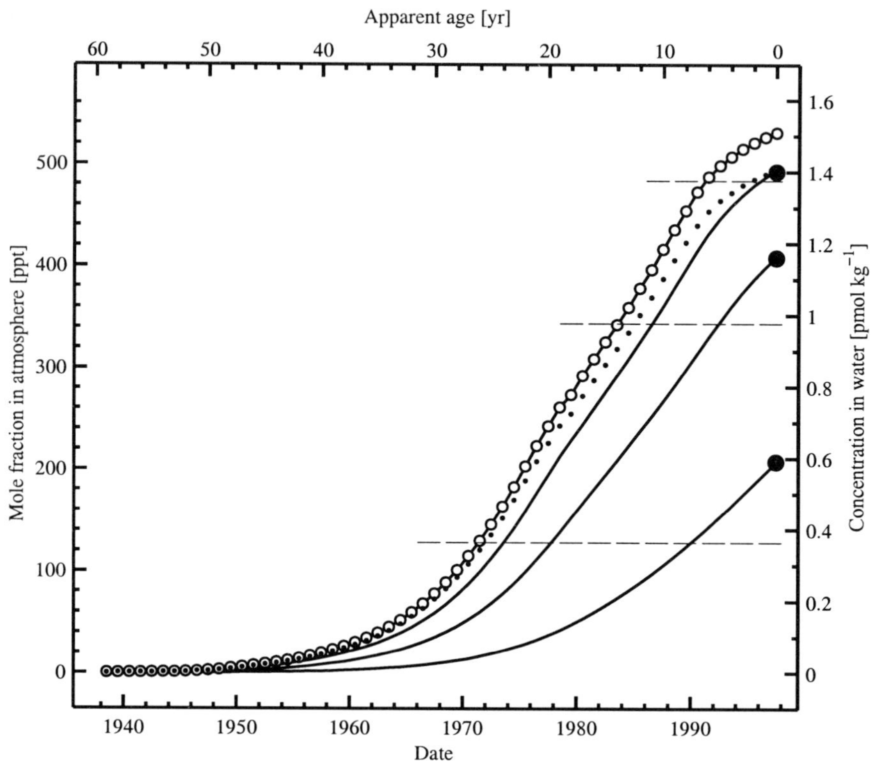


*Figure 5.* CFC-12 concentration reconstruction for the atmosphere and 3 box-layers in Lake Malawi/Nyasa. Atmospheric background concentrations vs. time (open circles) for CFC-12 in the Southern Hemisphere. Data are based on Walker *et al.* (2000) updated to the SIO-1998 calibration scale. The units are dry air mole fraction expressed as parts in $10^{12}$ (ppt), labeled on the left axis. The right axis gives CFC-12 concentrations in the water vs. time. The atmospheric record is equivalent to the epilimnion equilibrium concentration. The three solid lines ending in the volumetrically averaged layer concentrations (filled circles) for September 1997 are the results of our 3-box model. The dotted line was obtained by normalizing the atmospheric trend to a mean epilimnetic concentration of 1.40 pmol $kg^{-1}$ for September 1997 which corresponds to 93 % saturation at an estimated winter water temperature of 23.5°C. Dashed horizontal lines show the minimum CFC-12 concentration observed and the concentrations at the boundaries between the layers.

To make our results more directly comparable with those of Gonfiantini *et al.* (1979), the model calculations have also been performed for layers divided at 100 m and 250 m depths. The resulting volumetrically averaged CFC-12 concentrations were calculated as 1.41 pmol $kg^{-1}$, 1.13 pmol $kg^{-1}$, and 0.52 pmol $kg^{-1}$, for the epilimnion, metalimnion and hypolimnion, respectively. Following this earlier study we also assumed equal layer volumes. The resulting exchange times (Table 2) between the layers are 4.7 yr for $k_2^{-1}$ and 20.8 yr for $k_3^{-1}$ assuming no degradation, and 4.5 yr for $k_2^{-1}$ and 18.0 yr for $k_3^{-1}$ assuming $j$ = 0.012 $yr^{-1}$. Again these results are of similar magnitude for the two upper layers and significantly higher for the two lower layers compared to the earlier tritium-based result.

## 5. DISCUSSION

The model-derived air-water transfer velocity $k_p$ for CFC-12 when the layers are separated at 105 m and 220 m depths is calculated as 66 m $yr^{-1}$ assuming no CFC-12 degradation and 69 m $yr^{-1}$ including degradation. These results are based on the assumption of a constant year-round epilimnetic water temperature of 23.5°C and a mixed-layer thickness of 105 m, both of which are representative for the winter-time conditions. However gas exchange across the air-water interface is highly seasonal and effective only during part of the year. During spring and summer when the epilimnion is stratified, gas exchange is limited to a shallow surface mixed layer and, due to the warming during these periods, may result in supersaturation and subsequent outgassing of CFC-12. During this time the concentrations in the lower part of the epilimnion are expected to change little as mixing is greatly reduced. The important variables in the CFC-12 gas transfer are the onset of the seasonal cooling, the rate of the mixed layer deepening and the duration of a fully developed mixed layer before the next spring stratification begins. Our above assumption yields an overestimate of the term $c_{sat} - c_1$ (equation 1) which in turns results in an underestimate of the fitted mean annual transfer velocity $k_p$. If we assume gas exchange to be effective during a cooling period of about 6 months, an assumption based on multi-annual temperature observations (Patterson and Kachinjika, 1995), this would be equivalent to doubling our modeled $k_p$ to a value of about 130 m $yr^{-1}$. We compare this value to the empirical transfer velocity $k_p^*$ predicted by the air-sea gas exchange model of Wanninkhof (1992):

$$k_p^* = 87.6\left[0.39u^2\left(\frac{Sc}{660}\right)^{-1/2}\right] \tag{4}$$

where $u$ is the average winter wind speed, estimated as 2.5 m $sec^{-1}$ from the data of Patterson and Kachinjika (1995), and $Sc$ is the Schmidt number for CFC-12 in pure water (Zheng *et al.*, 1998), taken as 893 for the epilimnetic winter temperature of 23.5°C. This predicts a $k_p^*$ value of 184 m $yr^{-1}$ which is about one third larger than our estimated value. The modeled and predicted gas transfer rates agree surprisingly well considering that the Wanninkhof (1992) model is an empirical representation of open-ocean conditions with long wind fetch and breaking seawater (rather than freshwater) waves, and also considering the large uncertainties in our choice of a 6-month period of effective gas exchange.

*Table 2.* Exchange times for the Lake Malawi/Nyasa 3-box model under various assumptions. Numbers in parentheses are for layer depth boundaries used by Gonfiantini *et al.* (1979). The first-order CFC-12 degradation rate for anoxic water is given by *j*.

| | | $k_2^{-1}$ [yr] | | $k_3^{-1}$ [yr] | |
|---|---|---|---|---|---|
| This study | $j = 0\ yr^{-1}$ | 3.6 | (4.7) | 18.1 | (20.3) |
| This study | $j = 0.012\ yr^{-1}$ | 3.4 | (4.4) | 15.6 | (17.6) |
| Gonfiantini *et al.* (1979) | | | (4) | | (5) |

We have calculated the mean box model ages for the metalimnion and the hypolimnion. These ages are defined as the time elapsed since the water has last resided in the epilimnion and had therefore been in contact with the atmosphere. With this definition, the water age in the epilimnion is 0, and the mean box model ages of the water in the two subsurface boxes, $\tau_2$ and $\tau_3$, are related to the exchange times by the relationships:

$$\tau_2 = \left(1 + \frac{V_3}{V_2}\right)\frac{1}{k_2} \tag{5}$$

and,

$$\tau_3 = \tau_2 + \frac{1}{k_3} \tag{6}$$

resulting in mean box model ages (Figure 6) of $\bullet_2$ = 8.0 yr and $\bullet_3$ = 26.6 yr in the case of no CFC-12 degradation and $\bullet_2$ = 7.5 yr and $\bullet_3$ = 23.4 yr for $j$ = 0.012 yr.

We have also calculated the CFC-12 apparent ages for the individual water samples collected at the deep Station 940. The apparent age of a water parcel approximates the time elapsed since it last resided in the epilimnion and hence was in contact with the atmosphere through gas exchange. We explicitly define the apparent age as the sampling date minus the date at which the epilimnion had the same CFC-12 concentration as is measured in the sample. We have used the local water temperatures and salinities and the time-dependent epilimnetic CFC-12 saturation derived from our modeled epilimnetic CFC-12 concentration to calculate dates at which the epilimnetic CFC-12 concentration was equal to the measured concentration in a sample. In Figure 6 the Station 940 CFC-12 apparent age profile is shown along with the box model ages for our CFC-12 based calculations. The volumetrically averaged CFC-12 apparent ages of the metalimnion and the hypolimnion are 8.9 yr and 21.1 yr, respectively.

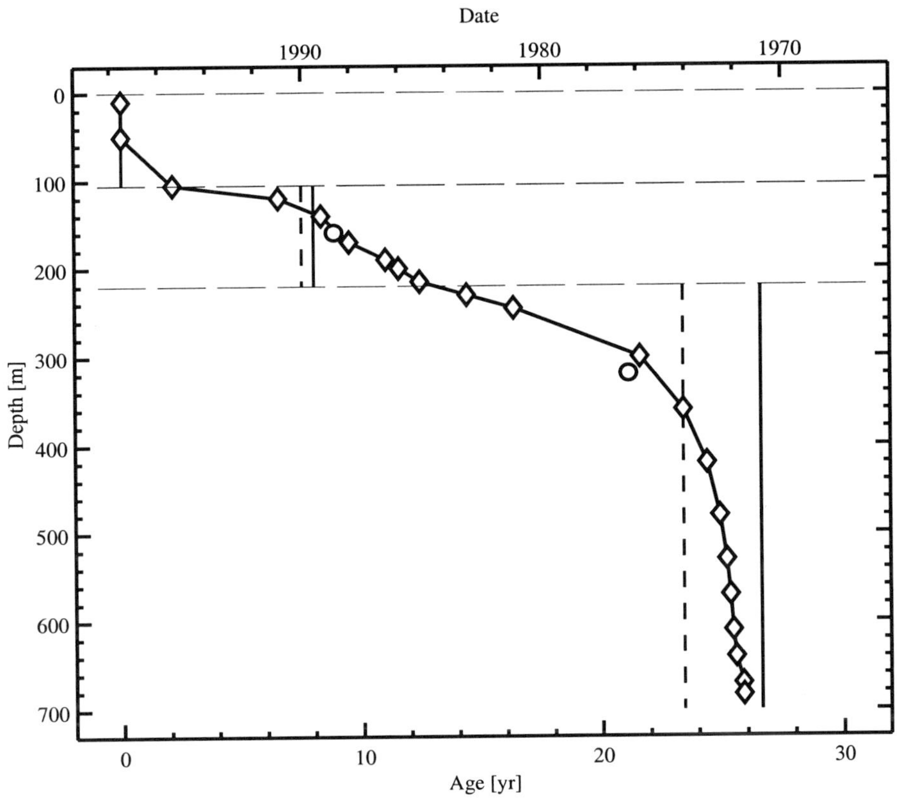


*Figure 6.* Age distribution for Lake Malawi/Nyasa water based on the CFC-12 concentration profile at Station 940. The CFC-12 apparent ages of the water samples (diamonds) and the box model ages for the case of no CFC-12 degradation (vertical solid lines) and a degradation rate of 0.012 $yr^{-1}$ (vertical dashed lines) are shown vs. depth. The volumetrically averaged apparent ages (circles) for the hypolimnion and metalimnion are shown at the vertical volumetric midpoints of these layers. Horizontal dashed lines denote the lake's surface and the boundaries between epilimnion and metalimnion at 105 m and between metalimnion and hypolimnion at 220 m.

The apparent age of a water sample may differ from its mixing age because there is subsurface mixing in Lake Malawi/Nyasa. Here the mixing age is defined as the volume-weighted average of the individual ages of the components that comprise a mixture. In the special case where the CFC-12 atmospheric source function has been effectively linear over the range of ages of a mixture's components, the apparent age approximates the mixing age (Weiss *et al.*, 1991). Otherwise, mixing components from an age range during which the CFC-12 atmospheric increase was accelerating with time (i.e. before the 1970s) will lead to an apparent age for the mixture which is less than the mixing age, while mixtures from an age range during which the atmospheric increase was decelerating with time (i.e. after the mid-1980s) will lead to an apparent age which is greater than the mixing age. The CFC-12 box model ages represent mean mixing ages, since mixing is included in the model. For the hypolimnion, the mean apparent age is lower than these box model ages and it is likely that the hypolimnetic apparent ages underestimate the mixing ages, since older (pre-1970s) individual water parcels undoubtedly contribute significantly to the mixtures in this part of the water column. For the metalimnion, where the mean apparent age is close to the box model ages, the contribution of young water might overestimate the mixing age and thus compensate for the age underestimation from older water which has upwelled into this layer. However it is impossible to quantify the mixing ages of the actual mixtures in the profile without a model which accurately represents the mixing history of the water column.

So far little attempt has been made to identify or quantify the physical processes involved with the ventilation of the deep water and this is not a goal of the present study. In the long-term, turbulent mixing across the pycnocline as simulated by the exchanges between the boxes in our model cannot be the only process of deep-water ventilation. To maintain the observed temperature gradient below the epilimnion, intrusions of cold water from the surface directly into the deeper layers must occur, thus compensating for heat diffusion from above and to a lesser extent for geothermal warming from the sediment. Such intrusions must result in an overall upwelling of the deep water, an hypothesis supported by the convex upward curvature of the vertical profiles of conservative properties such as potential temperature at depths below the influence of the seasonality of the upper water column.

It has long been recognized that ventilation of the Lake Malawi/Nyasa deep water may vary interannually. Beauchamp (1953) hypothesized that the deep water might warm and vertical temperature gradients decrease until an unusually cold `cool season' triggers strong mixing which brings the lake to uniform temperature. As an alternative scenario it is likely that during extreme winters sinking cold water, produced in localized areas, feeds the deep water but that the vertical temperature gradients throughout most of the lake persist. Using the mean of the observed temperatures in the lower two strata at Station 940, our box model calculations result in a yearly temperature increase of the hypolimnion of 0.025°C, which includes a small geothermal warming (Von Herzen and Vacquier, 1967) of about 0.0008 °C $yr^{-1}$. At steady state this warming must be compensated by cold water intrusions. Nevertheless the modeled warming of the hypolimnion is of similar magnitude as the observed temperature increases in the Lake Malawi/Nyasa deep water over the past six decades (Vollmer *et al.*, in preparation), suggesting that the hypolimnetic heat budget may not have been at steady state during this period.

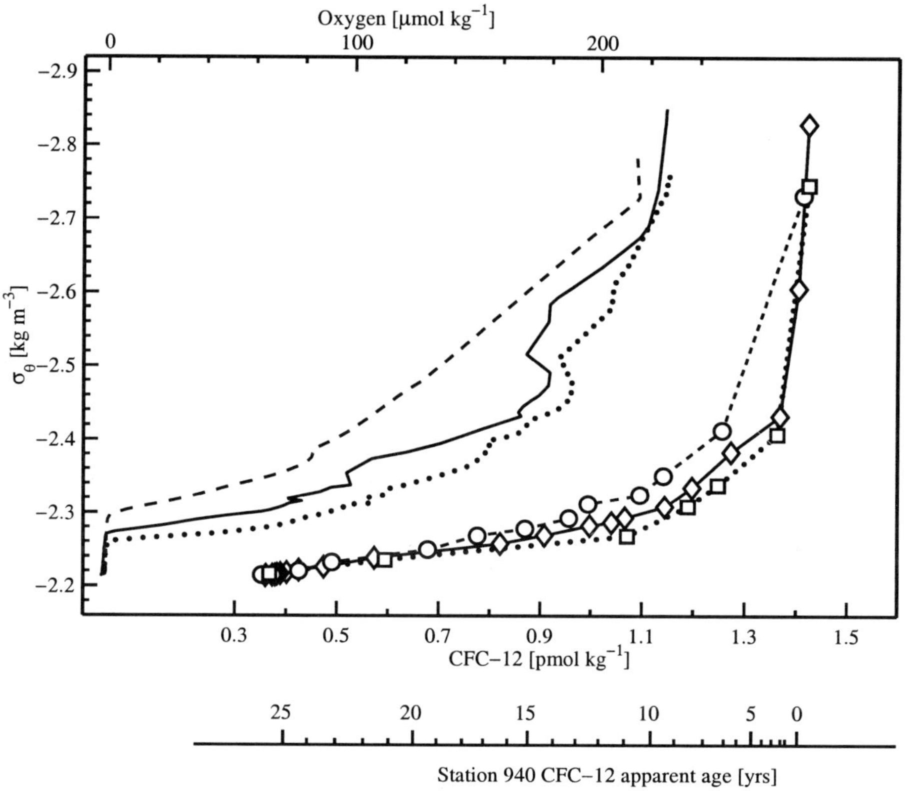


*Figure 7.* CFC-12 and dissolved oxygen concentrations plotted against potential density anomaly ($\sigma_\theta$) for Stations 915 (dotted lines), 940 (solid lines) and 924 (dashed lines). CFC-12 concentrations show a northward decrease for the metalimnion and the upper hypolimnion hence a northward increase in apparent age and therefore suggest that deep-water ventilation is dominant in the south. The northward decrease in dissolved oxygen on equipotential density layers supports this finding.

The modeled warming of the metalimnion depends strongly on the mean temperature assumed for the epilimnion. For a scenario in which the epilimnion has a mean temperature lower limit of 23.5°C the metalimnion would warm by 0.08 °C yr$^{-1}$ including geothermal warming of 0.002 °C yr$^{-1}$. This or any warming based on calculations using higher epilimnetic temperatures, is too large to be maintained over decades and it is therefore likely that this layer receives cold water intrusions more frequently than the hypolimnion. Interannual variability in ventilation rates is also a possible explanation for the differences obtained between this and the earlier tracer study (Gonfiantini *et al.*, 1979). It may well be that deep-water ventilation was stronger in the 1960's when atmospheric tritium concentrations were largest while ventilation has reduced during the last decade when atmospheric CFC-12 concentrations peaked.

Several mechanisms of cold water intrusion have been suggested, such as fluvial discharge, near-shore and open-water evaporative cooling during winter and turbidity currents (Halfman, 1993). The existence of such processes could be evidenced by a downward increase in CFC-12 concentrations near the lake floor which we do not find. However, without identifying specific processes, we find evidence that deep-water ventilation must primarily have occurred in the more southern parts of the lake in the past years. In Figure 7 we plot CFC-12 concentrations for our three stations against the potential density anomaly, rather than against depth, to eliminate effects of local vertical displacements of the water column by internal waves or upwelling. Significant differences in CFC-12 concentrations on isopycnals are observed in the metalimnion and upper hypolimnion with concentrations decreasing northward. The dissolved oxygen data also show northward concentration decreases and a corresponding shift of the oxic-anoxic boundary to lower density levels. These and the contoured oxygen data presented in Figure 3c support the CFC-12 based findings of older water in the north, although biases in the oxygen data can occur due to latitudinal changes in biological productivity and resulting changes in remineralization rates at depth.

The finding of a southern source of young deep water suggests that throughout the last decades deep plumes of fluvial discharge have not significantly contributed to the deep-water ventilation. Deep-water formation by river discharge is expected to be strongest in the northern part of the lake where the river inflows are larger and their water temperatures are lower than in the south (Kingdon *et al.*, 1999). This would lead to a northward decrease in deep-water ventilation age which is in contradiction to our findings from the CFC-12 observations.

## 6. CONCLUSION

Our 3-box model for CFC-12 in 1997 yields an exchange time between the hypolimnion and the metalimnion which is 2.2 to 2.7 times longer than that estimated by Gonfiantini *et al.* (1979) based on the tritium distribution they measured in 1976. These differences have significant effects in calculations of nutrient budgets and the residence times of compounds such as pollutants in the deep water. The reasons for this difference may be related to the incomplete representation of the actual mixing processes in this box model, including the transfer of these tracers from the atmosphere to the lake. These mixing rates must be treated with caution, since this 3-box model clearly oversimplifies the actual ventilation processes occurring in Lake Malawi/Nyasa. In fact, as noted earlier, this model cannot support steady-state temperature conditions because it does not allow for the direct intrusion of cold water from the surface into the hypolimnion. However it is likely that mixing conditions in Lake Malawi/Nyasa may experience variability on times scales of years to decades. A scenario of stagnant mixing conditions during the recent years when atmospheric CFC-12 concentrations were high or increased mixing during the early 1960's when atmospheric tritium concentrations were high, is a possible explanation for the different results obtained from these two studies. It is hoped that the results of the analyses of our helium-3 and tritium samples, collected simultaneously with our CFC-12 samples, as well as an improved model representation, will lead to a more realistic characterization of deep-water ventilation rates. In addition, better characterization and quantification of mixing processes in Lake Malawi/Nyasa, as well as continuing tracer observations, are necessary in order to improve our understanding of deep-water renewal and its long-term variability.

## ACKNOWLEDGMENTS

We thank the captain and the crew of R/V Usipa for their support of the expedition work, A. Krause for building the ampoule sampling devices, the group of W. Roether at the Institute of Environmental Physics in Bremen for their help with the ampoule technique, S. Walker for his help with the modeling calculations, and R. Hecky and the Canadian International Development Agency (CIDA) for their encouragement and financial support. This work has also been carried out under the auspices of the International Decade for the East African Lakes (IDEAL) program.

## REFERENCES

Aeschbach-Hertig W., Hofer M., Schmid M., Kipfer R., and Imboden D.M. (2001) The physical structure and dynamics of a deep, meromictic crater lake (Lac Pavin, France). *Hydrobio.*, in press.

Beauchamp R.S.A. (1953) Hydrological data from Lake Nyasa. *J. Ecol.* **41**, 226-239.

Bootsma H.A. and Hecky R.E. (1999a) Nutrient cycling in Lake Malawi/Nyasa, in Bootsma and Hecky (1999b), pp. 215-241.

Bootsma H.A. and Hecky R.E. (eds.) (1999b) Water Quality Report; Lake Malawi Biodiversity Conservation Project. Southern African Development Community (SADC), Global Environmental Facility (GEF).

Bullister J.L. and Lee B.S. (1995) Chlorofluorocarbon-11 removal in anoxic marine waters. *Geophys. Res. Lett.* **22**, 1893–1896.

Bullister J.L. and Weiss R.F. (1988) Determination of $CCl_3F$ and $CCl_2F_2$ in seawater and air. *Deep Sea Res.* **35**, 839–853.

Bulsiewicz K., Rose H., Klatt O., Putzka A., and Roether W. (1998) A capillary-column chromatographic system for efficient chlorofluorocarbon measurement in ocean waters. *J. Geophys. Res.* **103**, 15,959–15,970.

Busenberg E. and Plummer L.N. (1992) Use of chlorofluorocarbons ($CCl_3F$ and $CCl_2F_2$) as hydrologic tracers and age-dating tools: the alluvium and terrace system of Central Oklahoma. *Water Resour. Res.* **28**, 2257–2283.

Chen C.T.A. and Millero F.J. (1986) Precise thermodynamic properties for natural waters covering only the limnological range. *Limnol. Oceanogr.* **31**, 657–662.

Cook P.G., Solomon D.K., Plummer L.N., Busenberg E., and Schiff S.L. (1995) Chlorofluorocarbons as tracers of groundwater transport processes in a shallow, silty sand aquifer. *Water Resour. Res.* **31**, 425–434.

Doney S.C. and Bullister J.L. (1992) A chlorofluorocarbon section in the eastern North Atlantic. *Deep Sea Res.* **39**, 1857–1883.

Eccles D.H. (1974) An outline of the physical limnology of Lake Malawi (Lake Nyasa). *Limnol. Oceanogr.* **19**, 730–742.

Gammon R.H., Cline J., and Wisegarver D. (1982) Chlorofluoromethanes in the northeast Pacific Ocean: Measured vertical distributions and application as transient tracers of upper ocean mixing. *J. Geophys. Res.* **87**, 9441–9454.

Gonfiantini R., Zuppi G.M., Eccles D.H., and Ferro W. (1979) Isotope investigation of Lake Malawi, in *Isotopes in Lake Studies*, International Atomic Energy Agency, Vienna. pp. 195–207.

Halfman J.D. (1993) Water column characteristics from modern CTD data, Lake Malawi, Africa. *J. Great Lakes Res.* **19**, 512–520.

Hamblin P.F., Bootsma H.A., and Hecky R.E. (1999) Modeling nutrient upwelling in Lake Malawi/Nyasa, in Bootsma and Hecky (1999b), pp. 123–141.

Hutchinson G.E. (1957) *A Treatise on Limnology, Vol I.* Geography, Physics and Chemistry. Wiley, New York.

Kingdon M.J., Bootsma H.A., Mwita J., Mwichande B., and Hecky R.E. (1999) River discharge and water quality, in Bootsma and Hecky (1999b), pp. 29–84.

Klatt O. (1997) Entwicklungen am gaschromatographischen FCKW-Messsystem. Diploma thesis, University of Bremen.

Lovley D.R. and Woodward J.C. (1992) Consumption of Freons CFC-11 and CFC-12 by anaerobic sediments and soils. *Environ. Sci. Technol.* **26**, 925–929.

Oster H., Sonntag C., and Münnich K.O. (1996) Groundwater age dating with chlorofluorocarbons. *Water Resour. Res.* **32**, 2989–3001.

Patterson G. and Kachinjika O. (1995) Limnology and phytoplankton ecology, in A. Menz (ed.), *The Fishery Potential and Productivity of the Pelagic Zone of Lake Malawi/Niassa*, Natural Resource Institute, Chatham, UK. pp. 1–67.

Patterson G., Wooster M.J., and Sear C.B. (1997) Satellite-derived surface temperatures and the interpretation of the 3-dimensional structure of Lake Malawi, Africa: the presence of a profile-bound density current and the persistence of thermal stratification. *Verh. Internat. Verein. Limnol.* **26**, 252–255.

Pickart R.S., Hogg N.G., and Smethie W.M. (1989) Determining the strength of the deep western boundary current using the chlorofluoromethane ratio. *J. Phys. Oceanogr.* **19**, 940–951.

Plummer L.N., Busenberg E., Drenkard S., Schlosser P., Ekwurzel B., Weppernig R., McConnell J.B., and Michel R.L. (1998) Flow of river water into a karstic limestone aquifer —2. dating the young fraction in groundwater mixtures in the Upper Floridan aquifer near Valdosta, Georgia. *App. Geochem.* **13**, 1017–1043.

Prinn R.G., Weiss R.F., Fraser P.J., Simmonds P.G., Cunnold D.M., Alyea F.N., O'Doherty S., Salameh P., Miller B.R., Huang J., Wang R.H.J., Hartley D.E., Harth C., Steele L.P., Sturrock G., Midgley P.M., and McCulloch A. (2000) A history of chemically and radiatively important gases in air deduced from ALE/GAGE/AGAGE. *J. Geophys. Res.* **105**, 17,751–17,792.

Shapiro S.D., Schlosser P., Smethie Jr W.M., and Stute M. (1997) The use of $^3$H and tritiogenic $^3$He to determine CFC degradation and vertical mixing rates in Framvaren Fjord, Norway. *Mar. Chem.* **59**, 141–157.

Tanhua T. (1997) Halogenated substances as marine tracers. PhD thesis, Göteborg University, Göteborg.

Von Herzen R.P. and Vacquier V. (1967) Terrestrial heat flow in Lake Malawi, Africa. *J. Geophys. Res.* **72**, 4221–4226.

Walker S.J., Weiss R.F., and Salameh P.K. (2000) Reconstructed histories of the annual mean atmospheric mole fractions for the halocarbons CFC-11, CFC-12, CFC-113, and carbon tetrachloride. *J. Geophys. Res.* **105**, 14,285–14,296.

Wanninkhof R. (1992) Relationship between wind speed and gas exchange over the ocean. *J. Geophys. Res.* **97**, 7373–7382.

Warner M.J. and Weiss R.F. (1985) Solubilities of chlorofluorocarbons 11 and 12 in water and seawater. *Deep Sea Res.* **32**, 1485–1497.

Weiss R.F. (1970) The solubility of nitrogen, oxygen and argon in water and seawater. *Deep Sea Res.* **17**, 721–735.

Weiss R.F., Bullister J.L., Gammon R.H., and Warner M.J. (1985) Atmospheric chlorofluoromethanes in the deep equatorial Atlantic. *Nature* **314**, 608–610.

Weiss R.F., Carmack E.C., and Koropalov V.M. (1991) Deep-water renewal and biological production in Lake Baikal. *Nature* **349**, 665–669.

Wüest A., Piepke G., and Halfman J.D. (1996) Combined effects of dissolved solids and temperature on the density stratification of Lake Malawi, in T.C. Johnson and E.O. Odada (eds.), *The Limnology, Climatology and Paleoclimatology of the East African Lakes*, Gordon and Breach, Amsterdam. pp. 183–202.

Zheng M., DeBruyn W.J., and Saltzman E.S. (1998) Measurements of the diffusion coefficients of CFC-11 and CFC-12 in pure water and seawater. *J. Geophys. Res.* **103**, 1375–1379.

# APPLICATION OF SATELLITE AVHRR TO WATER BALANCE, MIXING DYNAMICS, AND THE CHEMISTRY OF LAKE EDWARD, EAST AFRICA

JOHN T. LEHMAN

*Department of Biology and Center for Great Lakes and Aquatic Sciences, University of Michigan, Ann Arbor, MI 48109-1048, USA*

**ABSTRACT**

Surface heat balance and evaporation rates for Lake Edward were calculated as part of a model for mixed layer dynamics. Evaporation rates were combined with rainfall and fluvial income to estimate water balance. Heat and water balance were subjected to sensitivity analyses to evaluate the relative effects of variations in minimum and maximum air temperature, humidity, wind speed, cloud cover, and rainfall. Cloud cover and wind speed emerged as the most influential factors for lake temperature, evaporation, and mixing dynamics. Historical variations in ionic strength of Lake Edward surface water underscore sensitivity of the lake to climate. Major ion composition of the lake is incongruous with chemical inputs reported for streams descending from the Ruwenzoris and the highlands of southwestern Uganda. Mass balance calculations reveal that major sources of many elements are unaccounted for by present measurements, and point to the dominance of as yet unstudied stream sources entering the lake from the Virunga volcano region at the south. Based on model calculations, the chemical composition of water sources from the Virunga region is predicted a priori. Lake Edward is remarkable because despite episodic richness of biogenic Si in its sediments, diatoms are not common at present, although they are more prevalent in the south than in the north. One reason for episodic success by diatoms may be linked to changing patterns of lake mixing. Seasonal mixing dynamics of the lake were inferred from weather records and radiation flux measured from Advanced Very High Resolution Radiometer (AVHRR)



*E.O. Odada and D.O. Olago (eds.),*
*The East African Great Lakes: Limnology, Palaeolimnology and Biodiversity,* 235–260.

satellite data. SSTs inferred from AVHRR imagery were used to test one hypothesis for the diatom distribution- namely that increased water column stability in the north and a shoaling mixed layer exaggerate the loss rates by sinking and depress the supply of nutrients from below. Variations in water balance for the lake are consistent with observations of significant changes in lake water chemistry during the 20$^{th}$ century.

# 1. INTRODUCTION

The Great Lakes of East Africa are rich sources of information about climatic variations in recent times as well as over long periods of human history and prehistory (Johnson and Odada, 1996). Paleolimnology has revealed that Lake Victoria, a major source of the White Nile, was dry during the last glacial period, and began to overflow comparatively recently (Johnson *et al.,* 1996; Johnson *et al.,* 1998). The great African lakes are sensitive indicators of climate, and several of them offer long sedimentary records. One such lake is Edward, a half graben on the equator in the Western Rift at the border of Uganda and the Republic of the Congo. Despite potentially rich geological history, it is little studied and remains largely unknown in many respects.

Water balance and mass budgets for chemical constituents provide essential framework for climate studies. No such framework has previously been assembled for Lake Edward, but there are indications that the lake is dynamic and responsive to climate variations. Surveys of its water chemistry during the past 40 years have identified two-fold changes in total ionic composition (Lehman *et al.,* 1998). The lake is noteworthy for its relatively high ionic strength despite being one of the headwater sources of the White Nile. The other headwater source, Lake Victoria, is of markedly lower ionic strength. Talling and Talling (1965) remarked on its unusual ion ratios, particularly its high contents of Mg and K. They ascribed its unusual chemistry to basic volcanic source rocks in the catchment. Degens (1973) stated that Edward, and all of the lakes of the African Western Rift, were influenced by hydrothermal inputs. If hydrothermal influences are indeed pervasive, there could be anomalies in material and possibly even heat budgets for the lake.

This study assembles meteorological, hydrological, and chemical data for Lake Edward and its catchment and organizes them into a coherent model for the lake. The model is based on principles of mass and heat balance constrained by existing data. The relative influence of different climate variables on water balance and mixing conditions is assessed. The purpose of the study was to collect, review, and analyze existing data so as to identify gaps in existing knowledge, as well as to uncover potential inconsistencies in the existing data record.

# 2. STUDY SITE

Lake Edward (Figure 1) receives drainage from the southwestern Ruwenzoris to the north, the highlands of Rwanda and SW Uganda (Kigezi district) to the east, and the Virunga volcanoes to the south. Net inflow occurs through the Kazinga Channel from a shallow but optically deep and hypereutrophic northeastern extension, Lake George. Lake George and its surrounding catchments were studied during the 1960s as part of a Royal Society component of the International Biological Programme (IBP) (Burgis *et al.,* 1972). Lake Edward itself is anoxic below ca. 40 m to its maximum depth of ca. 120 m, but it is only weakly chemically stratified. The temperature difference between surface and deep waters of the lake is only about 1 °C maximum, with surface temperatures around 26 °C. There is thus reason to believe that the lake circulates seasonally, but no time series of observations exist to confirm the deduction. The stratigraphic record in its sediments extends for at least hundreds, and probably a few thousand, meters (Degens, 1973).

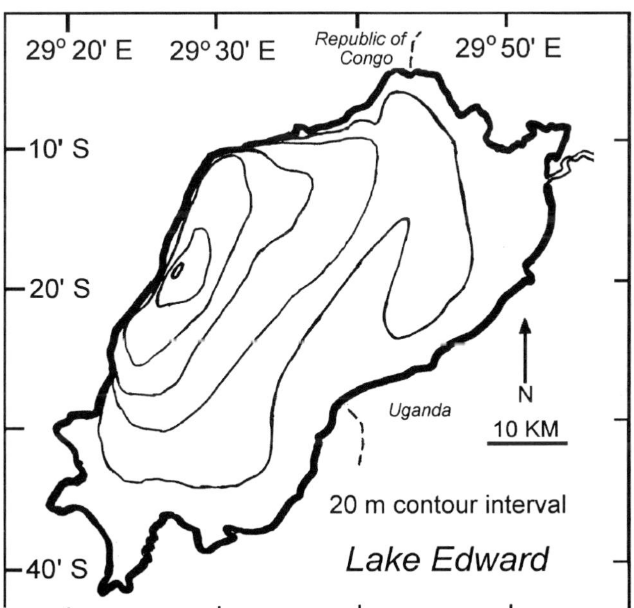


*Figure 1.* Lake Edward bathymetry, redrawn from original by T. Lærdal.

The sediments of Lake Edward are a rich diatom ooze, with biogenic silica varying between 5% and 50% by mass of the sediment (Johnson and Kelts, personal communication; J. Russell *et al.,* ms. in preparation). Authogenic calcite constitutes about 10% by mass on average, with marked temporal variation. Recently, the diatom flora has been weakly developed in the lake, particularly in the north (Hecky and Kling, 1987), although Worthington (1932) mentioned a conspicuous increase in diatoms along a transect from the Kazinga Channel into northern Lake Edward in 1931. Diatoms were nearly absent from northern lake sites sampled in 1995, consistent with diminished diatom representation in surface sediments (Lehman *et al.,* 1998). Dissolved silicate is abundant at present (150 μM SRSi, Lehman *et al.,* 1998), and is not being reduced to low levels by the biogenic precipitation typical of many lakes. The enormous concentrations and variability of fossil diatoms suggest that the present condition is aberrant, but not unique because wide variation in biogenic Si is a feature of the stratigraphic record.

## 3. DATA SOURCES

Analysis of radiation balance components requires information about insolation, cloud cover, and surface meteorology at appropriate geographic scales. Cloud cover and top-of-atmosphere short wave radiation were extracted from radiation budget estimates produced by the U.S. National Oceanic and Atmospheric Administration (NOAA) as part of its global radiation budget measurements. The radiation budget relies on sensor data from polar orbiting environmental satellites to compute short wave and long wave radiation flux from RADRET algorithms applied to Level 1b satellite data (I. Guch, NOAA NESDIS, pers. comm.). The radiation budget flux measurements are described at http://www2.ncdc.noaa.gov/docs/klm/html/c9/sec9-3.htm. Retrieval level information from NOAA-14 and NOAA-15 AVHRR sensors were obtained daily by special request for both ascending and descending orbital paths over East Africa (4° N to 15° S, 28° E to 38° E). This remote sensing scheme provided 4 scenes each day, 2 during day and 2 during night.

Data included geographically-referenced estimates for both available and absorbed short wave radiation (W $m^{-2}$), outgoing long wave radiation (OLR, W $m^{-2}$), as well as 12 daytime scene index values that incorporate cloudiness on a four point scale (Ruff and Gruber, 1988).

Data corresponding to the surface of Lake Edward were extracted from the larger set by selecting points within a prescribed search field with latitude and longitude boundaries that described a parallelogram. For Lake Edward, admissible Longitude was 29.3° E to 29.8° E. Within this range, admissible Latitude was constrained to be greater than (Longitude–30.2°) but less than (Longitude–29.7°). Binary data received from NOAA in sequential EDGEIS file format were extracted by software written by the author in Visual Basic.

Daily maximum and minimum air temperature, wind speed, barometric pressure, and dew point were extracted from Global Summary of Day climate observations from 1994 to 1998 for Kasese, Uganda (Stn 636740: 0° 11' N, 30° 6' E, elevation 961 m) archived by the National Climate Data Center (NCDC). Data may be accessed by way of http://www.ncdc.noaa.gov/cgi-bin/res40.pl?page=gsod.html. Mean values for each month were calculated for the reference period and the monthly means were used in model calculations.

Precipitation estimates for the geographic region around Lake Edward were obtained from NCDC archives in the Global Historical Climate Network version 2 (URL: http://www.ncdc.noaa.gov/cgi-bin/res40.pl?page=ghcn.html). Mean monthly precipitation estimates were obtained for ten stations ranging from 1.3° S to 0.8° N, and from 29.4° to 30.6° E, at elevations ranging from 920 to 2300 m.

## 3.1 Conceptual Overview

Changes in heat content and seasonal mixing depth of lakes and surface ocean waters are consequences of heat and momentum fluxes (Price, 1981; Monismith, 1985; Spigel *et al.,* 1986; Spigel, 1980; Spigel and Imberger, 1980; Imberger and Patterson, 1990). Heat gained from irradiance is lost to the atmosphere over time. The heat loss mechanisms are by long wave radiation, conduction, and evaporation; outflow losses of heat must be considered in some cases. Net heat loss by long wave radiation depends on the temperature difference between surface water and the sky as well as humidity and cloud cover (Keijman, 1974).

Variation in the water vapor or greenhouse gas content of the atmosphere can change the net gray body (i.e., emissivity is less than 1) long wavelength radiative heat flux from the lake surface. Heat loss by conduction is formulated as a linear function of wind speed and the temperature gradient from lake surface to air, with bulk transfer coefficients that vary according to stability, or the relative temperature of lake and air (Croley, 1989). Latent heat loss by evaporation is linked with conduction loss and other scalar fluxes by the concept of the Bowen ratio, using humidity differences in the air compared with a water saturated atmosphere at equilibrium with the lake surface temperature. Dalton is credited with being the first person to recognize that the rate of evaporation of water is proportional to the vapor pressure difference between the air and the saturated vapor layer at the water surface. Either decreased wind speed or increased relative humidity in the bulk atmosphere will decrease this heat flux. The effect of wind action is to replace the vapor-saturated boundary layer near the water surface with air that is not saturated with water vapor.

The heat balance of surface waters is fundamentally self-correcting; a temporary decrease in heat loss will be compensated for by elevated water temperature, with resultantly increased heat flux by all three heat loss mechanisms. When each mechanism is considered in isolation, if the wind slows down, or if the air temperature rises, or if the humidity increases, or if the sun shines more brightly, in each case the surface warms.

# 4.0 HYDROLOGY AND BALANCES FOR HEAT, WATER, AND CHEMICAL CONSTITUENTS

## 4.1 Heat Balance

Lake heat content is considered to be the result of a balance among short wave and long wave radiative fluxes, evaporation, and heat conduction. The energy balance model is

$$\Delta \text{ Lake Heat} = SW_{net} + LW_{net} - LH - SH \tag{1}$$

where SW is short wave radiation, LW is long wave radiation, LH is latent heat loss by evaporation and SH is sensible heat loss by free and forced convection. The scaling constants in equations were chosen to permit calculation of heat flux in cal $cm^{-2}$ $d^{-1}$. Net short wave radiation received into the lake water is calculated as insolation measured at the ground ($SW_0$) reduced by lake albedo ($\alpha$):

$$SW_{net} = (1 - \alpha)\, SW_0 \tag{2}$$

where $\alpha = 0.07$ (Yin and Nicholson, 1998). Incident $SW_0$ is calculated from satellite-derived measurements of available SW at the top of the atmosphere ($SW_{available}$) by using a correction for cloud effect (Yin and Nicholson, 1998):

$$SW_0 = SW_{available}\,(0.803 - 0.34\,C - 0.458\,C^2) \tag{3}$$

where C is the fraction of sky occluded by clouds. The NOAA radiation budget calculations generated mean daily $SW_{available}$.

Net heat loss by long wave radiation is estimated as gray body radiation upward from the lake surface combined with counter-radiation of LW radiation downward from the atmosphere, using the approach introduced by Keijman (1974) and subsequently adopted for modeling evaporation in the St. Lawrence Great Lakes (Croley, 1989):

$$LW_{net} = \sigma\, T_{Air}^{\,4}\,(0.53 + 0.065\, e_{Air}^{\,0.5})(p + (1 - p)(1 - C)) - \varepsilon\, \sigma\, T_L^{\,4} \tag{4}$$

where $\varepsilon$ is emissivity (= 0.97 Strub and Powell, 1987), $\sigma$ is the Stefan-Boltzmann constant, $T_L$ and $T_{Air}$ are lake surface and air temperature (K), $e_{Air}$ is atmosphere vapor pressure (mb), and p is a dimensionless parameter.

Heat losses by evaporation and conduction are computed from heat and vapor pressure gradients using bulk coefficients estimated for the turbulent atmospheric boundary layer over the lake. Two different methods were used to estimate the transfer coefficients. In one case (Model 1), bulk transfer coefficients were calculated by the iterative algorithm published by Croley (1989, his eqs. 1 to 10).

This method involved application of Monin and Obukhov's similarity hypothesis to wind speed and air temperature:

$$U = U^* / \kappa \, [\ln(Z/Z_L) - S_1] \tag{5}$$

$$T - T_L = T^*[\ln(Z/Z_L) - S_2] \tag{6}$$

Where U is mean wind speed at reference height Z above the surface, $U^*$ is friction velocity, $\kappa$ is von Karman's constant, $Z_L$ is roughness length, T is potential temperature at reference height, $T_L$ is potential temperature at $Z_L$, and $T^*$ is the scaling temperature defined as

$$T^* = Q_H /(r \, C_P \, \kappa \, U^*) \tag{7}$$

where $C_P$ is the specific heat of air at constant pressure, $Q_H$ is the turbulent heat flux to the surface, and r is air density.

The second approach (Model 2) was to approximate evaporation, or latent heat flux, by the bulk formula proposed by Maidment (1993):

$$\text{Evap (cm d}^{-1}) = 0.02909(\text{Lake Area, m}^2)^{-0.05} \, (e_L - e_{Air}) \tag{8}$$

Saturation vapor pressure at the temperature of the lake surface was defined (Maidment, 1993) as

$$e_L = 6.11 \exp(17.3 \, T_L/(T_L+237.3)) \tag{9}$$

Heat balance is thus a function of $SW_0$, $T_L$, $T_{Air}$, $e_{Air}$, C, and wind speed. Table 1 summarizes the key parameters and their definitions.

## 4.2 Theory of the Surface Mixed Layer

Modern synthesis of theory about the surface mixed layer is typically credited to Niiler and Kraus (1977). They drew on earlier studies to summarize how wind stress and buoyancy forces erode an existing thermocline. They are credited with solving the problem of the seasonal thermocline. Subsequent studies have extended the theory of the seasonal thermocline to the diurnal mixed layer.

Solar radiation that penetrates a lake surface is ultimately absorbed and converted to heat. Because the specific volume of water, and alternatively its density, are nonlinear functions of temperature, the absorption of heat causes buoyancy changes. Light attenuates differentially according to wavelength. The solar irradiance at longer wavelengths (IR) is absorbed within the top 2 meters or less by water itself. Light at shorter wavelengths, particularly 400-700 nm (PAR: photosynthetically active radiation), which is the photosynthetic and visible part of the spectrum, penetrates more deeply. Its attenuation rate varies strongly with dissolved and particulate matter in the water. In particular, PAR penetration varies inversely with algal biomass.

Heat loss occurs at the lake surface. The result is a cooled surface film. At temperatures greater than 4°C, the cooling generates convection by water that becomes denser than the water just below. The kinetic energy of resulting convection cells adds to kinetic energy produced by wind stress. The two sources of kinetic energy provide the turbulence velocity shears that work against any existing stable density gradient at the base of the turbulent mixed layer. During daytime, however, light absorption is a buoyancy generating mechanism that acts as a sink for this kinetic energy.

*Table 1.* Parameter definitions for heat balance model applied to Lake Edward.

| Parameter | Definition | Formula | Units |
|---|---|---|---|
| $T_L$ | lake surface temperature | | °C |
| $T_A$ | air temperature | | °C |
| $T_{DP}$ | dewpoint temperature | | °C |
| U | mean wind speed at 2 m | | $m\ s^{-1}$ |
| C | Fraction cloud cover | | none |
| $P_{STA}$ | Atmospheric pressure at lake | | mb |
| $\sigma$ | Stefan-Boltzmann constant | $8.26 \times 10^{-11}$ | $cal\ cm^{-2}\ min^{-1}$ |
| $\varepsilon$ | water emissivity | 0.97 | none |
| $\alpha$ | lake albedo | 0.07 | none |
| p | effect of clouds on $LW_{down}$ | 0.6 | none |
| $C_P$ | specific heat of dry air | 0.2403 | $cal\ g^{-1}\ K^{-1}$ |
| $SW_{available}$ | SW at top of atmosphere | | $cal\ cm^{-2}\ d^{-1}$ |
| A | lake surface area | $2325 \times 10^6$ | $m^2$ |
| $SW_0$ | Incident short wave radiation | $SW_{available}\ (0.803 - 0.34\ C - 0.458\ C^2)$ | $cal\ cm^{-2}\ d^{-1}$ |
| $e_A$ | air vapor pressure | $6.11\ \exp(17.3\ T_{DP}/(T_{DP}+237.3))$ | mb |
| $e_L$ | saturation vapor pressure at | $6.11\ \exp(17.3\ T_L/(T_L+237.3))$ | mb |
| LVT | Latent heat of vaporization at | $597.3 - 0.564\ T_L$ | $cal\ g^{-1}$ |
| B | Bowen ratio | $C_P(T_L-T_A)/LVT(e_L - e_A)$ | none |
| EVAP | Evaporation | $0.02909\ U\ A^{-0.05}(e_L - e_A)$ | $cm\ d^{-1}$ |
| LH | Latent heat flux | $LVT \times EVAP$ | $cal\ cm^{-2}\ d^{-1}$ |
| SH | sensible heat flux | $LH \times B$ | $cal\ cm^{-2}\ d^{-1}$ |
| $LW_{net}$ | net long wave radiation to | $\sigma(T_A+273.15)^4(0.53+0.065e_{Air}{}^{0.}$ | $cal\ cm^{-2}\ d^{-1}$ |
| $SW_{net}$ | net shortwave radiation to | $(1-\alpha)\ SW_0$ | $cal\ cm^{-2}\ d^{-1}$ |
| Balance | | SWnet + LWnet –LH – SH | $cal\ cm^{-2}\ d^{-1}$ |

Light attenuation is strongly nonlinear, such that more light is absorbed at the top half of any water layer than in the bottom half of the same layer. The result is continual production of positive buoyancy while the sun is shining. Turbulence kinetic energy is either partially or completely expended by working to mix the water in opposition to these buoyancy forces. As a result, during full sunlight hours the diurnal mixed layer shoals and mixing is contained much closer to the surface than during the night. Daytime mixed layer thickness can be predicted from irradiance, light attenuation, and the kinetic energy from wind and surface cooling.

Throughout the night, the mixed layer deepens as large convection cells, forced by the wind if it is present, entrain cool water at the base of the layer. The rate of layer deepening is set by the "generalized entrainment law" (Sherman *et al.,* 1978), which considers available kinetic energy, convection cell size, and the potential energy represented by stable density stratification that opposes the entrainment forces. It takes energy to pull cooler, dense water from the bottom of the mixed layer and lift it upward against gravity. As a result of the overnight process driven by surface cooling, the maximum extent of vertical mixing is typically reached at or around dawn.

## 4.3 Diffusion Processes Below the Mixed Layer

Mixing processes below the turbulent mixed layer were modeled for this study as eddy diffusion, using a constant diffusion coefficient of 0.02 $cm^2\ s^{-1}$. Imberger and Patterson (1990) cite this value as a lower limit of diffusion rates below the mixed layer, and it is probably most appropriate to the region of high gradient diffusive fluxes immediately below the surface layer. This coefficient probably underestimates eddy diffusion in the bulk of the hypolimnion, but is inconsequential to mixed layer dynamics and surface heat flux.

## 4.4 Sensitivity Analysis

Sensitivity analysis for the heat balance model was conducted by specifying a "standard" climate condition equal to empirical mean monthly conditions for air temperature, dew point, insolation, station barometric pressure, fraction cloud cover and wind speed. Day-of-month conditions were obtained by linear interpolation of the monthly means, assuming that the means represented conditions at mid-month. Simulations over an entire year were conducted with a horizontally averaged mixed-layer model using a time step of one hour.

Air temperature was simulated from the daily maximum and minimum records by assuming that minimum temperature occurred at sunrise, and that maximum temperature occurred midway between noon and sunset. Air temperatures at other times were obtained by linear interpolation. The initial condition for lake temperature was specified by running the model for one year under the “standard” climate scenario starting with an isothermal water column at 25.5 °C. The final calculated temperature distribution at the end of the year was stipulated as the initial state for all further simulations.

A reference, or Base simulation was conducted by running the model for one year under the standard mean climate conditions as specified in Table 2. Then, each climate variable was independently both increased and decreased by finite steps, and the response variables were recalculated by annual simulations for each perturbation. The magnitudes of the step perturbations were in proportion to the statistical variability of empirical climate data. Specifically, the standard deviation of each climate variable in the Kasese data set was calculated by month for data pooled across years. For each month and each climate variable, a day to day standard deviation ($SD_{i,m}$) was computed where i indicates the climate variable and m represents month. An index for variability of monthly means was defined as double the expected 30-day random sample standard error from the pooled data:

$$\text{Dev} = 2 \times SD_{i,m} / \sqrt{30} \tag{10}$$

This index, when either added or subtracted from the mean value, was taken to approximate a 95% confidence interval for year-to-year variation in mean climate values for any given month.

*Table 2.* Mean values by month for climate variables, and magnitudes of perturbations (Dev, from Eq. 10) used for sensitivity analysis.

| | Air $T_{MAX}$ °C | | Air $T_{MIN}$ °C | | $T_{DP}$ °C | | U m $s^{-1}$ | | Cloud fraction | |
|---|---|---|---|---|---|---|---|---|---|---|
| | Mean | Dev | Mean | Dev | Mean | Dev | Mean | Dev | Mean | Dev |
| Jan | 29.9 | 0.77 | 18.2 | 1.01 | 19.01 | 0.51 | 2.41 | 0.29 | 0.413 | 0.105 |
| Feb | 29.5 | 0.91 | 17.8 | 0.70 | 17.81 | 0.72 | 2.14 | 0.29 | 0.305 | 0.081 |
| Mar | 28.6 | 1.28 | 18.6 | 0.68 | 19.02 | 0.36 | 2.48 | 0.44 | 0.481 | 0.100 |
| Apr | 28.7 | 0.96 | 19.2 | 0.95 | 19.73 | 0.47 | 2.14 | 0.25 | 0.333 | 0.082 |
| May | 29.1 | 0.93 | 18.1 | 0.78 | 19.66 | 0.40 | 2.00 | 0.63 | 0.257 | 0.078 |
| Jun | 28.3 | 0.90 | 18.0 | 0.75 | 19.00 | 0.35 | 1.65 | 0.48 | 0.292 | 0.079 |
| Jul | 28.0 | 1.04 | 19.1 | 1.55 | 18.15 | 0.54 | 1.66 | 0.24 | 0.318 | 0.084 |
| Aug | 29.8 | 0.78 | 17.6 | 0.88 | 17.44 | 0.54 | 2.22 | 0.60 | 0.494 | 0.106 |
| Sep | 29.8 | 0.97 | 17.9 | 0.80 | 18.72 | 0.32 | 2.68 | 0.62 | 0.353 | 0.093 |
| Oct | 28.9 | 0.79 | 17.8 | 0.69 | 19.13 | 0.28 | 2.67 | 0.40 | 0.370 | 0.099 |
| Nov | 28.1 | 0.82 | 18.6 | 0.76 | 19.40 | 0.28 | 2.33 | 0.31 | 0.517 | 0.100 |
| Dec | 28.6 | 0.73 | 17.8 | 0.88 | 19.15 | 0.38 | 2.33 | 0.39 | 0.420 | 0.093 |

The differential effects of each independent climate factor (e.g., air temperature, dew point temperature, wind speed, or fractional cloud cover) on each response variable (e.g., lake surface temperature, mixing depth, evaporation rate) were then compared against the standard reference conditions. The variations defined in Eq. 10 are tabulated by month for each independent climate variable in Table 2. Top of atmosphere short wave radiation and barometric pressure were tabulated by month as mean values only in Table 3; insufficient data existed to identify the magnitudes of interannual variation in these two latter variables.

*Table 3.* Monthly means for top-of-atmosphere short wave radiation and station barometric pressure.

| | $SW_{available}$ | $P_{STA}$ |
|---|---|---|
| | W m$^{-2}$ | mb |
| Jan | 421.8 | 903.6 |
| Feb | 435.2 | 903.5 |
| Mar | 438.6 | 903.4 |
| Apr | 423.6 | 904.7 |
| May | 400.6 | 905.7 |
| Jun | 386.1 | 904.9 |
| Jul | 391.2 | 905.6 |
| Aug | 411.7 | 905.1 |
| Sep | 428.8 | 905.1 |
| Oct | 432.5 | 904.3 |
| Nov | 422.6 | 903.8 |
| Dec | 415.0 | 904.3 |

## 4.5 Hydrologic Model for Lake Edward

Hydrologic characteristics of Lake Edward and its catchment are assembled from data provided by Viner and Smith (1973), Bugenyi (1982), Hecky and Kling (1987), ILEC (2000), and Lærdal *et al.* (2001). Different sources are not in agreement about such fundamental points as lake mean depth (and volume) or catchment area. The modern bathymetry measured by Lærdal *et al.* (2001) was adopted and values are reported in Table 4. Drainage basin boundaries for Africa were taken from HYDRO1k, a hydrologically correct 1-km resolution version of the GTOPO30 global 30 arc-second Digital Elevation Model produced by the USGS EROS Data Center (Figure 2). The data and derivations are described at http://edcdaac.usgs.gov/gtopo30/hydro/readme.html. Water balance for Lake Edward was calculated as

$$\text{Outflow} = \text{Input from catchment} + \text{direct precipitation} - \text{evaporation} \quad (11)$$

Direct precipitation on the lake surface was taken to be the mean of the annual average precipitation at 10 stations around Lake Edward (1214 mm yr$^{-1}$). Six of the stations are located inside the catchment boundary, and the others are close by (Figure 2). The data represented 470 station-years of record ranging from 19 to 87 years per station. Precipitation did not vary significantly with elevation among these stations (Figure 3). Moreover, annual precipitation was not well correlated among the 10 stations. In 44 pairwise correlations the mean correlation was 0.25 (range = -0.23 to 0.76), and only 8 station pairs exhibited correlation greater than 0.5. The estimated precipitation reported by Viner and Smith (1973) for Lake George is 30% lower than the average value of 1214 mm $yr^{-1}$ found here, but it is within the range of variation of the data among stations.

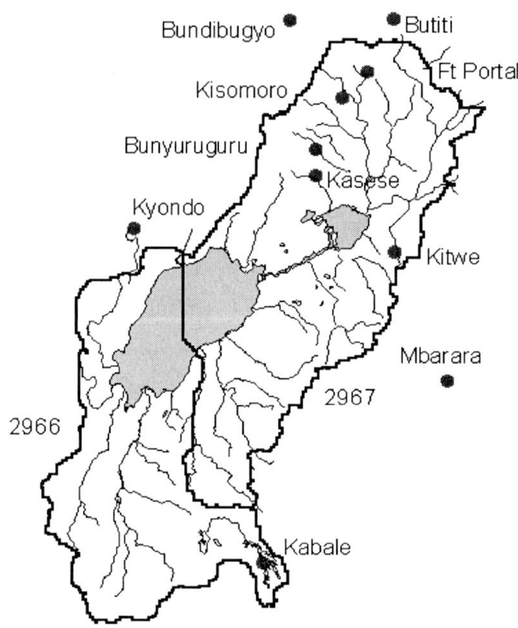


*Figure 2.* HYDRO1k drainage basin boundaries with Pfafstetter Level 4 codes, stream drainage (DCW, 1993), and GHCN version 2 climate stations.

Water input from the catchment was estimated by the method of Viner and Smith (1973). Total runoff was estimated from measurements of runoff yield by the Nyamuganasani River, which drains from the Ruwenzoris into northern Lake Edward. From measured discharge and river catchment area, Viner and Smith (1973) report 0.518 m $yr^{-1}$ in surface runoff per $m^2$ of land surface. Thus, for the terrestrial catchment of Lake Edward (20374 $km^2$ exclusive of lake surface area), the expected mean annual water income by direct proportions is $1.06 \times 10^{10}$ $m^3$. Because the relationship between mean precipitation and mean fluvial runoff is not known for Lake Edward catchments, the runoff yield may contain systematic bias when applied to the entire catchment area and to long term temporal averages. Hence, sensitivity analyses were used to infer the potential importance of these different assumptions.

*Table 4.* Morphometry and hydrology of Lake Edward.

| | |
|---|---|
| Surface Elevation | 912 m |
| Lake area | 2325 $km^2$ |
| mean depth | 33 m |
| Lake Volume | $7.67 \times 10^{10}$ $m^3$ |
| Catchment Area (less lake) | 20374 $km^2$ |
| Kazinga Channel discharge | $1.70 \times 10^9$ $m^3$ $yr^{-1}$ |
| Nyamugansani River discharge | $0.26 \times 10^9$ $m^3$ $yr^{-1}$ |
| direct rainfall | $2.82 \times 10^9$ $m^3$ $yr^{-1}$ |
| Other fluvial income | $8.59 \times 10^9$ $m^3$ $yr^{-1}$ |
| Outflow at Semliki River | $1.08 \times 10^{10}$ $m^3$ $yr^{-1}$ |
| Evaporation | $2.59 \times 10^9$ $m^3$ $yr^{-1}$ |
| Lake Volume/Outflow | 7 yr |
| Lake Volume/Evaporation | 30 yr |
| Theoretical concentration factor for conservative solutes | 1.2 |

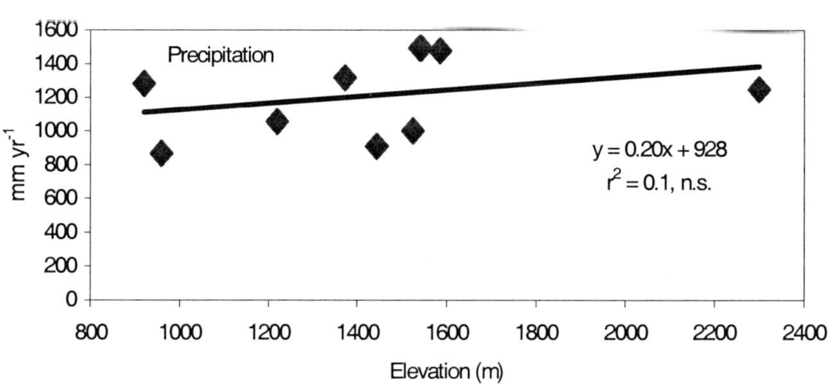


*Figure 3.* Mean annual precipitation versus elevation for 10 climate stations near Lake Edward.

Historical measurements were used for discharge of the Nyamuganasani River and Kazinga Channel into northern Lake Edward (Viner and Smith, 1973). Residual hydrologic income was assumed to derive from streams visible in Figure 2 including (1) the Lubilia River forming the international boundary at the north of the lake and also draining from the Ruwenzoris, (2) the Ishasha forming the international boundary at the south of the lake, and together with the Ntungwe, Mehuera, Ruampuno and Niamweru draining from the Kigezi and Ankole highlands of SW Uganda into the southeastern shore of the lake, (3) the Ruchuru and other unnamed streams draining the rift north of the Virunga volcanoes, and (4) miscellaneous unnamed mountain torrents which cascade from the Congo escarpment at the west. The HYDRO1k data set identifies two major sub-catchments for Lake Edward. Catchment 2967 includes drainage from the Ruwenzoris, Lake George, and the Uganda highlands. Catchment 2966 is dominated by drainage from the Virunga district but also includes the Congo escarpment.

Sensitivity analysis was conducted to evaluate the potential impact of regional variations in rainfall. Annual rainfall at ten stations around Lake Edward (GHCN version 2, National Climate Data Center archives: http://www.ncdc.noaa.gov/cgi-bin/res40.pl?page=ghcn.html) exhibits a mean CV (Coefficient of Variation = ratio of Standard Deviation to Mean) of 0.188 (SD = 0.056). Approximation of a 95% probability range for annual regional precipitation was assumed to be the mean (1214 mm) plus or minus double this CV multiplied by the mean. The same CV (i.e., 0.188) was assumed representative of total surface runoff.

## 4.6 Chemical Balance

The ion composition of hydrologic sources to Lake Edward was reconstructed from historical measurements and literature reports. The chemical composition of the Kazinga Channel discharge was assumed to be the same as that of Lake George during rainy season conditions (Viner, 1969). During the dry season there is little or no net discharge into Lake Edward from the Kazinga (Worthington, 1932). The chemical composition of the Nyamuganasani River was assumed to be that reported for streams from the Edward drainage of the Ruwenzoris (Visser, 1974). Initially, all other fluvial inputs were assumed to have the composition of streams draining the Kigezi highlands (Visser, 1974). Rainwater was assumed to have the composition reported by Visser (1961). By multiplying each source concentration by the mean hydrologic fluxes, bulk chemical inputs were obtained. Mass flux out of the lake was calculated as the composition of Lake Edward surface water reported by Lehman *et al.* (1998) multiplied by Semliki River discharge volume calculated from the hydrologic model.

Sediments of Lake Edward consist of rich siliceous ooze with about 40% $SiO_2$ by weight on average, 1.2% sulfur, and at least 15% calcite; most of the carbonates are $CaCO_3$, but about 12% are $MgCO_3$ (J. Russell *et al.,* ms. in preparation). Sediment accumulation rates are roughly 0.1 cm $yr^{-1}$. At sediment density of 2.5 g $cm^{-3}$, and porosity of 0.9, the mineral content and sedimentation rates indicate about 0.33 mol

$CaCO_3$, 0.05 mol $MgCO_3$, 0.09 mol S, and 1.7 mol $SiO_2$ $m^{-2}$ $yr^{-1}$ accumulate in the sediments. Removal rates by burial must be included in mass balance calculations.

## 4.7 Remote Sensing of Lake Edward Surface Temperature

Surface temperatures of Lake Edward were calculated from archival AVHRR (Advanced Very High Resolution Radiometer) data sets maintained by NOAA's Satellite Active Archive. The on-line archive available at www.saa.noaa.gov was searched for relatively cloud-free images of Lake Edward during 1995, 1996, and 1999. Suitable scenes at Local Area Coverage (LAC) resolution, nominally 1.1 km at satellite nadir, were downloaded in 10-bit Level 1B format. Spectral channel data were extracted from the packed binary files by XV-HRPT Data Acquisition software, version 4.3. The software was purchased from Dr. A. Mazurov, Space Research Institute RAS, Moscow (http://smis.iki.rssi.ru). Surface lake temperature in degrees C was calculated by multichannel split (MCSST split) algorithms for day or night images using the brightness temperatures (°K) recorded in AVHRR channels 4 and 5, and the calculated satellite zenith angles. Equations and calibration coefficients for NOAA-12 and NOAA-14 polar orbiting environmental satellites (POES) were used directly from values published by NOAA at http://psbsgi1.nesdis.noaa.gov:8080/EBB/ml/nicsst.html. It was not possible to calibrate the equations specifically for Lake Edward owing to the absence of contemporaneous lake surface temperature measurements.

The calculated SSTs were encoded as 8-bit scalar representations of their numerical values, and were projected and visualized as gray-scale images at 1-km resolution on a rectangular coordinate system by XV-HRPT software, using the latitude and longitude geo-reference data embedded in the AVHRR scan lines. Then, rectangular regions ranging from 36 to 289 $km^2$ were visually selected in cloud free offshore regions within the northern half and the southern half of the lake. Mean surface temperature and standard deviation were calculated for each region. The size of these analytical regions was chosen to provide good sampling density at the lake surface, but to be small enough to be located well offshore while avoiding any localized cloud cover.

Twenty-four satellite images were examined for this report. One pair of lake surface temperatures was sampled for each date.

# 5. RESULTS

## 5.1 Heat Balance Model and Sensitivity Analysis

Summary results are reported in Table 5 for the two alternative models used to calculate evaporation rates. Tables 6 and 7 report responses achieved from perturbations of the reference values during sensitivity analyses. Total mixing duration reports the total number of days during the year when the lake is calculated to mix to its full depth.

Table 5. Model results for reference (mean) climate conditions listed in Tables 2 and 3. Results are for two alternative models of evaporation described in the text.

| Response Variable | Mo<br>del 1 | Mo<br>del 2 |
|---|---|---|
| EVAP mm $yr^{-1}$ | 103<br>8 | 111<br>3 |
| $T_L$ annual mean | 27.9<br>3 | 27.0<br>4 |
| $T_L$ maximum for year | 29.3<br>7 | 28.6<br>5 |
| $T_L$ minimum for year | 26.9<br>2 | 26.0<br>0 |
| Total mixing duration, days | 167 | 106 |

Sensitivity calculations treated climate variables as independent and uncorrelated. Assumption of independence is supported by climate data from Kasese, as there are no correlations among daily maximum or minimum air temperature, dew point temperature, mean wind speed, and precipitation that exceed 0.36 in pair-wise comparisons. Sensitivity calculations reveal that cloud cover and wind speed are by far the most influential climate factors, within the limits of observed natural variation. Magnitudes of perturbations used in the analyses were reported in Table 2.

*Table 6.* Results of sensitivity analysis for Lake Edward heat balance with evaporation Model 1.

| Independent Variable | Perturbation | Δ EVAP (mm $d^{-1}$) | $\Delta T_L$ (°C) mean | $\Delta T_L$ (°C) max | $\Delta T_L$ (°C) min | Δ mixing period (d) |
|---|---|---|---|---|---|---|
| $T_{MAX}$ | Increase | -3 | +0.22 | +0.26 | +0.14 | -32 |
| | Decrease | -1 | -0.21 | -0.16 | -0.23 | +18 |
| $T_{MIN}$ | Increase | +15 | +0.15 | +0.17 | +0.11 | -7 |
| | Decrease | -14 | -0.14 | -0.15 | -0.15 | +22 |
| $T_{DP}$ | Increase | -7 | +0.15 | +0.17 | +0.10 | -20 |
| | Decrease | +7 | -0.14 | -0.10 | -0.12 | +21 |
| U | Increase | +77 | -0.60 | -1.03 | -0.58 | +30 |
| | Decrease | -93 | 0.74 | +1.48 | +0.25 | -71 |
| Cloud | Increase | -284 | -1.35 | -1.26 | -1.69 | +31 |
| Fraction | Decrease | +239 | +1.08 | +1.18 | +0.17 | -56 |

*Table 7.* Results of sensitivity analysis for Lake Edward heat balance using evaporation Model 2.

| Independent Variable | perturbation | Δ EVAP (mm $d^{-1}$) | $\Delta T_L$ (°C) mean | $\Delta T_L$ (°C) max | $\Delta T_L$ (°C) min | Δ mixing period (d) |
|---|---|---|---|---|---|---|
| $T_{MAX}$ | Increase | +22 | +0.13 | +0.19 | +0.14 | -2 |
| | Decrease | -22 | -0.13 | -0.18 | -0.14 | +3 |
| $T_{MIN}$ | Increase | +20 | +0.12 | +0.14 | +0.13 | -6 |
| | Decrease | -21 | -0.13 | -0.14 | -0.13 | +19 |
| $T_{DP}$ | Increase | -15 | 0.19 | +0.24 | +0.17 | -2 |
| | Decrease | +14 | -0.19 | -0.20 | -0.18 | +4 |
| U | Increase | +91 | -0.61 | -1.04 | -0.56 | +83 |
| | Decrease | -106 | +0.78 | +1.57 | +0.21 | -25 |
| Cloud | Increase | -229 | -1.41 | -1.39 | -1.83 | +90 |
| Fraction | Decrease | +222 | +1.27 | +1.30 | +0.36 | -33 |

## 5.2 Water Balance

Semliki River discharge from Lake Edward was calculated by difference from total surface runoff, direct precipitation, and evaporation. Historical direct measurements from the IBP period exist for hydrologic inputs from the Kazinga Channel, the Nyamuganasani River, and rainfall. Monthly precipitation and pan evaporation at Mbarara from 1969 to 1974 (Hydromet, 1991) exhibit no significant correlation between the two variables (R = -0.162, n = 68). Consequently, the effects of variability in precipitation was simulated independently from variations in evaporation rate.

The estimated Semliki outflow in Table 4 is about double the $5 \times 10^9$ $m^3$ $yr^{-1}$ calculated by Hurst (1927), about triple the equivalent annualized flow rate measured by Worthington (1932) during dry season conditions in June 1931 (ca. $3.3 \times 10^9$ $m^3$ $yr^{-1}$), and about 5 times the mean discharge reported by Talbot *et al.* (2000). Hurst had estimated the catchment area at 12000 $km^2$, much smaller than the modern DEM now indicates (Figure 2).

Mean annual outflow is about 80% of total water income to Lake Edward, leading to a theoretical concentration factor of 1.2 for conservative solutes. Surface runoff was scaled proportionally to variations in regional precipitation for sensitivity analysis. The range of variation in theoretical concentration factor for the ranges examined (approximating a 95% Confidence Interval) was 1.1 to 1.6 (Table 8). Results are shown in Table 8 for evaporative flux Model 1; results from Model 2 were essentially the same. Again, cloud cover emerged as the most influential climate variable other than variation in precipitation itself. These results imply variations in ionic strength of about 40%, which is less than has actually been observed for Lake Edward surface water over several decades (Lehman *et al.*, 1998). Ionic strength has varied by nearly a factor of two. Evaporation can produce only a few percent increase in salt concentrations of the lake during a single dry season. In addition to climate-induced variation, the cause of different salt concentrations measured in Lake Edward over the years may have to be sought in seasonal variations of the fluvial inputs to the lake.

*Table 8.* Theoretical concentration factors from sensitivity analysis. Results for Model 1; minimum and maximum values in bold print.

| Independent Variable | Climate perturbation | Concentration factor for specified rainfall perturbation reference (mean) | rainfall increase | rainfall decrease |
|---|---|---|---|---|
| Reference | None | 1.22 | 1.16 | 1.45 |
| $T_{DP}$ | Increase | 1.22 | 1.16 | 1.44 |
| | Decrease | 1.22 | 1.17 | 1.46 |
| $T_{MAX}$ | Increase | 1.22 | 1.17 | 1.46 |
| | Decrease | 1.22 | 1.16 | 1.44 |
| $T_{MIN}$ | Increase | 1.22 | 1.17 | 1.46 |
| | Decrease | 1.22 | 1.16 | 1.44 |
| U | Increase | 1.24 | 1.18 | 1.50 |
| | Decrease | 1.20 | 1.15 | 1.39 |
| Cloud | Increase | 1.15 | **1.13** | 1.33 |
| Fraction | Decrease | 1.29 | 1.20 | **1.59** |

## 5.3 Chemical Constituents and Mass Balance

Historical measurements of water chemistry in rain, Lake George, Lake Edward, Ruwenzori drainage, and streams of the Kigezi highlands (Visser, 1961; Talling and Talling, 1965; Viner, 1969) are combined in Table 9 to estimate the major element balance for the lake. The relatively conservative solutes Na, K, and Cl are far more concentrated in Lake Edward than the theoretical expectation for a conservative solute (Table 4), based on existing river and rain chemistry.

Hydrology and the bulk chemistry summarized in Table 9 underscore a geochemical anomaly about Lake Edward. The lake is far out of chemical mass balance based on existing measurements. Table 10 reports mass balance of major constituents computed from Tables 4 and 9. Massive sources are required for Na, K, Mg and bicarbonate to balance the mass budgets. There are not enough inputs of Ca, $SO_4$ and $SiO_2$ to account for measured sedimentation rates. Instead, there must be additional weathering sources of Ca, Si, as well as bicarbonate and other cations to close the water column and sediment budgets. The required ion composition and stoichiometry of the weathering product suggests an alkaline rock source material dominated by soluble carbonate salts. Absence of strong salt gradients in the lake argues against the likelihood that the salts are injected from groundwater sources.

*Table 9*. Mean (or rainy season) chemical composition (μM) of the main hydrologic sources to Lake Edward. "Mean Input" reports concentrations weighted by magnitudes of hydrologic fluxes in Table 4. For this calculation, all unmeasured inputs are assumed to have the chemistry measured for the Kigezi district.

| Component | Ruwenzori | Kigezi | L. George | Rain | Mean Input | L. Edward | Conc factor |
|---|---|---|---|---|---|---|---|
| Na | 534 | 478 | 890 | 74 | 410 | 3750 | 8.7 |
| K | 63 | 81 | 114 | 43 | 72 | 1620 | 21.7 |
| Ca | 389 | 215 | 361 | 1 | 166 | 310 | 1.7 |
| Mg | 206 | 216 | 300 | 0 | 141 | 1630 | 9.9 |
| Cl | 132 | 232 | 215 | 25 | 130 | 580 | 3.6 |
| $SO_4$ | 208 | 204 | 188 | 19 | 117 | 250 | 1.7 |
| $HCO_3$ | 1287 | 775 | 1250 | 0 | 577 | 7870 | 12.3 |
| $SiO_2$ | 273 | 270 | 366 | 0 | 175 | 150 | 0.7 |

Kilham and Hecky (1973) classed Lake Edward with other "sodium-potassium-magnesium-bicarbonate" waters, including Lakes Kivu and Tanganyika, and considered their common chemistry to be influenced by the volcanic rocks of the Virunga district. Talling and Talling (1965) as well as Kilham and Hecky (1973) alluded to the likely importance of the basic volcanic rocks in this region to the ion balance of Lake Edward. Virtually no direct measurements are available for the water chemistry of rivers draining the Virunga mountains into Lake Edward. Hurst (1927) measured the titration alkalinity of the Ruchuru River, which drains into the south of Lake Edward from the Virunga district. He reported its alkalinity as equivalent to 17.2 mEq. Hurst remarked (p. 23) that [the Ruchuru] "is the largest stream entering the lake in the dry season and probably the most important feeder of the lake." He estimated its dry season discharge at its mouth during August 1926 at 40 $m^3 s^{-1}$, with estimated rainy season flood discharge of 130 $m^3 s^{-1}$. A median value of 85 $m^3 s^{-1}$ for average discharge amounts to $2.7 \times 10^9$ $m^3 y^{-1}$, or 31% of all the "other fluvial income" identified in Table 4.

Hurst's measurement of alkalinity permits alternative estimation of the missing river sources by inverse methods. It would take 54% of the water income from "other fluvial" sources (Table 4) to balance the alkalinity budget of the lake at a mean annual supply concentration of 17.2 mEq. The requisite flow is greater than the estimated median discharge of the Ruchuru; it represents the composite discharge from the Virunga district. In point of fact, the sub-catchment south of Lake Edward (Figure 2) represents 6000 $km^2$, or 55% of the residual area remaining after drainage from the Ruwenzoris and Lake George/Kazinga Channel are subtracted from the total catchment area. The model result is thus completely feasible.

River discharge with the mean annual composition identified in the rightmost column of Table 10 would close the bulk geochemical budgets for Lake Edward. This model-generated composition is reminiscent of the mixolimnion chemistry of Lake Kivu, a lake that is influenced by drainage from the Virunga district. The real chemistry of the Ruchuru River and others that flow across the broad southern plain into Lake Edward is essentially unknown and unmeasured (Kilham, 1984). The figures presented in Table 10 thus constitute an explicit, model-based, *a priori* and quantitative hypothesis about the geochemistry of hydrologic sources to Lake Edward. They help explain the decrease in cations and carbonate from southwest to northeast in the lake that is evident in chemical analyses published by Worthington (1932, his table 3). Seasonal variations in these rather high predicted annual mean concentrations may also help to explain the two-fold range in the ionic strength of Lake Edward that has been recorded by different investigators.

*Table 10.* Mass balance anomalies for major chemical constituents in Lake Edward, and inverse model prediction of water chemistry for rivers entering from the Virunga district.

| | Input-Outflow anomaly mol $m^{-2}$ $yr^{-1}$ | sediment accumulation mol $m^{-2}$ $yr^{-1}$ | Total required sources mol $m^{-2}$ $yr^{-1}$ | Hypothetical composition of missing inputs mM |
|---|---|---|---|---|
| Na | -14.83 | | 14.83 | 7.9 |
| K | -7.08 | | 7.08 | 3.6 |
| Ca | -0.33 | 0.33 | 0.66 | 0.6 |
| Mg | -6.52 | 0.05 | 6.58 | 3.5 |
| Cl | -1.63 | | 1.63 | 1.0 |
| $SO_4$ | -0.22 | 0.09 | 0.32 | 0.4 |
| $HCO_3$ | -32.59 | 0.38 | 32.98 | 17.2 |
| $SiO_2$ | 0.60 | 1.67 | 1.07 | 0.8 |

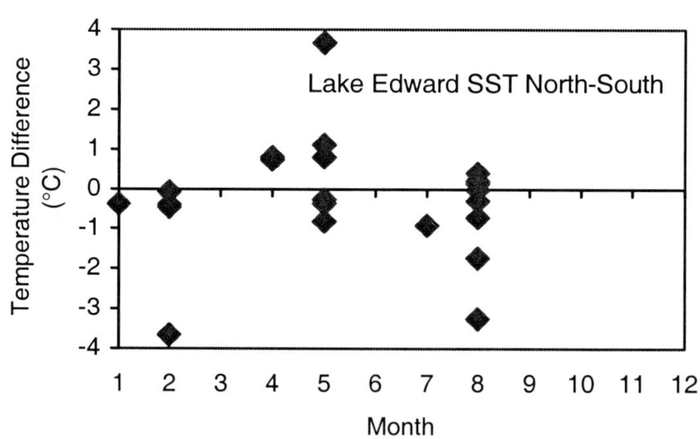


*Figure 4.* Temperature differences between northern and southern Lake Edward inferred from AVHRR brightness temperatures for different dates in 1995 and 1996. Calculated as TNORTH – TSOUTH.

## 5.4 Lake Temperature Gradients by Remote Sensing

In the absence of specific calibration of calculated SSTs to Lake Edward, analyses were confined to assessment of temperature differences within the lake rather than to temperature in absolute terms. Temperature differences between north central Lake Edward and south central Lake Edward are plotted in Figure 4. Discernible and statistically significant temperature differences exist on most dates, but there is not a consistent gradient between north and south throughout the year. Instead, there is a tendency for the lake to be warmer in the south during the dry seasons and to be warmer in the north during rainy season conditions. Temperature differences of 1°C or greater are not uncommon.

# 6. DISCUSSION

Lake Edward has experienced striking variation in dissolved salt concentration during the last 35 years. These variations are partially climate related, based on the water balance model, and are consistent with the generally warmer, wetter conditions of the present high elevation tropics compared with conditions 40 years ago (Lehman, 1998). Chemical data are presented in Table 11 along a gradient of increasing conductivity and ionic strength, from measurements by Kilham (1972), Lehman *et al.* (1998), and Talling and Talling (1965). Ion activities were calculated from ionic strength by the Güntelberg approximation (Stumm and Morgan, 1970), using temperature coefficients at 25 °C (Stumm and Morgan, 1970, table 5.1), and ion activity products (IAP) were calculated. Chloride is used as a conservative reference solute for comparisons. Several points can be made that have implications for the sedimentary record of Lake Edward.:

1. Ca/Cl ratios decrease with evaporative concentration. Ca is differentially lost, probably as carbonate to the sediments.
2. Ca concentrations remain relatively constant, but Ca/Mg ratios fall with progressive salt concentration.
3. During the last 40 years Lake Edward has varied from being at the saturation point with respect to calcite, to being 18-fold supersaturated.

The pH and Ca/Mg ratios of Lake Edward indicate that dolomite should be the thermodynamically stable phase for the solid minerals (Stumm and Morgan, 1970, Figures 5-11), as is also true of sea water. In fact, as in sea water the precipitating mineral is mainly calcite, but the carbonate should have a variable composition of $MgCO_3$. The range of supersaturation reported in Table 11 is compatible with up to about 10% $MgCO_3$ in the carbonate solid phase (Stumm and Morgan, 1970, Figures 5-11); percentage composition should increase with conductivity and ionic strength of the lake water. Consequently, the observation of about 12% Mg content in the sedimentary calcite of Lake Edward is consistent with the high levels of supersaturation observed earlier in the 20$^{th}$ century rather than later. Ca/Mg ratios of the precipitating carbonates may provide a proxy signal of changes in water balance through evaporative concentration and related changes. Sedimentary calcite is the result of (1) evaporative concentration of water already at the precipitation point, (2) elevated water temperature, or (3) increased rates of photosynthesis. Not all algal taxa can equally use bicarbonate as a carbon source (Prins and Elzenga, 1989; Riebesell *et al.*, 1993). Thus, the types of species that are stimulated under alternating conditions of water column stability or mixing may control some of the seasonality and isotopic composition of calcite deposition in Lake Edward.

*Table 11.* Chemical conditions affecting carbonate equilibria across a range of dilution states for Lake Edward. Ksp refers to calcite at 25 °C (Stumm & Morgan, 1970).

| Analyst: | Kilham (1972) | Lehman *et al.* (1998) | Talling & Talling (1965) |
|---|---|---|---|
| Year | 1969 | 1995 | 1961 |
| Ionic Strength | 0.007 | 0.012 | 0.015 |
| Ca, μM | 335 | 310 | 310 |
| Ca/Cl | 0.73 | 0.53 | 0.31 |
| Ca/Mg | 0.37 | 0.19 | 0.16 |
| IAP/Ksp | 0.95 | 10.2 | 17.9 |

Related arguments apply to the deposition of biogenic Si. Diatoms are not abundant at the surface sediments nor in the modern plankton. The dominant fossil diatom in the sediments of Lake Edward is *Cyclostephanos damasii*, a taxon formerly classified as *Stephanodiscus damasii* (Gasse *et al.*, 1983). This unicellular centric diatom has been identified as a low Si/P ratio species, with favorable stoichiometries near 1:1 by moles (Kilham *et al.*, 1986). Present day Lake Edward is very rich in SRSi (soluble reactive silicon), and the Si/P ratio is extremely high (ca. 200:1). These are not the typical conditions associated with success by *Cyclostephanos*/*Stephanodiscus*. Hence, its paucity in the plankton and the surface sediment is consistent with the chemistry, but the lake must have exhibited alternate states under perhaps altered climate conditions and mixing regimes that promoted vast production of these diatoms in the recent past.

One hypothesis for the paucity of diatoms in Lake Edward at present, particularly in the north, is that increased water column stability enhances diatom loss rates by sinking (E. Ralph, pers. comm.). The presumed mechanism would be a warmer surface layer resulting from the warm discharge of Lake George through the Kazinga channel during the rainy season. Water temperatures are in fact occasionally warmer in the north than in the south during the rainy months of April and May (Figure 4), but at other times the surface water is warmer in the south. The spatial pattern reported by Hecky and Kling (1987) was observed in samples collected in mid-March 1972, a month that can exhibit the start of rainy season conditions.

Sensitivity analyses which predict the responses of lake surface temperature and evaporation rates to climate changes also document predicted changes in lake mixing dynamics. Changes in wind speed, cloud cover, and other climate factors can theoretically increase or diminish the period of deep mixing in Lake Edward by weeks or months. It is plausible that altered mixing regimes under alternate climate scenarios have been the cause of the great variations in diatom success within the lake, as recorded in its sediments.

The heat balance model for Lake Edward and existing meteorological data indicate that temporal variations in wind speed and radiation balance owing to cloud cover are the dominating sources of variation in both lake temperature and evaporation (Tables 5 and 6). Model simulations were based on independent perturbation of individual climate variables. Any empirical positive correlations between wind speed and either air temperature or irradiance would tend to subdue the predicted effects on lake temperature, but would exaggerate effects on evaporation. Irrespective of these secondary effects which may amplify the variations in evaporation rate, it is clear that Lake Edward has exhibited wide variations in water turnover and evaporative concentration of dissolved salts in this century alone.

The methods used to estimate temperature and evaporation for Lake Edward may have wide application to tropical lakes in general, and in particular to the great lakes of East Africa. Similarly, sensitivity analyses based on these solution methods, using empirical variations in input variables as reference, should permit quantitative prediction of lake response to specific climate change scenarios.

# 7. CONCLUSIONS AND RECOMMENDATIONS

Lake Edward resonates strongly to climate change signals. Its water detention time is relatively short compared with other great lakes of the region, but the unusual geochemical composition of its inputs from the south interact with water balance to produce striking changes in chemistry and ionic strength. The lake is of adequate size to be investigated with existing satellite technology, and it exhibits temporal and spatial patterns that appear to be correlated with climate. The data sets and models developed for this study successfully brought into focus issues that require further field work, remote sensing, and additional development of theory. These are:

1. Improved elevation maps are needed to corroborate or refine the existing DEM for the region, to define more precisely the total area of drainage, and to partition the relative areas of sub-catchments with differing stream chemistry.
2. Water samples must be collected seasonally from the Ruchuru River and other streams entering Lake Edward from the south. Chemical analyses of these waters should be used to test the *a priori* hypothesis of their composition and their potential for altering the water chemistry of Lake Edward.
3. The discharge flow of the Semliki River at its source must be measured in order to test the validity of predicted water balance (Table 4). Furthermore, measurements are needed for the discharge flow of the southern rivers.

4. Ca/Mg ratios of sedimentary carbonates should be investigated as an index of lake water balance.
5. More precise estimates of cloud cover and wind speed over Lake Edward are required, and any cloudiness differences between day and night should be quantified.
6. Additional AVHRR scenes of Lake Edward from multiple years should be assembled to test the generality of surface temperature differences between lake regions in different seasons. An effort should be made to assess the accuracy of the temperature estimates in absolute terms. Then the heat balance model for the lake can be tested using remotely sensed temperatures.

## ACKNOWLEDGMENTS

This research was funded in part by grants from the National Geographic Society and the U. S. National Science Foundation. D. A. Lehman assembled and organized the data sets on which this study was based. T. Lærdal provided unpublished bathymetric maps of Lake Edward and made valuable comments on the content of the paper. J. Russell provided unpublished information about sediment composition and accumulation rates. D. Gesch provided information about the USGS digital elevation model and its derivative products.

## REFERENCES

Bugenyi, F.W.B. (1982) Copper pollution studies in Lakes George and Edward, Uganda: The distribution of Cu, Cd and Fe in the water and sediments. *Environmental Pollution (Series B)* **3**, 129-138.

Burgis, M. J., Dunn, L. G., Ganf, G. G., McGowan, L. M., and Viner, A. B. (1972) Lake George, Uganda. Studies on a tropical freshwater ecosystem, in K. Kajak and A. Hillbricht-Ilkowska (eds.), *Productivity Problems in Freshwaters*, Polish Scientific Publications, Warszawa-Krakow, pp. 301-309.

Croley, T.E. (1989) Verifiable evaporation modeling on the Laurentian Great Lakes. *Water Resources Research* **25**, 781-792.

DCW (1993) *Digital Chart of the World 1993 version on-line*, Pennsylvania State University Libraries, www.maproom.psu.edu/dcw/.

Degens, E.T. (1973) Hydrothermal origin of metals in some East African rift lakes. *Mineralium Deposita* **8**, 388-404.

Gasse, F., Talling, J.F., and Kilham, P. (1983) Diatom assemblages in East Africa: classification, distribution and ecology. *Revue d'Hydrobiologie Tropicale* **16**, 3-34.

Hecky, R.E., and Kling, H.J. (1987) Phytoplankton ecology of the Great Lakes in the rift valleys of central Africa. *Archiv für Hydrobiologie Ergebisse Limnologie* **25**, 197-228.

Hurst, H.E. (1927) *The lake plateau basin of the Nile*. Egyptian Ministry of Public Works. Physical Department paper No. 23, 66 p.

Hydromet (1991) *Hydrometeorological survey of the catchments of Lakes Victoria, Kyoga and Albert. Summaries of meteorological data observed over the project area.* Hydromet Survey Project Report, Entebbe.

ILEC (2000) *Lake Edward,* World Lakes Database, International Lake Environment Committee Foundation. http://www.ilec.or.jp/

Imberger, J. and Patterson, J.C. (1990) Physical Limnology, in J.W. Hutchinson and T.Y. Wu (eds.), *Advances in Applied Mechanics* **27**, 303-475.

Johnson, T.C. and 8 others. (1996) Late Pleistocene desiccation of Lake Victoria and rapid evolution of cichlid fishes. *Science* **273**, 1091-1093.

Johnson, T.C. and Odada, E. (1996) *The limnology, climatology, and paleoclimatology of the East African lakes,* Gordon and Breach.

Johnson, T.C., Chan, Y., Beuning, K.R.M., Kelts, K., Ngobi, G., and Verschuren, D. (1998) Biogenic silica profiles in Holocene cores from Lake Victoria: Implications for lake level history and initiation of the Victoria Nile, in J. T. Lehman (ed.), *Environmental change and response in East African lakes,* Kluwer Academic Publishers, Dordrecht, pp. 75-88.

Kilham, P. (1972) *Biogeochemistry of African lakes and rivers,* Ph. D. thesis, Duke University, 197 p.

Kilham, P. and Hecky, R.E. (1973) Fluoride: geochemical and ecological significance in East African waters and sediments. *Limnology and Oceanography* **18**, 932-945.

Kilham, P. (1984) Sulfate in African inland waters: Sulfate to chloride ratios. *Verhandlungen der Internationale Vereinigung für Limnology* **22**, 296-302.

Kilham, P., Kilham, S.S., and Hecky, R.E. (1986) Hypothesized resource relationships among African planktonic diatoms. *Limnology and Oceanography* **31**, 1169-1181.

Keijman, J.Q. (1974) The estimation of the energy balance of a lake from simple weather data. *Boundary-Layer Meteorology* **7**, 399-407.

Lærdal, T., Talbot, M.R. and Russell, J. (2001) Late Quaternary Sedimentation and Climate in the Lakes Edward and George area, Uganda-Congo, in E. Odada and D. Olago (eds.), *The East African Great Lakes: Limnology, Paleolimnology, and Biodiversity,* Kluwer Academic Publishers, Dordrecht.

Lehman, J.T. (1998) The role of climate change in the modern condition of Lake Victoria. *Theoretical and Applied Climatology* **61**, 29-37.

Lehman, J.T., Litt, A.H., Mugidde, R., and Lehman, D.A. (1998) Nutrients and plankton biomass in the rift lake sources of the White Nile: Lakes Albert and Edward, in J.T. Lehman (ed.), *Environmental Change and Response in East African Lakes,* Kluwer Academic Publishers, Dordrecht, pp. 157-172.

Maidment, D.R. (editor) (1993) *Handbook of hydrology,* McGraw-Hill.

Monismith, S.G. (1985) Wind-forced motions in stratified lakes and their effect on mixed-layer shear. *Limnology and Oceanography* **30,** 771-783.

Niiler, P.P., and Kraus, E.B. (1977) One-dimensional models of the upper ocean, in E.B. Kraus (ed.), *Modelling and prediction of the upper layers of the ocean,* Pergamon, pp. 143-172.

Price, J.F. (1981) Upper ocean response to a hurricane. *Journal of Physical Oceanography* **11**, 153-175.

Prins, H.B.A. and Elzenga, J.T.M. (1989) Bicarbonate utilization: function and mechanism. *Aquatic Botany* **34**, 59-83.

Riebesell, U., Wolf-Gladrow, D. A. and Smetacek, V. (1993) Carbon dioxide limitation of marine phytoplankton growth rates. *Nature* **361**, 249-251.

Ruff, I. and A. Gruber, 1988: General Determination of Earth Surface Type and Cloud Amount Using Multispectral AVHRR Data. *NOAA Technical Report NESDIS 39*, U. S. Dept. Comm., NOAA, NESDIS, Washington, D. C. 30 p.

Sherman, F.S., Imberger, J., and Corcos, G.M. (1978) Turbulence and mixing in stably stratified waters. *Annual Review of Fluid Mechanics* **10**, 267-288.

Spigel, R.H. (1980) Coupling of internal wave motion with entrainment at the density interface of a two-layer lake. *Journal of Physical Oceanography* **10**, 144-155.

Spigel, R.H., and Imberger, J. (1980) The classification of mixed-layer dynamics in lakes of small to medium size. *Journal of Physical Oceanography* **10**, 1104-1121.

Spigel, R.H., Imberger, J., and Rayner, K.N. (1986) Modeling the diurnal mixed layer. *Limnology and Oceanography* **31**, 533-556.

Strub, P.T. and Powell, T.M. (1987) The exchange coefficients for latent heat and sensible heat flux over lakes: dependence upon atmospheric stability. *Boundary-Layer Meteorology* **40**, 349-361.

Stumm, W. and Morgan, J. (1970) *Aquatic Chemistry*, Wiley.

Talbot, M. R., Williams, M. A. J. and Adamson, D. A. (2000) Strontium isotope evidence for late Pleistocene reestablishment of an integrated Nile drainage network. *Geology* **28**, 343-346.

Talling, J. F. and Talling, I. B. (1965) The chemical composition of African lake waters. *Internationale Revue gesamten Hydrobiologie* **50**, 421-463.

Viner, A. B. (1969) The chemistry of the water of Lake George, Uganda. *Verhandlungen der Internationale Vereinigung für Limnology* **17**, 289-296.

Viner, A.B. and Smith, I.R. (1973) Geographical, historical and physical aspects of Lake George. *Proceedings of the Royal Society of London, Series B* **184**, 235-270.

Visser, S. (1961) Chemical composition of rain water in Kampala, Uganda, and its relation to meteorological and topographical conditions. *Journal of Geophysical Research* **66**, 3759-3765.

Visser, S.A. (1974) Composition of waters of lakes and rivers in East and West Africa. *African Journal of Tropical Hydrobiology and Fisheries* **3**, 43-59.

Worthington, E.B. (1932) *A report on the fisheries of Uganda investigated by the Cambridge Expedition to the East African lakes, 1930-31. Part 1. Lakes Edward and George and the Kazinga Channel*, Zoological Laboratory, Cambridge, June 1932.

Yin, X. and Nicholson, S.E. (1998) The water balance of Lake Victoria. *Hydrological Sciences Journal* **43**, 789-811.

# LAKE - GROUNDWATER RELATIONSHIPS, OXYGEN ISOTOPE BALANCE AND CLIMATE SENSITIVITY OF THE BISHOFTU CRATER LAKES, ETHIOPIA

SEIFU KEBEDE[1], HENRY LAMB[2], RICHARD TELFORD[3], MELANIE LENG[4] and MOHAMMED UMER[1]

[1]*Department of Geology and Geophysics, Addis Ababa University, Ethiopia*

[2]*Institute of Geography and Earth Sciences, University of Wales, Aberystwyth, Wales, UK*

[3]*Department of Geography, University of Lancaster, Lancaster, UK*

[4]*NERC Isotope Geosciences Laboratory, Keyworth, Nottingham, UK*

**ABSTRACT**

Oxygen isotope data from groundwater, streams and crater lakes in central Ethiopia provide a basis for modelling lake hydrological and isotopic budgets. The environmental parameters $\delta_A$ (isotopic composition of vapor) and $\varepsilon^*$ (equilibrium fractionation factor) were determined by a semi-empirical approach using one lake as a terminal index lake. The models show that lake oxygen-isotope composition is more sensitive to rainfall-related parameters (humidity and $\delta_A$) than to the isotopic composition of inflow ($\delta_{in}$). The isotopic composition of the lacustrine sedimentary carbonates should therefore provide a record of past rainfall variation.



*E.O. Odada and D.O. Olago (eds.),*
*The East African Great Lakes: Limnology, Palaeolimnology and Biodiversity,* 261–275.

# 1. INTRODUCTION

Oxygen isotope analysis of lacustrine carbonates can provide quantitative reconstruction of past climate (Ricketts and Johnson, 1996). This approach requires an understanding lake water budgets, and their sensitivity to environmental change. Small lakes often have a substantial groundwater influence, but relatively few hydrological studies of lake-watershed systems emphasise the role of groundwater (Crowe & Schwartz, 1981; Crowe, 1990; Almendinger, 1990; Ayenew, 1998). Using hydrochemical (Winter, 1978; Kebede *et al.*, in press) and environmental isotope techniques (Dincer, 1968; Gat and Levy, 1978; IAEA, 1981; Krabbenhoft *et al.*, 1990; Krabbenhoft *et al.*, 1994), coupled with standard hydrological investigation, net groundwater flux can be separated into inflow and outflow components. The distribution of stable isotopes in groundwater near lakes may reveal useful information about flow directions and mixing processes.

Two of the Bishoftu crater lakes, at Debre Zeit, Ethiopia, contain laminated carbonate sediments (Lamb *et al.*, this volume), and thus have the potential to provide a high-resolution record of climatic change. This paper explores lake-groundwater relationships of the Bishoftu lakes, determines their water budgets using an isotope mass balance approach, and tests the sensitivity of their isotopic composition to hydrological and climatic parameters.

# 2. THE STUDY SITES

## 2.1 Lake Morphometry, Geology and Hydrogeology

The Bishoftu crater lakes are located at Debre Zeit on the western escarpment of the Main Ethiopian Rift, 45km southeast of Addis Ababa, at 1800 to 2000m elevation (Figure 1). The five permanent lakes are Hora (also known as Beite Mengist), Babogaya (Pawlo, Bishoftu Guda), Bishoftu, Kilole (Kilotes), and Arenguade. Artificial lakes and ponds in the area include Lake Kuriftu, a reservoir that fills an originally dry crater depression following the diversion of a tributary of the nearby Awash river. Lake Cheleleka is a large, shallow swamp that has been present since the early 1970's. Lake Balbala is a reservoir constructed on one of the tributaries of the Awash river in the early 1980's. The craters are maars, volcanic collapse structures above zones of fractured rock that extend to an igneous dike at depth (Lorenz, 1986). They are roughly circular in shape, with areas between 0.6 and 1 $km^2$. The lakes range in depth from 6.4m (Lake Kilole) to 87m (Lake Bishoftu; Prosser *et al.*, 1968; Table 1). The lakes have no perennial surface inlets or outlets, and are fed by direct precipitation, surface runoff from the crater walls, and by groundwater.

The bedrock of the area is composed of 9 Ma-old basalts and 1-4 Ma-old acid volcanics (Gasparon *et al.*, 1993). The maars, cinder cones, and lava flows represent more recent (10ka) volcanic activity. Ali (1999) determined the transmissivity of the older basaltic aquifer as ranging between 389 and 21600 $m^2$/day. The younger basic pyroclastic rocks interbedded with minor acidic products make up the largest part of the study area and have transmissivities ranging from 1100 to 18000 $m^2$/day. The scoria cones and the acid volcanic domes are believed to be the major zones of groundwater recharge. The static groundwater level is well above lake levels, implying the importance of groundwater inflow to the lakes.

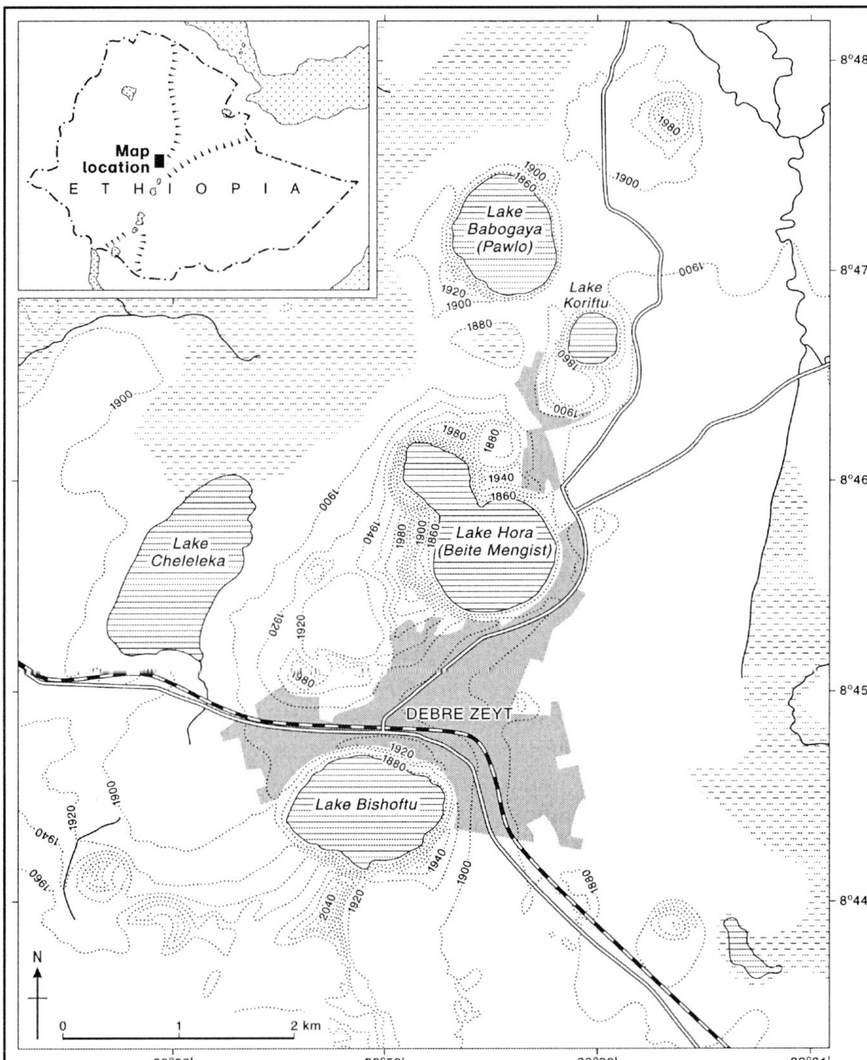


*Figure 1.* Location of the Bishoftu crater lakes.

*Table 1.* Morphology of the Bishoftu crater lakes, modified after Prosser *et al.* (1968).

| | *Kilole* | *Arenguade* | *Babogaya* | *Hora* | *Bishoftu* |
|---|---|---|---|---|---|
| Altitude(m asl) | 2000 | 1900 | 1870 | 1850 | 1870 |
| Lake area ($km^2$) | 0.771 | 0.541 | 0.579 | 1.029 | 0.929 |
| Catchment area ($km^2$) | 3.103 | 0.808 | 0.293 | 0.628 | 0.368 |
| Max. depth (m) | 6.4 | 32 | 68 | 38 | 87 |
| Ave. depth (m) | 2.6 | 18.5 | 38.0 | 17.5 | 55.0 |
| Volume ($10^6 m^3$) | 2 | 10 | 22 | 18 | 52 |

## 2.2 Climate

The study site is characterised by a semi-arid climate, with a mean rainfall of 830mm, and large seasonal and inter-annual rainfall variability (Gemechu, 1977). 'Little Rains' from March to May result from southeasterly airflow from the Indian Ocean. From mid June to mid September heavy rainfall (the 'Big Rains') originating from the Atlantic Ocean dominates. The dry season occurs between October and February. Temperature shows large diurnal but small seasonal changes with an annual average of 19ºC. The average annual humidity is 0.60, varying from 0.53 to 0.70. Penman (1948), energy balance and pan evaporation methods were combined to give an annual average open water evaporation rate of 1710 mm (Kebede, 1999).

## 2.3 Isotopic Composition of Rainfall

The long-term (1961-1996) mean weighted values for $\delta^{18}O$ and $\delta D$ for Addis Ababa (2360 m elevation) are -1.3‰ and 1.8‰ respectively (Rozanski *et al.*, 1996). The data for Addis Ababa station define a local meteoric water line (LMWL) $\delta D = (6.95 \pm 0.22)\ \delta^{18}O + 11.5 \pm 10.58$ (Figure 2). A single rainfall sample collected from the study site had an isotopic composition of -3.13‰ for $\delta^{18}O$ and -13‰ for $\delta D$, and lies close to the LMWL. There is a poor correlation between isotopic composition of meteoric waters and altitude in the Ethiopian Rift zone (Kebede, 1999).

## 2.4 Lake Chemistry and Chloride Mass Balance

Sodium ($Na^+$) and bicarbonate ($HCO_3^-$) are the dominant ions in all the crater lakes, coupled with low amounts of calcium ($Ca^{++}$) and magnesium ($Mg^{++}$). A chloride mass balance model shows that the lakes vary widely in their hydrologic properties (Lamb *et al*, this volume). The rate of groundwater outflow is greatest for Lake Babogaya, and least for Lake Hora (less than 3% of the total water loss) which can therefore be considered as a terminal lake.

## 3. METHODS

Water samples for isotope analysis were collected from January 1998 to January 1999 in leak-tight polyethylene bottles, additionally sealed with PVC tape. Samples were stored in a refrigerator prior to laboratory analysis at the NERC Isotope Geosciences Laboratory, Keyworth. Waters were treated with zinc turnings at 500°C to generate hydrogen for the D/H analysis, and equilibrated with $CO_2$ for $^{18}O/^{16}O$ analysis. Mass spectrometry was performed on a VG SIRA + ISOPREP 18 in conjunction with laboratory standards calibrated against VPDB and VSMOW-SLAP scales. Results are reported in the  notation in per mil versus these standards; reproducibility is better than 0.02 and 2‰ for $\delta^{18}O$ and $\delta D$ respectively.

## 4. RESULTS

The results of isotopic analyses are given in Figure 2 and Table 2. The lakes, and the northern and southern groundwater samples have distinctive isotopic signatures. The mean oxygen isotopic composition of lakes Hora, Babogaya, and Bishoftu is +7.2‰, +6.5‰ and +6.9‰ respectively. The oxygen isotopic composition of groundwater ranges from -3.3‰ to +4.0‰. Water from swamps, reservoirs and streams generally shows isotopic compositions intermediate in value between groundwater and lakes.

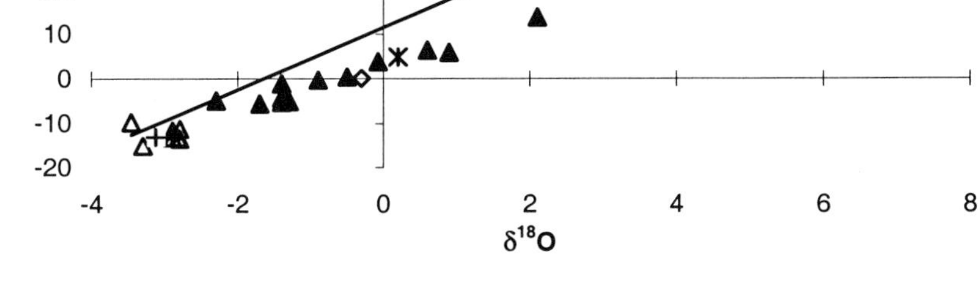


*Figure 2.* Isotopic composition of waters in the study area. Black diamonds = lakes; open diamonds = streams; black triangles = groundwaters from S, SW and SE; open triangles = groundwaters from N, NW, and NE; * = swamps, ponds and artificial lakes; + = rainfall. Also shown is the local meteoric water line for Addis Ababa (Rozanski *et al.*, 1996).

*Table 2.* Oxygen and deuterium isotopic composition of waters from Debre Zeit.

| *No* | *Code* | *Date* | Site | $\delta^{18}O$ | $\delta D$ |
|---|---|---|---|---|---|
| 1 | BAB98-1 | Jan-98 | Babogaya, margin | 6.5 | 38.9 |
| 2 | BAB98-2 | Jan-98 | Babogaya, 60m depth | 6.7 | 42.5 |
| 3 | BAB98-4A | Jan-98 | Babogaya, suface, center | 6.7 | 38.9 |
| 4 | BAB98-3 | Jan-98 | Babogaya,10m depth | 6.5 | 39.6 |
| 5 | HBM98-1 | Jan-98 | Hora, marginal (weeds) | 7.2 | 37.6 |
| 6 | HBM98-2 | Jan-98 | Hora, marginal openwater | 7.1 | 37.0 |
| 7 | HBM98-3 | Jan-98 | Hora, lake bottom (30m) | 7.2 | 37.6 |
| 8 | HBM98-4 | Jan-98 | Hora, 20m depth | 7.1 | 38.4 |
| 9 | HBM98-5 | Jan-98 | Hora, 10m depth | 7.2 | 37.4 |
| 10 | HBM98-6 | Jan-98 | Hora, surface lake center | 7.1 | 37.8 |
| 11 | B98-1 | Jan-98 | Bishoftu, margin | 6.9 | 42.9 |
| | | | ***Mean, lakes*** | ***6.9*** | ***39.0*** |
| 12 | CH98-1 | Jan-98 | Cheleleka, south shore | 7.0 | 40.6 |
| 13 | CH98-2 | Jan-98 | Cheleleka, east shore | 5.8 | 33.4 |
| 14 | KU98-1 | Jan-98 | Kuriftu, margin | 5.4 | 29.5 |
| 15 | KU98-2 | Jan-98 | Kuriftu, center | 5.3 | 31.4 |
| 16 | K98-1 | Jan-98 | Kilole, margin | 0.2 | 4.9 |
| | | | ***Mean, swamp and reservoir*** | ***4.7*** | ***28.0*** |
| 17 | MOJO98-1 | Jan-98 | Mojo river | -0.3 | 0.1 |
| 18 | WR98-2 | Jan-98 | Wedecha river | 4.2 | 28.4 |
| 19 | WT98-1 | Jan-98 | Wedecha tributary | 4.1 | 27.5 |
| | | | ***Mean, rivers*** | ***2.7*** | ***18.7*** |
| 20 | DZ98-3 | Jan-98 | Shimbra meda well field | -2.8 | -13.5 |
| 21 | DZ98-2 | Jan-98 | Management institute well | -2.9 | -11.6 |
| 22 | 66 | Jan-99 | Dragados DZ well | -3.3 | -15.0 |
| 23 | 63 | Jan-99 | Management institute well | -2.9 | -13.1 |
| 24 | DZ98-1 | Jan-98 | Shimbrameda well/tap | -2.9 | -13.4 |
| 25 | 60 | Jan-99 | Agri. Res center well 2 | -2.8 | -11.5 |
| 26 | 57 | Jan-99 | Elfora | -3.4 | -9.7 |
| | | | ***Mean, northern groundwater*** | ***-3.0*** | ***-12.5*** |
| 27 | 58 | Jan-99 | Hora Tannary well | -1.7 | -5.6 |
| 28 | 69 | Jan-99 | Almaz Doro well | -1.4 | -5.2 |
| 29 | 70 | Jan-99 | Slaughter House well | -1.4 | -4.3 |
| 30 | 62 | Jan-99 | Agri. Res. Inst well 1 | -1.4 | -1.1 |
| 31 | 61 | Jan-99 | N. Franco well | -0.9 | -0.2 |
| 32 | 67 | Jan-99 | B.Nile Plastic Factory well | -0.5 | 0.5 |
| 33 | 65 | Jan-99 | Defence Eng. well 1 | 0.9 | 6.0 |
| 34 | 59 | Jan-99 | Ada Pasta Factory well | 0.6 | 6.5 |

| | | | | | |
|---|---|---|---|---|---|
| 35 | 68 | Jan-99 | Hora Agro Industry well | 2.1 | 13.8 |
| 36 | 64 | Jan-99 | Defence Eng. well 2 | 0.7 | 22.2 |
| 37 | CHW98-2 | Jan-98 | Royal Hotel hand dug well | -1.3 | -5.0 |
| 38 | DZ98-1 | Jan-98 | Kebele hand dug well | -0.1 | 3.9 |
| 39 | 73 | Jan-99 | Ato Gizaw hand dug well | -2.3 | -4.9 |
| | | | ***Mean, southern groundwater*** | ***-0.5*** | ***2.0*** |

## 4.1 Lake-Groundwater Relationships

The groundwaters from north (samples 20, 24), northwest (22, 26) and northeast (21, 23, 25) of the lakes plot close to the meteoric water line (Figure 2), ruling out secondary processes, such as evaporation prior to infiltration, or isotopic exchange with aquifer rocks and the lakes. The groundwaters are probably recharged rapidly through fractures, faults, joints and scoria cones with little influence of evaporation before recharge. The isotopic composition of wells found in south, southwest (29, 32, 33, 36, 34, 37), and southeast (38) are enriched and plot on a groundwater-lake mixing line, indicating that groundwater is mixed with evaporated surface water from lakes, swamps and reservoirs. The enrichment of wells between lakes Babogaya and Hora (27, 31, 28 and 35) indicates the presence of hydraulic links between the lakes. Darling (1996) reported similar lake-groundwater mixing in the Bishoftu area, facilitated by fracture zones and faults that underlie the maars. The data show that groundwater flow is generally from north to south, an observation that is confirmed by static water level mapping.

Isotopic enrichment of the lakes (samples 1-11) relative to precipitation and groundwater results from evaporation during the 6-10 year lakewater residence times (Lamb *et al.*, this volume). Seasonal and depth variation in the oxygen isotopic composition of lakes Babogaya and Hora is negligible (Lamb *et al.*, this volume), indicating that the lakes are isotopically well mixed, despite wet-season stratification. Lakes Kilole (16) and Kuriftu (14 and 15) show relatively depleted values, which may be due to flushing by isotopically depleted river water or shorter water residence times. Water samples from the perennial streams Mojo and Wedecha and a tributary stream (17,18,19) show isotopic enrichment relative to groundwater and rainfall, probably because these stream waters are held in Balabala reservoir upstream of the collection point. The isotopic composition of the evaporated water bodies and waters influenced by them fall on an evaporation line defined by the equation $\delta D = 4.7\delta^{18}O + 6.3$ (Figure 2) which is close to the East African Lakes Evaporation Line (Craig *et al.*, 1977). The LMWL intersects with this local evaporation line at a point close to -2.8‰ for $\delta^{18}O$ and -13‰ for $\delta D$. The $\delta^{18}O$ intersection value may be used to estimate the weighted average isotopic composition of lake inflow (Dincer, 1968).

## 4.2 Lake Isotope Budgets

### 4.2.1 Theoretical Considerations

In hydrologic and isotopic steady state, hydrologic and isotope balance equations can be combined to separate the net groundwater flux into inflow and outflow components. The general water budget equation for a closed lake at hydrologic steady state is:

$$dV/dt = P + R + Gi - E - Go \qquad (1)$$

where $P$, $R$, and $E$ are precipitation, runoff and evaporation respectively. $Gi$ and $Go$ are groundwater inflow and outflow. $V$ is the volume of water in the lake. Equation 1 can be modified by substituting $R$ with $kP$, where $k$ is a runoff coefficient, to overcome the lack of measurements of surface runoff.

Similarly, the expression for the isotope mass balance of a lake at steady state is given by

$$d(\delta_L V)/dt = \delta_p P + \delta_p kP + \delta_{gi} Gi - \delta_{go} Go - \delta_E E \quad \text{(Pearson and Coplen, 1978)} \qquad (2)$$

where each term in equation 1 has been multiplied by its respective isotopic composition expressed in delta notation in units of per mil (‰) relative to SMOW. With complete lake mixing, $\delta_{go} = \delta_L$, so equation 2 can be further simplified as:

$$d(\delta_L V)/dt = \delta_p P + \delta_p kP + \delta_{gi} Gi - \delta_L Go - \delta_E E \qquad (3)$$

Combining equations 1 and 3:

$$Gi = [P(1+k)(\delta_L - \delta_p) + E(\delta_E - \delta_L)] / [\delta_{gi} - \delta_L] \qquad (4)$$

Equation 4 describes a lake budget in which the groundwater inflow can be determined independent of the groundwater outflow. However, difficulties arise because $\delta_E$, isotopic composition of evaporating water, is difficult to measure. However, $\delta_E$ can be obtained from Craig and Gordon's (1965) relation:

$$\delta_E = [\alpha^* \delta_L - h\delta_A - \varepsilon] / [1 - h + 10^{-3} \Delta\varepsilon] \qquad (5)$$

Where $\alpha^*$ is the equilibrium isotopic fractionation factor at the kinetic temperature of the evaporating surface ($1/\alpha^*$ is equal to 1.0098 for $\delta^{18}O$ at 25°C, Krabbenhoft *et al*, 1990), $h$ is the relative humidity normalized to water surface temperature, $\delta_A$ is the isotopic composition of local atmospheric vapor, $\varepsilon$ is the total fractionation factor, and $\Delta\varepsilon$ is the diffusion controlled or kinetic fractionation factor. The total fractionation factor is the sum of equilibrium and kinetic fractionation factors:

$$\varepsilon = \varepsilon^* + \Delta\varepsilon \qquad \text{(Krabbenhoft } et\ al.\text{, 1990)} \qquad (6)$$

where $\varepsilon^*$ is a convenient way to express the equilibrium fractionation factor in parts per thousand, $\varepsilon^*=1000\,(1-\alpha^*)$. The kinetic fractionation factor is given by the relation

$$\Delta\varepsilon = K\,(1-h) \qquad (7)$$

where $K$ has been empirically determined to be 14.3λ and 12.5λ for $\delta^{18}O$ and δD respectively (Gonfiantini, 1986).

Although Craig and Gordon (1965) formulated a suitable basis to relate the composition of $\delta_E$ to measurable system parameters through equation 5, the use of the isotope mass balance (i.e. equation 4) is still limited. This is because equation 5 introduces a new unknown, the atmospheric composition of local vapor $\delta_A$. It is the lack of measurement of $\delta_A$ that poses problems in the use of the isotope balance method. The method is suitable either where measurements of $\delta_A$ are available, or where extrapolation can be made from a lake whose water balance has been carefully determined - the index lake method. This method, whereby the isotope mass balance can be determined without reverting to the measurement of the environmental parameter ($\delta_A$), has been used here and in other studies (Dincer, 1968; Gat and Levy, 1978). Terminal lakes have been used as an index lakes in isotope balance studies (IAEA, 1981).

#### 4.2.2 The Index Lake Method

For the Bishoftu crater lakes, an overall isotopic and hydrologic steady state is assumed. The varied hydrology of the lakes provides a unique opportunity to solve the isotope balance without directly measuring the ambient parameters $\delta_A$. The chloride mass balance shows that Lake Hora may be considered as an index lake, because its water loss is dominated by evaporation, with negligible groundwater outflow (Lamb *et al*, this volume), and because its salinity is not high enough to influence isotopic fractionation processes.

The isotopic composition of a terminal lake with inflows equaling evaporative loss will approach a steady state value, so that

$$\delta_E = \delta_{in} \qquad \text{(IAEA, 1981)} \qquad (8)$$

where $\delta_{in}$ is the weighted average isotopic composition of all inflow to the lake. The intersection point of LMWL and Local Evaporation Line (LEL) of a given site may be used to estimate the weighted average isotopic composition of inflow ($\delta_{in}$) (Dincer, 1968). The intersection point is -2.8λ for $\delta^{18}O$ for the Bishoftu crater lakes, as shown above.

Using the terminal lake Hora as an index lake and substituting $\delta_E = \delta_{in} = -2.8\lambda$ into equation 5, the unknown $\delta_A$ is determined to be -12λ. This value is reasonably close to that of -15λ determined for the Afar region by Gonfiantini *et al.* (1973), and that of -9λ determined for West Africa by Turner *et al.* (1994). As a further check on the derived value of $\delta_A$, it can be used to calculate the limiting isotopic composition $\delta^*$ for the region. This limiting isotopic composition is the isotopic enrichment value approached by residual waters of a shrinking water body. It is given as:

$$\delta^* = \delta_A + \varepsilon / h \qquad \text{IAEA (1981)} \qquad (9)$$

The relative humidity ($h$), normalized to surface (kinetic) temperature is 0.6; $\varepsilon$ and $\Delta\varepsilon$ are determined from the empirical relations of equations 6 and 7. The limiting isotopic composition $\delta^*$ for the region is calculated to be +14.2λ. This value is slightly higher than the maximum oxygen isotopic composition reported for the lakes of the Ethiopian Rift which is +10.7λ (Craig *et al.*, 1977).

Further simplification of equations 4 and 5 can be made by removing $\alpha^*$ because it is close to unity and therefore does not alter $\delta_L$, and by removing $10^{-3} \times \Delta\varepsilon$ because it is close to zero and therefore does not contribute much to the denominator. This more easily manageable equation, originally derived by Gat and Levy (1978), is given as:

$$\delta_{Lss} = \frac{\left[\left(\frac{\varepsilon}{h} + \delta_A\right) + \frac{I}{E}\left(\frac{1-h}{h}\right)\delta_{in}\right]}{\left[1 + \frac{I}{E}\left(\frac{1-h}{h}\right)\right]} \qquad (10)$$

where $\delta_{Lss}$ is the steady state isotopic composition of the lake and $I$ is the total water inflow ($Gi + P + (R = kP)$).

Equation 10 can also be written as:

$$\frac{I}{E} = \frac{h}{1-h}\left(\frac{\delta^* - \delta_{Lss}}{\delta_{Lss} - \delta_{in}}\right) \qquad (11)$$

### 4.2.3 Lake Water Budgets from the Isotopic Mass Balance

For Lake Hora, the water balance is solved by directly substituting $Go = 0$ in equation 1 and calculating the unknown $G_i$. The solution of this equation is given in

Table 3. For lakes Babogaya and Bishoftu, total inflow (*I*) obtained by substituting $\delta_A$ = -12λ, $h$ = *0.6*, $E$=*1710*, $\delta^*$ =14.2λ, and the average oxygen isotopic compositions of lake Babogaya (6.5λ) and Bishoftu (6.9λ) into equation 11. The total inflow (*I*) is further separated into the unknown groundwater inflow and the known *P* ($P=0.83A_L$) and *R* ($R=0.83\kappa PA_c$). A runoff coefficient of *0.5* is assumed for the calculation. Groundwater outflow is obtained from the difference between *I* and *E*. The water budget of the two lakes is given in units of $m^3$/yr., and as the percentage each component of the water budget (Table 3).

*Table 3.* Water budgets of the Bishoftu crater lakes. Units are given in $m^3$/yr and as a percentage of the total hydrologic inflow or outflow.

| | *P* | *Gi* | *R* | *E* | *Go* |
|---|---|---|---|---|---|
| Hora | 854070 | 552355 | 353165 | 1759590 | 0 |
| | 49% | 31% | 20% | 100% | 0% |
| Babogaya | 480570 | 587514 | 121595 | 990090 | 199590 |
| | 40% | 50% | 10% | 83% | 17% |
| Bishoftu | 771070 | 804586 | 160190 | 1588590 | 147256 |
| | 44% | 46% | 9% | 92% | 8% |

The results show that Lake Babogaya differs from the other lakes in having a significant amount of groundwater outflow and inflow. Lake Bishoftu has groundwater flux and evaporative flux rates that are intermediate between the terminal lake Hora and the through-flow lake Babogaya. The accuracy of the water budget estimate for lakes Babogaya and Bishoftu depends on the validity of the assumption of the index-lake method. One way to assess the accuracy of the water budget estimates computed above is to compare them to estimates from other independent methods. There is a good agreement between water budgets determined by chloride mass balance (Kebede *et al*., in press) and isotopic approaches (Table 4).

*Table 4*. Comparison of groundwater inflow rates in $Mm^3$.

| Lake | Chloride method | Isotope method |
|---|---|---|
| Babogaya | 0.7 | 0.6 |
| Bishoftu | 0.7 | 0.8 |
| Hora | 0.7 | 0.6 |

The slight difference that exists between the two approaches may be attributed to the degree of precision in calculating $\delta_E$ using the terminal lake approach and the validity of considering lake Hora as a terminal lake which has 3% groundwater outflow. The similarity between the chloride mass balance and isotope budget calculation shows that the index lake approach is a valid means of determining the water budget of the Bishoftu crater lakes.

## 4.3 Climate Sensitivity of Lakes Hora and Babogaya

We used an isotopic model of lakes Hora and Babogaya to test the sensitivity of their oxygen isotope composition to climatic and hydrological variation. This approach may support quantitative paleoclimatic and paleohydrological interpretation of isotopic data from limnic carbonates (Turner *et al.*, 1994).

Equation 3, the isotopic balance of a well-mixed lake can be rewritten as:

$$\delta_{Lss} = \frac{I\delta_{in} - Q_o\delta_{Go} - E\delta_E}{\Delta V}\Delta t \qquad (12)$$

By substituting equation 5 into 12, and after further simplification, the following model can be formulated for lakes at isotopic steady state

$$\delta_{Lss} = \frac{\left[I\delta_{in}(1-h) + E(h\delta_A + 14.3 - 15h + \alpha^*)\right]}{(1-h)\dfrac{\Delta V}{\Delta t} + (1-h)Q_{go} + \alpha E} \qquad (13)$$

where $\delta_{Lss}$ is the steady state isotopic composition of the lake.

To demonstrate the sensitivity of the model to the changes in $h, \delta_{in}$, and $\delta_A$, the water balance components $I$, $E$, and $G_o$ were held constant, and $h, \delta_{in}$, and $\delta_A$ varied for each lake.

### 4.3.1 Model output

The model output, shown in Figure 3, indicates that the isotopic composition of the lake water is more sensitive to changes in the climate-controlled parameters $h$ and $\delta_A$, and less sensitive to the isotopic composition of inflow ($\delta_{in}$). For example, increasing h and $\delta_A$ by 10 units shifts the isotopic composition by 20 and 7 units respectively. Shifting $\delta_{in}$ by 10 units changes the isotopic composition by no more than 4 units.

Lake Hora is slightly more sensitive than Lake Babogaya to changes in humidity and isotopic composition of atmospheric vapor. These atmospheric moisture parameters $h$ and $\delta_A$, which control the stable isotopic composition of the lake water, are negatively correlated. $\delta_A$ is positively correlated with isotopic composition of rainfall and negatively correlated with relative humidity, so these climate factors will mutually reinforce one another to produce a lower $\delta_L$ under higher rainfall conditions.

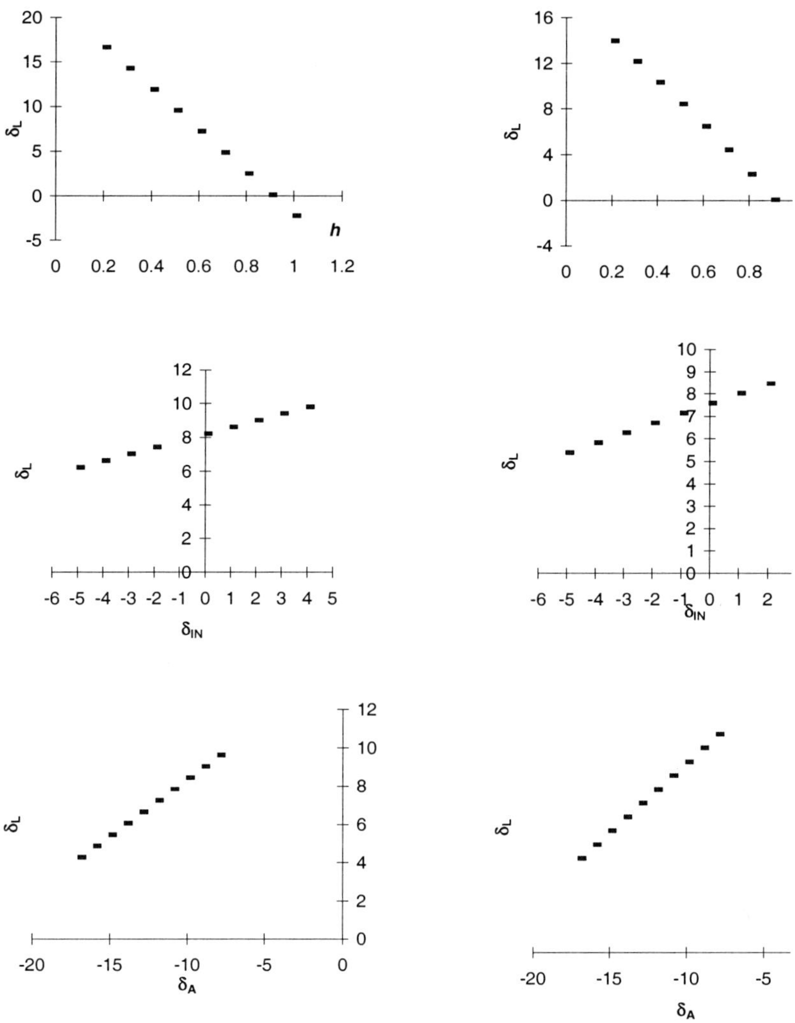


*Figure 3*. Sensitivity of oxygen isotopic composition of Lake Hora (left column) and Lake Babogaya (right column) to changes in humidity (top row), isotopic composition of inflow (middle row), and isotopic composition of water vapour (bottom row).

# 5. SUMMARY AND CONCLUSIONS

The isotopic approach demonstrated here appears to be a promising tool for solving the water balance of groundwater-fed lakes. In the case of the Bishoftu crater lakes, the aquifer is rapidly recharged by rainfall, principally during the summer wet season, without prior surface evaporation. Seepage from the evaporatively enriched lakes mixes with groundwater to the south of the lakes. The average isotopic composition of lake inflow, determined from the intersection of the local evaporation and meteoric water lines, allows an estimate of the isotopic composition of ambient moisture ($\delta_A$), by treating Lake Hora as a terminal index lake. The estimated value of $\delta_A$ is -12λ, which compares well with data previously obtained for the region, can then be used to estimate groundwater fluxes from the isotopic mass balance equation, and confirms that Lake Babogaya is most influenced by groundwater throughflow. The isotope balance can also be used to test the sensitivity of the lakewater isotopic composition to changing humidity, composition of atmospheric water vapor, and composition of inflow. The results suggest that lake oxygen-isotope composition is more sensitive to humidity and isotopic composition of ambient vapor (factors that are related to rainfall amount) than to the isotopic composition of inflow, and thus provide a basis for the quantitative interpretation of stratigraphic variation in the oxygen-isotope composition of authigenic carbonates in terms of past rainfall variation.

# REFERENCES

Ali, A. (1999) *River-Groundwater Interactions in the Sekelo-Akaki Basin*. Unpublished M.Sc thesis, Addis Ababa University.

Almendinger, J.A. (1990). Groundwater control of closed-basin lake levels under steady state conditions. *Journal of Hydrology* **112,** 293-318.

Ayenew, T. (1998) *The Hydrogeological System of the Lake District Basin, Ethiopia*. Ph.D. thesis, ITC, Enschede, The Netherlands.

Crowe, A.S. (1990). Numerical modelling of the groundwater contribution to the hydrological budget of lakes, in E.S. Simpson, E.S. and J.M. Sharp (eds.), *Selected Papers on Hydrogeology* IAH pub. **1**, 283-294.

Crowe, A.S. and Schwartz, F.W. (1981) Simulation of lake watershed systems. I. Description and sensitivity of the model. *Journal of Hydrology* **52**, 71-105.

Craig, H., and Gordon, L.I. (1965) Deuterium and oxygen-18 variations in the ocean and marine atmosphere, in E. Tongiorgi (ed.) *Stable Isotopes in Oceanographic Studies and Paleo-temperatures, Spoleto 1965,* Consiglio Nazionale della Richerche, Pisa, pp. 9-130.

Craig, H., Lupton, J.E., and Horowiff, R.M. (1977) Isotope Geochemistry and Hydrology of Geothermal Waters in the Ethiopian Rift Valley. *Scripps Inst. of Oceanogr. Univ. of California report,* 160pp.

Darling, W.G. (1996) *The Geochemistry of Fluid Processes in the Eastern Branch of the East African Rift System*. Ph.D. thesis, British Geological Survey, 235pp.

Dincer, T. (1968) The use of oxygen 18 and deuterium concentrations in the water balance of lakes. *Water Resources Research* **4**, 1289-1306.

Gasparon, M., Innocenti, F., Manetti, P., Peccerillo, A., and Tsegaye, A (1993) Genesis of Pliocene to Recent mafic-felsic volcanism in the Debre Zeyt area, central Ethiopia: volcanological and geochemical constraints. *Journal of African Earth Sciences* **17**, 145-165.

Gat, J.R., and Levy, Y. (1978) Isotope hydrology of inland sabkhas in the Bardawil area, Sinai. *Limnology and Oceanography* **23**, 841-850.

Gemechu, D. (1977) *Aspects of Climate and Water Budget in Ethiopia*. Addis Ababa University Press. Addis Ababa.

Gonfiantini, R., Borsi, S., Ferrara, G., and Panichi, C. (1973) Isotopic composition of waters from the Danakil depression (Ethiopia). *Earth and Planetary Science Letters* **18**, 13-21.

Gonfiantini, R. (1986) Environmental isotopes in lakes studies, *in* Fritz and Fontes (eds.), *Handbook of Environmental Isotope Geochemistry*, Elsevier.

IAEA, (1981) Stable isotope hydrology: deuterium and oxygen-18 in the water cycle, technical reports series no.201, Vienna, 203-221.

Kebede S., Lamb, H. F., Mohammed, U. and Telford, R. (in press) The relation between hydrology and hydrochemistry of Bishoftu crater lakes, Ethiopia. *SINET: Ethiopian Journal of Science*

Kebede, S. (1999) *The Hydrology and Hydrochemistry of Bishoftu Crater Lakes. Hydrological and Isotope Modeling*. Unpublished M.Sc thesis, Addis Ababa University, 125pp.

Krabbenhoft, D.P., Bowser, C.J., Anderson, M.Y., and Valley, J.W. (1990) Estimating Groundwater Exchange with lakes, the stable isotope mass balance method. *Water Resources Research* **26**, 2445-2453.

Krabbenhoft, D.P., Bowser, C.J., Kendall, C., and Gat, J. (1994) Use of oxygen-18 and deuterium to asses the hydrology of groundwater-lake systems. *In* Environmental chemistry of lakes and reservoirs, American Chemical Society, USA.

Lamb, H., Kebede, S., Leng, M., Ricketts, D., Telford, R. and Mohammed, U. (this volume) Origin and isotopic composition of aragonite laminae in an Ethiopian crater lake.

Lorenz, V. (1986) On the growth of maars and diatremes and its relevance to the formation of tuff rings. *J. Volcanology* **48**, 265-274.

Pearson, F.J., and Coplen, T.B. (1978) Stable isotope studies of lakes, in A. Lerman (ed.), *Lakes: Geology, Chemistry, Physics*, Springer.

Penman, H.L. (1948) Natural evaporation from open water, bare soil, and grass. *Royal Soc. London Proceed*, Series A, **193**, 120-145.

Prosser, M.V., Wood, R.B., and Baxter, R.M. (1968) The Bishoftu Crater Lakes: A bathymetric and chemical study. *Archiv. Hydrobiol.* **65**, 309-324.

Ricketts, R.D. and Johnson, T.C. (1996) Climate change in the Turkana basin as deduced from a 4000 year long $^{18}O$ record. *Earth and Planetary Science Letters* **142**, 7-17.

Rozanski, K., Araguas-Araguas, L., and Gonfiantini, R. (1996) Isotope patterns of precipitation in East Africa, in T.C. Johnson and E.O. Odada (eds.), *The Limnology, Climatology and Paleoclimatology of the East African Lakes,* Gordon and Breach, Amsterdam, pp 79-93.

Turner, B.F., Gardner, L.R., Sharp, W.E., and Blood, E.R. (1994) The geochemistry of lake Bosumtiwi, a hydrologically closed basin in the humid zone of tropical Ghana. *Limnology and Oceanography* **41**, 1415-1424.

Winter, T.C. (1978). Groundwater component of lake water and nutrient budgets. *Verh. Int. Ver. Limnol.* **20**, 438-444.

# A REVIEW OF SEDIMENT GAS CYCLING IN LAKES WITH REFERENCE TO LAKE VICTORIA AND SEDIMENT GAS MEASUREMENTS IN LAKE TANGANYIKA

DONALD D. ADAMS[1] AND SAMUEL O. OCHOLA[2]

[1]*Center for Earth and Environmental Science, State University of New York, Plattsburgh, NY 12901 U.S.A.*

[2]*Department of Geology, University of Nairobi, PO Box 30197, Nairobi, Kenya*

**ABSTRACT**

Little is known about gases in the sediments or the water column of African lakes. Considering the importance of lakes as life-supporting resources, these are critically neglected areas of limnologic investigation. The cycling of nutrients for food web requirements and maintenance of upper trophic levels, i.e. fish and fisheries, and a basic understanding of ecosystem processes requires a fundamental knowledge of the C and N cycles. As examples, nitrogen seems to be a limiting nutrient in these tropical lakes, likely a result of denitrification and $N_2$ export to the atmosphere; $CH_4$ loss from the sediments can represent a critical pathway for bacterial C production and lake deoxygenation. Sediment gases ($CH_4$, $CO_2$ and $N_2$) in the central basin of Lake Tanganyika were measured, probably for the first time. Total headspace concentrations (mM $\pm$ 1$\sigma$, range) were 0.47$\pm$0.18, 0.14-0.88 $CH_4$, 0.37$\pm$0.18, 0.12-0.75 $CO_2$ and 0.97$\pm$0.50, 0.1-1.97 $N_2$ for 33 sediment samples, with carbon gases increasing with depth to 35 cm. The high sediment carbon (TOC) and low C gas concentrations suggest the importation of recalcitrant C, coupled with extended preservation under continual anoxia at these Kalya Horst sites. It is unknown whether the sediments are present day sources of reduced gases and $CO_2$ to the overlying water column. High concentrations of $CO_2$ and $H_2S$ in deep waters suggest other sources, such as the possibility of geothermal emissions coupled with their long term accumulation in this permanently stratified system.



*E.O. Odada and D.O. Olago (eds.),*
*The East African Great Lakes: Limnology, Palaeolimnology and Biodiversity,* 277–305.

# 1. INTRODUCTION

The importance of gases as part of the carbon and nitrogen budgets of the African lakes has been addressed in only a few publications. With exception of dissolved oxygen, information concerning dissolved gases in the water column is fragmentary and scarce: Degens *et al.* (1973), Deuser *et al.* (1973), Hecky and Degens (1973), Jannasch (1975) and Tietze *et al.* (1980) for gas-charged Lake Kivu in East Africa, Rudd (1980) and Edmond *et al.* (1993) for Lake Tanganyika and Kling *et al.* (1987, 1991) for Lakes Nyos and Monoun in Cameroon. Kling *et al.* (1991) also provided water column gas data for other Cameroon lakes. In contrast to these listed references, there seems to be an even greater scarcity of published information for gases in African lake sediments, with no information concerning concentrations of major sediment gases, such as methane and carbon dioxide. There are a few publications which relate to the presence of sediment gases: Cerling (1996) discussed the processes of methanogenesis in sediment cores from Lake Turkana where $^{13}C$-enriched carbonate nodules were observed, De Batist *et al.* (1996) described a gas-enhanced reflector in the northern Livingston Basin sediments of Lake Malawi, most likely associated with recently-deposited deltaic sediments, and Scholz *et al.* (1990) evaluated the importance of sediment gases, possibly in gas-charged muds, in forming an acoustic Type B basement during echo-soundings taken in Lakes Victoria and Turkana. Johnson (1996) suggested that substantial amounts of gas, likely methane, caused disturbance and diffuse reflectivity in seismic profiles; cores from these areas contain high quantities of gas. He hypothesized that methanogenesis could be an important agent in sediment disturbance and would likely contribute to relief instability and downslope sediment transport. It seems, however, that none of these authors measured the sediment gas concentrations or composition, so much of their findings are conjectured. From previous studies, it is likely that gases of biogenic origin should be present in high abundance in some of the African Great Lake sediments, as observed in other lakes undergoing various conditions of eutrophication (Adams 1992a; Adams and Naguib, 1999). Even though there are little to no data because of the difficulties in collecting and analyzing sediment gases, their measurements are important in understanding many limnologic processes, such as hypolimnetic deoxygenation (Hecky and Bugenyi, 1992; Ochumba, 1996), cycling of carbon and nitrogen as described in the African lake systems (Livingstone and Melack, 1984; Johnson and Odada, 1996), and nutrient limitation phenomenon (Hecky *et al.,* 1996), to list a few. For example, the flux of methane from sediments to the overlying water column can be a major contributor to lake deoxygenation (Adams *et al.,* 1982; Adams, 1999). It is also expected that nitrogen limitation for many of these lakes (Livingston and Melack, 1984; Hecky and Bugenyi, 1992) could be related to $N_2$ loss during denitrification (Hecky *et al.,* 1996), where the focii of N-reduction normally occurs within low-$O_2$ waters in the vicinity of oxic-anoxic boundaries in the water column (Downs, 1988; Hecky *et al.,* 1996) or sediments (Fendinger and Adams, 1987). Examples of denitrified-N losses are: $N_2$ fluxes averaged 35% of total N for Narraganset Bay (Seitzinger *et al.,* 1984), 11-60% for 12 lakes in Denmark, Sweden and the U.S. (Tirén, 1977) and 65-88% for two eutrophic

Swiss lakes (Mengis *et al.,* 1997). As the rates of water eutrophication increase, as observed in the African Great Lakes (Cohen *et al.,* 1996; Hecky *et al.,* 1996), it is expected that the processes of denitrification will become even more important (Seitzinger and Nixon, 1985). The loss of nitrogen from these lake systems, along with increased phosphorus loading, will promote $N_2$-fixing, heterocystous cyanobacterium blooms; these microorganisms compensate for water column N-limitation by extracting $N_2$ directly from the water or atmosphere.

Even though little is known in the literature about the concentrations of naturally-occurring, dissolved biogenic gases and their cycling in African lakes, especially within bottom deposits, information from and references to other aquatic environments could provide assistance to those wishing to study sediment gases. Therefore, sampling techniques and some data from other sedimentary environments will be applied, when possible, to conditions for African lakes and especially Lakes Victoria and Tanganyika, at least with respect to the occurrence of natural biogenic gases. Significant differences in water column, sediment and geologic characteristics and climatic setting could make these comparisons questionable, yet it is hoped that this information would provide an initial step towards sediment gas research in the African lakes. Results from a recent study (July 2001) in the central basin of Lake Tanganyika, as part of the Nyanza Project, are included as a comparison to studies on other lakes; as far as the authors are aware, this is the first data set for sediment gases in East African lakes.

## 2. SEDIMENT SAMPLING AND GAS ANALYSIS

Sediment gases are routinely collected by two techniques - coring followed by sediment sampling and *in situ* deployment of sediment pore water equilibration devices. There are other methods being developed: microelectrodes which are capable of directly measuring some gases ($O_2$, $H_2$, $N_2O$, $H_2S$, etc.) and gas diffusion probes (silicon-covered tubing) for collecting pore water gases, with direct coupling to measuring devices (see Rothfuss *et al.,* 1994; Kühl *et al.*, 1998). Methane in anoxic paddy soils was measured by the latter technique (Rothfuss and Conrad, 1998). The coring technique lacks the capabilities for fine structure (mm) measurements at the sediment-water interface, but it is robust and has been tested under numerous field conditions. The *in situ* equilibration technique [called peeper, Hesslein (1976) or memocell, Van Eck and Smits (1986)] is useful for measuring the interface but the equilibrator needs to remain at a given site for at least two weeks. These devices are normally deployed with scuba divers, but other techniques are available for deeper placements. With the exception of microelectrodes, all of these techniques require instrumentation to measure gases in the collected sediment pore water sample.

For routine work, bottom sediments can be collected with gravity corers. Depending on the weight of the corer, consistency of the bottom deposits, and distance of the "free fall", 50-100 cm penetration is typical (Mudroch and

MacKnight, 1994). If lowered slowly into the bottom, methane concentrations in gravity core-collected sediments compare favorably with other sampling techniques, such as coring with the aid of scuba divers (Adams and Baudo 2001) and subsampling of box cores (Adams, 1991). Comparisons with *in situ* equilibrators suggest problems of gas loss during sub-sampling of the equilibration devices after their recovery (Adams *et al.,* in prep.). In the case of gravity coring for sediment gases, the time period between sampling and sediment processing should be minimized and the cores kept cold to inhibit further methanogenesis (Adams, 1994). Also, vibration and agitation of the cores should be avoided to lower bubble disruption and gas loss. Sediment cores are extruded horizontally within a glove bag ($I^2R$, Cheltenham, PA) flushed with helium, or another inert gas, within a few hours after collection. Horizontal processing avoids gas bubble migration while atmospheric contamination during processing is minimized by using an inert gas. Helium should be used if sediment $N_2$ and argon are to be measured. Sediments can be transferred at 1-2 cm depth intervals directly into tared 25-ml Sarstedt (Sarstedt, Numbrecht-Rommelsdorf, Germany) syringes and glass scintillation vials, using a core adapter syringe sampling (CASS) system (Adams, 1994). The internal air of the glove bag should be kept at about 0.2 ppm $O_2$, or lower, during sediment processing (Fendinger and Adams, 1986). In the field, sediment gas syringes are immediately placed in helium-filled, heavy-duty freezer bags and kept on ice. These bags should be flushed daily with helium or submerged in cold sediments to avoid oxygen contamination during storage. Glass scintillation vials, used for water content measurements, should be tightly closed and kept on ice to minimize water loss. These techniques for processing cored sediments lowers the possibility of introducing atmospheric contamination so that Ar and $N_2$ gases could be measured in sediment pore waters (Adams, 1994). A simpler method for collecting and measuring sediment methane is described in Casper (1992a).

Analysis of sediment gases should take place no later than 1-2 days after coring. At this time sediment gas syringes are submerged in a water bath at constant temperature (20°C) and 5.0 ml helium is added to each syringe; manipulations should be under water to avoid atmospheric contamination. Extraction of sediment gases into the helium headspace is facilitated by vibration for 4-5 minutes (with the sediment syringe kept in a helium-filled bag) followed by removal of most of the headspace (at least 3-4 ml) into a gas-tight syringe. About 1.5-2 ml of headspace is flushed through a sampling valve containing a 1.0 ml loop for direct injection into a gas chromatograph (GC). The following gases are measured: headspace extractable argon (Ar), oxygen ($O_2$), nitrogen ($N_2$), methane ($CH_4$) and carbon dioxide ($CO_2$). These gases, especially that last three, represent the major sediment pore water dissolved gases. The GC is equipped with a flame ionization detector (FID) for measuring carbon gases (with a C-H bond) and an in-line methanizer (operated at 380°C with $H_2$) to convert $CO_2$ to $CH_4$ for measurement with the FID and to remove $O_2$ for Ar analyses. The methanizer with FID is used for analyzing low levels of $CH_4$ and $CO_2$, for example, in the surface sediments and overlying water. Caution must be employed not to allow $O_2$ or $H_2S$ to pass through the methanizer; the catalyst will eventually be destroyed. An in-line thermal conductivity detector (TCD), also used with the methanizer, is employed to measure Ar (with or without $O_2$), $N_2$, $CH_4$ and $CO_2$. Since Ar and $O_2$ are measured at the same retention time by these GC techniques, the methanizer is switched into the gas flow (to remove $O_2$ for Ar analysis) or switched out of the gas flow (to assess the presence of $O_2$ in the injected sample); with two injections, 3-4 ml of headspace sample are required for proper flushing of the gas sampling loop. Measurement of low levels of $O_2$ in the sample suggests atmospheric contamination; in this case either the gas data are not reported or corrections are made for air contamination to the Ar and $N_2$ values. Argon is used as an inert tracer for calculating gas-to-Ar ratios to normalize gas data and improve accuracy for: 1) gas extraction temperatures effecting Bunsen coefficients (Weiss, 1970, 1974; Wiesenburg and Guinasso, 1979), 2) assessing possible air contamination, and 3) calculations of potential denitrification ($N_2$/Ar ratio; Nishio *et al.*, 1981; Fendinger and Adams, 1986).

Cores from the central basin of Lake Tanganyika's Moba-Kalya Ridge area were collected by subsampling one core of a multi-coring, 4-core sampling system. The two sites were: Kalya slope (NP01-MC01, 303 m depth; 6.55295 S. latitude, 29.97565 E. longitude) and Kalya horst (NPO1-MC02, 613 m depth; 6.6805 S., 29.8650 E.). Sediment gas analyses were conducted as described above, except difficulties were experienced with the ship's power, resulting in inefficient GC and methanizer performance. The carrier gases supplied for GC-operations were also of questionable quality, which led to unstable and erratic fluctuations in recorder baselines.

## 3. RESULTS AND DISCUSSION

Most of the biogeochemical processes involved with the internal cycling of C and N are inferred from temperate lakes which undergo seasonal temperature changes and water column mixing. This is not the case in tropical waters, where Hecky (1993) pointed out that these aquatic systems might suffer even greater effects of cultural eutrophication because of their endless summers and warm bottom waters. Precipitation can also represent one of the dominant water budget pathways. Under these conditions of poor circulation and high carbon loading, it is expected that methanogenesis will become an important year-round process resulting in rapid recycling of the labile portion of the C pool. The higher temperatures, occurrence of anaerobic conditions in the deeper portions of the water column, and rapid, sometimes diurnal, deoxygenation could also result in substantial denitrification at the oxic-anoxic boundary with subsequent $N_2$ loss to the atmosphere, as observed by Reddy and Patrick (1975) in flooded soils and Adams and Naguib (1999) in Lake Plußsee, northern Germany. The oxic-anoxic boundary, normally found within the sediments or at the sediment-water interface in oligotrophic lakes, would move upwards into the water column and can remain there continuously if thermal stratification persists. This boundary would represent both the zone of methanotrophy (oxidation of $CH_4$ diffusing upwards from the sediments) and denitrification (reduction of $NO_3$ being advected downwards, which could be reduced to $N_2$ gas). These two gas cycles, normally most ubiquitous near the sediment-water interface, but limited by diffusional processes, would be located at the oxic-anoxic boundary in the water column where greater advective mixing occurs.

## 3.1 Lake Victoria

The water volume of Lake Victoria is the third largest in Africa, yet it is the world's second largest freshwater body in terms of surface area (Herdendorf, 1990), suggesting a relatively shallow aquatic system with a mean depth of 40 m. This monomictic lake has a maximum depth of 80 m (IDEAL, 1990) and a calculated flushing time of about 140 years. The modern lake area is 68,800 $km^2$ with a catchment area of 184,000 $km^2$ (Hecky and Bugenyi, 1992) and a shoreline length of 3,440 km (Balirwa, 1995).

Since almost 90% of the water entering Lake Victoria come from precipitation (Hastenrath and Kutzbach, 1983), the hydrochemistry suggests a dilute aquatic system with probably the lowest sulfate concentrations of any large water body on earth (Hecky and Bugenyi, 1992). Sulfate for East African rain was 7.8 μM (Bootsma *et al.,* 1996), while in northern Germany's Brandenburg lake district during 1993-1999 it averaged 24 μM (range of 10-75 μM; Federal Environmental Agency 2000). The water column normally mixes throughout its entire depth at least once annually during the June to August period (Hecky, 1993). Thermal data suggested a warmer and more stable lake during 1990-1991 than in previous years (Hecky, 1993), with rapid depletion of hypolimnetic oxygen after the onset of stratification, as described earlier by Talling (1966). Presently, anoxia affects up to 50% of the lake's benthos for prolonged periods even though Hecky and Bugenyi (1992) listed 100% of the sediments as being oxygenated, compared to 45% for Lake Malawi and 20% for Lakes Kivu and Tanganyika during the full circulation period (Lehman, 1996). Some inshore areas undergo diurnal deoxygenation in their shallow water column, suggesting sites for intense denitrification. It is expected that the presence of warm, tropical temperatures (continuously in excess of 22°C; Hecky and Bugenyi, 1992) during the entire year would result in rapid decomposition of detrital organic material and an increase in internal recycling of nutritive micro- and macro-elements as compared to temperate lakes (Kilham and Kilham, 1989). It is believed that vertical mixing and stratification control the annual nutrient cycles within Lake Victoria (Hecky and Kling, 1987). As productivity in the surface waters increase and stratification worsens, it is likely that anoxia in the bottom waters will intensify. As with northern German lakes, the intensity of methane production could be related to the length of stagnation when the hypolimnion is anoxic (Casper, 1992b). This period will promote greater methanogenesis in both bottom sediments and within the anoxic water column.

Lake Victoria is one of the few tropical lakes where the annual cycles of plankton productivity, nutrient cycling and stratification were evaluated (Hecky and Bugenyi, 1992). A comparison of historic data (Talling, 1965, 1966) to recent measurements at offshore sites (Johnson *et al.,* 1992, Hecky, 1993, Mugidde, 1993) illustrates the rapid eutrophication of Lake Victoria: sedimentation rates changed from 57 to 90 g $m^{-2}$ $yr^{-1}$ (dry weight), dissolved silica decreased from 80 to 10 μM, the mixed layer thickness changed from about 40-50 to 30-40 m, and there has been a 5-fold decrease in transparency and 2- to 10-fold increase in chlorophyll. Higher photosynthesis and oxygen saturation occur in surface waters, where the biomass is now 3-5 times greater than observed in the 1960s (Hecky, 1993). Mugidde (1993) reported a doubling of the phytoplankton productivity compared to measurements three decades ago. The dominant phytoplankton group has changed from diatoms (mainly *Melosira*) to $N_2$-fixing blue-green cyanobacteria, perhaps as a result of the decrease in silicon and a change in the N:P ratio from 16:1 in the open lake mixed layer to less than 8:1 in the hypolimnion (Hecky, 1993). The macronutrient considered most likely to limit phytoplankton growth in East African lakes is nitrogen (Hecky and Bugenyi, 1992). Hecky (1993) reported that total N to total P varies from 16:1 (Redfield ratio) in the mixed layer to <8:1 in the hypolimnion; $N_2$-fixation by cyanobacteria maintain the Redfield ratio near the lake surface while metalimnetic N loss is likely from denitrification (Hecky *et al.*, 1991). The disappearance of *Melosira* from the lake and surficial deposits indicates that Lake Victoria is presently much different than it has been in the past 10,000 years (Hecky, 1993). Even though high denitrification rates are considered the prime reason for low N:P ratios and the abundance of heterocystous cyanobacteria, measurements of denitrification ($N_2$ or $N_2O$ gas production, and losses of these gases from the lake) have not been made in the African Great Lakes (Hecky *et al.,* 1991). Calculations by Hecky *et al.* (1996) of nitrate flux to the zone of denitrification provide an indirect measurement. These values of denitrification should be higher, because of vertical mixing, than found in the sediments (Table 1). Thus, movement of the oxic-anoxic boundary from the sediments into the water column would increase denitrification by introducing advective mixing and increasing the replenishment of electron acceptors, as compared to the much slower diffusional processes within the surface sediments (Hecky, 1993).

Knowledge about the lake's water is much greater than that of the sediments. There are some publications describing the recent sediments of Lake Victoria, yet nothing about their gas contents. The IDEAL (1990) report listed an approximate sediment accumulation rate of 0.2-0.5 mm $yr^{-1}$ (200-500 mm/1,000 yr) with no visible laminations. Relationships between benthic macroinvertebrate taxa, water depth, and sediment pH and elemental chemistry were described by Mothersill *et al.* (1980), while further data concerning the mineralogy and geochemistry at these same nearshore sites in northwestern Lake Victoria were provided by Mothersill (1976). The organic carbon content for surface sediments ranged from 4.1% to 20.6% dry weight (average of 8.3% for 32 samples; Mothersill, 1976). These values are about half the 16-19% carbon content reported by Lipiatou *et al.* (1996; northern depositional basin) or the uniform 16% reported by Ochumba and Kibaara (1989; 56-m deep site in northeastern basin), both for the 0-14 cm sediment depth interval. Further information about the sediments can be found in Fish (1956), Hesse (1958),

Newell (1960), Kendall (1969), Richardson and Richardson (1972), Stager (1984), Scholz *et al.* (1990) and Hecky (1993). Carbon and nitrogen budgets are given in Livingstone and Melack (1984) and Hecky and Bugenyi (1992).

*Table 1.* Theoretical Fickian diffusion rates (mM $m^{-2}$ $d^{-1}$) for two gases ($N_2$ and $CH_4$) across the sediment-water interfaces of Lake Taupo, New Zealand, Lakes Sempach and Baldegg, Switzerland, Lake Søbygaard, Denmark and Lago d'Orta, Italy (before and after liming). $N_2$ loss is expressed as molecular N so that comparisons with $NO_3$-N flux can be made. Estimated nitrate flux to the zones of denitrification in East African Lakes (taken from Hecky *et al.,* 1996) are given as a comparison. Even though considerable precautions were taken, high N flux calculations (from sediment $N_2$ pore water gradients) could be an artifact of atmospheric contamination during sediment $N_2$ measurements.

<table>
<tr><th>SITE</th><th>DINITROGEN (as $N_2$-N)</th><th>METHANE ($CH_4$)</th></tr>
<tr><td colspan="3">SEDIMENT-to-WATER DIFFUSE FLUXES FOR MOLECULAR N AND METHANE</td></tr>
<tr><td>LAKE TAUPO, NEW ZEALAND<br>(oligotrophic) Adams 1992a<br>Central Basin (155 m)<br>Taupo Bay (90 m)</td><td><br><br>15 - 22<br>6.3</td><td><br><br>0.2 - 1.3<br>0.4</td></tr>
<tr><td>LAKE SEMPACH<br>(artificially aerated eutrophic Swiss)<br>Höhener and Gächter, 1994<br>Urban et al., 1997</td><td></td><td>Littoral 1.4<br>Profundal 3.1</td></tr>
<tr><td>LAKE BALDEGG (66m)<br>(artificially aerated eutrophic Swiss)<br>Mengis et al., 1997</td><td>0.9 - 1.2 (acetylene block)<br>4.3 (isotope pairing)<br>6.1 (whole lake mass balance)</td><td></td></tr>
<tr><td>LAKE SØBYGAARD (hypertrophic Danish)<br>Risgaard-Petersen et al., 1999</td><td>1.3 (May - isotope pairing)<br>3.6 (Sept. - isotope pairing)</td><td></td></tr>
<tr><td>LAGO d'ORTA, ITALY<br>(Adams and Baudo 2001)<br>All Stations (1989)<br>Southern Basin (1989)<br>Southern Basin (1994)<br>Central Basin (1989)<br>Northern Basin (1989)<br>Omegna Basin (1989) (clayey sediments near lake outlet)</td><td><br><br>6.8 $\pm$ 6 (n = 16)<br>4.4 $\pm$ 3 (n = 7)<br>21 $\pm$ 4.6 (n = 4)<br>11.4<br>5.6<br>2.3 $\pm$ 0.7 (n=3)</td><td><br><br>1.7 $\pm$ 2.4 (n = 13)<br>0.13 $\pm$ 0.07 (n = 5)<br>2.12 $\pm$ 1.1 (n = 3)<br>6.6<br>7.4<br>0.17 $\pm$ 0.07 (n=3)</td></tr>
<tr><td colspan="3">(1989 – acidic lake water, pH ca. 3.8-4 1994 – four years after liming)</td></tr>
<tr><td colspan="3">WATER COLUMN DENITRIFICATION RATE ESTIMATES IN AFRICAN LAKES [Hecky et al., 1996; H. Bootsma (pers. comm.) considered that these estimates are high]</td></tr>
<tr><td>Lake Malawi</td><td>3.4</td><td></td></tr>
<tr><td>Lake Tanganyika</td><td>8.7</td><td></td></tr>
<tr><td>Lake Victoria</td><td>55</td><td></td></tr>
<tr><td colspan="3">(denitrification estimates were calculated by Hecky et al. 1996 for $NO_3$-N flux to the zones of nitrate depletion in the water columns of each lake; corrections were made to express fluxes in units of molecular N, i.e. 1.7 $NO_3$ = 1 $N_2$ (see section 3.3, last paragraph), to allow for comparisons, either as $NO_3$ or $N_2$)</td></tr>
</table>

Even though there is considerable information about oxygen in Lake Victoria (Hecky *et al.*, 1994), dissolved C and N gas data are non-existent. Data for other African lakes are more abundant. Carbon dioxide and methane were periodically measured in Lakes Kivu's water column (Degens *et al.*, 1973; Deuser *et al.*, 1973; Hecky and Degens, 1973 and Jannasch, 1975), while $CH_4$ (only to 360 m depth, Rudd, 1980) and $H_2S$, $NH_3$ and $\Sigma CO_2$ (Edmond *et al.*, 1993) were reported for Lake Tanganyika. The latter authors suggest the possibility of hydrothermal contributions to the $NH_3$ budget. Precise data for maximum $CH_4$ concentrations in deep waters are inadequate and contradictory: Rudd (1980) reported <2 mM (in text but not included as his own measurements) while Edmond *et al.* (1993) provided values from H. Craig (unpubl. data) of ~0.215 mM (Kigoma Basin) and ~0.125 mM (Kipili Basin). Many of the reduced gases have accumulated at depth, mainly in anoxic waters, to concentrations exceeding saturation at atmospheric surface pressure (Livingstone and Melack, 1984). This is especially the case for the Cameroon lakes (Kling *et al.*, 1991), where most gases are less than saturated (16-67%) at *in situ* depth but would exceed saturation at the surface; upward displacement would result in total gas saturation at 1 to 49 m below the surface of these Cameroon lakes. However, gas data for the sediments of these lakes also seem not to be available. As reported earlier, Scholz *et al.* (1990), Cerling (1996), De Batist *et al.* (1996) and Johnson (1996) describe sediment features created by methanogenesis and gas bubble generation, such as differences in reflectivity of echo-sounder profiles. Johnson (1996) stated that sediment cores from the large river deltas of Lake Malawi, and likely other African lakes, contain abundant gases, but these were not measured. There seems to be no continuous records, and little historical information, concerning these important, biologically active gases in any of the East African lake sediments. The rapid eutrophication and documented severe changes within Lake Victoria in the past few decades (Hecky, 1993) only highlights the urgency to develop an understanding of fundamental C and N biogeochemical pathways. A recent ICRAF (2001) report has illustrated the severe watershed degradation in the Nyando River Basin draining into Lake Victoria, identified problems and suggested possible remedial intervention strategies. Lake Victoria is considered one of the critically important water and fisheries resource in East Africa (Lowe-McConnell *et al.*, 1992), providing life-support requirements to more than 30 million people (Balirwa, 1995).

## 3.2 Lake Tanganyika, East African Rift

Detailed information is provided about Lake Tanganyika in Coulter (1991). It is not the purpose of this chapter to elaborate on this information. However, the limnologic and geologic setting surrounding the multi-coring sites occupied in July 2001 in the lake's central basin is necessary. These may be the first sediment gas data for an East African Rift Valley lake.

Lake Tanganyika is the second largest lake in the world, with a length of ~650 km, 50 km average width, a maximum depth of 1410 m and volume of ~18,940 $km^3$

(compared to Lake Baikal with 23,000 km$^3$; Edmond *et al.*, 1993). The central, or Kungwe Basin, has a maximum depth of 885 m and is separated by two sills, 655 m deep to the north and 700 m deep to the south. Sediment infilling (approximate rate of about ~290 mm/1000 yr) dates back to the Miocene, or about 13-20 m.y. (Tiercelin and Mondequer, 1991). Deep basin sediments are mostly autochthonous in origin, composed of dark-green or black muds of homogenous nature. Permanent anoxia allows for preservation of the "algal rain" in the deep basin sediments (average 4.3% TOC), forming biogenic oozes containing a hydrogen richness factor suggesting sediments of low to high petroleum potential (Tiercelin and Mondequer, 1991). Sediment carbon values at central basin sites were much higher, averaging 6.9% TOC with a range of 5.8-9.7% at the Kalya slope site in 2000, and 7.8% average with a range of 5-11.8% at the horst site (Zilifi and Eagle, 2000). Sedimentation rates (calculated from only one AMS $^{14}C$ date for each core) are ~460 mm/1000 yr at the slope and ~110 mm/1000 yr at the horst site; thus accumulation rates are 4 times faster at the slope of the horst than on the top of this relief feature.

The boundary between the mixolimnion and monimolimnion can vary somewhat but is usually located around 150-200 m deep. Even though oxygen was reported as deep as 240 m, Hecky *et al.* (1991) set the transition depth at 200 m, where the deeper water is continuously warm (>23 °C) and anoxic. Lake Tanganyika has the second largest volume of anoxic water in the world (the first being the Black Sea) with less that 25% of its benthos exposed to oxygen (Hecky, 1991). Methane in the bottom waters was reported, from secondary sources, to be 0.1-0.2 mM (Edmond *et al.*, 1993) or <2 mM (Rudd, 1980), while $\Sigma CO_2$ measured 6.7 mM at 1,174 m in both the Kigoma (northern) and Kipili (southern) Basins. Two hydrothermal sites at Pemba and Cape Banza were described by Tiercelin and Mondequer (1991) who provided a photograph of gases, likely $CH_4$ and $H_2S$, bubbling out of basement fractures located in shallow water. Because of geologic rifting, it is likely that geothermal fluids and gases are venting at deeper locations in the lake. The permanently anoxic hypolimnion is rich in $CH_4$, $H_2S$ and excess He (Tiercelin and Mondequer, 1991).

The two multi-core locations were positioned on the slope and top of the Kalya Horst, a previously uninvestigated relief, also called the Moba block, located south of the Mahali Mountains in the central basin. Its high topographic relief, too deep to be exposed during most low stands of lake level, separates the central and southern basins of the lake (Lezzar 2000). AMS $^{14}C$ dates for nearby locations, cored in 2000, were ~3000 years BP at 140 cm depth and ~9300 years BP at 109 cm depth (Zilifi and Eagle 2000) in the sediments near the 2001 coring sites at the Kalya slope and horst, respectively. Total organic and inorganic carbon at the slope (303 m) site in 2001 were 5.5-9.5% and >0.5%, respectively, with laminated light and dark bundles (thickness of bundles 1.5-8 cm). At the horst (613 m) site total organic carbon was higher (6-11.5%) while inorganic carbon was slightly lower (>0.4%, with a spike to 1% at 2 cm depth) with no observable stratification (Lezzar, pers. comm.). The cored horst site in 2000 (608 m) was different in that it had distinct laminations between yellow-grey diatom rich-oozes and dark grey clay bundles (Zilifi and Eagle, 2000).

The diatom stratigraphy for these two sites, cored in 2000, was described by Michelo and Chororoka (2000).

Water content for the two cores, taken from the same depth as the sediment gases, is shown in Figure 1. There is a rapid decrease in sediment water at 5 cm depth at the horst site and 8 cm in the slope core, while water content decreased again at ~25 cm depth (the $\%H_2O$ increase at the bottom of both cores could be an artifact of coring). Total methane ($CH_4$) and carbon dioxide ($CO_2$) concentrations (mM) in the sediment porewater (including both headspace and porewater phases) at the two coring sites (● = slope, ■ = horst) are given in Figure 2. Both $CH_4$ and $CO_2$ increased with depth in the sediments, which is typical for other lakes (Adams 1992a; Adams and Naguib, 1999; and Adams and Baudo 2001). There seemed to be unusual changes in the gas distributions within the 20-25 cm depth interval at both sites (except $CO_2$ at the horst site; Figure 2).

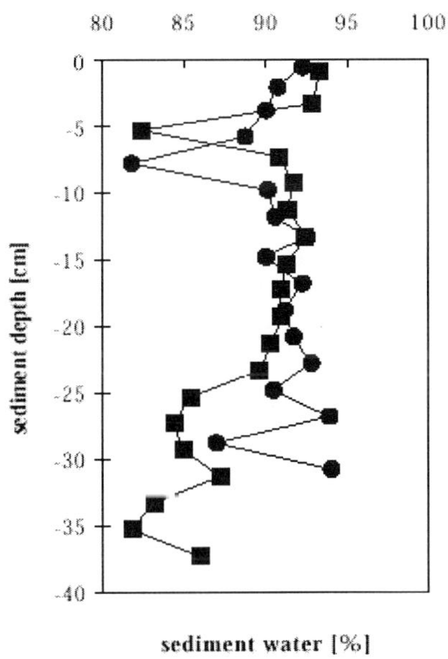


*Figure 1.* The vertical distribution of the sediment water content (%) in two cores collected from the slope (•) and top (■) of the Kalya Horst, central basin of Lake Tanganyika, July 2001, as part of the Nyanza Program. The depth extends from 0 to 35 cm below the sediment-water interface.

It was noted during visual observations of the cores, before extrusion into a helium-filled glove bag, that the sediments were homogenous for the 0-23 cm depth with laminae starting at 23 cm for the slope core, while at the horst site the sediments changed from dark black to brownish grey at 21-22 cm. Anoxic decomposition usually results in the equal proportions of $CO_2$ and $CH_4$; methane was slightly higher than $CO_2$ with $CH_4$:$CO_2$ ratios of $1.4 \pm 0.5$ (0.6-2.6, n=17) for the slope and $1.6 \pm 0.5$ (0.8-2.7, n=15) for the horst.

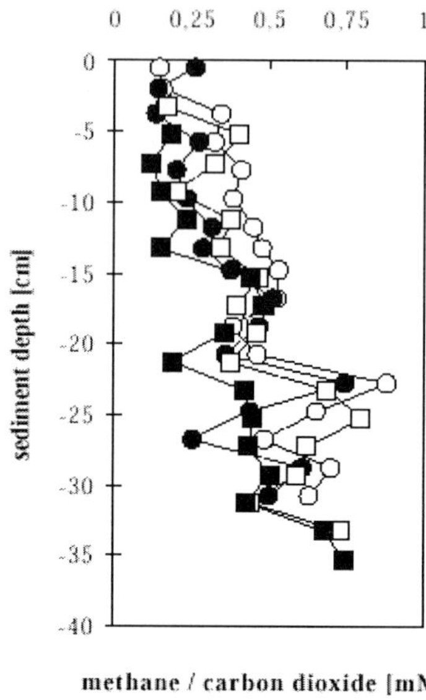


*Figure 2.* The vertical distributions of sediment methane (open symbols, ○ □) and carbon dioxide (closed symbols, ● ■) at two coring sites in the central basin of Lake Tanganyika, July 2001, as part of the Nyanza Program. Gas concentrations are millimolar (mM). Both locations are at the Kalya Horst: slope (○, ●) at 303 m depth and top of the horst (□, ■) at 613 m depth. $CH_4$ and $CO_2$ were averaged values from two detectors (FID and TCD) measuring the same gas sample.

Variability in $CO_2$ and $CH_4$ between the two detectors (FID and TCD) for the same gas samples at both sites is given in Figure 3. The three major gases at the two coring sites, including dinitrogen ($N_2$) which averaged 0.9±0.5 mM (0.5-2, n=15) for the slope and 1.0±0.5 mM (0.1-1.9, n=15) for the horst, are shown in Figure 4. Surface porewater $N_2$ measured 0.47 mM (0.5 cm depth at the slope site) and 0.71 mM (3 cm depth at the horst); both concentrations are close to equilibrium saturation at 24°C (=0.5 mM). Because of water column pressure, *in situ* $N_2$ gas pressures averaged 4.1±1.9 and 2.5±1.3 percent of total saturation for the slope (613 m) and horst (303 m) sites, respectively. Total *in situ* gas saturation values at the two locations are shown in Figure 4. Saturation at these depths was low at 5.7±2.6 percent for the slope, 3.2±1.4 percent for the horst. Expressed as total gas pressure, this was only 1.6± 0.9 atm at the slope to 1.9±0.9 atm at the horst. If these water parcels were moved towards the surface, full saturation would be achieved between 3 to 36 m and 1.5 to 32 m depth at the slope and horst sites, respectively. So, gases are supersaturated with respect to the surface of Lake Tanganyika. Averages for both detectors (TCD and FID) are used for $CH_4$ and $CO_2$ values in figures 2 and 4.

Normally, anoxic sediments are the source of reduced gases and $CO_2$ to the overlying water column. This is the case for most lakes. If one uses the data from Rudd (1980; <2 mM for "maximum" water column $CH_4$), then it would appear that the sediments are not a source of methane to overlying waters. However, values of

0.13-0.22 mM for bottom waters from Edmond *et al.* (1993) would be similar to surface pore water concentrations ($CH_4$ 0.14-0.17 mM) in Figure 3. It is likely that most of the "algal rain" is decomposed within the water column during sinking as observed in the 40-m deep Groot Rug reservoir in the Netherlands where 80% of the C was aerobically recycled (Adams and van Eck, 1988). The remaining C should be preserved in the sediments by the continual anoxic conditions. Deep or "maximum" water column data were cited as "unpublished", or from unnamed sources, obtained from different (northern and southern basins) or other unknown locations, and are 1-2+ decades ago. It appears that the deeper water column has a constant composition and long residence time, i.e., bottom water $CO_2$ and $H_2S$ in both Kigoma and Kipili Basins were indistinguishable from each other (Edmund *et al.*, 1993). Further studies would be required to confirm the sources of reduced gases to the water column since geothermal venting is likely a feature of Lake Tanganyika.

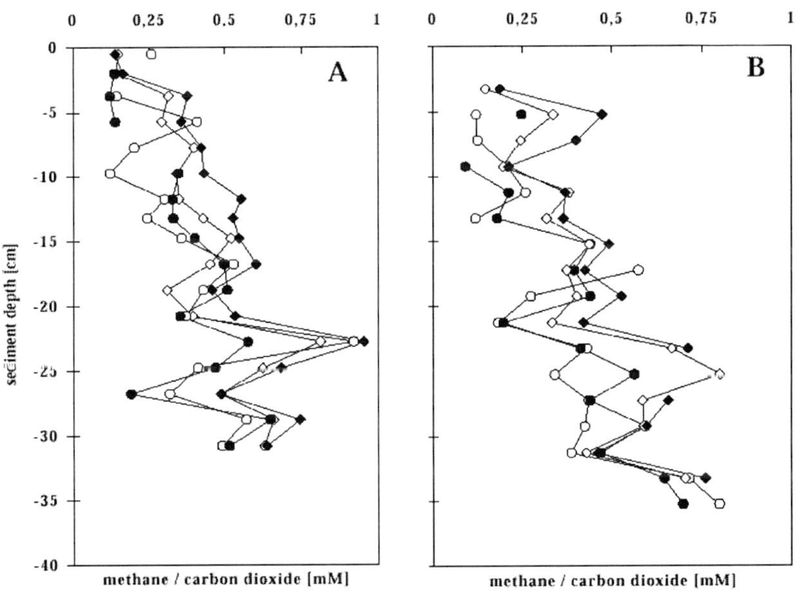


*Figure 3*. The vertical distributions of sediment methane (◇◆) and carbon dioxide (○ ●) at two coring sites in the central basin of Lake Tanganyika, July 2001, as part of the Nyanza Program. Gas concentrations are millimolar (mM). Both locations are at the Kalya Horst: slope (A) at 303 m depth and top of the horst (B) at 613 m depth. The flame ionization detector (FID) has filled (◆ ●) symbols while the thermal conductivity detector (TCD) has open (◇ ○) symbols.

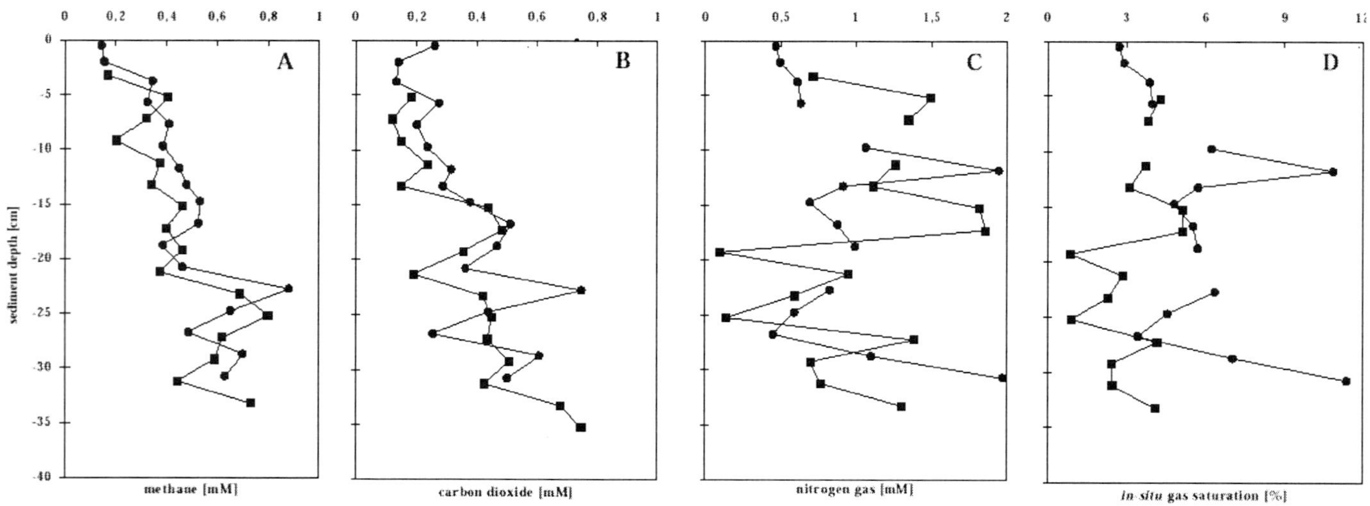


*Figure 4*. Vertical distributions of sediment gases (mM) and percent gas saturation (%) at two coring sites (slope of horst = ● and top of horst = ■) in the central basin of Lake Tanganyika, July 2001, as part of the Nyanza Program. Both locations are at the Kalya Horst, separating the central and southern basins of the lake. The figures are methane (A), carbon dioxide (B), nitrogen gas (C) and total *in situ* gas saturation (D). Gas concentrations are millimolar (mM). $CH_4$ and $CO_2$ were averaged values from two detectors (FID and TCD) measuring the same gas sample. Breaks in the profiles for $N_2$ represent samples which were likely contaminated with air.

## 3.3 Biogeochemical Cycling of Gases

The microorganisms involved in the cycling of trace gases as controllers of atmospheric composition were reviewed by Conrad (1996), while Oremland (1988) described the biogeochemistry of methanogens. Within the aquatic system, organic matter mineralization and sediment-water exchange are critical to the cycling of gases and nutritive elements between sediments and overlying waters. In fact, Klump (1987) stated that organic matter loading is probably the most important factor effecting the role sediments have in lake biogeochemical cycles, and, globally, lakes are accumulating about 42 Tg of organic carbon each year (Dean and Gorham, 1998). Exchange across the sediment-water boundary is considered one of the rate-determining steps in the biogeochemical cycling of nutritive elements back to the water column. These processes play an even more important role in shallow, warm environments such as Lake Victoria because sinking particulate matter will have less time to decompose in the aerobic portion of the water column and will thus provide additional C and N to sediment bacterial processes (Adams and van Eck, 1988). Livingstone and Melack (1984) stated that even though the understanding of nutrient control is primitive for East African lakes, the major supply of these chemicals and gases must come from recycling in surface waters and regeneration from depth. Rapid internal recycling and nutrient supply in excess of demand are perhaps one of the reasons for the relationship reported by Kilham and Kilham (1990), who stated that annual primary productivity was the same in each of the investigated tropical African lakes regardless of size. As well as productivity, understanding the cycling and coupling of bio-limiting nutrients between the sediments and overlying water column are essential in assessing oxygen budgets. The sedimentation rate and quality of the organic matter will determine the redox conditions of the sediments and the nature of the benthos. Even as early as four decades ago, Newell (1960) reported a common, thick (2-3 m) layer of loose flocculent material at the bottom of Lake Victoria, and Fish (1956) showed increased nutrients and sharp declines in oxygen content near the sediments.

It is necessary to calculate the flux of dissolved constituents, including gases, from the sediments to overlying waters for accurate C and N budgets. This can be done either indirectly from pore water gradients using Fick's first law of diffusion (and a knowledge of the sediment porosity and tortuosity) or directly with a benthic flux chamber. The magnitude of the flux is dependent upon the rates at which this material is decomposed and accumulated in sediment pore water, i.e., the magnitude of the produced concentration gradient, as well as the mixing mechanisms of sediment-water exchange. Devol (1987) found agreement to within 25% between flux calculations using pore water gradients and the increase in overlying water concentrations collected from *in situ* flux chambers at two locations (65 and 610 m depth) in oceanic areas where overlying waters were devoid of oxygen (thus no faunal populations which would cause bioturbation). However, these flux measurements with a chamber situated at the sediment-water interface (design shown in Hargrave and Connolly, 1978) suffer from numerous problems, in particular the simulation of turbulence at the diffuse boundary layer. Resistance to chemical transport across the benthic boundary layer was discussed by Santschi *et al.* (1983), while Hammond *et al.* (1985) provided an evaluation of flux chambers and sediment pore water profiles. Since the sediments of Lake Victoria are not oxygenated for much of the time, it is possible that asphyxiation (Rutgers van der Loeff *et al.*, 1984) would not be required to remove the effects of biologically-mediated transport at the sediment-water interface. The influence of porosity and tortuosity on molecular diffusion in the calculation of Fick's first law were discussed in Ullman and Aller (1982), Sweerts *et al.* (1991) and Iversen and Jorgensen (1993).

Extensive memocell (similar to peepers, or *in situ* equilibrators; Hesslein, 1976) studies and sediment gas coring in a shallow freshwater reservoir in the Netherlands showed that about 80% of the carbon (300 gC $m^{-2}$ $y^{-1}$ primary productivity) was recycled in overlying waters (Adams and van Eck, 1988); for that portion of the organic C which reached the sediments, 25% was permanently buried, 60% was decomposed aerobically while the remainder (15%) fueled anaerobic metabolism (divided equally between sulfate reduction and methanogenesis). Coring and sediment peeper studies in Lake Erie showed that deposited carbon was quickly mineralized within the surface 1-2 cm sediments with about 7-45% (high variability between locations) of the deposited carbon lost by diffuse flux of carbon gases (mostly $CO_2$; Adams and Fendinger (1986). The calculated diffusive flux of $CH_4$ to the sediment-water interface accounted for 30% of the oxygen demand measured with *in situ* chambers (Adams *et al.*, 1982); this was thought to be the cause of rapid dexoygenation of hypolimnetic Chilean reservoir waters (Adams *et al.*, 2000).

Nitrogen is an element of great importance in the African lakes. Its biogeochemical cycling, especially the processes of denitrification, were described earlier. It is likely a limiting nutrient. Hecky *et al.* (1996) provided a review of N cycling in the African Great Lakes. There have been a substantial number of publications concerning microbial nitrification and denitrification (Stouthamer, 1988, Tiedje, 1988 and Robertson and Kuenen, 1991, to cite a few). In an earlier review on aerobic denitrification, Robertson and Kuenen (1984) stated that it should not be surprising that anaerobic nitrification would also be found in nature under certain conditions. The acetylene inhibition technique adapted for sediments (Sorensen, 1978, Kasper, 1982) has frequently been used to estimate *in situ* denitrification rates. However, recent studies suggested that reduced sulfur reverses the acetylene block, and denitrification rates could be greatly underestimated if applied to intact sediment cores, i.e., systems with high sulfide or low nitrate content (Seitzinger *et al.*, 1984; Rudolph *et al.*, 1991). Since the terminal product of denitrification is dinitrogen gas ($N_2$,), Seitzinger *et al.* (1984) developed a special gas-tight system to measure the changes in gas phase $N_2$ after the aqueous phase overlying the sediments was replaced with low-$N_2$ water. From these measurements they calculated that 35% of the mineralized sediment nitrogen was converted to $N_2$ and lost from Narraganset Bay. This percentage is not much different than values reported by Tirén (11-60%; 1977) for loss of $N_2$ by denitrification from 12 lakes in Denmark, Sweden and the United States. $N_2$ flux was shown to be an important N-sink for U.S. estuarine and shelf waters (Nowicki *et al.*, 1997); for example, 43-54% of total nitrogen input was calculated as denitrified-N lost from Delaware Bay (Nixon *et al.*, 1996). Even higher values of N removal were reported for two eutrophic Swiss lakes by Mengis *et al.* (1997), where from mass budget calculations, denitrification accounted for 88% of the N loss from Lake Baldegg and 65% for Lake Zugg. Because of the high supersaturation of $N_2$ in the surficial sediments of Lake Erie, diffusive loss was calculated to represent ~20-32% of nitrogen deposited to the sediments (Fendinger and Adams, 1987). In four other lakes $N_2$/Ar ratios were used to calculate Fickian $N_2$-diffusive fluxes (see Table 1) to the sediment-water interface (Adams, 1992a; Abe *et al.*, 2000; Adams and Baudo 2001; Casper *et al.*, submitted). Sweerts *et al.* (1990) also reported complete denitrification of a $^{15}NO_3^-$ tracer by a freshwater mat consisting of *Beggiatoa* spp. Downes and Adams (unpublished data) found $N_2O$ and $N_2$ to be highly supersaturated below the sediment-water interface (0-10 mm) in Lake Taupo, New Zealand, suggesting active denitrification in surface sediments (within the zone of rapid deoxygenation, Adams, 1992b), as proposed earlier by Seitzinger (1988). In incubated freshwater samples containing $NO_3^-$ and glucose, Koyama (1990) found that most of the nitrate was reduced to $N_2$ gas; however, when $NO_3^-$ and $NO_2^-$ were exhausted and a small amount of glucose remained, denitrified-$N_2$ was utilized. Kasper *et al.* (1985) also found that denitrification accounted for 82-100% of nitrate reduction (3-75% of N mineralization) for marine sediments.

There is no easy way to calculate the reduction of nitrate to $N_2$ during sediment denitrification. Hecky *et al.* (1996) have done this by determining the downward flux of nitrate to the chemolimnion (oxic-anoxic boundary) in three East African lakes (Table 1). The amount of $N_2$ produced from nitrate reduction would depend on the quality of the organic matter. Stigg (1986) described the stochiometry of freshwater biomass as $(CH_2O)_{113}(NH_3)_{15}(H_3PO_4)$, which is similar to the Redfield ratio of 116C:16N:1P. If this organic matter is similar to that observed in African lakes, then one can assume the following stochiometry for decomposition through the denitrification pathway:

$$(CH_2O)_{113}(NH_3)_{15}(H_3PO_4) + 99.4\ HNO_3 \rightarrow 113\ CO_2 + 57.2\ N_2 + 185.2\ H_2O + H_3PO_4$$

thus, a molar N ratio of approximately 99.4 $NO_3$ to 57.2 $N_2$, or 1.7:1 during denitrification would be obtained. This relationship is used in Table 1 to compare denitrification rates through nitrate flux calculations (Hecky *et al.*, 1996) or theoretical diffuse flux of $N_2$ at the sediment-water interface. Song and Müller (1999) provide a review of these processes at the sediment-water interface.

## 3.4 Lake Surface Gas Emissions

Even though Tyler (1991) did not consider lakes in his global budget inventory, Aselman and Crutzen (1989) list them as a minor portion (0.3-4%) for their tropospheric methane budget inventory, while Brasseur and Chatfield (1991) calculated 1-27 Tg $yr^{-1}$. There is, however, mounting new evidence (Casper *et al.*, 2000) that these surfaces might be even more important. Further information is reviewed in Adams (1996). If reservoirs are included, a recent estimate (St. Louis *et al.*, 2000) places emissions from freshwater reservoir surfaces as high as 18% of all anthropogenic methane emissions (and 7% of all anthropogenic $CO_2$-equivalent C emissions for global warming potential), yet these aquatic systems are not included in IPCC (1996) inventories. Worldwide lake surface emissions of 700 mg $CO_2$ $m^{-1}$ $day^{-1}$ and 9 mg $CH_4$ $m^{-2}$ $day^{-1}$ are also provided in St. Louis *et al.* (2000); these data were computed by averaging over a 365-day period and assuming a 1.5 million $km^2$ global lake surface area. Casper *et al.* (2000) reported total methane surface flux rates from different aquatic freshwater systems to range from 0.001 to 69.6 mM $m^{-2}$ $day^{-1}$. For hypereutrophic lake Priest Pot in the lake district of the U.K., Casper *et al.* (2000) apportioned the $CH_4$ flux into 0.4 diffusion and 12 mM $m^{-2}$ $day^{-1}$ bubble ebullition. Duchemin *et al.* (1995) showed that emissions of C gases at the surfaces of two reservoirs in the boreal region of Canada were similar or lower than benthic fluxes. Even though there is considerable evidence that freshwater lakes could be important sources of greenhouse gases (Adams, 1996), emissions of C gases at the water-atmosphere interface of African lakes and reservoirs have not been investigated. A program needs to be initiated to study greenhouse gas emissions from various African biotopes of ecological significance.

Internal cycling of $CH_4$ is also important to the overall carbon (C) balance of aquatic freshwater ecosystems. Rudd and Hamilton (1978) reported production of methane ($CH_4$) and its subsequent oxidation represents 55% and 36%, respectively, of the total C budget for a eutrophic Canadian shield lake. It is likely that the difference, or 19%, is lost to the atmosphere. There are also numerous publications on the importance of $CH_4$ in the cycling of carbon, where values range from 7% of the sedimented C being decomposed to $CH_4$ gas (oligotrophic freshwater reservoir in the Netherlands; Adams and van Eck, 1988) to as high as 65% for eutrophic Lake Suwa in Japan (Koyama, 1990). Because of methane oxidation in the aerobic waters of lakes (Cicerone and Oremland, 1988), at the oxic-anoxic chemocline in the water column of eutrophic, stratified lakes (Casper *et al.*, 2000), or at the same boundary near the sediment-water interface (Frenzel *et al.*, 1990; Adams, 1992a, 1992b), Hessen and Nygaard (1992) reported that bacterial cellular production by methanotrophs can be as important a carbon source to pelagic food webs as bacterial secondary production. These organisms represent the "bacterial filter" which lower $CH_4$ emissions to the troposphere (by 60-90%) from inundated terrestrial surfaces (Galchenko *et al.*, 1989). So, internal cycling of gases, and their production and consumption, are important ecosystem processes whereby carbon is returned to the water column, and eventually the atmosphere, of freshwater lakes.

In the case of nitrogen, internal budgets and cycling of N components, especially the gases, become even more complex as described earlier. According to Cicerone (1989), recent attempts to obtain balanced atmospheric budgets have not been successful for $N_2O$, an intermediate gas in the denitrification and nitrification cycle. Only rudimentary budgets were attempted, mostly for terrestrial ecosystems (Bowden, 1986). In their construction of global and tropical (30 $^{O}$S - 30 $^{O}$N) $N_2O$

budgets, Seiler and Conrad (1987) did not list freshwater lakes as important sources for $N_2O$. Lakes are also not listed in the IPCC (1996) report. However, considerably more information is now known about internal N cycling within lakes (Yoh *et al.*, 1988; Downes, 1988; Terai and Yoh, 1996; Abe *et al.*, 2000, to list a few). Loss of $N_2$ by bubble ebullition from the sediments of one lake in northern Germany was reported as 6 ml $m^{-2}$ $day^{-1}$, or 25-36% of the collected gas, the remainder being $CH_4$ (61-74%) with a very small proportion (1%) as $CO_2$ (Adams and Naguib, 1999). It is expected that much of the sedimented inorganic N is lost as $N_2$ gas because of denitrification within the water column or near the sediment-water interface. As aquatic ecosystems become more eutrophic, denitrification rates will likely increase (Kaplan *et al.*, 1978, Seitzinger and Nixon, 1985, Mengis *et al.*, 1997) resulting in greater internal N cycling. Lake Victoria is presently undergoing rapid eutrophication; as shown by Hecky *et al.* (1996), rapid recycling of inorganic N and $N_2$ gas loss to the atmosphere is a major reason for nitrogen limitation in these ecosystems. Even though the diffuse flux of $N_2$ at the water-atmosphere interface would be difficult to measure, the loss of $N_2O$ could be attempted. The oxidation of reduced inorganic-N products, such as $NH_4^+$, to nitrate, followed by intense denitrification to $N_2$ in low-$O_2$ waters, will likely continue to make nitrogen a limiting nutrient in the East African lakes.

# 4. CONCLUSIONS

The cycling of gases in the water column, and especially the sediments, are important ecosystem processes. As lakes become more eutrophic, they tend to accumulate greater amounts of organic materials in their bottom deposits. These substances fuel organic matter decomposition processes which consume oxygen and produce $CO_2$ and reduced substances, such as methane, ammonium, sulfides and other elements (reduced forms of manganese and iron). The reduced substances diffuse into the overlying water and further consume oxygen. Eventually the oxic-anoxic boundary moves out of the sediments into the water column. Ammonium is oxidized to nitrate at the oxycline. The produced nitrate is reduced in the low-oxygen waters above the oxycline; these processes of nitrification and denitrification removes nitrogen from the water column as gaseous $N_2$. Eventually the water column becomes deficient in N, which will limit primary production. Special organisms, such as heterocystous cyanobacteria capable of fixing $N_2$ directly from the atmosphere, will likely flourish under these conditions. The hypolimnetic bottom waters continue along the course of deoxygenation, which further fuels methanogenesis and denitrification above the anoxic-oxic boundary. These are natural gas cycles, yet they become unwanted by producing gaseous end-products ($NH_3$, $H_2S$, mercaptans, amines, etc.) which are foul smelling, toxic and disruptive to normal oxygenated ecosystems.

As a first step in understanding gas cycling in the sediments of an East African lake, an elevated horst in the Central Basin of Lake Tanganyika was sampled as part of the Nyanza Project. Results indicated that the pore waters were undersaturated with gases at depth (300-600 m) but reached full saturation below the lake surface if the sediment-water parcel were transported upwards. Thus, the gases are supersaturated with respect to atmospheric pressure. It is suspected that carbon decomposition would have taken place in the water column during sinking, resulting in recalcitrant carbon reaching the bottom sediments. Continual anoxia would aid in preserving the deposited organic matter. This has led to high sediment TOC but low gaseous end-products. From present information, the sources of gases to the lake's bottom waters are unknown. As a rift valley lake, one likely source of gases in Lake Tanganyika would be geothermal emissions, yet further studies would be required to evaluate bottom water composition and fluxes and to answer the question whether the sediments are a source or sink for these gases.

It is obviously important to understand the biogeochemical cycles of gases, especially in the African lakes where there is limited or sporadic information. These lakes provide important life-supporting resources for the local populations. As shown in this review, there is little known about the gaseous bio-active compounds which are critical components of the pathways for biogeochemical cycling of C and N in aquatic ecosystems. There have been only a few studies of gases in the water column of some of the African lakes, but most of these were 1-3 decades ago. With the exception of the sediment gas data reported here for Lake Tanganyika, there seems to be no other information for the East African Great Lakes. Perhaps some data are available in internal reports, but as far as the authors are aware there have been no publications concerning gases in the sediments of these lakes. Yet, knowledge of these gases and their fluxes, both in the water column and sediments, is required for a better scientific understanding of these ecosystems and to insure appropriate management of these highly-utilized resources.

## ACKNOWLEDGEMENTS

D. Adams would like to thank the Fulbright Senior Scholar Program, the Council for International Exchange of Scholars and the U.S. Department of State for awarding a Fulbright Research fellowship for greenhouse gas studies in Kenya and Tanzania during 2001. Dr. Eric O. Odada and the PASS program are thanked for the invitation to attend the 2nd International Symposium in Malawi and in continuing their support of these climate change research efforts and capacity building in Africa. Colleagues at the Institut für Gewässerökologie und Binnenfischerei (IGB), Berlin, Germany, are thanked for providing the stimulus and atmosphere for initiating this article. Enormous gratitude is expressed by D. Adams to many colleagues around the world who have participated in coring expeditions, gas sampling, and numerous fruitful discussions: Renato Baudo, Paul Boon, Don Brown, Peter Casper, Christian Dinkel, Malcolm Downes, Nicholas Fendinger, André Furtado, Rene Gächter, Martin Mengis, Alena Mudroch, Monir Naguib, Yoshida Naohiro, Alla Nozhevnikova, Waldemar Ohle, Vladimir Samarkin, Jean-Pierre Sweerts, Hisayoshi Terai, Irma Vila, Bert van Eck, Martin Vollmer, Eitaro Wada, Bernhard Wehrli and Denis

Wiesenburg. The Max Planck Institute in Plön, Germany, and especially Jürgen Overbeck and Monir Naguib, are thanked for providing help with the initial development of the CASS system for sediment gas sampling. We would also like to thank Daniel O. Olago for inviting us to write this chapter, Harvey Bootsma for advice with an initial draft and Peter Casper for continuing to provide expert advice and critical review until the final draft. The Lake Tanganyika sediment gas study would not have been possible without the generous help of Andy Cohen, financial support from NSF Grant #ATM9619458 (The Nyanza Project), a fellowship to S. Ochola from the World Wildlife Fund and the cooperation of captain Mariage and crew of the *M/V Maman Benita*. Thanks are also extended to Kiram Lezzar of the Nyanza Project for his help as party chief and to Issa Petit for solving the extraordinary electrical problems on the cruise. Both authors would like to thank Gerlinde Adams for her enthusiasm and invaluable help throughout the research and for her concern with our welfare.

## REFERENCES

Abe, D.S., Kato, K., Adams, D.D., Terai, H. and Tundisi, J.G. (2000) Contribution of free-living and attached bacteria to denitrification in the hypolimnion of a mesotrophic Japanese lake. *Microbes and Environments* **15**, 93-101

Adams, D.D. (1991) Gas composition of Hamilton Harbour sediments – Changes during the 1980-1990 decade. Canada Centre for Inland Waters, National Water Res. Institute, Burlington, Ontario, Canada. Internal unpublished report, 38 p.

Adams, D.D. (1992a) Sediment pore water geochemistry of Taupo Volcanic Zone lakes with special reference to carbon and nitrogen gases. *Taupo Research Laboratory File Report* **135**, 24 p.

Adams, D.D. (1992b) Oxygen consumption at the sediment-water interface of Lake Taupo, New Zealand. *Taupo Research Laboratory File Report* **134** (now NIWAR), Christchurch, New Zealand), 34 p.

Adams, D.D. (1994) Sediment pore water sampling, in A. Mudroch and S.D. MacKnight (eds.), *CRC Handbook of Techniques for Aquatic Sediments Sampling,* CRC Press, Boca Raton, FL, pp. 171-202.

Adams, D.D. (1996) Aquatic cycling and hydrosphere to troposphere transport of reduced trace gases - A review, in D.D. Adams, S.P. Seitzinger and P.M. Crill (eds.), *Cycling of Reduced Gases in the Hydrosphere*, Schweizerbart'sche Verlagsbuchhandlungen, Stuttgart, Germany, pp. 1-13.

Adams, D.D. (1999) Methane, carbon dioxide and nitrogen gases in the surficial sediments of two Chilean reservoirs – diffusive fluxes at the sediment water interface, in L.P. Rosa and A.dos Santos (eds.), *Dams and Climate Change - Proceedings of an International Workshop on Hydrodams, Lakes and Greenhouse Gas Emissions*, Rio de Janeiro, Brazil, pp. 50-77.

Adams, D.D. and Baudo, R. (2001) Gases ($CH_4$, $CO_2$ and $N_2$) and pore water chemistry in the surface sediments of Lake Orta, Italy: Acidification effects on C and N gas cycling. *J. Limnol.* **60**(1), 79-90.

Adams, D.D. and Fendinger, N.J. (1986) Early diagenesis of organic matter in the Recent sediments of Lake Erie and Hamilton Harbour, in P.G. Sly (ed.) *Sediments and Water Interactions*. Springer-Verlag, New York, NY, pp. 305-318.

Adams, D.D. and Naguib, M. (1999) Carbon gas cycling in the sediments of Plußsee, a northern German eutrophic lake, and 16 nearby water bodies of Schleswig-Holstein, in K.-P. Witzel and T. Berman (eds.), *New Vistas in Aquatic Microbial Ecology, Archiv Hydrobiol. Spec. Issues: Advances Limnol.* **54**, 91-104. Schweizerbart'sche Verlagsbuchhandlung, Stuttgart, Germany.

Adams, D.D. and van Eck, G.Th.M. (1988) Biogeochemical cycling of organic carbon in the sediments of the Grote Rug reservoir. *Arch. Hydrobiol. Beih. Ergebn. Limnol.* **31**, 319-330.

Adams, D.D., Matisoff, G. and Snodgrass, W.J. (1982) Flux of reduced chemical constituents ($Fe^{2+}$, $Mn^{2+}$, $NH_4$ and $CH_4$) and sediment oxygen demand in Lake Erie. *Hydrobiologia* **92**, 405-414.

Adams, D.D., Vila, I., Pizarro, J. and Salazar, C. (2000) Gases in the sediments of two eutrophic Chilean reservoirs: potential sediment oxygen demand and sediment-water flux of $CH_4$ and $CO_2$ before and after an El Niño event. *Verh. Internat. Verein. Limnol.* **27**, 1376-1381.

Adams, D.D., Dinkel, C., Mengis, M. and Wehrli, B. (in prep.) Sediment pore water methane – A comparison of two sampling and measuring techniques. For submission to *Environ.. Science and Technology.*

Aselmann, I. and Crutzen, P.S. (1989) Global distribution of natural freshwater wetlands and rice paddies, their net productivity, seasonality and possible methane emissions. *J. Atmos. Chemistry* **8**, 307-358.

Balirwa, J.S. (1995) The Lake Victoria environment: Its fisheries and wetlands – a review. *Wetlands Ecology and Management* **3**, 209-224.

Bootsma, H.A., Bootsma, M.J. and Hecky, R.E. (1996) The chemical composition of precipitation and its significance to the nutrient budget of Lake Malawi, in T.C. Johnson and E.O. Odada (eds.), *The Limnology, Climatology and Paleoclimatology of the East African Lakes*, Gordon and Breach Publ., pp. 251-265.

Bowden, W.B. (1986) Gaseous nitrogen emmissions (sic) from undisturbed terrestrial ecosystems: An assessment of their impacts on local and global nitrogen budgets. *Biogeochemistry* **2**, 249-279.

Brasseur, G.P. and Chatfield, R.B. (1991) The fate of biogenic trace gases in the atmosphere, in T.D. Sharkey, E.A. Holland and H.A. Mooney (eds.), *Trace Gas Emissions by Plants*, Academic Press, San Diego, pp. 1-27.

Casper, P. (1992a) Methanhaushalt in Baltischen Seen unterschiedlichter Trophiestufe (Methane metabolism in Baltic Lakes of different trophic state). *Limnologica* **22**, 121-128.

Casper, P. (1992b) Methane production in lakes of different trophic state. *Arch. Hydrobiol. Beih. Ergebn. Limnol.* **37**, 149-154.

Casper, P., Maberly, S.C., Hall, G.H. and Finlay, B.J. (2000) Fluxes of methane and carbon dioxide from a small productive lake to the atmosphere. *Biogeochemistry* **49**, 1-19.

Casper, P., Furtado, A. and Adams, D.D. (Submitted) Biogeochemistry and diffuse fluxes of greenhouse gases (methane and carbon dioxide) and dinitrogen from the sediments in oligotrophic Lake Stechlin. For submission to *Lake Stechlin – An Approach to Understanding an Oligotrophic Lowland Lake* in Advances in Limnology (Archiv für Hydrobiologie) series, E. Schweizerbart'sche Verlagsbuchhandlung, Germany.

Cerling, T.E. (1996) Pore water chemistry of an alkaline lake: Lake Turkana, Kenya, in T.C. Johnson and E.O. Odada (eds.), *The Limnology, Climatology and Paleoclimatology of the East African Lakes*, Gordon and Breach Publ., pp. 225-240.

Cicerone, R.J. (1989) Analysis of sources and sinks of atmospheric nitrous oxide ($N_2O$). *J. Geophys. Res.* **94**, 18,265-18,271.

Cicerone, R.J. and Oremland, R.S. (1988) Biogeochemical aspects of atmospheric methane. *Global Biogeochem. Cycles* **2**, 299-327.

Cohen, A.S., Kaufman, L. and Ogutu-Ohwayo, R. (1996) Antropogenic threats, impacts and conservation strategies in the African Great Lakes: A review, in T.C. Johnson and E.O. Odada (eds.), *The Limnology, Climatology and Paleoclimatology of the East African Lakes*, Gordon and Breach Publ., pp. 575-624.

Conrad, R. (1996) Soil microorganisms as controllers of atmospheric trace gases ($H_2$, CO, $CH_4$, OCS, $N_2O$, and NO). *Microbiol. Rev.* **60**, 609-640.

Coulter, G.W. (1991) *Lake Tanganyika and its Life*. Natural History Museum Publications and Oxford Univ. Press, London.

Dean, W.E. and Gorham, E. (1998) Magnitude and significance of carbon burial in lakes, reservoirs, and peatlands. *Geology* **26**(6), 535-538.

De Batist, M., van Rensbergen, P., Back, S. and Klerkx, J. (1996) Structural framework, sequence stratigraphy and lake level variations in the Livingston Basin (northern Lake Malawi): First results of a high-resolution reflection seismic study, in T.C. Johnson and E.O. Odada (eds.), *The Limnology, Climatology and Paleoclimatology of the East African Lakes*, Gordon and Breach Publ., Amsterdam. pp. 509-521.

Degens, E.T., von Herzen, R.P., Wong, H-K., Deuser, W.G. and Jannasch, H.W. (1973) Lake Kivu: Structure, chemistry and biology of an East African rift lake. *Geologische Rundschau* **62**, 245-277.

Deuser, W.G., Degens, E.T., Harvey, G.R. and Rubin, M. (1973) Methane in Lake Kivu: New data bearing on its origin. *Science* **181**, 51-54.

Devol, A.H. (1987) Verification of flux measurements made with *in situ* benthic chambers. *Deep-Sea Res.* **34**, 1007-1026.

Downes, M.T. (1988) Aquatic nitrogen transformations at low oxygen concentrations. *Appl. Environ. Microbiol.* **54**, 172-175.

Duchemin, É., Lucotte, M., Canuel, R., and Chamberland, A. (1995) Production of the greenhouse gases $CH_4$ and $CO_2$ by hydroelectric reservoirs of the boreal region. *Global Biogeochemical Cycles* **9**, 529-540.

Edmond, J.M., Stallard, J.M., Craig, H., Craig, V. and Weiss, R.F. (1993) Nutrient chemistry of the water column of Lake Tanganyika. *Limnol. Oceanogr.* **38**, 725-738.

Fendinger, N.J. and Adams, D.D. (1986) A headspace equilibration technique for measurement of dissolved gases in sediment pore water. *Intern. J. Environ. Anal. Chem.* **23**, 253-265.

Fendinger, N.J. and Adams, D.D. (1987) Nitrogen gas supersaturation in the Recent sediments of Lake Erie and two polluted harbors. *Water Research* **21**, 1371-1374.

Federal Environmental Agency, Germany (2000) *Annual Report 1999 of the Monitoring Network.* Texte **58**, 89-102.

Fish, G.R. (1956) Chemical factors limiting growth of phytoplankton in Lake Victoria. E. Afr. Agric. J. 21, 152-158, cited from D.A. Livingstone and J.M. Melack (1984) *Some lakes of Subsaharan Africa*, pp. 467-497, in F.B. Taub (ed.) *Ecosystems of the World*, Vol. 23, *Lakes and Reservoirs*, Elsevier, Amsterdam.

Frenzel, P., Thebrath, B. and Conrad, R. (1990) Oxidation of methane in the oxic surface layer of a deep lake sediment (Lake Constance). *FEMS Microbiol. Ecol.* **73**, 149-158.

Galchenko, V.F., Lein, A. and Ivanov, M. (1989) Biological sinks of methane, in M.O. Andrea and D.S. Schimel (eds.), *Exchange of Trace Gases between Terrestrial Ecosystems and the Atmosphere*. John Wiley and Sons, New York, NY., pp. 59-71.

Hammond, D.E., Fuller, C., Harmon, D., Hartman, B., Korosec, M., Miller, L.G., Rea, R., Warren, S., Berelson, W. and Hager, S.W. (1985) Benthic fluxes in San Francisco Bay. *Hydrobiologia* **129**, 69-90.

Hastenrath, S. and Kutzbach, J.E. (1983) Paleoclimatic estimates from water and energy budgets of East African lakes. *Quaternary Res.* **19**, 141-153.

Hargrave, B.T. and Connolly, G.F. (1978) A device to collect supernatant water for measurement of the flux of dissolved compounds across sediment surfaces. *Limnol. Oceanogr.* **23**, 1005-1010.

Hecky, R.E. (1991) The pelagic ecosystem, in G.W. Coulter (ed.), *Lake Tanganyika and its Life*. Natural History Museum Publications and Oxford Univ. Press, London.

Hecky, R.E. (1993) The eutrophication of Lake Victoria. *Mitt. Internat. Verein. Limnol.* **25**, 39-48.

Hecky, R.E. and Bugenyi, F.W.B. (1992) Hydrology and chemistry of the African Great Lakes and water quality issues: Problems and solutions. *Mitt. Internat. Verein. Limnol.* **23**, 45-54.

Hecky, R.E. and Degens, E.T. (1973) *Late Pleistocene-Holocene Chemical Stratigraphy and Paleolimnology of the Rift Valley Lakes of Central Africa.* WHOI-73-28, Woods Hole Oceanographic Institution, Woods Hole, MA. 129 p.

Hecky, R.E. and H.J. Kling (1987) Phytoplankton of the great lakes in the western rift valley of Africa. *Arch. Hydrobiol. Beih. Ergebn. Limnol.* **25**, 197-228.

Hecky, R.E., Coulter, G.W. and Spigel, R.H. (1991) The nutrient regime, chapter 4, in G.W. Coulter, ed., *Lake Tanganyika and its Life*. Natural History Museum Publications and Oxford Univ. Press, London.

Hecky, R.E., Bootsma, H.A., Mugidde, R.M. and Bugenyi, F.W.B. (1996) Phosphorus pumps, nitrogen sinks, and silicon drains: Plumbing nutrients in the African Great Lakes, in T.C. Johnson and E.O. Odada (eds.), *The Limnology, Climatology and Paleoclimatology of the East African Lakes*, Gordon and Breach Publ., pp. 205-224.

Hecky, R.E., Bugenyi, F.W.B., Ochumba, P., Gophen, M., Mugidde, R. and Kaufman, L. (1994) Deoxygenation of the deep water of Lake Victoria. *Limnol. Oceanogr.* **39**, 1476-1480.

Herdendorf, C.E. (1990) Distribution of the world's largest lakes, in M.M. Tilzer and C. Serruya (eds.), *Large Lakes: Ecological Structure and Function*, Springer Verlag, Berlin, pp. 3-38.

Hesse, P.R. (1958) The distribution of sulphur in the muds, water and vegetation of Lake Victoria. *Hydrobiologia* **11**, 29-39.

Hessen, D. and Nygaard, K. (1992) Bacterial transfer of methane and detritus; implications for the pelagic carbon budget and gaseous release. *Arch. Hydrobiol. Beih. Ergebn. Limnol.* **37**, 139-1487.

Hesslein, R.H. (1976) An *in situ* sampler for close interval pore water studies. *Limnol. Oceanogr.* **21**, 912-914.

Höhener, P. and Gächter, R. (1994) Nitrogen cycling across the sediment-water interface in a eutrophic, artificially oxygenated lake. *Aquatic Sci.* **56**, 115-132.

ICRAF (2001) *Improved Land Management in the Lake Victoria Basin: Linking Land & Lake, Research & Extension, Catchment & Basin.* Final Technical Report. Start-up Phase, July 1999 to June 2000. International Centre for Research in Agroforestry, Nairobi, Kenya.

IDEAL (1990) *An International Decade for the East African Lakes.* Workshop Report 1 on the Paleoclimatology of African Rift Lakes, Bern, Switzerland, March 29-31, 1990. Tech. Report, Duke Univ. Marine Laboratory, Beaufort, NC. 39 p.

IPCC (1996) *Climate Change 1995: The Science of Climate Change.* Houghton, J.T., Meira Filho, L.G., Callander, B.A., Harris, N., Kattenberg, A. and Maskell, K. (eds.), Cambridge Univ. Press, U.K.

Iversen, N. and Jorgensen, B.B. (1993) Diffusion coefficients of sulfate and methane in marine sediments: Influence of porosity. *Geochimica Cosmochimica Acta* **57**, 571-578.

Jannasch, H.W. (1975) Methane oxidation in Lake Kivu (Central Africa). *Limnol. Oceanogr.* **20**, 860-864.

Johnson, T.C. (1996) Sedimentary processes and signals of past climatic changes in the large lakes of the East African Rift Valley, in Johnson, T.C. and Odada, E.O. (eds.), *The Limnology, Climatology and Paleoclimatology of the East African Lakes*, Gordon and Breach Publishers, pp. 367-412.

Johnson, T.C. and Odada, E.O., eds. (1996), *The Limnology, Climatology and Paleoclimatology of the East African Lakes*, Gordon and Breach Publishers.

Johnson, T.C., Kelts, K. and Scholz, C.S. (1992) *IDEAL - Paleoclimate Studies on Lake Victoria.* Funded NSF proposal.

Kaplan, W.A., Elkins, J.W., Kolb, C.E., McElroy, M.B., Wofsy, S.E. and Duran, A.P. (1978) Nitrous oxide in fresh waters: An estimate of the yield of atmospheric $N_2O$ associated with disposal of human waste. *Pure Appl. Geophys.* **116**, 423-438.

Kasper, H.F. (1982) Denitrification in marine sediment: Measurement of capacity and estimate of *in situ* rate. *Appl. Environ. Microbiol.* **43**, 522-527.

Kasper, H.F., Asher, R.A. and Boyer, I.C. (1985) Microbial nitrogen transformations in sediments and inorganic nitrogen fluxes across the sediment/water interface on the South Island West Coast, New Zealand. *Estuarine, Coastal Shelf Science* **21**, 245-255.

Kendall, R.L. (1969) An ecological history of the Lake Victoria basin. *Ecol. Monogr.* **39**, 121-176.

Kilham, P. and Kilham, S.S. (1989) Endless summer: internal loading processes dominate nutrient cycling in tropical lakes. *Freshwater Biology* **23**, 379-389.

Kilham, S.S. and Kilham, P. (1990) Tropical limnology: Do African lakes violate the "first law" of limnology? *Verh. Internat. Verein. Limnol.* **24**, 68-72.

Kling, G.W., Evans, W.C. and Tuttle, M.L. (1991) A comparative view of Lakes Nyos and Monoun, Cameroon, West Africa. *Verh. Internat. Verein. Limnol.* **24**, 1102-1105.

Kling, G.W., Clark, M.A., Compton, H.R., Devine, J.D., Evans, W.C., Humprey, A.M., Koenigsberg, E.J., Lockwood, J.P., Tuttle, M.L. and Wagner, G.N. (1987) Lake Nyos gas disaster in Cameroon, West Africa. *Science* **236**, 169-175.

Klump, J.V. (1987) Biogeochemical processes and chemical fluxes, in *Basic Issues in Great Lakes Research*, Special Report No. 123, Univ. Michigan, Ann Arbor, MI, pp. 86-116.

Koyama, T. (1990) Gases in lakes, their production mechanism and degassing ($CH_4$ and $H_2$) of the earth, in Durrance, E.M., E.M. Galimov, M.E. Hinkle, G.M. Reimer, R. Sugisaki and S.S. Augustithis (eds.), *Geochemistry of Gaseous Elements and Compounds.* Theophrastus Publ., Athens, Greece, pp. 271-335.

Kühl, M., Steuckart, C., Eickert, G. and Jeroschewski, P. (1998) A $H_2S$ microsensor for profiling biofilms and sediments: Application in an acidic lake sediment. *Aquat. Microb. Ecol.* **15**, 201-209.

Lehman, J.T. (1996) Pelagic food webs of the East African Great Lakes, in T.C. Johnson and E.O. Odada (eds.), *The Limnology, Climatology and Paleoclimatology of the East African Lakes*, Gordon and Breach Publ., pp. 281-301.

Lezzar. K.E. (2000) Limnology mentor's report, in A. S. Cohen (ed.) *The Nyanza Project 2000 Annual Report,* Univ. Arizona, Tucson, AZ, pp. 6-9.

Lipiatou, E., Hecky, R.E., Eisenreich, S.J., Lockhart, L., Muir, D. and Wilkinson, P. (1996) Recent ecosystem changes in Lake Victoria reflected in sedimentary natural and anthropogenic organic compounds, in T.C. Johnson and E.O. Odada (eds.), *The Limnology, Climatology and Paleoclimatology of the East African Lakes*, Gordon and Breach Publ., pp. 523-541.

Livingstone, D.A. and Melack, J.M. (1984) Some lakes of Subsaharan Africa, in F.B. Taub (ed.) *Ecosystems of the World*, Vol. **23**, *Lakes and Reservoirs*, Elsevier, Amsterdam, pp. 467-497.

Lowe-McConnell, R.H., Crul, R.C.M. and Roest, F.C. (1992) Symposium on resource use and conservation of the African Great Lakes, Bujumbura 1989. *Mitt. Internat. Verein. Limnol.* **23**, 1-128.

Mengis, M., Gächter, R. and Wehrli, B. (1997) Nitrogen elimination in two deep eutrophic lakes. *Limnol. Oceanogr.* **42**, 1530-1543.

Michelo, V. and Chororoka, K. (2000) Analytical study of a finely laminated core in central Lake Tanganyika Basin (Kalya slope): Variation of annual laminae-couplets and distribution of diatom populations in A. S. Cohen (ed.), *The Nyanza Project 2000 Annual Report*, Univ. Arizona, Tucson, AZ. Student report, pp. 24-26.

Mothersill, J.S. (1976) The mineralogy and geochemistry of the sediments of northwestern Lake Victoria. *Sedimentology* **23**, 553-565.

Mothersill, J.S., Freitag, R. and Barnes, B. (1980) Benthic macro-invertebrates of northwestern Lake Victoria, East Africa: Abundance, distribution, intra-phyletic relationships and relationships between taxa and selected element concentrations in the lake-bottom sediments. *Hydrobiologia* **74**, 215-224.

Mudroch, A. and MacKnight, S.D., eds. (1994), *CRC Handbook of Techniques for Aquatic Sediments Sampling, $2^{nd}$ Ed.* CRC Press, Boca Raton, FL.

Mugidde. R. (1993) The increase in phytoplankton primary productivity and biomass in Lake Victoria (Uganda). *Verh. Internat. Verein. Limnol.* **25**, 846-849.

Newell, B.S. (1960) The hydrology of Lake Victoria. *Hydrobiologia* **15**, 363-383.

Nishio, T., Koike, I. and Hattori, A. (1981) $N_2$/Ar and denitrification in Tama estuary sediments. *Geomicrobiol. J.* **2**, 193-206.

Nixon, S.W., Ammerman, J., Atkinson, L., Berounsky, V., Billen, G., Boicourt, W., Boynton, W., Church, T., DiToro, D., Elmgren, R., Garber, J., Giblin, A., Jahnke, R., Owens, N., Pilson, M.E.Q. and Seitzinger, S. (1996) The fate of nitrogen and phosphorus at the land-sea margin of the North Atlantic Ocean. *Biogeochemistry* **35**, 141-180.

Nowicki, B.L., Requintina, E., Van Keuren, D. and Kelly, J.R. (1997) Nitrogen losses through sediment denitrification in Boston Harbor and Massachusetts Bay. *Estuaries* **20**, 626-639.

Ochumba, P.B.O. (1996) Measurement of water currents, temperature, dissolved oxygen and winds on the Kenyan Lake Victoria, in T.C. Johnson and E.O. Odada (eds.), *The Limnology, Climatology and Paleoclimatology of the East African Lakes*, Gordon and Breach Publ., pp. 155-167.

Ochumba, P.B.O. and Kibaara, D.I. (1989) Observations on blue-green algal blooms in the open waters of Lake Victoria, Kenya. *Afr. J. Ecol.* 27, 23-34, cited from R.E. Hecky (1992) The eutrophication of Lake Victoria. *Verh. Internat. Verein. Limnol.* **25**, 39-48.

Oremland, R.S. (1988) Biogeochemistry of methanogenic bacteria, in A.J.B. Zehnder (ed.), *Biology of Anaerobic Microorganisms*. John Wiley and Sons, N.Y., pp. 641-705.

Reddy, K.R. and Patrick, W.H. (1975) Effect of alternate aerobic and anaerobic conditions on redox potential, organic matter decomposition and nitrogen loss in a flooded soil. *Soil Biol. Biochem.* **7**, 87-94.

Richardson, J.L. and Richardson, A.E. (1972) History of an African rift lake and its climatic implications. *Ecol. Monographs* **42**, 499-534.

Risgaard-Petersen, N., Skårup, S. and Nielsen, L.P. (1999) Denitrification in a soft bottom lake: evaluation of laboratory incubations. *Aquatic Microbial Ecology* **17**, 279-287.

Robertson, L.A. and Kuenen, J.G. (1984) Aerobic denitrification - old wine in new bottles? *Antonie van Leeuwenhoek* **50**, 525-544.

Robertson, L.A. and Kuenen, J.G. (1991) Physiology of nitrifying and denitrifying bacteria, in J.E. Rogers and W.B. Whitman (eds.), *Microbial Production and Consumption of Greenhouse Gases: Methane. Nitrogen Oxides, and Halomethanes*. Amer. Society Microbiology, Washington, D.C., pp. 189-199.

Rothfuss, F. and Conrad, R. (1998) Effect of gas bubbles on the diffusive flux of methane in anoxic paddy soils. *Limnol. Oceanogr.* **43**, 1511-1518.

Rothfuss, F., Frenzel, P. and Conrad, R. (1994) Gas diffusion probe for measurement of $CH_4$ gradients, in L.J. Stal and P. Caumette (eds.), *Microbial Mats. Structure, Development and Environmental Significance*. Springer-Verlag, Germany, pp. 167-172.

Rudd, J.W.M. (1980) Methane oxidation in Lake Tanganyika (East Africa). *Limnol. Oceanogr.* **25**, 958-963.

Rudd, J.W.M. and Hamilton, R.D. (1978) Methane cycling in a eutrophic shield lake and its effects on whole lake metabolism. *Limnol. Oceanogr.* **23**, 337-348.

Rudolph, J., Frenzel, P. and Pfennig, N. (1991) Acetylene inhibition technique underestimates *in situ* denitrification rates in intact cores of freshwater sediment. *FEMS Microbiol. Ecology* **85**, 101-106.

Rutgers van der Loeff, M.M., Anderson, L.G., Hall, P.O.J., Iverfeldt, A., Josefson, A.B., Sundby, B. and Westerlund, S.F.G. (1984) The asphyxiation technique: An approach to distinguishing between molecular diffusion and biological mediated transport at the sediment-water interface. *Limnol. Oceanogr.* **29**, 675-686.

Santschi, P.H., Bower, P., Nyffeler, U.P., Azevedo, A. and Broecker, W.S. (1983) Estimates of the resistance to chemical transport posed by the deep-sea boundary layer. *Limnol. Oceanogr.* **28**, 899-912.

Scholz, C.A., Rosendahl, B.R., Versfelt, J.W. and Rach, N. (1990) Results of high-resolution echo-sounding of Lake Victoria. *J. Afr. Earth Science* **11**, 25-32.

Seiler, W. and Conrad, R. (1987) Contribution of tropical ecosystems to the global budgets of trace gases, especially $CH_4$, $H_2$, CO, and $N_2O$, in R.E. Dickenson (ed.), *The Geophysiology of Amazonia, Vegetation and Climate Interactions*. John-Wiley, NY, pp. 1333-1362.

Seitzinger, S.P. (1988) Denitrification in freshwater and coastal marine ecosystems: Ecological and geochemical significance. *Limnol. Oceanogr.* **33**, 702-724.

Seitzinger, S.P. and Nixon, S.W. (1985) Eutrophication and the rate of denitrification and $N_2O$ production in coastal marine sediments. *Limnol. Oceanogr.* **30**, 1332-1339.

Seitzinger, S.P., Nixon, S.W. and Pilson, M.E.Q. (1984) Denitrification and nitrous oxide production in a coastal marine ecosystem. *Limnol. Oceanogr.* **29**, 73-83.

Song, Y. and Müller, G. (1999) *Sediment-Water Interactions in Anoxic Freshwater Sediments – Mobility of Heavy Metals and Nutrients*. Lecture notes in Earth Sciences Vol. **81**, Springer-Verlag, Heidelberg, Germany. 111 p.

Sorensen, J. (1978) Denitrification rates in a marine sediment as measured by the acetylene inhibition technique. *Appl. Environ. Microbiol.* **36**, 139-143.

Stager, J.C. (1984) The diatom record of Lake Victoria (East Africa): the last 17,OOO years, in D.G. Mann (ed.), *Proceedings of the Seventh International Diatom Symposium 1982*, Strauss and Cramer GmbH, Hirschberg, Germany, pp. 455-476.

Stigg, L. (1986) Metal transfer mechanisms in lakes; the role of settling particles, in W. Stumm, (ed.), *Chemical Processes in Lakes*, John Wiley and Sons, New York.

St. Louis, V.L., Kelly, C.A., Duchemin, É., Rudd, J.W.M. and Rosenberg, D.M. (2000) Reservoir surfaces as sources of greenhouse gases to the atmosphere: a global estimate. *Bioscience* **50**, 766-775.

Stouthamer. A.H. (1988) Dissimilatory reduction of oxidized nitrogen compounds, in A.J.B. Zehnder (ed.) *Biology of Anaerobic Microorganisms*. John Wiley, New York, NY, pp. 245-303.

Sweerts, J-P.R.A., DeBeer, D., Nielsen, L.P., Verdouw, H., Van den Heuvel, J.C., Cohen, Y. and Cappenberg, T.E. (1990) Denitrification by sulphur oxidizing *Beggiatoa* spp. mats on freshwater sediments. *Nature* **344**, 762-763.

Sweerts, J-P.R.A., Kelly, C.A., Rudd, J.W.M., Hesslein, R. and Cappenberg, T.E. (1991) Similarity of whole-sediment molecular diffusion coefficients in freshwater sediments of low and high porosity. *Limnol. Oceanogr.* **36**, 335-342.

Talling, J.F. (1965) The photosynthetic activity of phytoplankton in East African lakes. *Int. Revue ges. Hydrobiol.* **50**, 1-32.

Talling, J.F. (1966) The annual cycle of stratification and phytoplankton growth in Lake Victoria (East Africa). *Int. Revue ges. Hydrobiol.* **51**, 545-621.

Terai, H. and Yoh, M. (1996) Denitrification and nitrous oxide production in Lake Kizaki, in D.D. Adams, S.P. Seitzinger and P.M. Crill (eds.), *Cycling of Reduced Gases in the Hydrosphere*. E. Schweizerbart'sche Verlagsbuchhandlung, Stuttgart, Germany, pp. 97-104.

Tiedje, J.M. (1988) Ecology of denitrification and dissimilatory nitrate reduction to ammonium, in A.J.B. Zehnder (ed.) *Biology of Anaerobic Microorganisms*. John Wiley, New York, pp. 179-244.

Tiercelin, J-J. and Mondeguer, A. (1991) The geology of the Tanganyika Trough, in G.W. Coulter (ed.) *Lake Tanganyika and its Life*. Natural History Museum Publications and Oxford Univ. Press, London.

Tietze, K., Geyh, M., Müller, H., Schröder, L., Stahl, W. and Wehner, H. (1980) The genesis of methane in Lake Kivu (Central Africa). *Geologische Rundschau* **69**, 452-472.

Tirén, T. (1977) Denitrification in sediment-water systems of various types of lakes, in H.L. Golterman (ed.) *SIL-UNESCO Symposium on Interactions between Sediments and Freshwater*, Amsterdam, 1976. PUDOC, Wageningen, The Netherlands, pp. 353-369.

Tyler, S.C. (1991) The global methane budget, in J.E. Rogers and W.B. Whitman (eds.), *Microbial Production and Consumption of Greenhouse Gases: Methane, Nitrogen Oxides, and Halomethanes*. Amer. Society Microbiology, Washington, D.C., pp. 7-38.

Ullman, W.J. and Aller, R.C. (1982) Diffusion coefficients in nearshore marine sediments. *Limnol. Oceanogr.* **27**, 552-556.

Urban, N.R., Dinkel, C. and Wehrli, B. (1997) Solute transfer across the sediment surface of a eutrophic lake: I. Porewater profiles from dialysis samplers. *Aquatic Sci.* **59**, 1-25.

Van Eck, G.T.M. and Smits, J.G.C. (1986) Calculation of nutrient fluxes across the sediment-water interface in shallow lakes, in P.G. Sly (ed.), *Sediments and Water Interactions*. Springer-Verlag, New York, N.Y., pp. 289-301.

Weiss, R.F. (1970) The solubility of nitrogen, oxygen and argon in water and seawater. *Deep-Sea Res.* **17**, 721-735.

Weiss, R.F. (1974) Carbon dioxide in water and seawater: The solubility of a non-ideal gas. *Mar. Chem.* **2**, 203-215.

Wiesenburg, D.A. and Guinasso, N.L. (1979) Equilibrium solubilities of methane, carbon monoxide, and hydrogen in water and sea water. *J. Chem. and Engineering Data* **24**, 356-360.

Yoh, M., Terai, H. and Saijo, Y. (1988) Nitrous oxide in freshwater lakes. *Arch. Hydrobiol.* **113**, 273-294.

Zilifi D. and Eagle, M. (2000) Sediment cores from the Kalya platform, slope and horst: High resolution and long-time span paleolimnology and paleoclimate records, in A. S. Cohen (ed.) *The Nyanza Project 2000 Annual Report*, Univ. Arizona, Tucson, AZ. Student report, pages 20-23.

# Biodiversity, Food Webs and Fisheries

# REDUNDANCY AND ECOSYSTEM STABILITY IN THE FLUCTUATING ENVIRONMENTS OF LONG-LIVED LAKES

KOEN MARTENS
*Royal Belgian Institute of Natural Sciences, Freshwater Biology, Vautierstraat 29, B1000 Brussels, Belgium*

**ABSTRACT**

Ancient lakes like Tanganyika and Baikal are hot spots of megadiversity and endemicity. Although extant knowledge on the autecology of the more than 2000 species in each of these lakes is still limited, there seem to be a considerable number of redundant species in these ecosystems. In a more colloquial way, one could say that these lakes have too many species. However, redundancy can be a buffering mechanism against species loss in (more or less predictable) fluctuating environments. Cyclic changes in physical limnology and lake levels at Milankovich time scales have been documented for both Lake Tanganyika and Baikal. Such environmental alterations are bound to affect lacustrine ecosystems. A large number of species per functional guild will allow species loss without ecosystem-collapse. This effect of biodiversity on ecosystem stability presents a paradox, as in times of stasis, selection will favour higher efficiency of energy turn-over within the ecosystem, not mere survival. Competitive exclusion should than reduce diversity. However, post-stress (after extinction) speciation and immigration in ancient lakes increases high diversity levels. These levels of high diversity are maintained during periods of stasis by a combination of niche diversification, with mutualistic species interactions as a special case, and metapopulation dynamics. Conservation programs should thus not focus exclusively on keystone taxa and function, but should rather manage for redundancy as a buffer for ecosystem resilience to both climatic or human induced disturbances.



*E.O. Odada and D.O. Olago (eds.),*
*The East African Great Lakes: Limnology, Palaeolimnology and Biodiversity,* 309–319.

# 1. INTRODUCTION

Ancient lakes like Tanganyika (12-20 Ma) and Baikal (25-30 Ma) have often been cited as hotspots of megadiversity and endemicity. Most studies dealing with the faunas of these lakes have focussed on processes related to the dynamic aspects of these levels of diversity: immigration, extinction and, especially, speciation (reviewed in Martens *et al.,* 1994; Martens, 1997). Other investigations concentrated on species interactions in extant communities in order to explain the maintenance of these highly diverse communities (summary in Kawanabe *et al.,* 1997), but only few studies include long-term effects into such discussions (Cohen, 1995). None have thus far used ancient lakes as model ecosystems to assess effects of biodiversity on ecosystem functioning and, specifically and most importantly, on ecosystem stability.

The present contribution challenges the concept of redundancy, which is derived from a static view of ecosystems. If large time-scale dimensions are included, the redundancy concept is no longer tenable. Speciosity within functional groups (see below) provides a buffering mechanism to maintain ecosystem integrity in spite of species loss during periods of (predictable) environmental fluctuations (Solbrig, 1994; Lawton and Brown, 1994). I will address the apparent paradoxical maintenance of such high levels of diversity during periods of relative stability, when ruling theory would predict decreasing diversity levels through competitive exclusion. Both short and long time frames are necessary for understanding the dynamics of biodiversity in ancient lakes. IDEAL can play a significant role in such studies, especially in documenting long-term patterns and processes.

# 2. ANCIENT LAKES HAVE TOO MANY SPECIES

Standing diversity levels in ancient lakes are significantly higher than would be predicted, for example by species-area equations. Moreover, diversity (as well as density) often declines exponentially with depth in most lakes, so that these high diversities occur on even more limited surfaces. This situation is exemplified in the extreme in the East African lakes Tanganyika and Malawi, where waters below 100-150 are completely anoxic, so that all benthic (or benthic-associated) life is limited to a so-called bath tub rim, which constitutes a fraction of the total lake surface. Furthermore, surface-induced diversity limitations have a relative value only, as surface area derives from the equilibrium theory of island biogeography. This theory is severly limited in non-equilibrium, hyper-diversity states, such as found in ancient lakes.

This is further supported by other considerations. For a long time, it has been accepted that hyper-specialisation and resulting niche diversification allowed so many similar species to coexist in the lakes. Niche diversification, however, mostly requires a degree of segregation, be it seasonal, geographical (bathymetric, latitudinal) or otherwise. As will be argued in full below, some degree of niche diversification indeed exists. This is a logical consequence of the linkage of pattern and process: speciation is indeed often linked with a degree of ecological divergence and this should be most apparent in intra-lacustrine neo-endemics, i.e. endemic taxa which evolved in the lake itself (as opposed to palaeo-endemics, which are intra-lacustrine relics of previous, much wider distributions). However, the sympatric occurrence of dozens of species in a seemingly uniform habitat-patch defies this convenient concept. Mazepova (1990) reported 17 species of non-marine ostracod in a square meter of homogenous sediment. Cohen (1995) found a continuous diversity of about 30 ostracod species in one meter cores from Lake Tanganyika. Accepting some degree of micro-scale segregation (for example by vertical distribution of living specimens in the sediment) can still not explain such high levels of standing diversity. Tanganyikan benthic ostracods have no known trophic specialisation, are all of similar size and are mostly continuously crawling on and into the top millimeters of the sediment. Except when some hidden form of biochemical diversification occurs, for example different enzymatic exploitation of the same food resources, which has never been reported in this group, the high specific diversities occurring sympatrically cannot be explained by niche diversification alone. As an estimated 200 species of ostracods occur in both Baikal and Tanganyika (Martens, 1994), it is difficult to imagine an equal number of different niches occupied by these species. Similar high specific diversities, not supported by an equally high functional diversity, occur in other groups, notably in cichlid fish. Fryer (in litt.) addressed this in his *peaceful condominium* hypothesis.

Both ostracods and fish seem to have high redundancy in these ancient lake ecosystems, while it is at the same time very difficult to identify genuine keystone taxa (sensu Mooney *et al.*, 1995) in these lineages. Although it is uncertain as to how many species would be 'normal ' in an ecosystem as old and as complex as Tanganyika and Baikal, a high degree of redundancy seems apparent. Mooney *et al.* (1995) state that species diversity mostly affects ecosystem functioning between 1 and 100 species and although it is also not sure where this relationship saturates, a standing diversity of more than 2000 animal taxa appears to be very close to the asymptotic value.

## 3. BIODIVERSITY AND ECOSYSTEM STABILITY

Effects of the loss of one single species on ecosystem functioning depends on the functional identity of the species (Symstad *et al.,* 1998), i.e. whether it is a so-called keystone species which on its own constitutes a functional guild (mostly predators) or a redundant or compensatory species, of which there are several in a functional guild. However, it also depends on overall composition of the ecosystem. More in particular, it depends on how many species with similar (or identical) function are present (Solbrig, 1994). Correlations between diversity levels on the one hand and efficiency of ecosystem process rates (Johnson and Malmqvist, 2000) on the other, leading to community stability (Tilman, 1999) have been demonstrated, mostly through work on terrestrial grassland systems.

High levels of redundancy are thus functional, as they provide a buffering effect against species loss during environmental changes (Solbrig, 1994; and other papers in Schulze and Mooney, 1994; Johnson, 2000). Even when a relatively high number of species becomes extinct, for example through lake level fluctuations and resulting changes in the physical lake environment (oxygen, salinity), the remaining taxa will ensure the persistence of a skeletal structure of the ecosystem and its functioning.

The buffering effect would be especially useful in situations were such disturbances are cyclic, or at least to some extent predictable. Palaeohydrology of the East African lakes is still insufficiently known, especially in large and deep lakes such as Tanganyika because of the logistical difficulties in obtaining long cores. Nevertheless, recent reports demonstrate that there is a degree of cyclicity, and thus of predictability, in lake level fluctuations. For example, there seems to be a degree of congruence between the alteration of arid and pluvial periods in Africa with glacial and interglacial period in the European Pleistocene (Gasse *et al.,* 1980). Although such cyclicity is often not completely regular (Kerr, 1999), the deep drilling program on Lake Baikal has revealed cyclicities of 20, 40 and 100,000 years in physical variables and parameters; these are all related to the three Milankovitch parameters of solar insolation (Kashiwaya *et al.,* 1998, 1999).

Such cyclic (or predictable) events could be considered as selection forces operating at the ecosystem level. This would be highly controversial, although Rose (2000) considers ecosystems a perfectly valid level of selection. Most likely, however, ecosystem resilience through redundancy evolved as a side effect of selection processes at lower levels, i.e. populations or species. The level of predictability could nevertheless fuel such adaptation in both cases.

The above hypothesis, species redundancy as a strategy to maintain ecosystem integrity during periods of environmental fluctuations, has a paradoxical consequence. During periods of stasis, specific interactions will normally lead to competitive exclusion of species with similar niches and thus to a decrease of extant diversity. Nevertheless, we observe an increase of diversity in post-stress situations (Geary, 1990), while during stasis high diversity is maintained for long periods of time. Both mechanisms are discussed below.

# 4. SPECIATION AND IMMIGRATION IN POST-STRESS SITUATIONS

Increase of extant diversity is the result of immigration and/or of speciation. Martens (1997) has stressed the fact that ancient lakes, in spite of their high levels of endemicity, are not completely isolated from their surroundings and that surrounding water bodies (rivers, associated lagoons, other lakes) are highly influenced by (and themselves significantly influence) faunal dynamics in long-lived lakes. Fryer (1997) noted that cichlids of Lake Victoria might have survived recent desiccation events in that lake in surrounding refugia. Diversity levels in ancient lakes might thus (partly) be re-established in the post-stress phase by immigration of taxa from such refugia. Alternatively, other taxa can immigrate from surrounding water bodies in such conditions and give rise to new species flocks. The extant gammarid fauna of Lake Baikal is assumed to be the result of multiple invasions (Sherbakov *et al.*, 1999). The ostracod fauna of Lake Baikal consists of 3-4 lineages (Mazepova, 1990), that of Tanganyika of at least six lineages (Martens, 1994) and these radiations thus are a result of at least as many invasions.

The understanding of intralacustrine speciation has always been problematic to some extent, as most speciation was thought to occur in allopatry. Speciation within one lake basin must occur sympatrically or, at best, parapatrically along a geographical or an ecological cline (Martens *et al.*, 1994). Major lake level fluctuations, dividing the large Lake Tanganyika into 3 isolated basins, seemed to provide opportunities to maintain allopatric speciation as the major force causing high extant specific diversities in this lake. However, recent evolutionary theory provides support for the fact that speciation can be effective, even (or especially) when some degree of gene flow in contact zones is still possible, through mechanisms such as character displacement etc. In addition to this, Geary (1990) has shown that rapid speciation in molluscs occurs when water levels are rising again, i.e. when newly available habitats can be colonised without competition with extant communities, and not during the actual phase of isolation. Note that rising lake levels do not necessarily increase the available surface where life is possible; the combined available littoral surface of the three basins at low levels is equal to, or might even be larger than, the single bath rim of the lake at high levels. Nevertheless, the above observations are in accordance with, and even explain, the observed increasing levels of diversity in post-stress phases of ecosystems.

## 5. MAINTENANCE OF HIGH DIVERSITY LEVELS DURING STASIS

One could argue that littoral habitats in ancient lakes will never be in stasis, as they will always show fluctuations, however minor (Coulter, 1994). Sheldon (1993) presented a model, linking two extreme modes of speciation with different levels of habitat fluctuation, respectively punctuated speciation in widely fluctuating habitats and gradual speciation in weakly changing habitats. Martens *et al.* (1994) applied this model to ancient lakes, postulating that the first type could be congruent with littoral conditions, the second type with the pelagic processes. However, at least the first type of speciation will only occur if environmental variables will break out of so-called reflecting boundaries, i.e. imaginary values not reached by most of the oscillations. The importance of this model for the present discussion is that it shows that periods of stasis are indeed a biological reality, as systems apparently have buffers against minor fluctuations. The Sheldon model, which he named *Plus ça change*, applies to morphological stasis in species; it is here putatively expanded to ecosystems.

Thousands of years of nearly continuous stasis should lead to decreasing diversity through competitive exclusion (Johnson, 2000). Two different mechanisms buffering the system against species-loss during stasis could be relevant here. The first, niche diversification, is of the old ecological equilibrium paradigm. The second, metapopulation dynamics, fits more elegantly in the new ecological flux paradigm.

## 5.1 Niche Diversification

Although I reject the notion that each ancient lake species has a different niche, niche diversification does exist amongst a number of species in ancient lakes and contributes to the maintenance of high diversity levels during periods of stasis. Although redundancy implies large overlap of, or even identical, niches for many species, a difference must further be made between ecosystem functioning during periods of stasis and during periods of environmental fluctuation. During stasis, ecosystem functioning evolves towards a high efficiency of resource turn-over (energy flux) (Schultz and Mooney, 1994; Symstad *et al.*, 1998). Niches will then be defined in this light and in such conditions can be identical or nearly so for many species. During environmental fluctuations, however, species are subjected to different circumstances and other aspects of specific tolerances can then become important. During periods of stress, otherwise redundant species might show different reactions to different types of variables and it is worth preserving such genetic variability. An example of this can be found in the trophic adaptation amongst East African cichlids. Several authors have stressed the trophic adaptation in jaw and tooth structure in several cichlids. These examples of so-called adaptive radiation have been incorporated into several handbooks on evolution (Skelton, 1993; Ridley, 1996; Schluter 2000). Piscivores, algal grazers, scavengers, and even highly specialised types like scale eaters have been described. However, Reinthal (1990) has demonstrated that the presumed trophic adaptation is not always reflected in the actual diet of a fish, i.e. highly specialised species have a much wider diet than was previously assumed. These adaptations probably originated during periods of stress, and high competition for food resources. During the present periods with ample food resources, the specialisations can persist without actually serving their original purpose, because they have become secondarily functional in other ways, such as mate recognition and sexual selection. In summary, the statement that each species has its own niche requires the occurrence of hyper specialisation; this prediction is not corroborated in ancient lake biota.

A special type of niche diversification has recently (Lévêque, 1997) been cited as a mechanism to maintain high diversities. Years of behavioural studies on Tanganyikan cichlids by Kawanabe *et al.* (1997) have indeed demonstrated a wide range of mutualistic species interaction which do not lead to competitive exclusion, but rather contribute to the persistence of species-rich communities.

## 5.2 Metapopulation Dynamics

A final mechanism applies metapopulation dynamics to diversity of species communities in ancient lakes. The concept postulates that high diversities depend on non-equilibrium interactions between patchily distributed species and accepts a highly stochastic sequence of local extinctions and recolonisations, allowing for the sympatric occurrence of a higher number of species than can be explained by the available, and detectable, niches. The metapopulation structure of communities in lakes Tanganyika and Baikal is self-evident, as the size and shape of these lakes (*ca.* 660 km long) and the alternation of different types of sediment and depths impedes a continuous gene flow expected in a single population. Cohen (1995), investigating a number of short cores from Lake Tanganyika, provided evidence for stochastic changes in specific densities, including extinctions and re-appearances, in ostracod communities with otherwise stable overall diversity. Further research on the dynamics of these population patches is necessary before the actual buffering impact of this mechanisms against lacustrine extinctions can be estimated, but it is already established that these metapopulation dynamics at least contribute to the maintenance of high diversity levels.

# 6. GENETIC DIVERSITY

The above scenario on specific redundancy can also be applied to an intraspecific level, as persistence of species during periods of environmental fluctuation largely depends on the variability of their gene pool on which selection can act. Surviving genotypes will be those most fit under a particular set of environmental conditions and will ensure persistence of the species. Such strong selective pressures might cause the surviving populations to pass through so-called bottle necks of variability. When environmental conditions (e.g. lake levels) are restored to 'normal', recolonisation of newly available habitats will again increase extant variability.

Stochastic processes can influence intraspecific, genetic variability during periods of stasis as well as during periods of environmental fluctuations. However, local extinctions of subpopulations or haplotypes within these subpopulations will not lead to a loss of genetic diversity, if patches of this species exist elsewhere in the lake and if there is gene flow between the different localities. Metapopulation structures (see above) will be most significant in this context. Thus, the persistence of species during times of environmental change will depend mainly on redundancy of haplotypes during periods of stasis.

# 7. LAKE VICTORIA, A TEST CASE

The apparent extinction of hundreds of cichlid species in Lake Victoria was caused by the combined effects of the introduction of an alien predator (the Nile perch) and increasing eutrophication (Wanink and Witte, 2000). This ecosystem crash would at first glance thus constitute a negative test for the present hypothesis, but a number of remaining uncertainties hamper correct interpretation of the events to date.

On the one hand, this disaster could be an example of an ecosystem crash, caused by an unpredictable set of events where the buffering capacity provided by redundancy was insufficient to maintain ecosystem integrity. This scenario is strengthened by the fact that the large redundancy in the cichlid fauna was most likely not matched by similar diversities in other groups, more notably invertebrate grazers and scavengers. Redundancy in those functional groups was therefore limited, and this unbalanced state of the ecosystem could lead to a more limited buffering capacity. On the other hand, there are reports of recent findings of presumed extinct species. These would thus be remnants of the original ecosystem structure, which would constitute proof of successful buffering, i.e. ecosystem resilience. Only continuous monitoring of this lake will show which of the above scenario's holds truth.

## 8. REDUNDANCY AND CONSERVATION

Walker (1992, 1995) and Petchey (2000) argue that species conservation programs should concentrate on the protection of "keystone" species and functions, for example by concentrating efforts on functional groups with little or no apparent redundancy. Such approaches are clearly ignoring long-term benefits of functional redundancy, especially in long-lived habitats such as ancient lakes. Johnson (2000) argues that management for redundancy, as a driver for ecosystem resilience to disturbance, is much to be preferred. The present essay supports this view. From a practical point of view, managers could reject this approach. Policy makers would indeed find the argument, that present day redundancy must be preserved because it could be functional in ten thousand years, positively ludicrous. However, it is quite likely, that the buffering strategy is also functional at much shorter time frames, for example over several hundreds of years or even less. Moreover, present-day human impact is bound to cause dramatic environmental changes followed by significant biodiversity loss over very short time frames, less than a few dozens of years, requiring the buffering effect at this very moment.

## 9. CONCLUSIONS

Ancient lakes such as Baikal and Tanganyika have specific diversities higher than predicted. From a functional point of view, extant diversities are higher than deemed necessary. Following the hypothesis of Solbrig (1994) and others, such redundancy can ensure ecosystem integrity in spite of species loss through (predictable) environmental fluctuations. This hypothesis presents the paradox that competitive exclusion would cause biodiversity decrease during periods of stasis. Two major mechanisms providing a buffer against species extinctions through competition during periods of stasis might be relevant here: a certain degree of niche diversification, with mutualistic species interactions as a special case, and metapopulation dynamics. Although endemic species might be highly vulnerable to extinction, ancient lake ecosystems might be more robust than previously thought. Also, diversity in such buffered systems is not necessarily a good indicators of ecosystem function or long-term stability. The apparent ecosystem shift in Lake Victoria can be seen as a natural experiment testing the present hypothesis, and detailed monitoring of its recovery is therefore a priority.

## ACKNOWLEDGMENTS

I thank the organisers of the second IDEAL meeting, held at the shores of Lake Malawi early 2000, for inviting me to present this communication and the editors of the present volume for their patience. Dr Isa Schön and two anonymous referees read the manuscript and suggested important improvements.

## REFERENCES

Cohen, A.S. (1995) Paleoecological approaches to the conservation biology of benthos in ancient lakes: a case study from lake Tanganyika. *Journal of the North American Benthological Society* **14**, 654-668

Coulter, G. (1994) Speciation and fluctuating environments, with reference to ancient East African Lakes, in K. Martens, B. Goddeeris and G. Coulter (eds.), Speciation in Ancient Lakes, *Advances in Limnology* **44**, 127-137.

Fryer, G. (1997) Biological implications of suggested Lake Pleistocene desiccation of Lake Victoria. *Hydrobiologia* **354**,177-182.

Gasse, F. Prognon, P. and Street, F.A. (1980) Quaternary history of the Afar and Ethiopian Rift lakes, in M.A.J. Williams and H. Faure (eds.), *The Sahara and the Nile, Quaternary Environments and Prehistoric Occupation in Northern Africa*, A.A. Balkema, Rotterdam, pp. 361-400.

Geary, D.H. (1990) Patterns of evolutionary tempo and mode in the radiation of *Melanopsis* (Gastropoda, Melanopsidae). *Palaeobiology* **16**, 492-511.

Johnson, K.H. (2000) Trophic-dynamic considerations in relating species diversity to ecosystem resilience. *Biological Reviews* **75**, 347-376.

Johnson, M. and Malmqvist, B. (2000) Ecosystem process rate increase with animal species richness: evidence from leaf-eating, aquatic insects. *Oikos* **89**, 519-52.

Kashiwaya, K., Ryugo, M., Sakai, H. and Kawai, T. (1998) Long-term climato-limnological oscillation during the past 2.5 million years printed in Lake Baikal sediments. *Geophysical Research Letters* **25**, 659-662.

Kashiwaya, K., Ryugo, M., Horii, M., Sakai, H., Nakamura, T. and Kawai, T. (1999) Climato-limnological signals during the past 260 000 years in physical properties of bottom sediments from Lake Baikal. *Journal of Palaeolimnology* **21**, 143-150.

Kawanabe, H., Hori, M. and Nagoshi, M. (eds.) (1997) *Fish Communities in Lake Tanganyika*. Kyoto Univ. press. 298 pp.

Kerr, R.A. (1999) Why the Ice Ages don't keep time. *Science* **285**, 503-504.

Lawton, J.H. and Brown, V.K. (1994) Redundancy in ecosystems, in E.-D. Schulze and H.A. Mooney (eds), *Biodiversity and Ecosystem Function*, pp. 255-270.

Lévêque, C. (1997) *Biodiversity Dynamics and Conservation. The Freshwater Fishes of Tropical Africa*. Cambridge Univ. Press, Cambridge, 438 pp.

Martens, K., Coulter, G. and Goddeeris, B. (1994) Speciation in Ancient Lakes - 40 years after J.L. Brooks, in K. Martens, B. Goddeeris and G. Coulter (eds.), Speciation in Ancient Lakes, *Advances in Limnology* **44**, 75-96.

Martens, K. (1994) Ostracod speciation in ancient lakes: a review - in K. Martens, B. Goddeeris and G. Coulter (eds.), Speciation in Ancient Lakes, *Advances in Limnology* **44**, 203-222.

Martens, K. (1997) Review: Speciation in Ancient Lakes. *Trends in Ecology and Evolution* **12**, 177-182.

Mooney, H., Lubchenco, J., Dirzo, R. and Sala, O.E. (1995) Biodiversity and ecosystem functioning; basic principles, in V.H. Heywood and R.T. Watson (eds.) *Global Biodiversity Assessment*, Cambridge Univ. Press, Cambridge, pp. 275-326.

Mazepova, G. (1990) *Rakushkovye ratchki (Ostracoda) Baikala*. Sib. otdel. Akad. Nauk SSSR, Novosibirsk, 470 pp.

Reinthal, P. (1990) The feeding habits of a group of tropical herbivorous rock-dwelling cichlid fishes (Cichlidae: Perciformes) from Lake Malawi, Africa. *Environmental Biology of Fishes* **27**, 215-233.

Ridley, M. (1996) *Evolution*, 2nd ed. Blackwell Science, Cambridge, 719 pp.

Petchey, O.L. (2000) Species diversity, species extinction and ecosystem function. *The American Naturalist* **155**, 696-702.

Rose, S. (2000) Escaping Evolutionary Psychology, in H. Rose and S. Rose (eds.), *Alas, Poor Darwin. Arguments Against Evolutionary Psychology*, Jonathan Cape, London, pp. 247-265.

Schluter, D. (2000) *The Ecology of Adaptive Radiation*. Oxford Series in Ecology and Evolution, Oxford Univ. press, Oxford, 288 pp.

Schulze, E.-D. and Mooney, H.A. (eds) (1994) *Biodiversity and Ecosystem Function*. Springer Verlag, Berlin, 525 pp.

Sheldon, P.R. (1993) Making sense of micro-evolutionary patterns, in D.R. Lees and D. Edwards (eds.), *Evolutionary Patterns and Processes*. Linn. Soc. Symp. Ser. **14**, 19-31.

Sherbakov, D. Yu. (1999) Molecular phylogenetic studies on the origin of biodiversity in Lake Baikal. *Trends in Ecology and Evolution* **14**, 92-95.

Skelton, P. (1993) Evolution. *A Biological and Palaeontological Approach*. Addison-Wesley Publ. Cie, Wokingham, 1064 pp.

Solbrig, O.T. (1994) Plant Traits and adaptive strategies: their role in Ecosystem Function, in E.-D. Schulze and H.A. Mooney (eds), *Biodiversity and Ecosystem Function*, pp. 97-116.

Symstad, A.J., Tilman, D., Willson, J. and Knops, J.M.H. (1998) Species loss and ecosystem functioning: effects of species identity and community composition. *Oikos* **81**, 389-397.

Tilman, D. (1999) The ecological consequences of changes in biodiversity: a search for general principles. *Ecology* **80**, 1455-1474.

Walker, B. (1992) Biodiversity and Ecological Redundancy. *Conservation Biology* **6**, 18-23.

Walker, B. (1995) Conserving biological diversity through ecosystem resilience. *Conservation Biology* **9**, 747-752.

Wanink, J.H. and Witte, F. (2000) The use of perturbation as a natural experiment: effects of predator introduction on the community structure of zooplanktivorous fish in Lake Victoria, in A. Rossiter and H. Kawanabe (eds.), *Ancient Lakes: Biodiversity, Ecology and Evolution*. Advances in Ecological Research **31**, Acad. Press, San Diego, pp. 553-570.

# INVASION OF LAKE VICTORIA BY THE LARGE BODIED HERBIVOROUS CLADOCERAN *DAPHNIA MAGNA*

RYAN JONNA AND JOHN T. LEHMAN
*Department of Biology and Center for Great Lakes and Aquatic Sciences,*
*University of Michigan, Ann Arbor, MI 48109-1048, USA*

## ABSTRACT

Zooplankton collections from offshore Lake Victoria during the early to mid 1990s contained a *Ctenodaphnia* species that is morphologically similar to *Daphnia monacha (D. lumholtzi* var. *monacha*) which is present in Lakes Edward and Albert. Direct comparison of the Lake Victoria specimens with representative *D. monacha* from the Rift Valley lakes reveals that the taxa are not the same, however. Instead, the Lake Victoria animals conform more closely to the morphology of *Daphnia magna*. The large body size of *D. magna* makes it unlikely that it was overlooked by previous investigators of the lake. Instead, it appears that food web alterations of Lake Victoria created occasional conditions of relaxed planktivory by fish to such a degree that an herbivore of unusually large size could establish viable populations. The observation runs counter to prevailing paradigms about the small size structure of tropical zooplankton communities compared with those in temperate zones. The observation is consistent with claims that biological interactions rather than physical and chemical factors are the typical determinants of zooplankton size structure in the tropics.



*E.O. Odada and D.O. Olago (eds.),*
*The East African Great Lakes: Limnology, Palaeolimnology and Biodiversity,* 321–333.

# 1. INTRODUCTION

Predation is a strong organizing force in zooplankton communities. Size structure and species composition are consequences of complex interplay between visual orienting predators such as fish and mechanical or tactile predators such as *Chaoborus* and carnivorous copepods. Evidence for the role of predation comes from both temperate and tropical locations. Intentional or accidental introductions of planktivorous fish into African lakes and reservoirs have provided striking demonstrations of the power of visual planktivory (e.g., Gliwicz, 1985; Dumont, 1986; Marshall, 1991), either through direct predation or through a variety of indirect effects (Gliwicz, 1994).

One of the key studies in the development of understanding about visual predation and zooplankton community structure concerned tropical Cladocera in the Rift Lake sources of the White Nile (Green, 1967, 1971). Green demonstrated that a large bodied *Ctenodaphnia* known as *Daphnia lumholtzi* var. *monacha* (or *D. monacha* in earlier work) was confined to offshore waters of Lakes Albert and Edward, owing to relaxed magnitudes of predation by fish in those regions. The older terminology "D. *monacha*" is adopted in this paper for the purpose of clearly delineating two distinct morphological forms. Nearshore, *D. monacha* virtually vanished from the plankton and was replaced by *Daphnia lumholtzi*. *D. lumholtzi* has a smaller core body size and a smaller eye, making it less visually conspicuous. It also possesses exuberant tail and helmet spines which amplify its effective size to "gape-limited" predators (Zaret, 1980) and thereby help to frustrate their attempted attacks. Tollrian (1994) demonstrated that postembryonic elongation of the spines of *D. lumholtzi* is stimulated by kairomones released by fish. Experimental studies have demonstrated that the spines inhibit feeding by juvenile Centrachidae (Swaffer and O'Brien, 1996; Kolar and Wahl, 1998), in part by increasing the required handling time for spined prey.

*Daphnia lumholtzi* has a broad regional distribution in East Africa, south and Southeast Asia, and Australia (Benzie, 1988). Recently it has been introduced into North America and has colonized habitats northward along the Mississippi Valley into the Great Lakes. In Africa it occurs not only in the rift lakes of the White Nile, but in Lake Victoria as well. It co-occurs in Lake Victoria with one member of the subgenus *Daphnia*, *Daphnia longispina*, as well as with a variety of other Cladocera. Crustacean zooplankton of Lake Victoria have been the object of repeated study during the 20$^{th}$ century (Worthington, 1931; Worthington and Ricardo, 1936; Rzoska, 1957, 1975; Mavuti and Litterick, 1991). The lake is distinguished from other African Great Lakes in part by its greater species richness of diaptomid copepods (Lehman, 1996).

Diaptomidae are more abundant offshore than nearshore in Lake Victoria (Branstrator *et al.*, 1996), which is consistent with a gradient of decreasing planktivory by fish away from shore. Diaptomids and cyclopoid copepods are better at evading capture by teleost fish than are most Cladocera. Relative vulnerability of these two Copepoda suborders cannot be easily generalized, because responses can be both system and species specific (e.g., Drenner *et al.*, 1978; Mookerji *et al.*, 1998;

Nassal *et al.*, 1998; Viitasalo *et al.*, 1998), and are influenced by prey size and fecundity as well as taxonomy. Experience with planktivore introductions in Africa, however, has been that diaptomids are differentially reduced in environments of high predation intensity (e.g. Dumont, 1986; Marshall, 1991). The plankton community of Lake Victoria is consistent with the hypothesis that planktivory by fish is greater nearshore than offshore.

The food web of Lake Victoria has been in a state of flux throughout the 1980s and 1990s owing to changes in climate, land use, and fisheries (Lehman, 1998a, 1998b). Fish community changes included reductions in both herbivorous and zooplanktivorous species (Lowe-McConnell, 1997; Witte *et al.*, 1992). Not surprisingly, the zooplankton community has experienced changes, as well (Mwebaza-Ndawula, 1994).

Zooplankton collections obtained in 1992 from offshore Lake Victoria contained a large new daphnid that had not previously been a reported part of the community. The species was a *Ctenodaphnia*, possessing large cephalic fornices and conforming generally with published descriptions of the *Daphnia monacha* in Lake Albert. Branstrator *et al.* (1996) reported the presence of the novel species in collections made during April and May, 1992.

Subsequently, D. Verschuren (personal communication) pointed to the presence of ephippial eggs characteristic of *Daphnia magna* in surface sediments of Lake Victoria. The discovery called into question the identity assigned to the large *Ctenodaphnia* first seen in 1992. In order to resolve the issue, we inspected an archival collection of Lake Victoria zooplankton samples spanning the years 1991 to 1995, and compared the *Ctenodaphnia* contained within them with specimens of *D. monacha* collected from Lakes Edward and Albert in March and April of 1995.

# 2. METHODS

## 2.1 Sample Sites and Collection Methods

All of the zooplankton collections in this study were obtained with 0.5-m diameter nets constructed of 100-mm aperture Nitex that were hauled by hand from designated depths to the surface at ca. 0.5 m sec $^{-1}$ while the research vessel lay at anchor. Nets were manufactured by Research Nets, Inc. of Bothell, Washington, U.S.A., and had an aspect ratio of 1:5 diameter:length. Typically four replicate vertical hauls were combined in the field and preserved with 4% sucrose-formalin. Most samples were split quantitatively in Jinja, Uganda at the Fisheries Research Institute (FIRI), with one-half of the collection deposited at the University of Michigan in Ann Arbor and the remainder deposited at FIRI.

## 2.2 Microscopy and Documentation

Archival samples in Ann Arbor were stained with Eosin Y and diluted to total volumes ranging from 300 ml to 1200 ml. Successive quantitative 5 ml subsamples were withdrawn for enumeration. All *Daphnia* present in the aliquots were identified to species. Multiple subsamples were drawn until the number of *Daphnia* encountered exceeded 100 individuals. Larger volumes of the collections were searched for presence of large bodied *Ctenodaphnia*, in order to establish their temporal pattern of occurrence.

Similarity between *D. magna* and *D. monacha* called for detailed comparison of several key characteristics. These features were photographed and are reproduced here to document both similarities and differences among the taxa. Eighteen samples were inspected, several of which had been previously part of the report by Branstrator *et al.* (1996). Large *Ctenodaphnia* were distinguished by presence of cephalic fornices; no males or ephippia were encountered in the collections we inspected. Specimens were examined as whole mounts and in dissected detail.

Representative specimens were withdrawn from the collections and placed in glycerin for detailed examination and dissection. Morphological features were distinguished with the aid of an Olympus BHA compound microscope, using interference contrast optics at magnifications up to 1000X. Digital images of key distinguishing features were recorded for documentary purposes.

Size frequencies of *Daphnia* in some samples were measured by ocular micrometer measuring from the midpoint of the eye to the base of the tail spine. Clutch presence was recorded for ovigerous females in order to estimate the size at first reproduction. Clutch sizes could not be estimated reliably because the samples contained numerous loose eggs.

# 3. RESULTS

Morphological differences permit discrimination of three *Daphnia* species from Lake Victoria, none of which are identical with *D. monacha* from Lakes Albert and Edward. The Lake Victoria samples included two members of the subgenus *Ctenodaphnia*, identified as *D. lumholtzi* (typical spined form) and *D. magna*. The most common daphnid in Lake Victoria is a member of the subgenus *Daphnia*: *D. longispina*.

## 3.1 Carapace Valve Margins

Figure 1 shows ventral carapace margins of four taxa, three of which are members of the subgenus *Ctenodaphnia*. *D. lumholtzi* are distinguished by stout and widely spaced spines along the ventral margin of the valves. Setae are more densely arrayed along the ventral valve margin for *D. magna* from Lake Victoria compared with *D. monacha* from Lake Albert. In fact, *D. magna* have double the number of setae per

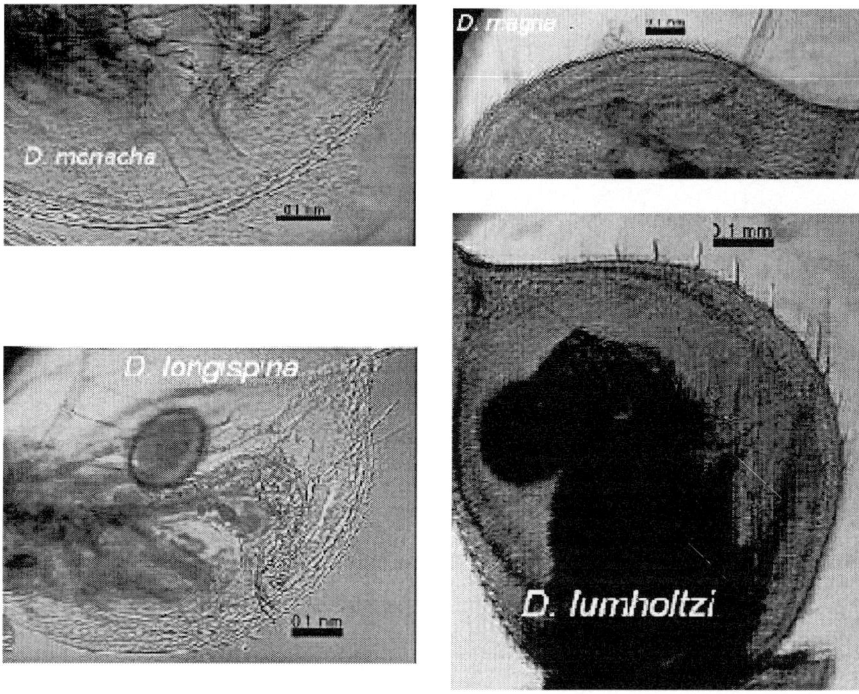


*Figure 1*. Carapace valve margins of *Daphnia monacha* (Lake Albert), and *D. longispina*, *D. magna*, and *D. lumholtzi* (Lake Victoria). All scale bars are 0.1 mm.

unit length compared with *D. monacha*. The valve margin of *D. monacha* is most similar to that of *D. longispina* in that the setae appear more conspicuous toward the posterior ventral margin of the valves.

## 3.2 Postabdomen

Figure 2 compares postabdomens among *Daphnia*. *D. longispina* is distinguished by its linear margin, diminishing length of anal spines from distal to proximal, and though not illustrated, distinctive brownish pigmentation of the terminal claw. All three *Ctenodaphnia* exhibit non-linear margins that are sinusoidal to varying degrees, least so with *D. lumholtzi*. The anal spines of *D. lumholtzi* are robust, short, and they taper in length proximally. They are also fewer in number, often as few as 10, compared with the other species. The postabdomens of *D. monacha* and *D. magna* are superficially similar, but the sinusoidal shape is more strongly expressed in *D. magna*, and *D. magna* bear a greater number of anal spines (typically 16) than do *D.*

*monacha* (typically 13). The most noticeable difference, however, is a gap in the anal spines at the anus of *D. magna*, but not so with *D. monacha*.

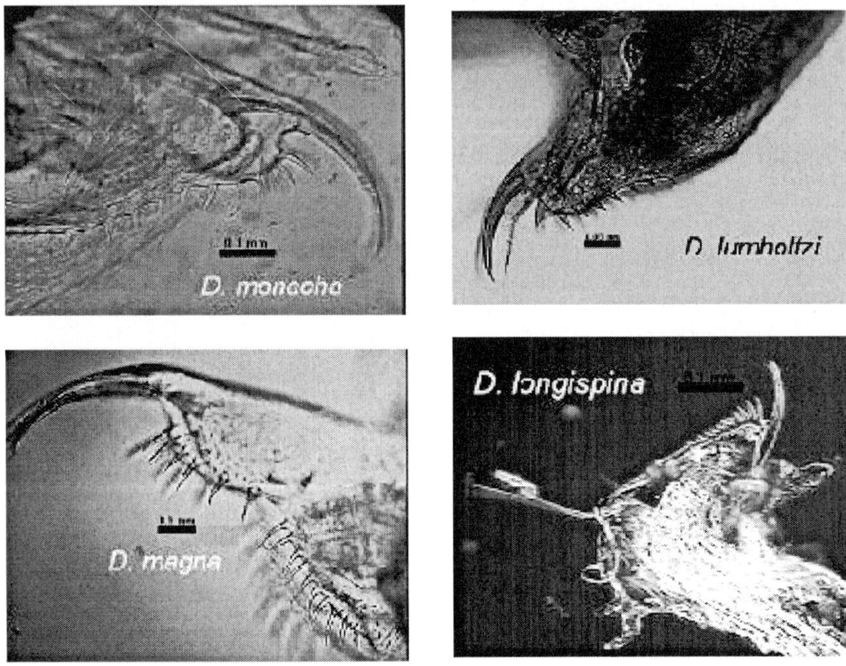


*Figure 2*. Postabdomens of *Daphnia* from Lakes Albert and Victoria, as identified in Figure 1.

## 3.3 Postabdomenal Claw

Figure 3 demonstrates the postabdomenal claws of the *Ctenodaphnia*. *D. lumholtzi* has a basal comb with dense elongated teeth, distinguished from uniformly short serrations that constitute the two distal combs of the claw. Both *D. monacha* and *D. magna* exhibit three distinct combs along the claw, with the middle teeth longest, and distal teeth shortest. The claws are not distinguishable by shape alone.

## 3.4 Fornices

The three *Ctenodaphnia (D. lumholtzi*, *D. monacha*, and *D. magna*) bear distinctive fornices; *Daphnia longispina* lacks fornices. *D. monacha* fornices have more acute apices than those of *D. magna*, though not nearly so acute or elongate as in the case of *D. lumholtzi* (Figure 4). The differences in shape between *D. monacha* and D. *magna* are subtle but consistent.

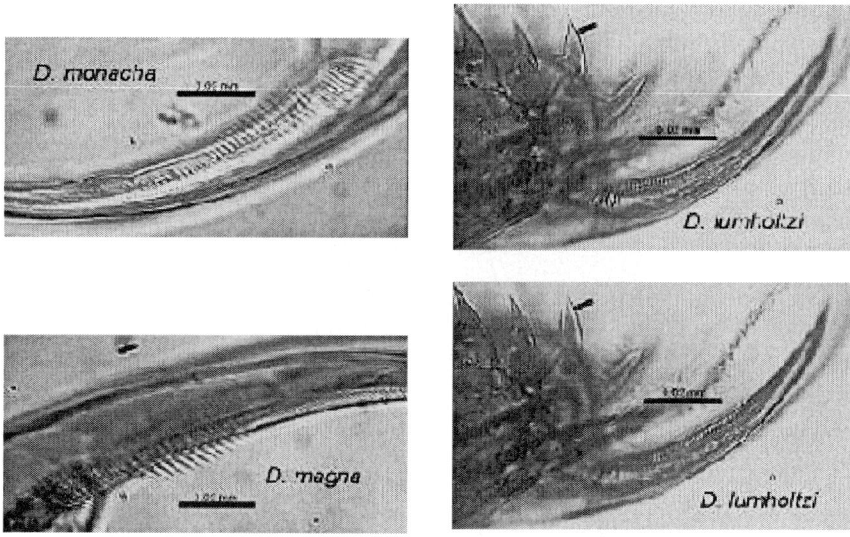


*Figure 3*. Postabdomenal claws from *Daphnia* specimens identified in Figures 1 and 2. *D. lumholtzi* (right) is illustrated at two different focal planes. All scale bars are 0.02 mm.

## 3.5 Eye and Rostrum

Differences also occur in the eye, rostrum, and general shape of the heads (Figure 5). *D. monacha* and *D. magna* differ by number and arrangement of lenses around the eye. *D. monacha* has more lenses than *D. magna* and they typically surround the entire eye. *D. lumholtzi* also has more lenses than *D. magna* but they do not surround the entire eye. *D. longispina* has the same arrangement as *D. magna* with a row of lenses along only the crown of the eye. Eye size also varies among specimens, but as Green (1971) noted, there is a positive correlation with body size. The rostrums of *D. monacha* and *D. magna* are similar, and the ventral margins of the heads are slightly arcuate compared with *D. lumholtz*i. The head of *D. monacha* is typically more broadly rounded than that of *D. magn*a. Ocelli are conspicuous for all taxa.

## 3.6 Size Distributions

In addition to distinguishing morphological characteristics, there are overt differences in the size distributions of the various *Daphnia* species. *D. magna* in Lake Victoria attain lengths exceeding 2 mm that are considerably larger than any specimens of *D. monacha* found in Lake Albert (Figure 6). *D. longispina* and *D.*

*monacha* produce neonates of about 0.5 mm length, but size of first reproduction is about 0.75 mm for *D. longispina*, and 1.1 mm for *D. monacha*. The smallest *D. magna* recorded with eggs was 1.0 mm in length.

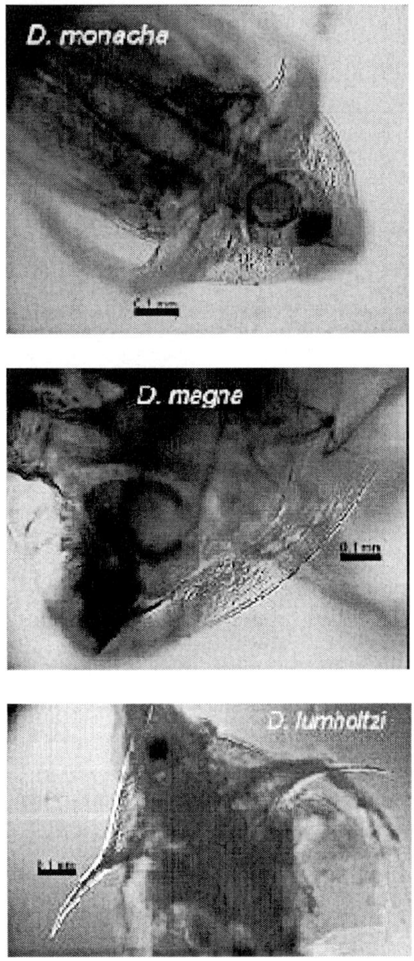


*Figure 4*. Fornices of *Daphnia* species from Lakes Albert and Victoria.

## 3.7 Seasonal Occurrence

*D. longispina* is a perennial member of the offshore zooplankton in Lake Victoria (Figure 7), but D. *magna* occurs episodically and inconsistently. The first known recorded occurrence of *D. magna* in the lake, April 1992, also seems to have been one of the most abundant. The peak abundances recorded for *D. magna* occurred in April 1992 during the rainy season but also in January 1994 during the dry season.

No simple pattern of seasonality is evident. This sporadic occurrence of *D. magna* may be a consequence of its high susceptibility to fish predation, and it may reflect opportunistic successes governed mainly by local planktivore abundance.

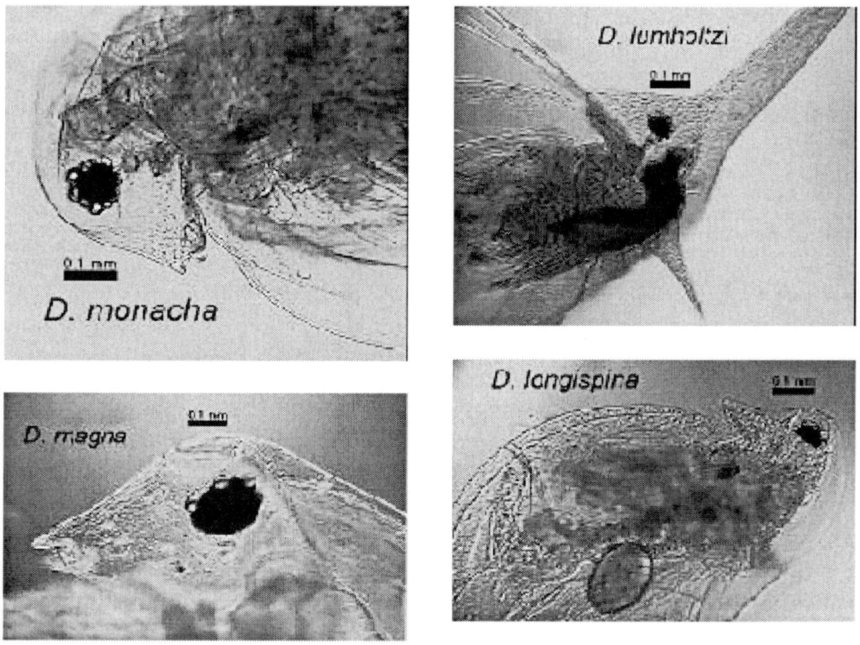


*Figure 5.* Eye and rostrum of *Daphnia* from Lakes Albert and Victoria. All scale bars are 0.1 mm.

## 4. DISCUSSION AND CONCLUSION

Based on review of existing plankton collections, there is no doubt that the large *Ctenodaphnia* first reported in Lake Victoria by Branstrator *et al.* (1996) was in fact *Daphnia magna*. There are sufficient similarities between *D. magna* and *D. monacha* that the taxa could easily be confused in the absence of reference specimens from the Rift Valley lakes. In the course of this study, we inspected *Daphnia* present in the equivalent of more than 100 $m^3$ of water from the offshore region of Lake Victoria.

Among the *Ctenodaphnia* present, not one could be reliably identified as *D. monacha*. Taxonomic nuances notwithstanding, the ecological significance of finding such a large-bodied cladoceran in the offshore waters of Lake Victoria must signal profound changes in the planktivory levels there. The explosive growth of introduced Nile perch during the 1990s is known to have decimated native stocks of haplochromines, including zooplanktivorous species. The ecological vacuum created

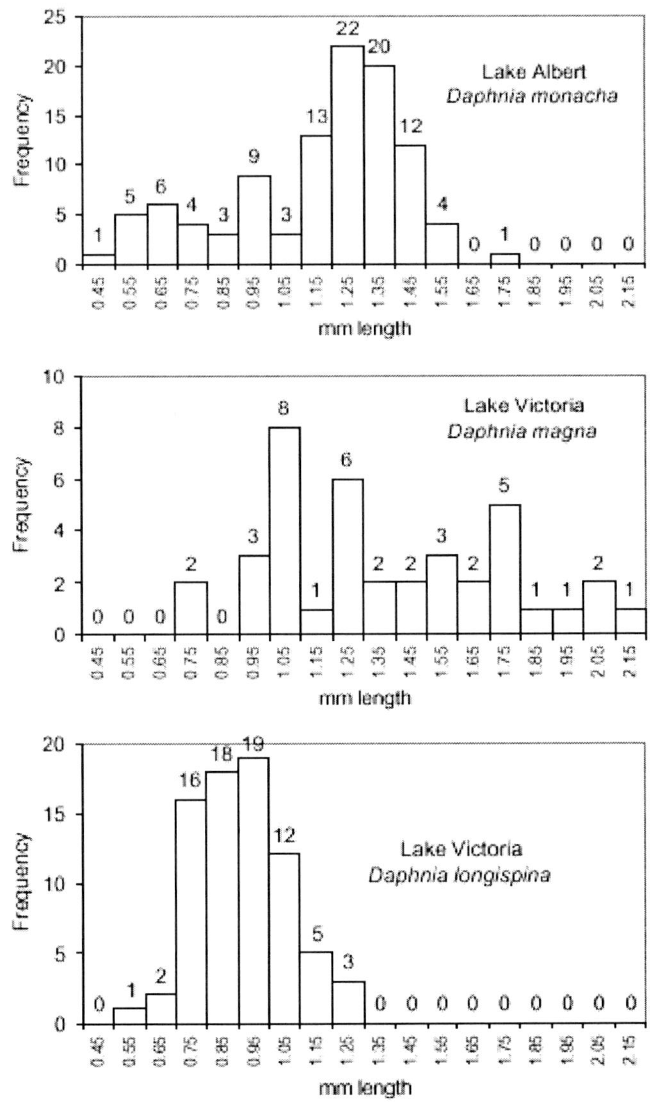


*Figure 6*. Size frequency for *D. monacha* from 29 March 1995 Lake Albert (upper); *D. magna* (middle), and *D. longispina* (lower) from 16 April 1992 Lake Victoria.

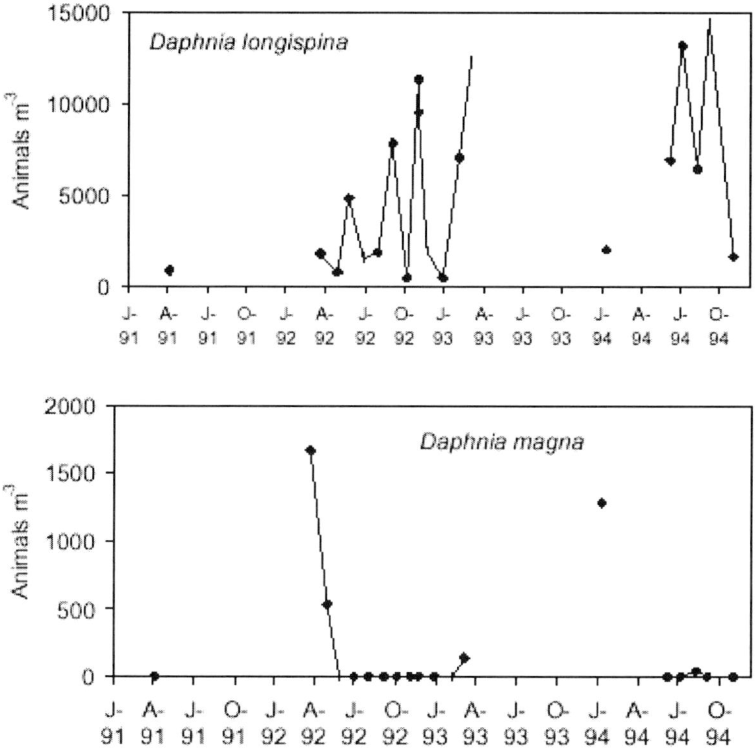


*Figure 7*. Occurrence of *D. longispina* (lower) and *D. magna* (upper) in Lake Victoria.

by collapse of a resident fish community seems to have created the opportunity for successful colonization of the lake by an atypical and unexpected herbivore.

One additional consequence of this study has been the documentation of striking and consistent morphological differences between *D. lumholtzi* and the taxon labeled *D. lumholtzi* var. *monacha* by Wagler in 1936. It seems untenable to us that these are polymorphisms of the species, and we therefore reverted to the older name *D. monacha* in this report. At this point, we can confirm the presence of *D. lumholtzi* in Lakes Victoria, Edward, and Albert. *D. monacha*, however, seems still to be absent from Lake Victoria.

## ACKNOWLEDGMENTS

We thank E. Ekdahl, E. Himrod, D. Lee, N. Malaibari, N. Paschka, and K. Vaughn for their help with preliminary searches of Lake Victoria samples for *Ctenodaphnia*.

# REFERENCES

Benzie, J. A. H. (1988) The systematics of Australian Daphnia (Cladocera: Daphniidae). Species descriptions and keys. Hydrobiologia **166**, 95-161.

Branstrator, D. K., Ndawula, L. M., and Lehman, J.T. (1996) Zooplankton dynamics in Lake Victoria, in T. C. Johnson and E. O. Odada (eds.), *The Limnology, Climatology and Paleoclimatology of the East African Lakes*. Gordon and Breach Publishers, pp. 337-355.

Brehm, V. (1959) Cladocera. Exploration hydrobiologique des Lacs Kivu, Edouard et Albert (1952-1954) **3**(3), 69-100.

Drenner, R. W., Strickler, J.R. and O'Brien., W.J. (1978) Capture probability: the role of zooplankter escape in the selective feeding of planktivorous fish. Journal of the Fisheries Research Board of Canada **35**, 1370-1373.

Dumont, H. J. (1986) The Tanganyika sardine in Lake Kivu: another ecodisaster for Africa? Environmental Conservation **13**, 143-148.

Gliwicz, Z. M. (1985) Predation or food limitation: an ultimate reason for extinction of planktonic cladoceran species. Archiv Hydrobiologie Beih. Ergebnisse Limnologie **21**, 419-430.

Gliwicz, Z. M. (1994) Relative significance of direct and indirect effects of predation by planktivorous fish on zooplankton. Hydrobiologia **272**, 201-210.

Green, J. (1967) The distribution and variation of *Daphnia lumholtzi* (Crustacea: Cladocera) in relation to fish predation in Lake Albert, East Africa. Journal of Zoology **151**, 181-197.

Green, J. (1971) Associations of Cladocera in the zooplankton of the lake sources of the White Nile. Journal of Zoology **165**, 373-414.

Kolar, C. S. and Wahl, D.H. (1998) Daphnid morphology deters fish predators. Oecologia **116**, 556-564.

Lehman, J. T. (1996) Pelagic food webs of the African Great Lakes, in T. C. Johnson and E. O. Odada (eds.), *The Limnology, Climatology and Paleoclimatology of the East African Lakes*. Gordon and Breach Publishers, pp. 281-301.

Lehman, J. T. (1998a) Role of climate in the modern condition of Lake Victoria. Theoretical and Applied Climatology **61**, 29-37.

Lehman, J. T. (ed.) (1998b) *Environmental Change and Response in East African Lakes*. Kluwer.

Lowe-Mcconnell R. (1997) EAFRO and after: A guide to key events affecting fish communities in Lake Victoria. South African Journal of Science **93**, 570-574.

Marshall, B. E. (1991) The impact of the introduced sardine *Limnothrissa miodon* on the ecology of Lake Kariba. Biological Conservation **55**, 151-165.

Mavuti, K. and Litterick, M.R.. (1991) Composition, distribution and ecological role of zooplankton community in Lake Victoria, Kenya waters. Verhandlungen Internationale Vereinigung Limnologie **24**, 1117-1122.

Mookerji, N., Heller, C., Meng, H.J., Buergi, H.R. and Mueller. R. (1998) Diel and seasonal patterns of food intake and prey selection by *Coregonus* sp. in re-oligotrophicated Lake Lucerne, Switzerland. Journal of Fish Biology **52**, 443-457.

Mwebaza-Ndawula, L. (1994) Changes in relative abundance of zooplankton in northern Lake Victoria, East Africa. Hydrobiologia **272**, 259-264.

Nassal, B., Burghard, W. and Maier, G. (1998) Predation by juvenile roach on the calanoid copepod *Eudiaptomus gracilis* and the cyclopoid copepod *Cyclops vicinus*: A laboratory investigation with mixed and single prey. Aquatic Ecology **32**, 335-340.

Rzoska, J. (1957) Notes on the crustacean plankton of Lake Victoria. Proc. Linnean Soc. London **168**, 116-125.

Rzoska, J. (1975) Zooplankton of the Nile system, in J. Rzoska (ed.) *The Nile, Biology of an Ancient River*. Junk, pp. 333-344.

Swaffar, S. M. and O'Brien, W.J. (1996) Spines of *Daphnia lumholtzi* create feeding difficulties for juvenile bluegill sunfish *(Lepomis macrochirus*). Journal of Plankton Research **18**, 1055-1061.

Tollrian, R. (1994) Fish-kairomone induced morphological changes in Daphnia lumholtzi (sars). Archiv für Hydrobiologia **130**, 69-75.

Viitasalo, M., T. Kiorboe, T., Flinkman, J., Pedersen, L.W. and Visser, A.W. (1998) Predation vulnerability of planktonic copepods: Consequences of predator foraging strategies and prey sensory abilities. Marine Ecology-Progress Series **175**, 129-142.

Wagler, E. (1936) Die Systematik und geographische Verbreitung des Genus *Daphnia* O. F. Müller mit besonderer Berücksichtigung der südafrikanischen Arten. Archiv für Hydrobiologie **30**, 505-556.

Witte, F., Goldschmidt, T., Wanink, J., Van Oijen, M., Goudswaard, K., Witte Maas, E. and Bouton, N. (1992) The destruction of an endemic species flock: Quantitative data on the decline of the haplochromine cichlids of Lake Victoria. Environmental Biology of Fishes **34**, 1-28.

Worthington, E. B. (1931) Vertical movements of freshwater zooplankton. Internationale Revue gesamten Hydrobiologie **25**, 394-436.

Worthington, E. B. and Ricardo, C.K. (1936) Scientific results of the Cambridge expedition to the East African lakes, 1930-1. No. 17. The vertical distribution and movements of the plankton in lakes Rudolf, Naivasha, Edward, and Bunyoni. J. Linnean Society (Zoology) **40**, 33-69.

Zaret, T. (1980) *Predation and Freshwater Communities*. Yale Univ. Press, 187 p.

# EFFECTS OF CLIMATE AND HUMAN ACTIVITIES ON THE ECOSYSTEM OF LAKE BARINGO, KENYA

P. A. ALOO
*Department of Zoology, Kenyatta University,*
*P.O. Box 43844, Nairobi, Kenya*

**ABSTRACT**

Since the beginning of the last century, the Lake Baringo ecosystem has undergone several ecological changes. Most of the changes are a result of human activities within the catchment basin and changes in climatic conditions. Currently, the most significant limnological feature of the lake is its extreme turbidity with an average secchi disc reading of 9.5 cm. The lake water is brownish and muddy because of considerable siltation resulting from high rates of soil erosion caused by overgrazing by livestock and deforestation of the surrounding hills. Due to increased siltation, the bed of the lake in the open waters is virtually devoid of invertebrate life. Primary production in the open waters is very low as the only phytoplankton present are positively buoyant species such as *Microcystis aeruginosa, Melosira granulata and Anabaena carcinalis*. The depth of the lake has a significant effect on the transparency of the water ($P<0.001$), where transparency decrease with depth. Under normal conditions, Lake Baringo has low alkalinity, but with less rain, the lake has become more saline with an average conductivity of 660 $\mu S\ cm^{-1}$ while the pH varies between 8.9 and 10.5. Five species of fish have been reported to occur in Lake Baringo: *Oreochromis niloticus* (Trewavas, 1983), *Protopterus aethiopicus* (Heckel, 1851), *Clarias gariepinnus* (Butchell, 1852), *Barbus intermedius* (Ruppell, 1836) and *Labeo cylindricus* (Peters, 1852). Presently, the species composition is dominated by *Oreochromis* (80.04%), *Clarias* (9.8%), and *Protopterus* (7.95%). *Barbus* rarely appear in the fisherman's catches while *Labeo* has almost disappeared from the lake since the damming of the inflowing rivers which interfered with its



*E.O. Odada and D.O. Olago (eds.),*
*The East African Great Lakes: Limnology, Palaeolimnology and Biodiversity,* 335–347.

breeding habits. Moreover statistical analysis has shown a significant relationship between fish yield and lake level changes ($P<0.001$).

# 1. INTRODUCTION

## 1.1 Study Area

Lake Baringo (Figure 1) is situated between latitudes 0°30'N and 0° 34' N and longitudes 36° 00' E and 36 10' E. It has an area of about 130 km$^2$ and lies at an altitude of 965 m. above sea level. The lake is shallow with a mean depth of 3.5 m with the deepest part in the northern zone measuring 3.8m. (Ssentongo, 1996). The depth of the lake has decreased from a maximum of 9m and a mean of 5.6m in 1972 to the current maximum of 4m and a mean of 2.5m due to the increasing magnitude of siltation, evaporation, damming and diversion of inflowing rivers.

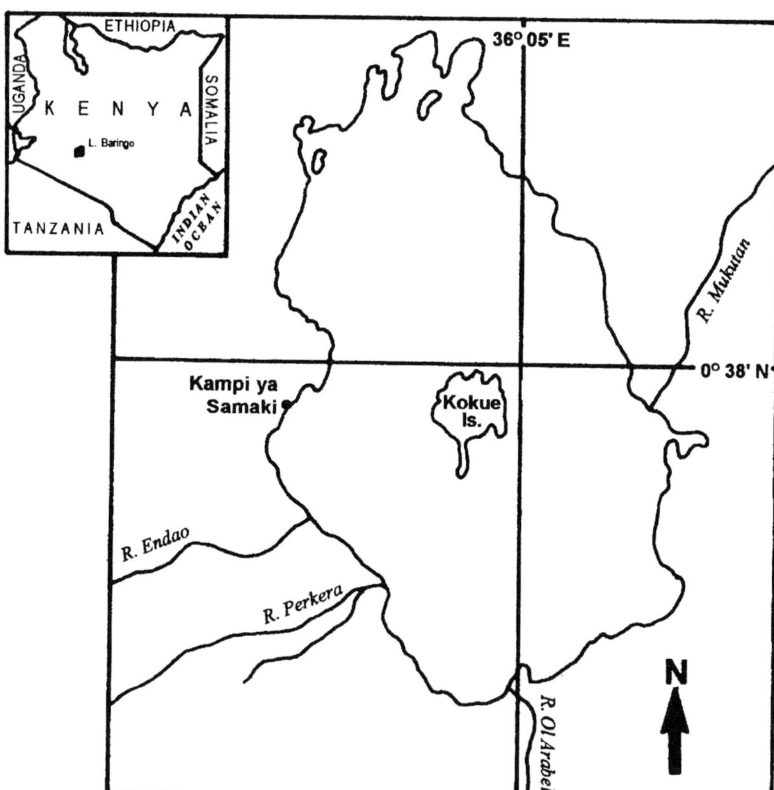


*Figure 1.* Lake Baringo showing its location.

## 1.2 Hydrographic Conditions

Lake Baringo is fed from the South by the Rivers Endao, Perkera, Molo, Ol Arabel and Mukutan (Figure 1). Other small rivers include Sandalo, Bargera, Tigeri and Marigato. All these rivers with the exception of Molo, Endao and Perkera are seasonal and have dried up during the last two years. The lake water appears brown in colour and muddy due to considerable siltation. Copley (1948) described the waters of Lake Baringo as being greenish in colour due to the vast quantities of blue-green algae present in the water. Under normal climatic conditions, Lake Baringo has a low alkalinity, but with less rain and more evaporation, the lake is steadily becoming saline (Talling and Talling, 1965).Therefore, the increase in the salinity of the lake since 1965 can be explained by evaporative concentration (Table 1).

*Table 1.* Variation in the water chemistry of Lake Baringo. Source: Talling and Talling 1965; Kallviqst 1978; Kenya Marine and Fisheries Research Institute, Annual Data 1998.

| Year | PH | Maximum Depth (m) | Conductivity ($\mu s\ cm^{-1}$) | Dissolved Oxygen (mg/l) | Transparency (cm) |
|---|---|---|---|---|---|
| 1965 | 7.9 | 3.8 | 416 | 7.02 | 54.0 |
| 1978 | 8.9 | 8.6 | 450 | 7.00 | 48.0 |
| 1998 | 8.8 | 6.0 | 898 | 6.80 | 10.8 |
| 2000 | 8.66 | 4.0 | 1195 | 6.40 | 11.5 |

*Table 2.* Variation in secchi disc transparency between 1990 and 2000. Source: Kenya Marine and Fisheries Research Institute, Annual Statistics, 2000.

| Year | Mean Transparency (cm) |
|---|---|
| 1990 | 11.0 |
| 1991 | 8.0 |
| 1992 | 6.4 |
| 1993 | 4.6 |
| 1994 | 4.9 |
| 1995 | 5.8 |
| 1996 | 5.4 |
| 1997 | 5.6 |
| 1998 | 10.8 |
| 1999 | 9.5 |
| 2000 | 11.5 |

The outstanding limnological feature of the lake is its extreme turbidity (mean transparency = 9.5cm), caused by the high rate of soil erosion resulting from overgrazing by livestock and deforestation of the catchment area. Okorie (1973) reported a secchi disc reading of 3 cm during his 1972-1973 survey of the lake while

Tompkins (1976) noted a slight improvement in the secchi disc reading of about 4.6 cm. However, between 1999 and the year 2000, the secchi disc reading has increased from 9.5cm to 11.5cm (Table 2). Primary production in the open waters is very low due to the turbid nature of the lake. Thus phytoplankton population is limited to only the positively buoyant species including *Microcystis aeruginosa, Melosira granulata* and *Anabaena carcinalis.* Others which were reported to occur in low numbers include the green algae *Pediastrum simplex and P. duplex* (Kallqvist, 1976; Ssentongo, 1996). The lake bed is virtually devoid of any invertebrate life except for molluscs occurring nearshore within aquatic macrophytes in sheltered offshore areas, this is due to the high rate of siltation.

## 2. FACTORS CONTRIBUTING TO FLUCTUATIONS IN THE LAKE BARINGO WATER LEVELS

Lake Baringo water levels have gone down over the years leading to changes in the physico-chemical parameters of the lake (Table 1). Two main factors have contributed to the fluctuations in lake level. These are: Siltation and damming.

### 2.1 Siltation

Siltation of Lake Baringo occur at a rate of about 315-350 tonnes per year (Fisheries Department, Annual Report, 2000). Several factors (both natural and human) have led to this high rate of siltation. The lake lies in a depression with loose soils which are easily washed into the lake during rains. Furthermore during dry season, the areas surrounding this lake experience wind erosion characterised by the dust clouds which eventually end up in the lake. Communities around this lake are pastoralists who keep large numbers of livestock, especially goats and these animals graze around the lake causing soil degradation which results in soil erosion. These people also fell trees to clear land for cultivation or to harvest wood for fuel. The indiscrimate cutting of trees around the lake region has left the catchment area bare and exposed to soil erosion. Whenever it rains around the lake there are flush-floods from inflowing rivers due to poor farming practices on the Turgen hills surrounding the lake. These rivers carry enormous amounts of soils into the lake (Olilo, 1993).

### 2.2 Damming

River Endao, one of the main rivers draining into Lake Baringo has been dammed to create the Chemeron dam whose water is being used for irrigating the surrounding farmlands where maize, onions, tomatoes and bananas are being grown. The Perkera and Molo rivers have not been dammed but they have been diverted for agricultural purposes around Marigat and Bogoria areas (Ssentongo, 1996). River Ol Arabel and other small rivers are seasonal and have dried again since the last heavy El Nino rains. (Gichuru, pers. comm.). The fact that damming and river diversion are the main factors that pose a serious threat to the survival of Lake Baringo can further be explained using the relationship between lake level and rainfall. Statistical analysis showed that the amount of rainfall had no significant effect on the lake level ($t_8$=0.014; P>0.05) (Table 3). However, a negligible positive relationship between rainfall amount and lake level was noticed (b=0.0000270). The damming and diversion of the rivers has not only caused changes in the water chemistry of the lake but have also interfered with the entire lake ecosystem. For example, two fish species *Barbus intermedius* and *Labeo cylindricus* which migrate to spawn upstream are today near extinction in the lake because their breeding habits have been interfered with.

*Table 3*. The relationship between rainfall and lake levels between 1989 and 1998. Source: Kenya Marine and Fisheries Research Institute Annual Report, 2000.

| Year | Lake Level (m, asl) | Rainfall (mm) |
|---|---|---|
| 1989 | 968.0 | 738 |
| 1990 | 969.3 | 530 |
| 1991 | 967.6 | 478 |
| 1992 | 967.0 | 450 |
| 1993 | 966.8 | 522 |
| 1994 | 966.4 | 618 |
| 1995 | 966.7 | 597 |
| 1996 | 966.6 | 576 |
| 1997 | 967.3 | 978 |
| 1998 | 967.5 | 538 |

# 3. FISH SPECIES COMPOSITION

The first report on the fish species composition of Lake Baringo was that of the Cambridge Expedition to the East African Lakes between 1930 and 1931 (Worthington and Riccardo, 1936). The above expedition recorded the following species: *Tilapia nilotica* (= *Oreochromis niloticus*. (Linnaeus), *Clarias mossambicus* (= *Clarias gariepinus*) (Peters), *Barbus gregorii* (= *Barbus intermedius)* (Boulenger) and *Labeo cylindricus* (Peters). During the expedition, *O. niloticus* was noted to be the most important commercial fish on which the pilot fish processing plant established in 1960's depended. This factory closed down in the 1980's due to the decline in both the numbers and the size of tilapia. Two other species *Aplocheilichthyes* sp. and *Barbus lineomaculatus,* which occurred in the lake were last recorded in 1969 (Okorie, 1975). Other reports of the fish species of Lake Baringo include those of Mann (1971) on taxonomic notes on the fishes and Ssentongo (1974) on the fish and the fisheries of the lake. Presently only five fish species are reported to occur in Lake Baringo: *Oreochromis niloticus, Barbus intermedius, Clarias gariepinus, Labeo cylindricus* and *Protopterus aethiopicus* which was introduced into the lake in 1974 from Lake Victoria (Lake Baringo Fisheries Department Annual Report, 1999). Also reported to be present in the lake but very rare is the guppy, *Poecilia reticulata* (Peters, 1859) (De Vos, pers. comm.).

Reports have indicated that *O. niloticus* contributing 80.04%, *C. gariepinus* (9.8%) and *P. aethiopicus* (7.95%) are the only commercially important species in the lake (Ssentongo, 1996). However, recent statistics have indicated an increase in *Protopterus* and *Clarias* numbers because the two species are adapted to the turbid, muddy waters of the lake. Although *O. niloticus* is still the most abundant fish in the lake, statistics have shown that its numbers are going down, due probably to the high rate of siltation which restricts the photic zone for primary production thus limiting plankton proliferation. *Barbus* and *Labeo* are near extinction not only due to changes in the lake water quality and the damming and diversion of the inflowing rivers but also probably due to the increased predation pressure from *Protopterus* (Aloo *et al.*, in press). Reports have indicated that introduced species usually have an effect on the distribution and densities of their prey species, but there is no unequivocal evidence that introductions can shift the whole nature of the ecosystem, although such an effect has been advocated for Lake Victoria (Pitcher, 1995).

## 4. THE LAKE BARINGO FOOD WEB

*Oreochromis niloticus* which occur abundantly in sheltered littoral zone feed mainly on phytoplankton and zooplankton while *C. gariepinnus* adults which feed mostly on the juveniles of *Oreochromis* sp. also occur abundantly in the littoral zones.

*Table 4.* Fish Distribution and Abundance in Lake Baringo. Source: Worthington and Riccardo, (1936); Aloo *et al.* in press; Fisheries Department Annual Report, 2000.

| Fish Species | Habitat | Abundance |
|---|---|---|
| *Oreochromis niloticus* | - Sheltered areas in the littoral zone<br>- Visits inshore reed areas principally to breed | Most abundant |
| *Protopterus aethiopicus* | - Occur both in deep waters and shallow waters | Moderately abundant |
| *Clarias gariepinus* | - Ubiquitous distribution throughout all over the lake but abundant in littoral zone<br>- During flood times large numbers occur in shallow waters | Not abundant |
| *Barbus intermedius* | - Distributed all over the lake especially in shallow areas<br>- Young fish occur abundantly along river mouths. | Rare |
| *Labeo cylindricus* | - Occurred abundantly in shallow macrophyte beds and river mouths. | Very rare |

Although *P. aethiopicus* occurs offshore, it is also abundant in macrophyte beds where their staple food, molluscs, are found (Aloo *et al.,* in press). The fish furthermore preys on the juveniles of other fish species. *Labeo* and *Barbus* are mainly bottom feeders on algae and insect larvae (Worthington and Riccardo,1936). The two species occurred abundantly all over the lake (Table 4) but more so at the inlet of river mouths and rocky areas. Stomach content analysis revealed the presence of zooplankton and phytoplankton in both species, the planktonic foods being prominent in the food web of the lake (Figure 2).

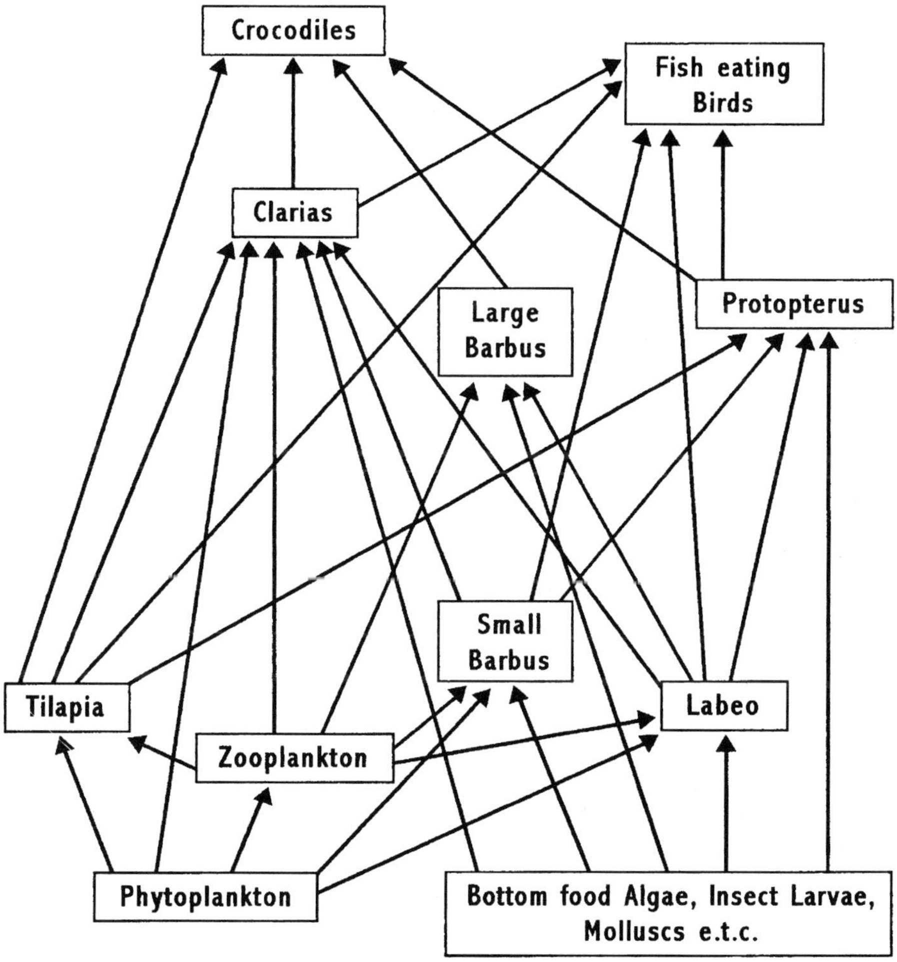


*Figure 2.* Food web in Lake Baringo (modified from Worthington and Riccardo, 1936).

## 5. LAKE FISHERY

Presently, only three species are commercially exploited in the lake: *P. aethiopicus, O. niloticus* and *C. gariepinus.* All three species are collected with the aid of gillnets but *P. aethiopicus* is also caught using hook and line. Fishing regulations allow fishermen to use only gillnets of 4" and 5" but most fishermen use 2½" mesh size gillnets. This has resulted in small sized tilapias being landed thus resulting in low production. Fishermen in this lake use simple small canoes made from stems of a special buoyant tree species *Aeshynomena elephroxylon.* The canoes are only used by one fisherman and propelled by hand using small paddles. There are about 20 small canoes in the lake, and 4-5 wooden boats and 4 fibre glass boats which are motorised and mainly used for transporting tourists around the lake and across to Island camp which is one of the major tourist attractions (Fisheries Department, Annual Report, 2000).

## 6. LAKE LEVEL FLUCTUATIONS AND FISH PRODUCTION

Lake level fluctuations affect fish production through effects on food, breeding grounds and predator-prey relationships. This may be partly due to the effects of reduced transparency at low lake level on food quality (for Planktivorous fish) or feeding efficiency (for zooplanktivorous fish). The fluctuations also affect breeding grounds and predator-prey relationship. However Food has been reported not to be a limiting factor for the fishes in Lake Baringo (Worthington and Ricardo, 1936; Okorie, 1975)( Figure 2).

*Oreochromis niloticus* numbers have gone down in recent times probably due to the decreasing water levels. Lowe-McConnell (1987) listed factors such as habitat drying and flooding as controllers of tilapia numbers in tropical fish communities. During periods of high water levels, fish populations increase in the shallow nearshore areas where they are easily accessed and caught by the artisanal fishermen and this is reflected in the high production during high water levels ( Muchiri, 1990) (Figure 3).

Due to the high rate of evaporation, low rain-fall and the damming and diversion of the inflowing rivers, it is predictable that Lake Baringo water levels will continue to decline ( Tompkins, 1976). This will not only affect the lake ecosystem but also the communities around the lake specifically dependent on fish production from the lake.

Between 1992 and 1994 there was severe drought in the region around Lake Baringo which caused drastic reduction in the water level. Subsequently the lake was closed down and no gillnet fishing was allowed except hook and line fishing permitted for a few fishermen. Compared to the 1980s, catches in 1990's have further gone down due to seasonality of the rivers, the damming and diversion of the rivers and increased siltation this is with the exception of the El Nino effect of 1998 (Figure 4).

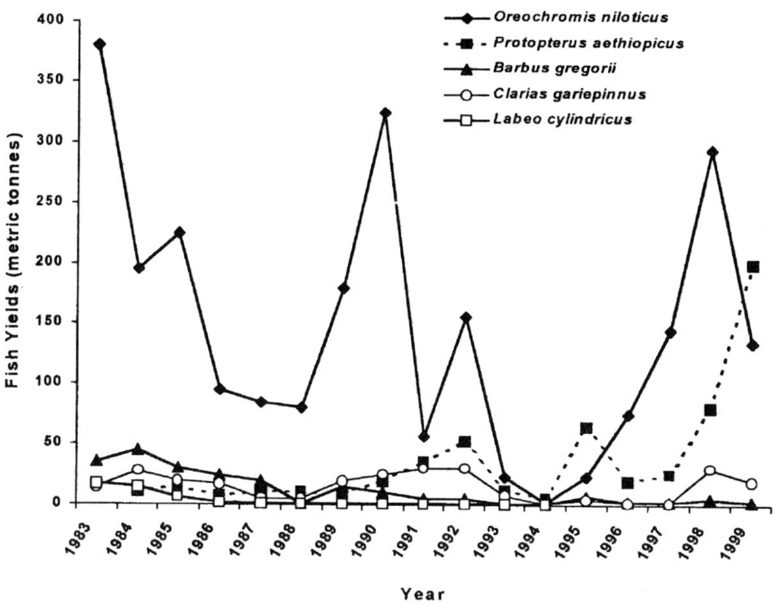


*Figure 3.* Lake Baring fish production and lake level changes (1989-1998).

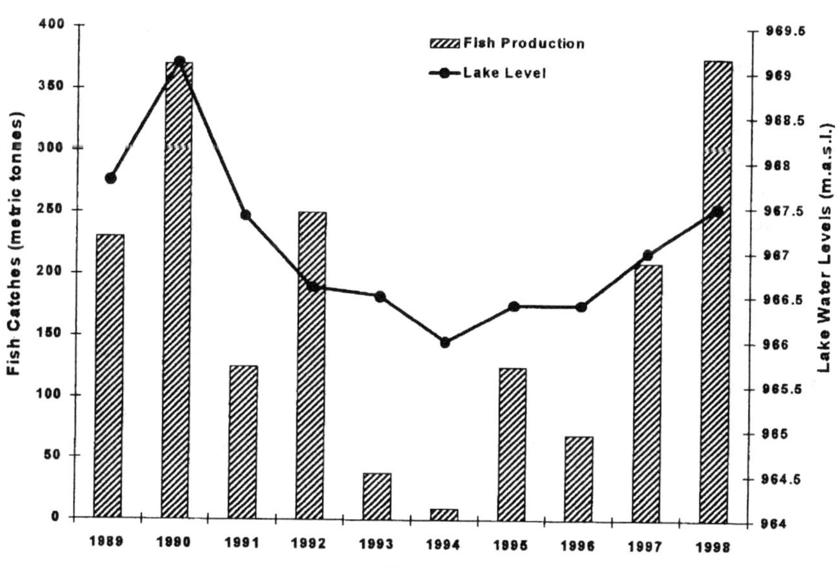


*Figure 4.* Lake Baring fish yields between 1983 and 1999 (government of Kenya, Fisheries Department annual statistics, 2000).

## 7. OTHER FAUNA ASSOCIATED WITH THE LAKE

Besides the fish and other small aquatic animals, there are a number of animals associated with this lake. Reports have indicated that there are between 40 –60 hippopotami in the lake. During the dry season, the hippopotami (*Hippopotamus amphibius*) are usually seen in the evening grazing on the green grass around Lake Baringo Club. Crocodiles (*Crocodilus niloticus*) numbering about 20, also occur in the lake. These reptiles feed mainly on fish including those caught in fishermen's nets leading to the destruction of the nets. During drought the crocodiles attack goats due to lack of fish. In the mid-morning hours, basking crocodiles are a common feature around the shores of Lake Baringo.

Other animals associated with the lake include the monitor lizards (the Nile Monitor and the Savannah Monitor). The Nile Monitor is most common around the lake and is always associated with the water. It is carnivorous, feeding mainly on fish and other small aquatic organisms. Also common around the lake are vervet monkeys, baboons, dikdiks, a few Grant's gazelles, zebras, hyenas, mongoose, ground squirrels, porcupines hares and the bat eared fox (Lewis, 1998).

Lake Baringo is an important tourist destination because of its diverse and numerous avifauna. The lake is home to over 500 species of birds which are residents around the lake. However, during winter in Europe an additional 300-500 migrant species occur around the lake (National Museums, Ornithology Department Bird Count report, 1998). These birds depend on the lake ecosystem in one way or another. For example, piscivorous birds such as the fish eagles, kingfishers and pelicans depend on this lake as a source of food. Weaver birds and other species construct their nests on the acacia trees which are closely associated with the lake.

## 8. HUMAN POPULATION AROUND THE LAKE

Between 8,000 and 10,000 people depend on the lake water for domestic use, fishing, livestock and agriculture. Many people also depend on this lake as a source of employment, for example beach boys who depend on tourists that visit the lake as their only source of income.

Kampi ya Samaki which is the main fishing village on the shores of Lake Baringo has a population of about 1,500 people. (Fisheries Department Estimates). Most of these people depend on the lake for their daily livelihood as a source of food and water for domestic use. The lake is also an important source of protein to the communities around.

## 9. DISCUSSION

Fish species inhabiting Lake Baringo were reported to have possibly come from the East African rivers. This explanation is supported by the fact that all the species present in the lake occur in one or other of the East African rivers (Worthington and

Riccardo, 1936). Only five species presently occur in the lake, and two are close to becoming extinct.

The silting of Lake Baringo, coupled with the decline in the lakes water level fluctuations over the last 35 years, make the disappearance of Lake Baringo a possibility. Also the high silt concentration prevents photosynthesis from taking place below the first 40 cm of the water column, this has led to the reduction of the photic zone resulting in low level of primary production and reduced fish numbers (Tompkins, 1976). Although lake level will continue to rise and fall in response to long-term rainfall variability (such as El Nino events), reduced water inflow due to damming and river diversion will increase the frequency of low stand periods during which fisheries are affected by the ecological consequences of bottom instability and high turbidity. If present trends of water extraction continue, it is conceivable that long-term survival of the Lake Baringo freshwater ecosystem (an internationally important breeding and over wintering ground for birds) will be threatened by periodic desiccation.

The precarious hydrologic balance of many African lakes make them vulnerable to disturbances caused by water diversion or impoundment projects. Elimination of swamps through water diversion or damming is likely to cause declines in diversity of indigenous fish through habitat loss, destruction of refugia and faunal mixing in many lakes (Chapman *et al*,. 1992) .

The decline in commercial catch during the dry seasons is probably caused by changes in the behavioural patterns of the fish species especially that of *Oreochromis,* thus making the fish less susceptible to capture (Van Someren, 1949). In the present review, the relationship between the lake level fluctuations and fish production strongly suggests that it is the environment more than the fishery which act as the dominant agent of change in the Lake Baringo ecosystem, confirming the findings of Kolding (1995). The existing relationships between lake level, water transparency and fish yields suggests that any negative effect of soil erosion on transparency will also negatively affect aquatic food web dynamics and the fisheries.

From the foregoing account, it is evident that Lake Baringo is an important ecosystem both ecologically and anthropogenically. Therefore, there is an urgent need to save it from the increased siltation and declining water levels which already have profound effects on the biota and the human populace around this lake.

## 10. CONCLUSIONS AND RECOMMENDATIONS

From the information given in this paper, and in agreement with Ssentongo (1996) the management of the lake's living resources, water quality and environment, as well as, improvement of the economic status of the communities depending on the lake would require an integrated multidisciplinary approach. The following activities can be undertaken to solve some of the problems facing Lake Baringo:

1. Creating public awareness among the lakeshore communities on the importance of conserving the lake.

2. Abatement of soil erosion by afforestation and reafforestation.
3. Conservation of the few wetlands still remaining around the lake e.g. Kichertel swamp.
4. Introduction of drought resistant and fast growing cash crops and food crops to reduce over dependency on the lake all the year round.
5. Control of any pollution and other sources of environmental degradation.
6. Appropriate agricultural practices to reduce soil erosion especially on the Turgen Hills.
7. Control of the damming and diversion of the inflowing rivers.
8. Lake patrols should be on a daily basis, if possible, to ensure compliance with fishing regulations (such as using the recommended mesh size nets).
9. The Fisheries Department should take tougher measures on those found using undersized nets.
10. The participation of fishermen and fish traders in decisions regarding the management of the lake should be encouraged.
11. Fisheries Department should improve the reliability of their catch statistics by employing qualified statisticians to collect data at the landing beaches so as to gain better control of the fisheries.
12. There is need for the formation of a national Task Force for the management of Lake Baringo to formulate guidelines for appropriate management, strategies and actions.

## ACKNOWLEDGEMENTS

The author wishes to acknowledge Kenya Marine and Fisheries Research Institute (KMFRI) Baringo field station and the Fisheries Department for providing data on fish catches and lake level changes. I am particularly grateful to the KMFRI centre Director, Mr. N. Gichuru for availing useful information, Mr. C. Olilo, Mr. P. Alela and Mr.G. Aondo for assisting in sampling. Mr S. Wambua of the Fisheries Department contributed a lot of information on the lake ecology, I wish to sincerely thank him. My special thanks go to my colleagues in the Department of Zoology at Kenyatta University especially the Chairman Dr. N.O. Oguge for their encouragement. I would also wish to thank Prof. R.O. Okelo for assisting with statistical analysis. Mr. J.N. Mwangi assisted in numerous ways during data collection. This paper has been developed from an on going research project on Lake Baringo ecosystem.

## REFERENCES

Aloo, P.A., Olilo C.O., Alela P., Mwangi J.N, Aondo, G. and Kilonzi J. (in press) The foods and feeding habits of Lake Baringo fish species.

Chapman, L., Chapman C., Kaufman, L., and Liem K. (1992) The role of papyrus swamps as barriers of fish dispersal. A case study and implications for fish diversity in lake Victoria basin. In Biodiversity, Fisheries and the Future of Lake Victoria Conference Abstracts p. 6.

Copley, H. (1948) *The Lakes and Rivers of Kenya.* Highway Press, Longman's Green and Company, Nairobi.

Fisheries Department Annual Report (2000). Lake Baringo Fish Statistics, Unpublished Report.

Kallviqst, (1976) A Study of Lake Baringo Water Chemistry. A report submitted to the Fisheries Department, 53 p.

Kenya Marine and Fisheries Research Institute (2000). Annual Report.

Kolding, J. (1995) Changes in species composition and abundance of fish populations in Lake Turkana, Kenya, in T.J. Pitcher and H.B. Hart (eds.) *The Impact of Species Changes in African Lakes,* pp. 336-363.

Lewis, M.N. (1998) *A Guide to Lake Baringo and Lake Bogoria.* Horizon Books, 161p.

Lowe McConell, R.H. (1987) *Ecological Studies in Tropical Fish Communities.* Cambridge Tropical Biology Series, 382 p.

Mann, M.J. (1971) Some taxonomical notes on the fish fauna of the Baringo area. *African Journal of Tropical Hydrobiology and Fisheries* **1**, 25-34.

National Museums of Kenya (1998) Annual Bird Count Report.

Okorie, O.O (1973) On the bionomics and population structure of non cichlid fishes in Lake Baringo. Report of East African Freshwater Research Organisation (EAFRO) Jinja, Uganda.

Okorie, O.O. (1975) On the bionomics and population structure of *Tilapia nilotica* (Linnaeus 1775) in Lake Baringo, Kenya. *African Journal of Tropical Hydrobiology and Fisheries* **4** (2), 192-218.

Olilo, C.O. (1993) Changes in Fish Populations and Effects of Fishing Gear in Lake Baringo Fishery Management. Discussion Paper No. I.

Pitcher, T.J. (1995) Species changes and fisheries in African lakes: Outline of the issue, in T.J. Pitcher and H.B.Hart (eds.) *The Impact of Species Changes in African Lakes.* Chapman and Hall, pp 1-16.

Ssentongo, G.W. (1974) On the fish and fisheries of Lake Baringo. *African Journal of Tropical Hydrobiology and Fisheries* **3**(1), 95-105.

Ssentongo, G.W. (1996) Report on the Present Fisheries Situation of Lake Baringo, Kenya, FAO Fisheries Department Publication, 10 p.

Talling, J.F. and Talling, T.B. (1965) The chemical composition of African lakes. *Internationale . Revue der Gesampten Hydrobiologie* **50**, 421-463.

Van Someren, V.D. (1949) Report on Lake Baringo Fisheries October 1994. Kenya Game Department files.

Worthington, E.B. and Riccardo, C. (1936). Scientific Results of the Cambridge Expedition to the East African Lakes (1030-1931) No.15. The fish of Lake Rudolf and Lake Baringo. *Zoological Journal of Linnean Society* **39**, 353-389.

# LIMNOLOGICAL PROFILES AND THEIR VARIABILITY IN LAKE TANGANYIKA

P.-D PLISNIER
Royal Museum for Central Africa, Leuvensesteenweg, 13
B-3080 Tervuren, Belgium

**ABSTRACT**

Results of frequent limnological profiles in the upper 0-300 m layer are compared for two years of sampling at three stations of Lake Tanganyika (Bujumbura, Kigoma and Mpulungu) in 1993-94 and 1994-95. Profiles of water temperature, pH, conductivity, nitrate, nitrite, amonnia and turbidity are compared for average values as well as for variability of observations. Differences in limnological profiles are discussed in relation with change in weather conditions between the two years and hypothesis on variable hydrodynamics states of the lake are presented. There exist few sets of data of frequent samplings at different stations over several years. Understanding present relationships between limnological environment and weather is important to interpret recent and past environmental changes in the Great Lakes in relation with climate or biotic changes such as algae blooms, fish kills or fisheries changes.

## 1. INTRODUCTION

East Africans lakes show signs of environmental changes (Lehman, 1998; Hecky *et al.*, 1994). One of the possible reasons could be related to their high sensitivity to climate variability (Johnson and Odada, 1996). There are indications of recent lake



*E.O. Odada and D.O. Olago (eds.),*
*The East African Great Lakes: Limnology, Palaeolimnology and Biodiversity,* 349–366.

warming at Lake Victoria (Hecky *et al.*, 1994) and at Lake Tanganyika (Plisnier, in press). The density stratification and stability of tropical lakes fluctuate widely at high water temperatures (Lewis, 1996). For this reason, lake hydrodynamic, which heavily depend on weather conditions, govern nutrient distribution in the water column, thereby influencing phytoplankton and food web productivity. It is interesting in this respect to investigate the interannual changes in limnological variables of the lakes. Although long time series of limnological variables measured are generally not available, interannual differences based on a great number of measurements may provide interesting background information to understand better the natural variability of these fragile ecosystems.

During a recent FAO/FINNIDA project "Research for the Management of the Fisheries in Lake Tanganyika (LTR)" frequent limnological samplings were carried out at three stations of Lake Tanganyika. Some results of the 1993-94 limnological annual cycle were presented earlier (Plisnier *et al.*, 1999). A preliminary comparison with data obtained during the second year of sampling of that project (1994-95) and possible hypotheses for the observed variations are presented here.

## 2. METHODOLOGY

Several sites were chosen for sampling at three stations in Lake Tanganyika (Bujumbura, Kigoma and Mpulungu) in the two periods for which data are presented: August 1993 to July 1994 (=year 1: 1993-94) and August 1994 to July 1995 (=year 2:1994-95) (Figure 1 and Table 1).

*Table 1*. Coordinates of sampling sites as recorded by GPS.

| Location | Sampling Sites | Latitude | Longitude |
|---|---|---|---|
| Bujumbura | Ab | 3°28.00' S | 29°17.00'E |
| | Bb | 3°45.00' S | 29°15.00 E |
| | Vb | 3°28.68' S | 29°16.18 E |
| Kigoma | Ak | 4°51.26' S | 29°35.54'E |
| | Bk | 4°50.69' S | 29°34.65'E |
| | Vk | 4°51.50' S | 29°34.50'E |
| Mpulungu | Am | 8°43.98' S | 31°02.43'E |
| | Bm | 8°34.45' S | 30°50.10'E |
| | Vm | 8°41.22' S | 30°59.80'E |

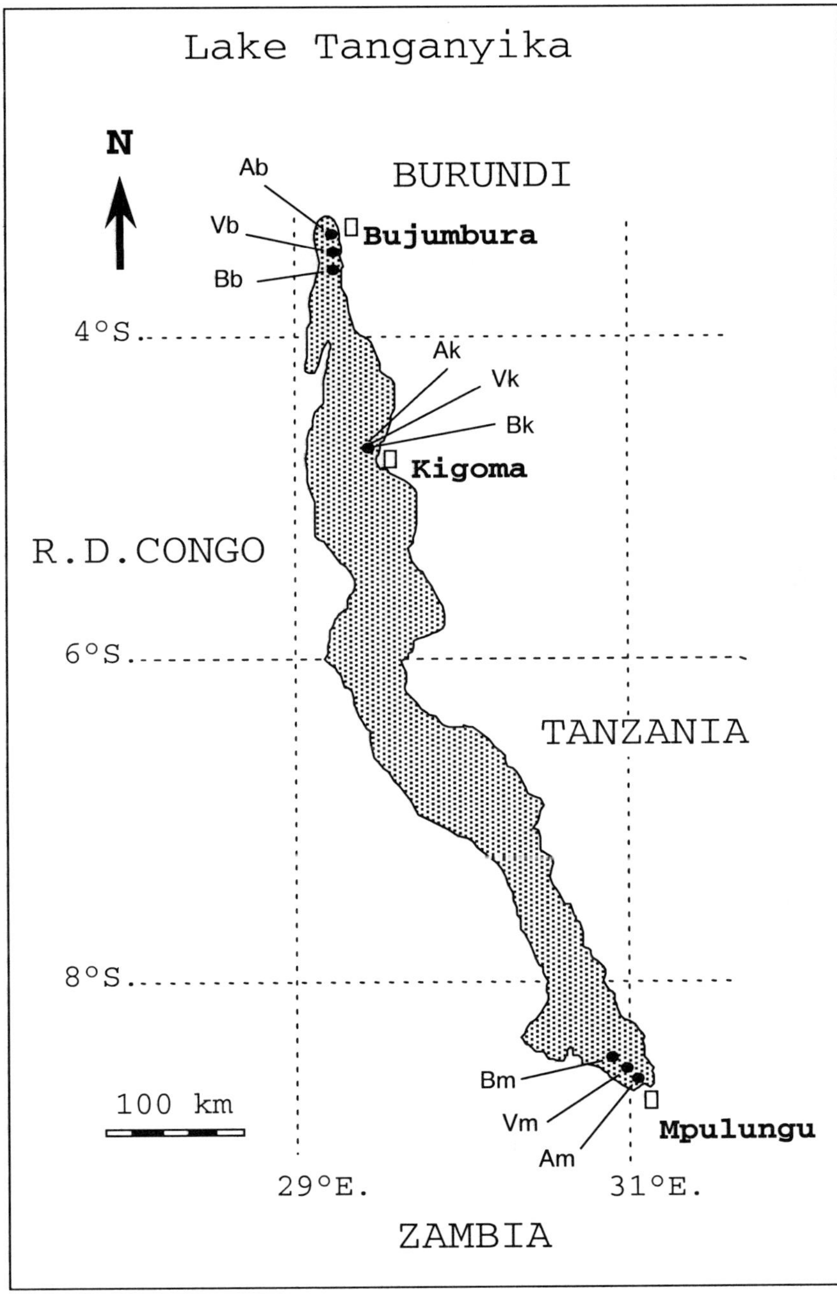


*Figure 1.* Map of Lake Tanganyika showing sites Ab, Ak, Am; Vb, Vk, Vm and Bb, Bk, Bm at the stations of Bujumbura, Kigoma and Mpulungu.

Vertical sampling was carried out every week or every two weeks down to 100 m (every 20 m) at sites Ab (Bujumbura), Ak (Kigoma) and Am (Mpulungu) during the two years. One measurement was done at each depth. However, measurements were repeated once or twice to check values out of the usual range.

At sites Bb, Bk and Bm, sampling was done 4 times during 24h cycles (6h, 12h, 18h and 24h) down to 300 m every 6 weeks in 1993-94 and every 8 weeks in 1994-95. Results of the measurements every 6 hours at each date were averaged. The 4 samplings were thus considered as replicates for each depth for the present analysis.

An additional site was sampled in 1994-95 once per month down to 140 m at Bujumbura and Kigoma and down to 160 m at Mpulungu (sites Vb, Vk and Vm). These sites are situated not far from sites A and data have been pooled together with those of sites A in the present paper. These sites, and the sampling frequency, had been chosen for practical reasons and distances from field stations. Sampling at sites Ab, Ak, Am and Vb, Vk, Vm was always in the morning around 9-11h (10-12 h in Tanzania).

Water temperature (accuracy $\pm$0.1°C) was measured with a thermometer placed inside the water sampler and, from April 1994 onwards, by *in situ* measurements down to 80 m using a digital thermometer made by Yellow Springs Instrument Co. with the same accuracy. pH readings were taken with a portable Hach pH meter, model 43800-00 (Precision $\pm$0.01). For conductivity (in $\mu S\ cm^{-1}$), a Hach conductivity meter, model 44600, was used. The instrument automatically compensates for temperature deviation from $25^{\circ}C$. Turbidity measurements (nephelometric turbid unit, NTU) were made with a Hach turbidimeter model 2100A (precision $\pm$0.01 NTU). All the above instruments were regularly calibrated at each station.

$NH_4^+$ and $NO_3^-$ were respectively expressed as $NH_4$-N and $NO_3$-N. They were measured using the spectrophotometric Nessler and cadmium reduction methods respectively (precision: 0.01 mg l-1). $NO_2^-$ was expressed as $NO_2$-N. It was measured with the diazotization method (precision: 0.001 mg $l^{-1}$). Total phosphorus (TP) and soluble reactive phosphorus (SRP) were measured with the acid persulphate and ascorbic acid methods respectively (precision: 0.01 mg $l^{-1}$). Results for phosphorus were preliminary due to possible unidentified interference. Dissolved oxygen data were not complete during the first year of sampling and comparison could not be done for the present analysis. More details on limnological methods used were presented earlier (Plisnier *et al.*, 1996b, 1999).

Median values are graphically compared using the "box and whiskers" representation (Tuckey, 1977). Statistical comparisons were realised using the t-tests and ANOVA procedures (Sokal and Rohlf, 1995).

Weather data were collected in each station. Methods and data have been described earlier (Verburg *et al.*, 1997).

# 3. RESULTS

## 3.1 Mean Values in the Upper 0 - 100 m Layer

Mean values in the upper 0 - 100 m layer in 1993-94 were presented earlier (Plisnier *et al.*, 1996 a). Data for 1994-95 are presented below.

Median temperature of the upper 0-100 m layer at Bujumbura (25.8°C) was close to that at Kigoma (25.7°C) but higher than at Mpulungu (24.7°C). Those values were generally higher than in 1993-94.

The thermocline depth from July to September showed some differences at Bujumbura (*ca.* 70 m in 1994-95 compared to 74 m in 1993-94). At Kigoma, a marked difference was noted between the two years: the average thermocline depth between July and September was 83 m in 1993 but only 70 m in 1994.

Secchi disk depth was lower at Bujumbura (8.2 m) than at Mpulungu (11.2 m) and Kigoma (13.0 m) in 1994-95. The relative differences are similar to that of the previous year but transparency was higher in 1993-94 by about 1 m in each station. Median pH for the layer from 0-100 m was similar to that of previous observations: Bujumbura and Mpulungu (9.0) and Kigoma (8.8). Minimal values for each station were generally higher than in 1993-94 for each station. Variation was similar at the Bujumbura and Kigoma stations (standard deviation was *ca.* 0.2) and slightly higher at Mpulungu (0.3). Conductivity was 662 $\mu$S/cm at Bujumbura and Kigoma and 659 $\mu$S/cm at Mpulungu. Those median values were comparable to 1993-94 for the upper 100 m layer. Turbidity was higher at Bujumbura and Mpulungu, 0.42 and 0.35 NTU respectively than at Kigoma, 0.26 NTU in 1994-95. This order was similar to that of the previous year but turbidity was lower in 1993-94 by about 0.01 NTU in each station, an average value not significantly different for the upper 0-100 m layer. Total phosphorus was 0.07 at Bujumbura, 0.05 at Kigoma and 0.05 mg/l $PO_4^{3-}$ at Mpulungu. Soluble reactive phosphorus (SRP) was 0.05 at Bujumbura, 0.04 at Kigoma and 0.04 mg/l $PO_4^{3-}$ at Mpulungu. Ammonia had a median value of 0.01 at Bujumbura and 0.00 at the other stations. As in 1993-94, individual measurements were sometimes high. They could occasionally reach 0.95 at Bujumbura, 0.15 at Kigoma and 0.06 mg/l $NH_4$-N at Mpulungu. Nitrates were similar to 1993-94: 0.05 at Bujumbura, 0.08 at Kigoma and 0.07 mg/l $NO_3$-N at Mpulungu. At Bujumbura and Kigoma, median values of nitrite were the same as in 1993-94: 0.002 and 0.004 mg/l $NO_2$-N. Nitrite was slightly higher at Mpulungu, 0.003 mg/l $NO_2$-N for the 0-100 m layer.

## 3.2. Average Vertical Profiles as compared between Two Years of Sampling

The results of the weekly or two-weekly measurements during the two years of sampling show relatively similar average vertical profiles. There are, however, differences between the two years that are statistically significant (Tables 2 and 3).

Temperature of surface water was generally higher in 1994-95 (Figure 2 and Tables 2 and 3). The difference is not statistically significant however (excepted at 40 and 60 m at Mpulungu at site Bm).

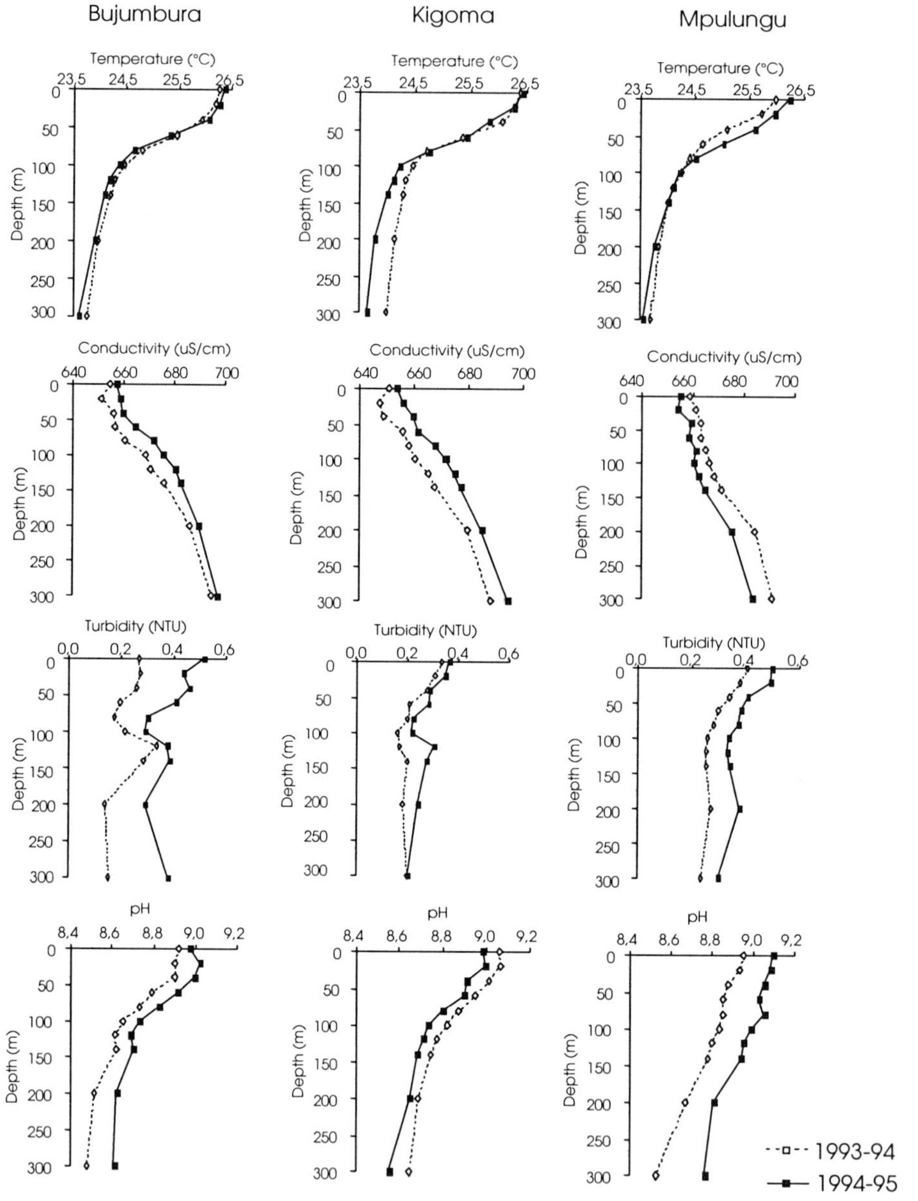


*Figure 2.* Average profiles of temperature, conductivity, turbidity and pH in Lake Tanganyika at sites Bb, Bk and Bm (0 to 300 m) at Bujumbura, Kigoma or Mpulungu, in 1993-94 compared to 1994-95 (period from August to July).

*Table 2.* Average temperature, conductivity, turbidity and pH measurements at each depth in 1993-94 compared with 1994-95 at sites B. Values of T observed, DF= degree of freedom, NS= non significant, * P<0.05, ** P<0.01, *** P<0.001.

| Station | Depth | Temperature (°C) | | | | | Conductivity (uS/cm) | | | | | Turbidity (NTU) | | | | |
|---|---|---|---|---|---|---|---|---|---|---|---|---|---|---|---|---|
| | | 1993-94 | 1994-95 | T obs | DF | Signif. | Average 1993-94 | Average 1994-95 | *T obs* | *DF* | Signif. | Average 1993-94 | Average 1994-95 | *T obs* | *DF* | |
| Bujumbura | 0 | 26.3 | 26.4 | 0.74 | 48 | NS | 655 | 657 | *0.46* | *50* | NS | 0.27 | 0.51 | *3.91* | *50* | *** |
| | 20 | 26.2 | 26.3 | 0.54 | 48 | NS | 652 | 659 | *2.04* | *50* | * | 0.27 | 0.43 | *7.37* | *49* | *** |
| | 40 | 25.9 | 26.0 | 0.85 | 47 | NS | 656 | 660 | *1.02* | *50* | NS | 0.26 | 0.46 | *4.29* | *49* | *** |
| | 60 | 25.5 | 25.3 | 0.73 | 48 | NS | 657 | 664 | *2.91* | *50* | ** | 0.20 | 0.40 | *7.71* | *49* | *** |
| | 80 | 24.8 | 24.6 | 1.25 | 48 | NS | 661 | 672 | *2.89* | *50* | ** | 0.17 | 0.30 | *6.05* | *49* | *** |
| | 100 | 24.5 | 24.3 | 1.03 | 48 | NS | 669 | 676 | *2.51* | *50* | * | 0.21 | 0.29 | *2.67* | *50* | * |
| | 120 | 24.3 | 24.2 | 1.48 | 49 | NS | 671 | 680 | *2.71* | *51* | * | 0.33 | 0.38 | *1.15* | *50* | NS |
| | 140 | 24.2 | 24.1 | 1.80 | 49 | NS | 676 | 682 | *2.28* | *51* | * | 0.28 | 0.38 | *2.26* | *50* | * |
| | 200 | 24.0 | 23.9 | 1.30 | 48 | NS | 686 | 689 | *0.86* | *50* | NS | 0.14 | 0.29 | *9.32* | *50* | *** |
| | 300 | 23.8 | 23.6 | 2.28 | 30 | * | 695 | 697 | *0.74* | *32* | NS | 0.15 | 0.38 | *4.19* | *32* | *** |
| Kigoma | 0 | 26.4 | 26.4 | 0.38 | 53 | NS | 651 | 653 | *0.70* | *53* | NS | 0.34 | 0.36 | *0.90* | *53* | NS |
| | 20 | 26.3 | 26.3 | 0.13 | 53 | NS | 648 | 656 | *2.16* | *53* | * | 0.31 | 0.35 | *1.62* | *53* | NS |
| | 40 | 26.0 | 25.8 | 1.56 | 53 | NS | 649 | 660 | *3.30* | *53* | ** | 0.28 | 0.29 | *0.36* | *53* | NS |
| | 60 | 25.3 | 25.4 | 0.33 | 53 | NS | 656 | 661 | *2.48* | *53* | * | 0.21 | 0.28 | *2.96* | *52* | ** |
| | 80 | 24.7 | 24.8 | 0.34 | 53 | NS | 658 | 668 | *2.44* | *53* | * | 0.20 | 0.22 | *1.07* | *52* | NS |
| | 100 | 24.4 | 24.2 | 1.30 | 53 | NS | 661 | 671 | *2.19* | *53* | * | 0.16 | 0.22 | *3.63* | *50* | *** |
| | 120 | 24.3 | 24.1 | 3.44 | 53 | ** | 665 | 675 | *2.36* | *53* | * | 0.17 | 0.30 | *3.79* | *50* | *** |
| | 140 | 24.3 | 24.0 | 4.99 | 53 | *** | 667 | 677 | *2.30* | *53* | * | 0.20 | 0.28 | *1.88* | *52* | NS |
| | 200 | 24.1 | 23.8 | 5.52 | 52 | *** | 680 | 685 | *1.61* | *52* | NS | 0.18 | 0.24 | *1.66* | *50* | NS |
| | 300 | 24.0 | 23.6 | 5.03 | 51 | *** | 688 | 695 | *1.88* | *52* | NS | 0.20 | 0.20 | *0.05* | *50* | NS |
| Mpulungu | 0 | 26.0 | 26.2 | 0.79 | 56 | NS | 659 | 655 | *1.07* | *52* | NS | 0.41 | 0.50 | *1.70* | *56* | NS |
| | 20 | 25.7 | 26.0 | 0.65 | 56 | NS | 661 | 654 | *2.75* | *52* | ** | 0.38 | 0.49 | *2.32* | *56* | * |
| | 40 | 25.1 | 25.6 | 2.00 | 56 | * | 663 | 659 | *2.46* | *51* | * | 0.34 | 0.41 | *1.54* | *55* | * |
| | 60 | 24.6 | 25.0 | 2.18 | 56 | * | 663 | 658 | *2.35* | *51* | * | 0.30 | 0.38 | *2.16* | *55* | * |
| | 80 | 24.4 | 24.5 | 1.02 | 56 | NS | 665 | 661 | *2.02* | *51* | NS | 0.28 | 0.37 | *2.42* | *55* | * |
| | 100 | 24.2 | 24.2 | 0.66 | 55 | NS | 666 | 660 | *3.40* | *51* | ** | 0.26 | 0.34 | *2.08* | *55* | * |
| | 120 | 24.1 | 24.1 | 0.11 | 54 | NS | 668 | 662 | *2.75* | *51* | ** | 0.25 | 0.33 | *2.02* | *55* | * |
| | 140 | 24.0 | 24.0 | 1.04 | 51 | NS | 671 | 664 | *2.61* | *50* | * | 0.26 | 0.34 | *1.87* | *54* | NS |
| | 200 | 23.8 | 23.7 | 1.89 | 52 | NS | 684 | 675 | *2.99* | *48* | ** | 0.27 | 0.38 | *2.50* | *54* | * |
| | 300 | 23.7 | 23.5 | 2.79 | 49 | * | 691 | 683 | *1.57* | *46* | NS | 0.24 | 0.30 | *1.75* | *50* | NS |

*Table 3*. Average temperature, conductivity, turbidity, pHn ammonia, nitrates and nitrites measurements at each depth in 1993-94 compared with 1994-95 at sites A and V. Values of T observed, DF= degree of freedom, NS= non significant, * P<0.05, ** P<0.01, *** P<0.001.

| | | Temperature (°C) | | | | | Conductivity (uS/cm) | | | | | Turbidity (NTU) | | | | | pH | |
|---|---|---|---|---|---|---|---|---|---|---|---|---|---|---|---|---|---|---|
| | | Average | | T obs | DF | Signif. | Average | | T obs | DF | Signif. | Average | | T obs | DF | Signif. | Average | |
| station | depth | 93-94 | 94-95 | | | | 93-94 | 94-95 | | | | 93-94 | 94-95 | | | | 93-94 | 94-95 |
| Buj./Uvira | 0 | 26.3 | 26.4 | 0.54 | 59 | NS | 649 | 659 | 2.61 | 61 | * | 0.40 | 0.56 | 2.63 | 61 | * | 9.013 | 9.016 |
| | 20 | 26.2 | 26.2 | 0.14 | 57 | NS | 648 | 659 | 2.15 | 55 | * | 0.38 | 0.47 | 2.03 | 56 | * | 9.033 | 9.054 |
| | 40 | 26.0 | 26.0 | 0.25 | 59 | NS | 655 | 660 | 1.58 | 57 | NS | 0.43 | 0.62 | 1.65 | 58 | NS | 8.979 | 9.044 |
| | 60 | 25.3 | 25.5 | 1.10 | 59 | NS | 659 | 665 | 2.00 | 57 | * | 0.32 | 0.72 | 1.61 | 58 | | 8.863 | 8.972 |
| | 80 | 24.7 | 24.6 | 0.90 | 59 | NS | 670 | 673 | 0.74 | 56 | NS | 0.31 | 0.54 | 2.26 | 57 | * | 8.79 | 8.826 |
| | 100 | 24.3 | 24.2 | 1.58 | 53 | NS | 675 | 678 | 1.00 | 57 | NS | 0.39 | 0.52 | 0.75 | 56 | NS | 8.664 | 8.737 |
| Kigoma | 0 | 26.6 | 26.6 | 0.41 | 62 | NS | 641 | 654 | 3.67 | 60 | *** | 0.32 | 0.35 | 1.03 | 61 | NS | 9.061 | 9.095 |
| | 20 | 26.4 | 26.3 | 0.64 | 62 | NS | 645 | 658 | 5.03 | 60 | *** | 0.30 | 0.34 | 2.78 | 61 | ** | 9.056 | 8.975 |
| | 40 | 26.1 | 26.0 | 0.40 | 62 | NS | 642 | 659 | 3.51 | 60 | *** | 0.27 | 0.30 | 2.14 | 61 | * | 9.031 | 8.923 |
| | 60 | 25.4 | 25.4 | 0.08 | 62 | NS | 650 | 662 | 3.29 | 60 | *** | 0.22 | 0.26 | 2.78 | 61 | ** | 8.957 | 8.881 |
| | 80 | 24.9 | 24.7 | 1.37 | 62 | NS | 661 | 668 | 3.73 | 60 | *** | 0.20 | 0.23 | 1.31 | 61 | NS | 8.863 | 8.788 |
| | 100 | 24.5 | 24.3 | 1.23 | 60 | NS | 666 | 673 | 3.29 | 59 | *** | 0.17 | 0.23 | 3.76 | 60 | *** | 8.799 | 8.716 |
| Mpulungu | 0 | 26.4 | 26.7 | 1.15 | 62 | NS | 647 | 645 | 0.54 | 63 | NS | 0.44 | 0.52 | 1.71 | 62 | NS | 9.019 | 8.958 |
| | 20 | 26.1 | 26.3 | 0.61 | 62 | NS | 653 | 651 | 0.73 | 62 | NS | 0.41 | 0.43 | 0.52 | 62 | NS | 8.978 | 9.027 |
| | 40 | 25.3 | 25.5 | 0.97 | 62 | NS | 658 | 658 | 0.15 | 62 | NS | 0.35 | 0.37 | 0.77 | 62 | NS | 8.931 | 8.938 |
| | 60 | 24.7 | 24.8 | 0.71 | 62 | NS | 662 | 659 | 2.02 | 62 | * | 0.31 | 0.33 | 0.73 | 62 | NS | 8.902 | 8.885 |
| | 80 | 24.3 | 24.3 | 0.70 | 62 | NS | 664 | 660 | 1.88 | 62 | NS | 0.29 | 0.31 | 1.11 | 62 | NS | 8.88 | 8.907 |
| | 100 | 24.1 | 24.2 | 1.01 | 60 | NS | 666 | 661 | 2.62 | 62 | * | 0.28 | 0.32 | 1.77 | 61 | NS | 8.865 | 8.887 |

| | | NH3-N (mg/l) | | | | | NO3-N (mg/l) | | | | | NO2-N (mg/l) | | | | |
|---|---|---|---|---|---|---|---|---|---|---|---|---|---|---|---|---|
| | | Average | | T obs | DF | Signif. | Average | | T obs | DF | Signif. | Average | | T obs | DF | Signif. |
| | depth | 93-94 | 94-95 | | | | 93-94 | 94-95 | | | | 93-94 | 94-95 | | | |
| Buj.-Uvi. | 0 | 0.02 | 0.02 | 0.30 | 57 | NS | 0.05 | 0.04 | 1.35 | 59 | NS | 0.002 | 0.001 | 2.83 | 59 | * |
| | 20 | 0.02 | 0.02 | 0.18 | 37 | NS | 0.05 | 0.04 | 1.48 | 37 | NS | 0.003 | 0.001 | 2.13 | 37 | * |
| | 40 | 0.03 | 0.02 | 0.86 | 37 | NS | 0.06 | 0.05 | 0.60 | 37 | NS | 0.002 | 0.002 | 1.32 | 37 | NS |
| | 60 | 0.03 | 0.03 | 0.29 | 36 | NS | 0.10 | 0.08 | 1.16 | 37 | NS | 0.003 | 0.002 | 0.58 | 37 | NS |
| | 80 | 0.04 | 0.03 | 0.49 | 37 | NS | 0.08 | 0.10 | 1.61 | 37 | NS | 0.005 | 0.003 | 1.12 | 37 | NS |
| | 100 | 0.10 | 0.05 | 1.63 | 37 | NS | 0.05 | 0.07 | 2.04 | 37 | * | 0.003 | 0.003 | 0.52 | 37 | NS |
| Kigoma | 0 | 0.02 | 0.02 | 0.25 | 60 | NS | 0.06 | 0.07 | 3.25 | 58 | ** | 0.003 | 0.005 | 3.07 | 58 | ** |
| | 20 | 0.03 | 0.03 | 0.05 | 39 | NS | 0.06 | 0.06 | 0.82 | 37 | NS | 0.003 | 0.007 | 2.29 | 37 | * |
| | 40 | 0.02 | 0.02 | 0.05 | 39 | NS | 0.06 | 0.08 | 1.43 | 37 | NS | 0.003 | 0.008 | 2.93 | 37 | ** |
| | 60 | 0.02 | 0.02 | 0.04 | 39 | NS | 0.09 | 0.10 | 0.89 | 37 | NS | 0.003 | 0.003 | 1.29 | 37 | NS |
| | 80 | 0.02 | 0.02 | 0.01 | 39 | NS | 0.12 | 0.17 | 2.32 | 37 | * | 0.003 | 0.004 | 1.08 | 37 | NS |
| | 100 | 0.02 | 0.02 | 0.15 | 38 | NS | 0.11 | 0.14 | 1.61 | 37 | NS | 0.004 | 0.006 | 2.6 | 37 | * |
| Mpulungu | 0 | 0.00 | 0.00 | 0.73 | 58 | NS | 0.05 | 0.05 | 0.33 | 62 | NS | 0.002 | 0.003 | 1.49 | 62 | NS |
| | 20 | 0.02 | 0.00 | 0.53 | 38 | NS | 0.05 | 0.04 | 1.69 | 42 | NS | 0.002 | 0.003 | 1.29 | 42 | NS |
| | 40 | 0.02 | 0.00 | 0.51 | 37 | NS | 0.07 | 0.08 | 0.64 | 42 | NS | 0.002 | 0.003 | 1.87 | 42 | NS |
| | 60 | 0.03 | 0.00 | 0.71 | 37 | NS | 0.09 | 0.10 | 1.26 | 42 | NS | 0.003 | 0.004 | 1.53 | 42 | NS |
| | 80 | 0.02 | 0.00 | 0.69 | 37 | NS | 0.09 | 0.11 | 1.22 | 42 | NS | 0.002 | 0.004 | 2.53 | 42 | * |
| | 100 | 0.01 | 0.01 | 0.51 | 39 | NS | 0.10 | 0.13 | 2.83 | 42 | ** | 0.003 | 0.003 | 0.16 | 42 | NS |

Inversely, below the upper metalimnion, a temperature decrease was observed at each station in 1994-95 that is highly significant ($p<0.01$) below 100 m depth at Kigoma (site Bk) and significant ($p<0.05$) at 300 m at Bujumbura (site Bb) and Mpulungu (site Bm) (Table 2). The same observations were made for profiles at sites Ab/Vb, Ak/V/k and Am/Vm around 80-100 m although the differences were not significant in those cases (Table 3).

The northern stations of Bujumbura and Kigoma had higher values of conductivity in 1994-95 (Figure 2). Those differences are generally significant in the upper layers at each site depending on the depths (Tables 2 and 3). Inversely, in the South, a significant decrease of mean conductivity is noted especially at site Bm (Table 2).

Turbidity shows some peaks in the water column. A first peak was observed at around 50 m depth in Bujumbura and 60 m in Kigoma. A second peak occurs clearly at around 120-140 m in Bujumbura each year and at 140 m in Kigoma. In Mpulungu, a small peak was observed around 200 m (Figure 2). Comparisons between years show that turbidity was higher at each station in 1994-95 compared to 1993-94. This is particularly significant in the two northern stations (Tables 2 and 3).

The profiles of pH show higher values at both ends of the lake during the second year. However, the difference between the two years is statistically significant only at site B of Mpulungu. In Kigoma, a decrease of the average values of pH for each depth in the 0-300 m layer in 1994-95 is statistically significant at several depths (Table 2).

Ammonia profiles show lower average values in 1994-95 at both ends of the lake (Figure 3). However, due to the high variance of the measurements, the differences are not statistically significant (Table 2).

Nitrate profiles show a peak situated on average at 60-80 m depth. The peak is well observed at Kigoma at a similar depth during the second year. Most of the interannual differences for this variable are not statistically significant (Table 2).

The nitrite profiles show some differences from year to year. An increase is observed at 20 and 40 m particularly well at Kigoma (Figure 3).

## 3.3 Variability of Observations

Variability around the mean was generally lower for most limnological variables in 1994-95 in the northern stations compared to the previous year. This can be observed at Figures 4, 5 and 6. The principle points of note are:

1. Temperature (Figure 4) showed relatively similar variability at each depth in 1994-95 compared to the previous year.
2. The variability in conductivity measurements was much lower in the northern stations, particularly in 1994-95 (Figure 4).
3. Turbidity variability was about the same in both years in the northern stations. Higher variability of turbidity is, however, observed in Mpulungu in 1994-95 (Figure 5).

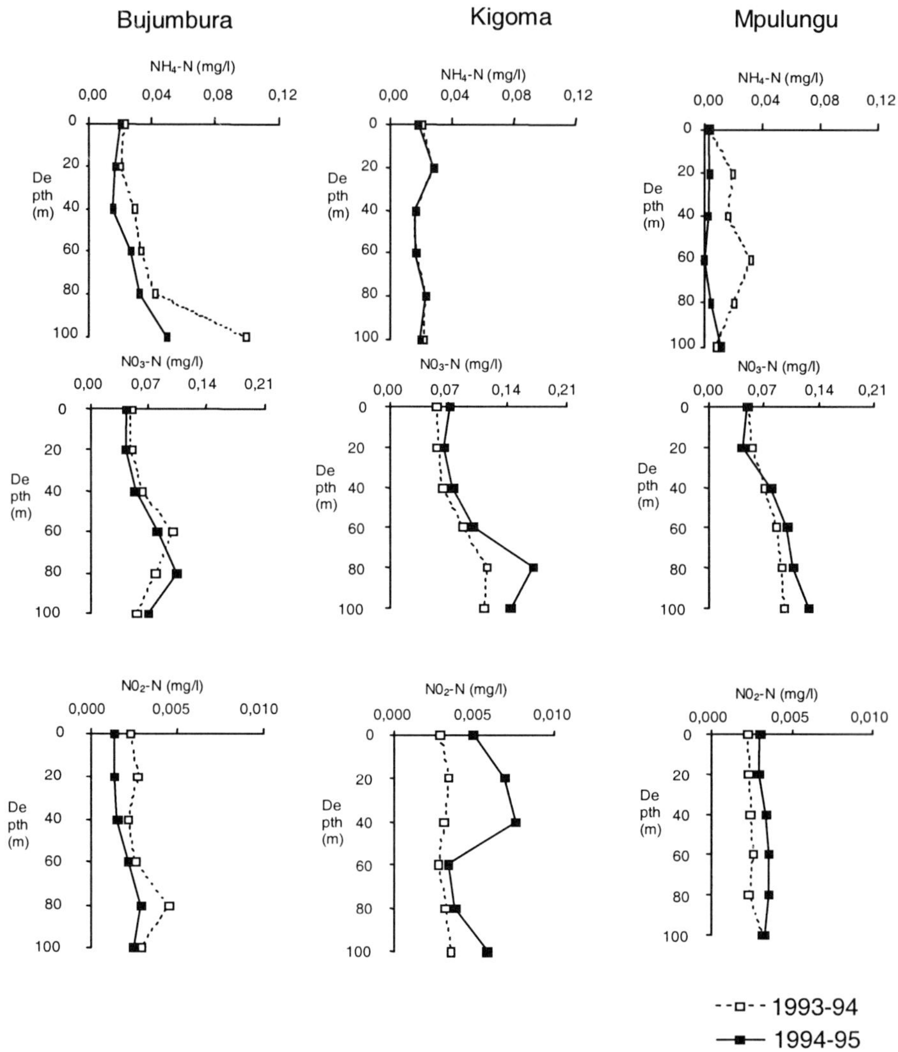


*Figure 3.* Average profiles of ammonia ($NH_4^+$ expressed as N), nitrites ($NO_2^-$ expressed as N) and nitrates ($NO_3^-$ expressed as N) in Lake Tanganyika at sites Ab/Vb, Ak/Vk and Am/Vm (0 to 100 m) at Bujumbura, Kigoma and Mpulungu, in 1993-94 compared to 1994-95 (period from August to July).

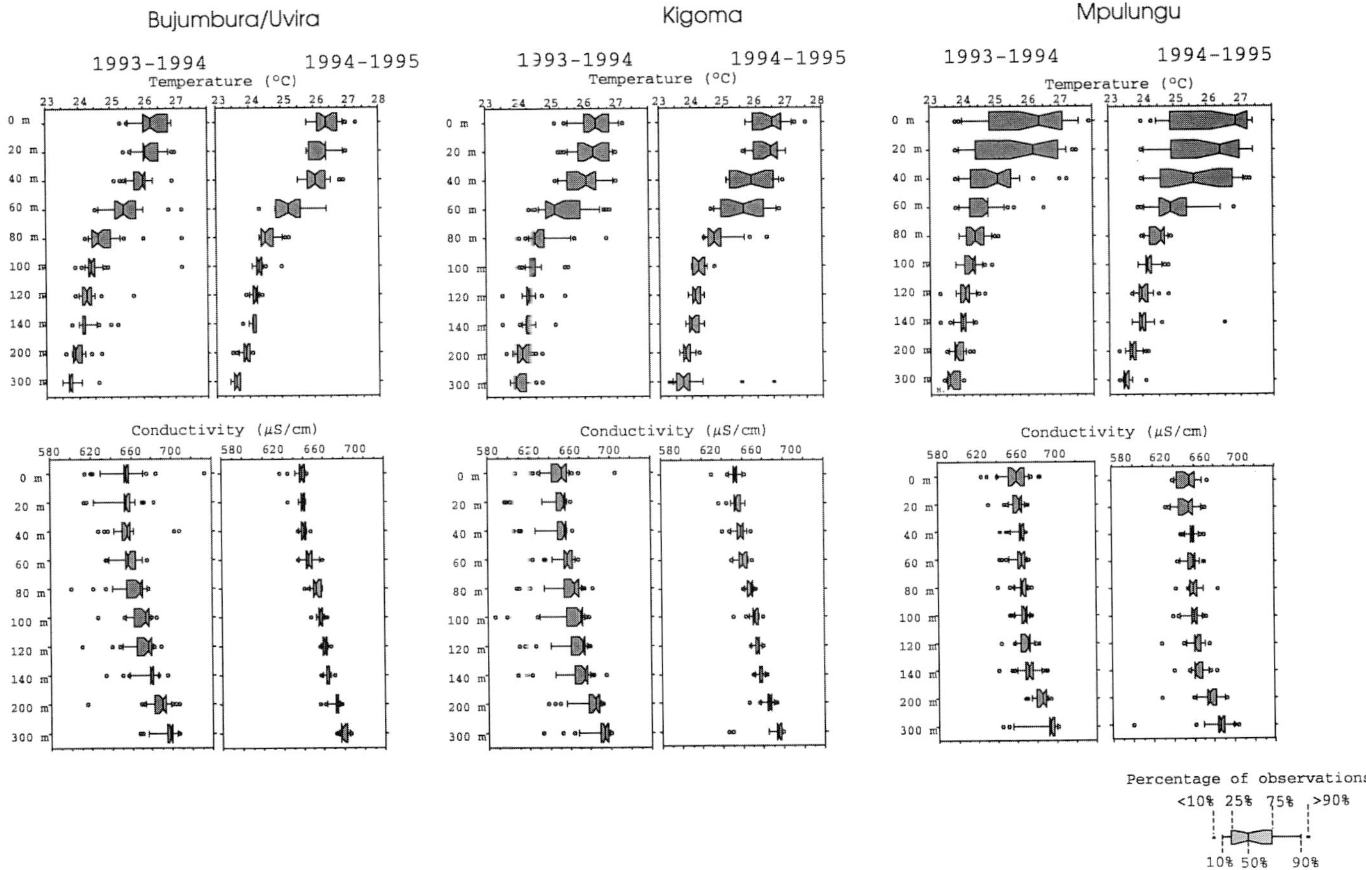


*Figure 4.* Median and percentile values of temperature and conductivity measured in Lake Tanganyika at sites Bb,Bk and Bm (0 to 300 m) at Bujumbura, Kigoma and Mpulungu in 1993-94 compared with 1994-95.

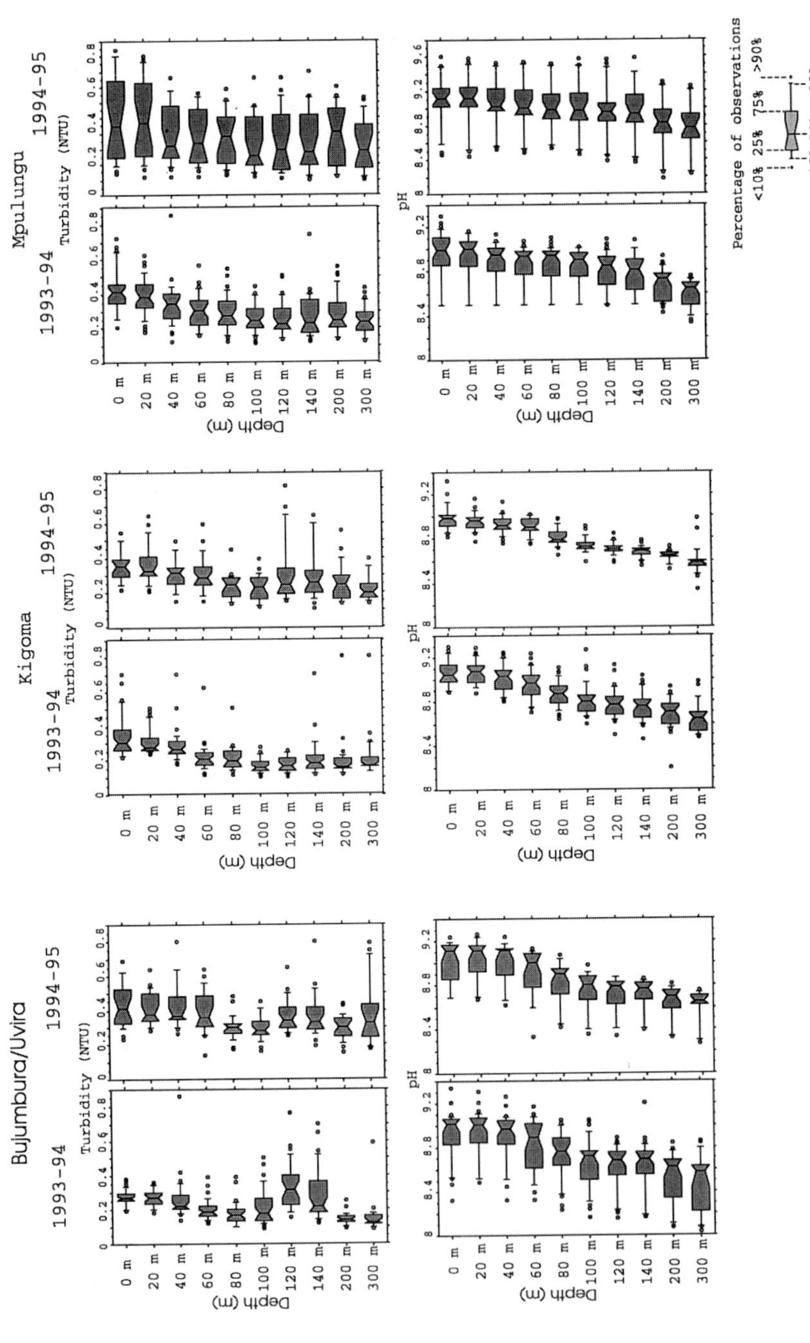


*Figure 5.* Median and percentile values of turbidity and pH measured in Lake Tanganyika at sites Bb,Bk and Bm (0 to 300 m) at Bujumbura, Kigoma and Mpulungu in 1993-94 compared with 1994-95.

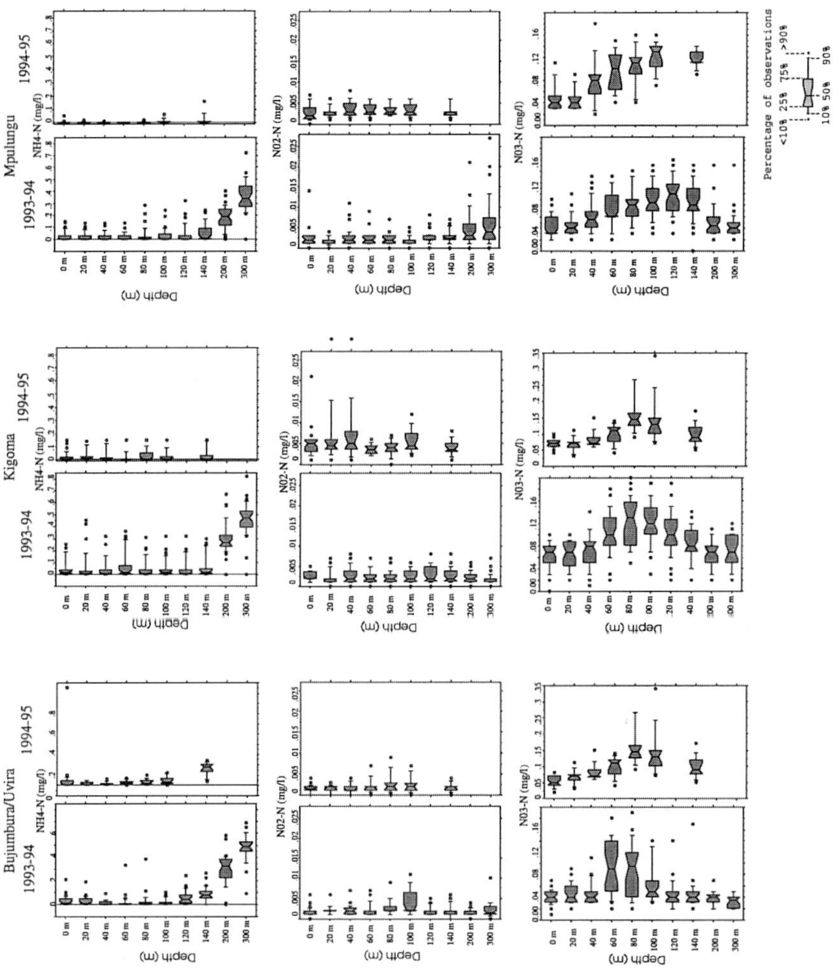


*Figure 6.* Median and percentile values of ammonia, nitrites and nitrates (0 to 300 m) in Lake Tanganyika at Bujumbura, Kigoma and Mpulungu in 1993-94 compared with 1994-95 at sites Bb.Bk.Bm and Vb.Vk.Vm.

4. The pH measurements showed more variability between stations than between years. This can be seen in Figure 5 where variability was higher at both ends of the lake compared to Kigoma. Variation around the mean was slightly more important at Mpulungu in 1994-95 as it was for turbidity. This is however not significant.
5. Values of ammonia were less variable in 1994-95 at each station, as it can be observed from the observations down to 140-160 m (Figure 6).
6. The variability of nitrate values was reduced generally at each depth for each station in 1994-95. For nitrite, it was quite similar except for Kigoma where a few high values were recorded during the second year (Figure 6).

# 4. DISCUSSION

"Average" layers may be identified from temperature profiles: an epilimnion between 0 and 35 m at each station, the "upper-metalimnion" (including the thermocline) between 35 and 90 m at Bujumbura and Kigoma and between 35 and 70 m at Mpulungu. The "lower metalimnion" ranges down to 250-300 m. The hypolimnion starts below 300 m. This is of course a rather subjective delimitation. The hypolimnion is often also considered to include "the lower metalimnion" (although temperature still decreases relatively faster there) and thus the classical three layers may be identified for comparative purposes with other lakes. The depths of those layers vary over the year, because of various factors linked to seasonal changes and direct meteorological conditions, hydrodynamics such as internal waves, tilting of the thermocline linked to wind forcing etc. (Coulter, 1991; Plisnier and Coenen, 2001).

Although the difference was not statistically significant, water temperature was higher near the surface in 1994-95 compared to the previous year. Inversely, measurements at deeper stations showed a statistically significant decrease of temperature during the second year. This slight change in the thermal stratification of the water column could be related to different weather conditions. Warmer and generally less windy weather was recorded in 1994-95 although the observations showed variable trends depending of the wind direction components considered (Verburg *et al.,* 1997). A regular survey of limnology and weather over several consecutive years should clarify the amplitude of interannual temperature variation vs. the apparent warming trend in the recent decades at Lake Tanganyika (Plisnier, in press).

The increase of conductivity in the surface layer of the northern stations during the second year is difficult to interpret. Increased mixing of deep water (more salty) toward the surface does not match with the apparent change in weather conditions. But a shallower thermocline in 1994-95 could explain this. In the South, a slight decrease of conductivity is noted. This could correspond to a decreased upwelling intensity in 1994-95 that may have resulted in less advection of deep waters, characterised by higher conductivity values, toward the surface, an hypothesis coherent with the preliminary observations on weather differences between the two years.

A turbidity peak at several depths in the water column each year may correspond to the presence of layers of micro-organisms (bacteria, protozoa, phytoplankton) there whose presence is directly or indirectly linked to the oxycline and reduced material in the lower metalimnion. This is an interesting field of investigation for future research, as migration of deep water organisms toward the surface during the night might partially originate from those layers. It was shown earlier (Plisnier *et al.*, 1996a) that these turbidity peaks are better observed when the lake conditions are calmer. The increase of turbidity in 1994-95, particularly in the North, could correspond to higher values of planktonic biomass. Turbidity seems to be a good indicator for planktonic abundance in the pelagic area (Coulter, 1991).

The profiles of ammonia, nitrates and nitrites show changes that are coherent with a reduction in intensity of hydrodynamic processes in 1994-95 compared with 1993-94. It was noted earlier (Plisnier *et al.*, 1996a) that occasional high values of ammonia occurred in the epilimnion when internal wave activity is high (such as between October and December for example). Previous observations have also shown that nitrate peaks were better observed during calm periods during an annual cycle (Plisnier *et al.*, 1996a, 1999). In the northern stations, this seems to be the case in 1994-95. Nitrites profiles correspond with nitrate profiles at each station as they show a similar decrease or increase (depending on the stations) for those variables simultaneously (Figure 3).

Although it was suggested that biological control dominates nutrients cycling in the epilimnion in tropical lakes (Kilham and Kilham, 1990), the changes in nitrogen concentration during the year in relation with the hydrodynamic cycle in Lake Tanganyika (Coulter, 1991; Plisnier *et al.*, 1999) and observations of turbulent events during 24h cycles during the FAO/FINNIDA project suggest that the physical environment (turbulence, stability, upwellings, currents…) plays an important role acting upon the nitrogen concentration in the epilimnion probably largely controlled by replenishment by N rich hypolimnion waters toward the surface.

A difference in variability of the measurements was observed from one year to the next at each station for many limnological variables (for example: conductivity). This difference was particularly important in the North. Since vertical profiles show significant differences for all the variables (ANOVA, depth factor: $p<0.01$ or 0.001), it is suggested that change in the extreme values recorded, from one year to the next, could reflect changes in hydrodynamics. Frequency of mixing between waters of different depths are probably modified depending on the hydrodynamic state of the lake, which is itself dependent upon climatic conditions.

A possible hypothesis is that the dynamics of the system had decreased during the second year. Lower velocities of SE winds, as noted in 1994-95 (Verburg *et al.*, 1997) have probably driven less warm epilimnion water to the North in the dry season (June-September). This agrees with a decreased thermocline depth for the two northern stations. A decreased amount of epilimnion water "piled up" in the North could correspond to less potential energy to redistribute through movements of water in 1994-95 (less oscillations of layers toward equilibrium once the monsoon has changed from SE to NE, each year around September) (Plisnier *et al.*, 1999). A higher surface water temperature in the South would suggest that upwelling was less important in 1994-95. Values of ammonia were less variable than in 1993-94 at each station. This may have resulted from decreased turbulence in the water column with less deep water mixing in the epilimnion.

Variations of turbidity and pH in the South were, however, more important in 1994-95. The differences in the bathymetry of the various basins and complex regional winds do not allow a unique explanation for all changes noted between years at each station. The East component of the wind may be tied, for example, with transversal tilting of epilimnion waters and changes in the monsoon direction may be accompanied by transversal oscillations of water layers with different amplitude in each basin. Hydrodynamics and ecological modelling, the study of ecological processes (such as primary or bacterial production), and a high frequency of limnological and weather measurements over several years, are necessary to better understand the ecosystem of Lake Tanganyika.

## 5. CONCLUSIONS

Limnological profiles from frequent measurements in 1993-94 and 1994-95 show that although the profiles remain very comparable, some significant differences, greater than the instrumental error, were observed for several variables. Variability of measurements and different weather conditions (warmer and less windy in 1994-95) agree with a probable decrease of hydrodynamic activity and less water mixing. This would explain the generally more stable observations for the variables at each depth sampled in the second year.

The sequence of a limnological cycle in 1994-95 appeared similar to the previous year (Plisnier *et al.*, 1999) but with a decreased intensity. A preliminary hypothesis links a decreased tilting of the epilimnion by weaker SE winds during the dry season with a weaker upwelling intensity in the South, decreased internal wave amplitude and turbulence leading to less changes of the variables at each depth during the second year. However, a secondary, northern upwelling could not be detected in 1994-95 and some higher fluctuations of pH and turbidity in the South during the second year of sampling may not yet be explained.

Understanding of limnology of Lake Tanganyika has been recently improved (Sarvala *et al.*, 1999; Coulter, 1991). However, many questions remain unsolved. Several years of continuous observations would be required to better understand differences in the limnological environment in relation to climate. This should enable improved interpretation of changes in communities of organisms, their recording in the sediments and subsequent paleoclimatic significance. More observations on limnological variability and its causes would help to improve understanding of important features of the East African Lakes such as eutrophication, fish kills, plankton blooms and recent fishery changes.

## ACKNOWLEDGEMENTS

Thanks are addressed to many researchers of the FAO/FINNIDA project involved in the sampling and analysis. At Bujumbura, we thank particularly E. Coenen, V. Langenberg, J.M. Tumba, K. Tshibangu, E. Gahungu, P. Paffen, E. Nikomeze, B. Ndimunzigo, S. Nyamushahu, S. Kimbadi, R. Varyianis, I. Rugoreka and P. Sinanka. At Kigoma, sincere thanks are addressed to D. Chitamwembwa, P. Mannini, E.W. Lyoba, D. Lyoba, N. Challe, H. Kurki, M. Chatta, S. Muhoza, A. Kalangali, E. Kadula, O. Kashushu, W. Kisisiwe and M. Agustino. Special thanks in Mpulungu are addressed to L. Mwape, E. Bosma, C. Lukwessa, R. Shapola, L. Makasa, I. Zulu, E. Chipulu, S. Sichivu, K. Kaoma, P. Verburg, J. Chimanga, B. Kassikila and C. Sichamba. We also thank G. Hanek, J.F. Craig J. Kapetsky, D. Blessich, M. Doeff, M. Mann, A. Bakun, as well as other officers of FAO involved with LTR for their support. Acknowledgement is made to O. Lindqvist and H. Mölsä, Kuopio University, Finland; J.-P. Descy and J.-C Micha, University of Namur, Belgium and K. Maertens of the Royal Institute for Natural Sciences, Belgium for help at various occasions. Thanks are addressed to the reviewers of this paper for their constructive comments. Invitation to participate to the IDEAL/GEF-STAP meetings in January 2000 was possible thanks to Prof. E. Odada and funding from Start/Pages/GEF and Mac Arthur Foundation. The analysis work for this paper was supported by the Federal Office for Scientific, Technical and Cultural Affairs, Belgium. Financial support was also provided by U.S.National Science Foundation Grant # ATM-9619458 (The Nyanza Project) and by the Lake Tanganyika Biodiversity Project (GEF/UNDP).

## REFERENCES

Coulter, G.W. (ed.) (1991) *Lake Tanganyika and its Life*. Oxford University Press, London, 354 p.

Hecky, R.E., Bugenyi, F.W.B., Ochumba, P., Talling, J.F., Mugidde R., Gophen, M., and Kaufman, L. (1994) Deoxygenation of the deep water of Lake Victoria, East Africa. *Limnol.Oceanogr.* **39**(6), 1476-1481.

Johnson, T.C., and Odada, E.O. (eds.) (1996) *The Limnology, Climatology and Paleoclimatology of the East African Lakes*. Gordon and Breach Publ., Amsterdam.

Kilham, S.S. and Kilham, P. (1990) Tropical limnology: Do African lakes violate the "first law" of limnology ? *Verh. Internat. Verein. Limnol.* **24**, 68-72.

Lehman, J.T. (ed.) (1998) *Environmental Change and Response in East African Lakes*. Kluwer Academic Publishers, Dordrecht. Monographiae Biologicae. **79**.

Lewis, W.M. (1996) Tropical lakes: how latitude makes a difference, in F. Schiemer and K.T. Boland (eds.), *Perspectives in Tropical Limnology*. SPB Academic Publishers, Amsterdam, pp.43-65.

Plisnier, P.-D., Langenberg, V., Mwape L., Chitamwembwa, D. and Coenen, E. (1996a) Limnological sampling during an annual cycle at three stations of Lake Tanganyika (1993 - 1994). FAO/FINNIDA Research for the Management of the Fisheries on Lake Tanganyika. GCP/RAF/271/FIN-TD/46 (En), 136 p.

Plisnier, P.-D. (1996b) Limnological sampling during a second annual cycle (1994-1995) and some comparisons with year one on Lake Tanganyika FAO/FINNIDA Research for the Management of the Fisheries on Lake Tanganyika. GCP/RAF/271/FIN-TD/56 (En), 60 p.

Plisnier, P.-D., Chitamwebwa, D., Mwape, L., Tshibangu, K., Langenberg, V. and Coenen, E.C. (1999) Limnological annual cycle inferred from physical-chemical fluctuations at three stations of Lake Tanganyika. *Hydrobiologia* **407**, 45-58

Plisnier, P.-D., and Coenen, E.C. (2001) Pulsed and dampened annual limnological fluctuations in Lake Tanganyika, in M. Munawar and R. Hecky (eds) *The Great Lakes of the World (GLOW): Food-web, health and integrity,* Ecovision World Monograph Series, Backhuys Publ., Leiden, The Netherlands, pp. 83-96.

Plisnier, P.D (in press) Recent climate and limnology changes in Lake Tanganyika, Verhandlungen Internationale Vereinigung für theoretische und angewandte Limnologie.

Sarvala, J., Salonen, K., Järvinen, M., Aro, E., Huttula, T., Kotilainen, P., Kurki, H., Langenberg, V., Mannini, P., Peltonen, A., Plisnier, P.-D., Vuorinen, I., Mölsä, H. and Lindqvist, O. V. (1999) Trophic structure of Lake Tanganyika: carbon flows in the pelagic food web. *Hydrobiologia*, **407**, 155-179.

Sokal, R.R. and Rohlf, F.J. (1995) *Biometry*. W.H. Freeman and Cie.

Tuckey, I.W. (1977) Exploratory data analysis. Addison Wesley Publ. Co., Massachusetts.

Verburg, P., Huttula, T., Kakogozo, B., Kihakwi, A., Kotilainen, P., Makasa L. and Peltonen, A. (1997) Hydrodynamics of Lake Tanganyika and meteorological results. FAO/FINNIDA Research for the Management of Fisheries on Lake Tanganyika. GCP/RAF/271/FIN-TD/59 (En), 54 p.

# Sedimentary Processes, Paleoclimate and Paleoenvironment

# SEDIMENTOLOGY AND GEOCHRONOLOGY OF LATE PLEISTOCENE AND HOLOCENE SEDIMENTS FROM NORTHERN LAKE MALAWI

SYLVIA L. BARRY[1]*, M. L. FILIPPI [2]#, MICHAEL R. TALBOT [2] AND THOMAS C. JOHNSON [1]

*[1] Large Lakes Observatory, University of Minnesota, 10 University Drive, Duluth, MN 55812*

*[2] Geological Institute, University of Bergen, Allégt. 41, 5007 Bergen, Norway*

** Current address: Harvard Forest, PO Box 68, Petersham, MA, 01366, USA*

*# Current address: Centro di Studio per il Quaternario l"evoluzione Ambientale, c/o Dipartimento di Scienze della Terra Universita "La Sapienza", Piazzale Aldo Moro, 5-00185, Roma, Italy*

**ABSTRACT**

Six ca. 9 m long piston cores and 16 multi-cores up to ca. 50 cm long were recovered from the North Basin of Lake Malawi, East Africa. Five of these cores were selected for palaeoenvironmental studies. Five stratigraphic units are recognised, which together with volcanic ash beds and magnetic susceptibility profiles, were used to correlate the cores. A geochronology for the cores has been established by means of $^{210}$Pb assays of the uppermost part of the record, radiocarbon dating and varve counts. Radiocarbon age determinations on bulk organic matter spanning the last ~24,000 years are complicated by the presence of variable quantities of reworked organic debris and a calibration correction of 450 years has been provisionally assumed. Three facies: laminites (varves), homogenites, and unlaminated clays were recognised and used to interpret the depositional environment.

The varved sections provide the first high-resolution varve chronology from the southern tropics. The uppermost varved section provides an outstanding basis for elucidating Lake Malawi's response to historically documented climatic, volcanic and seismic events in the region. Beds of homogeneous clay or



*E.O. Odada and D.O. Olago (eds.),*
*The East African Great Lakes: Limnology, Palaeolimnology and Biodiversity,* 369–391.

silty clay (homogenites) were probably deposited from muddy turbidity flows, and are particularly abundant in the section spanning the Pleistocene-Holocene boundary, when the lake level of Lake Malawi fell dramatically. Homogenite deposition at this time was probably due to reworking of newly exposed, unconsolidated muddy sediments. Some younger homogenites can be correlated across large distances and may have originated from turbidity flows triggered by seismic shocks.

The new sedimentary record provided by these cores is difficult to reconcile with current understanding of lake-level history. Although our new data are consistent with previously inferred lowstand conditions in the Early Holocene, our interpretation of the oldest section recovered in the cores also suggests lowstand rather than highstand conditions at around the time of the Last Glacial Maximum.

# 1. INTRODUCTION

Lake Malawi is the most southerly and second largest (by volume) of the East African Rift Valley lakes. Despite the tectonic origin of the lake, its sedimentary record preserves substantial palaeoclimatic information. Because evaporation plays a major role in its hydrological budget, the lake is sensitive to changing climate conditions that affect lake-level and its status as an open or closed basin. Lake Malawi's location, between the equatorial and subtropical rainfall zones that tend to vary inversely with each other (Nicholson, 1998), makes it particularly suitable for unravelling the interactions of tropical and Southern Hemisphere palaeoclimates. Indeed, lowstands dated at about 6-10 ka (Finney and Johnson, 1991), and 28-37 ka (Finney *et al.*, 1996), show that some lake-level variations in Malawi were out of phase with all other large East African lakes (Johnson, 1996). Recent lake-level records from Lake Malawi (Birkett *et al.*, 1999), on the other hand, are in phase with other East African lakes and appear to correlate with climatic events in the Indian Ocean.

The lake is permanently anoxic below a water depth of ca. 220 m (Vollmer *et al.*, in press, this volume), a condition that favours the preservation of organic matter and primary sedimentary structures. Finely laminated muds and silts (annual varves, see Pilskaln and Johnson, 1991) form in the northern basin, which is influenced by several large rivers and is proximal to the Rungwe Volcanic Province. A fast sedimentation rate (approximately 1.4 mm/yr.; Barry, 2001) and lack of biological disturbance contribute to the preservation of the laminae. Time-series analysis of varve thickness suggests the presence of El Niño - Southern Oscillation (ENSO)-scale periodicities in some sequences (Rosenmeier, 1998), thus strengthening the probability that the Lake Malawi sedimentary record is in part forced by global climatic variations.

In order to investigate the intriguing potential of this varved record, six piston cores (8-9 m long) and sixteen multi-cores (less than 60 cm long) were taken at different depths from several sites in the northern basin of Lake Malawi during an IDEAL (International Decade of East African Lakes) cruise in 1998. The cores have been studied with a multiproxy approach, including sedimentological analysis, physical properties (water content, magnetic susceptibility), various dating methods (radiocarbon, $^{210}$Pb, varve counting), and geochemical analyses (TOC, $N_{tot}$, biogenic silica, Hydrogen Index, C and N stable isotope ratios in organic matter). The diatom assemblages in one core (M98-2P) have also been studied (Gasse *et al.*, in press, this volume). The aim of this paper is to present the chronological and sedimentological framework for these cores, in order to furnish a common stratigraphic base for future work. We address some of the problems in assigning an accurate chronology to the cores, and in the interpretation of their sedimentary record.

## 2. GEOLOGIC SETTING AND HYDROLOGIC REGIME

Lake Malawi (9°S-14°S, 475 m a.s.l.) lies in the southernmost section of the western arm of the East African Rift Valley (Figure 1-A). The rift is composed of a series of half-grabens (Rosendahl, 1987), with the highest topography formed by the Livingstone Mountains along the eastern side of the northern basin (North-West of the Ruhuhu river in Figure 1-B). The catchment is primarily composed of Precambrian Basement Complex rocks, with Cenozoic pyroclastic rocks of the Rungwe volcanic field occurring along the Songwe River valley in Tanzania, adjacent to the northern basin (Harkin, 1982). The lake is about 570 km long, 40 km wide and nearly 700 m deep, with a catchment area that is ca. 3 times the area of the lake. It is meromictic, with only the upper 100 to 250 m of water column mixing during periods of low stratification during the cool, windy months (April through October). The waters are anoxic and nearly isothermal at about 22.6° C below 220 m (Eccles, 1974). Permanent stratification occurs due to small thermal and chemical differences between the epilimnion and the hypolimnion. (Vollmer, *et al.*, in press, this volume).

Regional climate is dominated by the African monsoon, which is controlled by the passage of the Inter-tropical Convergence Zone (ITCZ). Lake Malawi is at the ITCZ's southern limit, and therefore experiences only one rainy season per year, in austral summer (November through March), in contrast to the large rift lakes to the north that experience two wet seasons per year. There is differential rainfall along the length of the lake, with ~250 cm/yr falling in the northern basin, decreasing to ~60 cm/yr in the south. In the austral winter the ITCZ migrates northward, allowing strong, dry, southerly tradewinds to dominate the lake basin.

Evaporation controls the hydrology of the basin, with outflow at the Shire River accounting for only about 20% of the total water loss (Beadle, 1981). At times lake level drops below the outlet, creating closed-basin conditions. Ample evidence exists for low lake stands, including submerged wave-cut terraces and beach sands (Johnson and Ng'ang'a, 1990, Finney and Johnson, 1991), buried lowstand deltas (Scholz, 1995), and isotopic data from offshore sediment sequences bearing authigenic calcite (Ricketts and Johnson, 1996).

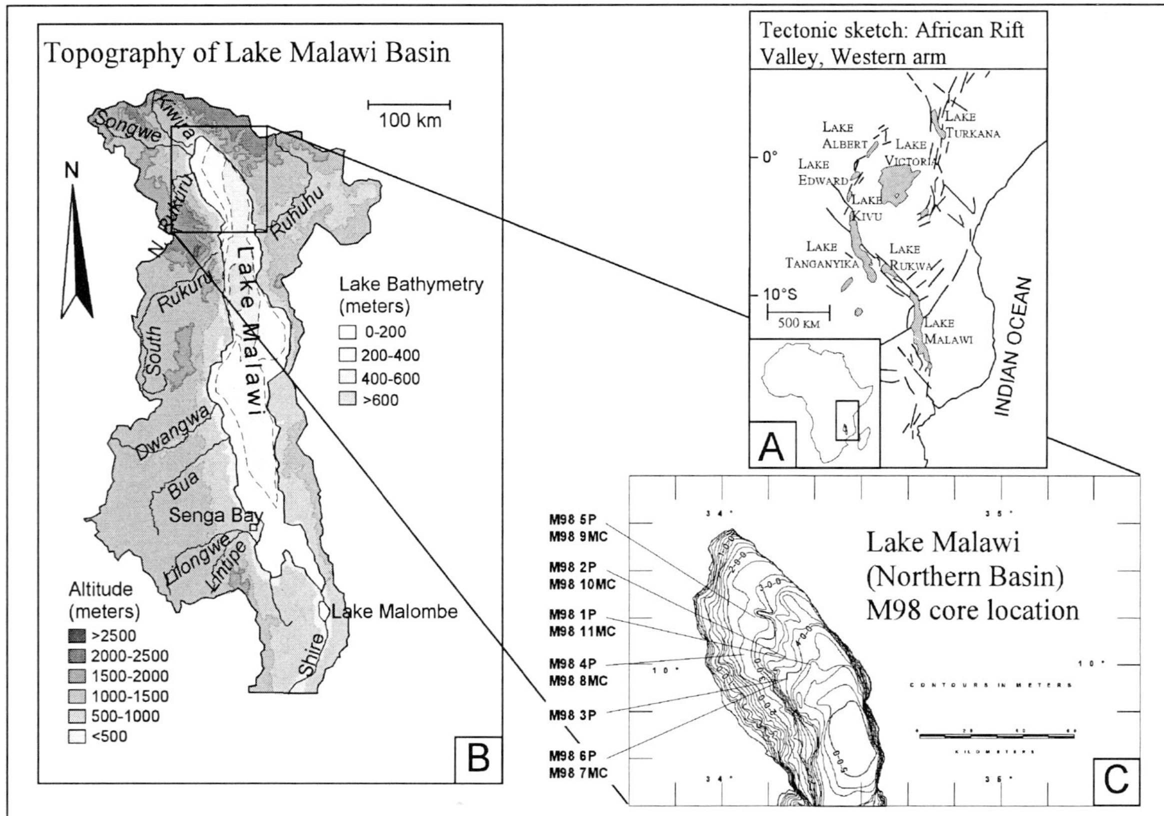


*Figure 1*. A) Tectonic sketch of the East African rift system, Western Arm (redrawn after Rosendahl et al., 1992), B) Topography of Lake Malawi basin (from Crossley, 1984, modified); C) M98 core location

## 3. METHODS

Coring locations were selected using high-resolution seismic profiles obtained during Duke University's PRoto-Ocean Basin Evolution (Project PROBE) expedition in 1986-1987 (Scholz and Rosendahl, 1988) and its sequel (SEPRO) in 1990 (Johnson *et al.* 1995). As high-resolution palaeoclimatic records were the primary goal of our cruise, sites were chosen where we judged the disruptive effects of turbidites and downslope mass movement were likely to be minimal. Core site positions were determined by satellite global positioning system (GPS). During the 1998 IDEAL cruise on Lake Malawi, six Kullenberg piston cores (ca 8-9 m long) with associated trigger cores (ca 0.5-1 m long) and 16 short multicores (ca 0.5 m long, 2-4 at each site) were taken from five locations (Figure 1-C, Table 1) in the northern basin. One piston core (core M98-13P) and 4 multicores (cores M98-12MC A to D) were taken from 604 m in the deepest part of the lake, south of the northern basin (Table 1).

The piston cores were taken with a modified Kullenberg corer (Kelts *et al.*, 1986), a deep-water piston operated gravity corer with polycarbonate liners. The load (typically 250 kg) influences penetration and it tends to over-penetrate the sediment-water interface by as much as 1-m. A short gravity "trigger" core is used as a trip for the assembly and collects approximately one meter of sediment (labelled PG in Table 1). The short cores were recovered with an Ocean Instruments Multi-Corer that retrieves up to four cores simultaneously (labelled A-D in Table 1). It is hydraulically damped and collects a high-quality sample of the sediment-water interface. During descent, the position of the coring apparatus was followed on the ship's echosounder. Five of the multi-cores were extruded on board (Table 1) and sliced horizontally for later analysis. The Kullenberg cores were cut into 1.5 m-long sections, sealed and shipped by airfreight to the Large Lakes Observatory in Duluth, Minnesota where they are held in a cold locker at 4° C. All of the piston cores and their accompanying trigger cores were scanned on a Bartington multi-sensor track logging system for non-destructive high-resolution (1cm) measurement of whole core magnetic susceptibility at the Limnological Research Center (LRC), University of Minnesota, Minneapolis. They were then split longitudinally, visually described for sedimentary structures; colour was noted using a Munsell colour chart (Munsell, 1954). Half of each core was archived and the other half was designated as the working half. The archived halves were photographed and the working halves were sampled every 10 cm for water content, biogenic silica and smear-slide analysis, and geochemical characteristics. Smear slides were taken not only from the 10-cm sampled intervals but also from intervening horizons of interest, such as zones with high magnetic susceptibility peaks. Digital imagery of the cores was accomplished using a Pixera Professional Digital Camera with a Fujinon-TB 1:1.4 macro lens. Images were recorded on CD at a resolution of 600 dpi. Varves were counted by eye in every core using the magnified digital image. The light-coloured (diatomaceous) lamination of each varve couplet was assigned a number from the top of the unit and counted downcore using the software program Adobe Photoshop. The tops of the cores were correlated using distinct marker beds and thin (1-2 cm thick) muddy homogenites (unlaminated clay) which interrupt the varves. Age assignments are assumed to be equal to the maximum number of varves counted between each marker bed in the least disturbed cores, and were combined to create a master stratigraphy.

*Table 1*. Lake Malawi core sites. P= Piston core; PG= Piston gravity core (trigger core associated to piston core); MC= Multi-cores; (OF): denotes cores that were opened and discarded in the field; (EX): denotes cores that were extruded in the field. Latitude and Longitude are in decimal degrees.

| Core | Latitude | Longitude | Water Depth (m) | Core Length (m) |
|---|---|---|---|---|
| M98-1P | 10°00.278' S | 34°17.233' E | 403 | 7.70 |
| M98 1PG | 10°00.278' S | 34°17.233' E | 404 | 0.50 |
| M98-2P | 09°58.664' S | 34°13.898' E | 363 | 9.01 |
| M98 2PG | 09°58.664' S | 34°13.898' E | 363 | 0.94 |
| M98-3P | 10°03.219' S | 34°13.625' E | 392 | 8.81 |
| M98-3PG (OF) | 10°03.219' S | 34°13.625' E | 392 | N/d |
| M98-4P | 09°57.974' S | 34°04.875' E | 261 | 8.30 |
| M98 4PG | 09°57.974' S | 34°04.875' E | 261 | 0.95 |
| M98-5P | 09°50.091' S | 34°07.163' E | 268 | 8.11 |
| M98 5PG | 09°50.091' S | 34°07.163' E | 268 | 0.96 |
| M98-6P | 10°03.201' S | 34°13.688' E | 394 | 9.04 |
| M98 6PG | 10°03.201' S | 34°13.688' E | 394 | 0.99 |
| M98-7MC-A (EX) | 10°03.312' S | 34°13.677' E | 392 | 0.41 |
| M98-7MC-B | 10°03.312' S | 34°13.677' E | 392 | 0.48 |
| M98-7MC-C | 10°03.312' S | 34°13.677' E | 392 | 0.44 |
| M98-8MC-A | 09°58.021' S | 34°04.817' E | 260 | 0.18 |
| M98-8MC-B | 09°58.021' S | 34°04.817' E | 260 | 0.15 |
| M98-9MC-A | 09°49.951' S | 34° 07.022' E | 270 | 0.48 |
| M98-9MC-B | 09°49.951' S | 34° 07.022' E | 270 | 0.50 |
| M98-9MC-C | 09°49.951' S | 34° 07.022' E | 270 | 0.49 |
| M98-9MC-D (EX) | 09°49.951' S | 34° 07.022' E | 270 | 0.50 |
| M98-10MC-A | 09°58.598' S | 34°13.913' E | 367 | 0.49 |
| M98-10MC-B | 09°58.598' S | 34°13.913' E | 367 | 0.52 |
| M98-10MC-C (EX) | 09°58.598' S | 34°13.913' E | 367 | 0.32 |
| M98-11MC-A | 10°00.246' S | 34°17.271' E | 404 | 0.52 |
| M98-11MC-B | 10°00.246' S | 34°17.271' E | 404 | 0.48 |
| M98-11MC-C (EX) | 10°00.246' S | 34°17.271' E | 404 | 0.33 |
| M98-12MC-A | 11°16.210' S | 34°26.116' E | 606 | 0.48 |
| M98-12MC-B | 11°16.210' S | 34°26.116' E | 606 | 0.48 |
| M98-12MC-C (EX) | 11°16.210' S | 34°26.116' E | 606 | 0.51 |
| M98-13P | 11°15.990' S | 34°26.108' E | 604 | 7.92 |

Thirty samples of approximately 1cc were also selected for AMS $^{14}C$ dating, which was carried out in two groups. The first set of 20 samples, including one blind duplicate, was submitted to the National Ocean Sciences Accelerator Mass Spectrometry Facility (NOSAMS), Woods Hole Oceanographic Institution, Massachusetts. The second group of 10 samples was sent to the Isotope Geochemistry Laboratory, Department of Geosciences, University of Arizona (Lab. code AA in Table 2). The first set of bulk samples was treated with 10% HCl to remove carbonates and 10% NaOH to remove humic material. The second group was treated in the same manner, but in addition, two duplicate but untreated samples were also submitted for dating (T= treated and U= untreated in Table 2). There were no significant differences in the results for treated and untreated samples from the same core interval. Samples of organic matter and wood charcoal (unidentified wood fragments) were retrieved and cleaned with a brush and de-ionised water under a microscope. Bulk sediment samples should have a mean total organic carbon (TOC) content of 3±1%, corresponding to the mean TOC calculated over all analysed bulk sediment Malawi samples. All dates are expressed with respect to a half-life of 5570 years, and are calibrated using Calib v.4.2 from the University of Washington Quaternary Isotope Lab (Stuiver and Reimer, 1993). The CALIB datasets are summarised in Stuiver *et al.* (1998a, 1998b).

The uppermost 20 cm from two extruded multi-cores, M98-10MC and M98-11MC, were sent to Dr. Paul Wilkinson of the Freshwater Institute, Winnipeg, Canada, for $^{210}Pb$ analysis by $\alpha$-counting spectrometry.

## 4. LITHOSTRATIGRAPHY

The piston cores all show a generally well defined sequence of sedimentary facies. We present here our results from piston cores M98-1P to 6P, except for core M98-5P that was disturbed during the coring process and will not be considered in this paper. Data from cores M98-13P and M98 12MC, from the central part of the lake, will be presented elsewhere.

Five major sedimentary units are defined based on detailed stratigraphic analysis of the cores (Figure 2), described from bottom to top:

The basal unit (Unit V) is composed of stiff dark brown clay that exhibits occasional zones with a lighter brown/tan colour. The colour changes generally have sharp basal contacts and grade upward from tan to dark brown, and are at some levels associated with small magnetic susceptibility peaks. The sediments of Unit V stand out from those above in lacking distinct lamination. A few vague laminae are visible at some levels and there are also traces of mottling, which may be a result of bioturbation. At approximately 650 cm there is a sharply defined marker bed that consists of ~1 cm of almost pure fine, sub-angular volcanic glass fragments. This tephra, designated A5, is found in four of the six piston cores.

*Table 2.* Radiocarbon results. Calibrated 14C ages using Calib 4.2 (Stuiver and Reimer 1993); calibration dataset as summarised in Stuiver *et al.* (1998a). Ages are corrected by 450 yr before calibration (see text). SCYS= Sample Calendar Year Span option in the Calib 4.2 program. Calibrated ages are rounded to the nearest 10 years; B= bulk sediment; C= Charcoal; O= plant/wood; r= blind replicate; T= pre-treatment; U= untreated

| Lab code | Sample ID | Notes | depth (cm) | $^{14}$C Age | 2sigma maximum (cal BP) | Cal age BP intercepts | 2sigma minimum (cal BP) | SCYS* (years) |
|---|---|---|---|---|---|---|---|---|
| 'AA32949' | M98 11 MC 4-5 | B, T | 4.50 | 475±55 | 260 | 0 | 0 | 10 |
| 'AA32950' | M98 11 MC 4-5 | B, U | 4.50 | 470±50 | 260 | 0 | 0 | 10 |
| 'AA32951' | M98 11 MC 46-47 | B, T | 46.50 | 820±65 | 530 | 460 | 290 | 10 |
| 'AA32952' | M98 11 MC 46-47 | B, U | 46.50 | 740±50 | 480 | 310 | 10 | 10 |
| 'AA32943' | M98 1P B 20-21 | B | 55.50 | 1880±55 | 1410 | 1310 | 1270 | 20 |
| 'AA32944' | M98 1P C 10-11 | B | 167.00 | 4015±60 | 4070 | 3840 | 3690 | 20 |
| 'AA32945' | M98 1P C 150-151 | B | 307.00 | 6270±65 | 6790 | 6660 | 6480 | 20 |
| 'AA32946' | M98 1P D 150-151 | B | 460.50 | 9010±85 | 9700 | 9540 | 9430 | 20 |
| 'AA32947' | M98 1P F 10-11 | B | 627.50 | 13330±100 | 15840 | 15510 | 14520 | 40 |
| 'AA32948' | M98 1P F 148-149 | B | 765.50 | 19010±270 | 23000 | 22200 | 21460 | 40 |
| '18271' | M98 2P A 139-140 | B | 139.50 | 2510±50 | 2150 | 2000 | 1890 | 20 |
| '18272' | M98 2P B 100-101 | B | 248.50 | 4020±65 | 4080 | 3860 | 3690 | 20 |
| '18566' | M98 2P C 80-81 | B | 379.50 | 6260±65 | 6750 | 6640 | 6450 | 20 |
| '18692' | M98 2P D 49-50 | B | 500.50 | 8820±110 | 9540 | 9460, 9430 | 9030 | 20 |
| '18565' | M98 2P D 86-87 | B | 537.50 | 9550±120 | 10560 | 10230 | 9920 | 20 |
| '18567' | M98 2P E 46-47 | B | 648.00 | 11450±85 | 13160 | 13000 | 12680 | 20 |
| '18561' | M98 2P E 142,5 | C | 744.00 | 14450±100 | 17790 | 17360 | 16950 | 20 |
| '18696' | M98 2P E 142-143 | B | 744.00 | 15150±130 | 18240 | 17840 | 17460 | 40 |
| '16698' | M98 2P E 142-143 | B, r | 744.00 | 15100±120 | 18170 | 17790 | 17410 | 40 |
| '18693' | M98 2P F 147-148 | B | 900.00 | 21000±240 | | out of range | | 40 |
| '18147' | M98 3P A 85,5 | O | 85.50 | 5950±50 | 6890 | 6750 | 6670 | 20 |
| '19504' | M98 3P B 104 | C | 221.00 | modern | | | | |
| '18562' | M98 3P C 94 | C | 366.00 | 6890±50 | 7800 | 7680 | 7620 | 20 |
| '18695' | M98 4P A 2-3 | B | 2.50 | 1500±90 | 1170 | 950 | 760 | 20 |
| '18699' | M98 4P B 47-48 | B | 134.50 | 2550±65 | 2310 | 2060 | 1900 | 20 |
| '18563' | M98 4P C 20 | O | 257.50 | 3650±40 | 4090 | 3970, 3940 | 3940 | 20 |
| '18697' | M98 4P C 20-21 | B | 257.50 | 4420±80 | 4790 | 4420 | 4160 | 20 |
| '18694' | M98 4PG 2-3 | B | 2.50 | 1140±95 | 790 | 660 | 520 | 10 |
| '18148' | M98 6P D 16 | O | 387.00 | 7550±60 | 8420 | 8370 | 8190 | 20 |
| '19495' | M98 6P E 2,5 | C | 517.50 | 9020±530 | 11640 | 10200 | 8780 | 20 |
| '18564' | M98 6P F 17,5 | O | 670.50 | 6690±50 | 7660 | 7570 | 7470 | 20 |

Unit IV is a short sequence (approximately 1m thick), consisting of alternating brown/green clay or silty clay homogenites and discontinuous, eroded packets of laminites. The contact between Units V and IV is gradational over several centimetres and in some cores is difficult to define. Laminite packets within this unit are generally 2-4 cm thick, with eroded tops, and several are deformed. Each couplet is composed of two sub-mm laminae, a light-coloured diatomaceous (primarily *Aulacoseira* and *Stephanodiscus cyclostephanus*) lamina and a thinner, dark clay layer. The homogenites are 5-10 cm thick, and more numerous at the bottom of the sequence.

Unit III is a finely laminated sequence of about 1-2 m consisting of up to 2400 sub-mm scale regular tan (diatomaceous) and brown (clay) laminae with few or no homogenites. A diffuse ash layer, designated A4, can be correlated between three of the six piston cores using magnetic susceptibility peaks. The top and bottom stratigraphic contacts for Unit III are abrupt and easy to identify.

Unit II consists of alternating homogenites and laminite packets for an interval of approximately 2-3 m. Each homogenite is a clay layer one to several cm thick resting on the eroded top of a laminite packet. The latter are generally a few cm thick and consist of horizontal mm-scale laminae. Many varve packets have an eroded top and a few show signs of soft-sediment deformation. A major ash layer, up to 30 cm thick, and with pumice clasts in some cores, occurs in Unit II in all of the northern basin piston cores. This prominent marker bed is labelled A3. Radiocarbon dates bracketing this major tephra constrain its age to approximately 4000 yr BP.

The contact between Units I and II is gradational over a few centimetres but is nevertheless fairly well defined. Unit I is approximately 1m thick and composed of well-defined, mm-scale couplets of light-coloured diatomaceous and dark-coloured clay laminae. These laminae have been demonstrated to correspond to a seasonal signal, indicating that the light-dark couplets are varves (Pilskaln and Johnson, 1991, see Section on Laminites below). Five to seven homogeneous layers of brown to olive-brown silty clay, approximately 1-3 cm thick (homogenites), are interspersed among the laminae and may be correlated between the cores. These layers have erosive basal contacts but no apparent internal structure. Grain-size analysis by laser particle counter reveals no significant size gradation between the top and bottom of a homogenite. Two 1-cm thick ash layers (A1 and A2) and one distinctively coloured lamina (C1) may be correlated between each core within Unit I. The multi-cores and trigger cores are composed entirely of Unit I sediments.

## 4.1 Stratigraphic Correlations

The depths of stratigraphic boundaries in the cores are listed in Table 3. The piston cores were correlated visually using Units I-V and distinctive marker beds, as well as by magnetic susceptibility peaks. Smear-slide examination reveals that the latter typically represent tephra deposits or zones of coarser grain size within the silty

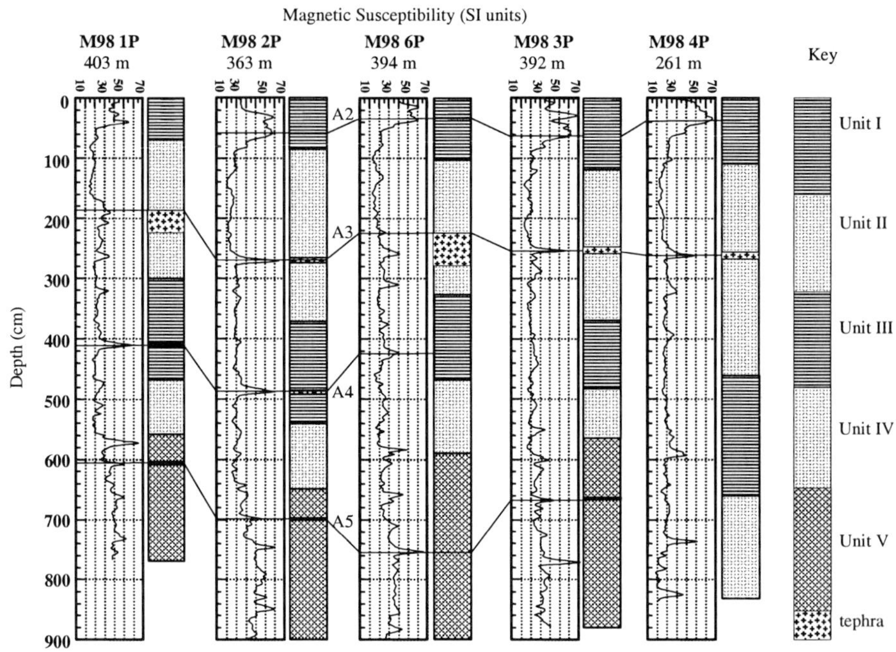


*Figure 2*. Stratigraphic logs and correlation by magnetic susceptibility peaks on 5 piston cores from the northern basin of Lake Malawi. Unit I :mm-scale annual laminations of gray-brown diatomaceous ooze and dark clay with organic matter with few homogenites. Unit II : fine grained brown homogenites regularly interbedded with discontinuous, basal-eroded lamination packets. Unit III: very fine diatomaceous/clay laminations with few homogenites. Unit IV: fine-grained brown homogenites interbedded with discontinuous, eroded lamination packets. Unit V: fine grained homogeneous brown clay. Scale is depth in core in cm; water depth in meters.

*Table 3*: Depths and estimated ages of unit boundaries and marker beds in piston cores. Depths are expressed in centimetres from the top of the piston core. Ages were calculated by mean linear interpolation between uncalibrated, uncorrected radiocarbon dates form cores M98-1P and 2P. Interpolated ages are rounded to the nearest 10 years. x= not present.

| Unit limits and Marker beds | M98-1P Depth (cm) | M98-2P Depth (cm) | M98-3P Depth (cm) | M98-4P Depth (cm) | M98-6P Depth (cm) | Interpolated mean age (yr. BP) | Estimated mean error (yr. BP) |
|---|---|---|---|---|---|---|---|
| Unit I-II | 68 | 121 | 116 | 109 | 100 | 2190 | ±65 |
| Unit II-III | 296 | 381 | 368.5 | 458.5 | 324 | 6190 | ±100 |
| Unit III-IV | 463 | 552 | 483 | 658 | 482 | 9440 | ±400 |
| Unit IV-V | 558 | 661 | 566 | x | 591 | 11620 | ±100 |
| A3 | 185.5 | 269 | 253 | 263.5 | 224 | 4350 | ±25 |
| A4 | 408 | 488 | x | x | 432 | 8350 | ±250 |
| A5 | 606 | 699 | 666.5 | x | 757 | 12950 | ±200 |

*Table 4.1.* Depth and varve number of marker beds in Unit I. "Varve number" represents the number of varves from the top of the core. "Corrected varve number" represents the total number of varves from the sediment water interface. See text for marker beds description.

| Cores | | Marker beds | | | | | |
|---|---|---|---|---|---|---|---|
| | | H1 | H2 | H3 | A1 | C1 | A2 |
| 10 MCA | depth (cm) | 5 | 18 | 35 | 40 | | |
| | varve number | 43 | 133 | 253 | 293 | | |
| 10 MCB | depth (cm) | 7.5 | 21.5 | 42 | 52 | | |
| | varve number | 41 | 131 | 252 | 290 | | |
| 11 MCA | depth (cm) | 8.5 | 23 | 40 | 47 | | |
| | varve number | 45 | 125 | 228 | 264 | | |
| 11MCB | depth (cm) | 9.5 | 22 | 38 | 47 | | |
| | varve number | 38 | 133 | 245 | 299 | | |
| 7MCB | depth (cm) | 8 | 22 | 42.5 | | | |
| | varve number | 61 | 144 | 267 | | | |
| 1PG | depth (cm) | | | | 9 | 37 | |
| | varve number | | | | 6 | 220 | |
| | corrected varve number | | | | 290 | 504 | |
| 1P | depth (cm) | | | | 15 | | |
| | varve number | | | | 23 | | |
| | corrected varve number | | | | 298 | | |
| 2PG | depth (cm) | | 7.5 | 32 | 38.5 | 70.5 | 88 |
| | varve number | | 29 | 171 | 211 | 441 | 578 |
| | corrected varve number | | 112 | 254 | 294 | 524 | 661 |
| 2P | depth (cm) | | | | 5 | 35.5 | 53 |
| | varve number | | | | 0 | 158 | 304 |
| | corrected varve number | | | | 294 | 524 | 670 |
| 3P | depth (cm) | | | | 12 | 36 | 56.5 |
| | varve number | | | | 42 | 242 | 364 |
| | corrected varve number | | | | | | |
| 4PG | depth (cm) | | 21.5 | 53 | 70 | | |
| | varve number | | 122 | 252 | 314 | | |
| | corrected varve number | | | | 314 | | |
| 4P | depth (cm) | | | | | | 25.5 |
| | varve number | | | | | | 118 |
| 6PG | depth (cm) | | | | 36 | 62,5 | 79 |
| | varve number | | | | 167 | 379 | 480 |
| 6P | depth (cm) | | | | 1.5 | 14.5 | 30 |
| | varve number | | | | 0 | 103 | 225 |
| | corrected varve number | | | | | 379 | 501 |

clays. The piston cores were correlated by means of four tephras (labelled A2, A3, A4, A5), as well as by other tephra or distinct, diatom-rich beds (Figure 2, Table 3).

Correlation of the marker beds in Unit I is unambiguous, as distinctively coloured laminae may be found at the same stratigraphic position in all of the cores. Similarly, the volcanic ash A3 in Unit II is a reliable marker bed found in every core. However, the thickness of A3, as well as the amount it disturbed the underlying sediment, varies between cores. Fewer correlations can be made below A3, which occurs at approximately 200-cm depth. An exception is the A5 marker bed, which is composed of clean, well-sorted sub-angular volcanic glass, is associated with a high magnetic susceptibility peak, and provides an unequivocal tie between the cores. In one case (core M98-3P) comparison of geochemical trends (Filippi and Talbot, in prep.) helped to correlate and define unit boundaries.

The multi-core, trigger core and top of the piston core from each location were also correlated visually by stratigraphic marker beds within Unit I (Table 4.1 and 4.2). These include an ash layer (designated A1), a distinctively coloured diatomaceous lamina (C1), and a thin ash layer associated with a magnetic susceptibility peak (A2).

*Table 4.2*. Estimate age of marker beds in unit I.

| | MARKER BEDS | | | | | |
|---|---|---|---|---|---|---|
| | H1 | H2 | H3 | A1 | C1 | A2 |
| estimated age | 45 | 131 | 261 | 323 | 552 | 698 |
| varve year AD | 1952 | 1866 | 1736 | 1674 | 1445 | 1299 |
| estimated error | ±9 | ±8 | ±13 | ±13 | ±61 | ±78 |

The multi-cores contain three homogenites (H1, H2, and H3) at similar varve intervals that can also be correlated. The trigger cores typically contain two of the three homogenites, A1 and at least one of the other two marker beds. The piston cores over-penetrated the lake floor but can be correlated to the trigger cores by the two bottom marker beds (C1 and A2). The A1 ash layer was often disturbed during the coring process, as it typically occurs at the depth where the multi-cores end and the trigger cores and piston cores start.

# 5. CHRONOLOGICAL FRAMEWORK

## 5.1 Varve Counts

Each varve within Unit I has been assigned an age based on $^{210}Pb$ dating and our best estimate of varve years from the sediment-water interface. While the small homogenites, 1-2 cm thick, found within the laminated core units prevent an uninterrupted varve count, they can be correlated between multi-cores and top sections of the piston cores, based on their stratigraphic location and the number of varves between each one. Varve couplet counts agree well with sediment age based on $^{210}Pb$, with an uncertainty of ±8 yr. By comparing numbers of varves between marker beds in sediments lower than the depth of unsupported $^{210}Pb$ activity, we conclude that the erosional events associated with the homogenites removed ca. 20 years of sedimentary record from Unit I.

## 5.2 $^{210}$Pb Dating

$^{210}Pb$ analysis was used to date the core-top sediments in short cores 10MC and 11MC. Insufficient sample material (20 cm) was extruded from M98-10MC to extend the record beyond the excess $^{210}Pb$ to the depth of unsupported activity. However, since it was within a few cm of the end of the unsupported $^{210}Pb$ and the decay curve was linear, the excess $^{210}Pb$ integral was simulated before running the model (P. Wilkinson, personal communication).

The data obtained with the $^{210}Pb$ method and the varve counts from the same core proves to be reassuringly consistent (Figure 3), especially with the constant rate of supply (CRS) method. The CRS method assumes that flux of $^{210}Pb$ to the sediment surface is constant through time. It differs from the linear or C.I.C. model, which assumes that the $^{210}Pb$ activity in the uppermost layer of the sediment is constant. In the CRS method, if the accumulation rate of bulk sediment changes, then sedimentary $^{210}Pb$ will vary (Appleby and Oldfield, 1983, Binford, 1990). Although it is doubtful that effects of anthropogenic sediment loading can be seen in the off-shore sediments of Lake Malawi, the several small event deposits (homogenites) may have influenced the lead flux, making the CRS method the more appropriate model.

## 5.3 Radiocarbon Correction

Comparison of the results from varve counting and the corresponding calibrated $^{14}C$ dates reveals significant differences, varying from ca 400 to 1500 yr, in the calculated age of the samples. $^{14}C$ dates are consistently older than the age estimated from varve counting. This difference is larger in the piston core samples (ca. 880 and 1500 yr.) than in the multi-cores (ca. 420-540 yr.), and it is variable in time and space (Table 5). Where charcoal or plant/wood organic matter have been dated from the same horizon as a bulk sediment sample, the bulk sediment is between 650 and 770 years older than the organic material. Samples of lacustrine bulk organic matter yield less reliable $^{14}C$ dates than those from terrestrial material (Wohlfarth, 1996, Oldfield *et al.*, 1997). Smear slide analysis indicates that the organic matter is a mixture of terrestrial and aquatic plant remains.

These observations lead to the conclusion that the bulk sediment $^{14}C$ dates are affected more by the presence of a variable amounts of reworked organic matter within the dated samples than by a reservoir effect *sensu strictu*. In this environment, an undoubtedly variable amount of old, reworked particulate carbon is mixed with the new carbon fixed from dissolved atmospheric carbon, which increases the apparent age of the sediment. In the piston cores, varve counting may underestimate the age due to erosion of varves by small turbidites or due to lack of varve sedimentation. Moreover, the varved sequence is discontinuous in the lower part of the cores. In the youngest sediments (those sampled by the multi-cores) the varve counting is very reliable, as confirmed by the $^{210}$Pb dating. We therefore assume that our $^{14}$C dates are affected by a contribution from reworked organic matter, giving them an apparent age that is older by ca. 450 yr. Although we know this contribution may be variable in time and space, our current data set is not sufficient to model this variability. For now we thus assume the effect as constant. Calibrated ages for $^{14}$C age determinations corrected for this 450 yr reworked organic matter effect are listed in Table 2, together with the uncalibrated ages and constraints used for calibration.

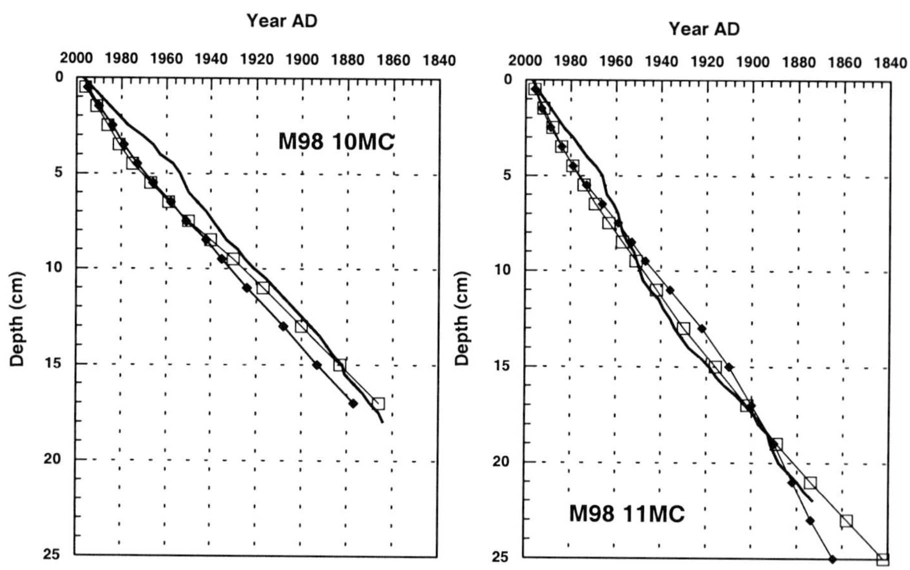


*Figure 3.* Varve counts from M98 10MCA and 11 MCA compared with the CRS (solid squares) and linear method (open squares) for $^{210}$Pb interpretation. Solid line represents varve count at that depth.

*Table 5*: Calculated difference in age between calibrated radiocarbon dates and varve dates. T: Treated bulk sediment samples; U: Untreated bulk sediment samples. * M98 4PG varve date is underestimated due to core top disturbance.

| Core ID | Depth cm | Reported yBP | Error | Cal yr before 1998 | Varve yr | Difference in years |
|---|---|---|---|---|---|---|
| M98-11MCA 4-5T | 4.5 | 475 | 55 | 561 | 21 | 540 |
| M98-11MCA 4-5U | 4.5 | 470 | 50 | 560 | 21 | 539 |
| M98-11MCA 46-47T | 46.5 | 820 | 65 | 776 | 293 | 483 |
| M98-11MCA 46-47U | 46.5 | 740 | 50 | 717 | 293 | 424 |
| M98-1PB20-21 U | 55.5 | 1880 | 55 | 1869 | 578 | 1291 |
| M98-2PA 139-140 T | 139.5 | 2510 | 50 | 2667 | 1275 | 1392 |
| M98 4PG 2-3 T* | 2.5 | 1140 | 95 | 1072 | 2 | 1070 |
| M98 4P-A 2-3 T | 2.5 | 1500 | 90 | 1405 | 552 | 853 |

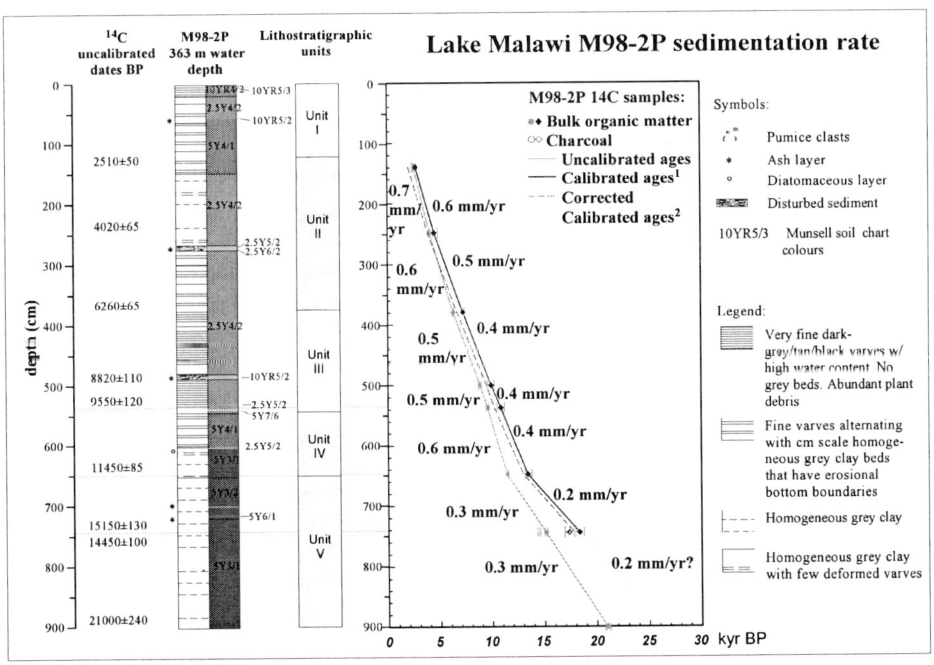


*Figure 4*. Sedimentation rate for core M98-2P with uncalibrated, calibrated, and calibrated corrected $^{14}C$ ages. (1) Radiocarbon dates were calibrated using Calib v. 4.2 (Stuiver and Reimer, 1993, calibration data set Stuiver, *et al.*, 1998a). A sample calendar year span option = 20 was used for all samples except for bulk samples older than 15000, where the option was set = 40. (2) Ages are corrected by 450 years before calibration (see text for explanation).

## 5.4 Sedimentation Rate

Sedimentation rates based on varve counts vary from ~ 0.15 cm/yr in Unit I to ~0.07 cm/yr in Unit III. While the $^{210}$Pb-sedimentation rate is in good agreement with our varve analysis for Unit I (Figure 3), there is a discrepancy between the calculated calibrated $^{14}$C-sedimentation rate and the sedimentation rate calculated from varve counts between the same horizons in Unit III.

Using the uncalibrated radiocarbon data from cores M98-1P and 2P (6 and 9 dates respectively), it is possible to calculate a sedimentation rate for each core. Ash layer A1, with a varve-determined age of 1675 AD ±20 (275 BP ±20) was used to calculate the sedimentation rate in the upper part of the cores. If calibrated $^{14}$C dates are used, the sedimentation rates are slightly lower, as is apparent in M98-2P (Figure 4).

# 6. DISCUSSION

The cores discussed here were all taken from the long, SE-oriented slope that links the flexural western margin of the North Basin to the deep basin plain, located close to the Livingstone Mountain border fault (Figure 1C). As indicated earlier, core sites were selected to minimise the influence of turbidity currents and downslope mass movement. In this we seem to have been largely successful, despite the slope setting. No major turbidites interrupt the cored sequences and the generally good between-core correlation suggests that their stratigraphies record a basin-wide sedimentary history. Several large rivers enter the northwest corner of Lake Malawi (Figure 1B) and have formed major deltas which are the principal source of terrigenous sediment to the North Basin. From these deltas, sediment is transferred downslope, and ultimately to the deep basin, by a variety of sedimentation processes (Johnson and Ng'ang'a, 1990, Scholz, 1995, Wells *et al.*, 1999). Several major, deep-water distributary channel systems have been identified on seismic profiles, many of which seem to be structurally controlled (Johnson and Ng'ang'a, 1990, Scholz, 1995, De Batist *et al.*, 1996). Lateral movement of such channels seems to be limited and most of the turbidity flows they carry are probably confined to the channel and immediate overbank areas. The platform and mid-slope sites where our cores come from are thus probably out of reach of the normal, channelised flows; sedimentation in these settings is dominated by pelagic and hemipelagic processes. We recognise three principal facies, laminites, homogenites, and unlaminated clays within the Late Pleistocene - Holocene deposits found on offshore slope and platform areas in the North Basin of Lake Malawi.

## 6.1 Laminites (Varves)

These are a major component of all but Unit V and thus an important Late Pleistocene - Holocene facies in the North Basin. Their characteristic occurrence is as couplets of alternating light, diatomaceous, and dark, clastic laminae containing relatively abundant terrestrial plant debris, as recognised by means of smear slides analysis. Insights into the probable mode of formation of these couplets is provided by suspended sediment studies in modern Lake Malawi. Pilskaln and Johnson (1991) used a sediment trap to monitor settling particles in the deep offshore waters of the Northern Basin southeast of the 1998 coring sites. They found that diatoms dominate the sediment flux at the time of peak productivity, during the dry, windy season (April through November), while terrigenous clastic particles and degraded plant material characterise wet-season deposition. The flux of material that forms the dark laminae was, however, lower than would have been predicted from their rate of accumulation on the underlying lake floor (Pilskaln and Johnson, 1991, Halfman and Scholz, 1993). Although data from a single site should be interpreted with caution, it is possible that the additional silt- and clay-grade clastic sediment on the lake floor is supplied by dilute density inter- or underflows. Halfman and Scholz (1993) recorded a rainy season turbidity maximum between depths of 30-65 m up to 10 km off the mouth of the Ruhuhu River, which drains the eastern shore of Lake Malawi. Tropical lakes are generally favourable waterbodies for the generation of such flows, and by spreading along density interfaces these may distribute sediment over large areas (Talbot and Allen, 1996). It is thus probable that density flows supply significant quantities of fine-grained clastic sediment to the offshore areas of the lake, and the sediment trap may have been moored too far above the lake bottom to sample underflows.

Although the contrast between dry and wet-season sedimentation seems to be responsible for the formation of the couplets, we are currently uncertain as to what combination of factors has led to the preservation of this seasonal signature in some parts of the sedimentary record, but not others. One possibility is, of course, that well-marked seasonality has not been a constant feature of the regional climate, but at present we have no means of assessing the nature of any putative annual climatic cycle in the non-varved sections of the cores. What does seem clear is that varve formation is not directly dependent upon lake level. Of the two most consistently varved sections, Units I and III, the former accumulated during the last ca. 2000 years, when the lake has been generally high and periodically overflowing. During Unit III time, on the other hand, there is evidence that the lake was perhaps as much as 150 m lower than its present level (Finney and Johnson, 1991).

## 6.2 Homogenites

These silty layers (0.5 - 4 cm thick), which typically interrupt and erode the laminites, have been a characteristic, although episodic feature of offshore sedimentation in the North Basin throughout the terminal Pleistocene and Holocene. Three, not necessarily mutually exclusive processes could have been responsible for producing these deposits.

1) Exceptional sediment-laden inter- or underflows similar to those responsible for the clastic laminae in the varved sections. Such an origin seems unlikely. The erosive basal contact to most of the homogenites eliminates suspension fallout from interflows as the dominant depositional process. Although centimetres-thick silty underflow laminae are a relatively common feature of some pro-delta lacustrine environments (e.g. Born, 1972, Ashley, 1975, Talbot and Allen, 1996), the sites sampled by our cores are probably too far from the lake margin to be directly influenced by deltas.
2) Overflow of exceptionally thick, channelised turbidity flows. Well-developed levées along some of the major sublacustrine channels in the North Basin (Johnson and Ng'ang'a, 1990, Scholz *et al.*, 1993, Scholz, 1995) indicate that large flows frequently overtop the channel margins. However, most core sites are too far from, and too high above any of these channels for deposition from overbank flows to be a feasible explanation for the homogenites.
3) Local, non-channelised flows originating upslope. Resuspension of muddy sediments and their subsequent transport as turbidity flows of local origin seems, on present evidence, to be the most likely explanation for the homogenites. Beds of this type are particularly prominent in Unit IV. The emplacement of numerous homogenites is presumably the reason for the high sedimentation rates. Their coincidence with a period of falling lake level suggests that the high terrigenous sediment yields and increased incidence of turbidity flows were probably due to reworking of newly exposed, unconsolidated muddy sediments upslope from the core sites. Although some homogenites may be related to lake-level fall, the fact that others (e.g. in Unit I) can be correlated across several core sites suggests that mobilisation of the sediments that produced them may have been due to basin-wide events, even if they were produced by local erosion and slope-failure processes. The most likely mechanism for triggering geographically extensive flows is seismic shock, raising the intriguing possibility that some of the homogenites are in fact seismites and as such preserve a record of the frequency of large earthquakes in and around northern Lake Malawi.

## 6.3 Unlaminated Clays (Unit V)

Unit V differs markedly from the overlying deposits. Varves and distinct turbidites are absent, and the generally homogeneous clays lack lamination, displaying only subtle colour changes. At present we have no completely convincing explanation for these deposits. The colour differences may be due to the presence of variable amounts of pyroclastic debris, but we have so far been unable to confirm this hypothesis. In view of their generally homogeneous appearance, it is possible that the silty clays represent a series of dilute turbidites, similar to those that produced the homogenites, but if so, the basal erosive contact with underlying sediments, which characterises many of the homogenites, is rare in Unit V. In addition, the sedimentation rate of the latter is generally lower than the overlying sediments

(Figure 2), which does not suggest that incursions of turbidity flows were so frequent that they swamped the pelagic and hemipelagic sediment accumulation.

We have noted possible evidence of bioturbation at some horizons in Unit V. Intense burrowing certainly provides an alternative explanation for the homogeneous nature of the sediments and is also compatible with low sedimentation rates. Extensive bioturbation would imply relatively prolonged periods of bottom-water oxygenation. Complete mixing of the water column because of lower lake levels is one way in which such a situation could be achieved. Possible evidence of lowstand conditions during some of the period of accumulation of Unit V also comes from studies of the diatom assemblages in Core 2P (Gasse *et al.*, this volume). In addition, both diatom and geochemical evidence (Filippi and Talbot, unpublished) suggest a major rise in lake level between 17 and 14 ka BP. Organic matter preservation, on the other hand, is generally good throughout Unit V, as demonstrated by smear slides analysis, and higher than normal TOC content and Hydrogen Index values (Filippi and Talbot, unpublished), and is not consistent with permanently well-oxygenated bottom waters. However, it is possible that oxic conditions and colonisation of the sediments by burrowing organisms was intermittent, occurring only during periods of intense overturn.

Evidence of a lowstand in Lake Malawi at the time of the LGM are at variance with previous interpretations of the state of the lake during this period, which suggest relatively highstand conditions (Finney *et al.*, 1996). They are, however, in accord with recent reinterpretations of the palaeoclimate of northern South Africa where arid conditions seem to have characterised the LGM (Partridge, 1997).

## 7. CONCLUSIONS

Sedimentation in offshore areas of Lake Malawi is clearly complex, reflecting the interplay of climate, a variety of sedimentary processes, and tectonics in a large dynamic lake. Nevertheless, the suite of cores collected during the 1998 cruise in the North Basin demonstrates that there is a consistent, basin-wide stratigraphy in areas remote from the effects of large-scale turbidity current and mass-flow influence (cf., for example, Soreghan *et al.*, 1999, Wells, *et al.*, 1999). We thus feel reasonably confident that the cores provide an excellent framework upon which to build a detailed understanding of the palaeoenvironmental history of northern Lake Malawi over the last ca. 25 ka years (e.g. Gasse, *et al.*, this volume). Other features to note from our preliminary study of the cores are:

1. Although a significant portion of the last ca. 25 ka years of the offshore sedimentary record in the North Basin is varved, it seems unlikely that there are any continuously varved successions in this part of the lake. The varved sections have great potential to provide very high-resolution chronologies (albeit a floating chronology for Unit III) and palaeoclimatic records. The existence of a probable ENSO signal has already been demonstrated for Unit I (Rosenmeier 1998) and

spectral analysis of lamina thickness in Unit III may reveal the degree to which ENSO forcing influenced the climate of southern Africa during the early Holocene. The most recent varved section (<100 yr) provides a temporal uncertainty of ±8 yr based on the comparison of varve counts with $^{210}Pb$ activity; this value lower to ~20 yr in the older Unit I sediments where several homogenites are present.

2. Beds of clay or silty clay (homogenites), typically one to several cm thick, interrupt the varve packets. These are event beds, formed from muddy turbidity flows. Some homogenites can be correlated between several cores, across varied lake floor topography, suggesting that the flows that formed these were probably triggered by seismic shocks. Other homogenites, particularly those in the section spanning the Pleistocene-Holocene boundary, seem to have been formed at times of falling lake level and may thus result from the reworking of newly exposed, unconsolidated sediment.

3. A number of ash horizons, the largest up to 30 cm thick, have been identified. These provide valuable marker beds within the cored sediments.

4. The oldest sediments encountered (Unit V) differ from those above in lacking any clear laminations. Their mode of deposition is at present uncertain. Independent geochemical evidence (Filippi and Talbot, unpublished) and diatom evidence (Gasse *et al.* this volume) suggest low lake levels for at least some of the period of Unit V accumulation, so the homogeneous texture could be due to bioturbation during times of increased oxygen supply to the lake bottom. Moreover, the low sedimentation rate could be compatible with a drier climate and a reduced detrital input by rivers. Such an environment is, however, at variance with geochemical evidence (TOC content and Hydrogen Index values, Filippi and Talbot, unpublished) of favourable conditions for organic matter preservation.

5. Dating the cores has been problematic. Radiocarbon dates yield bulk sediment ages that are 500 - 1000 years too old, primarily due to downslope reworking and the influx of allochthonous organic matter to the lake. As volcanic ash layers are common in the lake, Ar/Ar dating may eventually yield a reliable tephrochronology, which can then be used together with the lithological correlations to construct a more reliable chronostratigraphic framework for the North Basin of Lake Malawi.

## ACKNOWLEDGEMENTS

We acknowledge the support of the Lake Malawi Biodiversity Conservation Project and extend our thanks to Drs. Tony Ribbink, Director, and Harvey Bootsma, Chief Scientist, for their assistance during the 1998 field season. Funding for this research was obtained through grants from the National Science Foundation, the Norwegian Research Council and the European Union's Marie Curie Research Training Program (TMR contract n° ERBFMBICT 972029). Captain Mark Day and the crew of the R/V *Usipa* gave invaluable field assistance. Yvonne Chan and Erick Hallie provided lab assistance.

# REFERENCES

Appleby, P. G. and Oldfield, F. (1983) Assessment of $^{210}Pb$ from sites with varying sediment accumulation rates, *Hydrobiologia* **103**, 29-35.

Ashley, G. M. (1975) Rhythmic sedimentation in glacial Lake Hitchcock, Massachusetts, Connecticut. *SEPM Special Publication* **23**, 304-320.

Barry, S. L. (2001) Stratigraphic Correlation and Geochronology of Varved Sediments from Lake Malawi, East Africa. Masters Thesis, University of Minnesota, 135 pp.

Beadle, L. C. (1981) *The Inland Waters of Tropical Africa. An Introduction to Tropical Limnology.* Longman, London, 468 pp.

Binford, M. W. (1990) Calculation and uncertainty analysis of 210 Pb dates for PIRLA project lake sediment cores. *Journal of Paleolimnology* **3**, 253-267.

Birkett, C., Murtugudde, R. and Allan, T. (1999) Indian Ocean climate event brings floods to East Africa's lakes and the Sudd Marsh. *Geophysical Research Letters* **26**, 1031-1034.

Born, S. M. (1972) *Late Quaternary history, deltaic sedimentation and mudlump formation at Pyramid Lake, Nevada.* Center for Water Resources Research, Desert Research Institute, University of Nevada, 1-97, Reno.

Crossley, R. (1984) Controls of sedimentation in the Malawi Rift Valley, Central Africa, in P. F. Burollet and A. C. Grant (eds.), *Basin Analysis: Principles and Applications*, Sedimentary Geology, 40, pp. 33-50.

De Batist, M., Van Rensbergen, P., Back, S. and Klerkx, J. (1996) Structural Framework, Sequence Stratigraphy and Lake Level Variations in the Livingstone Basin (Northern Lake Malawi): First Results of a High-Resolution Reflection Seismic Study. In T. C. Johnson and E. O. Odada (eds.), *The Limnology, Climatology and Paleoclimatology of the East African Lakes*, Gordon and Breach, Toronto, pp. 509-521.

Eccles, D. H. (1974) An outline of the physical limnology of Lake Malawi (Lake Nyasa). *Limnology and Oceanography* **19**,(5), 730-742.

Finney, B. P. and Johnson, T. C. (1991) Sedimentation in Lake Malawi (East Africa) during the past 10,000 years: a continuous paleoclimatic record from the southern tropic. *Palaeogeography, Palaeoclimatology, Palaeoecology* **85**, 351-366.

Finney, B. P., Scholz, C. A., Johnson, T. C. and Trumbore, S. (1996) Late Quaternary Lake-level Changes of Lake Malawi, in T. C. Johnson and E. O. Odada (eds.), *The Limnology, Climatology and Paleoclimatology of the East African Lakes*, Gordon and Breach, Toronto, pp. 495-508.

Gasse, F., Barker, P. and Johnson, T. (in press) A 24,600 yr diatom record from the Northern Basin of Lake Malawi, *This volume*.

Halfman, J. D. and Scholz, C. A. (1993) Suspended Sediments in Lake Malawi, Africa: A Reconnaissance Study. *Journal of Great Lakes Research* **19**(3), 499-511.

Harkin, D. A. (1982) The Rungwe volcanics at the northern end of Lake Nyasa. In A. M. Quennell (ed), *Rift Valleys Afro-Arabian*, Hutchinson Ross Publishing Co., Stroudsburg, Pennsylvania.

Johnson, T. C. (1996) Sedimentary Processes and Signals of Past Climatic Change in the Large Lakes of the East African Rift Valley, in T. C. Johnson and E. O. Odada (eds.), *The Limnology, Climatology and Paleoclimatology of the East African Lakes*, Gordon and Breach, Toronto, pp. 367-412.

Johnson, T. C., Wells, J. D. and Scholz, C. A. (1995). Deltaic sedimentation in a modern rift lake. *GSA Bull*, **107**, 812-829.

Johnson, T. C. and Ng'ang'a, P. (1990) Reflections on a Rift Lake. In B. J. Katz (ed), *Lacustrine basin exploration: case studies and modern analogs*, Memoir 50, American Association of Petroleum Geologists, Tulsa, pp. 113-135.

Kelts, K., Briegel, U., Ghilardi, K. and Hsü, K. (1986) The limnogeology-ETH coring system. *Schweizerische Zeitschrift für Hydrologie* **48**(1), 105-115.

Munsell (1954) *Soil Color Charts*. Machbeth, Kollmorgen Ins.Corp, Newburgh.

Nicholson, S. E. (1998) Fluctuations of Rift Valley Lakes Malawi and Chilwa during historical times: a synthesis of geological, archeological and historical information, in J. T. Lehman (ed.), *Environmental Change and Response in East African Lakes*, Monographiae Biologicae, 79, Kluwer Academic, Dordrecht, pp. 207-231.

Oldfield, F., Crooks, P. R. J., Harkness, D. D. and Petterson, G. (1997) AMS radiocarbon dating of organic fractions from varved lake sediments; an empirical test of reliability. *Journal of Paleolimnology* **18**(1), 87-91.

Partridge, T. C. (1997) Cenozoic environmental change in southern Africa, with special emphasis on the last 200000 years. *Progress in Physical Geography* **21**, 3-22.

Pilskaln, C. H. and Johnson, T. J. (1991) Seasonal signals in Lake Malawi sediments. *Limnology and Oceanography* **36**(3), 544-557.

Ricketts, R. D. and Johnson, T. (1996) Early Holocene Changes in Lake Level and Productivity in Lake Malawi as Interpreted from Oxygen and Carbon Isotopic Measurements of Authigenic Carbonates, in T. C. Johnson and E. O. Odada (eds.), *The Limnology, Climatology and Paleoclimatology of the East African Lakes*, Gordon and Breach, Toronto, pp. 475-493.

Rosendahl, B. R. (1987) Architecture of continental rifts with special reference to East Africa. *Ann. rev. Earth Planet. Sci.* **15**(445-503), 235-256.

Rosendahl, B. R., Kilembe, E. and Kaczmarick, K. (1992) Comparison of the Tanganyika, Malawi, Rukwa and Turkana rift zones from analyses of seismic reflection data. *Tectonophysics* **213**,(1-2), 235-256.

Rosenmeier, M. A. (1998) Annual-to Centennial-scale Climate Variability in the Tropics: An Introduction to the Limnologic and Climatic Setting of Lake Malawi, East Africa and the Significance of Late Holocene Environmental Change. Unpublished Master Thesis, University of Minnesota, Duluth.

Scholz, C. A. and Rosendahl, B. R. (1988) Low lake stands in Lakes Malawi and Tanganyika, East Africa, delineated with multifold seismic data. *Science* **240**, 1645-1648.

Scholz, C. A. (1995) Seismic stratigraphy of an accomodation-zone margin rift-lake delta, Lake Malawi, Africa. In J. J. Lambiase (ed), *Hydrocarbon Habitat in Rift Basins*, Geological Society Special Publication, 80, London, pp. 183-195.

Scholz, C. A., Johnson, T. C. and McGill, J. W. (1993) Deltaic sedimentation in a rift valley lake: New seismic reflection data from Lake Malawi (Nyasa), East Africa. *Geology* **21**, 395-398.

Soreghan, M. J., Scholz, C. A. and Wells, J. T. (1999) Coarse-grained, deep-water sedimentation along a border fault margin of Lake Malawi, Africa: Seismic stratigraphic analysis. *Journal of Sedimentary Research* **69**(4), 832-846.

Stuiver, M. and Reimer, P. J. (1993) Extended C-14 data-base and revised Calib 3.0 C-14 age calibration program. *Radiocarbon* **35**(1), 215-230.

Stuiver, M., Reimer, P. J., Bard, E., Beck, J. W., Burr, G. S., Hughen, K. A., Kromer, B., McCormac, G., VanderPlicht, J. and Spurk, M. (1998a) INTCAL98 radiocarbon age calibration, 24,000-0 cal BP. *Radiocarbon* **40**(3), 1041-1083.

Stuiver, M., Reimer, P. J. and Braziunas, T. F. (1998b) High-precision radiocarbon age calibration for terrestrial and marine samples. *Radiocarbon* **40**, 1127-1151.

Talbot, M. R. and Allen, P. A. (1996) Lakes, in H. G. Reading (ed), *Sedimentary environments : processes, facies and stratigraphy.* Blackwell Science, Oxford, pp. 83-124.

Vollmer, M. K., Weiss, R. F. and Bootsma, H. A. (in press) Ventilation of Lake Malawi/Nyasa. *This volume.*

Wells, J. T., Scholz, C. A. and Soreghan, M. J. (1999) Processes of sedimentation on a lacustrine border-fault margin: Interpretation of cores from Lake Malawi, East Africa. *Journal of Sedimentary Research* **69**(4), 816-831.

Wohlfarth, B. (1996) The chronology of the last termination: a review of radiocarbon-dated, high-resolution terrestrial stratigraphies. *Quaternary Science Reviews* **15**, 267-284.

# A 24,000 yr DIATOM RECORD FROM THE NORTHERN BASIN OF LAKE MALAWI

FRANÇOISE GASSE[1], PHILIP BARKER[2], THOMAS C. JOHNSON[3]

[1]*CEREGE, UMR 6635, CNRS-Université d'Aix-Marseille III, BP. 80, 13545 Aix-en-Provence, Cedex 04, France. gasse@cerege.fr*

[2]*Hysed, Department of Geography, Institute of Environmental and Natural Sciences, Lancaster University, Lancaster LA1 4YB, UK. p.barker@lancs.ac.uk*

[3]*Large Lakes Observatory, University of Minnesota - Duluth, 10 University Drive, Duluth, MN, 55812, USA. tcj@d.umn.edu*

**ABSTRACT**

Diatom analysis of core M98-2P from the northern basin of Lake Malawi shows changes in lake level and mixing during the last 24,000 calendar years. Fluctuations between the major planktonic diatoms (*Aulacoseira nyassensis, Stephanodiscus* spp. and *Cyclostephanos* spp.) are interpreted as reflecting different degrees of nutrient upwelling whereas the percentages, concentrations and influxes of periphytic diatoms are primarily related to variations in water level. Total diatom biovolume is used to give a first-order approximation of relative diatom productivity at the core site. The diatom results suggest periods of low lake level at the LGM and for much of the Late Holocene period. Short-lived regressions are also suggested during the YD, at 10,600 and 8500-8200 years BP. Conversely the lake was high and probably overflowing 15,700-13,000 and 7500-6600 years BP. A low lake level during the LGM contradicts the interpretation of this period made by previous studies from Lake Malawi. However, intermediate lake levels in the Early Holocene support earlier work that suggested that Lake Malawi was lower than northern East African lakes at this time.



*E.O. Odada and D.O. Olago (eds.),*
*The East African Great Lakes: Limnology, Palaeolimnology and Biodiversity,* 393–414.

## 1. INTRODUCTION

Lake Malawi (Figure 1) extends over 5° of latitude in southern East Africa (10-15 °S). It has a strongly seasonal tropical climate regime with rain occurring during November to April and a dry season characterised by southerly winds from May to October. It is Africa's second largest lake with a length of 650 km, a width of 40 km and a depth of about 700 m. Lake Malawi is divided into three principal regions, the northern basin is most clearly delimited and has the smallest surface area of these three and a depth >500 m, the central sector is deepest (>700 m) and is connected to the southern basin which has an extensive area but is also the shallowest region of the lake. Malawi has an epilimnetic salinity intermediate between Victoria and the other rift valley lakes (0.211 ‰ ; Wüest *et al.*, 1996). Lake Malawi waters are richer in Ca (0.90 meq. $l^{-1}$) than the more northern lakes, Tanganyika, Kivu, Edward and Albert (< 0.62 meq. $l^{-1}$; Beadle, 1974; Hecky and Kling, 1987). It is also subject to larger seasonal temperature fluctuations due to its southern location and has lower mean annual productivity (700 mg C $m^{-2}$ $day^{-1}$) than in the other large lakes of East Africa (Hecky and Kling, 1987). The water balance is dominated by direct precipitation and evaporation from the lake surface since fluvial inputs account for only 32% of total input and outflow through the Shire river (Figure 1) is only 20% of the water loss (Spigel and Coulter, 1996). The lake is meromictic with a permanent thermocline at 200-250 m in the northern basin. Strong vertical mixing occurs during the dry, windy season generating upwelling which brings nutrients into the epilimnion. This seasonal mixing regime is also reflected in strong variations in primary productivity (Hecky and Kling, 1987).

Lake Malawi's sediments have been the subject of palaeolimnological and palaeoclimatological study for the last two decades. Numerous long cores (average length of 10m) were taken in 1986 from the different basins by a Duke University expedition and provide an outline stratigraphy of the Late Pleistocene and Holocene sediments (Finney and Johnson, 1991; Finney *et al.*, 1996; Ricketts and Johnson, 1996). The results of these studies suggested that the lake was 200-300m lower than present 46,000-32,700 calendar years before present (hereafter years BP) and 11,500-6800 years BP. Palynological results from the same cores and using the same chronology also suggested that the catchment area was relatively wet during the Last Glacial Maximum (LGM) (DeBusk, 1998). These results differ from those from lakes to the north, eg., Lake Tanganyika (Gasse *et al.*, 1989) and Lake Rukwa (Barker *et al.*, in press; Haberyan, 1987) as well as to the south, eg., Pretoria Salt Pan (Partridge *et al.*, 1997) and Madagascar (Gasse and van Campo, 1998), and pollen studies from Namibia and Angola (Ning Shi *et al.*, 1998). Clearly the ambiguous position of the sedimentary records from Lake Malawi must be explained.

Here we present the results of diatom analysis from a well-dated and stratigraphically-complete core taken from the northern basin in 1998 (Figure 1). We focus on the diatom ecology and palaeoecology together with its interpretation in palaeohydrological terms. This diatom study is part of a multi-proxy investigation whose results will be presented in this volume (Barry *et al.*, this volume) or elsewhere.

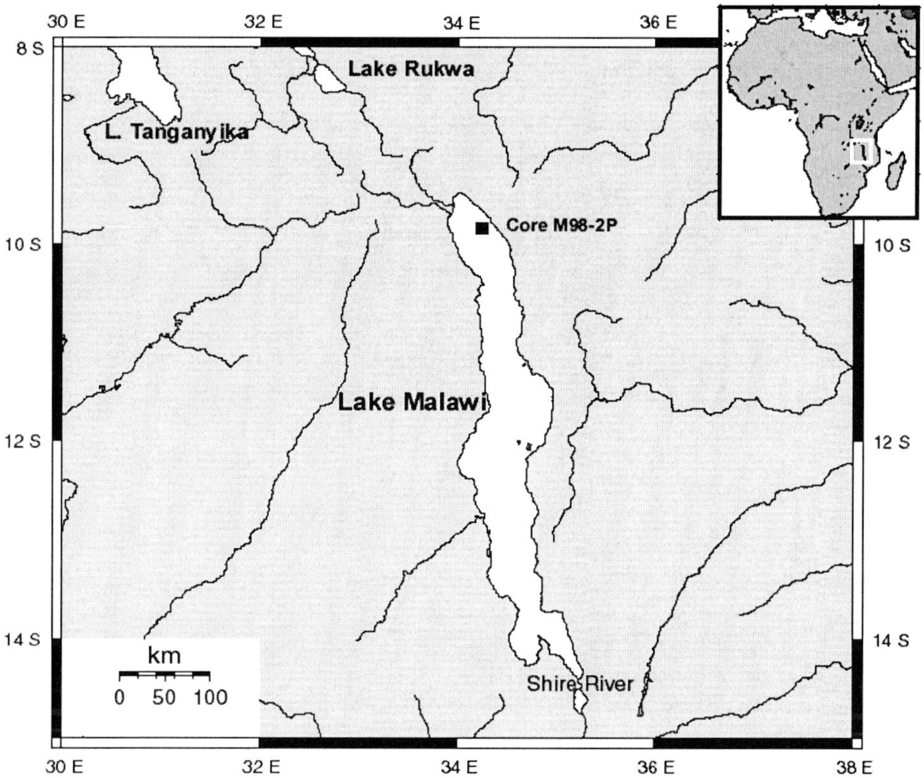


*Figure 1*. Location map of Lake Malawi in southern East Africa and the position of core M98-2P in the northern basin. This map was drawn online using software available at: http:\\aquarius.geomar.de\omc\make_map.html.

# 2. MATERIAL AND METHODS

## 2.1 Core Stratigraphy and Chronology

Six piston cores and five multi-cores were recovered from the northern basin of Lake Malawi by the Large Lakes Observatory of the University of Minnesota-Duluth and the University of Bergen in early 1998. Core sites were located by GPS satellite navigation. The cores were returned to the U.S. where they were first measured for magnetic susceptibility and then split open, described visually, and immediately sampled for water content and various chemical analyses. Stratigraphic correlations were established among the cores, based primarily on magnetic susceptibility, the presence of discrete layers of volcanic ash, and the abundance of biogenic silica

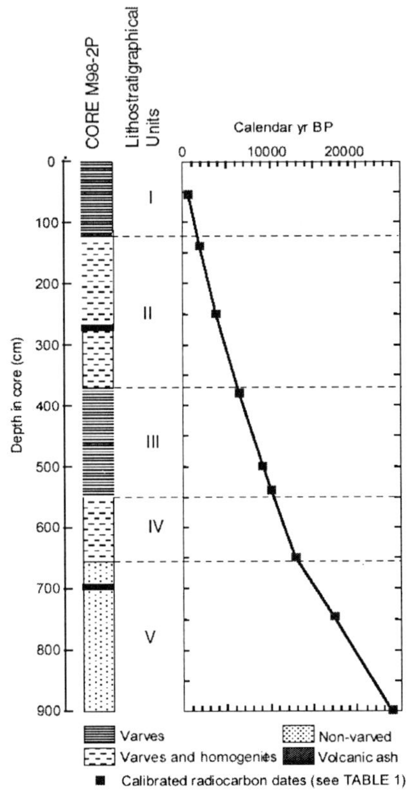


*Figure 2.* Lithostratigraphical units and calibrated radiocarbon ages versus depth in core MP98-2P.

(Barry *et al.*, this volume). The core selected for this study, M98-2P, is the most extensively dated of the piston cores and has a stratigraphy that is representative of all of the cores recovered. It was taken from a depth of 363 m.

The suite of piston cores show five main lithostratigraphic units that are all represented in M98-2P (Figure 2). The upper 121 cm of this core (Unit I) is laminated throughout. $^{210}$Pb dates on multi-cores from this part of the lake reveal that the laminations are annual varves composed of silt and clay (Johnson *et al.*, 2001). Varve counting (Table 1; Barry *et al.*, this volume) provided calendar ages from the core top dated at 699 years BP expressed from 1998 at 53 cm. The varves are interrupted by an occasional homogeneous layer 1-2 cm thick, that appears identical in composition to the varved silty clays. These 'homogenites' are muddy turbidites that can be correlated among cores taken as much as 40 km apart (Johnson *et al.*,

2000). The varved zone overlies a unit of alternating varves and homogenites that extends from 121 to about 381 cm depth in core (Unit II). The alternating packets of varves and homogenites are only a few cm thick. From 381 cm to about 552 cm depth (Unit III), the sediments are once again continuously varved, with the presence of only an occasional homogenite, as in the case of the youngest lithostratigraphic zone. The interval from 552 to 661 cm resembles Unit II, i.e., alternating packets of varves and homogenites (Unit IV). The bottom 240 cm of the core (Unit V) is a uniform clayey silt, displaying no evidence of varves. Major layers of volcanic ash, which can be correlated among several of the piston cores from the north basin, are found at about 270 cm and 700 cm depth in this core (Barry *et al.*, this volume).

*Table 1.* Chronology of core M98-2P. Varve counts after Barry *et al.* (this volume). Radiocarbon results. Calibrated $^{14}$C ages using Calib 4.2 (Stuiver and Reimer, 1993), and a polynomial equation (Bard *et al.* , 1998) for 21,000 $^{14}$C years BP. Radiocarbon ages were corrected by 450 years subtracted before calibration (see text). B = bulk sediment - C = Charcoal.

| Depth (cm) | Calendar yrs before 1998 | $^{14}$C age ($^{14}$C yr BP) | 2sigma maximum (cal yr BP) | Calibrated $^{14}$C age intercepts (cal yr BP) | 2sigma minimum (cal yr BP) | Laboratory code |
|---|---|---|---|---|---|---|
| 5 | 323 | - | | | | (varve count) |
| 35 | 552 | - | | | | (varve count) |
| 53 | 698 | - | | | | (varve count) |
| 139.5 | | 2510±50 | 2150 | 2000 | 1890 | NOSAMS 18271 |
| 248.5 | | 4020±65 | 4080 | 3860 | 3690 | NOSAMS 18272 |
| 379.5 | | 6260±65 | 6750 | 6640 | 6450 | NOSAMS 18566 |
| 500.5 | | 8820±110 | 9500 | 9460, 9430 | 9030 | NOSAMS 18692 |
| 537.5 | | 9550±120 | 10560 | 10230 | 9920 | NOSAMS 18567 |
| 648.0 | | 11450±85 | 13160 | 13000 | 12680 | NOSAMS 18561 |
| 744.0 | | 14450±100 | 17790 | 17360 | 16950 | NOSAMS 18561 |
| 900.0 | | 21000±240 | | 24170 | | NOSAMS 18693 |

AMS radiocarbon dates were obtained at 8 depth intervals on bulk organic matter, but one was obtained on wood charcoal (Table 1, Figure 2). Since bulk sediment samples result in dates on a combination of aquatic and reworked terrigenous organic matter, a. correction is believed to be necessary because of the finite amount of older organic matter that is eroded and redeposited in this distal, pro-deltaic environment. A value of 450 years was subtracted from the $^{14}$C ages before calibration as a correction factor based on dates we have obtained from sediment core tops, other cores and varve counts in this region of Lake Malawi (Barry *et al.*, this volume). Corrected radiocarbon dates were converted to calendar years before present by using the calibration program CALIB v4.2 (Stuiver and Reimer, 1993), except for the oldest one for which the polynomial equation established by Bard *et al.* (1998) was used :

$$y = -1005 + 1.2896x - 3.0126e\text{-}6x^{\wedge 2}$$

where x is the $^{14}C$ age in yr-BP corrected for the 450 years ageing,
y is the calendar age in cal-yr-BP (present is 1950 AD in all cases).

[For homogenization with varve chronology, 48 years were subtracted from the calendar ages. In the text below, ages are given in cal yr BP, that is before 1950. Ages of diatom samples were calculated by linear interpolation between two dated levels].

## 2.2 Diatom Analytical Methods

Diatom analysis of core M98-2P was performed at approximately 10 cm intervals. Standard cleaning procedures ($H_2O_2$, HCl) were followed, and the diatoms were mounted on slides in Naphrax. A known quantity of polystyrene microspheres was added to the suspension for quantitative estimates of the diatom content per unit of sediment weight (valves $mg^{-1}$) (Battarbee and Kneen, 1982) (Figure 4) and diatom influx to the lake bottom. Identification was performed using studies of East African diatoms by Müller (1903), Hustedt, (1949), Gasse, (1986), Cocquyt, (1998), Klee and Casper, (1992 and 1996), and general works (Krammer and Lange-Bertalot, 1986, 1988, 1991a, b). For individual samples, at least 300 diatom valves were counted to establish the fossil assemblage composition (percentage of individual taxa), except for one very diatom-poor level at 551.5 cm (112 valves). Elsewhere, the sediments are generally rich in diatoms; most of the diatom valves are heavy silicified and dissolution was not a major problem. However, in many samples, diatom valves were badly broken due to either turbulence and/or grazing. Breakage was a particular problem for large *Aulacoseira* and *Cyclostephanos* spp. So, as not to underestimate the diatom concentration, fragments have also been counted. Fragments greater than one third of the valve were recorded and then expressed as whole-valves.

Biovolume was also estimated by morphological measurements of the taxa having percentages greater than 2% in two or more samples (15 taxa). For each of these species, twenty individual valves were measured from at least four different samples. The measurements were used to calculate the mean biovolume of these species using the algorithms in the program BIOVOL 2.1 (Kirschtel, 1996). The influx of diatoms to the sediment was calculated in terms of total cell volume ($mm^3\ cm^{-2}\ yr^{-1}$) using the product of mean sample biovolume ($\mu m^3$) X diatom concentration (cells $g^{-1}$) X bulk density (g $cm^{-3}$) X accumulation rate (cm $yr^{-1}$). This provides a first order approximation of the relative diatom productivity at the core site, although it is recognised a single core cannot be used to infer productivity in a lake of this size.

We used three indices to enumerate the importance of periphytic species in the core sediments. Percentage abundance gives an indication of the importance of periphyton relative to other life-forms and is therefore also dependent on changes in the other groups. Absolute abundance or concentration is independent of changes in other diatom life-forms but not of other sedimentary constituents. The calculation of influx overcomes these problems but also reflects changes in lake productivity. All diatom studies (and most other palaeoecological indicators) assume the fossil assemblage is representative of the living community and has not been biased by sedimentary changes such as turbidity flows.

In order to test the stability of the diatom populations and the potential mixture of species having different habitats, the diversity of the diatom assemblage was estimated using the Shannon and Weiner diversity index cited in Magurran (1988). Interpretation of this index also must consider sedimentation rate since diversity is also a function of the time represented in each core slice (Smol, 1981).

## 3. RESULTS

### 3.1 Taxonomy

A total of 120 diatom taxa were identified. Many of these taxa are widespread in East Africa and described in the taxonomic literature, others are endemic to Lake Malawi or confined to the largest East African lakes. Some taxonomic uncertainty is associated with the centric planktonic forms belonging to the three genera *Aulacoseira, Cyclostephanos,* and *Stephanodiscus* that dominate the fossil diatom assemblage (Figures 3 and 4). Diatoms of the genus *Aulacoseira* are the most abundant and ubiquitous in the core samples, with *A. nyassensis* Müller as the principal taxon. Müller (1903) has described several subspecies in his work on diatoms from Lake Malawi. Here, we have recognised three morphological types of *A. nyassensis*, labelled A, B, and C, on the basis of the diameter/height ratio, size, shape, and density of areolae, and mantle thickness (Table 2; Figure 3). Other planktonic *Aulacoseira* (*A. granulata* and var. *angustissima*, *A. ambigua*, *A. agassizii,* and *A. muzzanensis*) are present but not found at levels >2% in the sediments.

The centric diatoms *Stephanodiscus* and *Cyclostephanos* (Thalassiosirales) are known to be an important component of the plankton in East African lakes including Lakes Tanganyika and Victoria but their taxonomy is still under revision. One large species of the Thalassiosirales group was defined by Hustedt (1949) as *Stephanodiscus damasii.* This taxon was later transferred to the genera *Cyclostephanos* by Stoermer and Håkansson (1983) based on SEM observations of specimens from Lake Edward and light microscope slides from Lake Tanganyika. However, Klee and Casper (1992) have observed specimens with the central rosette (rosulate central area) typical of *S. damasii* Hustedt from Lake Malawi which are clearly attributable to *Stephanodiscus* because of the structure of the internal marginal costae. They describe their taxon as a new species *S. mülleri* Klee and Casper, with a size range from 15-70 μm diameter but with most specimens between 35-50 μm. In our material we found one large diameter *Stephanodiscus* (29-56 μm) which is similar to *S. damasii* Hustedt and falls within the definition of *S. mülleri*

Klee and Casper. We have not found *Stephanodiscus* specimens <29$\mu$m diameter with a rosulate central area, but we have found a diatom resembling *S. medius* or what was previously called *S. astrea v. intermedia* before the revision of this species, that will be referred to as *S.* af. *medius* in this study (Figure 3). We also have found two smaller centric taxa that conform well to the descriptions of *S. nyassae* Klee and Casper and *C. malawiensis* Casper and Klee (Table 2, Figure 3) (Klee and Casper, 1992; Klee and Casper, 1996).

*Table 2.* Morphological characteristics of common centric diatoms. Measurements of height (H) and diameter (D) are based on 20 valves per taxon from at least 4 different samples. Cell height was not measured for *Stephanodiscus* and *Cyclostephanos* species and has been estimated in the biovolume calculation.

| Diatom species | Mean ($\mu$m) | Max ($\mu$m) | Min ($\mu$m) | Other characteristics |
|---|---|---|---|---|
| *A. nyassensis* type A | 19.1 | 22.5 | 13.7 | small, round, densely spaced areolae |
| | 23.7 | 30.4 | 16.7 | |
| *A. nyassensis* type B | 18.1 | 25.5 | 13.7 | large, round or irregularly shaped areolae and thick mantle |
| | 13.7 | 18.6 | 7.8 | |
| *A. nyassensis* type C | 18.9 | 21.6 | 14.7 | small, densely spaced, square or rectangular areolae |
| | 19.5 | 27.4 | 11.8 | |
| *S. mulleri* | 42.9 | 56.8 | 29.4 | rosulate central area |
| *S. nyassae* | 6.6 | 10.8 | 2.9 | |
| *S.* af. *medius* | 17.8 | 32.3 | 9.8 | undulating central area |
| *C. malawiensis* | 12.5 | 17.6 | 7.8 | undulating and rosulate central area |

The pennate diatoms are more diversified than the centrales, and are abundant in certain sections of the core material (Figure 4). Most are cosmopolitan in freshwater lakes. However, some taxa are regarded as endemic to East Africa. Amongst these are *Cymatopleura nyassansae*, *Fragilaria africana*, *Gomphonitzschia ungeri*, *Navicula nyassensis*, *Nitzschia lancettula*, *Rhopalodia* af. *vermicularis*, and *Rhopalodia* sp. We have also encountered a small *Fragilaria* which cannot be assigned to any described species known to us (Figure 3) and appears to be intermediate between *F. pinnata* and *F. africana*. It is lanceolate, 10 $\mu$m long and 2 $\mu$m wide, with short but coarse striae, 14 in 10 $\mu$m. This species is relatively abundant in some levels, and seems to be similar to a small *Fragilaria* found in Lake Victoria by Stager (1984).

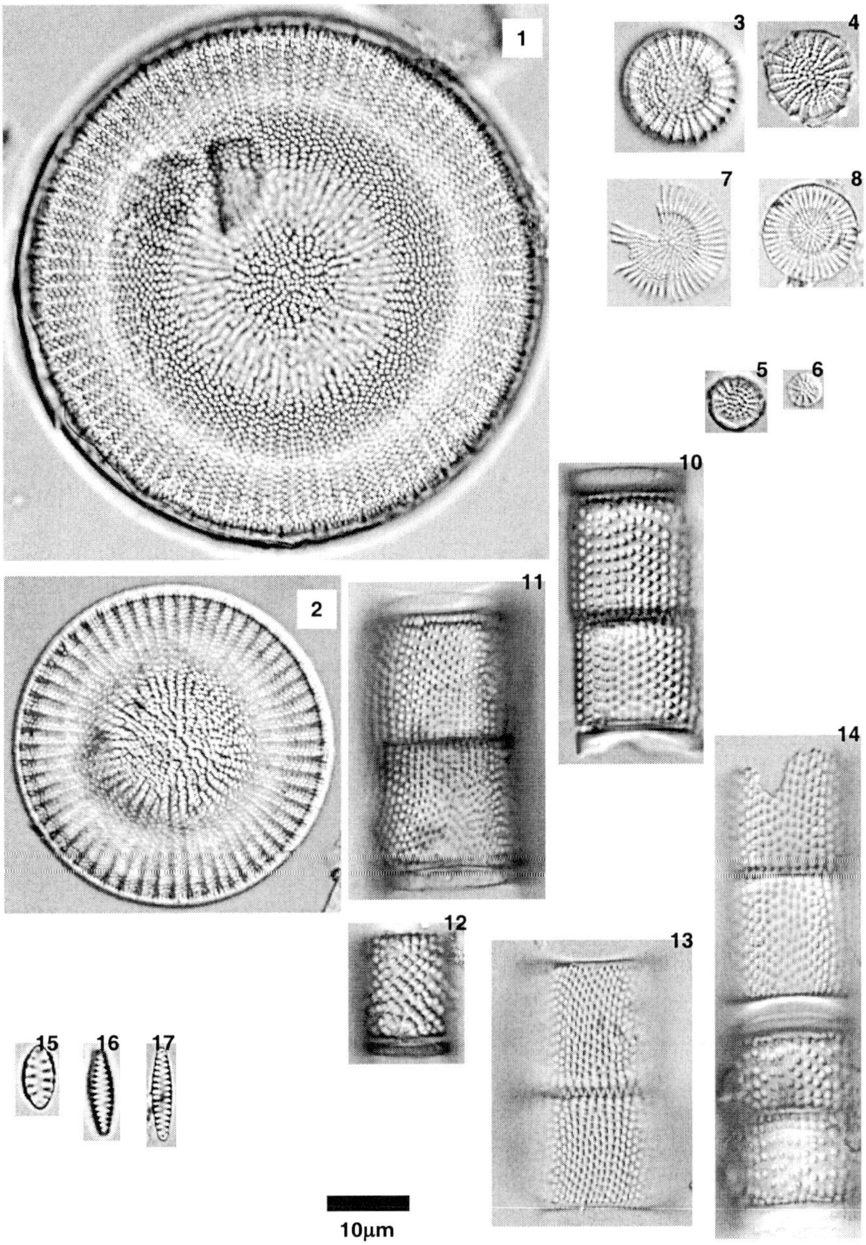


*Figure 3.* Common diatom taxa encountered in core M98-2P. Plates 1 and 2 *Stephanodiscus damasii*; 3 and 4 *Stephanodiscus* af. *medius*; 5 and 6 *Stephanodiscus nyassae*; 7 and 8 *Cyclostephanos malawiensis*; 10 and 11 *Aulacoseira nyassensis* type A; 12 *Aulacoseira nyassensis* type B; 13 *Aulacoseira nyassensis* type C; 14 Chain incorporating different morphological types of *Aulacoseira nyassensis*; 15 *Fragilaria pinnata*; 16 and 17 *Fragilaria* af. *pinnata* (small specimens).

## 3.2 Diatom Ecology and Palaeoecological Interpretation

Few long-term studies of phytoplankton ecology have been carried out in the large East African lakes and little is known of the auto-ecology of the diatom species present in the Lake Malawi sediment. Our interpretation is therefore based on the limited information available in comparative studies of the large East African lakes (Hecky and Kling, 1987; Talling, 1986) and hypotheses generated from eco-physiological experiments (Kilham *et al.*, 1986). In 1980, diatoms dominated the plankton of Lake Malawi during the well-mixed windy season and were prominent from April to September, cyanobacteria and chlorophytes were more important during the months of reduced mixing (Hecky and Kling, 1987). This strong seasonality in the phytoplankton succession that related to the mixing regime was also found in 1985-86 in the southern basin of Lake Malawi (Haberyan and Mhone, 1991).

At present the diatom plankton of Lake Malawi is dominated by *Aulacoseira nyassensis, Cyclostephanos* spp., *Stephanodiscus* spp. and *Nitzschia* spp. (Haberyan and Mhone, 1991; Hecky and Kling, 1987). A succession of *Stephanodiscus* (probably including *Cyclostephanos* spp.) – *Aulacoseira* – *Nitzschia* was found in 1985 by Haberyan and Mhone (1991) although *Stephanodiscus* occurred after *Nitzschia* in 1980 (Hecky and Kling, 1987). According to the data of Haberyan and Mhone (1991) *Stephanodiscus* appears slightly more important than *Aulacoseira* under stratified conditions but *Nitzschia* cf. *spiculum* was more important than either of these during the stratified period.

Spatial differences also occur in the diatom communities related to variability in mixing and nutrient supply (Owen and Crossley, 1992). Hecky and Kling (1987) found the southern part of the lake to be more productive with a greater abundance of diatoms. *Aulacoseira* was found to account for much of the higher diatom biomass at the southern end of the lake in April and May 1980 but was rare in their central sampling station. Surface sediments from 242 short cores confirm the spatial variability in diatom distribution within the lake (Owen and Crossley, 1992). This study showed the Northern section of the lake to be dominated by *A. nyassensis*, the central section to contain a mixed diatom assemblage of *A. nyassensis, Stephanodiscus, Cyclostephanos* and *Nitzschia* spp., the western Nkhotakota sector to be dominated by *Stephanodiscus,* and the Southern sector to contain a mixture of *A. nyassensis, A. agassizi* and *A. ambigua.* The spatial patterns are considered to be a result of differences in resource ratios brought about by the variability in mixing-induced nutrient supply (Owen and Crossley, 1992). The dominance of *Aulacoseira* in the northern and southern parts of the lake is associated with deep mixing bringing nutrients to the surface In contrast the *Stephanodiscus-Cyclostephanos* dominated Nkhotakota sector is thought to be less well mixed and to receive fewer nutrients from the deep waters (Owen and Crossley, 1992).

In the few phytoplankton studies completed in the large East African lakes, a clear relationship exists between diatoms and periods of mixing caused by their ability to withstand turbulence relative to other algae, a tolerance of low light conditions, and a requirement for strong nutrient cycling (Talling and Lemoalle,

1998). Observations on within-lake diatom community ecology are not of sufficient duration to enable relationships between environmental conditions and particular diatom genera to be firmly established. Nevertheless, comparative data from different lakes and ecophysiological studies do show relationships between the major planktonic diatoms and the interrelated variables of light availability, mixing regime and nutrient availability. Strong mixing promotes high nutrient concentrations (Si, P, N being most important for diatoms) in the epilimnion but reduces the depth of the euphotic zone by a factor of 4 in Lake Malawi (Hecky and Kling, 1987). According to experimental studies, *Aulacoseira* spp. require higher Si:P ratios than do *Stephanodiscus* species (Kilham *et al.*, 1986) and may also need a greater degree of turbulence than other planktonic diatoms to keep them in suspension (Lund, 1954). Little is known of the N requirements of East African diatoms but the supply of this nutrient may be important in limiting the diatoms to the periods of strong mixing and enables the N-fixing cyanobacteria to succeed the diatoms. Few diatoms are thought to fix N and some including *Nitzschia* spp. are obligate nitrogen heterotrophs taking their nutrient from cyanobacteria (Kilham *et al.*, 1986).

The planktonic diatom assemblages in the sediments of core M98- 2P shows up-core fluctuations in the ratio (expressed as $\log_{10}$ to downweight the dominance of *Aulacoseira*) of *Aulacoseira* / *Cyclostephanos* + *Stephanodiscus* ratio (A/C+S) (Figure 5). Based on the modern data available to us, we consider that when the sedimentary A/C+S ratio is high, the lake had a deep mixed layer with a strong nutrient flux from the hypolimnion and high Si:P ratios in the euphotic zone. Conversely, a low A/C+S ratio is interpreted as representative of a more stratified lake with a lower rate of nutrient recycling, lower Si:P ratio in the epilimnion and a strong nutrient loss to the hypolimnion and the sediments. Changes from well-mixed to stratified conditions and vice versa can be brought about by a number of factors including lake-level relative to outlet, the strength and direction of wind, degree of upwelling, water temperature and density gradients between epilimnion and hypolimnion. In Lake Malawi, evaporative cooling of surface water, seiches and wind-induced turbulence are the most important variables in promoting mixing (Hecky and Kling, 1987; Owen and Crossley, 1992). The A/C+S may therefore represent mean decadal changes in the length or intensity of the cooler, windy season between May and September with higher values A/C+S indicating cooler, windier conditions. According to the surface sediment studies of Owen and Crossley (1992) and also the sediment traps (François *et al.*, 1996) this ratio is independent of lake level.

In comparison to its importance in the phytoplankton studies, the abundance of *Nitzschia* spp. in the sediments of core M98-2P is low. The combined total of *Nitzschia amphibia* var. *pelagica*, *N. epiphytica*, and *N. lacuum*, reaches a maximum of just 15% (Figure 4). This may be due to spatial differences, limnological changes since the age of the core top or selective breakage and dissolution. Owen and Crossley (1992) found *Nitzschia* spp. (especially *N. nyassensis*) to be most important in the Central and Nkhotakota sectors of the lake, a distribution associated with fault scarps and river mouths where turbidity currents may promote mixing.

In our calculation of periphytic abundance (percentage and absolute abundance in valves mg$^{-1}$, Figure 5), we have excluded the small *Fragilaria* species and included only the diatoms which live attached to different substrates in the euphotic zone at the lake margin or in rivers. Among them, the best represented in the cores sediments are *Amphora pediculus*, *Cocconeis neothumensis, Achnanthes lanceolata, Rhopalodia* spp. and *Navicula* spp. The occurrence of periphyton in offshore sediment is a function of several factors including the proximity of the lake margin and/or a river mouth, productivity in the littoral zone and transport processes to the offshore zone. In a very large lake such as Lake Malawi, autochthonous diatom production is expected to greatly exceed allochthonous inputs from rivers. Furthermore, in a study of algal communities near Cape Maclear, southern Lake Malawi, Haberyan and Mhone (1991) showed that benthic taxa never made up more than 2% of cells in offshore tows. They concluded that an abundance of benthic taxa in sediment cores could be interpreted as lower lake level if sediment redistribution can be excluded. In some sections of core M98-2P, the periphytic taxa exceed 20% of the diatom assemblages, have high absolute abundance and influx. As there is no systematic association with these values and known turbidite layers, we interpret high quantities of periphytic taxa as indicating relatively low lake levels.

Furthermore, small chain forming *Fragilaria* are best regarded as facultative planktonic as they are tolerant of different habitats including many shallow lakes and the marginal zones of large lakes in East Africa (Gasse, 1986). Here we found that their frequency corresponds to that of the periphytic taxa, suggesting they are also indicative of low lake levels.

## 3.3 Diatom Assemblage Zones

Nine major diatom assemblages zones have been identified (Figures 4 and 5).

Zone 1 (901-868 cm; 24,200 – 22,800 years BP). This poorly diversified diatom assemblage (diversity index : 0.8-2.3) is dominated by *A. nyassensis* type A. This assemblage suggests a lake with a deep mixed layer favouring the large-celled *Aulacoseira* species. The periphyton content falls to a minimum at 883 cm (23,430 years BP) also indicating deep waters.

Zone 2 (868-737 cm; 22,800 – 17,050 years BP). This zone is characterised by the highest percentages (together with very high absolute abundance and influx) of periphyton (8-28%) and also of facultative planktonic *Fragilaria* (2-18%) (Figure 5). The diversity index increases markedly from the previous zone to a maximum value of 3.5 (Figure 5). Enhanced diversity is primarily attributed to the proximity of the littoral zone where the number of habitats is greater than offshore, and possibly to unstable, fluctuating limnological conditions. The plankton also comprises more species than in the previous zone (Figure 4) suggesting greater niche availability. In addition to the ubiquitous *A. nyassensis* type A, there is a peak in the more robust form *A. nyassensis* type B (33 % at 852 cm, ca. 22,070 years B.P.), followed by the development of *Stephanodiscus damasii* which reach 21% at 792 cm (ca. 19,460 yr

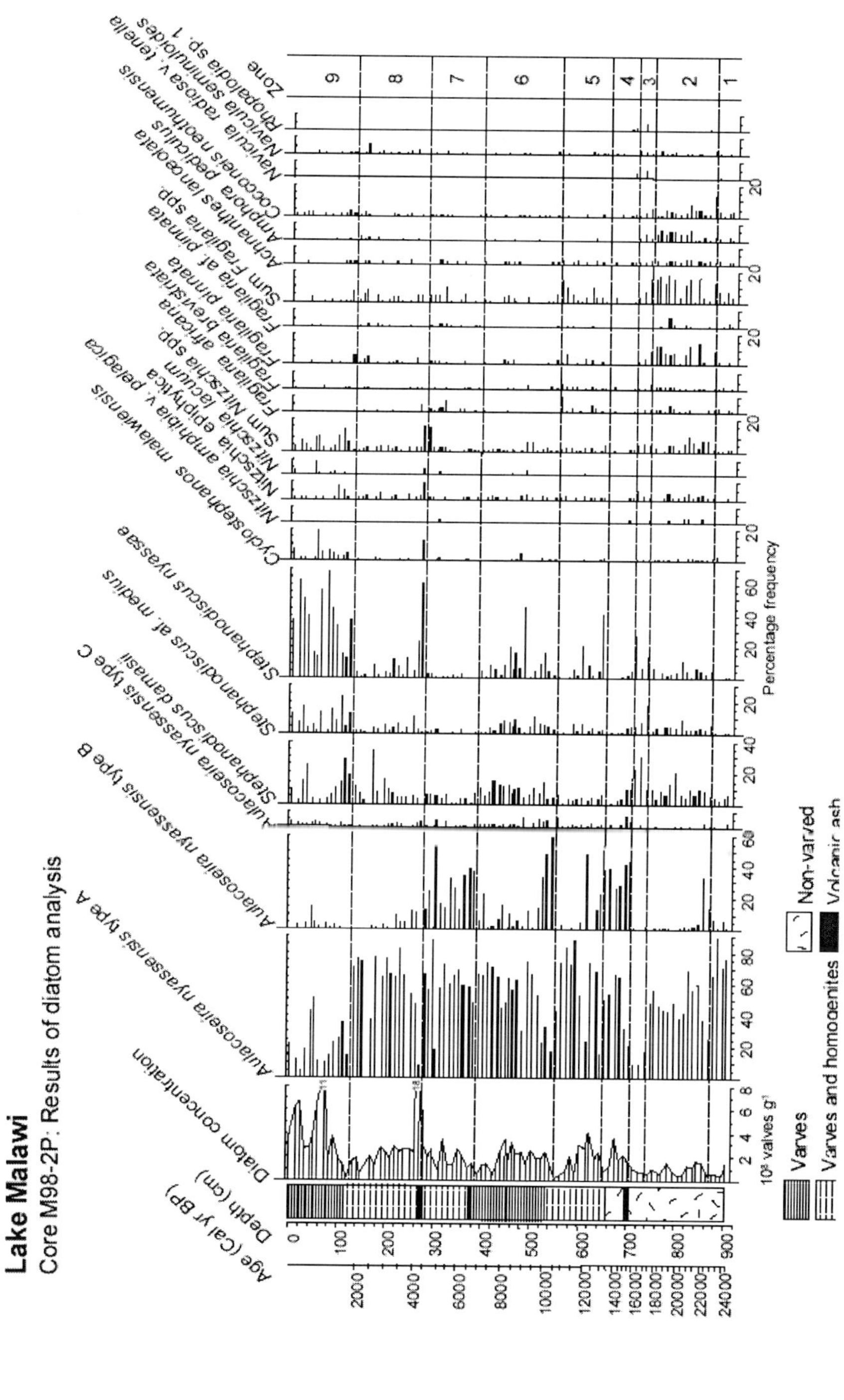


*Figure 4.* Summary of the diatom record from core M98-2P as a function of depth.

B.P.) and also of small *Stephanodiscus*. Other planktonic forms including small *Nitzschia* spp. (eg. *N. lacuum* and *N. amphibia* var. *pelagica*) reach significant percentages in certain samples (cumulatively >10% at 813 cm, ca. 20,370 years BP). Nevertheless, this time interval is one of complex environmental change, as shown by sharp fluctuations of several taxa.

Zone 3 (737-707 cm; 17,050 – 15,680 years BP). *Stephanodiscus damasii* and small *Stephanodiscus* are predominant in this zone (Figure 4) while the percentage of *A. nyassensis* falls drastically to a minimum value of 10% at 722-712 cm, around 16,000 years BP. As in the previous zone, periphytic taxa and small planktonic *Nitzschia* occur in relatively high numbers in the lower part of this zone, reaching 22% and 4% respectively, but then the percentages, absolute abundance and influx of periphyton and of *Fragilaria* spp. decrease markedly. This interval shows the highest diversity index observed for the whole core (4.0 at 722 cm, around 16,360 years BP) (Figure 5).

Zone 4 (707.5-647 cm; 15,680 – 12,980 years BP). An abrupt change is recorded by the sudden development of *A. nyassensis* at the beginning of this zone, which reaches 93.7% (sum of all forms) at 682 cm (14,540 years BP), and replaces *Cyclostephanos* and *Stephanodiscus* in the plankton (Figure 4). Valves of the highly silicified *A. nyassensis* type B are more important than type A at the beginning of this zone. Diatom biovolume (Figure 5) also increases significantly from the previous zone to a maximum at 672 cm (ca. 14,090 years BP). There is a marked negative shift in the diversity index which falls to 1.35 at this point, associated with the almost complete disappearance of all taxa from shallow water environments. The development of a quasi- monospecific assemblage of *A. nyassensis* (Figure 4) indicates the establishment of a deep lake, with a deep well-mixed epilimnion. Therefore, this interval is interpreted as a stage of high lake level.

Zone 5 (647-546 cm; 12,980 – 10,510 years BP). The A/C+S ratio reduces sharply in this zone (Figure 5), together due to the growth of small *Stephanodiscus* that represent up to 44% of the diatom assemblage (Figure 4). Diatom productivity reaches a maximum at 621.8 cm (12,270 years BP). *Fragilaria* and periphytic taxa increase substantially in this zone, the latter showing a sharp peak (0.31 x $10^8$ valves $g^{-1}$) in absolute abundance and influx at 621.8 cm (Figure 5), but reaching their maximum percentage (21%) in this zone at 551.5 cm (10,640 years BP). The last step, between samples 591 and 551.5 cm (11,610 – 10,640 years BP) contains diatom assemblages resembling those of Zone II, in respect to the plankton / periphyton ratio. This interval also coincides with a considerable decrease in total diatom concentration and influx with a minimum value occurring at 551.5 cm (Figure 4). This level is also characterised by very poor diatom preservation. Most valves are broken, not just the most vulnerable large-celled species, but also surprisingly including small *Fragilaria*. Even the robust valves of *Aulacoseira* are partly dissolved and most commonly reduced to their sulcus. Breakage and poor preservation is consistent with turbulent shallow water and even reworking of previously accumulated diatom ooze. In summary, this assemblage zone is interpreted as a period of environmental instability including a series of falls in lake

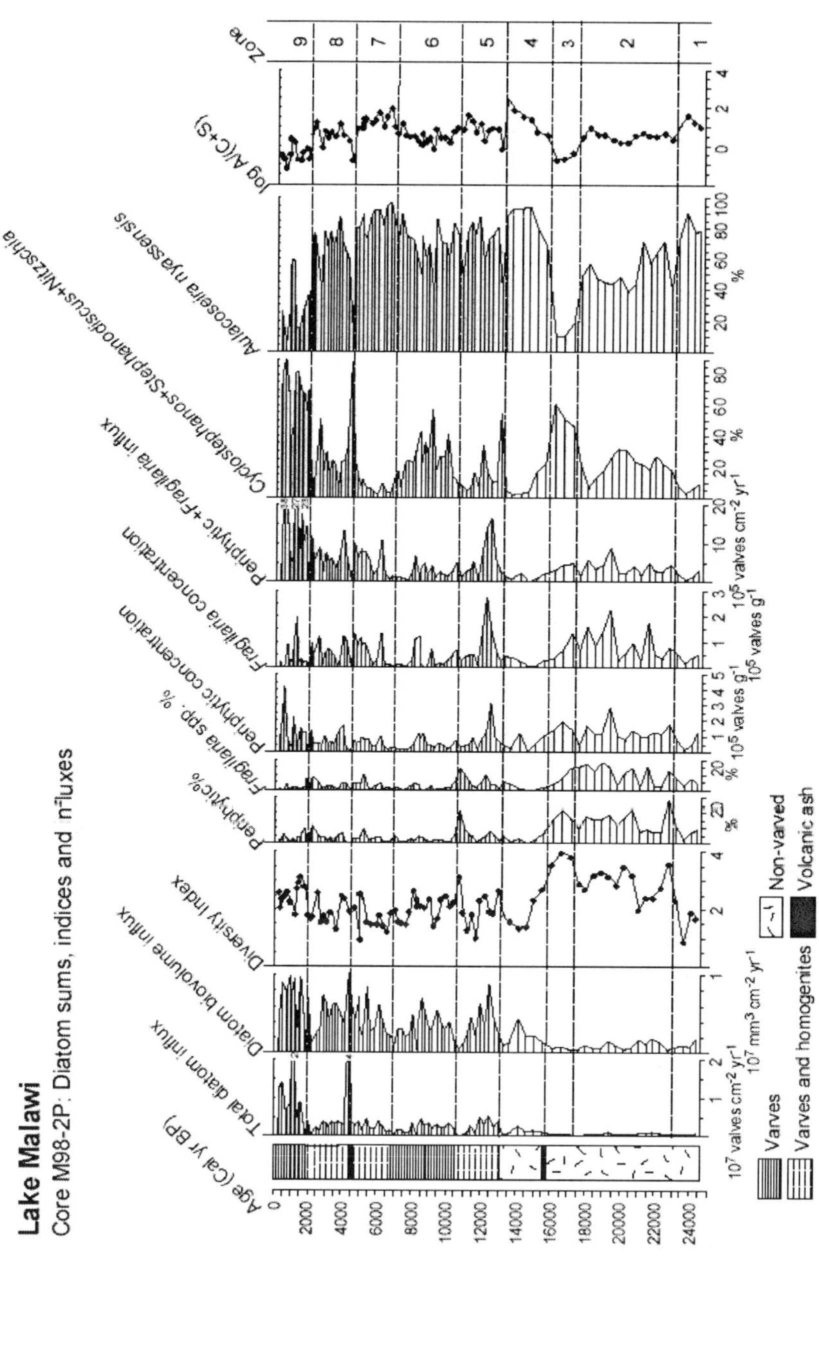


*Figure 5*. Diatom indices, sums and influx. The influx values for the zone 9 are not considered reliable because of difficulty in calculating the chronology of this section.

level and ending with a remarkable regressive event that ended suddenly after ca. 10,600 years BP.

Zone 6 (546-384 cm; 10,510 - 6740 years BP). Diatom content (Figure 4) and influx (Figure 5) show a progressive increase to reach their highest values in this zone between 531.5 and 439 cm (ca. 10,150- 8030 years BP) and then fall sharply to 389 cm (6860 years BP).The diversity index is intermediate to levels found elsewhere in the core and fluctuates between 1.4 and 2.6 (Figure 5). Planktonic diatoms in this zone are a mixture of *Stephanodiscus, Cyclostephanos* and *Aulacoseira* (Figure 4). *A. nyassensis* type B dominates the lower part of this zone after which it is replaced by type A. *S. minutulus* has a maximum percentage (45%) at 481 cm (9020 years BP) in phase with a maximum of small planktonic *Nitzschia* (6%) and a small increase in facultative planktonic *Fragilaria*. The periphytic taxa reach their maximum concentration and abundance in this zone at 461 and 439 cm (ca. 8550 - 8030 years BP). The upper part of this zone, 419-384 cm (ca. 7560-6740 years BP) has an increase in the percentage of *Aulacoseira nyassensis* associated with a decline in periphytic species, diatom diversity and productivity.

Zone 7 (384-273 cm; 6740 - 4380 years BP). This zone is characterized by the highest percentage of *A. nyassensis* (all types) observed in the whole record (97% at 370 cm, ca. 6440 years BP) (Figure 4). Type B is relatively abundant in the lower part of this zone where periphyton influx is at a minimum. As in Zone IV, this plankton-dominated assemblage has generally low diversity (Figure 5). Productivity increases progressively through this zone but shows some large fluctuations between samples. The A/C+S ratio declines due to an increase in *S. damasii* at the top of the zone. One sample at 309 cm (ca. 5140 years BP) contains a high percentage of periphyton and *Fragilaria* (9 and 10% respectively). Zone VII suggests a sustained deep-water phase that falls in two steps, firstly at 349.5 cm (6000 years BP) and then from 319.5 cm (5370 cal years BP).

Zone 8 (273-125 cm; 4380 – 1770 years BP). This zone starts with a very large peak of *S. minutulus* (61%) together with *S. hantzschii* (12%), and *Nitzschia epiphytica* (10%) in the sample at 268.5 cm (ca. 4280 years BP). Diatom abundance and influx reaches its maximum in this sample (17.9 x $10^8$ valves $g^{-1}$), although biovolume is only moderate due to the small cell volumes of the *Stephanodiscus* spp. This sample seems to represent a short lived event, represented by a bloom in *Stephanodiscus* productivity. It coincides with a 10-cm thick tephra layer (ca. 270 cm, 4320 years BP) and this may have been a stimulus to the diatom productivity. The effects of the event are also shown to a reduced extent in the following sample (258 cm, ca. 4060 years BP) where *Stephanodiscus* still account for 24% of the assemblage. Following these unusual diatom assemblages, *Aulacoseira* returns to dominate the rest of the zone with *S. damasii* as the sub-dominant species. The periphytic taxa and the *Fragilaria* represent together up to 11% at 248.5 cm (3860 years BP) and 20% at 148.5 cm (ca. 2150 years BP) indicating relatively dry periods. Influx of periphytic taxa is relatively high throughout this zone.

Zone 9 (125-0 cm; 1770 – 240 years BP). The sample at 120.5 cm (1700 years BP) at the base of this zone is almost devoid in diatoms. The concentration is just 1.8 x $10^5$ valves $g^{-1}$, almost 2 orders of magnitude less than other samples analysed

(Figure 4). Zone IX differs from the preceding zones by being dominated by *Stephanodiscus* which represent up to 77% of the total diatom assemblage, together with high frequencies of *S. damasii*, planktonic *Nitzschia* spp., and relative small percentages of periphytic taxa. However, as the abundance of diatoms in the rest of the zone is high, the concentration of periphyton in the sediment is greater than the percentages might suggest, and reaches a maximum of 0.42 x $10^8$ valves $g^{-1}$ at 20 cm (430 years BP), the highest in the core. This zone is also characterised by high diversity, although lower than during Zones II and III. The influx values for this zone are not considered reliable because of difficulties in calculating the accumulation rate at the water-rich core top.

## 4. DISCUSSION AND CONCLUSION

The diatom record of core M98-2P enables the reconstruction of major limnological changes in Lake Malawi over the past 24,200 years BP. In particular, the planktonic diatoms show important up-core variations in mixing and nutrient cycling. No clear relationship is evident between the A/C+S ratio and water level as shown by the distribution of these diatoms in the modern plankton and surface sediments (Owen and Crossley 1992; François *et al.*, 1996; Hecky and Kling, 1987). Moreover, changes in the A/C+S ratio in our core samples are not associated with the frequency of periphytic taxa and *Fragilaria* spp. (Figure 5). We conclude that the A/C+S shows mean decadal changes in the length or intensity of the mixing season with higher values of the ratio indicating cooler, windier periods. Low lake levels are indicated by the quantity of periphyton and of shallow-water *Fragilaria* spp. in the sediments as measured by percentage, concentration and influx (Figure 5). These taxa mark the proximity of the littoral zone to the core site assuming that sediment distribution processes have remained constant.

The core base indicates relatively deep and well-mixed water conditions preceding the LGM from 24,200-22,800 years BP. From ca. 22,800 to 17,000 years BP (including the LGM), the diatom record suggests that the lake experienced its lowest and most sustained low-stand represented by this core as shown by the percentage and concentration of shallow water diatoms. This was a period of moderate mixing and nutrient levels in the euphotic zone. This period ended at ca. 17,000 years BP. Poorly mixed conditions then established and lasted about 1300 years (17,000-15,700 years BP) when nutrients (especially Si) limited the growth of *Aulacoseira* in the northern basin. This phase with stratified water preceded the onset of very wet conditions. At ca. 15,700-15,000 years BP, the lake-level rose sharply and a maximum high stand with very deep mixing is registered over ca. 2700-2000 years (15,700-15,000 to13,000 years BP). Interestingly, Lake Tanganyika experienced a comparable evolution during the last deglaciation period (Gasse *et al.* 1989). In that lake, the re-establishment of a positive water balance after the LGM lowstand led first to planktonic assemblages of stratified waters (*Cyclostephanos, Stephanodiscus* and needle-shape *Nitzschia* spp.) from ca. 17,000 to 15,000 years

BP. Around 15,000 years BP, an abrupt shift to *Aulacoseira* assemblages was attributed to a major, sudden rise in water level and to the lake opening.

At Malawi, the shallow water diatom data suggest hydrological instability and regressive events of short duration during the late Glacial period. A marked rise in concentration and influx of periphytic diatoms began slowly after 13,000 years BP, in phase with an increase in diversity, and then rose most rapidly after 12,600 years BP to a maximum at 12,400 years BP (Figure 5). A near symmetrical fall in periphyton influx occurred after this point to minimum values at 11,600 years BP. This event may have begun shortly before the Younger Dryas chronozone (YD) as recorded in the northern polar ice records from Greenland (Johnsen *et al.*, 1992) during the Antarctic cold reversal (Jouzel *et al.*, 1995) but reached its maximum intensity during the YD. The termination of the YD was followed by a further regression centred upon 10,600 years BP. This is indicated by rises of the periphytic taxa and *Fragilaria* spp. percentages, by an increase of the diversity index, and by a dramatic shift in diatom concentration and biovolume influx (Figures 4 and 5).

From ca. 10,500 to 5400 years BP, the diatom flora suggests generally high or intermediate lake levels, but with different degrees of mixing and nutrient cycling and some regression events. The earlier part of this period (10,500-6700 years BP experienced less deep mixing and some nutrient limitation. A short-lived regression is found at 8500-8000, but this was less marked than at the LGM or around 10,600 years BP. The influx of periphyton began after 5400 and reached a peak at 4900 years BP before briefly declining at 4300 years BP.

For most of the last ca. 4000 years, the diatom record suggests that lake levels were generally lower than in the Early Holocene, and the water column was less well mixed, especially after 1770 years BP. This general trend toward lower lake-levels was not linear. Short-term phases of very low water level are suggested by some samples within the last two millennia but our chronology does not allow these to be dated precisely.

Our diatom results imply fluctuations in lake level that are not in agreement with earlier work on cores from Lake Malawi (Finney and Johnson, 1991; Finney *et al.*, 1996; Johnson, 1996; Ricketts and Johnson, 1996). These studies suggested that lake level was low prior to 32,700 (30,000 $^{14}$C yr BP) and from 6850-11,500 (6000 to 10,000 $^{14}$C yr BP), and high during the intervening period including the LGM. The high level at the LGM was assumed because of the absence of carbonates in the sediments in this age range. However their sampling was skewed since most of the cores they studied did not extend further back than 10,000 radiocarbon years BP. The two cores that had calcareous sediments older than 20,000 years (M86-18P and 27P) were taken from relatively shallow water, and may very well have hiatuses spanning the LGM, consistent with our evidence for a low stand at this period.

Earlier studies also proposed a lowstand during the early Holocene 11,500- 6850 (10,000-6000 $^{14}$C yr BP) due to the presence of endogenic calcite in sediments, in cores from water depths shallower than 300 m (Finney and Johnson, 1991; Finney *et al.*, 1996; Johnson, 1996; Ricketts and Johnson, 1996). Cores from below this depth do not contain calcite in sediments of any age, presumably because they have always been below the calcite compensation depth. Furthermore, nearshore sands are associated with the carbonate horizons in M86-12P (Finney and Johnson, 1991) that have recently been dated at ca. 10,600 years BP (ca. 9400 radiocarbon yr BP). Diatoms are rare and poorly preserved in the calcareous sediments of core M86-12P, but sponge spicules are relatively abundant. This lowstand at ca. 10,600 years BP is in good agreement with our periphytic diatom data in core MP98-2P. The new diatom data described in this paper suggest water levels during the carbonate interval to be higher than those that occurred at the LGM, during short periods between ca. 13,000 and 10,600 years BP and most of the Late Holocene, but lower than in the period 15,700-13,000 and 7500-6700 years BP. Since the lake is close to its outlet at present, it is likely it was closed for much of the Early Holocene as suggested in previous studies. Moreover, our results are consistent with significantly lower levels during parts of this period as indicated by peaks in periphyton influx and concentration, especially ca. 8500-8000 years BP (Figure 5).

In summary, our data suggest that Lake Malawi experienced a low stand during the LGM, like most other lakes in Africa including lakes in the southern tropics (Partridge *et al.*, 1997; Gasse, 2000). However, the lake also exhibits some Southern hemisphere traits that set it apart from African lakes to the north. Firstly, environmental conditions in Lake Malawi changed relatively early after the LGM (from ca. 17,300 years BP), when an increase in available moisture occurred in Lake Tanganyika (Gasse *et al.*, 1989), in Madagascar (Gasse and van Campo, 1998), and in western southern Africa (Ning Shi *et al.*, 1998). Secondly, Lake Malawi was well mixed and probably open with maximum lake levels during the interval 15,700-13,000 years BP, while lakes in the northern tropics remained relatively low until 11,500 years BP (Gasse, 2000). Thirdly, the lake was at an intermediate level in the early Holocene and may have been lower than the outlet during most of this period to allow for the deposition of carbonate. It was not characterised by the extremely wet early-mid Holocene conditions found in lakes Tanganyika (Gasse *et al.*, 1989), Victoria (Johnson, 1996), Albert (Beuning *et al.*, 1997), Turkana (Johnson, 1996) and the Ethiopian rift lakes (Gasse and Street, 1978). This may be linked to the minimum summer insolation (438 W $m^{-2}$) around 10,000 years BP in the southern tropics (Berger and Loutre, 1991) as suggested by Finney and Johnson, (1991) and Finney *et al.* (1996). The diatom record from the northern basin of the lake adds the very important suggestion that Lake Malawi was even lower during the LGM and parts of the Late Holocene when summer insolation receipt was close to its maximum (473 W $m^{-2}$ and 470 W $m^{-2}$ respectively at 15°S; Berger and Loutre (1991). Our results suggest that orbitally-induced forcing of rainfall is insufficient to fully explain the climate variability of East Africa's southern tropics during the last glacial cycle.

## ACKNOWLEDGEMENTS

The cores were taken under the auspices of the IDEAL programme funded by the National Science Foundation. Prof. Mike Talbot, University of Bergen, and Maria Letizia Filippi, CNR-Roma, provided helpful comments on early drafts. Philip

Barker wishes to thank the director and staff of the CEREGE who hosted his sabbatical visit.

## REFERENCES

Bard, E., Arnold, M., Hamelin, B., Tisnerat-Laborde, N., and Cabioch, G. (1998) Radiocarbon calibration by means of mass spectrometric $^{230}Th/^{234}U$ and $^{14}C$ ages of corals. An updated data base including samples from Barbados, Mururoa and Tahiti. *Radiocarbon* (40 (3), 1085-1092

Barker, P., Telford, R., Gasse, F. and Thevenon, R. (in press). Late Pleistocene and Holocene palaeohydrology of Lake Rukwa, Tanzania, inferred from diatom analysis. *Palaeogeography, Palaeoclimatology, Palaeoecology.*

Barry, S.L., Filippi, M.L., Talbot, M.R. and Johnson, T.C. (This volume ) A 20,000 year sedimentological record from Lake Malawi, East Africa: The Late Pleistocene -Holocene transition in the southern tropics.

Battarbee, R.W. and Kneen, M.J. (1982) The use of electronically counted microspheres in absolute diatom analysis. *Limnology and Oceanography* **27**, 184-188.

Beadle, L.C. (1974) *The Inland Waters of Tropical Africa*, Longman, New York.

Berger, A. and Loutre, M.F. (1991) Insolation values for the climate of the last 10 million years. *Quaternary Science Reviews* **10**, 297-317.

Beuning, K.R.M., Talbot, M.R. and Kelts, K. (1997) A revised 30,000-year paleoclimatic and paleohydrologic history of Lake Albert, East Africa. *Palaeogeography, Palaeoclimatology and Palaeoecology* **136** (1-4), 259-279.

Cocquyt, C. (1998) Diatoms from the N. Basin of Lake Tanganyika. *Bibliotheca Diatomologica* **39**, Cramer, Stuttgart.

DeBusk, G.H. (1998) A 37,500-year pollen record from Lake Malawi and implications for the biogeography of afromontane forests. *Journal of Biogeography* **25** (3), 479-500.

Finney, B.P. and Johnson, T.C. (1991) Sedimentation in Lake Malawi (East-Africa) during the past 10,000 Years - a continuous paleoclimatic record from the Southern tropics. *Palaeogeography, Palaeoclimatology, Palaeoecology* **85** (3-4), 351-366.

Finney, B.P., Sholz, C.A., Johnson, T.C. and Trumbore, S. (1996) Late Quaternary lake-level changes of Lake Malawi, in T.C. Johnson and E.O. Odada (eds.), *The Limnology, Climatology and Paleoclimatology of the East African Lakes*, Amsterdam, pp. 495-508.

François, R.H., Pilskaln, C.H. and Altabet, M.A. (1996) Seasonal variation in the nitrogen isotopic composition of sediment trap materials collected in lake Malawi, in T.C. Johnson and E.O. Odada (eds.), *The Limnology, Climatology and Paleoclimatology of the East African Lakes*, Gordon and Breach, Amsterdam, pp. 241-250.

Gasse, F. (1986) East African diatoms. Taxonomy, ecological distribution. *Bibliotheca Diatomologica* **11**, J. Cramer, Berlin.

Gasse, F. (2000) Hydrological changes in the African tropics since the Last Glacial Maximum. *Quaternary Science Reviews* **19**, 189-211.

Gasse, F., Ledée, V., Massault, M. and Fontes, J.Ch. (1989) Water-level fluctuations of Lake Tanganyika in phase with oceanic changes during the last glaciation and deglaciation. *Nature* **342**, 57-59.

Gasse, F. and Street, F.A. (1978) Late Quaternary lake level fluctuations and environments of the Northern Rift Valley and Afar Region (Ethiopia and Djibouti). *Palaeogeography, Palaeoclimatology, Palaeoecology* **24**, 279-325.

Gasse, F. and van Campo, E. (1998) A 40,000-yr pollen and diatom record from Lake Tritrivakely, Madagascar, in the southern tropics. *Quaternary Research* **49** (3), 299-311.

Haberyan, K.A. (1987) Fossil diatoms and the palaeolimnology of Lake Rukwa, S.W.Tanzania. *Freshwater Biology* **17**, 429-436.

Haberyan, K.A. and Mhone, O.K. (1991) Algal Communities Near Cape Maclear, Southern Lake Malawi, Africa. *Hydrobiologia* **215** (3), 175-188.

Hecky, R.E. and Kling, H.J. (1987) Phytoplankton ecology of the great lakes in the rift valley of central Africa. *Ergebnisse der Limnologie* **25**, 197-228.

Hustedt, F. (1949) *Susswasser diatomeen aus dem Albert-National Park Belgisch-Kongo, Exploration du Parc National Albert, Mission H. Dumas (1935-36),* Inst. des Parcs Nationaux au Congo Belge. Hayeg, Bruxelles.

Johnsen, S.J. *et al.* (1992) Irregular Glacial Interstadials recorded in a New Greenland Ice Core. *Nature* **359** (6393), 311-313.

Johnson, T.C., (1996) Sedimentary processes and signals of past climatic change in the large lakes of the East African Rift valley, in T.C. Johnson and E.O. Odada (eds.), *The Limnology, Climatology and Paleoclimatology of the East African Lakes*, Gordon and Breach, Amsterdam, pp. 367-412.

Johnson, T.C., Barry, S.L., Chan, Y. and Wilkinson, P. (2001) A decadal record of climate variability spanning the last 700 years in the southern tropics of East Africa. *Geology* **29**, 83-86.

Jouzel, J. *et al.* (1995) The 2-Step Shape and Timing of the Last Deglaciation in Antarctica. *Climate Dynamics* **11** (3), 151-161.

Kilham, P., Kilham, S.S. and Hecky, R.E. (1986) Hypothesized resource relationships among African planktonic diatoms. *Limnology and Oceanography*, **31** (6), 1169-1181.

Kirschtel, D.B. (1996) BIOVOL Ver. 2.1. available at http://www.uvm.edu/~dkirscht/biovol.html (last accessed on 27/7/99).

Klee, R. and Casper, S.J. (1992) New Centric Diatoms (Thalassiosirales) of Lake Malawi. *Archiv Fur Protistenkunde* **142** (3-4), 179-192.

Klee, R. and Casper, S.J. (1996) *Stephanodiscus nyassae* Klee et Casper nov spec, a new centric diatom. *Archiv Fur Protistenkunde* **146** (3-4), 293-297.

Krammer, K. and Lange-Bertalot, H. (1986) *Bacillariophyceae. I Teil Naviculaceae. Susswasserflora von mitteleuropa*, Gustav Fischer Verlag, Stuttgart.

Krammer, K. and Lange-Bertalot, H. (1988) *Bacillariophyceae. II Teil: Bacillariaceae, Epithemiaceae, Surirellaceae. Susswasserflora von mitteleuropa*, Gustav Fischer Verlag, Stuttgart.

Krammer, K. and Lange-Bertalot, H. (1991a) *Bacillariophyceae. III Teil : Centrales, Fragilariaceae, Eunotiaceae. Susswasserflora von mitteleuropa,* Gustav Fischer Verlag, Stuttgart.

Krammer, K. and Lange-Bertalot, H. (1991b) *Bacillariophyceae. IV Teil: Achnanthaceae, kritische ergänzungen zu Navicula (Lineolatae) und Gomphonema Gesamliteraturverzeichnis. Susswasserflora von mitteleuropa*, Gustav Fischer Verlag, Stuttgart.

Lund, J.W.G. (1954) The seasonal cycle of the planktonic diatom *Melosira italica* ssp. *subarctica* O. Müll. *Journal of Ecology* **42**, 151-179.

Magurran, A.E. (1988) *Ecological diversity and its measurement*, Princeton University Press.

Müller, O. (1903) Bacillariaceen aus dem Nyasalande und einigen benachbarten Gebeit. Engler's *Bot. Jahrb,* **34** (36, 45), 9-28, 255-301, 137-206, 69-122.

Owen, R.B. and Crossley, R. (1992) Spatial and temporal distribution of diatoms in sediments of Lake Malawi, Central Africa and ecological implications. *Journal of Paleolimnology* **7**, 55-71.

Partridge, T.C., DeMenocal, P.B., Lorentz, S.A., Paiker, M.J. and Vogel, J.C. (1997) Orbital forcing of climate over South Africa: A 200,000-year rainfall record from the Pretoria Saltpan. *Quaternary Science Reviews* **16** (10), 1125-1133.

Ricketts, R.D. and Johnson, T.C. (1996) Early Holocene changes in lake level and productivity in Lake Malawi as interpreted from oxygen and carbon isotope measurements of authigenic carbonates, in T.C. Johnson and E.O. Odada (eds.), *The Limnology, Climatology and Paleoclimatology of the East African Lakes*, Gordon and Breach, Amsterdam, pp. 475-493.

Ning Shi, Dupont, L.M., Beug, H.J. and Schneider, R. (1998) Vegetation and climate changes during the last 21000 years in SW Africa based on a marine pollen record. *Vegetation History and Archaeobotany* **7**(3), 127-140.

Smol, J.P. (1981) Problems associated with the use of 'Species Diversity' in paleolimnological studies. *Quaternary Research* **15**, 209-212.

Spigel, R.H. and Coulter, G.W. (1996) Comparison of hydrology and physical limnology of the East African Great Lakes: Tanganyika, Malawi, Victoria, Kivu and Turkana (with reference to some North American Great Lakes), in T.C. Johnson and E.O. Odada (eds.), *The Limnology, Climatology and Paleoclimatology of the East African Lakes*, Gordon and Breach, Amsterdam, pp. 103-140.

Stager, J.C. (1984) The diatom record of Lake Victoria (East Africa) : the last 17,000 years, in D. Mann (ed.), *7th International Diatom Symposium (1982),* Koeltz, Philadelphia, pp. 455-476.

Stoermer, E.F. and Håkansson, H. (1983) An interpretation of the morphological structure and taxonomic relationship of *Stephanodiscus damasii* Hustedt. *Bacillaria* **6**, 245-255.

Stuiver, M. and Reimer, P.J. (1993) Extended $^{14}C$ database and revised CALIB 3.0 $^{14}C$ age calibration program. *Radiocarbon* **35**, 215-230.

Talling, J.F. (1986) The seasonality of phytoplankton in African Lakes, *Hydrobiologia* 138, 139-160.

Talling, J.F. and Lemoalle, J. (1998) *Ecological Dynamics of Tropical Inland Waters*, Cambridge University Press, Cambridge.

Wüest, A., Piepke, G. and Halfman, J.D. (1996) Combined effects of dissolved solids and temperature on the density stratification of Lake Malawi, in T.C. Johnson and E.O. Odada (eds.), *The Limnology, Climatology and Paleoclimatology of the East African Lakes*, Gordon and Breach, Amsterdam, pp. 183-202.

# LAKE TANGANYIKA HOLOCENE RECORD ON VARIABILITY IN PRECIPITATION IN THE MALAGARASI CATCHMENT BASIN

ALFRED N. MUZUKA AND NTAHONDI NYANDWI

*University of Dar es Salaam, Institute of Marine Sciences,*
*P.O. Box 668, Zanzibar, Tanzania.*

**ABSTRACT**

The stable isotope composition of organic carbon (OC), abundance of OC and nitrogen, and C/N ratios for core T97-69V are used to document late Holocene variability in the sources of organic matter (OM) in the Malagarasi delta (Lake Tanganyika) and in precipitation in the Malagarasi catchment basin. Core T97-69V, located at latitude 5°12.92'S and longitude 29°40.50'E, was retrieved at a water depth of 60 m. The organic $\delta^{13}C$ values for this core ranges from -25.0λ to -19.8λ, and averages -23.0∀1.4λ. The $\delta^{13}C$ values decrease down-core to the base of the core with one spike of depleted $\delta^{13}C$ values interrupting this general trend. A similar trend of down-core decrease is also observable for the contents of OC and nitrogen whose abundance averages 3.75∀1.1% and 0.4∀0.1% respectively. In contrast, the C/N ratios with mean value of 11.45∀2.3 do not display any clear down-core trend. Furthermore, there is a sharp shift in the stable isotope compositions of OC to lower values at about 200 cm. An enrichment in $^{13}C$ in the upper 200 cm associated with higher contents of OC and nitrogen and relatively low C/N ratio values can probably be attributed to complete utilisation of available nutrients (high primary productivity) in conjunction with deposition of large proportion of $C_4$ material derived from the catchment areas. Low isotope values of up to 4λ in the lower 200 cm of the core suggest higher input of $C_3$ type of material derived from phytoplankton and terrestrial plants. Furthermore, a downcore decrease in $^{13}C$ suggests that the level of precipitation in the Malagarasi Basin has decreased since mid-Holocene, and most likely this



*E.O. Odada and D.O. Olago (eds.),*
*The East African Great Lakes: Limnology, Palaeolimnology and Biodiversity,* 415–428.

trend was associated with a decrease in the lake levels. Two spikes of low isotope values may suggest period of above normal precipitation that resulted into transportation of a significant quantity of terrestrial OM.

## 1. INTRODUCTION

East African climate is largely influenced by variability of the monsoon winds, which are considered to have been established during the middle Miocene (Léclaire, 1974; Kroon *et al.*, 1991). Since its establishment, the intensity of monsoon winds has changed considerably with stronger southwest (SW) north of the equator or southeast south of the equator ( SW/SE) during interglacial periods ( Peterson *et al.*, 1979; Kolla *et al.*, 1981; Prell and Van Campo, 1986; Pokras and Mix, 1985; Bigg, 1996). Variability in the monsoon winds has been observed to correlate well with precipitation, with stronger SW/SE (north/south of equator) monsoon being associated with higher precipitation (Pokras and Mix, 1985; Bigg, 1996; Karlén *et al.*, 1999). For example, at the beginning of the Holocene period (9000 years) the intensity of the monsoon winds was high and this led to high precipitation and deposition of sapropel in the Mediterranean Sea (Bigg, 1996).

Precipitation is an important measure of the climate change and a knowledge of its variability through time may help to predict future trends. Variability in precipitation during the Holocene has been documented through the use of palaeolake levels in many parts of the world (e.g. Hillaire-Marcel and Cassanova, 1987; Talbot and Livingstone, 1989; Ricketts and Johnson, 1996). Another way of documenting variability in precipitation is through the use of stable isotopes particularly those of OC. In the tropics, the vegetation tends to change from $C_4$ (mainly grassland) to $C_3$ types of vegetation (mainly high land trees) as precipitation increases. These two groups of plants have significantly different stable isotope composition, which can be utilised in documenting vegetation changes. The stable isotope composition for $C_4$ plants averages -12λ while that of $C_3$ plants averages -26λ (See Deines, 1980; Fry and Sherr, 1984; Gearing, 1988; Hillaire-Marcel *et al.*, 1989; Meyers, 1994; Muzuka, 1999).

Increased precipitation is also associated with a significant transport of soil to the basins of lakes and oceans. Allochthonous soils rich in terrestrial vascular material are expected to have higher C/N ratios than lacustrine phytoplanktonic organic material because of little diagenetic effect of OM degradation during reworking, transportation and post-deposition (Meyers, 1994). The C/N ratios in the OM of lake sediments have been observed to be variable depending on relative proportion between terrestrial and phytoplanktonic material (e.g. Tenzer *et al.*, 1997; Herczeg *et al.*, 2001). Generally, the C/N ratio of unaltered autochthonous lake OM ranges from 4 to 10 while that of terrestrial material is generally higher than 20 (Hedges *et al.*, 1986; Gearing, 1988; Meyers, 1994). Since sedimentation of particulate material in lakes includes soil material derived from the catchment basin, the use of C/N ratios in conjunction with the stable isotope compositions of organic carbon (OC) may help in establishing variability in rainfall in a particular region.

Climatic records retrieved from the ice cores in Greenland and Antarctica show significant climatic variability during the Holocene (e.g. Alley and Bender, 1998; Petit *et al.*, 1999). More dramatic signals of climate change have been obtained from marine and lacustrine environments in Europe, Africa, North America, North Atlantic Ocean and Pacific Ocean (e.g. Bond *et al.*, 1997; Bluemle *et al.*, 1999; Karlén *et al.*, 1999). In some areas (e.g. Mount Kenya) this variability has been associated with changes in monsoon wind intensity and precipitation (Karlén *et al.*, 1999). Few studies have been conducted in the East African region to document variability in palaeoprecipitation. Studies that have been conducted in the region have indicated that precipitation at the end of the last glacial period was low, and lake levels were at their lowest level, with shallower lakes like Lake Victoria being completely dry (Hillaire-Marcel and Cassanova, 1987; Johnson *et al.*, 1996; Stager *et al.*, 1997; Holmgren and Karlén, 1998). This phase was later followed by an increase in precipitation to a maximum level during the early Holocene that culminated at about 5 ka (Hillaire-Marcel and Cassanova, 1987; Bigg, 1996; Stager *et al.*, 1997; Machado *et al.*, 1998; Karlén *et al.*, 1999).

High resolution records of the Holocene variability in precipitation in East Africa, may be obtained from basins that have high sedimentation rates. Such basins include small crater lakes, deltas of large lakes, and some coastal marine areas. For this reason, a 5.96 m (Core T97-69V) long core recovered from the Malagarasi delta, Lake Tanganyika, was analysed for the stable isotope composition of OC and the abundance of OC and nitrogen. This work documents sources of OM and their relative contribution and Holocene variation in precipitation, and correlates the Malagarasi delta record with other available records in the region, particularly Lake Rukwa, Lake Victoria and peat bogs in Burundi (Talbot and Livingstone, 1989; Stager *et al.*, 1997).

## 2. MATERIALS AND METHODS

### 2.1 Study Area Description

Lake Tanganyika, which is the second deepest lake in the world and is located within the western branch of the East African rift system, is believed to have been formed during the Oligocene-Miocene period (Tiercelin and Mondeguer, 1991). There are several rivers draining into the lake including the Rusizi in the north, the Malagarasi in the central eastern part, and the Rugufu and Kalambo in the south. The Lukuga river, located on the central western shoreline, drains the waters of Lake Tanganyika into the Congo river system. Among these rivers, the Rusizi and Malagarasi are the largest and have the biggest deltas.

The modern Malagarasi catchment basin includes most of the western part of Tanzania (Figure 1), and the river forms the second largest delta after the Rusizi. The river runs across various lithologies ranging from Precambrian rocks to recent sediments. Seismic profiles show the sediment cover on the Malagarasi delta to be at least 500 m thick (Tiercelin and Mondeguer, 1991).

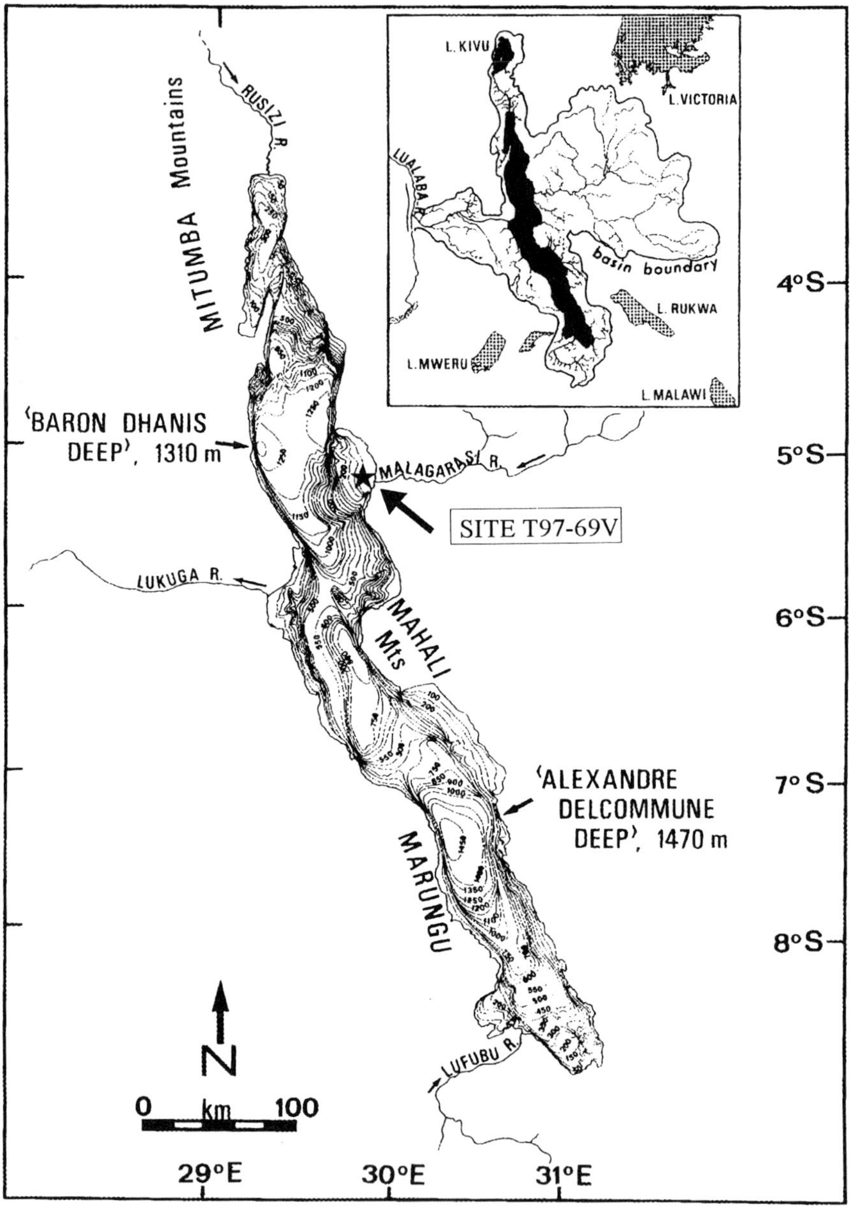


*Figure 1.* A bathymetric map showing the core site (star), major rivers entering and exiting Lake Tanganyika and the catchment basin of various rivers (insert) (modified after Tiercelin and Mondeguer, 1991).

## 2.2 Methods

Core T97-69V, which is a 5.96 m long vibro-core, was retrieved in 1997 by Prof. A. Cohen of the University of Arizona during the GEF (Global Environmental Facility) biodiversity campaign on the Malagarasi delta. The sampling site, which is located at a water depth of 60 m, was determined by GPS to be located at latitude 5°12.92'S and longitude 29°40.50'E. The core was sampled every 10 cm, and each sample represents a homogenate of 1 cm. All samples were dried at a temperature of 60°C and then ground to a fine powder. Total carbon and nitrogen measurements were performed on ground aliquots using a Carlo-Erba™ elemental analyzer at GEOTOP, University of Quebec at Montreal, Canada. Another aliquot was then acidified twice with HCl (1 M) to remove carbonates, washed, dried, and divided into two portions for stable isotope analysis and determination of residual OC content. OC and nitrogen contents are expressed in weight percent of the dry sediment. Uncertainties, as determined from replicate measurements of standard substances, are estimated to average ∀5% (relative) for OC and nitrogen.

Stable isotope compositions of OC were determined from the remaining aliquot using ISOPRIME™ instrument at GEOTOP, University of Quebec at Montreal. Isotope data of OC are reported in δλ values, after usual corrections (Craig, 1965), and with reference to V-PDB (Coplen, 1995). Uncertainties are lower than ∀0.1λ, as determined from routine replicate measurements of standards.

Although a good time control is important in documenting any palaeoclimatic variabilities, dating of this core has not been undertaken, and this weakens to some degree the interpretation presented here. To partially overcome this problem, age of the sediments deposited at this site is estimated through correlation of various published works in the Lake Tanganyika area. Although this age assignment procedure has a potential of high marginal error, it is unlikely that the actual sedimentation rates will be out of the upper (160 cm/ka) and lower (50 cm/ka) limits set using sedimentation rates that have been reported throughout the lake basin. A summary of sedimentation rates determined using $^{14}C$ data for the whole lake is given by Tiercelin and Mondeguer (1991). From this published information it is evident that the northern basin has the highest sedimentation rates which averages 160 cm/ka, and the highest sedimentation rate of about 500 cm/ka was observable in the Burton Bay (See Figure 10 of Tiercelin and Mondeguer, 1991). The southern basin has sedimentation rates that averages 100 cm/ka (see Figure 10 of Tiercelin and Mondeguer, 1991).

Data on the sedimentation rate for the central basin (Kigoma Basin) is scanty (see Figure 10 of Tiercelin and Mondeguer, 1991). Few of the available data shows that central basin has the lowest values of sedimentation that are equal or less than 50 cm/ka. Cores used to estimate sedimentation rates for the central basin were recovered from water depths in excess of 600 m, while cores for the other two basins were recovered in water depths less than 500 m (see Figure 10 of Tiercelin and Mondeguer, 1991). Three cores that were recovered in the central basin and located relatively closer to the study site are used as a base for estimating the sedimentation rate for core T97-69V. These cores include Sd 36 from the northern part, core 15 from western part, and core 19 from southern part (north of Mahale National Park 19) (see Figure 10 of Tiercelin and Mondeguer, 1991). The sedimentation rate for core Sd 36 that was recovered from a water depth of 1245 m, is about 44 cm/ka, while that of core 15, recovered at a water depth of 1180 m, is 50 cm/ka (Tiercelin and Mondeguer, 1991). The sedimentation rate for core 19, which was recovered at a depth of 660 m, is 0.7 cm/ka and is the lowest that has so far been recorded in the Lake Tanganyika (Tiercelin and Mondeguer, 1991). Since the present study site is located at 60 m, the sedimentation rate most likely is higher than the three sites surrounding it owing to the fact that sedimentation rate is inversely related to the water depth. But it is also unlikely that the rate is more than twice that of site 15, and most likely this site has a similar sedimentation rate as those observed in the southern basin. Thus, a sedimentation rate of 100 cm/ka, which is 2 times that of core 15 and equal to that of the southern basin seems to be reasonable. This sedimentation rate of 100 cm/ka is in agreement with core lithostratigraphy which shows that the sediments recovered at this site are composed of dark to grey silt-clay mud with some plant remains and laminations (Figure 2). These types of sediments correspond to previously reported lithostratigraphy of various cores collected from the lake and were deposited during the Holocene (Tiercelin and Mondeguer, 1991).

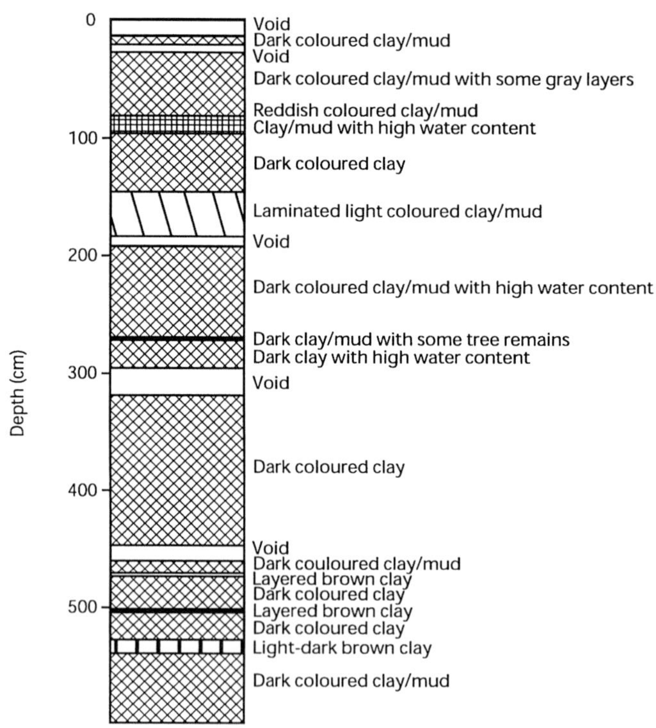


*Figure 2.* Stratigraphy for core T97-69V collected off the Malagarasi delta.

## 3. RESULTS

The stable isotopic composition of OC ranges from -25.0λ to -19.8λ and averages -23.0 ∀1.4λ. Values are relatively high in the upper 200 cm and then show a general down-core decrease to the base of the core (Figure 3). This general trend is interrupted at a depth of 370 cm by a zone that is depleted in $^{13}C$ (Figure 3). There is a sharp change in the stable isotope values of more than 3λ to lower values at a depth of 200 cm. The $^{13}C$ values show a positive correlation with the abundance of OC and nitrogen (Figs. 4a, b).

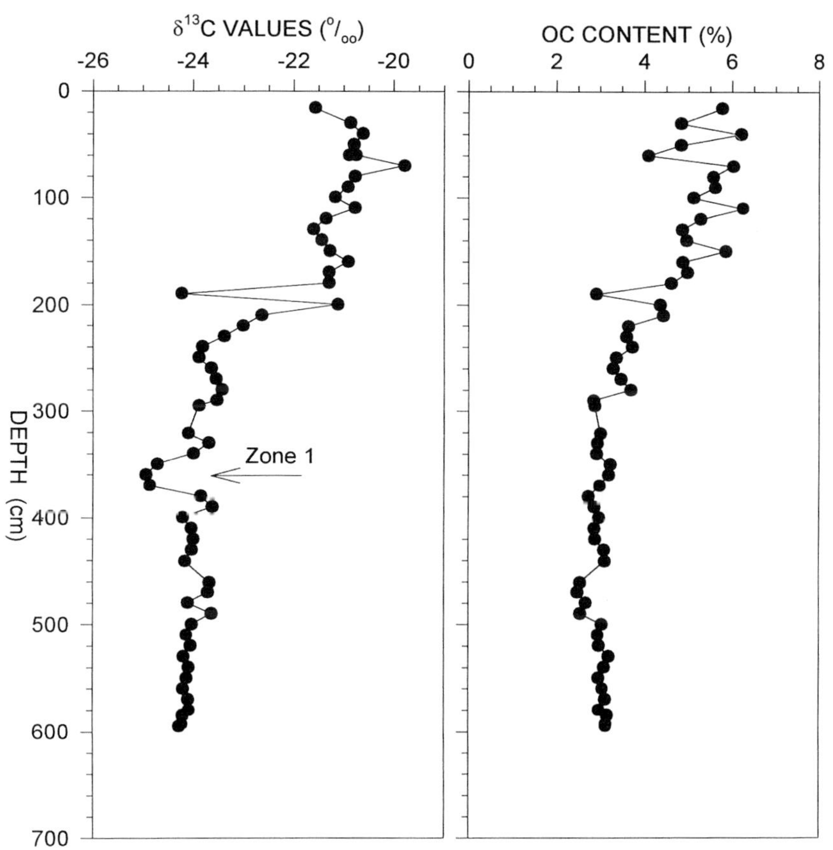


*Figure 3*. Downcore variations in the stable isotope compositions of organic carbon (left) and content of organic carbon (right) for a core collected in the eastern side of Lake Tanganyika.

The abundance of OC, ranges from 2.48% to 6.24% and averages 3.75∀1.1%. The OC decreases systematically down-core to the base of the core (Figure 3). The upper 200 cm has the highest concentration of OC, typically higher than 4% (Figure

3). Like the OC content, a higher concentration of nitrogen is observable in the upper 200 cm of the core (Figure 5). Below this depth nitrogen shows a down-core decrease to the base of the core (Figure 5). The nitrogen content ranges from 0.21% to 0.72% and averages 0.40∀0.1%. These two parameters, i.e. OC and nitrogen contents have a positive correlation (Figure 4).

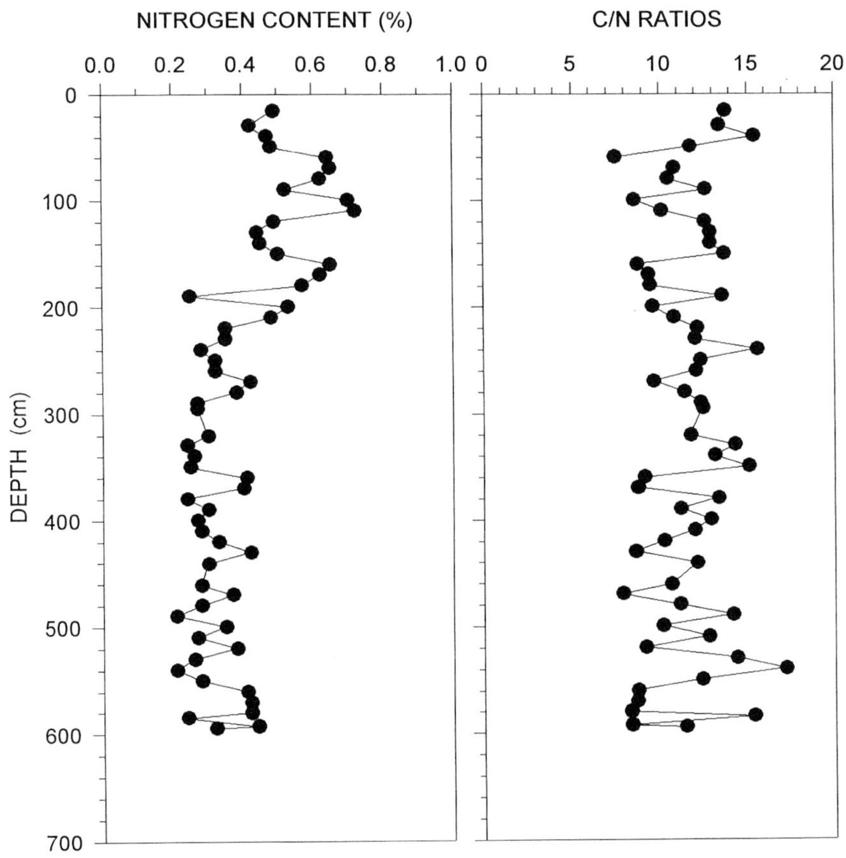


*Figure 4*. Temporal variability in the content of nitrogen (left) and C/N ratio (right) for a core recovered off the Malagarasi delta.

The C/N ratios, which range from 7.46 to 17.11 and average 11.45∀2.3, do not show any significant correlation with either $^{13}$C or OC content. This parameter shows a weak correlation (r = 0.53) with the nitrogen content, and lack any downcore trend of increase or decrease to the base of the core (Figure 5).

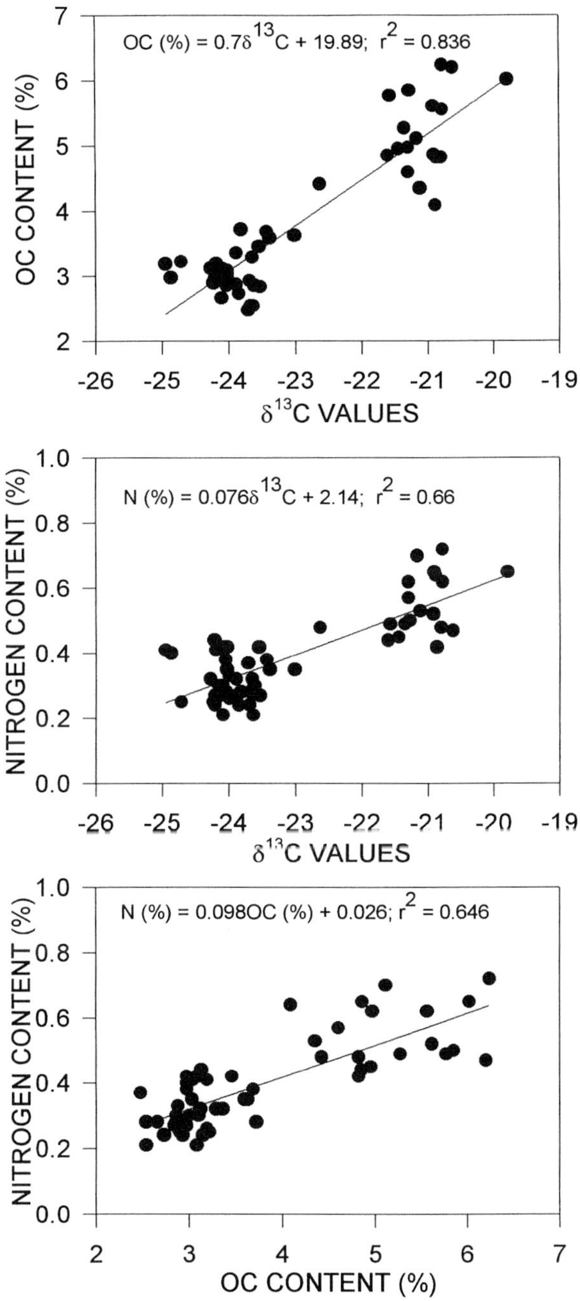


Figure 5. Bivariate plots showing correlation between $\delta^{13}C$ and OC content (top); $\delta^{13}C$ and nitrogen content (centre); contents of organic carbon and nitrogen (bottom).

## 4. DISCUSSION

The stable isotopic composition of $C_3$ plants from Burundi, Rwanda, coastal Tanzania and Kenya (Amboseli National Park) averages -28λ (Hillaire-Marcel *et al.*, 1989; Koch *et al.*, 1991; Muzuka, 1999). Similarly, the stable isotopic composition of OC for lake phytoplanktonic material range from -16λ to -38λ with most of $\delta^{13}C$ values falling within and below the range of $\delta^{13}C$ values of $C_3$ plants (Deines, 1980; LaZerte and Szalados, 1982; Aravena *et al.*, 1992; Meyers, 1994; de Giorgio and France, 1996; Mitchell *et al.*, 1996; Tenzer *et al.*, 1997). Because of their overlapping range of isotopic composition, material derived from lacustrine phytoplankton and $C_3$ plants are isotopically indistinguishable. Thus, the $\delta^{13}C$ values as high as -20λ, particularly in the upper 200 cm, are likely to have resulted from either complete utilisation of available nutrients and/or mixing of OM derived from $C_4$ and $C_3$ terrestrial and lacustrine environments.

An enrichment in $^{13}C$ in phytoplanktonic material may occur when available nutrients are utilised owing to high primary productivity. This is because as demand for nutrients and $CO_2$ increases owing to increase in primary productivity preferential utilisation of $CO_2$ rich in $^{12}C$ is reduced, and leading to formation of OM that is enriched in $^{13}C$. Thus, enrichment in $^{13}C$ in the upper 200 cm my partly be attributed to increase in lake productivity associated with complete utilisation of available nutrients. High concentration of OC and nitrogen associated with relatively low C/N ratios that averages 11.5∀2.3 suggest that the Malagarasi delta has experienced high primary productivity in recent years. However this mechanism alone is not enough to cause a change in $^{13}C$ of at least 8λ. The most probable reason is the systematic deterioration in climatic condition that led to input of OM enriched in $^{13}C$ from land. A systematic change in climatic condition could have led to exposure of the part of the delta forming wetland (marshes) owing to a fall in lake level, and thus causing high input of nutrients and land material due to the sampling site being in close proximity to the land. The C/N ratio signature of land derived material could have easily been diluted by phytoplanktonic material.

Therefore an up-core increase in the $\delta^{13}C$ values observed in this study can also be attributed to a progressive increase in the relative proportions of $C_4$ derived OC with time. This inference means that there has been a general vegetation change from a dominance of $C_3$ to $C_4$ type of plants. This change was most likely in response to a progressive decrease in precipitation in the Malagarasi catchment basin since the middle Holocene. A decrease in precipitation is supported by a diatom record from Lake Victoria which has indicated that the conditions in precipitation were more seasonal between 7200-2000 yr BP and were drier from 2200 yr BP to the present (Stager *et al.*, 1997). Even a record from Lake Bosumtwi, Ghana shows that climatic conditions were drier between 3 ka and present when compared to the early-middle Holocene (Talbot and Johannessen, 1992). Another line of evidence on a decline in precipitation is provided by a down-core depletion in $^{18}O$ of authigenic carbonate that was deposited in Lake Turkana owing to a decrease in the input of freshwater since the middle Holocene (Ricketts and Johnson, 1996).

Since precipitation in the East African region is largely regulated by the intensity of the monsoon winds, the inferred decline in precipitation could have been a response to a decline in the intensity of the African monsoon. A decline in the intensity of the Asian monsoon winds since the middle Holocene has been documented through the use of planktonic foraminifera in the Arabian Sea (Naidu and Malmgren, 1996), suggesting that the monsoon has weakened on both continents. Weaker SE monsoons over the Indian Ocean have also been inferred from $\delta^{18}O$ in planktonic foraminifera and in the abundance of dolomite, and $CaCO_3$ in the Arabian Sea (Sirocko *et al.*, 1993). Glacier fluctuations on mount Kenya also support a weaker African monsoon in the late Holocene (Karlén *et al.*, 1999).

A progressive down-core decrease also suggests that the lake level in Lake Tanganyika has been declining since the middle Holocene. High lake levels in the region for the mid-Holocene has been also reported for Lake Victoria (Talbot and Johannessen, 1992; Johnson *et al.*, 1996; Stager *et al.*, 1997), Lake Rukwa (Talbot and Johannessen, 1992), and Lake Turkana (Ricketts and Johnson, 1996).

A zone depleted in the stable isotope compositions (Figure 2) may represent a separate event when precipitation was higher than normal. Although there has been a progressive decrease in lake levels throughout the region, thus low precipitation, fluctuations towards high lake levels superimposed on generally declining lake levels could have occurred as has been documented in the Afar region, Ethiopia (Machado *et al.*, 1997). Furthermore, it has been documented that wetter conditions persisted between 4.5-3.3, and 2.0-1.8 ka corresponding to periods of high precipitation on Mount Kenya (Karlén *et al.*, 1999). Although the present work lacks age control, this event probably records/represents one of the time intervals. If the estimated sedimentation rate of 100 cm/ka is correct, 370 cm corresponds to 3.7 ka, a period which is within time interval when Mt. Kenya has been inferred to have high precipitation. Our Lake Tanganyika record correlates well with other records that suggest that a decline in precipitation was wide spread in tropical East Africa.

Available isotope data from the southern Lake Tanganyika basin (core MPU 12 from Mpulungu basin, recovered at a water depth of 460 m) show a similar down-core trend as the one observed in this study for sediments deposited since early Holocene (Hillaire-Marcel *et al.*, 1989). However, the stable isotope values for core MPU 12 are lower relative to those of this core. A major difference in the stable isotope record between the southern and central basins of the Lake Tanganyika could be a result of differences in the proportion of $C_4$ derived OM between the two areas, with the Malagarasi delta having a high proportion of $C_4$ derived material. The result for the stable isotope composition of OC for core MPU 12 would suggest a high input of $C_3$ type of organic derived from plant and phytoplanktonic material in the area when compared to the sampling site off the Malagarasi delta. This is more likely because of the presence of the large Malagarasi river, while in the southern part only small rivers capable of transporting large quantity of allochthonous material are found there.

## 5. CONCLUSION

Enrichment in $^{13}C$ observed in the upper 200 cm of this core from the Malagarasi delta can be attributed to a high proportion of $C_4$ plant organic material resulting from reduction in $C_3$ type of vegetation in the Malagarasi catchment area. Moreover, an up-core increase in $^{13}C$ observed in this study can be attributed to a progressive increase in the relative proportion of $C_4$ type of OM. This trend is attributed to a progressive decreased in precipitation since the middle Holocene. A decrease in precipitation may have been accompanied by a decrease in lake levels. A reduction in precipitation since the middle Holocene was probably due to a decrease in the intensity of the monsoon winds. Although precipitation has been declining since the mid-Holocene, some time intervals experienced high precipitation as documented by one zone depleted in $^{13}C$.

## ACKNOWLEDGEMENT

We highly acknowledge the crew members and Prof. A. Cohen for the core. We acknowledge assistance accorded to us by Dr. Alphonce Dubi and Juliana Gwakahuzu for shipment of the core from Kigoma to Dar es Salaam. We are grateful for the laboratory assistance accorded to us by Dr. Guy Bilodeau, and Prof. Claude Hillaire-Marcel of the GEOTOP, University of Quebec at Montreal for allowing use to carry out analytical work at a reduced price.

## REFERENCES

Alley, R. B. and Bender, M. (1998) Greenland ice cores: Frozen in time. *Scientific American, February 1998*, 66-71.

Aravena, R., Warner, B., MacDonald, G. L. and Hanf, K. (1992) Carbon isotope compositions of lake sediments in relation to lake productivity and radiocarbon dating. *Quaternary Research* **37**, 333-345.

Bigg, G. R. (1996) *The Oceans and Climate*. Cambridge University Press, Cambridge.

Bluemle, J. P., Sabel, J. M. and Karlen, W. (1999) Rate and magnitude of past global climate changes. *Environmental Geosciences* **6**, 63-75.

Bond, G., Showers, W., Cheseby, M., Lotti, R., Almasi, P., deMenocal, P., Priore, P., Cullen, H., Hajdas, I. and Bonani, G. (1997) A pervasive millennial-scale cycle in North Atlantic Holocene and glacial climates. *Science* **278**, 1257-1266.

Coplen, T. B. (1995) Discontinuance of SMOW and PDB. *Nature* **375**, 285.

Craig, H. (1965) The measurement of oxygen isotope paleotemperatures. *Earth and Planetary Science letters* **5**, 837-858.

Deines, E. T. (1980) The isotopic composition of reduced organic carbon, in P. Fritz and J.Ch. Fontes (eds.) *Handbook of Environmental Isotope Geochemistry, Vol. 1A*. Elsevier, New York, pp. 329-406.

de Giorgio, P. A. and France, R. L. (1996) Ecosystem-specific patterns in the relationship between zooplankton and POM or microplankton $\delta^{13}C$. *Limnology and Oceanography* **41**, 359-365.

Fry, B. and Sherr, E. B. (1984) $\delta^{13}C$ measurements as indicators of carbon flow in marine and freshwater ecosystems. *Contributions to Marine Science* **27**, 13-47.

Gearing, J. N. (1988) The use of stable isotope ratios for tracing the nearshore-offshore exchange of organic matter, in B. O. Jansson (ed.) *Lecture Notes on Coastal-Offshore Ecosystem Studies, Coastal-Offshore Ecosystem Interactions, Vol. 22*. Springer-Verlag, Berlin, pp. 69-101.

Hedges, J. I., Clark, W. A., Quay, P. D., Richey, J. E., Devol, A. H. and Santos, U. M. (1986) Compositions and fluxes of particulate organic material in the Amazon River. *Limnology and Oceanography* **31**, 717-738.

Herczeg, A. L., Smith, A. K. and Dighton, J. C. (2001) A 120 year record of changes in nitrogen and carbon cycling in Lake Alexandria, South Australia: C:N, $\delta^{15}N$ and $\delta^{13}C$ in sediments. *Applied Geochemistry* **16**, 73-84.

Hillaire-Marcel, C. and Cassanova, J. (1987) Isotopic hydrology and palaeohydrology of the Magadi (Kenya)-Natron (Tanzania) basin during the late Quaternary. *Palaeogeography, Palaeoclimatology, Palaeoecology* **58**, 155-181.

Hillaire-Marcel, C., Aucour, A-M., Bonnefille, R., Rioldet, G., Vincens, A. and Williamson, D. (1989) $^{13}C$/Palynological evidence of differential residence times of organic carbon prior to its sedimentation in East African rift Lakes and Peat Bogs. *Quaternary Science Review* **8**, 207-212.

Holmgren, K. and Karlen, W. (1998) Late Quaternary changes in climate. Technical Report TR-98-13, Swedish Nuclear Fuel and Waste Management Co. (SKB), Stockholm.

Johnson, T. C., Scholz, C., Talbot, M. R., Kelts, K., Ricketts, R. D., Ngobi, G., Beuning, K., Ssemmanda, I. and McGill, J. W. (1996) Late Pleistocene desiccation of Lake Victoria and rapid evolution of Cichlid fishes. *Science* **273**, 1091-1093.

Karlén, W., Fastook, J. L., Holmgren, K., Malmström, M., Mathews, J. A., Odada, E., Risberg, J., Rosqvist, G., Sandgren, P., Shemesh, A. and Westerberg, L-O. (1999) Glacier fluctuations on mount Kenya since ~6000 Cal. years BP: Implications for Holocene climatic change in Africa. *Ambio* **28**, 409-418.

Koch, P. L., Behrensmeyer, A. K. and Fogel, M. (1991) The isotopic ecology of plants and animals in Amboseli National Park, Kenya. *Geophysical Laboratory, Carnegie Institution Yearbook 1990/91*, pp. 163-171.

Kolla, V., Kostecki, J. A., Robinson, F., Biscaye, P. E. and Ray, P. K. (1981) Distribution and origins of clay minerals and quartz in the surface sediments of the Arabian Sea. *Journal of Sedimentary Petrology* **51**, 563-569.

Kroon, D., Steens, T. N. F. and Troelstra, S. R. (1991) Onset of monsoonal related upwelling in the western Arabian Sea as revealed by planktonic foraminifers, in W. L. Prell and M. Niitsuma, *et al.*, *Proceedings of the Ocean Drilling Program, Scientific Results, Vol. 117*, College Station, Texas, pp. 257-263.

LaZerte, B. D. and Szalados, E. (1982) Stable isotope ratio of submerged freshwater macrophytes. *Limnology and Oceanography* **27**, 413-418.

Léclaire, L. (1974) Late Cretaceous and Cenozoic pelagic deposits - paleoenvironment and paleoceanography of the central Western Indian Ocean, in E. S. W. Simpson, R. Schlich *et al.*, (eds.) *Initial Report of the Deep Sea Drilling Project Vol. 25*, Government Printing Office, Washington D.C., pp. 481-512.

Machado, M. J., Perez-Gonzalez, A. and Benito, G. (1997) Paleoenvironmental changes during the last 4000 yr in the Tigray, Northern Ethiopia. *Quaternary Research* **49**, 312-321.

Meyers, P. A. (1994) Preservation of elemental and isotopic source identification of sedimentary organic matter. *Chemical Geology* **114**, 289-302.

Mitchell, M. J., Mills, E. L., Idrisi, N. and Michener, R. (1996) Stable isotopes of nitrogen and carbon in an aquatic food web recently invaded by *Dreissena polymorpha* (Pallas). *Canadian Journal of Fisheries and Aquatic Sciences* **53**, 1445-1450.

Muzuka, A. N. N. (1999) Isotopic compositions of tropical East African flora and their potential as source indicators of organic matter in coastal marine sediments. *Journal of African Earth Sciences* **28**, 757-766.

Naidu, P. D. and Malmgren, B. (1996) A high-resolution record of late Quaternary upwelling along the Oman Margin, Arabian Sea based on planktonic foraminifera. *Paleoceanography* **11**, 129-140.

Peterson, G. M., Webb, I. T., Kutzbach, J. E., van der Hammen, T., Wijmstra, T. A. and Street, F. A. (1979) The continental record of environmental conditions at 18000 years B.P: An initial evaluation. *Quaternary Research* **12**, 47-82.

Petit, J. R., Jouzel, J., Raynaud, D., Barkov, N. I., Barnola, J-M., Basile, I., Bender, M., Chappellaz, J., Davis, M., Delaygue, G., Delmotte, M., Kotlyakov, V. M., Legrand, M., Lipenkov, V. Y., Lorius, C., Pépin, L., Litz, C., Saltzman, E. and Stievenard, M. (1999) Climate and atmospheric history of the past 420,000 years from the Vostok ice core, Antarctica. *Nature* **399**, 429-435.

Pokras E. M. and Mix, A. C. (1985) Eolian evidence for spacial variability of late Quaternary climates in tropical Africa. *Quaternary Research* **24**, 137-149.

Prell, W. L. and Van Campo, E. (1986) Coherent response of Arabian Sea upwelling and pollen transport to late Quaternary monsoonal winds. *Nature* **323**, 526-528.

Ricketts, R. D. and Johnson, T. C. (1996) Climate change in the Turkana basin as deduced from a 4000 year long $\delta^{18}O$ record. *Earth and Planetary Science Letters* **142**, 7-17.

Sirocko, F., Sarnthein, M., Erlenkeuser, H., Lange, H., Arnold, M. and Duplessy, J. C. (1993) Century-scale events in monsoonal climate over the past 24,000 years. *Nature* **364**, 322-324.

Stager, J. C., Cumming, B. and Meeker, L. (1997) A high resolution 11,400-Yr diatom record from Lake Victoria, East Africa. *Quaternary Research* **47**, 81-89.

Talbot, M. R. and Johannessen, T. (1992) A high resolution paleoclimatic record for the last 27,500 years in tropical West Africa from the carbon and nitrogen isotopic composition of lacustrine organic matter. *Earth and Planetary Science Letters* **110**, 23-37.

Talbot, M. R. and Livingstone, D. A. (1989) Hydrogen index and carbon isotopes of lacustrine organic matter as lake level indicators. *Palaeogeography, Palaeoclimatology, Palaeoecology* **70**, 121-137.

Tenzer, G. E., Meyers, P. A. and Knoop, P. (1997) Sources and distribution of organic and carbonate carbon in surface sediments of Pyramid Lake, Nevada. *Journal of Sedimentary Research* **67**, 884-890.

Tiercelin, J-J. and Mondeguer, A. (1991) The geology of the Tanganyika trough, in G. W. Coulter (ed.) *Lake Tanganyika and its Life*. Oxford University Press, London, pp. 7-48.

# LATE QUATERNARY SEDIMENTATION AND CLIMATE IN THE LAKES EDWARD AND GEORGE AREA, UGANDA - CONGO

TINE LÆRDAL[1], MICHAEL R. TALBOT[1] AND JAMES M. RUSSELL[2]

[1]*Geological Institute, University of Bergen, Allégaten 41, N-5007 Bergen, Norway. E-mail: tine.lardal@geol.uib.no*

[2]*Limnological Research Centre, University of Minnesota, 220 Pillsbury Hall, 310 Pillsbury Dr. SE, Minneapolis, MN 55455-0219, USA*

**ABSTRACT**

Sedimentological and geochemical analyses of four cores from Lake Edward, East Africa, provide a detailed record of climate and lake-level changes during the Late Pleistocene and Holocene. Our record suggests that the lake was lower during the Late Pleistocene, in agreement with previous records of lake-level change in East Africa. Following an Early Holocene high stand, during which Lakes Edward and George were connected as one large water body, lake levels dropped as a result of a shift to drier climates, possibly combined with tectonic lowering of the Semliki outlet. This fall in lake level led to the desiccation of Lake George, was accompanied by tectonic activity in the basin, and exposed a >10 m fault scarp associated with the Kasindi Fault Zone. This fault zone divides the basin into a western and an eastern section and the exposed fault scarp caused damming of rivers that entered the basin from the north and northeast, leading to the creation of two separate water bodies. Low Lake Edward in the west, which was a closed lake and where extensive beaches formed along the eastern shore, and Lake Mweya located east of the fault zone. Lake Mweya was at a higher elevation than low Lake Edward and was apparently a well-flushed basin, receiving water from several rivers and possibly draining into low Lake Edward. Sediments deposited during this low stand have different geochemical characteristics,



*E.O. Odada and D.O. Olago (eds.),*
*The East African Great Lakes: Limnology, Palaeolimnology and Biodiversity,* 429–470.

suggesting that limnological conditions in the two basins were different. Lake levels began to rise around 4.5 ka BP, uniting the two lakes and creating modern Lake George some time before 3.6 ka BP. The transgression caused drowning of extensive swamps and marks a change to gradually more lacustrine conditions in the Edward and George basins. There are indications of a shift to a slightly more arid climate around 1.7 to 1.8 ka BP, correlating with climate records from other East African lakes.

# 1. INTRODUCTION

Lake Edward is located in the northern part of the western arm of the East African Rift System (EARS; Figure 1). The lake is about 40 km wide and 70 km long and occupies a half-graben basin formed during periods of Cenozoic rifting (Rosendahl, 1987; Ebinger, 1989). Volcanism in the area has been estimated to have initiated about 10 to 12 Ma years ago, suggesting a Late Tertiary age for rift initiation in this part of the EARS (Ebinger, 1989; Pasteels *et al.*, 1989).

The lake was subject to research by the Woods Hole Oceanographic Institute (WHOI) in the early 70's, at which time Hecky and Degens (1973) completed a detailed investigation of the chemical stratigraphy and paleolimnology of the Central African Rift Lakes. More recently, the lakes has been subject to attention through the IDEAL (International Decade for the East African Lakes) project, mainly as an archive of past climate changes, but also because of probable eutrophication, changes in fish stock, its relation to volcanism and as a possible modern example for syn-rift sedimentation and tectonics (Russell and Kelts, 1999; Russell and Kelts, 2000; Lehman, in press; Beuning, in prep; Russell *et al.*, in prep).

Sedimentological and geochemical analyses of organic material (OM) from four cores collected from Lake Edward in 1996 yield a high resolution record of the Holocene period, supporting earlier interpretations of a regional lake low stand in East Africa around 4.5 ka BP (~4000 $^{14}$C years ago) (Hecky and Degens, 1973; Viner and Smith, 1973; Talbot 1982; Street-Perrott and Harrison, 1984; Talbot and Livingstone, 1989; Bonnefille *et al.*, 1990; Beuning *et al.*, 1997; Marchant *et al.*, 1997). In addition, our records show the importance of combining an understanding of basin tectonics with the sedimentary history.

## 1.1 Geological Background

Lake Edward (912 m asl) is situated in the western arm of the EARS, on the border between Uganda and Congo (Bishop, 1969; Hecky and Degens, 1973; Livingstone and Melack, 1984; Musisi, 1991). The deepest part of the lake (>120 m) is located just a few km off the western shore and correlates to the main border fault of the basin (the Lubero Border Fault). From there the lake floor rises gradually towards the eastern shore (Figure 1).

The lake floor is affected by neotectonics, especially along the Kasindi Fault Zone (KFZ) located in the central part of the basin (Lærdal and Talbot, in press).

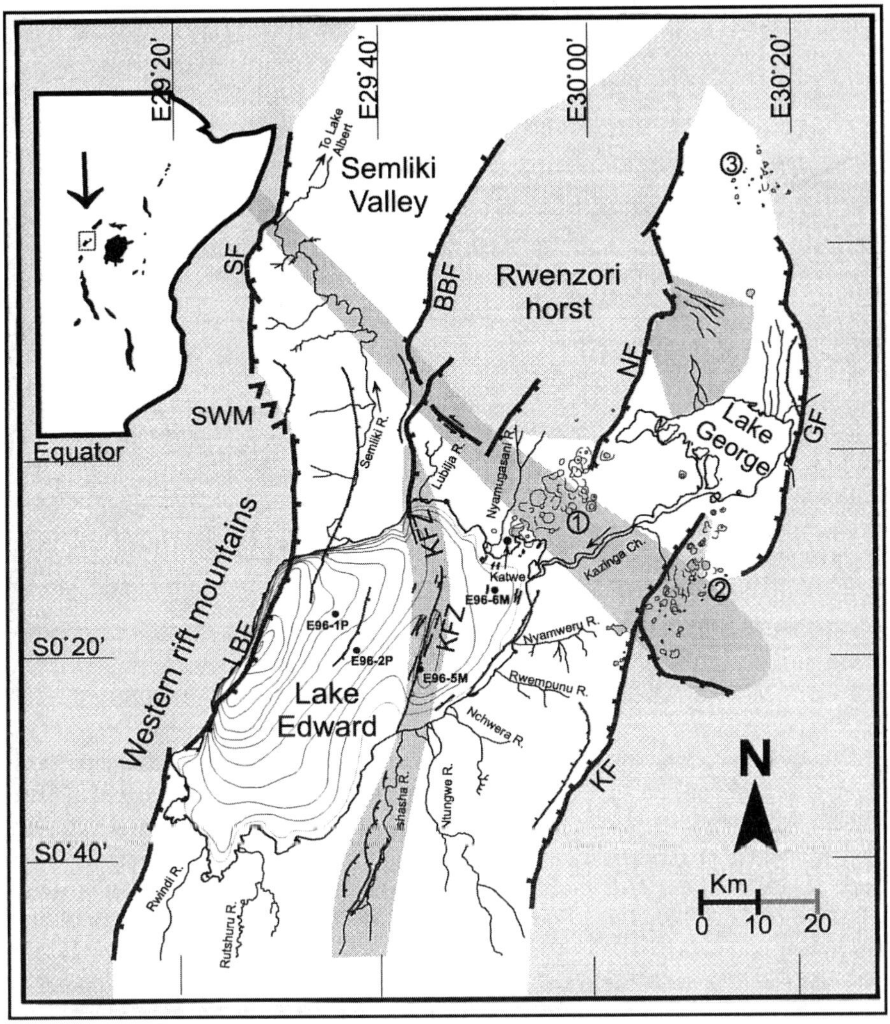


*Figure 1.* Map of the Lakes Edward and George area, showing the main structural features in the area, lake bathymetry and core sites in Lake Edward. SWM = Semliki Fault, BBF = Bwamba Border Fault, KFZ = Kasindi Fault Zone, KF = Kichwamba Fault, NF = Nyamwamba Fault, GF = George Fault. 1 - The Katwe-Kikorongo volcanic craters, 2 - the Bunyaruguru volcanic crater areas and 3 - the Fort Portal volcanic craters. Shaded NS and NW-SE oriented areas are related to accommodation zones in the basin, while the shaded area north of Lake George represents large alluvial fans building into the basin from the Ruwenzori mountains. Note how the N-S oriented accommodation zone related to the KFZ separates Lake Edward into a western and an eastern half, and creates a local deep "pool" just east of the fault zone. The structural limits of the western and eastern rift mountains and the Ruwenzori horst are indicated by the patterned areas. Inset map for location of the area within the East African Rift System (EARS).

The KFZ correlates to a ~N-S oriented, low-relief accommodation zone, which divides the Edward basin into a western and an eastern part (Figure 1). The western sub-basin is dominated by the western border fault, and is the deeper of the two. The eastern part is shallower and affected by small-scale faulting, associated with the flexural margin. The KFZ has been active throughout the Holocene period, and separates Holocene deposits in the western and eastern basins (Russell and Kelts, 2000; Lærdal and Talbot, in press: Russell *et al.*, in prep).

The Kazinga Channel connects Lake Edward to the much smaller and shallower (avg. 2.5 m deep) Lake George in the northeast (Figure 1). It runs across a slightly elevated area, related to a high-relief accommodation zone (Rosendahl, 1987; Ebinger, 1989; Lærdal and Talbot, in press). This zone is also associated with volcanism (Figure 1), represented by two fields around the Kazinga Channel, one to the north (Katwe-Kikorongo), the other, Bunyaruguru, to the south (Boven *et al.*, 1998; Kampunzu *et al.*, 1998). Volcanic activity in the Katwe-Kikorongo area has occurred in several phases during the Pliocene and Quaternary, with the most recent eruptions dated to about 8 to 6 ka BP (de Heinzelin, 1955; Brooks and Smith, 1987; Boven *et al.*, 1998). In addition, drowned volcanic craters have been identified within the northeastern corner of Lake Edward, just south of the town of Katwe (Livingstone, pers comm. 2000) and within the southwestern corner of Lake George. The course of the Kazinga Channel has been affected by faulting related to the accommodation zone located between Lake Edward and Lake George, and to river incision into unconsolidated lake sediments during a rapid fall in lake level (Bishop, 1969; Beadle, 1981; Musisi, 1991; Byakagaba, 1997; Lærdal and Talbot, in press). There is a weak flow of water from Lake George (914 m asl) to Lake Edward (912 m asl). Due to the shallowness of Lake George and the absence of a sill at the outflow, a lake level fall of >2.5 m of Lake Edward would lead to complete desiccation Lake George.

To the north, the Edward basin is limited by the Ruwenzori mountains, which formed as an uplifted, N-S oriented horst block during Miocene rifting (Maasha, 1975; Rosendahl, 1987; Ebinger, 1989), and to the south, by the Virunga Volcanic chain (Mohr and Wood, 1976; Kampunzu and Mohr, 1991). The high rift mountains in the west and in the east form additional structural and topographic boundaries between the basin and surrounding areas (Figure 1).

## 2. BASIN AND LAKE HYDROLOGY

### 2.1 Lake Edward

Lake Edward is an open lake, draining into Lake Albert through the Semliki River in the north. The lake receives water from several rivers, the most important being the southern inputs (dominated by the Rwindi, Rutshuru and Ishasha rivers) and the Nyamugasani and the Lubilja rivers in the north, sourced by the Rwenzori mountains. In addition, the Kazinga Channel draining the George basin, enters the lake from the NE (Figure 1). There are also several smaller rivers running off the western and eastern rift mountains, many of which are seasonal (Beadle, 1981; Musisi, 1991; Lehman, in press). See Lehman (in press) for a detailed presentation of the morphometry and hydrology of Lake Edward.

Based on numbers presented by Lehman (in press) precipitation (P) directly onto the lake surface accounts for as much as 21% of the total income of water to Lake Edward. Similarly about 19.4 % of the water leaves the system through evaporation (E) from the lake surface. As many of the rivers entering the Edward basin are intra-basinal (meaning that their drainage areas are located within the limits of the Edward-George basin), climatic change within the basin as a whole would result in variations in river discharge, thereby amplifying changes in lake level. Climatic change, resulting in variations in the P/E ratio and river discharge may therefore often lead to changes in the lake level (Street-Perrott and Harrison, 1984).

The lake is anoxic below ~40 m of water depth (Lehman, in press), but mixing to greater depths may occur during the windy season (Beadle, 1981; Livingstone and Melack, 1984). Anoxic bottom waters provide excellent conditions for OM preservation (Talbot, 1988). The lake floor, especially the shallower areas, may be subject to bottom currents, causing sediment reworking, erosion and re-deposition. Generally, the sediment cover in the shallower areas appear thinner than in the deeper and more protected areas of the lake (pers. observ., 1999). Lake Edward is eutrophic and sedimentation in the open waters is (presently) dominated by cyanophytes and chlorophytes (Hecky and Degens, 1973; Hecky and Kling, 1987). Near-shore sediments (i.e. around river mouths and deltas) are more sandy/silty and clastic sediments appear to be lacking in the deeper parts of the lake. Diatoms are at present not common in the lake (Lehman, in press), but have made a significant contribution to the sediments in the past.

## 2.2 Lake George

Extensive swamps occupy the northern part of the George basin and the lake shore. Several rivers enter the lake after draining the eastern slopes of the Rwenzori mountains and the eastern rift escarpment (Figure 1). River discharge closely follows the bimodal pattern of rainfall in the area, with two wet (March to May and September to November) and two dry seasons (Viner and Smith, 1973). Despite the seasonal changes, discharge in the rivers draining the almost continuously cloud-covered Rwenzori mountains are relatively high, even during the driest season. This maintains the relatively high level of both Lakes George and Edward, despite their location in a climatically rather dry part of the EARS. Lake George drains into Lake Edward through the Kazinga Channel in the west (Figure 1).

Large alluvial fans build into the George basin from the eastern slopes of the Rwenzori mountains (Figure 1), but most of the sediment is trapped in the fringing swamps. Only the finest grain sizes reach the open waters of Lake George, resulting in a high input of dissolved nutrients which sustain a very high algal production (Viner, 1969, 1977; Beadle, 1981; and pers. observ., 1996, 1999). Despite its shallow depth, the bottom waters may periodically become anoxic as a result of massive decay of OM accumulating on the lake floor. Lake stratification breaks down as the winds increase (generally the evening breeze) and/or temperature changes. Decomposition of this OM produces abundant biogenic methane, which has proved to be a problem during attempts to collect high-resolution seismic data from the lake floor (Lærdal and Talbot, in press).

## 2.3 Lake Level Variations

The level of the two lakes follow variations in P/E within the basin, but the system is also regulated by the sill at the Semliki outlet. A marked change to more humid conditions during a period on the early 60's only resulted in a ~1 m raise in the level of Lake George (Beadle, 1981; Livingstone and Melack, 1984), as opposed to several metres in many of the other East African lakes (i.e. Lakes Victoria and Albert; Beadle, 1981; Nicholson, 1996, 1997; Nicholson and Yin, 2000).

In a tectonically active region like the Edward-George basin, it may be difficult to separate lake-level variations induced by tectonics from those related to changes in climate. Active faulting may cause damming of water along fault scarps, and a tectonically controlled outlet may cause rapid lowering or raising of lake levels. As climate changes also appear to have caused rapid changes in lake level, separating the climatic from the tectonic signal left in the lake sediments may be very difficult. As a simplified rule, tectonics tend to cause more local changes (intra-basinal) in lake level, while climate changes affect a broader area. Studies suggest that many of the East African lakes followed a similar pattern of Late Pleistocene-Holocene variations in lake level, whether the lake is located in the western or the eastern branch of the EARS, or outside the main rift (Adamson *et al.*, 1980; Beadle, 1981; Hastenrath and Kutzbach, 1983; Street-Perrott and Harrison, 1984; Haberyan and Hecky, 1987; Johnson *et al.*, 1996; Beuning *et al.*, 1997; Lærdal, 1997; Delvaux *et al.*, 1998). These changes have therefore been related to changes in the regional climate rather than to tectonics.

# 3. MATERIALS AND METHODS

Two field seasons (1996 and 1999) were completed on Lakes Edward and George, during which two piston cores (E96-1P and E96-2P) and two Mackereth cores (E96-5M and E96-6M) were collected from Lake Edward (here referred to as 1P, 2P, 5M and 6M), in addition to many hundred km of high-resolution seismic data (Lærdal and Talbot, in press). Core 1P is 706 cm long and was collected in 60 m of water depth. 2P is 489 cm long and was collected in 46 m of water depth. 5M is 768.5 cm long and was collected in 29 m of water depth and the last core, 6M, is 426 cm long and was collected in only 12 m of water just off the mouth of the Kazinga Channel (Figures 1 and 2).

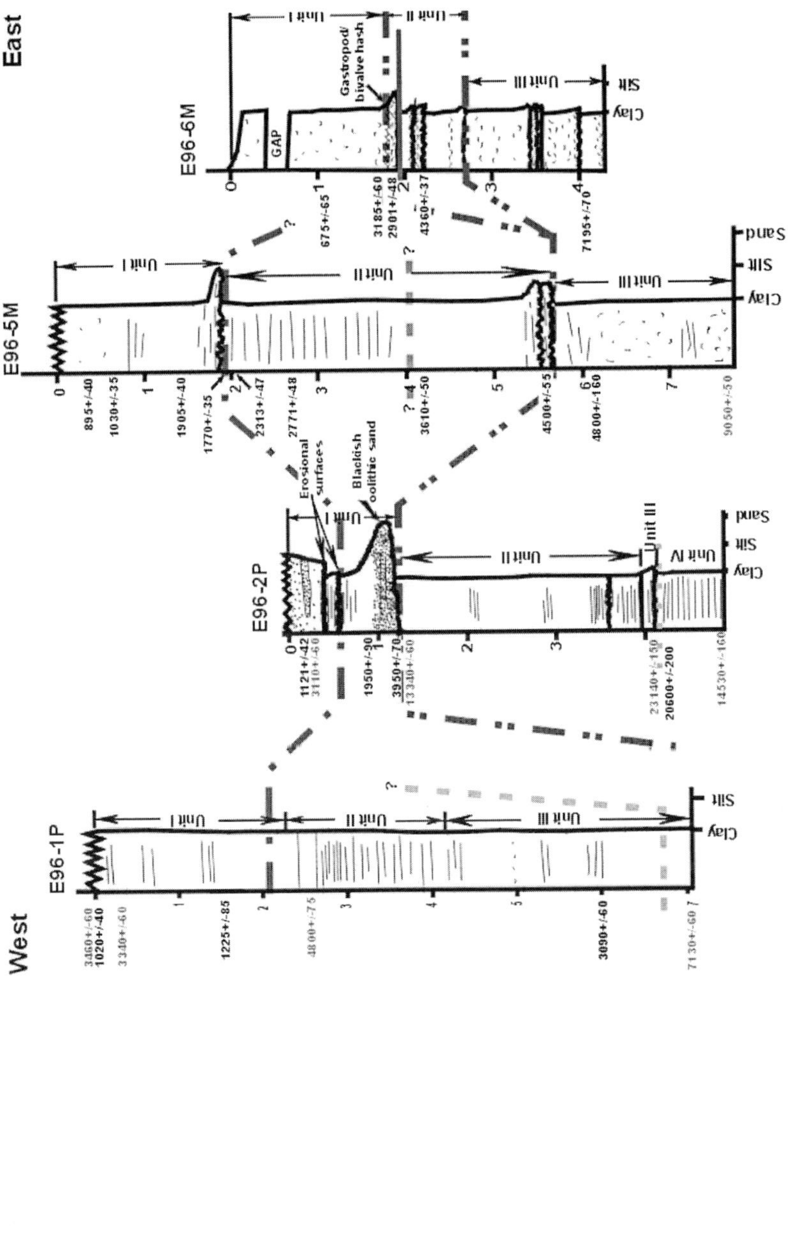


*Figure 2.* Core logs with dates and correlations. Dates in black are from charcoal and/or terrestrial macrofossils, while those in grey are from microflora concentrates (see Table 1 for more details). Correlations are based on core chronology, the geochemical records and sedimentology. The cores have been separated into units that are described in detail in the text.

The cores were cut into approximately one metre long sections in the field and sealed before immediate shipment to the Limnological Research Centre (LRC), University of Minnesota (UM), Minneapolis, where they are stored at 4°C. The cores were sampled at 10 cm intervals in 1996, re-sampled in 1999, and the sub-samples were prepared for analysis at the University of Bergen (UB).

Each sample was treated with 1M HCl to remove possible carbonate, before being washed in deionised water, dried at 45°C and pulverized. Appropriate amounts of prepared sediment were then weighed for the different analyses. The stable isotope analyses were performed at the GMS Laboratory, Geological Institute, UB. Between 11 and 49 mg were used for the carbon isotopic analyses. Elemental composition and gases for isotopic analyses were obtained simultaneously by combustion of the samples at 1020°C in a Carlo Erba 1500 element analyser, on line to a Finnegan Delta E mass spectrometer. Results are expressed in the conventional $\delta$ notation with respect to the PDB standard for carbon. Standard deviation for replicate runs are typically 0.05‰ (Talbot and Johannessen, 1992; Talbot and Lærdal, 2000). From the elemental data we have calculated the C/N ratio. The elemental nitrogen values were corrected after plotting the TN% against TOC% (Figure 3).

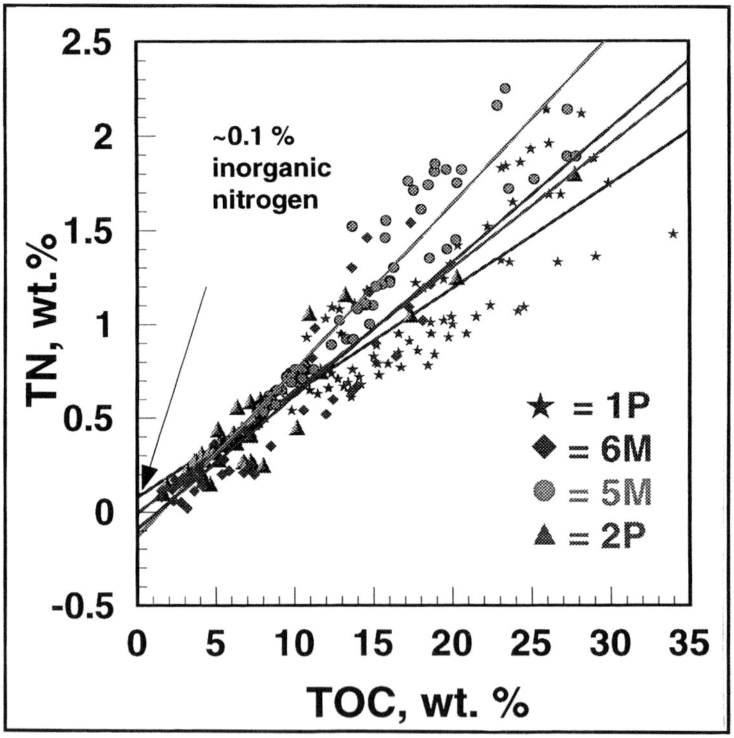


*Figure 3*. Plot of TN% vs TOC% for all four cores. Estimated ~0.1% of inorganic nitrogen present in the 1P core, based on where the regression line crosses the TN% axes. No inorganic nitrogen appears to be present in the three other cores.

A linear fit was computed from the data and the intersection point of the line with the TN% axis was determined. A value of >0 indicates the presence of inorganically bound nitrogen. The value for the mean amount of inorganic nitrogen in the cores was then subtracted from the original TN% values which was used to calculate a corrected C/N ratio. In core 1P an average of 0.1% inorganic nitrogen was assumed to be present. The other cores do not indicate the presence of any inorganic nitrogen (Figure 3). The 5M and 6M data show a linear fit where the regression line crosses the TN% axes below zero, indicating that some N-free, oxidized, possibly reworked OM is present in the cores. This also helps to explain the high C/N ratios in these cores.

Between 13 and 62 mg of powdered sediment were weighed for Rock-Eval© pyrolysis. The results are expressed as the conventional Hydrogen Index (HI), indicating the amount of milligrams of liberated hydrocarbon per gram of organic carbon. All pyrolysis analyses were performed at the Pyrolysis Laboratory at Norsk Hydro AS in Bergen, on a Rock-Eval 6 Turbo Version apparatus, which is calibrated with the IFP55000 standard. The Large Lakes Observatory (LLO) in Duluth, Minnesota provided biogenic silica (BSi%) determinations and the LRC in Minneapolis provided magnetic susceptibility and water content (WC%) (Figure 4) from the four cores. The reader is referred to Stager and Johnson (2000), and the LRC webpage (http://lrc.geo.umn.edu/Core_Facility) for methodology of these determinations.

All dates and details of the dated material are presented in Table 1. Their distribution in the cores are also shown in Figure 2. Dating of core material is still in progress and a more detailed chronology will be presented by Russell *et al.* (in prep).

## 3.1 Core Descriptions and Chronology

The cores are divided into units (Figure 2) based on changes in colour, occurrence of lamination, the relative abundance of diatoms and geochemical characteristics (Russell and Kelts, 1999).

### 3.1.1 Core E96-1P (706 cm long)

Unit I (0 - 240 cm): Dark, olive-green, poorly laminated calcareous sapropels and diatom oozes. A sample of fine-grained charcoal from 143 cm was dated to 1125+/-85 $^{14}$C years. Two $^{14}$C dates on microflora concentrates from 17 and 3 cm gave ages of 3340+/-60 and 3460+/-60 respectively, while fine-grained charcoal collected at 2 cm gave a $^{14}$C-age of 1020+/-40.

Unit II (240 - 401 cm): Well-laminated, light olive-green, slightly calcareous diatomaceous oozes and dark-olive calcareous sapropels. Laminae/beds vary in thickness, but are typically of mm scale, in places corresponding to marked changes in colour (light and dark alternations). A sample of microflora concentrates collected at 241 cm was dated to 4800+/-75 years.

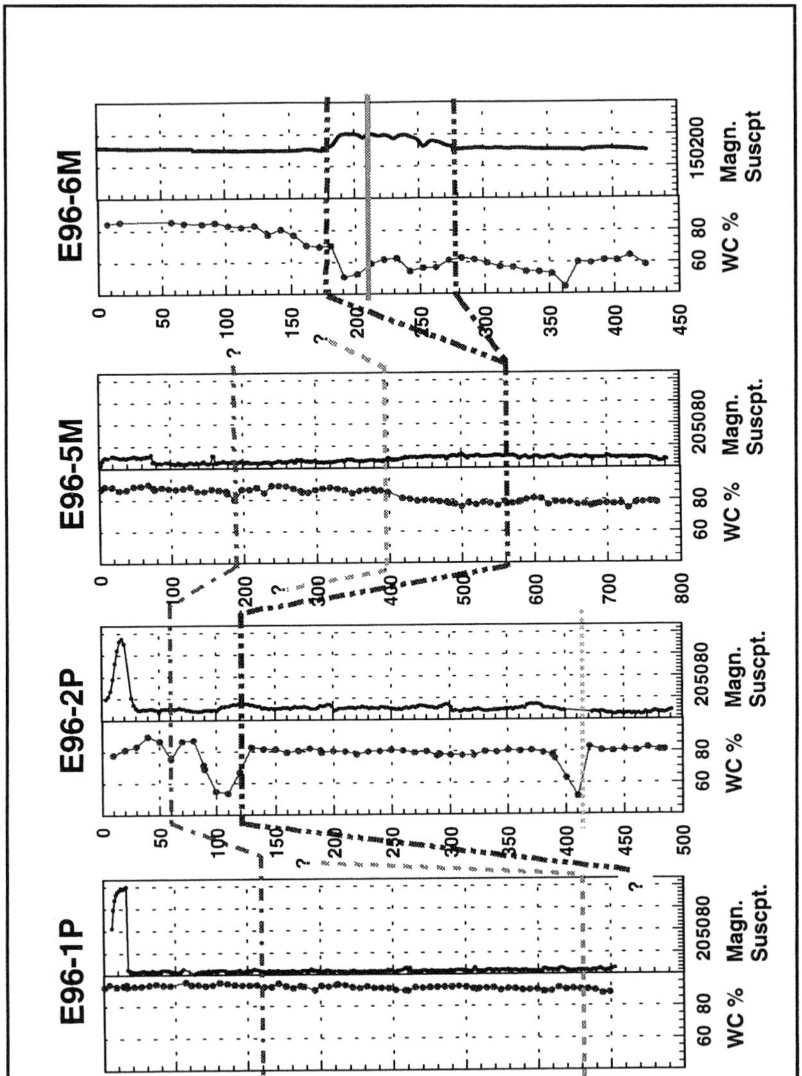


*Figure 4*. Water content (WC%) and magnetic susceptibility (Magn. Suscpt.) data for all four cores. Stippled lines are based on geochemical correlations from figures 5,6, 8 and 9. The"spike" in magnetic susceptibility values in the uppermost part of cores 1P and 2P correspond to the presence of aluminium foil in the core barrels, which was originally put into the core barrels in the field, to prevent sediment disturbance during transport.

*Table 1*. Dated samples from the Edward cores. Depth (cm) is from the top of the core section. Absolute depth (cm) is from top of core. Calendar years BP are calibrated ages using the CALIB 4.2 program developed by Stuiver and Reimer (1993). Only dates obtained from wood and charcoal fragments have been calibrated, due to the reservoir effect of the lake waters on aquatic organic matter. Calibrated ages with standard deviations in radiocarbon years greater than 50 yrs have been rounded to the nearest 10 yr.

| Core | Section | Depth | Absolute | $^{14}$C - Date | Calendar | Lab. Number | Dated material |
|---|---|---|---|---|---|---|---|
| E96-1P | I | 13 | 2 | 1020±40 | 932 | OS-26829 | charcoal |
| E96-1P | I | 15 | 3 | 3460±60 | | | pollen |
| E96-1P | I | 29 | 17 | 3340±60 | | | pollen |
| E96-1P | III | 53.5 | 143 | 1225±80 | 1170 | AA32873 | charcoal |
| E96-1P | IV | 49 | 241 | 4800±75 | | | pollen |
| E96-1P | VIII | 85 | 592.6 | 3090±60 | 3280 | AA32874 | charcoal |
| E96-1P | IX | 91 | 698.9 | 7130±60 | | | pollen |
| E96-2P | I | 12 | 12 | 1121±42 | 1030 | AA39442 | charcoal |
| E96-2P | I | 17 | 17 | 3110±60 | | | pollen |
| E96-2P | I | 86 | 86 | 1950±90 | 1910 | OS-26648 | charcoal |
| E96-2P | II | 35 | 124.7 | 3950±70 | 4420 | | charcoal |
| E96-2P | III | 37 | 126.7 | 13340±60 | | | pollen |
| E96-2P | V | 18 | 408.5 | 20600±200 | | OS-27760 | charcoal/ wood |
| E96-2P | V | 21 | 411.5 | 23140±150 | | | pollen |
| E96-2P | V | 93 | 483.3 | 14530±160 | | | pollen |
| E96-2P | core catcher | | 490.5 | 9800±60 | 11200 | OS-27768 | wood |
| E96-5M | I | 52 | 30 | 895±40 | 789 | AA37100 | charcoal |
| E96 5M | I | 82 | 60 | 1030±35 | 942 | OS-26830 | charcoal |
| E96-5M | II | 58 | 135 | 1905±40 | 1846 | AA37101 | charcoal |
| E96-5M | III | 6 | 184.2 | 1770±35 | 1640 | OS-22529 | charcoal |
| E96-5M | III | 17.5 | 196.3 | 2313±47 | 2343 | AA39457 | charcoal |
| E96-5M | III | 83.5 | 262.3 | 2771±48 | 2854 | AA39455 | charcoal |
| E96-5M | V | 16.8 | 396.1 | 3610±40 | 3899 | AA35871 | wood |
| E96-5M | VI | 74.5 | 554.2 | 4500±55 | 5170 | OS-19442 | wood |
| E96-5M | VII | 16.5 | 595.8 | 4800±160 | 5500 | OS-26646 | charcoal |
| E96-5M | VIII | 78 | 755 | 9050±50 | | | pollen |
| E96-6M | II | 15 | 90 | 675±65 | 650 | AA33218 | wood |
| E96-6M | II | 88 | 163 | 3185±60 | 3390 | AA33217 | wood |
| E96-6M | II | 99 | 174 | 2901±48 | 3052 | AA39443 | charcoal |
| E96-6M | III | 37 | 212 | 4360±37 | 4870 | AA33216 | wood |
| E96-6M | V | 14 | 388 | 7195±70 | 7990 | AA33215 | wood |
| *Mutela* | Ishango | | shore | 4130±55 | modern | T-14063 | Bivalve |
| *Caelatura* | Nyakasia | | shore | 3905±115 | modern | T-14064 | Bivalve |
| E96-3G | | 0-2 | 0-2 | 2140±50 | | | pollen |

Unit III (401 - 706 cm): Lighter coloured, diatomaceous muds and sapropels. A sample of microflora concentrates from 698.9 cm gave a $^{14}C$ date of 7130+/-60, while a piece of charcoal from 592.6 cm was dated to 3090+/-60 $^{14}C$ years.

### 3.1.2 Core E96-2P (489 cm long)

Unit I (0-125 cm): Blackish "oolitic" sand with abundant plant macro-fossils and scattered fish teeth, often with reddish-brown ferruginous coatings. The unit has an overall fining upwards trend. Erosional surfaces occur at 37.5 and 58 cm, and a lense of black "oolitic" sand is located between 28 and 31 cm. The base of the unit is marked by an undulating, erosional surface. A piece of charcoal from 124.7 was dated to 3950+/-70 $^{14}C$- years. A date on microflora concentrates from 17 cm gave a date of 3110+/-60 $^{14}C$- years, while two charcoal dates from 86 and 12 cm gave ages of 1950+/-90 and 1121+/-42 $^{14}C$-years respectively.

Unit II (125 - 396 cm): Light olive-green, diatomaceous mud, with scattered laminated intervals, especially well developed in the lower part. Microflora concentrates from 126.7 cm depth have been dated to 13340+/-60 $^{14}C$ years.

Unit III (396 - 409 cm): Calcareous reddish-grey clay. The base of the unit is associated with a slight change in grain size (Figure 2) and colour, and has been interpreted as erosional. There are faint laminations associated with this interval, and a weak fining-upwards trend. Pollen and microflora concentrates from 408.5 and 411.5 cm depth, gave $^{14}C$ ages of 20600+/-200 and 23140+/-150 years respectively.

Unit IV (409 - 489 cm): Dark olive-green/brown, faintly laminated, diatomaceous mud. A piece of wood collected from the core catcher (490.5 cm depth) gave a $^{14}C$ age of 9800+/-60, while a sample of microflora concentrates from 483.5 cm core depth was $^{14}C$ dated to 14530+/-160 years.

### 3.1.3 Core E96-5M (768,5 cm long)

Unit I (0 - 190 cm): Dark olive-brown, organic-rich and diatom-depleted mud. The bottom of the unit is marked by an erosional surface (Figure 2). Fine-grained charcoal from 184.8 cm gave a $^{14}C$ date of 1770+/-35 years, while a similar sample from 135 cm gave an age of 1905+/-40 $^{14}C$ years, suggesting possible reworking of sediment at this level. Fine-grained charcoal from 60 and 30 cm gave $^{14}C$ ages of 1030+/-35 and 895+/-40 years respectively. This could indicate much lower sedimentation rates during the last 1 ka years, or more likely, a lack of the most "recent" sediments in the core, possibly as a result of over-penetration during coring, or removal of surface sediments by bottom currents.

Unit II (190 - 564 cm): Dark olive-brown, organic-rich and diatom depleted mud. Horizontal gas expansion cracks in the upper part of the unit may be associated with lamination (although laminae are not visible). There is a slightly higher diatom content in this unit relative to the one above, especially in the lower part. Ostracod-, plant-, fish- and scattered shell remains occur throughout. A terrestrial plant fragment from 554.2 cm depth gave an age of 4500+/-55 $^{14}$C years and a piece of wood from 396.1 cm gave a date of 3610+/-40 $^{14}$C years, suggesting high sedimentation rates (Figure 2). Two samples of micro-grained charcoal from 196.3 and 262.3 cm of depth gave $^{14}$C-dates of 2771+/-48 and 2313+/-47 years respectively.

Unit III (564 - 768,5): Slightly darker than the units above. The sediment has a more crumbly appearance and a higher diatom content. Microflora concentrates from 755 cm depth was dated to 9050+/-50 $^{14}$C years, suggesting an Early Holocene age. Fine-grained charcoal from 595.8 cm depth was dated to 4800+/-160 $^{14}$C years.

### 3.1.4 Core E96-6M (426 cm long)

Unit I (0 - 177 cm): Reddish, olive-brown calcareous silt with abundant ostracod remains and few diatoms. Fine-grained charcoal from 174 cm of depth gave a $^{14}$C-age of 2901+/-48, while a plant fragment collected from 163 cm was dated to 3185+/-60. This could suggest that the sediments have been reworked. Another plant fragment from 90 cm was dated to 675+/-65 $^{14}$C years.

Unit II (177 - 277 cm): The upper ~10 cm is a gastropod/bivalve hash layer (~7-8 cm thick) with an erosive base, overlain by a thin (1-2 cm) peaty layer. This unit is slightly greener than the one above and consists of decreasingly calcareous clayey silts and sands. There are several erosional surfaces with associated shell layers (Figure 2). A piece of wood from 212 cm was dated to 4360+/-37 $^{14}$C years. This date was obtained ~20 cm's below the erosional surface associated with the bivalve/gastropod hash layer.

Unit III (277 - 426 cm): Greenish-brown clayey silt. Abundant shell and ostracod debris throughout, occasionally arranged in layers, and in places the mud has a slightly more peaty appearance. A wood fragment from 388 cm was dated to 7195+/-70 $^{14}$C years.

## 3.2 Problems Related to Dating

Investigations of the microflora concentrates that were submitted for dating have subsequently showed that some of them contained as much as 60% lake OM. This is especially a problem in productive lakes, such as Lake Edward, as many algae have sporopollenin walls. The ages obtained from microflora concentrates are therefore not true ages, but merely represent the age of a mixture of terrestrial and aquatic organic material (Beuning, in prep; Russell *et al.*, in prep). In general, dates on micro-flora concentrates give ages that are between 3 and 4 ka "older" than dates measured on fine-grained charcoal picked from the same or similar stratigraphic levels in the core. These results point towards a substantial amount of old carbon in the lake waters. Two of the topmost samples in the 1P core gave $^{14}$C dates (Table 1) from microflora concentrates that are about 2300 years older than a date on charcoal from the same stratigraphic level. This could suggest a reservoir age of >2300 years and/or a possible lack of the topmost 1 ka from the core. The more algal material present in the sample, the older the obtained $^{14}C$ date. The only reliable ages (at least for the time being) from the cores are therefore likely to be those obtained from charcoal and/or terrestrial macro-fossils (Meyers and Ishiwatari, 1995).

In addition to the discrepancy between dates on microflora concentrates and those on charcoal, dates on modern shell samples support the suggestion of a substantial old carbon reservoir in the lake waters (Table 1). Two shell samples provided by the Royal Museum for Central Africa in Belgium, and collected in the 1950's by De Heinzelin, one near Ishango on the northern lake shore (*Mutela nilotica*) and one near Nyakasia on the eastern lake shore (*Caelatura stuhlmanni*), yielded $^{14}C$ ages of 4130+/-55 and 3905+/-115 years respectively. (i.e. modern shells in Lake Edward have $^{14}C$ ages of the order of 4 ka; Table 1). These ages confirm a date of 3000+/-200 $^{14}C$-years from a shell collected from a beach near Ishango and reported by Rubin and Suess (1956). As the samples came from geographically separated localities, the effect is likely to be lake wide, rather than local. Additionally, a microflora concentrate from the top cm of a gravity core collected from Lake Edward in 1996 gave a $^{14}C$ age of 2140+/-50, suggesting that the microflora (originally believed to be pollen), was in fact mixed with autochthonous OM.

Old carbon may enter the lake waters either by degassing of $CO_2$ from drowned volcanic craters, fault planes or through hydrothermal springs (Rubin and Suess, 1956), or is possibly carried in rivers draining the Virunga Volcanoes in the south. Hydrothermal spring activity has been reported from the George basin (Twesigomwe, 1996).

### 3.2.1 Age Corrections

According to the carbon reservoir effect in the modern lake waters (based on the shell dates), it is tempting to subtract 3-4 ka from all dates on microflora concentrates. It may nevertheless be unreasonable to assume that the input of old carbon has been constant through time. The amount of carbon brought in through rivers draining the Virunga mountains is likely to have varied relative to variations in river discharge. Furthermore, the Toro Ankole volcanic province has been volcanically active for the last 50 ka (Bishop, 1969; Kampunzu and Mohr, 1991; Musisi, 1991; Boven *et al.*, 1998; Kampunzu *et al.*, 1998), and volcanic eruptions during this period have occurred in pulses. One of the last volcanic periods in the Edward area occurred around 8 ka BP, during which volcanoes were active in both the Toro Ankole and Virunga provinces (Beadle, 1981). Another volcanic period has been estimated to have occurred around 5-4 ka BP, based on a dated ash layer in a glacial moraine lake in the Rwenzori mountains (Livingstone, 1967). The input of old carbon to the waters of Lake Edward may thus have varied relative to volcanic activity in the nearby volcanic provinces.

In addition to the possible variations through time, the amount of terrestrial pollen versus aquatic algae in the microflora concentrates are likely to vary from sample to sample. The higher the percentage of pollen in the sample, the more accurate the date. Therefore, it would only be reasonable to subtract 3-4 ka (the reservoir age?) from dates obtained from 100% algal material. Most of our microflora concentrates are a mixture of the two (pollen and algae) and the "reservoir effect" on each sample would therefore vary.

In an attempt to avoid the complicating effects of "old" carbon in the lake, we have primarily relied upon $^{14}$C-dates from hand-picked or sieved charcoal and/or plant macro-fossils of a non-aquatic origin. These are likely to be little influenced (if at all) by the "old carbon" in the lake waters. Ages (from charcoal and wood samples only) have been calibrated into calendar years using the CALIB 4.2 program developed by Stuiver and Reimer (1993), and are presented in Table 1. For the calibrations of the terrestrial material we used "Dataset 1" which is based on the 1998 atmospheric decadal dataset (Stuiver *et al.*, 1998). Core correlations are based on both geochemical records and core chronology (Figure 2). All dates referred to in the text are in calendar years, unless otherwise stated.

## 4. RESULTS AND INTERPRETATIONS

The geochemical record of the four Edward cores show variations related to the core site's distance from shore. The best record of open water conditions is found in the 1P core, which was collected in the deepest water, and furthest offshore. Cores collected in shallower waters typically bear a stronger signal of fluctuating lake levels, such as erosion surfaces, deltaic- and beach-deposits, terrestrial plant material and drowned soils. Very high sedimentation rates in the open waters of the lake have resulted in a >7 m long sediment record (1P core) which appears to cover only the last ~3.9 ka. Our 10 cm sampling interval therefore offers an exceptionally detailed record of open-water conditions and palaeolimnology over the Late Holocene period in this part of East Africa. The three other cores are interrupted by erosional surfaces and hiati, which, together with very different sedimentary patterns and problematic dating, have made core-correlations difficult (Figure 2). The correlations presented here are mainly based on the calibrated $^{14}$C-dates, sedimentary patterns, major erosional surfaces and geochemical signals.

### 4.1 The 1P Core Record

Core 1P (collected in 60 m of water) contains no apparent erosional surfaces nor hiati (Figure 2) and the core record appear to span the last 3.9 ka (Figure 5). TOC% and TON% values fluctuate between 10 and 30% and 0.4% and 2% respectively. Shifts are more frequent in the upper third of the core. C/N values are high in the lower half of the core, averaging about 27, decreasing to generally low values (between 8 and 16) in the upper third of the core.

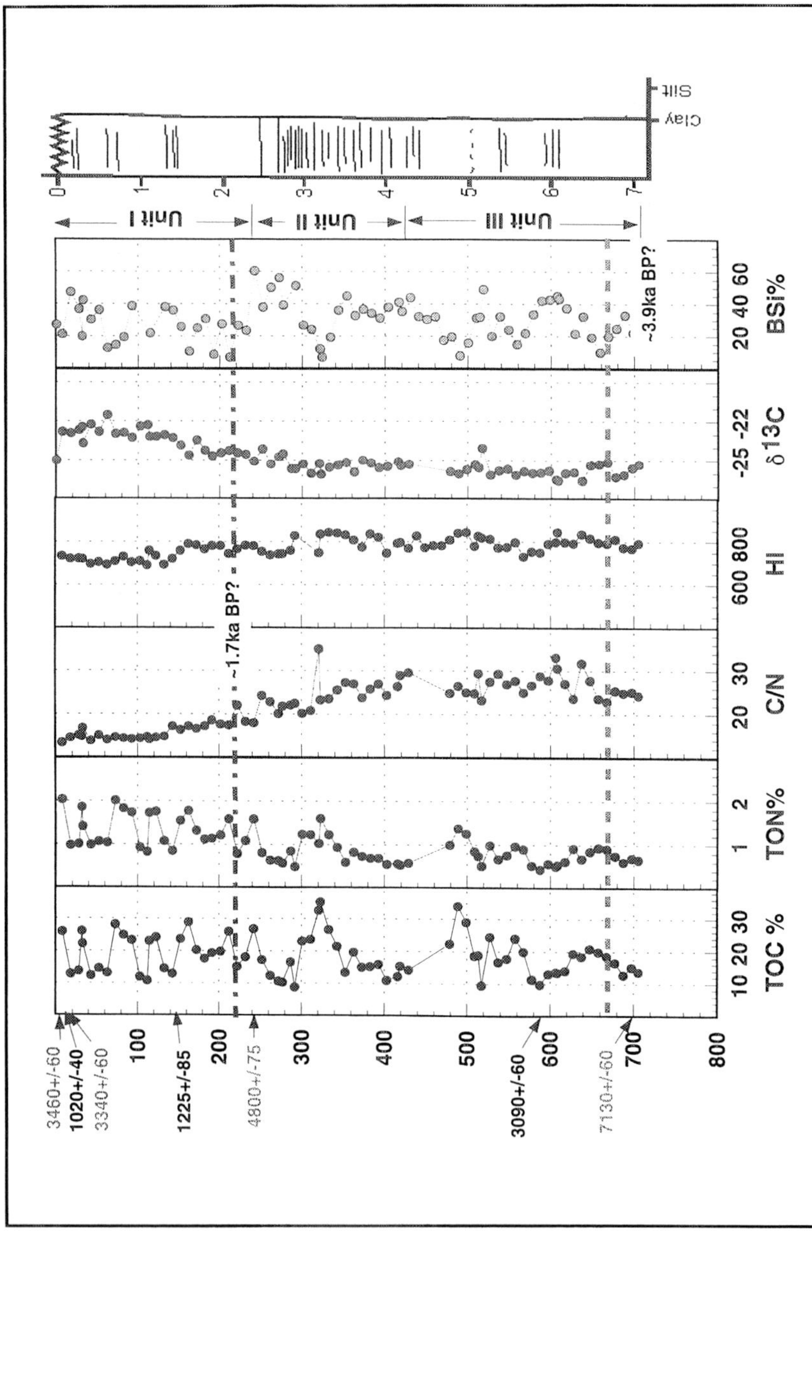


*Figure 5.* Geochemical data from the E96-1P core. Units discussed in the text and the core log are drawn to the right. Correlations are based on geochemical parameters and core chronology.

HI values are high, between 700 and 850, decreasing towards the top of the core and $\delta^{13}C$ values rise from about -27 ‰ near the base of the core to around -23 ‰ around the top, before dropping to lower values in the very upper sample of the core. Biogenic silica values fluctuate between 2 % and 60 %. In units II and III, periods with generally high values appear to be punctuated with short intervals with markedly lower values. Water content and magnetic susceptibility values are high throughout the core (Figure 4).

### 4.1.1 Interpretations

The overall rise in $\delta^{13}C$ values is interpreted to reflect the fractionation effect related to lake production in a stratified water column. This involves a gradual depletion of $^{12}C$ from the epilimnion as algae preferentially utilize the light carbon isotope during production, resulting in rising $\delta^{13}C$ values in the produced OM (McKenzie, 1985). On the other hand, if the lake is well-mixed, isotopically light $CO_2$ released to the bottom waters during degradation of OM on the lake floor, may be transported back into the epilimnion. OM produced in the open waters of the lake under such conditions will therefore have low $\delta^{13}C$ values. In addition, the $\delta^{13}C$ of phytoplankton OM is inversely related to the concentration of dissolved $CO_2$. A lake with a high $p[CO_2]$ will have OM with lower $\delta^{13}C$ values (Tyson, 1995; Meyers and Lallier-Vergès, 1999; Talbot and Lærdal, 2000). Low $\delta^{13}C$ values occasionally correspond to diatom-depleted intervals and high values of HI and TOC, suggesting possible periods of cyanobacteria-dominated lake production (Ariztegui *et al.*, in press).

As the high C/N values correspond to HI values around 800, they probably do not reflect a high amount of terrestrial plant material, but rather suggest that the OM formed in waters subject to severe nitrogen deficiency (Talbot and Lærdal, 2000). The overall decrease in C/N towards the present could suggest that nutrient replenishment is gradually improving. But, this does not appear to fit with our interpretation of the $\delta^{13}C$ record, suggesting a stratified water column during this period. One possible explanation could be that the decrease in C/N and increase in TON values reflect higher precipitation and runoff rates in the basin during this period, increasing the amount of nutrients brought into the basin. Higher precipitation rates would increase the flushing rate of Lake George and probably also the amount of nutrients washed into the basin from the surrounding mountains. The hypereutrophic Lake George may act as a nutrient source to Lake Edward and as the flushing time of the former decreases, due to sediment infilling of the basin (Viner, 1977), rate of supply of nutrients to Lake Edward may gradually have increased.

Shifts in biogenic silica in 1P probably record changes in lake production. Low values of biogenic silica correlate to peaks in TOC and HI (Figure 5). The higher TOC values are due to decreased dilution by biogenic silica, but the very high values of HI indicate excellent preservation of the OM, possibly related to periods of lake stratification and anoxic bottom waters at the core site. Low BSi% values possibly correspond to periods when lake production was dominated by other forms of algae than diatoms. In African rift lakes in general, a well-mixed water column favours diatoms, while stable stratification favours a dominance of cyanobacteria (Hecky and Kling, 1987). The periods with low BSi% values and high TOC % and HI values, could thus reflect periods when lake production was dominated by cyanobacteria. The high C/N values over this interval probably indicate OM grown under severe nitrogen deficiency, which could be a result of lake stratification and thus limited recycling of lake nutrients. Conversely, periods with biogenic silica peaks correspond to lower values of HI and TOC, possibly reflecting periods of better mixing.

A drop in biogenic silica and TOC has been estimated to have occurred ~1.7 ka BP (Figure 5). This change correlates with a shift to a more cyclical pattern and to generally lower BSi% values. There is also a general lowering of the HI values. The described shifts may correspond to a regional climate change to drier conditions and/or stronger winds (better lake mixing) that has been recorded in several of the East African lakes from this period (Talbot, 1982; Bonnefille and Mohammed, 1994).

We have counted 13 or 14 geochemical "cycles" (peaks and troughs) during the last 1.7 ka, occurring simultaneously in the TOC, TON and biogenic silica records, suggesting that the shifts reflect changes in the algal production of the lake. Over the same period HI values are high and relatively stable, and $\delta^{13}C$ values increase gradually. If the rising $\delta^{13}C$ values of the OM really do reflect a stratified water column, possible shifts in the algal production does not appear to be related to shifts in the mixing conditions in the lake.

Assuming that the shifts all occurred during the last 1700 years, this points towards periodicities of 120 to 130 years (largely depending on confirmed dates). This correlates to periodicities of lake level variations reported from the Turkana basin by Johnson et al (1991). Our sampling interval for the TOC, TON and BSi% is every ten cm. As the laminae are on a mm scale it is possible that even denser sampling would reveal even shorter periodicities. Based on this, we will delay any further discussions on the observed oscillations until a more detailed, high-resolution investigation of the cores has been performed.

## 4.2 The 2P Core Record

TOC% values are relatively low in units II and IV of this core, averaging around 5 to 8% (Figure 6), and slightly lower (avg. 2%) in unit III. There is a marked increase in unit I, with maximum values as high as 28%. TON% follows the same pattern with values in units IV and II averaging between 0.2 and 0.4%, and the lowest values occurring in unit III (average between 0.1 and 0.2%). Unit I have maximum TON% values of about 1.8 %.

C/N values reflect the pattern of the TOC% and TON% with relatively high values (~15) in the lower three units. Unit II has several peaks to relatively high values (>30), while unit I shows a weakly decreasing trend. HI values are generally

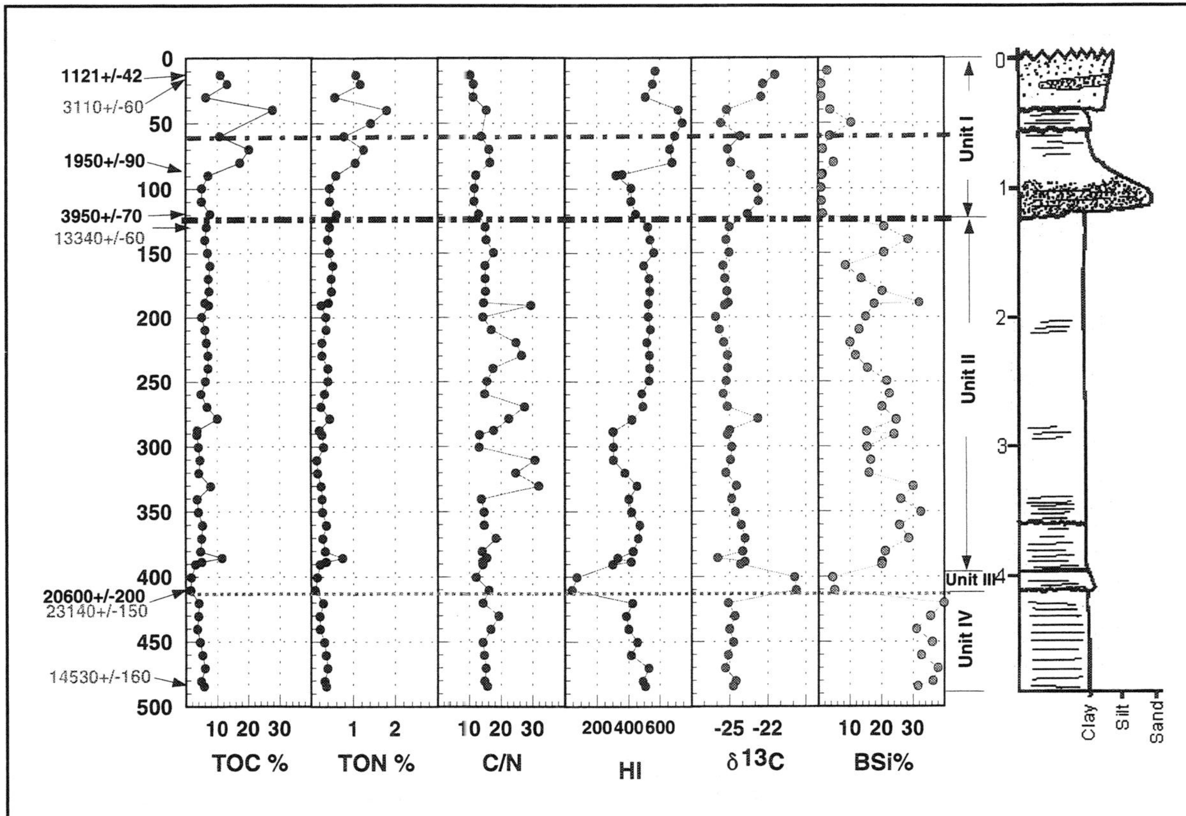


*Figure 6*. Geochemical data from the E96-2P core. Units discussed in the text and the core log are drawn to the right. Correlations are based on geochemical parameters and core chronology.

very high, averaging around 400 to 600 in units IV and II, while unit III has relatively low values (<100). Unit II has a more step-like pattern, with values around 450 between 390 and 330 cm, slightly lower values (around 360) between 330 and 280 cm, and values around 540 between 280 and 130 cm. Values first decrease in the lowermost part of unit I, but peak to high values (~700) between 380 and 35 cm. $\delta^{13}C$ values are relatively stable around -25 ‰ in units IV and II and markedly higher (around -20 ‰) in unit III. Unit I shows the opposite pattern of the HI record, first increasing, then decreasing and finally increasing towards the top of the core (Figure 6). BSi% values are high in unit IV (between 30 and 40 %) and drop to values around 5 % in unit III. Values increase in the lower part of unit II and then decrease towards the top of the unit. Unit I is marked by a rapid drop from the high values of unit II to generally low values throughout (avg. ~3 %).

### 4.2.1 Interpretations

The generally high HI values in units II and IV of the 2P core indicate to good preservation of the OM and suggest that it mainly consists of algal material. Very high HI values may be related to periodic stratification of the water column and/or high sedimentation rates, which could partly explain the high values in the lower part of the core and in the upper part of unit II. Lower values immediately above unit III suggest a period of poorer preservation of the OM, possibly related to better mixing of the water column. This could reflect increased wind activity in the basin, which could furthermore lead to increased input of terrestrial plant material into the lake waters. Increased amounts of terrestrial organic material in the lake sediments would lower the HI values and could furthermore explain the occasional peaks in C/N reported from this interval.

C/N values around 15 have generally been interpreted to reflect lake OM with a substantial contribution of terrestrial OM (Talbot and Johannessen, 1992; Tyson, 1995). This is probably not the case in the 2P core, as smear-slide analyses and high HI values show that the sediment is dominated by diatoms. Talbot and Lærdal (2000) described even higher C/N values from algal-dominated OM in Lake Victoria and related them to nitrogen starvation of the lake waters. Healy and Hendzel (1980) have demonstrated that OM produced in a lake that is starved of nitrogen, can lead to C/N values in the OM that are well above that expected for phytoplankton dominated OM. Hecky et al (1993) suggested that C/N values around 15 (as reported from unit IV, III and II) could indicate OM grown in an environment with severe N limitations. Hecky *et al.* (1993) reported similar high values from Lakes Albert and George, and related these to N deficiency in the waters. Preliminary nitrogen isotopic analysis of the OM show generally high $\delta^{15}N$ values (Talbot, and Lærdal, in prep), supporting our hypothesis of nitrogen starvation.

N is preferentially removed from the OM during early burial of the sediment and analyses of gravity cores from Lake Victoria indicate an increase of the C/N values by as much as 5 units during the early stages of burial as a result of this process (Talbot and Lærdal, 2000). Unfortunately we do not have a similar geochemical data record from Lake Edward, but a similar scenario is likely for this lake. The relatively high C/N values could indicate limited nitrogen supply to the lake waters during this period, due to high rates of OM production and burial.

The low $\delta^{13}C$ and high HI values suggest a dominance of autochthonous OM, probably related to high lake levels, during which the core site was subject to open water conditions. We have interpreted the low $\delta^{13}C$ values in units IV and II as evidence for a well-mixed lake during the Late Pleistocene-Holocene transition, allowing the light carbon released during degradation of OM on the lake floor to be transported back into the epilimnion of the lake. A slight increase in $\delta^{13}C$ values in the upper part of unit II corresponds to slightly elevated HI values (Figure 6) and this could indicate a period with lake stratification and gradual depletion of light carbon in the epilimnion (McKenzie, 1985).

The amount of biogenic silica reflects the relative abundance of diatoms in the sediment. Diatom-depleted intervals intervene with diatom-rich intervals (Russell *et al.*, in prep) and these shifts correlate with shifts in the C/N, TOC and HI records, suggesting that the shifts in the geochemical parameters may reflect changes in diatom production in the lake (Figure 6). In the intervals between 330 and 260 cm, low BSi% values correspond to slightly lower HI values and an increase in C/N values. This could suggest a period of increased input of terrestrial plant material and/or oxidizing conditions (Talbot and Livingstone, 1989; Tyson, 1995; Talbot and Johannessen, 1992; Talbot and Lærdal, 2000). Russell et al (in prep) document an increase in the clay content over the same interval, while Beuning (in prep) describes this interval as one with a marked drop in pollen concentrations. She also notes, that while most pollen concentrations decreased, the concentration of *Pteridophytes* remained constant, and she relates this to a possible lake-level lowering. The observed lowering of HI and increase in C/N values over the same interval, could support this theory.

Unit III has been interpreted to consist of in part reworked lacustrine sediments, based on the erosive character of the base of the unit, the presence of laminations and the obtained dates. The low HI values and the reddish colour of the sediments suggest that the OM has been oxidized and possibly exposed prior to deposition (Figure 6). The high $\delta^{13}C$ values may be related to high amounts of terrestrial plant material, or possibly reflecting C4 plants that grew in the lake basin. Silt-sized grains of calcite and feldspar occur in the sediments and Russell and Kelts (1999) interpreted these as reworked lake deposits that was washed into the lake as the lake level recovered from more arid Late Pleistocene climates.

The sediments are incorporated within sediments of a Late Pleistocene-Early Holocene age (units IV and II; Figure 2). We suggest that this (the Late Pleistocene - Early Holocene) might have been a period with fault activity along the Kasindi Fault Zone (KFZ; Figure 1) and relate the tilt of the sedimentary layers (including the sediments discussed here) observed on the seismic data from this area (Figure 7), to this fault activity (Lærdal and Talbot, in press). The sediments were probably tilted (to the present angle observed on the seismic data) after deposition, suggesting that the slope of the lake floor at the 2P core site was more gentle than at present during

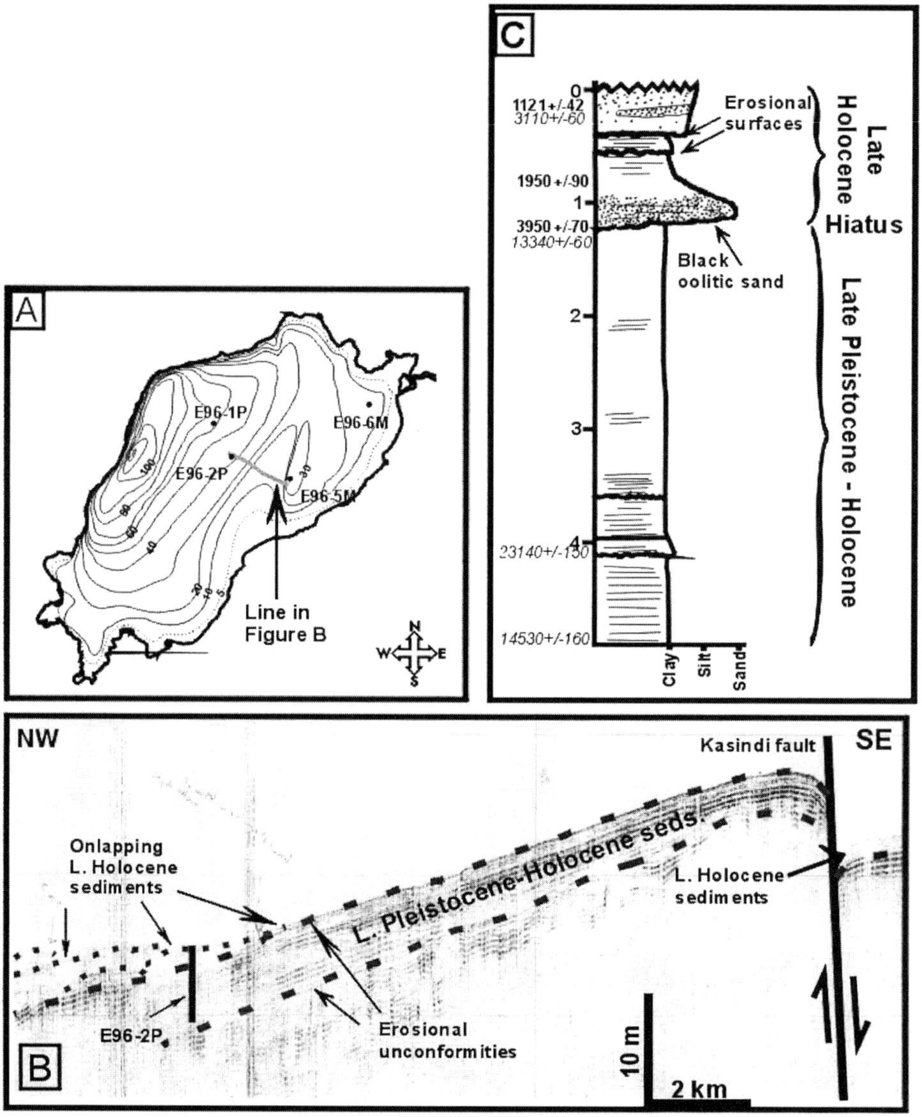


*Figure 7.* A map of Lake Edward with core sites and the location of the seismic line shown in Figure B. B – seismic line crossing the core site of 2P. Note the on-lapping sediments NW of the 2P core site which are interpreted as beach deposits. The two stippled lines represent erosional unconformities. The upper one interpreted to represent the ~4.5ka BP low-stand in the basin, and the lowermost possibly correlating to a Late Pleistocene low-stand. The tilt of the Late Pleistocene – Holocene sediments has been related to movement (uplift) on the Kasindi Fault in the East. C – Log of the 2P core related to the on-lapping beach sediments (Late Holocene) and the underlying tilted Late Pleistocene – Holocene sediments. The base of the black oolitic sands corresponds to the uppermost erosional unconformity on Figure B.

deposition of these sediments. Such a gently sloping lake floor would probably not be subject to gravity driven sediment-flows, and we therefore speculate whether the reworking and deposition of the sediments in unit III was triggered by increased wind and/or wave activity in the lake basin.

Units I and II are separated by an erosional surface (Figure 2) and a significant hiatus, and unit I is characterized by large shifts in the geochemical record. The presence of black "oolitic" sands and their geometry as revealed in seismic profiles (Figure 7), has led us to interpret these sediments as littoral deposits. On seismic profiles the deposits onlap older, tilted sediments (Figure 7) and can be traced at approximately the same depth (varies by a few metres due to later periods of tectonic uplift) in a N-S direction on several of the seismic lines (Lærdal and Talbot, in press). This suggests that the lake level was >46 m lower than today during deposition of the sands at the 2P core site, and that extensive beaches developed along the eastern shore of the lake. The sands may have been formed by reworking of Late Pleistocene deposits. Fe-rich oolitic deposits with a blackish coating have been described from Late Pleistocene outcrops in the Semliki Valley and from the northern shore of the lake (Bishop, 1969; Musisi, 1991). Similar iron-stained oolitic sands are presently forming on beaches on the eastern shore of Lake Albert (Tiercelin, pers. comm., 2001) and in the shallow waters of Lake George (pers. observ., 1999). Thick Late Cenozoic deposits have also been identified in the Albert rift (Tiercelin, pers. comm., 2001).

The lack of any preserved sediments in the 2P core from the period just prior to beach sedimentation, and the erosive character of the base of unit I, lead us to suggest that the core site may have been exposed or within a littoral, high-energy environment during this period. Due to the lack of sediments covering the period prior to beach sedimentation it is also impossible to determine for how long the shallow water conditions (or exposure?) persisted at the 2P core site. There are no signs of exposure in the upper part of unit II, but the large hiatus (covering the period between ~9 ka and 4.4 ka BP) suggests that any such signs may have been eroded by beach/shore-face processes.

Following this low stand, increasing TOC and HI values suggest a change to deeper waters. High $\delta^{13}C$ values in the lower part of unit I probably correlates to increased input of terrestrial plant material during the low-stand. The following decrease in values have been interpreted to reflect a change to more algal dominated material, and thus suggests a change to deeper water conditions. Increasing $\delta^{13}C$ values towards the upper part of the core may reflect a depletion of light carbon in the epilimnion of the lake, possibly reflecting stabilization of the water column and possible stratification. The very low BSi% values suggest that the planktonic algal community was dominated by something else than diatoms during this period. At present the lake is dominated by blue-green algae and it is thus possible that Unit I was deposited under limnological conditions similar to the present.

## 4.3 The 5M Core Record

The geochemical data record of the 5M core show much less variation than the 2P record (Figure 8). TOC % values increase gradually from the lowermost part of the core, from values just below 10%, to values around 25 % at the top of unit II. Values drop from unit II to I and are relatively stable, around 18-20 %, in the upper unit of the core. Similarly, TON % values show a gradually increasing trend from the bottom of the core (0.5%) towards the top (~2.5% at 60 cm). Values are slightly lower in the upper 50 cm. C/N values are stable throughout the core, with a weakly decreasing trend from values around 16 near the base of the core, to minimum values (~10) at the top. Hydrogen Index (HI) values are generally high throughout the core, increasing from the base (~500) towards the top of unit II (~740). Unit I has values around 620. $\delta^{13}C$ values are relatively stable around -24 ‰ throughout units III and II, increasing to a maximum of -21 ‰ in the upper part of the core. As for the C/N record, the BSi% record have generally stable values, decreasing in a step like manner through each unit towards the top of the core. Values average around 7 in Unit III, 4 in unit II and 2 in unit I.

### 4.3.1 Interpretation

The relatively invariant geochemical values and similarities with the lowermost part of the 2P core record, suggest that unit III in core 5M was deposited during a period of stable lake levels. High HI values indicate excellent preservation of the OM and relatively high TOC % values suggest high lake production. Low $\delta^{13}C$ and high HI values reflect a predominance of algal material in the OM.

Russell and Kelts (1999) noted a relatively high component of allochthonous material in the unit and related this to possibly slightly lower lake levels. For the same interval they reported an absence of calcite, which could suggest a relatively diluted lake (cf. Lehman, in press). We have interpreted this as a period with stable lake levels, although slightly lower than today's. The lake appears to have been well flushed, which may explain the high allochthonous input (several rivers enter the lake in the vicinity of the 5M core site) and apparent lack of calcite precipitates. Low values of biogenic silica could reflect a periodic lowering of diatom production as a result of limnological and/or climatic changes.

The geochemical record of unit II suggests a change to wetter climatic conditions and the rising $\delta^{13}C$ and HI values suggest a gradually deepening of the lake, possibly accompanied by a stabilization of the water column. This trend is accompanied by rising TOC and TON % values, suggesting increased production in the lake waters, possibly related to increased nutrient supply, as the rising lake transgressed formerly exposed and vegetated areas around the lake (Talbot and Lærdal, 2000). The low C/N values suggest limited supply of terrestrial plant material as a result of the rising lake levels.

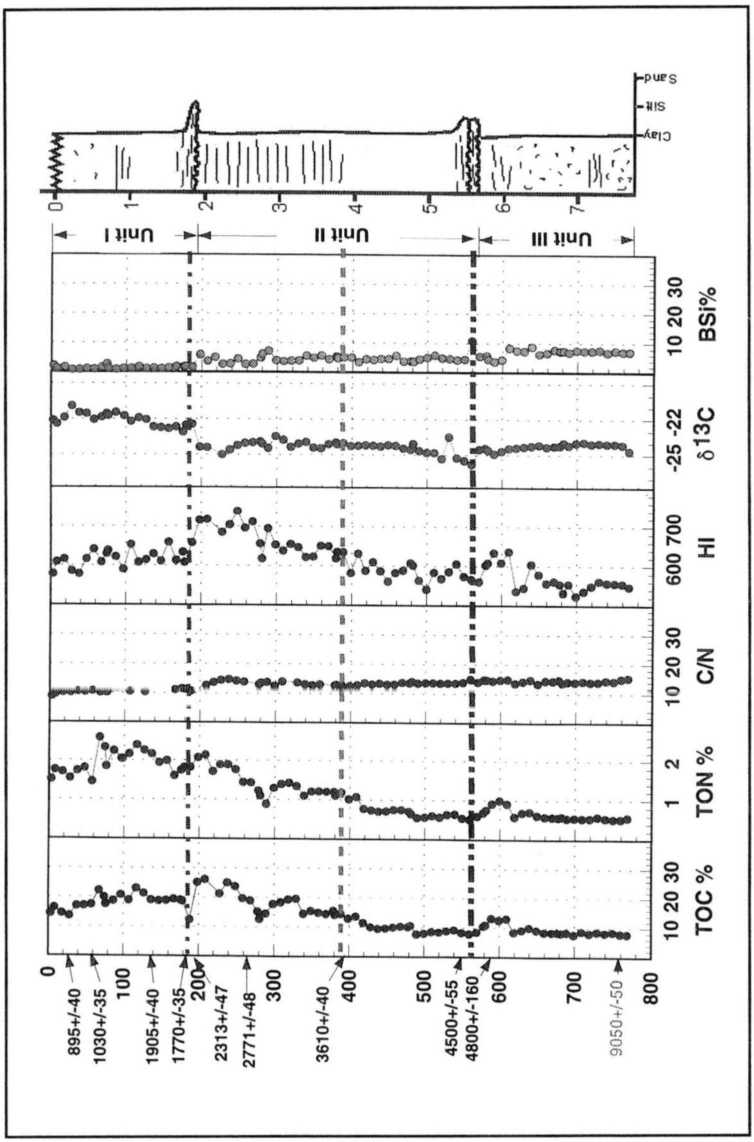


*Figure 8.* Geochemical data from the E96-5M core. Units discussed in the text and the core log are drawn to the right. Correlations are based on geochemical parameters and core chronology.

The base of unit I appear to be slightly erosional and this may be related to bottom currents due to intensified winds. The marked changes in the geochemical record in unit I suggest changes in lake production, possibly related to changes in lake water pH. Higher pH (>8.5) could lead to assimilation of bicarbonate from the lake waters, which could furthermore explain the increase in $\delta^{13}C$ values in this interval (Talbot and Johannessen, 1992). A similar increase in $\delta^{13}C$ could also be related to a change from a diatom- to cyanophyte-dominated production. Low BSi% values in the same unit could support this theory. However, the observed changes could also reflect poor preservation of diatom frustules due to alkaline bottom waters. Preliminary $\delta^{15}N$ data from the OM does not support a cyanobacteria-dominated lake at this time, as $\delta^{15}N$ values appear to be relatively high (averaging around +6‰) throughout this unit. We therefore suggest that the observed changes are related to changes in lake pH.

Calcite with a relatively high Mg/Ca ratio is present in units II and I (Russell *et al.*, in prep). This may indicate occasionally increased evaporation, lake temperatures and/or lake production (cf. Lehman, in press). Warm water enters in the northeastern corner of the lake from the shallow, less saline and slightly warmer Lake George (Viner and Smith, 1973; Lehman, in press). This could lead to a difference in water temperatures between the eastern and western parts of Lake Edward, favouring precipitation of calcite with a high Mg/Ca ratio in the warmer and shallower, eastern water body. As opposed to many of the other lakes in the area, that have a much higher proportion of sodium over calcium, Lake George is relatively richer in calcium (Viner, 1977; Lehman, in press). It is thus possible that Lake George is supplying Lake Edward (especially the eastern waters) with calcium.

## 4.4 The 6M Core Record

TOC % values are relatively low in units II and III, varying between 2 and 8 % (Figure 9). There is a marked increase in values in unit I, averaging around 15 %. TON % show a similar pattern to the TOC % record, with low values in units II and III (averaging ~0.3 %) and a marked increase in values in unit I, from about 0.5 % at the base of the unit to about 1.5 % near the top. C/N values are generally high in this core, averaging around 20 in the lower two units, and decreasing to about 10 at the top of the core. HI values are much lower in this core than in the three previous cores. The lower two units have HI values <200 and magnetic susceptibility is high over the same interval (Figure 4). There is a gradual increase in HI from the base of unit I and values of ~200 to maximum values of >400 at the top of the core. $\delta^{13}C$ values are relatively high, averaging around -22 to -23 ‰, in the lower two units, and increasing to about -20 ‰ in unit I.

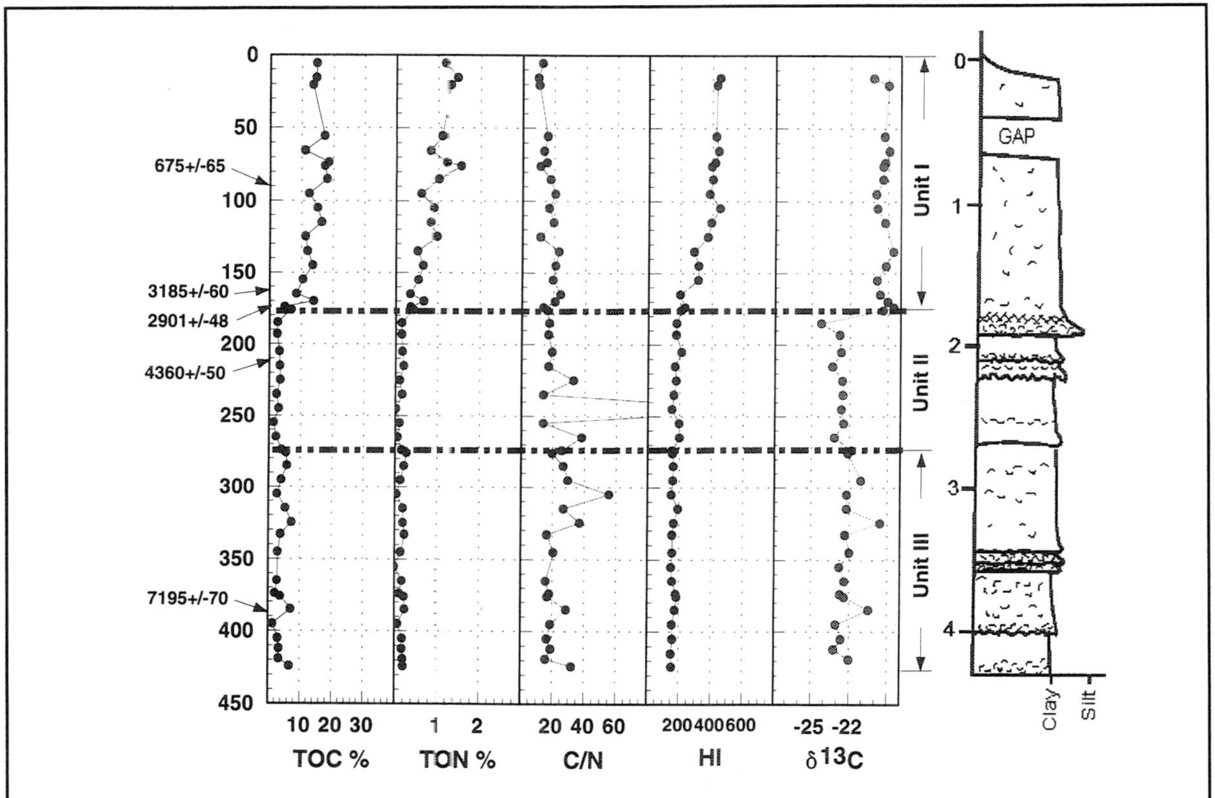


*Figure 9*. Geochemical data from the E96-5M core. Units discussed in the text and the core log are drawn to the right. Correlations are based on geochemical parameters and core chronology.

### 4.4.1 Interpretation

The geochemical parameters of units II and III all point towards a shallow-water depositional environment. Abundant shell material, erosional surfaces and well-preserved terrestrial plant fragments suggest a near-shore, littoral or fluvial environment. If so, this indicates a lake-level lowering of at least 12 m (the current depth of the core site) during deposition of the sediments. The depth at the 5M core site would then have been reduced to <17 m, at which stage Lake George would be desiccated. Under these conditions, much of the sediment brought into the George basin would most likely be transported by rivers across the exposed lake floor, through the Kazinga Channel, and deposited in the lower Lake Edward. The sediments may have been deposited in a fluvial/deltaic environment around the (lowered) lake. As the 6M core site is located just off the mouth of the Kazinga Channel (Figure 1) and just south of the Nyamugasani delta, it is possible that these rivers formed a broad fluvial/deltaic plain surrounding the 6M core site during this low-stand.

The high $\delta^{13}C$ and C/N values in units II and III may indicate nearby swamps with a C4-dominated vegetation, possibly associated with a delta. *Papyrus* swamps are common around the shores of Lake George (Viner, 1969) and also associated with the present Nyamugasani delta in the northeastern corner of Lake Edward. We suggest that similar swamps formed around the mouth of the Kazinga and Nyamugasani river mouths/deltas during this low stand. Much of the clastic sediment transported by these rivers may have been trapped in these swamps, possibly preventing it from reaching the 5M core site further south.

This gradual lowering of the lake level could suggest a change to a slightly drier climate. Nevertheless, the lake appears to have been relatively well flushed, suggesting that rivers entering the basin must have had a substantial discharge. Another possible explanation could be a lake level lowering related to a tectonic lowering of the Semliki outlet. The upper course of the Semliki river is controlled by a set of syn- and antithetic faults to the main boundary fault of the Edward basin (the LBF; Figure 1). Movement on one or several of these fault could result in a lowering of the sill of the Semliki river on the northern shore of Lake Edward. The present depth of the sill is only about 3 m deep (Beadle, 1981). Therefore, if a lake level lowering of >12 m during deposition of these sediments where to be explained by a tectonic lowering of the outlet, it also suggests that later tectonic activity have raised the level of the sill again (up to the present 3 m depth). This suggests a change from normal to reverse fault movement on the fault that control the outlet (or the main boundary fault) over a relatively short time period, which seems unrealistic in an extensional setting such as the EARS.

Unit II appears to represent a very short time interval (based on dates; see following section) around 4.8 ka BP (~4300 $^{14}C$-years), and has been related to a period of rising lake levels and high sedimentation rates at the core site. The shelly layers with erosive bases could represent switching delta lobes, - river channels and/or a fluctuating shoreline. We have interpreted this to reflect rising lake levels following the low-stand, forcing the deltas to migrate landwards, finally drowning the

6M core site. The final stage of drowning is believed to be represented by the bivalve/gastropod hash with generally well-preserved shells, and the overlying peaty layer (Figure 2). The erosional surface beneath the hash (solid line in Figure 2) may be related to erosion during this transgression, and probably represents a ravinement surface (Bruun, 1962; Swift, 1968). The trend of higher $\delta^{13}C$ values, increasing TOC and TON% values and high HI values as observed in the upper unit of core 5M can also be observed in the upper unit of this core, suggesting that the observed changes are related to a lake-wide event.

# 5. PALAEOCLIMATIC AND PALAEOLIMNOLOGIC RECONSTRUCTIONS

Based on the $^{14}C$-based chronology for the cores and the previously presented interpretations of the geochemical record, we have separated several different time periods of lake history (Figure 10).

## 5.1 The Late Pleistocene - Holocene

The three lower units (II, III and IV) of core 2P contain the oldest sediments encountered in the four cores. $^{14}C$ dates (Table 1) from unit III (Figures 2 and 6) suggest a Late Pleistocene age. Based on the calibrated age for the lowermost sample from the core (and subtracting ~3-4 ka for the reservoir effect on the microflora-dates) units II and IV appear to cover the Late Pleistocene-Holocene transition in the lake, possibly spanning the interval between 11.2 and 9 ka BP (Table 1 and Figure 2).

Unit III has been interpreted to consist of reworked lacustrine sediments. Being reworked, the sediments does not really leave any confident information about lake conditions during, or prior to, deposition. Russell and Kelts (2000) suggested that these sediments could indicate lowered lake levels during the Late Pleistocene, as has been reported from several of the other East African Lakes (Livingstone and Melack, 1984; Talbot and Livingstone, 1989; Johnson *et al.*, 1996; Lærdal, 1997; Beuning *et al.*, 1997; Talbot and Lærdal, 2000; Stager and Johnson, 2000). We find the evidence too speculative for such a conclusion to be drawn. Nevertheless, an erosional unconformity identified on seismic profiles just below the base of the cored sediments (Figure 7) could be related to exposure of the core site and thus support a theory for lowered lake levels during the Late Pleistocene. But, this unconformity needs to be cored and studied before any further conclusions can be drawn. Reworking of the sediments in Unit III appear to have occurred in the Late Pleistocene-Early Holocene. The sediments are incorporated within sediments of a Late Pleistocene-Early Holocene age (Figure 2) and we suggest that this was a period with fault activity along the Kasindi Fault Zone (KFZ).

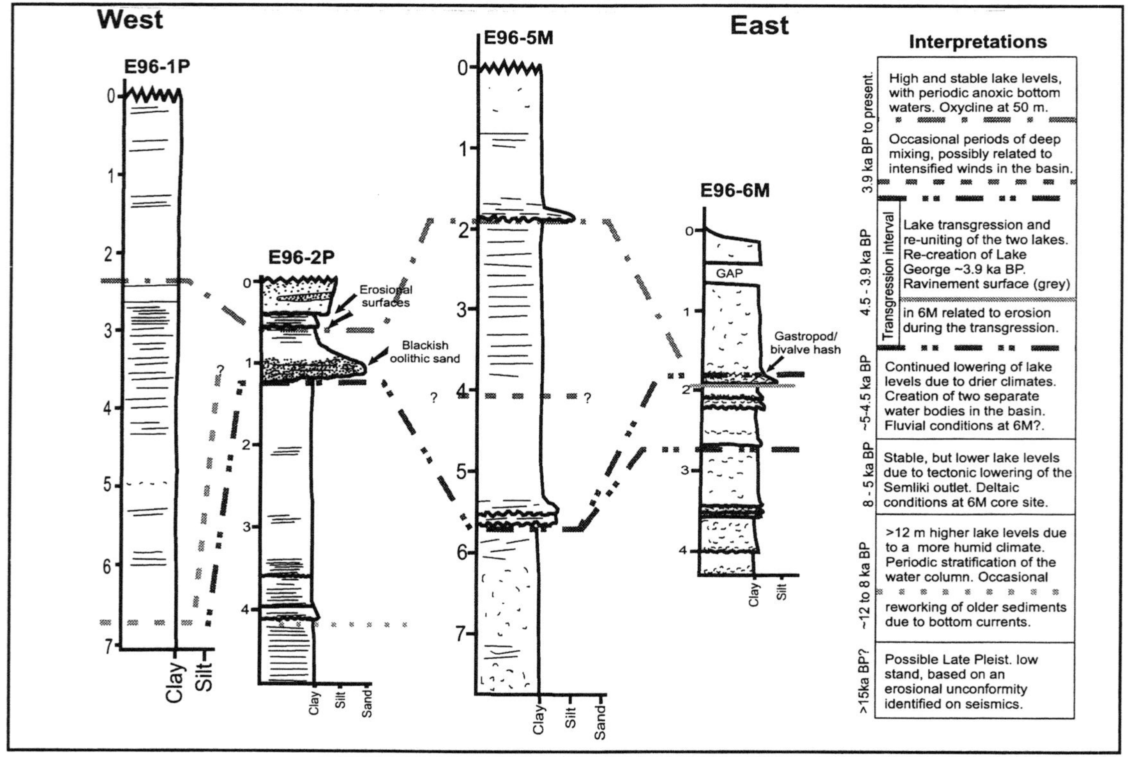


*Figure 10*. Logs of the Edward cores and the interpretation of the sedimentological and geochemical parameters. Indicated time periods are in calendar years BP (see Table 1).

Sediments of units IV and II are of an Early Holocene age. Lacustrine sediments of the same age have also been described from river gullies and ravines in the Kazinga Channel area (Bishop, 1969; Musisi, 1991), where they are located several metres above present lake level, suggesting high-stand conditions during their accumulation. Brooks and Smith (1987) described an ancient shoreline on the northern shores of Lake Edward located 12.5 m above present lake level and related it to a Late Pleistocene-Early Holocene high stand. Furthermore, hanging valleys on the western shore of Lake Edward are located some tens of metres above the present lake surface (pers. observ., 1996) and this could also be related to former periods of higher lake levels. As the Late Pleistocene-Holocene lacustrine high-stand sediments have been described both from the northern and eastern parts of the basin, we believe that they represent a humid climatic period, with lake levels >12 m above present level (Figure 11). The hanging valleys of the western rift mountains are probably related to a combination of a higher lake level and tectonic uplift. It is tempting to suggest that the high stand was associated with a somewhat higher level of the Semliki outlet, and that, following this high stand, movement on the western border fault led to a lowering of the outlet (and thus lake level).

We have furthermore suggested that the Early Holocene period was characterized by oscillations between periods with water column stability (stratification) and periods when it was (well) mixed. The reworked layer (unit III) located between units IV and II (Figures 2 and 6), may have been deposited as a result of bottom currents triggered by intensified winds.

## 5.2 The Mid-Holocene

None of the four cores cover the period between the Late Pleistocene-Holocene transition and the mid-Holocene (Figure 2). The next interval of lake history sampled in our cores is present at the bottom of 6M and 5M. In core 5M, the lowermost part of unit III has been dated on microflora concentrates to 9050 +/-50 $^{14}$C-years (~6 ka BP after corrections?). This has been interpreted as a period with stable lake conditions, although slightly lower than today, allowing for high lake production and good preservation of the OM. The lower part of unit III in the 6M core was deposited around 8 ka BP (Table 1) and is thus slightly older than the lower part of the 5M core. The unit (III of 6M) is believed to cover the period between 8 and 6 ka BP(?), and the geochemical and sedimentological records suggest lower lake levels (>12 m) than present. This suggests a possible fluvial/deltaic setting for the 6M core site.

Stager and Mayewski (1997) suggested drier climates around 8.2 ka BP from Lake Victoria. They correlated this to ice cores from the Arctic and Antarctica and suggested a possible global climatic transition in the Early-Mid Holocene period. Nevertheless, their chronology of the Victoria core is disputed (Talbot and Lærdal, 2000) and the reported signal may very well <u>not</u> be the 8.2 ka event. Our record does not give any clear evidence of drier climates, and the ~8 ka BP lowering could have been triggered by tectonics. (See earlier discussion).

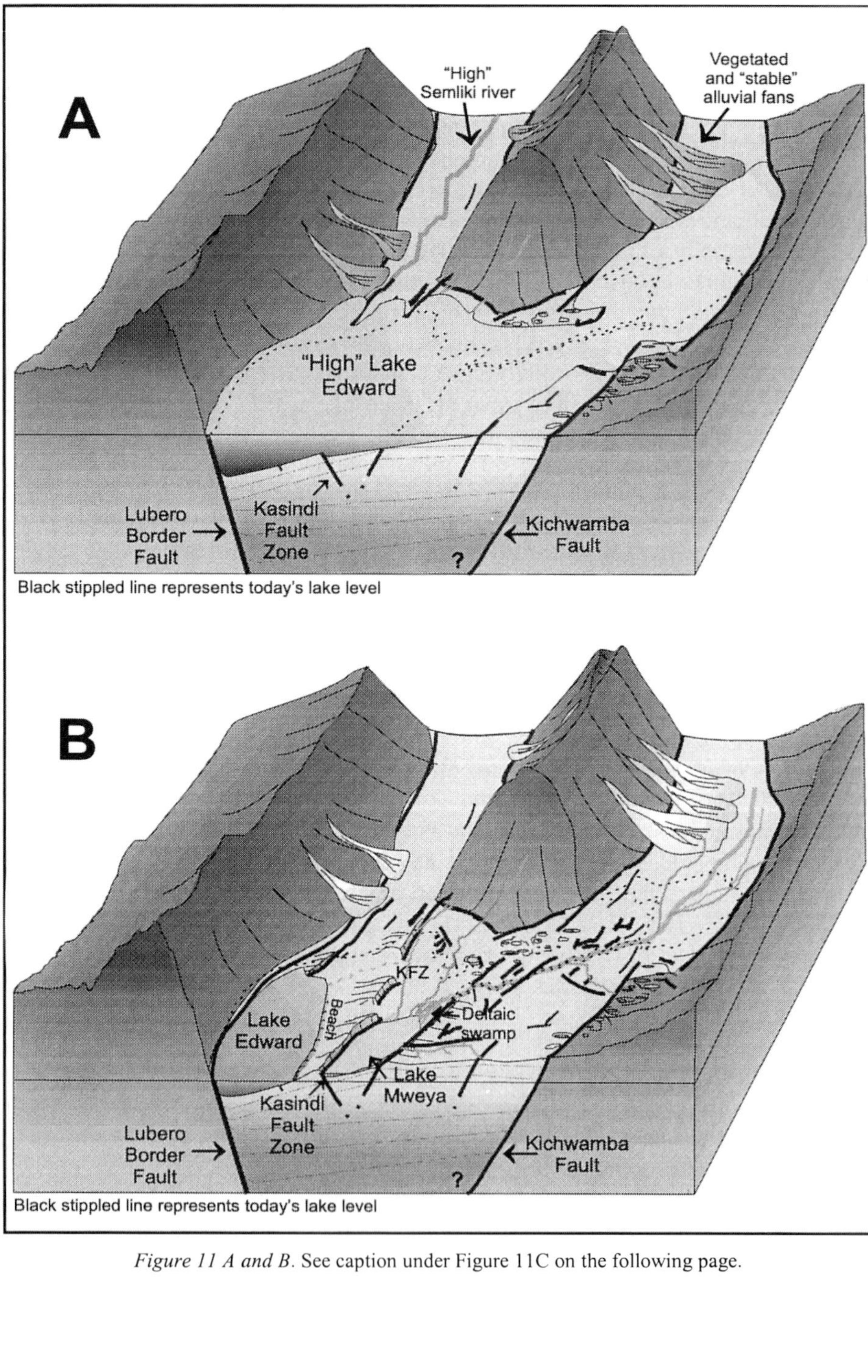


*Figure 11 A and B*. See caption under Figure 11C on the following page.

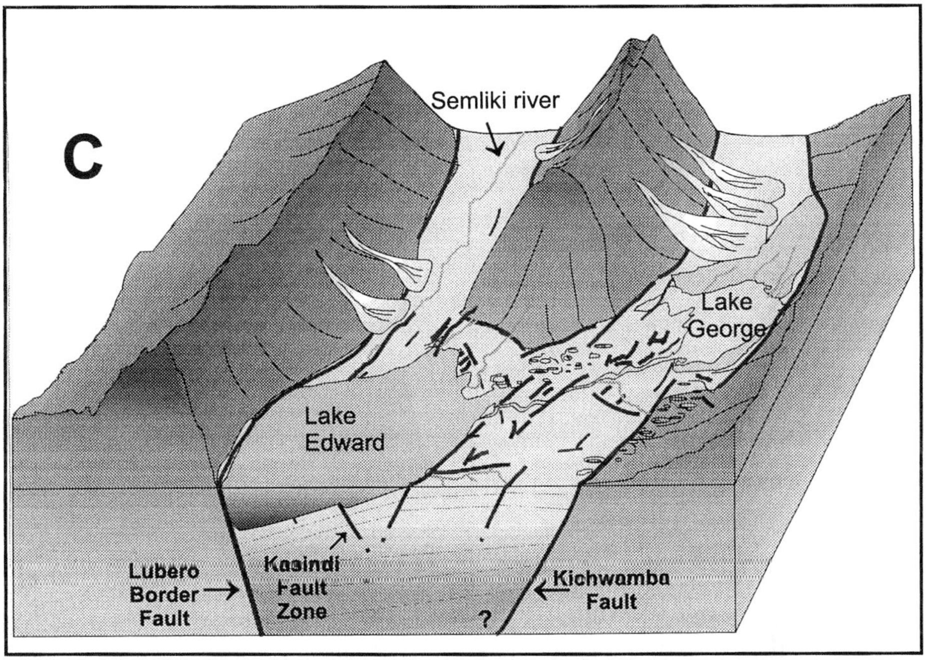


*Figure 11.* A possible palaeogeographical interpretation of the Edward-George basins during three time intervals (see previous page). A – The "high" Lake Edward conditions during the humid interval between 12 and 8 ka BP. B – The ~4.5 ka BP low-stand in the basin, related to drier climates, and the creation of low Lake Edward and Lake Mweya in the Edward basin. See text for more details. C – The present conditions in the basin.

Volcanic activity in the area has been dated to about 8 ka BP (Brooks and Smith, 1987; Boven *et al.*, 1998), and it is possible that this also corresponds to fault activity in the basin. This tectonically induced lake level low stand was then later followed by a change to gradually drier climates towards the Mid Holocene, gradually lowering the lake levels towards a minimum about 5 to 4.5 ka BP (about 4500 to 4000 $^{14}$C-years), finally leading to the desiccation of Lake George some time prior to 4 ka BP.

The 2P core was collected in 46 m of water depth, while the 5M core was collected from 29 m of water. If the 2P core site was exposed (or within shallow water) during this low stand, the 5M core site should also have been exposed. However, our data provide no evidence of exposure at this time. In fact, the relatively high HI values of the OM suggest a fairly stable water column, and possibly high lake productivity. We believe that the contrast between the two core sites is related to fault activity in the basin. As mentioned earlier, the Kasindi Fault Zone (KFZ) divides the Edward basin into a western and an eastern part. The 5M and 6M core sites are located to the east, while the 1P and 2P core sites lie west of this fault zone (Figure1). A N-S trending and east facing fault with a >12 m fault scarp has been identified just west of the 5M core site (Figure 7). Water depth to the top of the fault scarp just west of the 5M core site is around 20 m.

When regionally drier climates, and/or tectonics, some time prior to ~4.5 ka BP caused a lowering of the lake level, water entering the eastern half of the Edward basin was dammed against the uplifted fault scarp. As a result, two separate lakes were created, one east of the fault scarp with a surface elevation at a maximum of 892 m asl (912 - 20 m, which is today's elevation - depth to fault scarp); the other to the west of the fault scarp, with a surface elevation of maximum 865 m asl (912 m - 47 m, which is depth to the base of the black sand in core 2P). To simplify further discussions, we refer to the western water body as low Lake Edward and that in the east as Lake Mweya (Figure 11).

Lake Mweya was possibly an open lake (with several inflowing rivers),with an outlet in the west-northwest. The KFZ consists of several smaller faults linked in an en echelon fashion with offsets between the faults (Lærdal and Talbot, in press). It is thus possible that a river draining through one of these offset zones connected Lake Mweya to low Lake Edward, but this is only speculation (Figure 11). However, flushing rates in Lake Mweya appear to have been high enough as to limit the precipitation of calcite during these otherwise arid conditions.

The present depth of the sill at the Semliki outlet is 3 m (Beadle, 1981) and a lake level lowering of 47 m would lead to closed conditions in low Lake Edward. Seismic studies suggest that the eastern shore, located along the western side of the uplifted fault block, was dominated by beach sedimentation (Lærdal and Talbot, in press). Blackish oolitic sands were deposited in a N-S oriented belt along the palaeo-coastline (Figure 11).

During this low stand, Lake Mweya had a maximum depth of 15 m ((depth of 5M core site + 6 m of core sediments) - depth to fault scarp) while low Lake Edward would be >73 m deep (today's maximum depth of >120m(?) - 47 m of lake lowering). A lake level lowering of >20 m in Lake Mweya suggest that the 6M core site was located onshore, >6 m (>20 m - (depth to core site + 2 m of core sediment)) above the eastern water body. This points towards a deltaic/fluvial setting during deposition of unit III. The core site may have been located on a broad, fluvial flat, possibly with pools and swamps, where periodically more lacustrine conditions persisted, thereby preventing desiccation and/or exposure of the core site and sediments. With the elevation of the Rwenzori mountains, and a year-round supply of glacial melt water, rivers entering the Edward-George basin probably sustained a relatively high discharge even during this climatically dry period. Viner and Smith (1973) and Viner (1977) describe a core from Lake George. Based on the presence of a desiccation surface, they estimated the present lake to be no more than 3600 $^{14}$C-years old (~3.9 ka BP), supporting our interpretation of lower lake levels in the Edward-George basin prior to ~ 4.5 to 4 ka BP.

This lowering of lake levels in the Mid- to Late Holocene has been recorded in lakes throughout East Africa and correlates to a regional (possibly global) climate change at this time (Livingstone, 1975; Talbot, 1982; Street-Perrott and Harrison, 1984; Talbot and Livingstone, 1989; Bonnefille *et al.*, 1990; Talbot and Johannessen, 1992; Ricketts and Johnson, 1996; Beuning *et al.*, 1997; Marchant *et al.*, 1997; Le Turdu *et al.*, 1999).

## 5.3 The Late Holocene

Following the low stand, climate changed and wetter conditions returned to the Edward basin some time after 4.5 ka BP. Rising lake levels caused a decrease in the sediment grain size in the 2P and 6M cores and reflect a shift to open lake conditions. Rising lake levels furthermore caused drowning of the deltaic swamps in the 6M area and as the transgression proceeded the supply of terrestrial plant material decreases. The rise in $\delta^{13}C$ values in all four cores over the same interval could reflect an overall increase in the input of (C4-) terrestrial/deltaic material to the lake waters, as a result of gradually more humid climate and higher flushing rates in the rivers entering the basin (Meyers and Lallier-Vergès, 1999). The associated rise in HI corresponds to a rise in the TOC values (Figures 5, 6, 8 and 9), interpreted as a sign of better preservation of the OM as a result of increased water depth and a gradually more stable water column.

The erosional surfaces in unit I of the 2P core may be related to bottom currents, possibly triggered by increased wind stress (storms) on the lake surface. They may also reflect tectonic activity, as the 2P core site is located just west of the uplifted Kasindi Fault block (Figure 2). It is possible that movement along the fault could trigger mass flows and cause reworking of earlier deposited beach deposits. Unfortunately we lack HI samples from the upper part of 2P, but the TOC values decrease in the reworked sands in the upper part of unit I, suggesting that the sediments might previously have been exposed (Figure 6). Laminae in the 1P core are occasionally tilted which could indicate slumping, possibly related to the suggested fault activity.

HI and TOC values are generally higher at the 5M relative to the 2P core site in the period between ~4 and 2 ka BP, suggesting that OM preservation was better at the 5M site (in the east) than at the 2P site (in the west). This may seem odd as OM preservation generally improves with depth. Lake Edward is today seasonally anoxic below ~40 m and permanently anoxic below 80 m. As a result, OM preservation at the 2P site (46 m of water depth) should be better than at the 5M site (29 m of water depth), and not the other way around, as is observed. A possible explanation for this could be that the local deep "pool" (5M core site) just east of the Kasindi fault (Figure 1) is sheltered from lake currents and mixing in the western half of the lake. Thus, stratification may be locally preserved east of the fault at the same time as mixing occur at the 2P site (Figure 12).

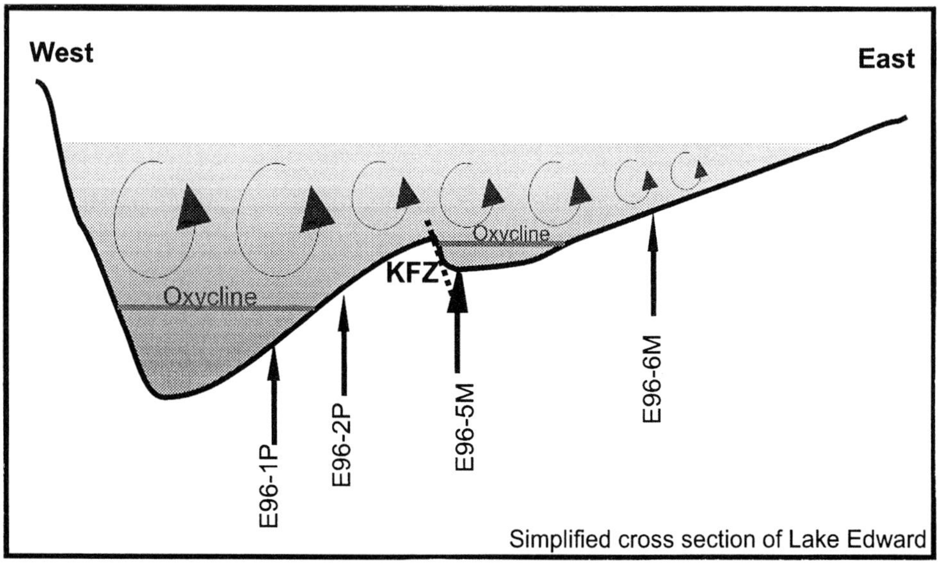


*Figure 12.* Simplified cross-section of the Edward basin illustrating the effect of the Kasindi Fault Zone (KFZ) on lake mixing. Core sites are indicated with arrows, as well as potential local oxyclines/chemoclines west and east of the KFZ. Large arrows illustrate depth of mixing in the basin, suggesting that this may be deeper west in the fault zone than in the east.

The relatively higher biogenic silica values in the deep water core (1P) relative to the shallower water cores (2P and 5M) could reflect differential preservation of the diatom frustules and/or differences in the diatom production between the open water and near-shore environments. The 1P core site is located well below the seasonal oxycline of the present lake (~40 m), while the 2P and 5M core sites are located in shallower waters, probably within the oxycline and/or well above it. A stratified lake with anoxic bottom waters could result in a low pH hypolimnion, as degradation of the OM proceeds, favouring preservation of biogenic silica (Talbot and Allen, 1996). More alkaline, shallower waters at the 2P and 5M core sites, on the other hand, could cause dissolution of diatom frustules. The very high BSi% values in the 1P core and low values in 2P could thus indicate that the chemocline was located somewhere between the two core sites (60 to 46 m of water depth) during this period.

The 1P core has the most detailed record of the last 2.5 ka BP of lake history (units II and I; Figure 4) and this has been interpreted as a period with high lake production and a stratified water column, possibly with periods of cyanobacteria-dominated phytoplankton. High C/N values observed during this period suggest that the OM formed in waters subject to severe nitrogen deficiency (Talbot and Lærdal, 2000).

At about 1.7 ka BP, a regional climate change to drier conditions and/or stronger winds (better lake mixing) was recorded in several of the East African lakes (Talbot, 1982; Bonnefille and Mohammed, 1994). This shift is also apparent from the 5M and 2P records (stippled upper line in figures 2 and 10), suggesting changes in lake production during this period, possibly related to shifts in lake water pH.

## 6. CONCLUSIONS

Our palaeoclimatic and palaeolimnologic record from Lake Edward (Figure 11) fits well with earlier interpretations of Late Pleistocene and Holocene palaeoclimate in East Africa (Livingstone, 1975; Viner, 1977; Talbot, 1982; Hastenrath and Kutzbach, 1983; Talbot and Livingstone, 1989; Johnson *et al.*, 1991; Johnson et al, 1996; Ricketts and Johnson, 1996; Beuning *et al.*, 1997; Lærdal, 1997; Le Turdu *et al.*, 1999; Stager and Johnson, 2000; Talbot and Lærdal, 2000). Our data suggest that the level of Lake Edward may have been lower during the Late Pleistocene arid interval (>13.5 ka BP), recorded as an erosional discontinuity in seismic profiles. This low-stand may also be represented by an erosional unconformity identified on a seismic line crossing the core site (Figure 7) and could indicate that the level of the lake fell below the depth of the 2P core site.

The Late Pleistocene - Early Holocene transition was a period characterized by a stable water column dominated by diatom production in the open waters of the lake. Some time around 8 ka BP lake levels dropped, probably as a result of a combination of tectonic lowering of the Semliki outlet, and a shift to drier climates. Tilting of the fault block associated with the KFZ may have occurred during this period, and possibly coincides with volcanism in the basin. These conditions culminated in the desiccation of Lake George some time prior to 4 ka BP. Exposure of a >12 m fault scarp related to the KFZ in the Edward basin (Figures 1 and 7) resulted in damming of west-flowing rivers in the eastern half of the Edward basin, and the formation of two separate water bodies in the Edward basin. The two lakes have informally been named low Lake Edward (in the west) and Lake Mweya (in the east). Lake Mweya was located at a slightly higher level than low Lake Edward and there might have been a fluvial connection between the two.

Lake levels began to rise after ~4.5 ka BP, drowning the uplifted fault block and connecting the two water bodies to form the present Lake Edward. The transgression caused drowning of formerly vegetated areas around the lake, triggering a massive release of terrestrial OM and nutrients to the lake water. Hecky and Degens (1973) suggested an increase in the input of nutrients from Lake Edward into Lake Albert at about 5,000 years BP. We believe that this corresponds to the rise in lake levels recorded in the Edward cores at about 4.4 ka BP. The discrepancy in dates may be related to different dating techniques on the Albert cores relative to the Edward cores and generally high uncertainties about $^{14}$C-age determinations in rift lakes.

The Late Holocene was a period with high lake production and a stratified water column, possibly with anoxic waters below 60 m. There are indications of climate oscillations during the last 1700 years, possibly corresponding to a shift to a more seasonal climate in the area.

## ACKNOWLEDGEMENTS

The work on Lakes Edward and Lake George has been funded by the Norwegian Research Council (NFR) and BP Amoco Norway AS. We have also benefited greatly from the support and resources of the International Decade for the East African Lakes (IDEAL) during our work on the lakes; through exchange of information and data obtained from the cores. Field work on Lake Edward in 1996 was largely supported by an NSF grant to Thomas C. Johnson (Large Lakes Observatory, University of Minnesota, Duluth), while work on the lake in 1999 benefited in part from an NSF grant to Christopher Scholz (University of Syracuse, New York). John Mothersill kindly donated his Mackereth cores to the IDEAL project. We are grateful for the help of the crew of the R/V Topi on Lakes Edward and George in both 1996 and 1999 and would also like to thank Mr. Honey Malinga, Mr. Reuben J. Kashambuzi, Mr. Charles Lukyamuzi, Mr. Aron Sserubiri Musore and Mr. Abdul B. Byakagaba of the Petroleum Exploration and Production Department (PEPD) of Entebbe (Uganda), for help with "everything" while working in Uganda. We wish also to thank the Ministry of Energy (Kinshasa, Dem. Rep. of Congo) for help during field work on Lake Edward in 1996. Nick Peters and Geoffrey Ellis from the University of Miami (USA) and Knut Inge Brendeland (University of Bergen) are thanked for help during field work in 1999, and everyone participating in the IDEAL expedition to Lake Edward in 1996. Warren Eversly, Sissel Kvernes and Tom Remi Ellingsen helped with sample preparations, and Odd Hansen of the University of Bergenran the stable isotope analyses. Marian Våge and Norsk Hydro are thanked for access to and help with Rock Eval analyses at the Norsk Hydro Research Centre, Sandsli, Bergen. The authors are grateful to D. Ariztegui and J-J. Tiercelin for constructive reviews on an earlier version of the manuscript.

## REFERENCES

Adamson, D. A., Gasse, F., Street, F. A. and Williams, M. A. J. (1980) Late Quaternary history of the Nile. *Nature* **288**, 50-55.

Ariztegui, D., Chondrogianni, C., Lami, A., Guilizzoni, P. and Lafargue, E. (in press) Lacustrine organic matter and the Holocene paleoenvironmental record of Lake Albano (central Italy). *Journal of Paleolimnology*

Beadle, L. C. (1981) *The Inland Waters of Tropical Africa*. Longman, London, pp. 433.

Beuning, K. R. M., Talbot, M. R. and Kelts, K. (1997) A revised 30,000-year paleoclimatic and paleohydrologic history of Lake Albert, East Africa. *Palaeogeography, Palaoeclimatology, Palaeoecology* **136**, 29-279.

Bishop, W. W. (1969) Pleistocene Stratigraphy of Uganda. *Geological Survey of Uganda Memoir*. X, pp. 128.

Bonnefille, R., Roeland, J. C. and Guiot, J. (1990) Temperature and rainfall estimates for the past 40,000 years in equatorial Africa. *Nature* **346**, 347-349.

Bonnefille, R. and Mohammed, U. (1994) Pollen-inferred climatic fluctuations in Ethiopia during the last 3000 years. *Palaeogeography, Palaoeclimatology, Palaeoecology* **109**, 331-343.

Boven, A., Pasteels, P., Punzalan, L. E., Yamba, T. K. and Musisi, J. H. (1998) Quaternary perpotassic magmatism in Uganda (Toro-Ankole Volcanic Province): age assessment and significance for magmatic evolution along the East African Rift. *Journal of African Earth Sciences* **26**(3), 463-476.

Brendeland, K. I. (2001) Sein pleistocene/holocene strontium isotop stratigrafi av Lake Edward og Lake Albert, Uganda/Kongo. *Unpublished Masters Thesis*, Department of Natural Science, Geology. Bergen, University of Bergen, pp. 91.

Brooks, A. S. and Smith, C. C. (1987) Ishango revisited: new age determinations and cultural interpretations. *The African Archaeological Review* **5**, 65-78.

Bruun, P. (1962) Sea level rise as a cause of shore erosion. *Journal of the Waterways and Harbours Division of the American Society of Civil Engineers* **88**, 117-130.

Byakagaba, A. B. (1997) Report on the Geological mapping of the Lakes Edward-George Basin. *Report of the Department of Petroleum Exploration and Production*, Uganda (September), pp. 1-55.

De Heinzelin, J. (1955) Le fosse tectonique sous le parallele d'Ishango. *Inst. Parc. nat. Congo Belge, Brussels* Fasc. 1.

Delvaux, D., Kervyn, F., Vittori, E., Kajara, R. S. A. and Kilembe, E. (1998) Late Quaternary tectonic activity and lake level change in the Rukwa Basin. *Journal of African Earth Sciences* **26**(3), 397-421.

Ebinger, C. J. (1989) Tectonic development of the western branch of the East African rift system. *Geological Society of America Bulletin* **101**, 885-903.

Haberyan, K. A. and Hecky, R. E. (1987) The Late Pleistocene and Holocene Stratigraphy and Paleolimnology of Lakes Kivu and Tanganyika. *Palaeogeography, Palaeoclimatology, Palaeoecology* **61**, 169-197.

Hastenrath, S. and Kutzbach, J. E. (1983) Paleoclimate Estimates from Water and Energy Budgets of East African Lakes. *Quaternary Research* **19**, 141-153.

Healy, F. P. and Hendzel, L. L. (1980) Physiological indicators of nutrient deficiency in lake phytoplankton. *Can. J. Fisheries Aquatic Sci.* **37**, 442-453.

Hecky, R. E. and Degens, E. T. (1973) Late Pleistocene - Holocene Chemical Stratigraphy and Paleolimnology of the Rift Valley Lakes of Central Africa. *Woods Hole Ocean. Inst. Tech. Report.* 73(28), pp. 93.

Hecky, R. E. and Kling, H. J. (1987) Phytoplankton ecology of the great lakes in the rift valleys of Central Africa. *Arch. Hydrobiol. Beih. Ergebn. Limnol.* **25**, 197-228.

Hecky, R. E., Campbell, P. and Hendzel, L. L. (1993) The stoichiometry of carbon, nitrogen and phosphorous in particulate matter of lakes and oceans. *Limnol. Oceanogr.* **38**, 709-724

Johnson, T. C., Halfman, J. D. and Showers, W. J. (1991) Paleoclimate of the past 4000 years at Lake Turkana, Kenya, based on the isotopic composition of authigenic calcite. *Palaeogeography, Palaoeclimatology, Palaeoecology* **85**, 189-198.

Johnson, T. C., Scholz, C. A., Talbot, M. R., Kelts, K., Ricketts, R. D., Ngobi, G., Beuning, K. R. M., Ssemmanda, I. and McGill, J. W. (1996) Late Pleistocene Desiccation of Lake Victoria and Rapid Evolution of Cichlid Fishes. *Science* **273**, 1091-1093.

Kampunzu, A. B. and Mohr, P. (1991) Magmatic Evolution and Petrogenesis in the East African Rift System. In A. B. Kampunzu and R. T. Lubala (eds.), *Magmatism in Extensional Structural Settings: The Phanerozoic African Plate.* Springer-Verlag, Berlin, pp. 85-136.

Kampunzu, A. B., Bonhomme, M. G. and Kanika, M. (1998) Geochronology of volcanic rocks and evolution of the Cenozoic Western Branch of the East African Rift System. *Journal of African Earth Sciences* **26**(3), 441-461.

Lærdal, T. (1997) A reconstruction of the Sedimentology and Paleolimnology of Lake Victoria during the last 15 thousand years, using stable isotopes (carbon and nitrogen) and Rock-Eval Pyrolysis. *Unpublished Masters Thesis*, Department of Natural Science, Geology. Bergen, University of Bergen, pp. 99.

Lærdal, T. and Talbot, M. R. (in press) Structure and Neotectonics in the Edward and George basins, Uganda-Congo, East Africa. *Palaeogeography, Palaeoclimatology, Palaeoecology*

Le Turdu, C., Tiercelin, J-J., Gibert, E., Travi, Y., Lezzar, K-E., Richert, J-P., Massault, M., Gasse, F., Bonnefille, R., Decobert, M., Gensous, B., Jeudy, V., Tamrat, E., Mohammed, M. U., Martens, K., Atnafu, B., Chernet, T., Williamson, D. and Taieb, M. (1999) The Ziway-Shala basin system, Main Ethiopian Rift: Influence of volcanism, tectonics, and climate forcing on basin formation and sedimentation. *Palaeogeography, Palaeoclimatology, Palaeoecology* **150**, 135-177.

Lehman, J. T. (this volume) Application of satellite AVHRR to water balance, mixing dynamics, and the chemistry of Lake Edward, East Africa..

Livingstone, D. A. (1967) Post-glacial vegetation on the Rwenzori Mountains in equatorial Africa. *Ecol. Mon.* **37**, 25-52.

Livingstone, D. A. (1975) Late Quaternary climatic change in Africa. *Annual Review of the Ecological Systems* **6**, 249-280.

Livingstone, D. A. and Melack, J. M. (1984) Some lakes of Sub-Saharan Africa, in F. B. Taub (ed.), *Lakes and Reservoirs.* Elsevier, Amsterdam, 467-497.

Maasha, N. (1975) The Seismicity of the Rwenzori Region in Uganda. *Journal of Geophysical Research* **80**(11), 1485-1496.

Marchant, R., Taylor, D. and Hamilton, A. (1997) Late Pleistocene and Holocene History at Mubwindi Swamp, Southwest Uganda. *Quaternary Research* **47**, 316-328.

McKenzie, J. A. (1985) Carbon isotopes and productivity in the lacustrine and marine environment, in W. Stumm (ed), *Chemical Processes in Lakes*, Wiley, New York, pp. 99-118.

Meyers, P. A. and Ishiwatari, R. (1995) Organic Matter Accumulation Records in Lake Sediments, in A. Lerman, D.M. Imboden and J. R. Gat (eds.), *Physics and Chemistry of Lakes.* Springer-Verlag, Berlin, 279-328.

Meyers, P. A. and Lallier-Vergès, E. (1999) Lacustrine sedimentary organic matter records of Late Quaternary paleoclimates. *Journal of Paleolimnology* **21**, 345-372.

Mohr, P. A. and Wood, C. A. (1976) Volcanic spacings and lithospheric attenuation in the Eastern Rift of Africa. *Earth and Planetary Science Letters* **33**, 126-144.

Musisi, J. H. (1991) The Neogene-Quaternary Geology of the Lake George-Edward Basin, Uganda. *Unpublished PhD Thesis*, Faculty of Science. Brussel, Vrije Universiteit, pp. 298.

Nicholson, S. E. (1996) Victoria lake-level modelling aims to predict basin rainfall. *IDEAL Bulletin*(spring), p. 5.

Nicholson, S. E. (1997) Extending the African lake record to the early 19th century. *IDEAL Bulletin*(winter), pp. 4-5.

Nicholson, S. E. and Yin, X. (2000) Inferring past rainfall conditions from lake-level fluctuations: an example based on the historical levels of Lake Victoria. *Abstract Volume, 2nd International IDEAL Symposium*, Malawi, p. 18.

Pasteels, P., Villeneuve, M., De Paepe, P. and Klerkx, J. (1989) Timing of the volcanism of the southern Kivu province: implications for the evolution of the western branch of the East African Rift system. *Earth and Planetary Science Letters* **94**, 353-363.

Ricketts, R. D. and Johnson, T. C. (1996) Climate change in the Turkana basin as deduced from a 4000 year long delta 18O record. *Earth and Planetary Science Letters* **142**, 7-17.

Rosendahl, B. R. (1987) Architecture of the continental rifts with special reference to East Africa. *Annual Review of Earth and Planetary Science* **15**, 445-503.

Rubin, M. and Suess, H. E. (1956) U. S. Geological Survey Radiocarbon Dates III. *Science* **123**, 442-448.

Russell, J. and Kelts, K. (1999) The Sedimentologic History of Lake Edward, Uganda. *IDEAL Bulletin* (Summer), pp. 1-3.

Russell, J. and Kelts, K. (2000) Sediment history based on piston cores from Lake Edward, East Africa. *Abstract Volume, 2nd International IDEAL Symposium*, Malawi, pp. 33-34.

Stager, J. C. and Johnson, T. C. (2000) A 12,400 $^{14}C$ yr offshore diatom record from east central Lake Victoria, East Africa. *Journal of Paleolimnology* **23**, 373-383.

Stager, J. C. and Mayewski, P. A. (1997) Abrupt early- to mid-Holocene climatic transition registered at the Equator and the Poles. *Science* **276**, 1834-1835.

Street-Perrott, F. A. and Harrison, S. P. (1984) Lake levels and climate reconstruction. In, A. D. Hecht (ed.), *Paleoclimate Analysis and Modelling*. Wiley, New York, pp. 291-340.

Stuiver, M. and Reimer, P. J. (1993) Extended 14C data base and revised CALIB3.0 14C age calibration program. *Radiocarbon* **35**(1), 215-230.

Stuiver, M., Reimer, P. J., Bard, E., Beck, J. W., Burr, G. S., Hugen, K. A., Kromer, B., McCormac, F. G., Plicht, J. and Spurk, M. (1998) High precision radiocarbon age calibration for terrestrial and marine samples. *Radiocarbon* **40**, 1041-1083.

Swift, D. J. P. (1968) Coastal erosion and transgressive stratigraphy. *Journal of Geology* **76**, 444-456.

Talbot, M. R. (1982) Holocene Chronostratigraphy of Tropical Africa. In, J. Mangerud, H. J. B. Birks and K. D. Jäger (eds.), *Chronostratigraphic subdivision of the Holocene*. Striae, Uppsala **16**, 17-20.

Talbot, M. R. (1988) The origins of lacustrine oil source rocks: evidence from the lakes of tropical Africa. In, A. J. Fleet, K. Kelts and M. R. Talbot (eds.), *Lacustrine Petroleum Source Rocks*. Geological Society Special Publications. **40**, 29-43.

Talbot, M. R. and Livingstone, D. A. (1989) Hydrogen Index and Carbon Isotopes of Lacustrine Organic Matter as Lake Level Indicators. *Palaeogeography, Palaeoclimatology, Palaeoecology* **70**, 121-137.

Talbot, M. R. and Johannessen, T. (1992) A high resolution paleoclimatic record for the last 27,500 years in tropical West Africa from the carbon and nitrogen isotopic composition of lacustrine organic matter. *Earth and Planetary Science Letters* **110**, 23-37.

Talbot, M. R. and P. A. Allen (1996) Lakes. In, H. G. Reading (ed.), *Sedimentary Environments,* Blackwell, Oxford, pp. 83-124.

Talbot, M. R. and Lærdal, T. (2000) The Late Pleistocene - Holocene paleolimnology of Lake Victoria, East Africa, based upon elemental and isotopic analyses of sedimentary organic matter. *Journal of Paleolimnology* **23**, 141-164.

Twesigomwe, E. M. (1996) Probabilistic Seismic Hazard Assessment of Uganda. *Unpublished Dr. Phil. thesis, Department of Physics*. Makerere University, Department of Physics, Uganda, pp. ~200.

Tyson, R. V. (1995) *Sedimentary Organic Matter*. Chapman & Hall, London, pp. 416.

Viner, A. B. (1969) The chemistry of the water of Lake George, Uganda. *Verh. Internat. Verein. Limnol.* **17**, 289-296.

Viner, A. B. (1977) The Sediments of Lake George (Uganda); Vertical Distribution of Chemical Features in Relation to Ecological History and Nutrient Recycling. *Archive fur Hydrobiologie* **80**, 40-69.

Viner, A. B. and Smith, I. R. (1973) Geographical, historical and physical aspects of Lake George. *Proc. R. Soc. Lond.* **184**, 235-270.

# PIGMENT ANALYSIS OF SHORT CORES FROM THE CENTRAL ETHIOPIAN RIFT VALLEY LAKES

MOHAMMED M.U.[1], BONNEFILLE, R.[2] AND SEIFU KEBEDE[1]

[1] *Department of Geology and Geophysics, Addis Ababa University, c/o P.O.Box 1176, Addis Ababa, Ethiopia*

[2] *CEREGE, BP. 80, 13545 Aix-en Provence, France*

## ABSTRACT

Pigment analysis (Chlorophyll Derivatives - CD, and Total Carotenoids - TC) from surface and core sediments of three lakes: Langano, Abijata and Shalla in the Central Ethiopian Rift Valley are presented. The results show that pigment concentration is very low in modern sediments with CD generally higher than TC. This is in accordance with the present low productivity of the lakes. Within laminated portions of the cores, collected from Lakes Langano and Shalla, the concentration of pigments is as much as 5 to 40 times higher than in the modern sediments. This is interpreted as resulting from persistent lake stratification due to higher lake levels. This allowed the conservation of both laminae and pigments in a reducing environment. The fact that CD and TC values increased contemporaneously could also imply a high terrestrial organic matter input into the lakes as well as a high nutrient supply, which might have led to enhanced lake productivity. This was more so when pigment concentration was the highest in cores from Shalla and Langano at around 2000 yr BP. The above interpretation is supported by data from lake level variations showing that there was a rising trend at around 2000 yr BP with inference of humid conditions coming from pollen and other proxies in the study area and in the region. Pigment concentration declined after about 1000 YR BP when the CD/TC ratio became high. This could result from a fall in lake productivity shown by a low TC value and a poor conservation condition shown by a more homogeneous sediment.



*E.O. Odada and D.O. Olago (eds.),*
*The East African Great Lakes: Limnology, Palaeolimnology and Biodiversity,* 471–485.

# 1. INTRODUCTION

Pigments preserved in lake sediments are indicators of lake history (Raybak and Raybak, 1985; Swain, 1985). However, the interpretation of the stratigraphy of the principal types of pigments (Chlorophylls and Carotenoids) is often difficult due to a number of factors governing the sedimentation and preservation processes (Likens and Devis, 1975; Daley *et al.*, 1977; Hutton and Tolonen, 1977; Tolonen, 1978). In spite of this difficulty, they have been used to understand the autochthonous or allochthonous origin of organic matter. Since the concentration of allochthonous pigments coming from litter and soil humus is very weak, due to the aerobic nature of the environment, a good proportion of the amount preserved in eutrophic lakes have been attributed to an autochthonous origin, resulting from the lake's productivity (Gorham and Sanger, 1975). In particular, the degradation of Carotenoids is faster than the pheopigments (chlorophyll a derivatives). Pigments in lake sediments have a higher chance of preservation than those in soils and litter, and are therefore characterized by a low proportion of CD/TC (derivatives of chlorophyll/carotenoids) (Sanger and Gorham, 1970). On the other hand, oligotrophic lakes contain a relatively high proportion of CD/TC because of a high concentration of oxygen (Gorham and Sanger, 1976) or because of an important contribution of allochthonous organic matter (Gorham and Sanger, 1975; 1976; Sanger and Gorham, 1972)

Swain (1985) stresses the effect of the amount of dissolved oxygen in lakes on pigment preservation emphasising that increased primary productivity leads to a high pigment concentration because of better conservation due to a high consumption of dissolved oxygen.

This work presents pigment measurements made on modern and core sediments from three lakes: Abijata, Langano and Shalla from the Ethiopian Rift Valley in order to show the origin and degree of preservation of organic matter during the last 2000-3000 years. Results from pollen and other proxies have previously been published from core Abijata A (Bonnefille *et al.*, 1986) and Langano C (Mohammed and Bonnefille, 1991). Two pigment types have been distinguished: CD or Chlorophyll a derivatives (Chlorophyll a and Pheophytin) and TC (Total Carotenoids). The results of the measurement of the two pigment types and the ratio CD/TC are presented here.

# 2. PRESENT ENVIRONMENT OF THE LAKES

Lakes Langano, Abijata and Shalla are located in the Central Ethiopian Rift Valley, in the Ziway-Shalla basin. They are found between about 7°30' and 8° N latitude and 38°30' and 39° E longitude (Figure 1). All the three lakes occur in a semi-arid type of climate with a mean annual rainfall of about 600 mm. The precipitation increases with altitude on the bordering escarpment to the east to reach its maximum value of 1600 mm around 2500 m altitude. The lakes are mainly fed by perennial rivers from the highlands. They are all tropical in character having surface-water temperatures averaging 24 - 25°C. Their water chemistry is dominantly of sodium bicarbonate-carbonate type.

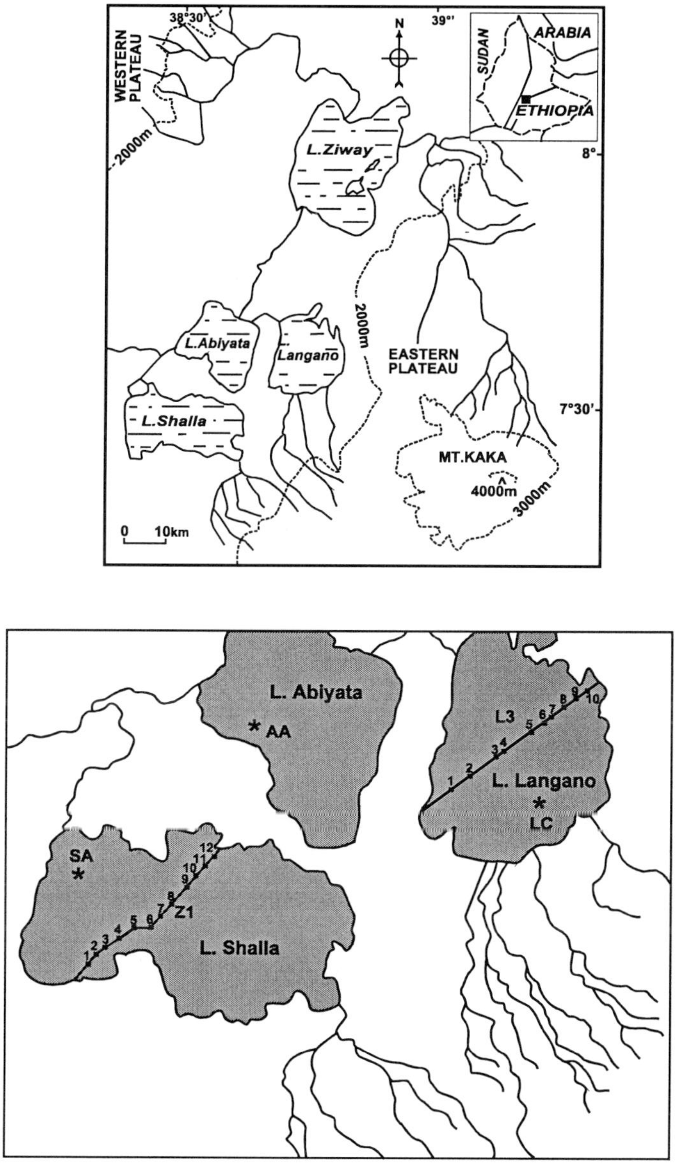


*Figure 1.* Map of the study area: location of core sites (asterisk) and surface sample transects of the studied lakes (Langano, Shalla and Abijata in the Central Ethiopian Rift Valley)

From a biological point of view the aquatic vegetation is poorly developed due to the high alkalinity and is limited to a discontinuous belt of cattail (*Typha cf. latifolia*), *Cynodon* and *Sporobolus* (Street, 1979; Mohammed and Bonnefille, 1991). The microflora is dominated by diatoms, green algae and blue green algae (Gasse, 1987).

Lake Abijata is found at 1578 m altitude with a maximum recent depth of 9 m. The lake is fed by the River Bulbula coming from Lake Ziway and River Horakello coming from Lake Langano. At present this lake is a terminal lake and it is assumed that the main water loss is through evaporation. The primary productivity of this lake is the highest of the three lakes. The Chlorophyll a content has been estimated to be 56.6 mg/m$^3$ (Gasse, 1987).

Lake Langano is found at 1582 m altitude. The lake is the nearest to the eastern escarpment of the Rift from where the rivers entering this lake come from. It has a maximum recorded depth of 48 m. Its water colour is reddish-brown due to the high concentration of suspended particles. The primary productivity of this lake is relatively low with a 7.1 mg/m$^3$ of Chlorophyll a .

Lake Shalla is a saline caldera lake with a bluish colour. It is located at 1558 m altitude. The maximum depth of the lake is 266 m. The primary productivity of this lake is very low with 5mg/m$^3$.

## 3. MATERIALS AND METHODS

Surface samples were taken by bottom dredging equipment while the cores measuring about 6 m long were taken by a Mackereth corer in 1981 by R. Bonnefille, J.J. Tiercelin and C. Barton (Figures 2 and 3). In total of 82 samples were analysed of which 10 were from the surface and 72 were from the cores. Sampling of the core sediment was done at approximately 25 cm intervals. The samples were analysed following methods adopted by Lorenzen (1967), Schultz (1988/89) and Züllig (1982,1985). Pigment extraction was done on 1cm$^3$ fresh sediment samples mixed in 15ml of 50:50 acetone (tech.) and ethanol (tech.) solution. The operation was done in the dark and under an atmosphere of nitrogen gas. After filtration, the residual sediment was dried at 80°C. The total quantity of main pigment types was calculated from measurements obtained in a photometer (1cm cuvette) at specific wavelengths (665 and 700 nm for chlorophyll, 450 nm for Crude Carotenoids) and the results are given in terms of the dry weight of the sediment. In this study Chlorophyll derivatives, CD (sum of Chlorophyll a and Pheophytin a) as well as total Carotenoids (TC) have been measured. The ratio CD/TC is usually used as a measure for the relative importance of allochthonous to autochthonous input. This ratio can also be considered as an index for the quality of pigment preservation since TC degrades more than CD (Engstrom *et al.*, 1985).

## 4. LITHOLOGY AND AGE OF THE CORES

The core sediments are characterized by clay, silty clay and silt either homogeneous or laminated. The cores were taken under water depths of 50 m, 40 m and 10.8 m for Shalla A, Langano C and Abijata A respectively. The laminae are well developed in Lake Shalla, the deepest of the three lakes. They are generally millimetric and show grey and dark couplets.

*Table 1.* $^{14}C$ dates from cores in the Rift Valley Lakes of Ethiopia. Dating was undertaken in the "Centre des Faibles Radioactivités'' de GIF sur Yvette (Delibrias).

| Core | Depth (cm) | $^{14}C$ dates (yr BP) | Lab No. (Giff/Yvettes) |
|---|---|---|---|
| Langano C | 105-115 | 1060±200 | Gif 6533 |
| (LC) | 367.5-380 | 2100±220 | 6534 |
| | 535-545 | 1850±220 | 6535 |
| | 590-592 | 2370±220 | 6483 |
| Abijata A | 162.5-172.5 | 1600±150 | 6205 |
| (AA) | 345-347.5 | 1540±200 | 5989 |
| | 543-545 | 1720±80 | 5988 |
| Shalla A | 217-225 | 2170±220 | 6204 |
| (SA) | 520-527 | 2700±230 | 6359 |

## 5. RESULTS OF PIGMENT ANALYSIS ON MODERN AND CORE SEDIMENTS

### 5.1. Lake Shalla

Five surface samples from Lake Shalla show variable amounts of pigment distribution along a SW-NE transect $Z_1$ (Figure 2a). Again CD is more important than TC. The highest values are encountered in sample $Z_1 4$ taken at a water depth similar to that of the core site. The shallowest and deepest samples again have low pigment concentrations. The general trend of CD/TC is similar to that of the pigment curves.

Core Shalla A has recorded higher CD than TC values with four peaks having values five to eight times greater for CD and twenty to five times greater for TC than the highest values measured in the modern samples. These correspond to the laminated parts of the core. Pigment content decreases towards the top part of the core where the CD/TC values increase (Figure 3a).

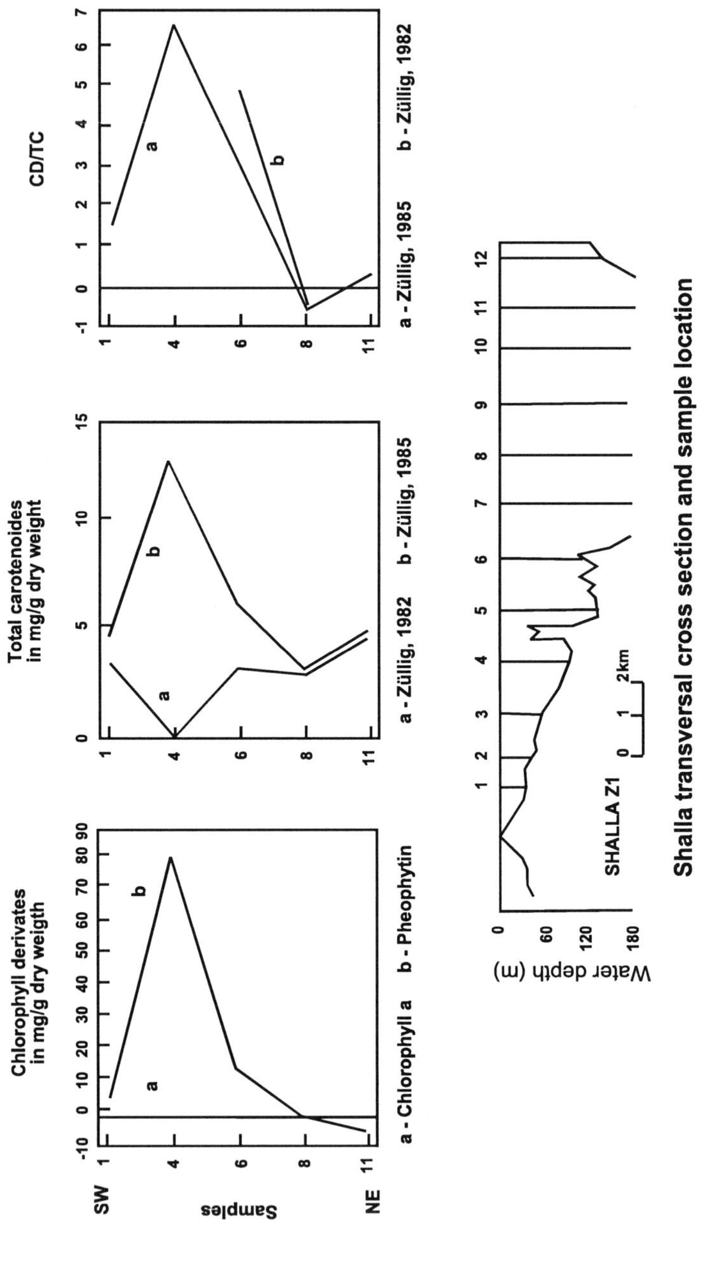


*Figure 2a.* Pigment curves from sub-modern lacustrine samples, along transect Z1 in Lake Shalla.

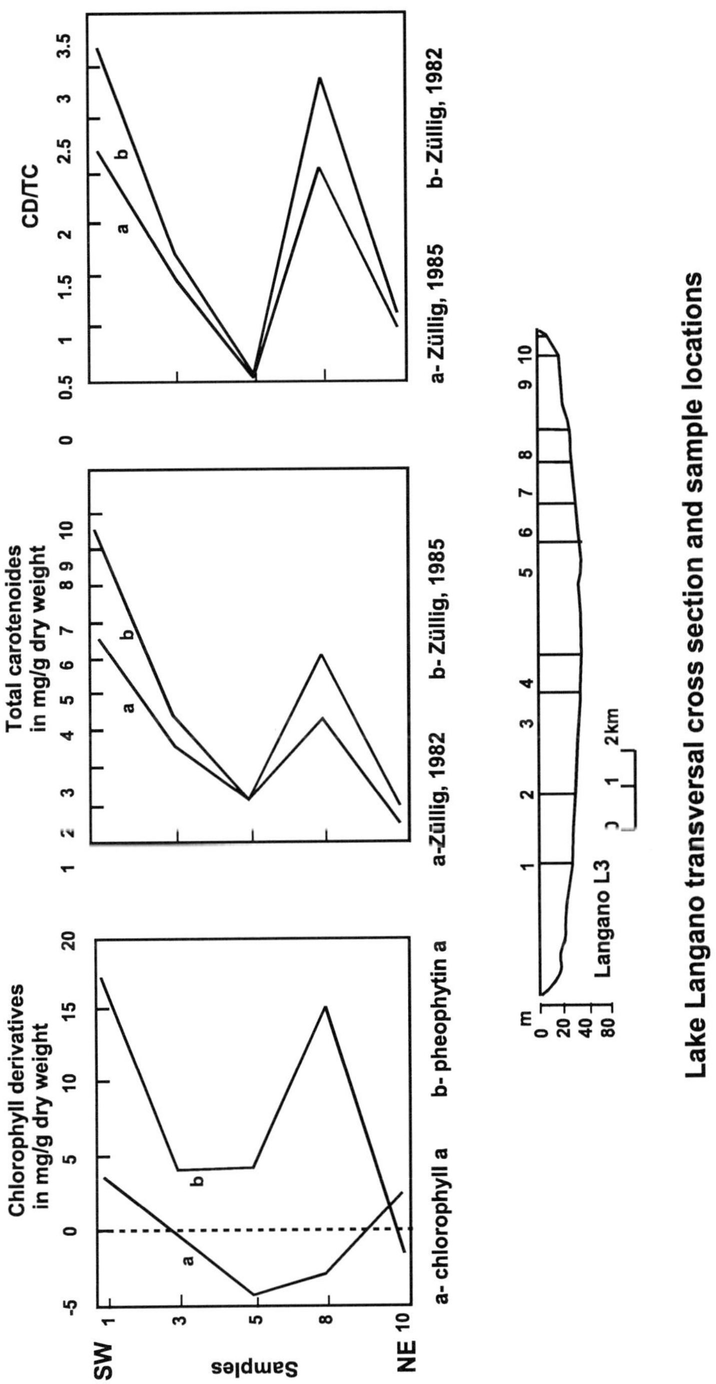


*Figure 2b.* Pigment curves from sub-modern lacustrine samples, along transect L3 in Lake Langano.

## 5.2 Lake Langano

As in Lake Shalla, the results of five modern samples of transect $L_3$ oriented SW-NE show a variable amount of pigment counts (Figure 2b) with more abundant CD than TC. Highest pigment values were recorded in samples under a water depth of about 30m from either side of the shore line (in samples $L_3$ 3 and $L_3$ 8). The samples from the deepest water, $L_3$ 5 and from the coast $L_3$ 10 have low pigment values. The CD/TC ratio follows the trend shown by both pigment types. Sample $L_3$ 5 comes from a water depth at which the core LC was taken.

Measurements from the core sediments from LC show an abundance of CD over TC as in the modern samples. Three peaks of high CD and TC pigment values can be seen, which correspond to laminated parts of the core. These values are nearly 40 times greater for CD and 15 times for TC than in the surface samples, particularly when compared to the surface sample $L_3$ 5 from the same water depth as the core. Pigment concentration falls towards the top part of the core while the CD/TC ratio rises, with the highest value attained in the core section with minimum pigment content (122.5-205 cm) (Figure 3b).

## 5.3 Lake Abijata

No surface samples have been collected from this lake during the field work. In the core, two pigment peaks could be observed. It can also be seen that the TC and CD peaks are out of phase. Again the CD/TC value increases towards the top part of the core (Figure 3c).

# 6. DISCUSSION

The above results show that modern samples (Figures 2a and b) from Langano and Shalla have recorded low pigment values with CD generally dominating over TC. This could be interpreted as resulting from low algal productivity as is the case for non-productive oligotrophic lakes, which are dominated by allochthonous detritus coming from the drainage basin. No linear relationship between water depth and pigment concentration exists. Following the Züllig (1985) formula, higher pigment values were measured in samples from intermediate depths (L1, L8 and Z4, Z6) as compared to the shallowest (L10 and Z1) and to the deepest ones (L3, L5 and, Z8, Z11). This could be due to a better conservation condition compared to samples from the shallowest depths and due to better productivity conditions (shown by higher TC) and/or higher allochthonous supply (shown by higher CD) compared to those from deepest ones.

The pigment curves from the cores (Figures 3 a, b and c) show the presence of peaks several times greater than in modern samples (about 40 times for CD and 15 times for TC in Langano and about 5-8 times for CD and 5-20 times in Shalla). In particular, they correspond to laminated parts of the cores. The increase in the more

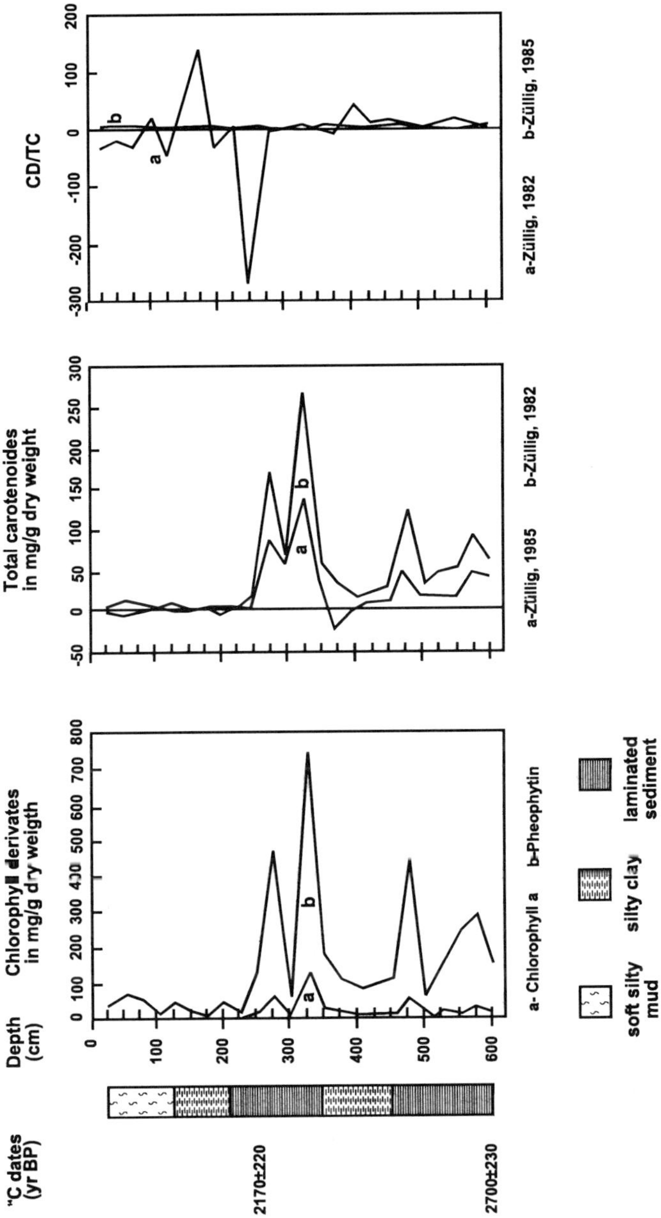


*Figure 3a.* Pigment curve from core SA (Lake Shalla, Central Ethiopian Rift Valley.

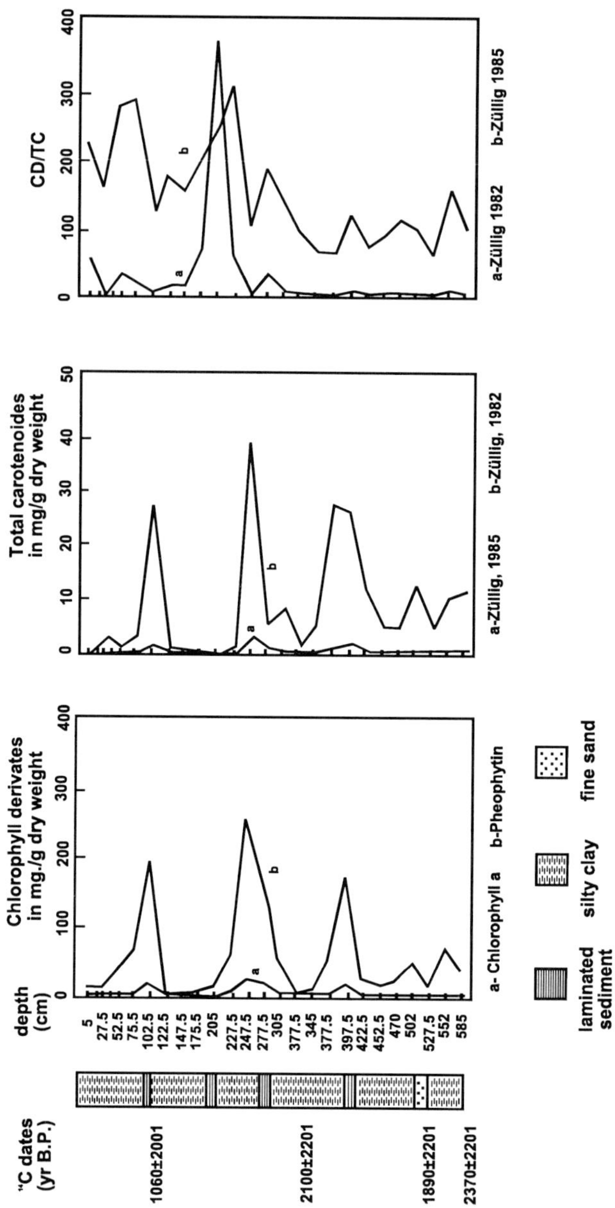


*Figure 3b.* Pigment curve from core LA (Lake Langano, Central Ethiopian Rift Valley).

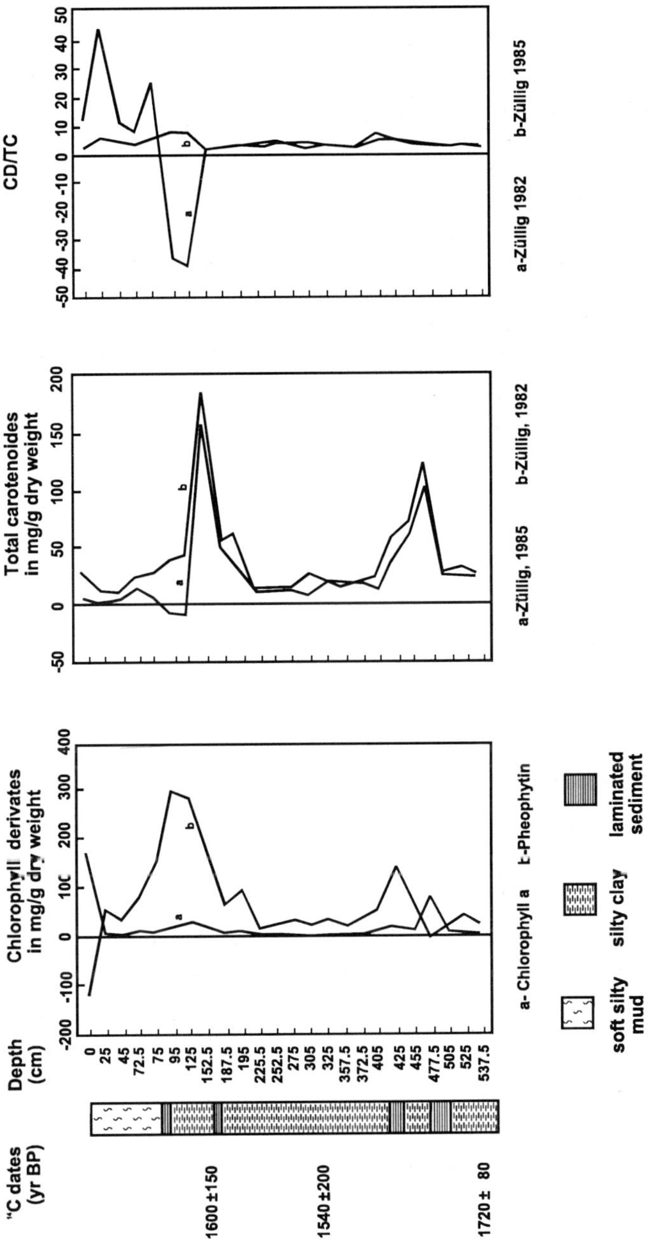


*Figure 3c.* Pigment curve from core AA (Lake Abijata, Central Ethiopian Rift Valley).

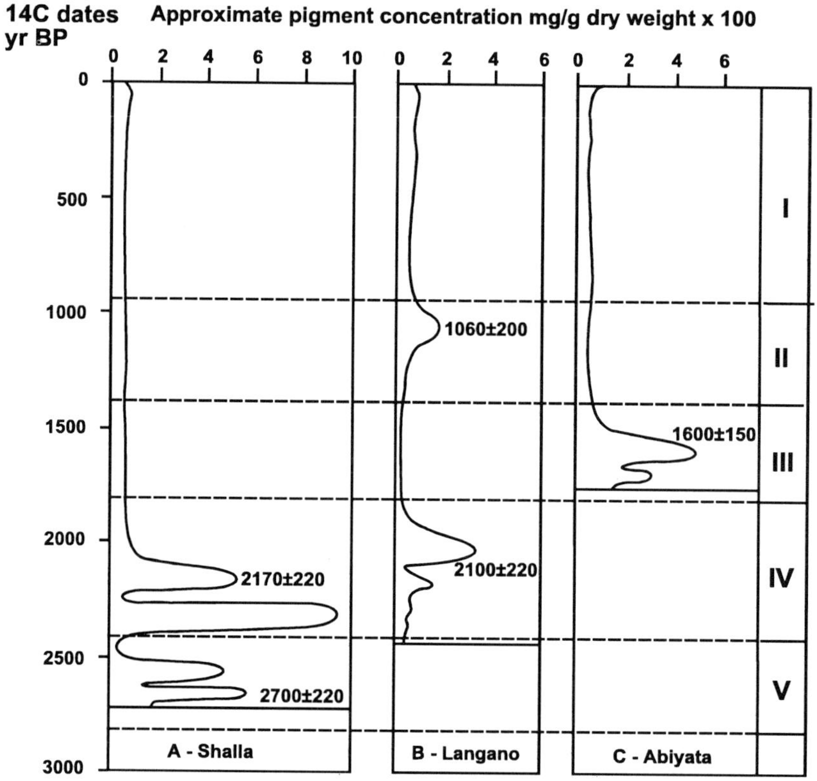


*Figure 4.* Synthetic pigment curve from Lakes Shalla, Langano and Abijata (Central Ethiopian Rift Valley). The chronology is from an interpolation of radiocarbon ages obtained on each core and considering the top of the cores as modern. A. Shalla, B. Langano, C. Abijata.

labile Carotenoids in the laminated portions could indicate high productivity and hence more eutrophication contributing to the autochthonous supply into the sediments (Engstrom *et al.*, 1985). Also, the formation of laminae could follow bottom anaerobic conditions due to persistent lake stratification which also means less bioturbation and mixing. This would have created better pigment preservation conditions.

The high values of CD in the laminated sediments could equally be due to more allochthonous input from terrestrial organic matter contemporaneous to the period favouring abundance in TC.

As has been argued for modern sediments, optimum depths at which conservation and organic matter supply coincide will favour the amount of pigment in sediments. However, factors influencing the fluctuations in the pigment curves of the cores should be those controlling the variability in abundance, supply and conservation of autochthonous and allochthonous organic matter.

Laminated sediment facies could correspond to the presence of stratified meromictic lakes which in tropical regions are mainly related to thermal stratification but reinforced by a chemical one. Such a sediment could also be the marker for a higher productivity resulting from increased nutrient input (Talbot, 1988). In the studied lakes, relatively persistent meromictic conditions have ceased to happen after 1000 years ago when homogeneous sediments dominated and pigment concentration declined.

Seasonal patterns show that deep lakes such as Langano and Shalla can produce stratification (thermal and also chemical, mainly of oxygen) much more easily than shallow lakes, such as Abijata, where the stratification is superficial and unstable (Kassahun, 1982). Observations conducted on the Debre Zeiet Crater lakes (North of the studied lakes), have shown that thermal stratification takes place during summer rainfall while mixing occurs during the cold and dry winter seasons (Baxter *et al.* 1965). The studies on Lakes Langano and Abijata indicated that the fall in oxygen in the deep stratified water also occurs during rainy seasons. For Lake Abijata, a fall in temperature by 1°C has been measured in 1981 and created thermal discontinuity which has been explained by descending cold water to produce stratification (Kassahun, 1982).

This observation indicates that the most important pigment peaks could have resulted from high rainfall conditions favouring lake stratification which led to the conservation of laminae and organic matter

More specifically the peaks at around 2000 YR BP in Shalla and Langano (Figure 4, zones V and IV) correspond to a period of Late Holocene rising lake levels uniting the three studied lakes (Street, 1979; Gillespsie *et al.,* 1983) and to inferences of humid conditions from pollen and other proxies studied in the area (Bonnefille *et al.*, 1986, Mohammed and Bonnefille, 1991) and in the region (Mohammed *et al.,* 1996). Pigment accumulation from about 1000 YR BP in the three lakes during which time they were separated to modern conditions. However the history of the last few hundred years may be lacking due to the absence of the top soft sediment in the core.

This study has shown that autochthonous and allochthonous organic matter supply have been declining in the lake sediment record after 1000 yr BP. High pigment concentration coincided with laminated portions of the cores which had durations of about 100 to 500 years and recurrence intervals of about 500 years in either one of the lakes. Such laminae are associated with more persistent lake stratifications than the current seasonal behaviour, perhaps following increased rainfall. Particularly that which occurred in Lakes Shalla and Langano between 2500 to 2000 years is contemporaneous with rising lake levels centred at about 2000 yrs ago.

## ACKNOWLEDGEMENTS

We are very much grateful to Dr. E. Schultz who provided laboratory facility for pigment analysis at the limnological institute, Mondsee, Austria. Our thanks go to J.J Tiercelin and C. Barton who did the coring. Funding from the French Ministry of Foreign Affairs have helped Mohammed Umer to do several research visits to

France. The Department of Geology (Addis Ababa University) and The Ethiopian Ministry of Mines and Energy have facilitated field work in Ethiopia when R. Bonnefille coordinated the coring campaign. J.P. CAZET and Dagnachew Legesse have improved the quality of the figures.

## REFERENCES

Bonnefille, R., Robert, C., Delibrias, G., Elenga, C., Herbin, J.P., Lézine, A.M., Perinet, G., Tiercelin, J.J. (1986) Paleoenvironement of Lake Abijata, Ethiopia during the past 2000 years. In: L. Frostick *et al.* (eds.), *Sedimentations in the African Rifts*. Geol. Soc. London Spec. Publ. **25**, 253-265.

Baxter, R.M., Posser, M.V., Talling, J.F., and Wood, R.B. (1965) Stratification in tropical African Lakes at moderate altitudes (1,500 to 2000m)- *Limnol. Oceanogr.* **10**, 510-520.

Daley, R.J., Brown, S.R. and McNeely, R.N. (1977) Chromatographic and SCDP measurements of fossil phorbins and the postglacial history of Little Round Lake, Ontario. *Limnology and Oceanography* **22**, 349-360.

Engstrom, D.R., Swain, E.B., Kingston, J.C. (1985) A paleolimnological record of human disturbance from Harvey's Lake, Vermont: geochemistry, pigments and diatoms. *Freshwater Biology,* 261-288.

Gasse, F. (1987) Ethiopie et République de Djibouti. Directory/Répertoir, in M.J. Burgis, and J.J. Symoens (eds.) *African Wetlands and Shallow Water Bodies/ Zones Humides et Lacs Peu Profonds d'Afrique*. ORSTOME. 300-331.

Gillepsie, R., Street-Perrott, F.A. and Switsur, R. (1983) Postglacial arid episodes in Ethiopia have implications for climatic prediction. *Nature* **306**, 680-683.

Gorham, E. (1960) Chlorophyll derivatives in surface muds from the English lakes. *Limnol.Oceanogr.* **5**, 29-33.

Gorham, E. and Sanger , J.E. (1967) Plant pigments in woodland soils. *Ecologia,* **48**, 306-308.

Gorham, E. and Sanger, J.E. (1975) Fossil pigments in Minnesota lake sediment and their bearing upon the balance between terrestrial and aquatic inputs to sedimentary organic matter. *Verh.int.Ver.limnol.* **19**, 2267-2273.

Gorham, E. and Sanger , J.E. (1976) Fossilised pigments as stratigraphic indicators of cultural eutrification in Shagawa lake, north eastern Minnesota. *Geol. Soc. Amer. Bull.* **87**, 1638-1642.

Huttunen, P. and Tolonen, K. (1977) Human Influence in the History of Lake Lovojarvi . S. Finland. *Finiskt Museum* 1975, pp. 68-105.

Kassahun, W. (1982) Comparative limnology of lake Abijata and Lake Langano in relation to primary and secondary production. Msc. Thesis. Addis Ababa university, School of Graduate studies.

Likens, G.E. and Davis, M.B. (1975) Post glacial history of Mirror Lake and its watershed in new hampshire. USA: an initial report. *Vereinigung für Theoretische und Angewandite Limnologie* **19**, 982-993.

Mohammed, M.U. and Bonnefille, R. (1991) The recent history of vegetation and climate around Lake Langano (Ethiopia). *Palaeoecology of Africa* **22**, 275-285.

Lorenzen, C.S. (1967) Determination of Chlorophyll and Pheopigmants: spectrophotometer equations. *Limnology and Oceanography* **12**, 343-346.

Mohammed, M.U., Bonnefille, R. and Johnson .T.C. (1996) Pollen and isotopic records of Late Holocene sediments from Lake Turkana, N. Kenya. *Palaeogeogr, Palaeoclimatol, Palaeoecol.* **119**(3-4), 371-383.

Rayback, I. and Rayback, M. (1985) Stratigraphy of fossil pigments as a guide to the environmental changes of the Kortwoskie Lake, Poland. *Acta Hydrochim. Hydrobiol.* **13**(3), 297-305.

Sanger, J.E. and Gorham, E. (1970) The diversity of pigments in lake sediments and its ecological significance. *Limnol.Oceanogr.* **15**, 59-69.

Sanger, J.E. and Gorham, E. (1972) Stratigraphy of fossil pigments as a guide to the postglacial history of kirchner marsh, Minnesota. *Limnology and Oceanography*, **23**, 1059-1066.

Schlutz,E. (1988/99) Falstudien zur Pläolimnologie. Geologija 31, 32, 437-516.

Street, F.A. (1979) Late Quaternary lakes in the Ziway-Shalla basin, Southern Ethiopia. Unpublished PhD Thesis, University of Cambridge, 457p.

Swain, E.B. (1985) Measurements and interpretation of sedimentary pigments. *Fresh Water Biology* **15**, 53-75.

Talbot, M.R. (1988) The origins of lacustrine oil source rocks: evidence from the lakes of tropical Africa, in A.J. Fleet, K. Kelts and M.R. Talbot (eds.) (1988) *Lacustrine Petrolium Source Rocks. Geol. Soc. Spec. Publ.* **40**, 29-43.

Tolonen, M. (1978) Palaeoecology of an annually laminated sediments in Lake Ahvenainen. S. Finland. III. Human influence in the lake development. *Annales Botanici Fennici*, **15**, 223-240.

Züllig, H. (1982). Untersuchungen Uber die Stratigraphie von Carotenoiden im geschichichteten Sediment von 10Schweizer Seen zur erkundung frûherer Phytoplankton -entfaltungen. *Schweiz Z. Hydro.* **44**(1), 1-98.

Züllig, H. (1985) Pigmente phototropher Bakterien in See Sedimenten und ihre Bedeutung f.r. die Seenforschung (mit Ergebnissen aus dem Lago Cadona, Rotsee und Lobsigensee). *Schweiz Z. Hydrol.* **47**, 87-126.

# ORIGIN AND ISOTOPIC COMPOSITION OF ARAGONITE LAMINAE IN AN ETHIOPIAN CRATER LAKE

HENRY LAMB[1], SEIFU KEBEDE[2], MELANIE LENG[3], DOUGLAS RICKETTS[4], RICHARD TELFORD[5] and MOHAMMED UMER[2]

[1]*Institute of Geography and Earth Sciences, University of Wales, Aberystwyth, Wales, UK*

[2]*Department of Geology and Geophysics, Addis Ababa University, Ethiopia*

[3]NERC Isotope Geosciences Laboratory, Keyworth, Nottingham, UK

[4]*Large Lakes Observatory, University of Minnesota, Duluth, Minnesota, USA*

[5]*Department of Geography, Newcastle University, Newcastle-upon-Tyne, UK*

**ABSTRACT**

Millimetre-scale white aragonite laminations alternating with dark diatom-rich organic layers are present in the uppermost sediments of the crater lakes Hora and Babogaya, at Debre Zeit, Ethiopia. The sediment accumulation rate calculated from lamina counts matches that estimated from a $^{210}$Pb chronology, indicating that the laminations were deposited annually. The oxygen-isotope composition of the surface white layer is equivalent to that of surface water, which shows that white-layer aragonite is formed in isotopic equilibrium with the lake water, and suggests that isotopic analyses of these layers may provide valid paleoclimatic information. Because the aragonite is probably precipitated during dry-season mixing, aragonite $\delta^{18}$O values for individual laminae reflect the composition of the entire lake, integrated over its water-residence time of about 10 years. The sedimentary record of oxygen-isotope variations should therefore be interpreted as a proxy-climate record with decadal rather than annual resolution. Comparisons between $\delta^{18}$O values for the laminae and climate data for equivalent years show no clear relationships, so calibration of the sedimentary record requires a more detailed understanding of the



*E.O. Odada and D.O. Olago (eds.),*
*The East African Great Lakes: Limnology, Palaeolimnology and Biodiversity,* 487–508.

climatic controls on the isotopic composition of these groundwater-fed lakes. An isotopic mass-balance model of the lake's response to rainfall variation shows (1) that the oxygen isotope composition of the lake waters varies by about 1 ‰, which is comparable to the range of $\delta^{18}O$ values determined from the individual laminae; and (2) that modelled lake level is a reasonable match to observed levels, confirming that climate changes can interpreted from the oxygen-isotope record.

## 1. INTRODUCTION

Annually-laminated sediments from mid-latitude lakes have proved valuable in providing high-resolution records of past environmental change (Peglar *et al.*, 1984; Lotter, 1989; Bradbury and Dean, 1993). Where laminae are present to the sediment surface, a varve chronology may allow calibration of sedimentary proxies against instrumental records of environmental change. In contrast, such biochemically varved sediments are rare in tropical lakes, and opportunities for calibrating them against instrumental records are rarer still. Here we report the presence of authigenic aragonite laminations in the recent sediments of Lake Hora and Lake Babogaya, two of the five crater lakes at Debre Zeit, Ethiopia, and examine the potential for calibrating the isotopic composition of individual year-assigned laminae against the 100-year climate record from Addis Ababa. This requires answers to two principal questions. First, are the laminations deposited annually? We approach this question by comparing the sediment accumulation rate estimated from $^{210}Pb$ determinations with that estimated from lamina counts. Second, can we obtain a high-resolution record of climate variability from oxygen isotopic composition of the laminae? We compare $\delta^{18}O$ values from the laminae to the instrumental record of climate during the last 30 years. We also examine the lake's sensitivity to rainfall variability by modelling the lake's isotopic hydrology, which requires data on the hydrochemistry, hydrology and isotopic composition of the lakes and their surface and groundwater inflows. If the isotopic composition of the lacustrine carbonates can be calibrated in this way, and the calibration applied to the pre-instrumental sedimentary record, long cores from the lakes may provide a high-resolution record of variations in the strength of the African monsoon, and ENSO-related climate variability.

## 2. STUDY AREA AND SITE DESCRIPTIONS

The Bishoftu lakes lie in a group of maars (volcanic explosion craters) on the western shoulder of the Ethiopian Rift Valley, at Debre Zeit (8°45'N 38°59'E; ~1900m altitude; see Figure 1 of Kebede *et al.*, this volume), 47 km southeast of Addis Ababa. Basaltic lavas and pyroclastics erupted in the area between one and four million years ago (Gasparon *et al.*, 1993). More recent activity, possibly in the early Holocene (Mohr, 1961), generated scoria cones, rhyolitic lava domes, and maars. Because of their proximity and common mode of origin, the lakes are subject to the same climatic, geological and land-use influences, and have broadly similar morphologies. Despite these shared characteristics, they differ in depth and water chemistry. Of the two lakes considered here, Lake Babogaya is deeper (65m) and more dilute (conductivity 800 μS.cm$^{-1}$) than Lake Hora, which has a maximum depth of 38m, and is relatively saline (2200 μS.cm$^{-1}$; Table 1). Lake Hora lies within a double crater, and a small submerged volcanic cone is present in Lake Babogaya.

## 2.1 Climate and Vegetation

The climate of the region is monsoonal with pronounced wet and dry seasons. The dry season extends from October to March with northeasterly airflow from the Arabian high, and ends in March or April with the onset of the spring rains, derived from airmasses of Indian Ocean origin. Slightly lower rainfall is recorded for May, during the transition from spring to summer rains. The main rainy season occurs from June to September, when frequent and heavy precipitation of Atlantic origin falls during passage of the inter-tropical convergence zone. July and August show the highest rainfall totals.

Rainfall data have been recorded at two stations in Debre Zeit: at the Airforce Base (AFB) since 1951 , and at the Agricultural Research Station (ARS) since 1953. Other meteorological data from Debre Zeit are available only for much shorter series, of ten years or less. The record from Addis Ababa is longer, from 1898 onwards. Data for the period 1951-1996 from AFB (1850m altitude) show that mean annual precipitation is 856mm, with large seasonal and inter-annual variability. Annual totals show a rising trend from ~600mm in 1953 to a maximum of 1200mm in 1966, followed by a more gradual decline to a minimum of 420mm in 1995. Mean April-September rainfall is 753mm, in contrast to mean dry-season (October - March) precipitation of 103mm. Rainfall data show marked local-scale spatial variability, because of the convective nature of tropical weather systems, so there is little or no correlation between sites only a few km apart. Penman, energy-balance and pan evaporation methods combined give an annual average open-water evaporation rate of 1710mm (Kebede, 1999). Mean annual air temperature is 19.5°C, with large diurnal but comparatively small seasonal variation. Lowest daily minimum temperatures occur during the dry season.

The natural vegetation of the area was probably *Acacia albida* savanna (Zerihun and Mesfin, 1990), with *Juniperus procera* forest at higher elevations. Today, the landscape is dominated by agriculture, principally by cultivation of tef, an indigenous Ethiopian cereal. Introduced trees and shrubs have been extensively planted around the lakes, including *Eucalyptus, Casuarina equisetifolia, Schinus molle* and *Opuntia.* The steep northern and western slopes of the Hora crater are covered by savanna vegetation of *Acacia albida, A. tortilis*, and tall grasses including *Hyparrhenia.*

## 2.2 Isotopic Composition of Rainfall

The oxygen isotopic composition of rainfall has been measured at Addis Ababa (2360m altitude), 47km to the northwest of Debre Zeit, in 24 of the 36 years between 1962 and 1997 (IAEA/WMO, 1999). Spring and summer rains have mean $\delta^{18}O$ values of +0.57‰ (-0.09‰ weighted mean) and -1.7‰ (-0.71‰ weighted mean) respectively, reflecting their different synoptic origins. Weighted mean annual values range from +0.23‰ in 1984 to -2.7‰ in 1975, with an overall weighted mean of –1.18 ‰ for the 24 years. These data define a local meteoric water line, $\delta D = (7.15 \pm 0.16)\delta^{18}O + (11.63 \pm 0.41)$ (*cf.* Rozanski *et al.*, 1996), shown in Figure 1. A few datapoints from the Rift Valley (Craig *et al.*, 1977) suggest that rainfall at Debre Zeit is probably similar in isotopic composition to that at Addis Ababa, despite the 460m difference in altitude.

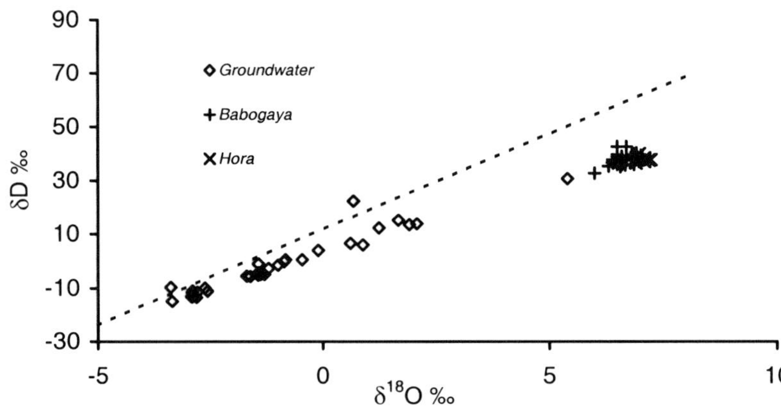


*Figure 1.* Local meteoric water line for rainfall at Addis Ababa, and isotopic composition of lakes and groundwater at Debre Zeit.

## 2.3 Limnology

Baxter *et al* (1965), Prosser *et al.* (1968), Wood *et al.* (1976, 1984), Rippey and Wood (1985), Wood and Talling (1988), and Zinabu (1994) have made detailed investigations of the limnology of the Bishoftu lakes. Lake Babogaya gradually develops thermal stratification during March to November, leading to the formation of indistinct thermoclines at 11-16m depth late in the summer wet season (Figure 2). The lake mixes from November to February, during the dry season, as a result of evaporative and night- time radiative cooling under conditions of low humidity and low cloud cover, finely balanced with solar inputs. Oxygen supersaturation develops in the epilimnion during the wet season, because of photosynthetic activity by the phytoplankton (Figure 3). Epilimnetic oxygen minima coincide with dry-season mixing, and occur occasionally during the wet season when the thermocline deepens, as in July-August 1966. The hypolimnion below 30m remains anoxic, and contains discernible amounts of hydrogen sulphide. Surface pH usually varies between 8.7 and 9.2, decreasing with depth to pH 8.4 in stratified conditions. Calcium concentration is lower in the epilimnion, because of photosynthetic precipitation of aragonite, but there is no significant depth increase of magnesium. Wood and Talling (1988) list only the cyanophyte *Microcystis aeruginosa* in Lake Babogaya, and also record it from Lakes Hora and Bishoftu. The similarity in limnological behaviour of Lakes Babogaya and Bishoftu points to climatic control (Wood *et al.*, 1984), and makes it reasonable to assume that Lake Hora shows comparable seasonal dynamics.

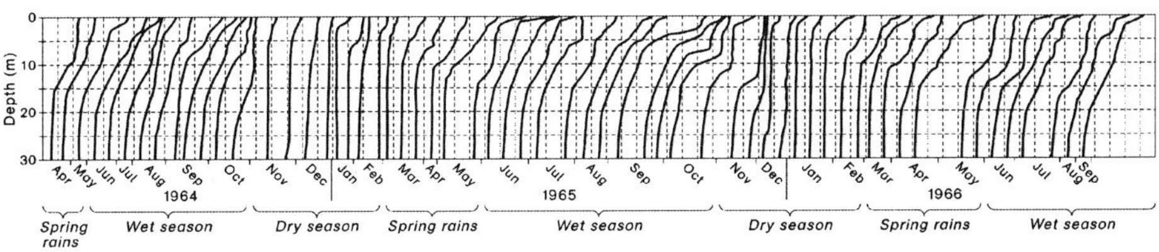


*Figure 2.* Temperature profiles for Lake Babogaya, April 1964 - September 1966. All profiles are between 19° and 20°C at 30m depth; vertical grid lines are spaced at 1°C intervals. (After Wood *et al.*, 1976).

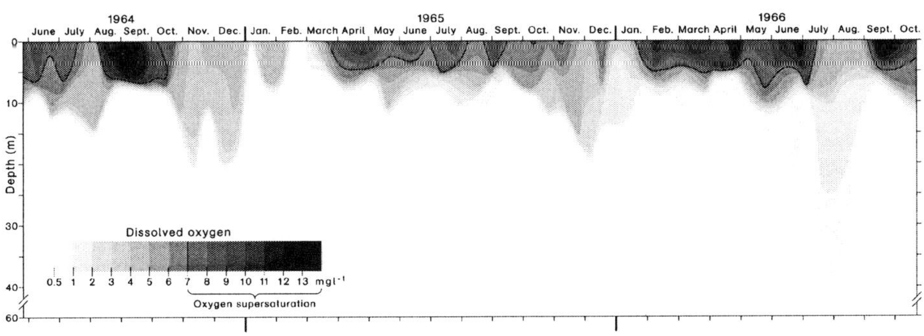


*Figure 3.* Dissolved oxygen content of Lake Babogaya water, May 1964 - October 1966 (After Wood *et al.*, 1984).

## 3. METHODS

We sampled the surface sediments of Lake Hora in January 1997 with a 45cm Glew corer. One core (HG97-1) was sub-sampled on site, for loss-on-ignition and diatom analysis (Pearson, 1997); the second (HG97-2) was returned intact to the laboratory. In January 1998, we again sampled the surface sediments with an Eckman Grab, from which we took a 10.4 cm mini-core within a capped perspex cylinder. This mini-core, HE98-1, was frozen by immersion in liquid nitrogen, sliced vertically into 1cm-thick slabs with a powered band saw, and freeze-dried. The dried slabs were photographed, and the laminations counted, labelled, and then individually removed with a scalpel for isotope analysis. Core HG97-2 was halved lengthwise, and one half freeze-dried, photographed, and the laminations labelled and counted, before sampling at 0.5cm intervals for $^{210}$Pb determinations. Samples at 0.5cm intervals from the other half were used for isotopic analyses. Additional samples for $^{210}$Pb analyses were obtained from core HE98-1, because lamina correlations showed that the Glew corer failed to take the top few cm of sediment.

Sediment samples were analysed for $^{210}$Pb, $^{226}$Ra and $^{137}$Cs by direct gamma assay in the Liverpool University Environmental Radioactivity Laboratory using Ortec HPGe GWL series well-type coaxial low background intrinsic germanium detectors (Appleby *et al.*, 1986). $^{210}$Pb was determined via its gamma emissions at 46.5keV, $^{226}$Ra by the 295keV and 352keV $\gamma$-rays emitted by its daughter isotope $^{214}$Pb following 3 weeks storage in sealed containers to allow radioactive equilibration. $^{137}$Cs was determined by its emissions at 662keV. The absolute efficiencies of the detectors were determined using calibrated sources and sediment samples of known activity. Corrections were made for the effect of self-absorption of low energy $\gamma$-rays within the sample (Appleby *et al.*, 1992).

For stable isotope analysis, sediment samples were disaggregated in 5% sodium hypochlorite solution (10% chlorox) for 24 hours to oxidise reactive organic material. Samples were then washed three times in distilled water and sieved at 85μ m. The <85μm fraction was filtered through quartz microfibre filter paper, dried at 40°C, and ground in agate. The isolated material was reacted with anhydrous phosphoric acid in vacuo overnight at a constant 25°C (McCrea, 1950). The $CO_2$ thus liberated was separated from water vapour and collected for analysis. Measurements were made on a VG Optima mass spectrometer. Overall analytical reproducibility is normally better than 0.1‰. $\delta^{18}$O for lake water at 20°C was calculated from the $\delta^{18}$O of aragonite using Kim and O'Neil's (1997) fractionation factor. Isotope values ($\delta^{13}$C, $\delta^{18}$O and $\delta$D) are reported as per mil (‰) deviations of the isotopic ratios ($^{13}$C/$^{12}$C, $^{18}$O/$^{16}$O and D/H) from standards (V-PDB for carbonates, VSMOW for water).

Water samples were collected from the surface, 10m and 60m depths in Lake Babogaya, on ten dates during 1998-1999. Groundwater samples were collected from 27 boreholes and wells. Total dissolved inorganic carbon (TDIC) for $\delta^{13}$C analysis was precipitated on site by addition of $BaCl_2$ + NaOH solution; untreated water was collected for $^{18}$O, D and major ion and trace element analysis. Samples were stored under refrigeration in leakproof nalgene bottles for up to 12 weeks before transport to Aberystwyth and Keyworth for laboratory analysis. $\delta^{18}$O and $\delta$D values for water were measured on a SIRA mass spectrometer, and are reported in per mil versus VSMOW. Reproducibility is better than 0.1 and 2‰ respectively. Temperature and conductivity readings were taken with a Hanna instrument water test meter. Measurements of pH were made with Hanna instrument HI-9024/5. Trace element analysis was carried out by ICP-MS, and major ions were determined with a Dionex DX-100 ion chromatograph. Alkalinity was calculated from the charge balance.

# 4. LAKE HYDROCHEMISTRY AND HYDROLOGY

## 4.1 Hydrochemistry

The groundwaters are dilute relative to the lakewaters, Ca being the only ion that is more concentrated in the groundwaters (Table 1). Some groundwater samples show anomalously high Cl values (0.73 -1.94 mequiv.$l^{-1}$), probably because they are affected by seepage from lakes. However, there is no clear spatial trend in Cl concentrations, indicating that seepage outflow is fault or fracture controlled, rather than diffuse. The conductivity of Lake Babogaya is only slightly higher than that of the groundwaters, whereas Lake Hora has conductivity values three times that of groundwater. Calcium (from weathering of plagioclase feldspars, pyroxenes, and calcic amphiboles in the basic volcanics) has lower concentrations in the lakes than in groundwater, since it is precipitated as carbonate from the lake surface waters. Mg/Ca ratios in the lakes are therefore 10 and 20 times that of groundwater. Calcium shows an increase with depth in lake Babogaya for all samples taken in 1998. At 10m and 60m depths, the calcium concentrations remained fairly constant (mean values 6.5, 9.4ppm respectively). Lowest calcium concentrations (1.8 and 3.6ppm) were recorded at the surface in April 98, increasing progressively to 5.8ppm in January 99. Depth differences in magnesium concentrations were less pronounced, ranging from 43 ppm at the surface to 52 ppm at 60m, so the Mg/Ca ratio is highest in surface waters (mean values 11.3, 7.4, and 5.7 at surface, 10m and 60m depths, respectively). In the dry season, chloride increases by 5-10ppm in the surface waters, despite lake mixing. Overall, hydrochemical data from the lakes are similar to those of Prosser *et al.* (1968). However, the concentration of chloride, and to a lesser extent of sodium, in lakes Babogaya, Hora and Bishoftu have dropped since about the 1960's, which apparently corresponds with the timing of a recent water-level rise in these lakes.

## 4.2 Hydrology

Groundwater influx ($G_i$) and outflow ($G_o$) for each lake can be estimated (Table 2), assuming steady state conditions, by solving the hydrological and solute budgets simultaneously. Open-water evaporation from the lakes is the mean value of estimates based on pan evaporation, Penman, and energy balance approaches (Kebede, 1999). Surface runoff, calculated here with an estimated coefficient of 0.25, is small, because the crater walls define a very restricted catchment area. The solute budget is based on chloride, described by $G_iC_{Gi}=G_oC_{lake}$, where $C_{Gi}$ = concentration of chloride in the inflowing groundwater and $C_{lake}$ = chloride concentration of the lake, assumed to be the same as that of the groundwater outflow. Chloride in precipitation, runoff and evaporation is assumed to be negligible.

The principal difference between the two lakes is in the extent of groundwater seepage, which makes up 21% of total water loss from Babogaya but only 3% from Hora (Table 2). Lake Hora receives 43% of its total inflow from groundwater, but almost all water loss is by evaporation (97%). By contrast, groundwater loss from Babogaya is 21% of the total. Lake Hora may therefore be regarded as a terminal lake, in comparison to Babogaya, a through-flow lake. Because the lakewater chloride concentrations were used to estimate the hydrological budgets, we cannot strictly use the estimates to explain the differences in lake chemistry. Nevertheless, there is little doubt that the salinity contrasts are due to the differences in lake hydrology. Loss of solutes via groundwater explains why Lake Babogaya is more dilute than Hora, which has a greater influence of evaporation. The residence time of water in the lakes, calculated as lake volume divided by water flux (evaporation plus groundwater outflow), is 15 years in Babogaya, and 10 years in Hora.

*Table 1.* Hydrochemistry of Debre Zeit groundwaters and lake surface waters, mean values 1998

| | *Conductivity* k20 $\mu S.cm^{-1}$ | *pH* | $HCO_3^-$ $meq.l^{-1}$ | $Cl^-$ $meq.l^{-1}$ | $Na^+$ $meq.l^{-1}$ | $Mg^{++}$ $meq.l^{-1}$ | $Ca^{++}$ $meq.l^{-1}$ | *Mg/Ca ratio* |
|---|---|---|---|---|---|---|---|---|
| *Groundwater* | 685 | 7.3 | 2.62 | 0.42 | 1.99 | 2.52 | 1.98 | 1.3 |
| *Babogaya* | 776 | 8.9 | 7.67 | 0.69 | 3.74 | 3.74 | 0.22 | 17 |
| *Hora* | 2235 | 8.4 | 15.3 | 6.83 | 18.2 | 5.95 | 0.3 | 19.8 |

*Table 2.* Hydrology of lakes Hora and Babogaya

| | *% Inflow* | | | *% Outflow* | | *Residence time, yrs* | |
|---|---|---|---|---|---|---|---|
| | *Direct Rainfall* | *Runoff* | *Ground-water* | *Evaporation* | *Ground-water* | *Water* | *Cl* |
| *Hora* | 47 | 10 | 43 | 97 | 3 | 6 | 500 |
| *Babogaya* | 38 | 6 | 56 | 79 | 21 | 10 | 81 |

## 4.3 Isotope Hydrology

The $\delta^{18}O$ and $\delta D$ composition of groundwater samples from boreholes and wells sampled at Debre Zeit between January 1998 and January 1999 is shown in Figure 1, and in Table 2 of Kebede *et al.* (this volume). Most values plot close to the Addis Ababa meteoric water line, indicating that the aquifer is fed by precipitation without any appreciable effects of either evaporation before infiltration, or of selective recharge by the more intense, isotopically light summer rains. Similarly, there is no evidence of isotopic exchange with the aquifer rocks. In some groundwater wells, there is clearly a considerable degree of mixing with seepage outflow from the lakes, as shown by the trend of enrichment towards values for lakewaters.

Oxygen isotope values for Hora and Babogaya lakewaters are enriched relative to rainfall and groundwater, reflecting evaporative enrichment during their 10 and 15 year water residence times. The values for Hora (mean 7.0‰) are only slightly more enriched than those for Babogaya (mean 6.7‰), reflecting a combination of higher percentage water loss by evaporation in Hora, its greater surface:volume ratio and the longer water residence time of Babogaya. $\delta^{18}O$ values from all three depths in Babogaya are similar for samples taken in January to June 1998, presumably as a result of mixing (Figure 4). Depletion of the surface waters in August probably results from dilution of the epilimnion by rainfall, when the deeper waters become isolated below the thermocline. The surface waters then show progressive $^{18}O$ enrichment from 6‰ to 6.6‰ between August and January, from the effects of evaporation and dry-season mixing. The deeper waters show slight but progressive $^{18}O$ enrichment by about 0.3‰ from April to November, for unknown reasons.

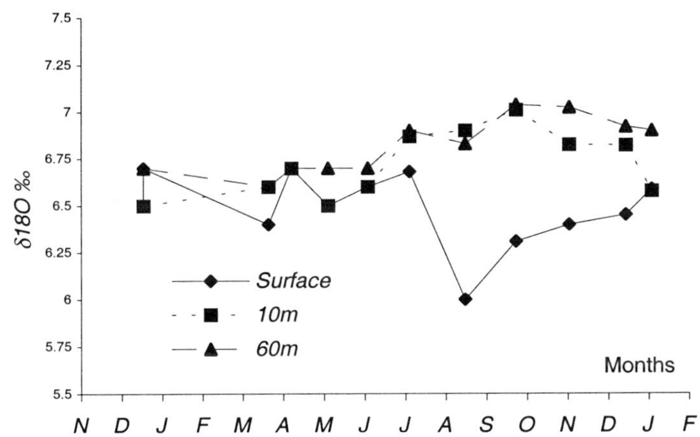


*Figure 4*. Variation in oxygen-isotope composition of Lake Babogaya surface, middle and deep water during 1998.

Groundwater $\delta^{13}C_{TDIC}$ values vary from -6.5 to -9.5‰, which may reflect equilibration with soil-respired carbon dioxide beneath a vegetation dominated by C4 grasses (typical values from C3-dominated western Europe range from -10‰ to -15‰; Andrews *et al.*, 1993). Dissolution of magmatic $CO_2$ in groundwater may also account for the values observed here. Lake Hora has unusually high $\delta^{13}C_{TDIC}$ values (+6.2‰ to +7.1‰). $\delta^{13}C$ values for Lake Babogaya TDIC range from +0.1‰ to +3.0‰ , with most positive values from the deeper waters. $^{13}C$ enrichment relative to groundwater results in part from isotopic fractionation associated with $CO_2$ outgassing from the $CO_2$-saturated water, and from preferential uptake of $^{12}C$ during algal photosynthesis in the surface waters. Anoxic decay of organic matter in the surface sediments with release of isotopically light methane would also lead to enriched TDIC (Curry *et al.*, 1997).

## 5. LAMINA COUNTS

At the times of sampling (January 1998, 1999) a white carbonate layer was visible on the surface of the sediment retained by the Eckman grab. Scanning electron microscopy shows that the white layers are composed of aragonite, the 5-10μm crystals identifiable by their characteristic linear shape and frequent cruciform twinning (Figure 5). Aragonite is typically precipitated by algal photosynthesis in waters with a Mg:Ca ratio greater than 3 (Folk, 1974; Kelts and Hsu, 1978), so its occurrence in Hora and Babogaya is not unexpected. In the cores, the aragonite forms discrete white laminae, less than 1 mm thick, separated by thicker (0.5 – 8mm) brown layers of algal detritus, with a low carbonate content (Figure 5). The white layers are sub-parallel, undulating irregularly at a fine scale as a result of deposition over an uneven sediment surface. These irregularities, and the pattern of spacing between laminae, allowed identification of individual laminae, and correlation between cores. In core HE98-1, which is 104 mm long, there are 40 white laminae, with an average interval of 2.6 mm between adjacent laminae. Eight laminae lie in close-set pairs ~0.5mm apart. The 42cm-long Glew core HG97-1 revealed 95 white laminations, of which layers 1-16 (counted from the top in core HE98-1) were missing.

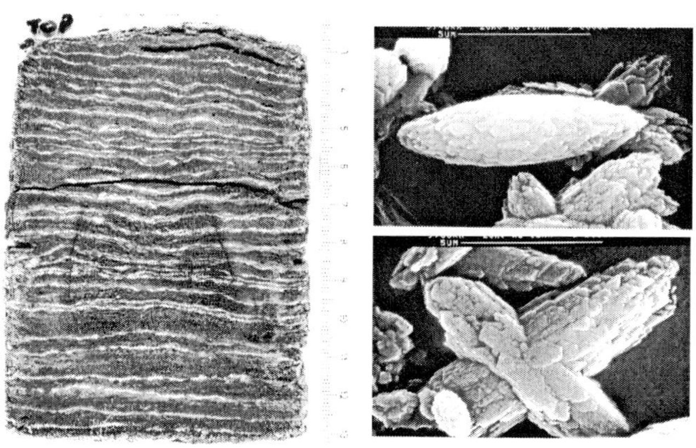


*Figure 5.* Aragonite laminations in freeze-dried section of Lake Hora core HG98-1 (left; scale in cm). Scanning electron microscope images of aragonite crystals from the core (right; 5μm scale at top).

Assuming, as a working hypothesis, that the laminae were deposited annually, we assigned a year of deposition for each carbonate lamina by counting down from the sediment surface. The four close-set lamina pairs were counted as single years, because they may represent years when mixing events also interrupted wet-season stratification, as recorded by Wood *et al.* (1984) in July-August 1966. Thus 104mm of sediment in core HE98-1 apparently accumulated from AD 1962 to 1997 at a rate of 2.9mm/yr. The 43cm-long Glew core HG97-2 contains ~96 laminae, giving a mean sediment accumulation rate of 4.5mm/yr for the period AD 1886 - 1982. However, the sequence is interrupted by a coarse reddish unit (14-9.5cm), and by several smaller unlaminated units, so we disregard the sediment accumulation rate estimated from core HG97-2.

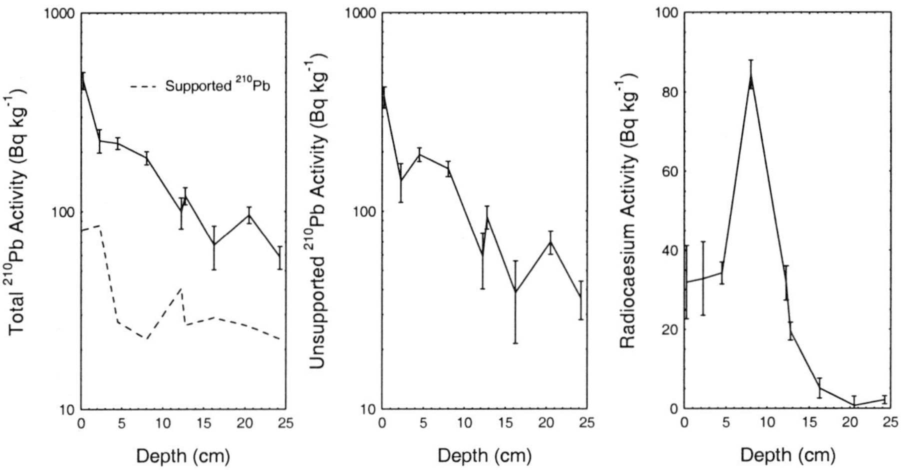


*Figure 6.* Fallout radionuclide concentrations versus depth in Lake Hora sediments showing (a) total and supported $^{210}$Pb (dashed line), (b) unsupported $^{210}$Pb, (c) $^{137}$Cs.

# 6. $^{210}$PB AND $^{137}$CS CHRONOLOGY

Fallout $^{210}$Pb and $^{137}$Cs concentrations in the sediments of cores HG97-2 and HE98-1 are shown in Figure 6. Radiometric dates calculated using the constant rate of supply (CRS) $^{210}$Pb dating model (Appleby and Oldfield, 1978) are shown in Figure 7 together with dates determined from the $^{137}$Cs stratigraphy. The final chronology for the core was established following an assessment of the data using procedures described in Appleby and Oldfield (1983) and Appleby (1998). Equilibrium between total $^{210}$Pb activity and the supporting $^{226}$Ra (corresponding to ~150 years) was reached at a depth of 30-35 cm (Figure 6a). The unsupported $^{210}$Pb activity versus depth profile (Figure 6b) has a number of small-scale irregularities that may be related to the apparent large fluctuations in dry bulk density of the sediment. However, the overall trend follows an exponential relationship, indicating that mean accumulation rates on a decadal time-scale have been relatively uniform. The $^{137}$Cs activity versus depth profile (Figure 6c) has a well-resolved peak at 7.5-8.5 cm that presumably records the 1963 fallout maximum from the atmospheric testing of nuclear weapons.

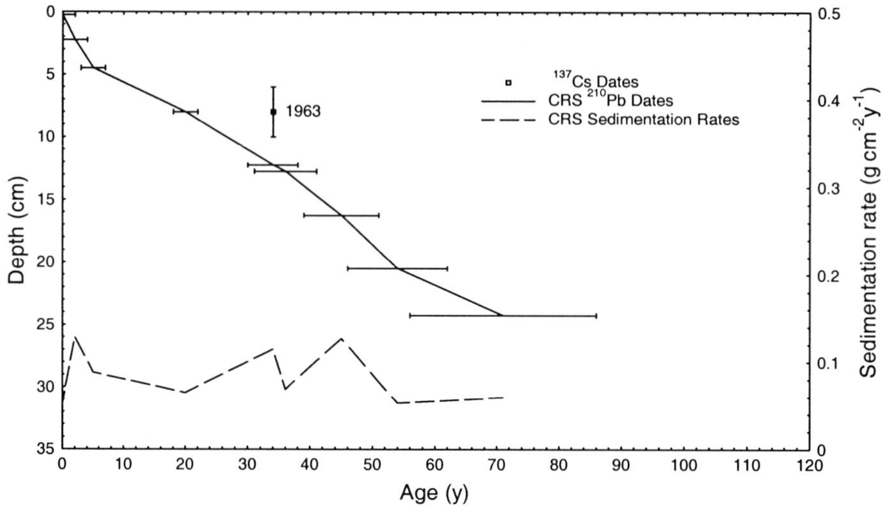


*Figure 7.* Radiometric chronology of Lake Hora core showing CRS model $^{210}$Pb dates together with an estimate of the 1963 depth determined from the $^{137}$Cs stratigraphy. Also shown are sedimentation rates calculated using the CRS model

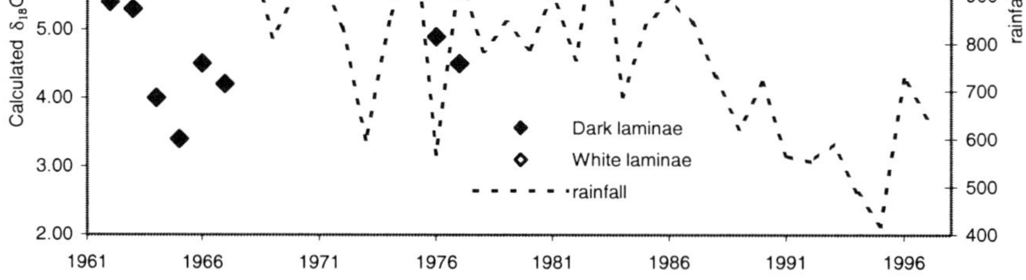


*Figure 8.* Calculated $\delta^{18}O_{lakewater}$ from aragonite in dark and light layers of Lake Hora sediments, plotted against year of deposition and compared to mean annual rainfall for Debre Zeit. $\delta^{18}O_{lakewater}$ (SMOW ) values were calculated for water temperature of 20ºC using the fractionation factor of Kim and O'Neil (1997).

*Table 3.* Oxygen and carbon isotope values (V-PDB) for aragonite from individual year-assigned laminae in the sediments of Lake Hora.

| *Year* | *Light layers* | | *Dark layers* | |
|---|---|---|---|---|
| | $\delta^{18}O$ | $\delta^{13}C$ | $\delta^{18}O$ | $\delta^{13}C$ |
| 1997 | 6.54 | 10.30 | | |
| 1996 | 6.30 | 10.93 | | |
| 1995 | 6.37 | 10.53 | | |
| 1994 | 6.38 | 10.32 | | |
| 1993 | 6.06 | 10.41 | | |
| 1992 | 6.78 | 10.83 | | |
| 1991 | 7.01 | 10.70 | | |
| 1990 | 7.06 | 10.86 | | |
| 1989 | 7.26 | 10.71 | | |
| 1988 | 6.89 | 10.16 | | |
| 1987 | 7.09 | 10.70 | 5.9 | 9.3 |
| 1986 | 6.49 | 10.73 | | |
| 1985 | 6.75 | 10.55 | | |
| 1984 | 6.65 | 10.87 | | |
| 1983 | 7.03 | 10.63 | | |
| 1982 | 7.42 | 10.65 | | |
| 1981 | 7.63 | 10.60 | 5.1 | 8.0 |
| 1980 | 7.21 | 10.57 | | |
| 1979 | 7.23 | 10.12 | 5.1 | 7.5 |
| 1978 | 6.87 | 10.41 | | |
| 1977 | 7.08 | 10.21 | 3.8 | 6.3 |
| 1976 | 6.99 | 10.17 | 4.2 | 6.3 |
| 1975 | 6.94 | 10.00 | 5.2 | 7.9 |
| 1974 | 6.76 | 10.34 | | |
| 1973 | 6.65 | 10.36 | | |
| 1972 | 7.26 | 10.12 | 5.7 | 8.1 |
| 1971 | 6.87 | 9.98 | 5.0 | 6.6 |
| 1970 | 7.15 | 9.99 | | |
| 1969 | 6.87 | 9.82 | | |
| 1968 | 7.11 | 9.68 | | |
| 1967 | 7.15 | 9.73 | 3.5 | 0.8 |
| 1966 | 6.99 | 9.72 | 3.8 | 5.5 |
| 1965 | 6.87 | 9.51 | 2.7 | 0.7 |
| 1964 | 6.72 | 9.66 | 3.3 | -0.7 |
| 1963 | 6.84 | 9.91 | 4.6 | 6.1 |
| 1962 | 7.43 | 9.47 | 4.7 | 6.3 |

Figure 8 shows the $^{210}Pb$ chronology calculated using the CRS model together with the 1963 depth suggested by the $^{137}Cs$ fallout record. Use of the constant initial concentration (CIC) model was precluded by the irregularities in the $^{210}$Pb record. There is a significant discrepancy between the $^{210}$Pb and $^{137}$Cs dates, due to an apparent episode of accelerated sedimentation during the past few decades. Since the samples in the top 3.5 cm mainly responsible for the discrepancy were from core HE98-1, they were excluded from the calculations. The $^{210}$Pb data between 4-20 cm indicate a mean sedimentation rate of 2.9 mm yr$^{-1}$. A corrected chronology was determined by applying this rate to the top 4 cm. The results of these calculations, shown in Figure 8, place 1963 at a depth of 9 cm, in better agreement with the $^{137}$Cs results.

## 7. ISOTOPIC COMPOSITION OF THE LAMINAE, AND COMPARISON WITH CLIMATE DATA

Measured $\delta^{18}O_{lakewater}$ values for February 1998 are identical to those calculated from $\delta^{18}O_{aragonite}$ of the surface layer of carbonate, suggesting that the white-layer aragonite was precipitated in isotopic equilibrium with the lake water. Dark-layer $\delta^{18}O_{lakewater}$ values (not all dark layer samples contained sufficient carbonate for analysis) are consistently lower than the white layer values. Four dark layers near the base of the sequence, representing years 1964-1967, have especially low values; these were years of exceptionally high rainfall (Figure 8). The depleted dark-layer values may represent dilution of the epilimnion by rainfall during wet-season stratification. Oxygen and carbon isotope data for the laminae are given in Table 3.

Calibration of the oxygen-isotope record against instrumental climatic data requires finding a relationship between $\delta^{18}O_{laminae}$ and rainfall or P-E at Debre Zeit. Linear regressions of $\delta^{18}O$ values for the white layers, each assigned to individual years by counting down from the surface, against rainfall, P-E and dry-season evaporation at Debre Zeit show no strong relationships. Lower $\delta^{18}O$ values since ~1993 (Figure 8) may be related to a recent lake-level rise, shown by the presence of dead trees standing in about 1m of water near the lake margins. The lake-level rise may be a response to increased mean annual rainfall since the 1995 minimum of 419mm, after almost 30 years of diminishing rainfall. Allowing for error in the lamina-count chronology, it appears that the $\delta^{18}O$ signal records lake response to the increase in rainfall.

## 8. MODELLING LAKE RESPONSE TO ENVIRONMENTAL CHANGE

A more complete understanding of Lake Hora's response to changing environmental conditions requires a model of lake isotope hydrology. The model used here involves solving an isotopic mass balance equation iteratively (Ricketts and Johnson, 1996; Ricketts and Anderson, 1998):

$$\frac{d(V\delta_L)}{dt} = G_i\delta_{Gi} + P\delta_P + S_i\delta_{Si} - G_o\delta_{Go} - E\delta_E \qquad (1)$$

where V is lake volume, L is the lake water, $G_i$ and $G_o$ are groundwater inflow and outflow respectively, P is direct precipitation to the lake, $S_i$ is surface inflow, E is evaporation from the lake surface and δ is the isotopic composition of each parameter. The isotopic composition of evaporation from the lake ($\delta_E$) was derived from the following equation (Benson and White, 1994; equation 20):

$$\delta_E = \left( \frac{\left(\frac{1-k}{\alpha_{eq}}\right)\left(1+\frac{\delta_L}{1000}\right) - (1-k)h\left(1+\frac{\delta_A}{1000}\right)}{1-h} - 1 \right) * 1000 \qquad (2)$$

where h is relative humidity, $\delta_{eq}$ is the equilibrium fractionation factor, k is the kinetic fractionation factor and $\delta_A$ is the isotopic composition of atmospheric moisture. The equilibrium fractionation factor between liquid water and water vapor is dependent on temperature (Friedman and O'Neil, 1977) and was calculated using a water temperature of 20ºC. The kinetic fractionation factor is dependent on wind speed and in most continental settings can be assumed to be 6.8‰ (Merlivat and Jouzel, 1979; Phillips *et al.*, 1986). The isotopic composition of atmospheric moisture ($\delta_A$) was assumed to be in equilibrium with the composition of local precipitation at the yearly average air temperature at Debre Zeit. Local precipitation ($\delta_P$) was estimated from the weighted mean isotopic composition of rainfall at Addis Ababa from the IAEA/WMO network (-1.18‰; IAEA/WMO, 1998). Relative humidity was set at 47%, which is the yearly average of daytime humidity at Debre Zeit (Kebede *et al.*, 1999).

We first modelled the isotopic response of the lake to changes in the amount of rainfall. To solve equation (1) we used:

a) An input of artificial sinusoidal rainfall variations with decadal periodicity. The input rainfall range (1210mm/yr to 450mm/yr) is similar to that experienced at Debre Zeit during the last 44 years.

b) Surface inflow was calculated from P using a runoff coefficient of 0.25.

c) The evaporation rate was held constant at 1710 mm/y (Kebede *et al.*, this volume).

d) Groundwater inflow was calculated using a cumulative departure from mean P-E over 5 years, that interval being an empirical estimate of the aquifer residence time. In effect, the modern P-E value was divided by the mean P-E value of the modelled year and the preceding four years. This departure from mean P-E was multiplied by the current estimate of groundwater inflow to the lake, thereby generating an inflow value ($G_i$) that incorporates an estimate of the aquifer residence time.

e) Groundwater outflow was calculated as 3% of total outputs.
f) Fluxes for P, $S_i$, E and $G_o$ were generated from rates using surface areas from topographic maps of the basin ($S_i$) and depth vs. surface area curves for Lake Hora (P, E and $G_o$).
g) The isotopic composition of surface inflow ($\delta_{Si}$) was set equal to $\delta_P$ and the isotopic composition of groundwater outflow ($\delta_{Go}$) was set equal to $\delta_L$.
h) The isotopic composition of groundwater inflow ($\delta_{Gi}$) was calculated from the average value found for the two closest wells to the lake (-0.9; Hora Tannery Well and the Kebele hand dug well; Table 2, Kebede *et al.*, this volume).

Figure 9 shows simulation results for Lake Hora, which demonstrate that $\delta^{18}O_{lakewater}$ fluctuates by about 1‰ for a 50-year model run, in response to the input of artificial sinusoidal rainfall variations with decadal periodicity. The lakewater $\delta^{18}O$ response is comparable to the range of $\delta^{18}O$ variation determined from carbonate laminations.

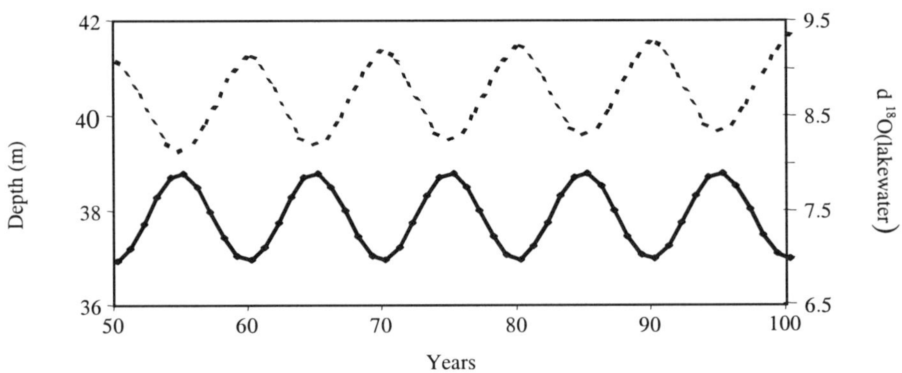


*Figure 9.* Isotope hydrology model simulation for lake Hora, demonstrating that depth (dotted line) and $\delta^{18}O_{(lakewater)}$ (solid line) should be sensitive to decadal rainfall variations.

Next, we used the isotopic composition of the aragonite laminations as a proxy for paleo-$\delta^{18}O_L$ composition (Table 3), and used equation (1) to solve for input fluxes (and therefore lake level) instead of $\delta O_L$. The same values for h, $\delta_{eq}$, k, and $\delta_A$ were used as in the previous model run with evaporation flux again held constant. To further simplify the equation, groundwater outflow was ignored, since it is only 3% of total outflow, and all input fluxes were combined. The $\delta^{18}O_{input}$ value was held constant at –1.06‰, which was calculated, using modern flux values for the lake (Table 2), assuming 57% of inputs had a composition of –1.18‰ (precipitation on the lake surface and runoff into the lake) and 43% of the inputs had a composition of –0.9‰ (groundwater inflow into the basin). Using this version of equation (1), and starting the calculations assuming that during 1963 the lake had a depth of 38meters (Prosser *et al.*, 1968), we are able to generate a paleo-lake-level curve (Figure 10)

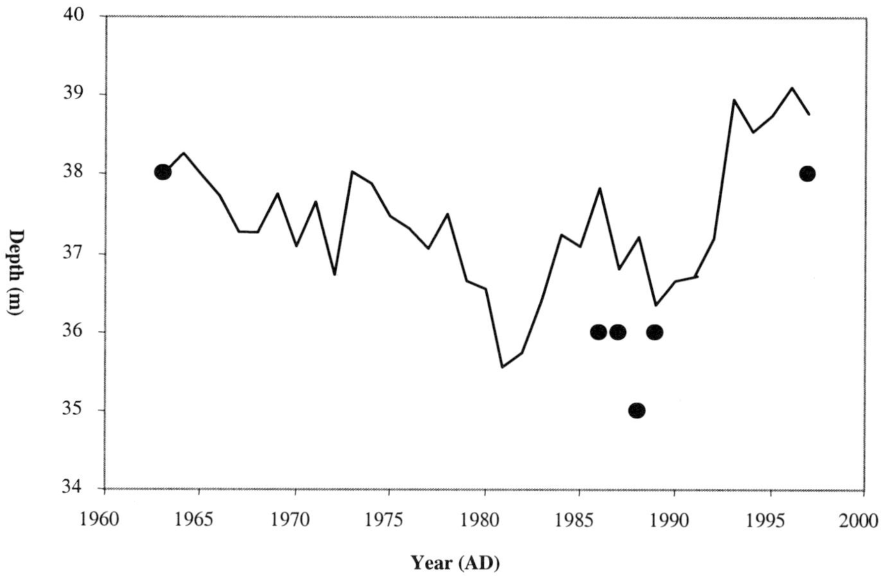


Figure 10: Comparison of actual lake-level measurements (●) Prosser *et al.*, 1968; Kebede, 1999) to a lake level curve (——) generated from a isotopic mass balance model using the isotopic composition of aragonite laminae as a proxy for paleo-$\delta^{18}O$ (lakewater).

and compare it to past observations of lake level (Prosser et al., 1968; Kebede, 1999). While there is not a perfect match between the observed lake level and the lake-level curve generated from the modelled $\delta^{18}O_L$(aragonite) values, the modelled lake-level curve is within 1-2 meters of the observed lake levels, which increases our confidence that the aragonite laminae do contain information about recent conditions in the Hora basin. As Kebede et al. (this volume) and earlier authors (e.g. Pearson and Coplen, 1978) have noted, isotopic mass-balance models of lacustrine systems are very sensitive to changes in input parameters, especially relative humidity (h) and the isotopic composition of atmospheric moisture ($\delta_A$). While this sensitivity hampers interpretation, in basins such as Lake Hora where modelling results can be compared to actual lake-level records, modelling can help interpret isotope data from carbonate laminations.

# 9. DISCUSSION

Carbonate laminations, also known as biochemical varves, are well known from temperate lakes (Peglar *et al.*, 1984; Lotter, 1989; Bradbury and Dean, 1993), but have not previously been reported from the tropics. Laminated sediments in other tropical lakes may be composed of diatoms, as in Lake Malawi (Barry, this volume), or formed from seasonal deposition of silts and clays, as in Lake Turkana (Muchane, 1996). Laminations in Lake Bosumtwi, Ghana, are composed of aragonite with a clastic component, and are not forming at the present-day sediment surface (Talbot *et al.*, 1984; Talbot and Johannessen, 1992). Lake Barrine, an upland maar lake in northeast Australia, has laminae that probably form by episodic redeposition of organic detritus from water shallower than 50m to the deeper meromictic zone during periods of mixing (Walker and Owen, 1999).

Carbonate laminae from eutrophic temperate-zone lakes usually consist of alternating light carbonate and dark organic layers, formed as a result of seasonal changes in productivity, closely linked to the seasonal cycle of temperature stratification. The light layers, formed in spring and early summer, are typically composed of authigenic calcite crystals and centric diatoms, the calcite being precipitated during photosynthetic uptake of dissolved $CO_2$. Surface-water temperatures in temperate lakes typically vary seasonally over a range of 20ºC. Tropical lakes, by contrast, experience much less pronounced seasonal temperature variations; at Debre Zeit the surface-water seasonal temperature range is only about 3ºC. Lake productivity and varve formation thus appear to be controlled by seasonal stratification due to small seasonal differences in the lakes' heat contents (Wood *et al.*, 1976) rather than by the dominant influence of temperature on stratification and productivity, as in temperate-zone lakes.

The close similarity between Lake Hora sediment accumulation rates estimated from lamina counts (2.9 mm $yr^{-1}$) and from the $^{210}Pb$ chronology (2.2 to 2.9 mm $yr^{-1}$) confirms that the laminae are deposited annually. Lamina formation must be linked to the annual cycle of lake stratification and mixing, which is well documented for Babogaya, and is probably similar in Hora. Small amounts of aragonite are present in the dark organic layers, suggesting that aragonite deposition takes place year round, but that its rate of precipitation increases markedly, and/or that deposition of organic material ceases, during part of the annual stratification cycle. Sediment traps located below the epilimnion retain dark organic material during summer and early autumn, suggesting that the white layers are not deposited during the summer wet season, despite high epilimnetic oxygen concentrations indicating maximum algal photosynthesis. Initial results from the sediment traps suggest that the white layers form during the November-January dry-season, when mixing may promote photosynthesis and aragonite precipitation by bringing calcium and nutrients from the hypolimnion to the surface. Mixing may also cause lighter algal detritus to be retained in suspension in the water column, while the denser aragonite particles fall to the lake floor. In some years, aragonite layers may also form during mixing events associated with thermocline deepening in the wet season, giving rise to the occasional pairs of close-interval laminae that can be seen in the core. In support of this hypothesis, algal densities appear to be highest in the dry season, and we observed an aragonite layer at the mud-water interface in January. The isotopic composition of this surface layer matches the calculated value for carbonate precipitated from surface-water samples collected at the same time as the sediments, again suggesting that the aragonite was precipitated during the dry season. The isotopic depletion of dark-layer aragonite also supports a wet-season dark, dry-season white model.

The seasonal timing of lamina deposition is relevant to the interpretation of oxygen isotope values determined from the white laminations (Figure 8). If the white layers are formed from aragonite precipitated in the epilimnion when the lake is stratified (February - October), they represent surface water conditions for the wet season, and could provide an annually resolved record of rainfall variation. If our preferred model of aragonite precipitation and deposition during mixed lake conditions (November-January) is correct, then aragonite $\delta^{18}O$ values for individual laminae reflect the composition of the entire lake, integrated over its water-residence time of about 10 years. The sedimentary record of oxygen-isotope variations may therefore be interpreted as a proxy P-E record with some degree of decadal smoothing.

## 10. SUMMARY AND CONCLUSIONS

Aragonite laminations in the uppermost sediments of Lake Hora appear to have been deposited annually, presenting an opportunity to investigate recent environmental change at high time resolution. They are probably deposited during the dry season, when lake mixing causes nutrient and calcium enrichment of the photic zone, thus enhancing the photosynthetic precipitation of aragonite. If this is the case, then the oxygen isotope composition of each lamina reflects lakewater composition of the mixed lake, which has a water residence time of about 10 years. The isotopic composition of each lamina may therefore represent lake conditions integrated over 10 years rather than a single year. Initial runs of a model of the lake's isotope hydrology suggest that the lakewater isotopic composition is sensitive to rainfall variation, but there is no clear relationship between rainfall at Debre Zeit and the isotopic composition of individual laminae deposited since about 1965. In contrast, Muchane (1996) found a good fit between rainfall and $\delta^{18}O$ of calcite in laminated sediments from Lake Turkana, a river-fed lake. Since rainfall records from tropical uplands show marked local-scale variability, comparison of the Lake Hora isotopic record with rainfall data integrated over the area of its groundwater catchment may provide a better calibration.

## ACKNOWLEDGEMENTS

We thank Robert Turberville for painstaking work on the sediments, and Emi Ito and Mike Talbot for helpful comments on an early draft of the manuscript.

# REFERENCES


Andrews J.E., Riding R., and Dennis, P.F. (1993) Stable isotopic compositions of recent fresh-water cyanobacterial carbonates from the British Isles - local and regional environmental controls. *Sedimentology* **40**, 303-314.

Appleby, P.G. (1998) Dating recent sediments by $^{210}$Pb: Problems and solutions. Proc. 2nd NKS/EKO-1 Seminar, Helsinki, 2-4 April 1997.

Appleby, P.G., Nolan, P.J., Gifford, D.W., Godfrey, M.J. , Oldfield, F., Anderson, N.J. and Battarbee, R.W. (1986) $^{210}$Pb dating by low background gamma counting. *Hydrobiologia* **141**, 21-27.

Appleby, P.G. and Oldfield, F. (1978) The calculation of $^{210}$Pb dates assuming a constant rate of supply of unsupported $^{210}$Pb to the sediment. *Catena* **5**, 1-8.

Appleby, P.G. and Oldfield, F. (1983) The assessment of $^{210}$Pb data from sites with varying sediment accumulation rates. *Hydrobiologia* **103**, 29-35.

Appleby, P.G., Richardson, N. and Nolan, P.J. (1992) Self-absorption corrections for well-type germanium detectors. *Nucl. Inst.and Methods B* **71**, 228-233.

Baxter, R.M., Prosser, M.V., Talling, J.F. and Wood, R.B. (1965) Stratification in tropical African lakes at moderate altitudes (1500 to 2000 m). *Limnology and Oceanography* **10**, 510-520.

Benson, L.V. and White, J.W.C (1994) Stable isotopes of oxygen and hydrogen in the Truckee River - Pyramid Lake surface water system. ***Limnology and Oceanography*** **32**, 745-751.

Bradbury, J.P. and Dean, W.E. (1993) Elk Lake Minnesota: evidence for rapid climate change in the north-central U.S. *Geological Society of America Special Paper* **276**, 336pp.

Craig, H., Lupton, J.E. and Horowitz, R.M. (1977) *Isotopic geochemistry and hydrology of geothermal waters in the Ethiopian Rift Valley*. La Jolla, California: Scripps Institute of Oceanography, SIO 77-14.

Curry, B.B., Anderson, T.F. and Lohmann, K.C. (1997) Unusual carbon and oxygen isotope ratios of ostracodal calcite from last interglacial (Sangamon episode) lacustrine sediment in Raymond Basin, Illinois, USA. *Journal of Paleolimnology* **17**, 421-435.

Emrich, K., Ehhalt, D. H., and Vogel, J. C. (1970) Carbon isotope fractionation during the precipitation of calcium carbonate. *Earth and Planetary Science Letters* **8**, 363-371.

Folk, R.L. (1974) The natural history of crystalline calcium carbonate: effect of magnesium content and salinity. *Journal of Sedimentary Petrology* **44**, 40-53.

Friedman, I, and O'Neil, J.R., (1977) Compilation of stable isotope fractionation factors of geochemical interest. In: *Data of Geochemistry*, M. Fleischer, ed., US Geol. Surv. Prof. Paper **440-KK**, 6$^{th}$ ed., 1-12.

Gasparon, M., Innocenti, F., Manetti, P., Peccerillo, A., and Tsegaye, A. (1993) Genesis of the Pliocene to recent bimodal mafic-felsic volcanism in the Debre Zeyt area, central Ethiopia. *Journal of Afican. Earth Sciences* **17**, 145-165.

IAEA/WMO (1998) Global Network for Isotopes in Precipitation. The GNIP Database. Release 3 October 1999. URL: http://www.iaea.org/programs/ri/gnip/gnipmain.htm

Kebede, S. (1999) *The hydrology and hydrochemistry of Bishoftu crater lakes. Hydrological and isotope modeling*. Unpublished M.Sc thesis, Addis Ababa University.

Kebede, S., Lamb, H., Telford, R., Leng, M., and Mohammed, U. (this volume) Lake - groundwater relationships, oxygen isotope balance and climate sensitivity of the Bishoftu crater lakes, Ethiopia.

Kelts, K. and Hsu, K.J. (1978) Freshwater carbonate sedimentation, in A. Lerman (ed.), *Lakes: geology, chemistry, physics,* Springer Verlag, New York, pp. 295-323.

Kim, S., and O'Neil, J.R., (1997) Equilibrium and nonequilibrium oxygen isotope effects in synthetic carbonates. *Geochimica et Cosmochimica Acta*, 61, 3461-3475.

Lotter, A.F. (1989) Evidence of annual layering in Holocene sediments of Soppensee, Switzerland. *Aquatic Sciences* **51**, 19 - 30.

McCrea, J.M. (1950) On the isotopic chemistry of carbonates and a palaeotemperature scale. *Journal of Chemical Physics* **18**, 849-853.

Merlivat, L., and Jouzel, J. (1979) Global climatic interpretation of the Deuterium-Oxygen 18 relationship for precipitation. *Journal of Geophysical Research* **84**, 5029-5033.

Mohr, P.A. (1961) The geology, structure and origin of the Bishoftu explosion craters. *Bulletin of the Geophysical Observatory Addis Ababa,* **2**, 65-101

Muchane, M.W. (1996) Comparison of the isotope record in micrite, Lake Turkana, with the historical weather record over the last century, in Johnson,T.C. and Odada, E.O. (eds.), *The Limnology, Climatology and Paleoclimatology of the East African Lakes* Gordon and Breach, Amsterdam, pp 431-441.

O'Neil J.R., Clayton, R.N., and Mayeda, T.K. (1969) Oxygen isotope fractionation in divalent metal carbonates. *Journal of Chemical Physics* **51**, 5547-5558.

Pearson, E.J. (1997) The validation and application of a merged African diatom-based transfer function for reconstructing hydrochemical characteristics at Lake Hora, an Ethiopian crater lake. Unpubl MSc thesis, University of London.

Pearson, F.J., and Coplen, T.B. (1978) Stable isotope studies of lakes. In: *Lakes: Chemistry, Geology, Physics*, A. Lerman, ed., Springer-Verlag, pp 325-339.

Peglar, S.M., Fritz, S.C., Alapieti, T., Saarnisto, M., and Birks, H.J.B. (1984) Composition and formation of laminated sediments in Diss Mere, Norfolk, England. *Boreas* **13**, 13-28.

Phillips, F.M., Person, M.A., and Muller, A.B. (1986) A numerical lumped-parameter model for simulating the isotopic evolution of closed-basin lakes. *Journal of Hydrology* **85**, 73-86.

Prosser, M.V., R.B. Wood and R.M. Baxter, (1968) The Bishoftu crater lakes: a bathymetric and chemical study. *Archiv Hydrobiologie* **65**, 309-324.

Ricketts, R.D. and Johnson, T.C. (1996) Climatic change in the Turkana basin as deduced from a 4000-yr long $\delta^{18}O$ record. *Earth and Planetary Science Letters* **142**, 7-17.

Ricketts, R.D. and Anderson,R.F. (1998) A direct comparison between the historical record of lake level and the $\delta^{18}O$ signal in carbonate sediments from Lake Turkana, Kenya. *Limnology and Oceanography* **43**, 811-822.

Rippey, B. and Wood, R.B. (1985) Trends in major ion composition of five Bishoftu crater lakes. *Ethiopian Journal of Science* **8**, 9-28.

Rozanski, K., Araguas-Araguas, L., and Gonfiantini, R. (1996) Isotope patterns of precipitation in East Africa, in Johnson,T.C. and Odada, E.O. (eds.), *The Limnology, Climatology and Paleoclimatology of the East African Lakes* Gordon and Breach, Amsterdam, pp 79-93.

Talbot, M.R. and Johannessen, T. (1992) A high-resolution palaeoclimatic record for the last 27,500 years in tropical West Africa from the carbon and nitrogen isotopic composition of lacustrine organic matter. *Earth and Planetary Sciences Letters* **110**, 23-37.

Talbot, M.R., Livingstone, D.A., Palmer, P.G., Maley, J., Melack, J.M., Delebrias, G. and Gulliksen, S. (1984) Preliminary-results from sediment cores from Lake Bosumtwi, Ghana. *Palaeoecology of Africa* **16**, 173-192.

Walker, D. and Owen, J.A.K. (1999) The characteristics and source of laminated mud ar Lake Barrine, Northeast Australia. *Quaternary Science Reviews* **18**, 1597-1624.

Wood, R.B. and Talling, J.F. (1988) Chemical and algal relationships in a salinity series of Ethiopian inland waters. *Hydrobiologia* **158**, 29-67.

Wood, R.B., Baxter, R.M. and Prosser, M.V. (1984) Seasonal and comparative aspects of chemical stratification in some tropical crater lakes, Ethiopia. *Freshwater Biology* **14**, 551-573.

Wood, R.B., Prosser, M.V. and Baxter, R.M. (1976) The seasonal pattern of thermal characteristics of four of the Bishoftu crater lakes, Ethiopia. *Freshwater Biology* **6**, 519-530.

Zerihun, W. and Mesfin, T. (1990) The status of the vegetation in the lakes region of the Rift Valley of Ethiopia and the possibility of its recovery. *Sinet: Ethiopian Journal of Science* **13**, 97-120.

Zinabu G.-M., (1994) Long term changes in indices of chemical and productive status of a group of tropical Ethiopian lakes with differing exposure to human influence. *Archiv Hydrobiologie* **132**, 115-125.

# VEGETATION CHANGES AND THEIR CLIMATIC IMPLICATIONS FOR THE LAKE VICTORIA REGION DURING THE LATE HOLOCENE

IMMACULATE SSEMMANDA[1] AND ANNIE VINCENS[2]

[1]Geology Department, Makerere University, Box 7062, Kampala, Uganda.

[2]CNRS, CEREGE BP 80, 13545 Aix-en-Provence, Cedex 04, France.

## ABSTRACT

The pollen sequence of the core V95-2P (00°58.67' S, 33°27.32' E, 67 m depth) from Lake Victoria mirrors larger extension of forests between *ca.* 6500 and *ca.* 4100 yr BP, implying a wetter climatic conditions than during the later period. Highest humidity was experienced in the region prior to *ca.* 6500 yr BP, before the semi-deciduous forest formations became widespread in the region. From *ca.* 5000 yr BP, the forests around Lake Victoria were mainly of semidecideous character with increasing abundance of *celtis* associated with *Holoptelea grandis*, mixed with some Guineo-Congolian elements such as *Tetrorchidium*. The period *ca.* 4100 to *ca.* 3000 yr BP shows a progressive decline of the semideciduous forest formations and the establishment of open vegetation with Capparidaceae and Gramineae attesting to a dry climate. After *ca.* 3000 yr BP, the pollen data particulary that from high altitude, evidence an amelioration of climate. The dry montane forest with *Podocarpus* and *Juniperus procera* underwent significant extension which reached a maximum at *ca.* 1700 yr BP. At low altitude, the extension of the semideciduous forests in relation to this improvement of climate, is evidenced mainly until *ca.* 2200 yr BP. During this sub-humid climatic phase, either the amount of precipitation was inadequate or the dry season was too long for a large development of evergreen forests in the Lake Victoria region. Pollen data from this core and from the other cores in the region indicates that the increase in rainfall during this period was larger and lasted for a longer duration in the high altitude sites than at low altitude. From *ca.* 1700 yr BP, the significant decrease in the abundance of the pollen of the regional taxa indicates a



*E.O. Odada and D.O. Olago (eds.),*
*The East African Great Lakes: Limnology, Palaeolimnology and Biodiversity,* 509–523.

decline of the *Juniperus-Podocarpus* dry montane forest or a reduction in precipitation at high altitude. During this dry period, both the montane and the semideciduous forests decline at the profit of the open grass dominated formations: open woodlands and probably savannas.

# 1. INTRODUCTION

In the recent past, human induced environmental degradation arising from deforestation of humid areas through cultivation and burning for better pasture, has become a major factor controlling the geographical extent of the forested areas in Uganda. The interaction between climate, anthropogenic deforestation and non-sustainable farming methods has led to poor agricultural yields. The latter, the eutrophication of Lake Victoria and the invasion of some of Ugandan water bodies by the water hyacinth (*Eichornia crassipes*) have caused concern for the Government and Non-Governmental Organizations. These factors have led the attention of the above organizations on environmental issues and human impact on the vegetation. Preservation and sustainable use of lakes and forests are imperative, for a large population in East Africa depends on these for water, food, transport and fuel. However, conservation and protection of the natural habitats require knowledge of the past environments, their changes, when and why they changed as well as the factors that controlled the changes. In this region archaeological and meteorological data is scarce, often incomplete and very recent. Understanding vegetation dynamics and the factors that have influenced the latter in the past necessitated coring and studying paleorecords in the lake sediments. Interdisciplinary research was the tool to apply to get solutions to the problems.

In the early 1990s, the International Decade for East African Lakes (IDEAL) undertook an interdisciplinary study to determine the biological, chemical and physical responses of the deepest African lakes to climatic forcing. Its aim was to acquire adequate data for simulation in the General Circulation Models (GCMs).

This palynological analysis and interpretation is one of the various disciplines contributing to the achievements of the goals of the IDEAL project on Lake Victoria. It may also contribute to knowledge of past environments necessary for planning of land use and management.

## 1.1 Location

Lake Victoria is located between longitudes 31° 39' and 34° 53' East and between latitudes 0° 20' North and 3° 00' South. Its water surface lies at 1134 m above sea level. Figure 1 shows the core site, the main tributaries of Lake Victoria and the outlet.

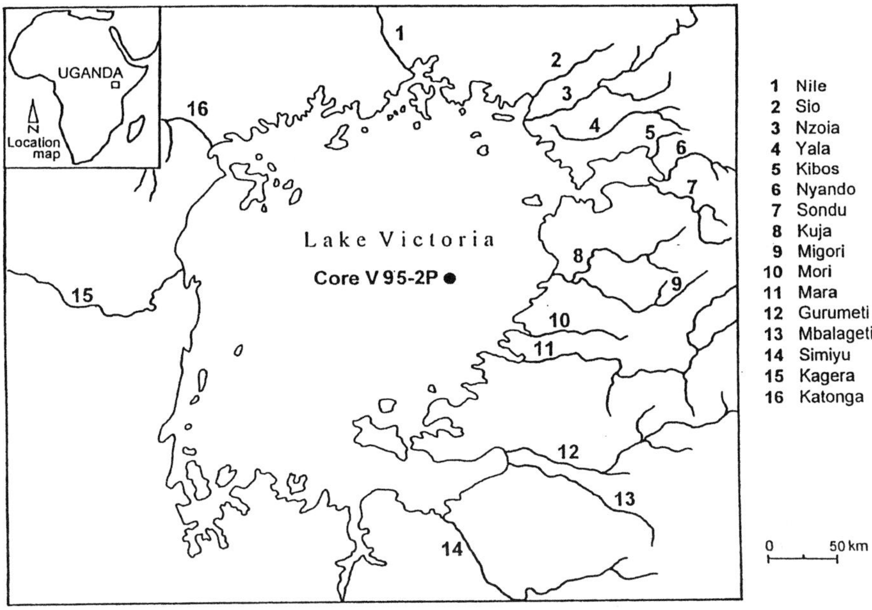


*Figure 1.* Lake Victoria's main tributaries, outlet and core site V95-2P.

## 1.2 Climate

The climate of the Lake Victoria region is part of the East African climate described by Nicholson (1996). MacCallum (1962), Griffiths (1972), Beltrando and Cadet (1990), Rosqvist (1990) and Edmond *et al.* (1993) have also contributed greatly to the knowledge of climate in this part of Africa. The surroundings of Lake Victoria correspond to region 3 (Victoria) as classified by Beltrando and Cadet (1990). The Congo humid air and the two relatively dry northeast and southeast monsoons constitute the wind patterns in the area (Nicholson, 1996). The northeast and southeast trade winds converge in the low-pressure zone over Lake Victoria. Temperatures and evaporation are high and rain condenses from the rising air masses (Kendall, 1969). Rainfall is mainly an afternoon phenomenon and is distributed throughout the year with two maxima in April-May and October-November. The amount of precipitation is highest on the lake (1780 mm/yr Edmond *et al.*, 1993; 2000 mm/yr Nicholson, 1996) and decreases west and north-westwards with a gradient of approximately 16 mm per km (MacCallum, 1962). Monthly rainfall is always higher than 50 mm. For this reason there is no real dry season. The inter-annual variability in rainfall is related to large-scale atmospheric and oceanic factors (Nicholson, 1996). During the ENSO (El-Nino Southern Oscillation) years, rainfall is enhanced by 15% to 25% in the Lake Victoria catchment area (Birkett *et al.*, 1999).

## 1.3 Vegetation

Langdale-Brown *et al.* (1964), Trapnell and Langdale-Brown (1972) and White (1983) describe the vegetation formations surrounding Lake Victoria. The lake is situated within a phytogeographical region known as the Lake Victoria mosaic (White, 1983). In this region five distinct flora meet. From the driest to the most humid these are Somali-Masai, Sudanian, Zambezian, Guineo-Congolian and Afromontane (Figure 2).

The Somali-Masai regional center of endemism is represented by the dry bushland and thicket occurring in the extreme northeast of Uganda, extending eastwards to northwest Kenya. The vegetation is characterized by presence of *Acacia*, *Commiphora*, Capparidaceae and *Grewia*. Succulents such as: *Euphorbia, Adenium obesum, Cissus*, Asclepiadaceae and Crassulaceae are abundant components (White, 1983).

The *Combretum* and *Terminalia* woodland, belonging to the Sudanian regional center of endemism, extends northwestwards from the plains to the north of Mount Elgon through Northern Uganda to Sudan. Occasionally, this vegetation is replaced by grass savanna with *Hyparrhenia* in the drier areas (Langdale-Brown *et al.*, 1964).

The Zambezian woodland dominated by Caesalpiniaceae (*Brachystegia spiciformis* and *Brachystegia boehmii*) is widespread to the south of the lake. The other common species include *Julbernardia* species, *Uapaca* species, *Pterocarpus angolensis, Albizia antunesiana, Pericopsis angolensis, Burkea africana, Erythrophloeum africanum and Parinari curatellifolia* (Trapnell and Langdale-Brown, 1972; White, 1983).

To the west of Lake Victoria, the Forest/Savanna mosaic is composed of forest remnants and grass savannas dominated by *Pennisetum purpurem*. In the mosaic, the forests of Guineo-Congolian affinity include 1. Medium Altitude Moist Semi-deciduous forests with the associations: *Celtis-Chrysophyllum, Albizia-Chlorophora, Albizia-Markhamia and Cynometra-Celtis* 2. Medium altitude moist evergreen forests with the associations: *Piptadeniastrum-Uapaca*, *Piptadeniastrum-Albizia-Celtis,* and forests with *Parinari* (Langdale-Brown *et al.*, 1964; Lind and Morrison, 1974; White, 1983; Osmaston, 1959).

To the northeast of Lake Victoria, Mount Elgon rises to alpine belt. On the southern slopes, the montane forest with *Prunus africana* and *Podocarpus milanjianus* pass into the bamboo zone. On the northern side, the dry montane forest with *Juniperus-Podocarpus* includes species such as *Juniperus procera, Podocarpus gracilior, Ilex mitis, Olea europaea* ssp. *africana, Euclea divinorum* and *Olinia usambarensis.* Above the bamboo zone, in the alpine belt, open woodland is dominated by *Senecio elgonensis* below 3900 m. Above this level, *Senecio barbatipes* dominates associated with communities of *Lobelia telekii* and *Euryops elgonensis* (Hedberg, 1951; Dale, 1940; Langdale-Brown *et al.*, 1964; Hamilton and Perrott, 1981).

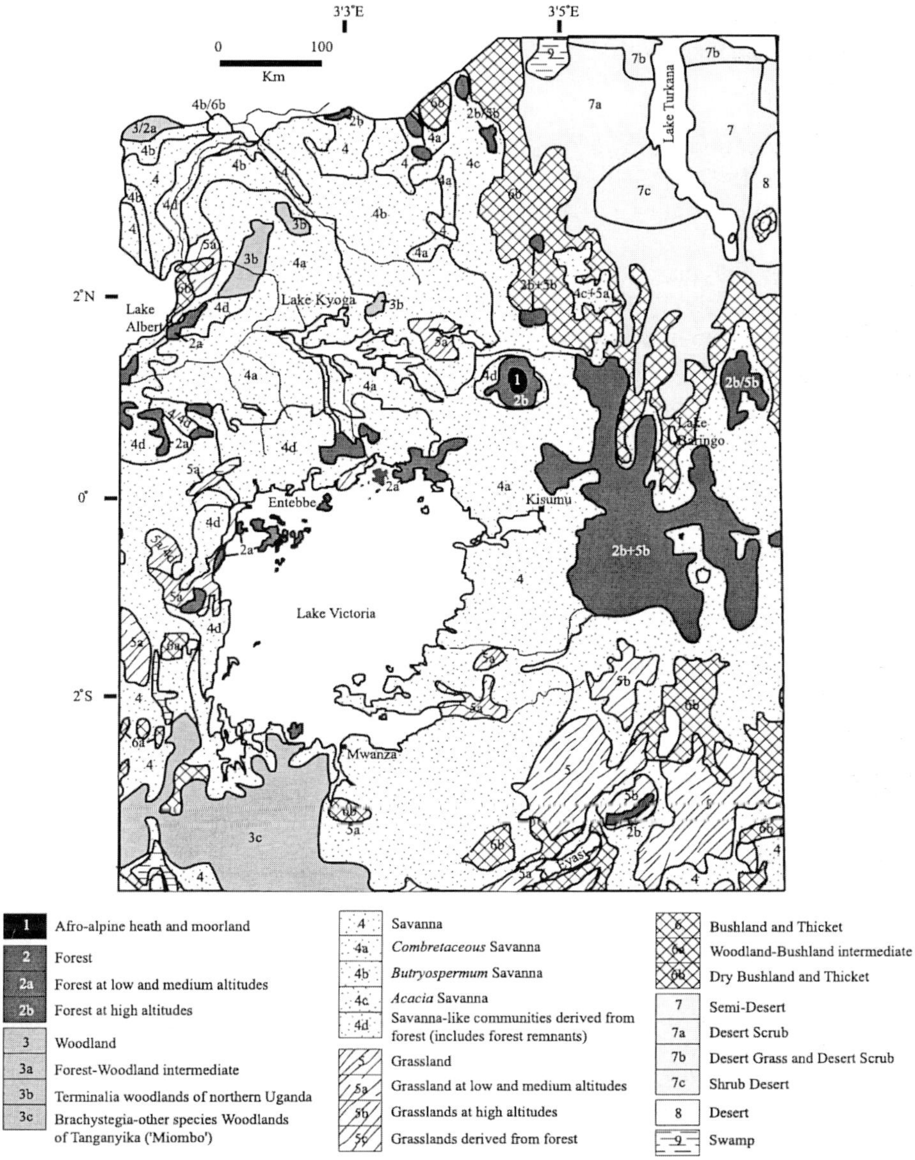


*Figure 2*. Vegetation map of Lake Victoria region.

## 2. MATERIAL STUDIED

The sedimentary sequence V95-2P was recovered from Lake Victoria at latitude 00° 58.67' South, longitude 033° 27.32' east and a water depth of 67 m (Figure 1).

The core is constituted of dark, organic-rich, diatomaceous mud. Various studies have been made on this core including water content, total carbon, total sulfur, magnetic susceptibility, hydrogen index, nitrogen isotopic analysis and pollen (Beuning *et al.* 1997; N'Gobi *et al.*, 1998; Ssemmanda and Vincens 1999; Talbot and Laerdal, (2000,). Beuning (1999) did pollen analysis on the lower section of this core.

This paper presents the results of pollen analysis of the upper 3.78 m of the sediments of the core V95-2P, the study of which was aimed at interpreting human activities and climate in the region.

Several carbon-14 dates have been obtained on the upper section of this core by Accelerator Mass Spectrometer (AMS) (Table 1). The dates were derived from treated sub-samples of 20-70 mm pollen-lignin-charcoal concentrate isolated from bulk sediment. The ages quoted for this core are published by Beuning *et al.* (1997), as uncorrected carbon dates. The same authors calculated a sedimentation rate of 0.66 m /ky for the upper section of the core analyzed here. The top soft sediments of this core were lost as a result of the impact due to the arrival of the piston.

*Table 1.* Radiocarbon dates for the upper section of the core V95-2P.

| Absolute Core Depth (cm) | Uncorrected Radiocarbon Age (yr BP) | CAMS # |
|---|---|---|
| 19 | 1450 ± 61 | 22963 |
| 141 | 3050 ± 51 | 22968 |
| 217 | 4110 ± 61 | 22964 |
| 330 | 6290 ± 61 | 22969 |
| 438 | 9260 ± 60 | 33998 |

Assuming a constant rate of sedimentation between the levels 3.30 and 4.38 m, the part of the sequence from 3.56 to 3.78 m, represents a period *ca.* 6500 to *ca.* 7100 yr BP.

The section of the core for which the results are presented in this paper covers the period *ca.* 7100 to *ca.* 1100 yr BP. Forty-one levels of the core were analyzed and 120 taxa were identified.

The identification of the fossil pollen grains was facilitated by the reference collection of modern pollen grains at Makerere University and pollen atlas: Bonnefille 1971a, 1971b and Bonnefille *et al.* 1980. For each level the fossil pollen grains counted exceeded 400 grains.

# 3. RESULTS

Figure 3 presents the principal pollen taxa encountered in the core V95-2P. The pollen sequence is divided into five zones.

## 3.1 Zone 1 (3.78 to 3.56 m)

In this zone dated between *ca.* 7100 and *ca.* 6500 yr BP, the tree taxa represented mainly by *Celtis, Olea, Alchornea, Podocarpus* and *Macaranga* constitute 29%. The Gramineae, *Artemisia* and the other non-arboreal taxa contribute 28%. The Moraceae/Urticaceae (diporate) (38-32%) constitutes the dominant taxa of the zone. *Acalypha,* Cyperaceae and the smooth fern spores of Pteridophyta (monolete) show low representation (less than 3%). Generally, the zone displays a low diversity of taxa.

## 3.2 Zone 2 (3.56 to 2.17 m)

From *ca.* 6500 to *ca.* 4100 yr BP, the arboreal taxa are abundant and diversified: *Celtis* (22%), *Olea* (7%), *Alchornea* (10%), *Podocarpus* (4%) and *Macaranga* 3%. Occasionally, *Rapanea melanophloeos, Nuxia/Ficalhoa,* and *Phoenix reclinata* are significant. The Combretaceae/Melastomataceae and *Holoptelea grandis* attain significant abundance (2.5%) in the upper part of the zone. The non-arboreal taxa still maintain low percentages. The Gramineae (21-27%), always a significant component, is occasionally associated with *Artemisia,* Amaranthaceae/ Chenopodiaceae and the Compositae Tubuliflorae. The Moraceae/Urticaceae (23-35%) is still the dominant taxa, associated with *Acalypha* (7%). The Cyperaceae contributes (4%) to the sum. The smooth monolete spores of Pteridophyta are significant (2.5%) in the lower part of the zone.

## 3.3 Zone 3 (2.17 to 1.4 m)

During the period *ca.* 4100 to *ca.* 3000 yr BP, the tree taxa particularly *Celtis and Macaranga* decrease abundance to the advantage of *Maerua*-type (2%), the Gramineae (51%) and *Acalypha* (12%). The Moraceae/Urticaceae reduce percentages from 23% to 10%. The *Podocarpus* (4%), *Olea* (5.5%) and *Alchornea* (2.5%) are always present occasionally associated with Combretaceae/ Melastomataceae, *Holoptelea grandis, Tetrorchidium* and *Trilepisium madagascariensis.*

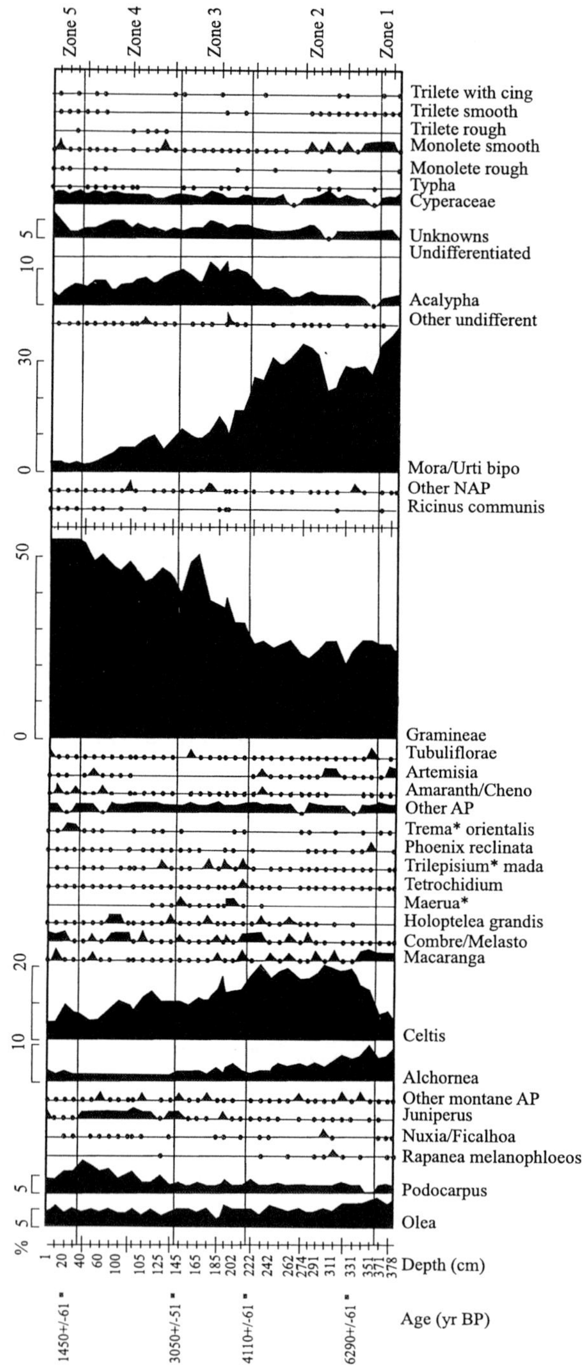


*Figure 3.* Pollen diagram for Lake Victoria (percentages are calculated using the sum of total identifiable pollen and spores).

## 3.4 Zone 4 (1.4 to 0.35 m)

The period *ca.* 3000 to *ca.* 1700 yr BP shows an increase in the abundance of the arboreal taxa particularly the *Podocarpus* (5% to 9%), *Juniperus procera* (2.5%) and *Celtis* (15%). The pollen of *Maerua*-type disappears from the pollen diagram. *Acalypha* and the Moraceae/Urticaceae decrease abundance. The percentages of the Gramineae reduce to 43% in the lower part of the zone. However, the abundance of this taxa increases to 51% in the upper part of the zone where it is associated with Amaranthaceae/Chenopodiacaeae and *Artemisia*. The Cyperaceae attain 4% while the smooth fern spores (monolete) are rare.

## 3.5 Zone 5 (0.35 to 0.0 m)

In this zone dated between *ca.* 1700 to *ca.* 1100 yr BP, the arboreal taxa decrease abundance. The *Podocarpus* reduce percentages from 9% to 4% and *Celtis* from 10% to 5%. *Juniperus procera* becomes generally insignificant. However, *Trema*-type *orientalis* becomes a significant representative of the vegetation. *Olea* and *Alchornea* are always present occasionally associated with *Holoptelea grandis,* Combretaceae/Melastomataceae and *Macaranga.* The Gramineae attains its highest abundance (55%) in the sequence in this zone. The Cyperaceae is always present (5%) while the smooth fern spores of Pteridophyta (monolete) are rare.

# 4. INTERPRETATION

## 4.1 *ca.* 7100 to *ca.* 6500 yr BP

During this period, the low abundance of the non-arboreal taxa, particularly the Gramineae, indicates that the vegetation around Lake Victoria was not of open type. The high abundance of Moraceae/Urticaceae signifies a humid environment. During this time, the taxa considered as the main indicators of semi-deciduous forests, *Holoptelea grandis* and *Celtis*, (Kendall, 1969) reflect low occurrence or are absent from the floristic composition of the forests. It is possible that much of the taxa in the vegetation were not good pollen producers. Considering the pollen data, a humid climate can be inferred for this period.

## 4.2 *ca.* 6500 to *ca.* 4100 yr BP

The pollen data mirrors a well developed a forest with *Celtis* and *Alchornea* associated with some *Macaranga* in the low-lying areas of the lake. At high altitude occurred arboreal formations with *Podocarpus, Olea, Nuxia/Ficalhoa* and *Rapanea melanophloeos.* The development of forests, concomitant with the high abundance of

Moraceae/Urticaceae reflects a humid climate. In the upper part of the zone, the significant abundance of the Combretaceae/Melastomataceae could signify a better representation of the Combretaceae woodland from *ca.* 5300 yr BP onwards than before. At *ca.* 5000 yr BP, the significant representation of *Holoptelea grandis* in the vegetation signifies a semi-deciduous character acquired by the arboreal formations in the neighborhood of the lake. The climate is humid. However, the semi-deciduous nature of the forests indicates either lower humidity or more seasonal climatic conditions than before.

## 4.3 *ca.* 4100 to *ca.* 3000 yr BP

After *ca.* 4100 yr BP, the abrupt decrease in the abundance of Moraceae/Urticaceae indicates a reduction in humidity. The progressive decrease of *Celtis* and the increase of Gramineae and *Acalypha* suggest a decline of forests as a result of progressive installation of dry climatic conditions in the region. The presence in significant abundance of *Maerua*-type, a taxon common in dry areas (Elffers *et al.*, 1964), confirms this interpretation. The pollen data permits to infer for this period, the development of vegetation types that are in equilibrium with climatic conditions that are drier than those of the previous period.

## 4.4 *ca.* 3000 to *ca.* 1700 yr BP

During this period, the increase in the abundance of *Podocarpus* and *Juniperus procera* indicates either an extension of the montane forest at the northeastern and eastern periphery of Lake Victoria or greater river discharge from these highland areas than today. More humid climatic conditions than before can be inferred. In the lower part of the zone, from *ca.* 3000 to *ca.* 2200 yr BP, the increase in *Celtis* and the decrease of the Gramineae suggest that forests were more widespread in the surroundings of the lake. The increase in humidity is reflected also by the reduction in the abundance of *Acalypha* and the disappearance from the pollen sequence of *Maerua*-type. The increase in abundance, shown by the tree taxa, reflects a sub-humid climate with probably a seasonal contrast as shown by the extension of dry montane forests.

## 4.5 *ca.* 1700 to *ca.* 1100 yr BP

The decrease in the abundance of *Juniperus procera* and *Podocarpus* mirrors a decline in the extension of the montane forest during this period. In the neighborhood of the lake, the reduction in the abundance of *Celtis* and Moraceae/Urticaceae, the high increase of grasses and the occurrence in significant abundance of *Trema*-type *orientalis* suggest an open vegetational environment. Such a decline of forest during this time is probably linked to drier climatic conditions but human deforestation can also be proposed to have influenced this decline of forests during this period or soon after.

## 5. DISCUSSION

The pollen data attests to the existence of forests in the Lake Victoria region, the extent and the floristic composition of which have varied in time and space mainly according to climate. The pollen sequence mirrors larger extension of forests between *ca.* 6500 and *ca.* 4100 yr BP implying a wetter period than during the later Holocene. Highest humidity was experienced in the region prior to *ca.* 6500 yr BP, before the semi-deciduous forest became widespread in the neighborhood of the lake. From *ca.* 5000 yr BP, the forests were mainly of semi-deciduous character mixed with some Guineo-Congolian elements such as *Tetrorchidium*. These results are in agreement with those previously obtained for the core recovered from Pilkington Bay (Kendall, 1969). During this same period, the forests around Lake Albert (Nakimera-Ssemmanda, 1991; Ssemmanda and Vincens, 1993), Muchoya site in Southwestern Uganda (2260 m) (Taylor, 1990) and Kuruyange site (2000 m) in Burundi (Bonnefille *et al.*, 1991) were also of semi-deciduous type. For these forests, the acquisition of the semi-deciduous character is probably a response to a climate characterized by more seasonal rainfall.

The period *ca.* 4100 to *ca.* 3000 yr BP shows a progressive decline of the semi-deciduous forest and the establishment of open vegetation with Capparidaceae (particularly *Maerua*-type) in the neighborhood of Lake Victoria. Farther west on Mount Rwenzori (Bwamba Pass site), climatic dryness is evidenced by the development in the montane forest, of *Rumex* and *Dodonaea viscosa* from *ca.* 3900 yr BP, followed by the decline of the montane forest at *ca.* 3300 yr BP (Nakimera, 2001). Dry climatic conditions were also experienced around this time on Mount Elgon (Hamilton, 1982). The decline of the semi-deciduous forests at this time is in agreement with the progressive lacustrine regression towards modern lake levels (Kendall, 1969; Stager *et al.*, 1997).

After *ca.* 3000 yr BP, the pollen data, particularly that from high altitude evidences an amelioration of climate. The dry montane forest underwent significant extension reaching a maximum at *ca.* 1700 yr BP At low altitude in the neighborhood of the lake, the extension of the semi-deciduous forest is evidenced mainly until *ca.* 2200 yr BP. During this climatic phase, either the amount of precipitation was inadequate or the dry season was too long for the development of extensive evergreen forests in the Lake Victoria region. A sub-humid climate is inferred from the pollen data. More evidence for humidity in the region after *ca.* 3000 yr BP, is provided by the development of forests on Mount Rwenzori (Bwamba Pass site) and in its neighborhood (Fort Portal site) (Nakimera, 2001). The pollen data for this period indicates that the increase in rainfall in these Ugandan sites was greater at higher altitudes than at low altitudes. A humid climatic phase post *ca.* 3000 yr BP has also been evidenced for Lake Turkana by use of high-resolution isotopic studies and pollen analysis. For the Turkana basin, the wet phase is dated between *ca.* 2000 to *ca.* 1800 yr BP (Johnson *et al.*, 1991; Mohammed, 1992).

From *ca.* 1700 yr BP, the significant decrease in the pollen of the regional taxa signifies either decline of the *Juniperus-Podocarpus* dry montane forest or a decrease in the water discharge of the rivers from high altitude areas. During this period, both the dry montane and the semi-deciduous forests decline at the profit of open grass dominated formations: open woodlands and probably savannas.

Comparison of the pollen data from this core and from the core 2P (Kendall, 1969), with the pollen data from cores in the neighboring region, shows that the vegetation in the neighborhood of Lake Victoria at times responds slowly to climate changes particularly those of small magnitudes such as the ones that occurred from *ca.* 4000 yr BP onwards. In the pollen sequence V95-2P, the dry phase *ca.* 4000 to *ca.* 3000 yr BP is marked by a progressive decrease in the abundance of the of *Celtis*, a small appearance of Capparidaceae and a little rise in the abundance of *Acalypha*. Similarly, the increase in humidity between *ca.* 3000 and *ca.* 1700 yr BP, is indicated by a small increase in the abundance of *Celtis*, a small reduction in the amount of *Acalypha* and a rise of the Gramineae. It is mainly the increase in the abundance of the *Podocarpus* and *Juniperus* originating from high altitude that provides a good signal for this sub-humid phase that is linked to the expansion of the dry montane forest or to an increase in rainfall at, and river discharge from high altitude. However, for the Lake Albert region, the core 3PC and 2PC show an abrupt decrease in the abundance of the arboreal taxa at *ca.* 3000 yr BP (Ssemmanda and Vincens, 1993; Nakimera-Ssemmanda, 1991). Also the core from Bwamba Pass site provides clear signals for the decline of the montane forest at *ca.* 3300 yr BP and a large development of the forest between *ca.* 3000 to *ca.* 1000 yr BP (Nakimera, 2001). The Lake Victoria basin seems to have a capacity to buffer climatic signals of small amplitude. This capacity could be due to the very wide catchment area supplying the paleobotanical data documented in the sediments, the geomorphology of gently sloping tectonic sag (particularly in the south and west) and the good hydrological network of river and swamps. This buffering capacity against climatic oscillations of small amplitude shown in the upper part of the pollen sequence V95-2P, was also reflected in the elemental and isotopic analyses of sedimentary organic matter of three cores from this lake (Talbot and Laerdal, 2000).

Among the pollen taxa present in this sequence, *Juniperus procera* has two possible sources: the first one is Northern Mount Elgon dry montane forest from where it can be transported to Lake Victoria by river Nzoia. The second source is the Mau escarpment from where it is transported by the rivers from the east. *Juniperus procera* grows on the Mau escarpment (S. M. Rucina pers. comm.). Pollen transport by wind is discussed by Ssemmanda and Vincens, 1999.

## 6. CONCLUSION

The pollen data for the core V95-2P indicates to humid climatic conditions prior to *ca.* 6500 yr BP followed by the spread of semi-deciduous forests between *ca.* 6500 and *ca.* 4100 yr BP, indicating probably lower humidity or greater seasonal contrasts in the rainfall distribution. The data also testifies to local and regional forest decline in response to drier climatic conditions from *ca.* 4100 to *ca.* 3000 yr BP, reflected by the significant representation of Capparidaceae and an increase in open grass dominated vegetation. After *ca.* 3000 yr BP, the extension of the *Juniperus-Podocarpus* dry montane forest or an increase in river discharge from high altitude, the absence of the Capparidaceae, the reduction in the abundance of *Acalypha* and an increase in *Celtis* indicate sub-humid period that lasted until *ca.* 1700 yr BP. The increase in rainfall during this time was greater and lasted for a longer duration at high altitude than at low altitude. The decline of the montane and semi-deciduous forests recorded for the period, *ca.* 1700 to *ca.* 1100 yr BP, testifies to a dry climate probably accompanied by human interference.

The geomorphology of the Lake Victoria basin and the hydrological network of swamps and rivers create a buffering capacity for the lake's water catchment so that the vegetation responds slowly and progressively to climatic oscillations of small amplitudes such as those that post-date *ca.* 4000 yr BP.

## ACKNOWLEDGEMENT

This research received funding from IDEAL (An International Decade for East African Lakes) and MEDIAS-FRANCE (INCO-DC PL 972473). We greatly acknowledge Prof. Thomas Johnson, Prof. Kerry Kelts, Dr. Michel Hoepffner, Dr. Anne-Marie Lezine and the technical assistance provided by Mr. Guy Riollet and Mr. Guillome Buchet.

## REFERENCES

Beltrando, G. and Cadet, L. (1990) Variabilité interannuelle de la petite saison des pluies en Afrique orientale: relations avec la circulation atmosphérique générale. *Veille Climatique Satellitaire* **33**, 19-36.

Beuning, K.M., Kelts, K., Ito, T. and Johnson, T.C. (1997) Paleohydrology of Lake Victoria, East Africa, inferred from $O^{18}/O^{16}$ ratios in sediment cellulose. *Geology* **25**, 1083-1086.

Beuning, K.M. (1999) A re-evaluation of the Late Glacial and Early-Holocene vegetation history of the Lake Victoria region, East Africa. *Palaeoecology of Africa* **26**, 115-136.

Birkett, C., Murtugudde, R. and Allan, T. (1999) Indian Ocean climate event brings floods to East Africa's lakes and the Sudd Marsh. *Geophys. Res. Letters* **26**(8), 1031-1034.

Bonnefille, R. (1971a) Atlas des pollens d'Ethiopie: pollens actuels de la basse vallée de l'Omo (Ethiopie). Récoltes botaniques 1968, Adansonia, 2, 3, 463-518.

Bonnefille, R. (1971b) Atlas des pollens d'Ethiopie: principales espèces des forêts de montagne. *Pollen et Spores* **13**(1), 15-72.

Bonnefille, R. and Riollet, G. (eds.) (1980) *Pollens des Savanes d'Afrique Orientale*. CNRS, Paris, 140 pp.

Bonnefille, R., Riollet, G., and Buchet, G. (1991) Nouvelle séquence pollinique d'une tourbière de la crête Zaire-Nil (Burundi). *Rev. Palaeobot. Palynol.* **67**, 315-330.

Dale, I.R. (1940) The forest types of Mount Elgon. *J. East Afr. and Uganda Nat. Hist. Soc.* **15**, 74-82.

Edmond, J.M., Stallard, R. F., Craig, V., Weiss, R. F. and Coulter, G. W. (1993) Nutrient chemistry of the water column of Lake Tanganyika. *Oceanogr.* **38**(4), 725-738.

Elffers, J., Graham, R.A. and Dewolf, G.P. (1964) Capparidaceae. Flora of Tropical East Africa, Hubbard C.E. and Milne-Redhead E. (eds.), 88 p.

Griffiths, J. F. (1972) *Climate,* in W.T.W. Morgan (ed.), *East Africa: Its Peoples and Resources,* Oxford Univ. Press, pp. 107-118.

Hamilton, A.C. (1982) *Environmental History of East Africa. A study of the Quaternary,* Academic Press, London, 328 pp.

Hamilton, A.C. and Perrott, R.A. (1981) A study of altitudinal zonation in the montane forest belt of Mount Elgon, Kenya/Uganda. *Vegetation* **45**, 107-125.

Hedberg, O. (1951) Vegetation belts of East African mountains. *Svensk Bot. Tidskr.* **45**, 140-202.

Johnson, T.C., Halfman, J.D. and Showers, W.J. (1991) Paleoclimate of the past years at Lake Turkana, Kenya, based on the isotopic composition of authigenic calcite. *Palaeogeogr. Palaeoclimatol. Palaeoecol.* **85**, 189-198.

Kendall, R.L. (1969) An ecological history of the Lake Victoria basin, *Ecol. Monogr.* **39**, 121-176.

Langdale-Brown, I., Osmaston, H. A. and Wilson, J.G. (1964) The vegetation of Uganda and its bearing on land use, Gov. of Uganda, 159 pp.

Lind, E.M. and Morrison, M.E.S. (1974) *East African vegetation*, Longman ed., London, 257 pp.

MacCallum, D. (1962) *Atlas of Uganda*, Dept. of Lands and Surveys, Uganda.

Mohammed, M.U. (1992) Paléoenvironnement et paléoclimatologie des derniers millénaires en Ethiopie, Contribution palynologique. Unpublished Thesis, Univ. Aix-Marseille III, 209 pp.

Nakimera-Ssemmanda, I. (1991) Histoire des végétations et du climat dans le Rift occidental ougandais depuis 13 000 ans BP. Etude palynologique de séquences sédimentaires des lacs Albert et Edouar., Diplôme de l'Ecole Pratique des Hautes Etudes, Bordeaux, 125 pp.

Nakimera, I. (2001) The impact of human activities and climate on the vegetation in the Lake Victoria region and on the Rwenzori mountain and its neighbourhood. Thesis, Makerere University, 339 pp.

N'Gobi, G.N., Kelts, K., Johnson, T.C. and Solheid, P.A. (1998) Environmental magnetism of the Late Pleistocene-Holocene sequences from Lake Victoria, East Africa, in J.T. Lehman (ed.) *Environmental Change and Response in East African Lakes,* Kluwer, Dordrecht, pp. 59-74.

Nicholson, S.E. (1996) A review of climate dynamics and climate variability in East Africa, in T.C. Johnson and E.O. Odada (eds.), *The Limnology, Climatology and Paleoclimatology of the East African Lakes*, Gordon and Breach Publishers, pp. 25-56.

Osmaston, H.A. (1959a) Working plan for the Kibale and Itwara forests, 1959-1965, Uganda Forest Dept., Entebbe.

Rosqvist, G. (1990) Quaternary glaciations in Africa. *Quat. Sci. Rev.* **9**, 281-297.

Ssemmanda, I. and Vincens, A. (1993) Végétation et climat dans le bassin du lac Albert (Ouganda, Zaire) depuis 13 000 ans BP.: Apport de la palynologie. *C. R. Acad. Sci. Paris* **316**(2), 561 - 567.

Ssemmanda, I. and Vincens, A. (1999) Preliminary pollen record from the deep waters of Lake Victoria (East Africa). *Palaeoecology of Africa* **26**, 137-145.

Stager, J.C., Cumming, B. and Meeker, L. (1997) A high resolution 11 400-yr diatom record from Lake Victoria, East Africa. *Quat. Res.* **47**, 81-89.

Talbot M.R. and Laerdal T. (2000) The late Pleistocene-Holocene palaeolimnology of Lake Victoria, East Africa, based upon elemental and isotopic analyses of sedimentary organic matter. *Journal of Paleolimnology* **23**, 141-164.

Taylor D. M. (1990) Late Quaternary pollen records from two Ugandan mires : evidence for environmental change in the Rukiga Highlands of southwest Uganda. *Palaeogeogr. Palaeoclimatol. Palaeoecol.* **80**, 283 - 300.

Trapnell C.G. and Langdale-Brown I. (1972) Natural vegetation, in W.T.W. Morgan (ed.) *East Africa: Its People and Resources.* Univ. Press, Oxford, 127-140.

White, F. (1983) The vegetation map of Africa, A descriptive memoir to accompany the UNESCO/AETFAT/UNSO vegetation map, UNESCO, Paris, 356 pp.

# ORGANIC CONTENT AND X-RAY DENSITY OF LACUSTRINE SEDIMENTS FROM HAUSBERG TARN, MOUNT KENYA

WIBJÖRN KARLÉN[1], ERIC ODADA[2] and OLA SVANERED[1]
[1]*Department of Physical Geography, Stockholm University, 106 91 Stockholm, Sweden*
[2]*University of Nairobi, Department of Geology, Chiromo Campus, Riverside Drive, P.O. Box 30197, Nairobi, Kenya*

**ABSTRACT**

The variations in the organic content of a sediment core from a pro-glacial lake, Hausberg Tarn, located on the NW slope of Mount Kenya has been determined. Two types of methods for determining organic content were compared using sediments from the same core. One type was direct measurement of organic carbon using a carbon analyser and the other was X-ray analysis. The results show that both techniques reveal changes in organic content, which are believed to reflect changes in the glacier size during the last 5700 years. However, the X-ray technique discloses a much more detailed view of the changes in the environment. In addition, the method is considerably faster than the methods of determining organic carbon content or weight loss on ignition.

## 1. INTRODUCTION

Lacustrine sediments can record changes in the environment of a drainage area. The relation between organic and minerogenic content has frequently been considered to indicate general changes in the environment. In addition, it has also



*E.O. Odada and D.O. Olago (eds.),*
*The East African Great Lakes: Limnology, Palaeolimnology and Biodiversity,* 525–533.

been used for the stratigraphic description of the sediments. Variations in the organic content of sediments, especially sediments from pro-glacial lakes (Karlén 1981; Lehman and Niessen, 1994), provide particularly useful information about climatic change. The organic content of lacustrine sediment has commonly been determined by a technique involving the heating of samples to a temperature of around 550ºC (loss on ignition - LOI). More recently, direct measurements of organic carbon have been made using a carbon analyser (e.g. an ELTRA Cs 500 equipment). Although both techniques yield quantitative information, the sample size required only permits a limited resolution; a sampling density of about 1 cm will produce a resolution somewhere between 20 and 100 years, depending on the rate of sedimentation. However, modern carbon analyser permits higher resolution.

In this paper we discuss the possibility of using X-ray density as a quantitative or semi-quantitative measure of organic content. With this method resolution can be increased and the time-consuming laboratory work of determining organic content by the weight loss at ignition method can be reduced. Because the relative variations are more important than the absolute values in most studies, a possible loss of precision in the determination of organic carbon content will be of limited importance. We use studies of pro-glacial lacustrine sediments from Hausberg Tarn, Mount Kenya, to illustrate the technique. Both the advantages and the disadvantages of the technique, based on our experience, are discussed. The X-ray technique has been discussed previously on a number of occasions (Axelsson, 1972, 1983; Karlén, 1976, 1981; Rosqvist, 1995; von Rad *et al.*, 1999; Brown *et al.*, 2000).

## 2. STUDY AREA

Hausberg Tarn is a pro-glacial lake on Mount Kenya, in East Africa. Mount Kenya, located on the equator with peaks reaching an altitude of 5200 m, has several now rapidly retreating glaciers on the slopes of its old volcanic neck. Hausberg Tarn is located in physical contact with the alpine glacier moraines formed by the advances of Cesar and Josef Glaciers, advances which have been dated to the early 1900s (Baker, 1967; Young and Hastenrath, 1991). The lake is located inside a moraine predating the Little Ice Age maximum position. The bottom organic sediments in Hausberg Tarn have been dated to 5700 cal. yr BP, as has the retreat after a major advance of two neighbouring glaciers (Karlén *et al.*, 1999). A glacier chronology spanning the last 5700 cal. years has been determined on the basis of variations in the organic content of the sediments in Hausberg Tarn (Karlén *et al.*, 1999). According to this study, a major glacier expansion before 5700 cal. yr. BP was followed by a relatively warm period around 5000 cal. yr. BP. This glacier retreat was followed by several advances; major advances are dated to 4500-3900, 3500-3300, 3200, 2300, 1300-1200 and 600-400 cal. yr. BP. In addition, advanced glacier front positions have been documented to the beginning of the 20$^{th}$ century (Young and Hastenrath, 1991).

## 3. METHODS

A modified Livingstone coring equipment was used for the Hausberg Tarn project. A core 290-cm long (including sediments from 43-333 cm below the sediment surface) was transported to the laboratory, where the sediments were sampled for X-ray density and analyses of organic content. The top section of the core was not included because the sediments were too soft for the equipment used and for transport. Plastic boxes, measuring 19.5-cm by 3 cm by 2 cm, were pushed into the centre of the core. These samples were used for X-ray studies. X-ray photographs were taken using a Hewlett Packard 43 80 5 N X-ray analyser and Kodak X-ray industrial films.

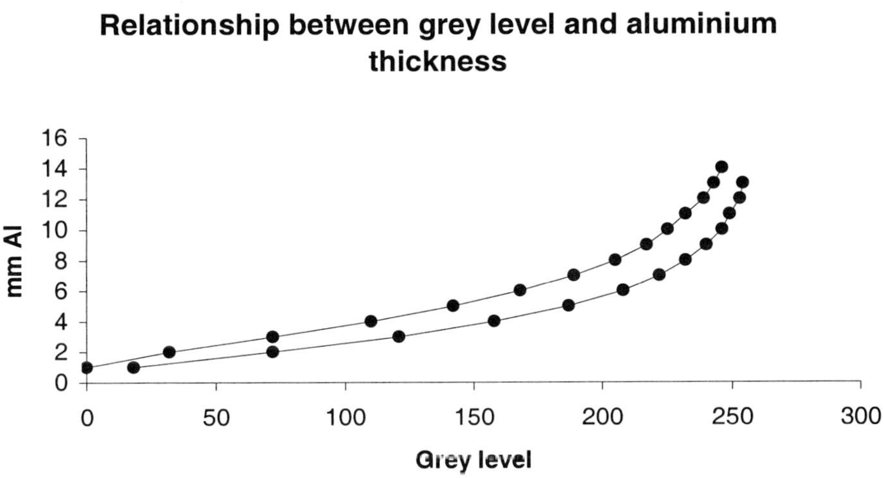


*Figure 1*. Two curves as an example of the nonlinear relationship between aluminium wedge steps and the grey level in an X-ray negative.

The focal distance was about 40 cm. Using between 35 and 50 KVP, an exposure time of around 10 minutes gave the best results. The X-ray negatives were digitised using an Agfa Arcus II-scanner. The scanning resolution was set to 70 dpi and the digital image format was set to 8-bits, where colour representation is expressed as grey levels with a range of 2^8 = 256 grey levels. Black is level 0 and white is level 255. A profile of the changes in grey level was then made using Image Tool, a free-ware computer program made for digital image processing and analysis (Department of Dental Diagnostic Science at the University of Texas Health Science Center, San Antonio, Texas). An aluminium wedge with a thickness ranging stepwise from 0.5 mm to 10 mm was exposed together with the sediment samples. This enable the calibration of the grey levels and to then transfer them into a standard scale expressed as millimetre aluminium (mm Al). All digitised sediment densities have been compared with this standard.

The relationship between the aluminium steps and the grey values are not linear (Figure 1). In order to explore the effect of this non-linearity, different calibration functions, both linear regression and cubic spline interpolation were tested (Figure 2).

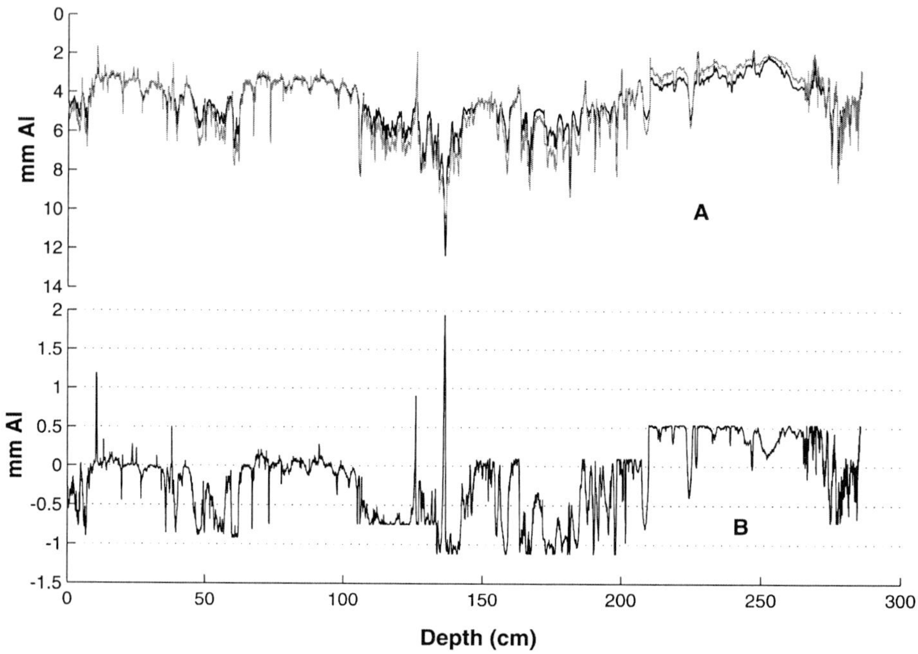


*Figure 2.* Comparison between linear regression and cubic spline interpolation of the X-ray density calibration curve. Diagram A shows the result of the different transfer functions (note, scales inverted). The black curve is cubic spline and the grey linear regression. Diagram B shows the difference between the curves in A.

In this paper we compare the results obtained by X-ray photographs with the results obtained by the direct measurement of organic carbon, using ELTRA Cs 500 equipment (precision about +/- 1 % of carbon present, minimum sample size 100-200 mg). Since the analysis of the organic carbon content yields a mean value for each 1 cm sample, the same sample width had to be obtained from the X-ray negatives. As mentioned earlier the scanning resolution for digitising the negatives was set to 70 dpi, which yields a digital picture resolution of 0.36 mm per picture element (pixel). The line profile drawn from the pictures were therefore subsampled by averaging whole centimetres.

# 4. RESULTS

The analysis of the organic carbon content yields a mean value for each 1-cm sample, which results in a resolution of about 20 years for the Hausberg Tarn core. The difference in organic carbon content varied between about 0.8 and 5.7 %.

These two techniques, one involving the direct measurement of organic carbon content and the other the scanning of X-ray films, give generally similar results (Figure 3). One problem however, is that the sampling intervals does not correspond exactly with each other but there are unsystematic displacements between the two series. That is probably one of the reasons that the correlation coefficient is no more than –0.77. The correlation is mainly caused by the similarities in trends.

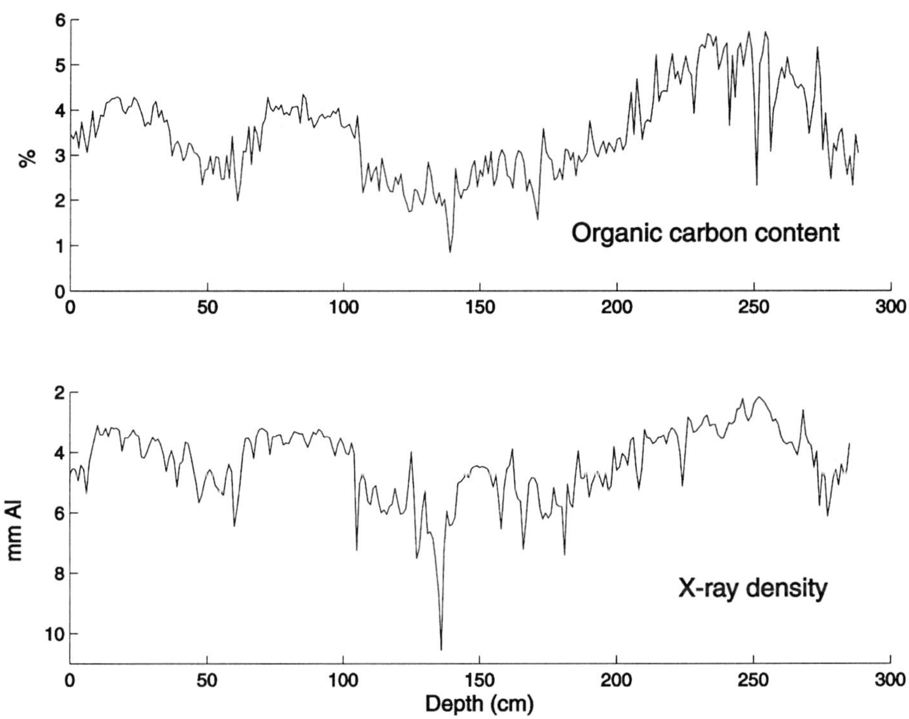


*Figure 3*. The upper panel shows organic carbon content and the lower panel shows X-ray density (subsampled every centimeter). The depth at which the organic carbon was located is given as the distance in cm from the core top (located at about 43 cm under the sediment surface).

In addition to variations on the order of 100s or 1000s of years, there are distinct, short-term deviations in organic carbon content. Because the sediments associated with these events display sharper boundaries, poorer sorting and coarser grain size than is usual for sediments typical of glacier-eroded rock flour, these short-term events can be caused by rapid slope processes at the surface of the glaciers.

The sub-sampled X-ray density record is similar to the organic content record obtained by direct measuring (Figure 3), but neither of them yield the information that the raw X-ray density record includes. Figure 2, based on the complete data set, shows that while some events with an increased flux of rock flour were limited to short periods (less than 5 years), others lasted 100 years or more. The sub-sampled version of the X-ray data and the record of directly measured organic content appear to give similar results concerning the duration of both the short and the long events. Because this type of event does not occur in a lake of similar size located a few hundred meters downstream, these short events are likely to be a result of processes at the glacier surface.

## 5. DISCUSSION

The scanning of lacustrine sediment X-radiographs yield a detailed record of the changes in the organic carbon content of sediments. The photographic image, which shows details about the variations in density as well as information about the structure of strata in the sediment sample and the disturbances resulting from sampling, also reveals the occurrence of large grains etc. (Figure 4).

The scanning of radiographs permits a relatively rapid quantitative measure of organic carbon content, which in our example from Hausberg Tarn appears to be well correlated with the results obtained on 1 cm samples using the ELTRA Cs 500 equipment. There are questions though. In the upper part of the series, from about 200 cm to 0 cm, the amplitude corresponds rather well between the series, but from 200 cm and down to the bottom the X-ray density curve has a lower amplitude. That is an anomaly that has to be investigated further. As a whole though, the results show that the scanning technique can be used to reduce the number of traditional organic carbon determinations.

The main purpose with the analysis of weight loss at ignition and the determination of organic carbon content is to identify the time at which changes occurred in the organic content. The relative change in organic carbon indicates a change in the environment. In general only this relative value is of importance because the organic carbon content will be basin-specific; the figure will depend on drainage basin size, morphology and other conditions in the basin. The X-ray technique can, in our opinion, be calibrated and the technique will then yield a much more detailed view of the changes in organic content than the traditional organic carbon determination will.

Results from the scanning of X-ray photographs (negatives) have not changed in any major way the interpretation of glacier expansions and contractions reviewed briefly under the section study area. However, the scanning has emphasised the frequent, short-term fluctuations in organic content. The unfiltered plot has to be read with caution. The scanning includes some grains, turbidites and other disturbances of unknown origin. In the data presented, these events have not been removed, but in an interpretation of glacier size variations, very short-term changes in density are not a measure of glacier size but only of inorganic influx. However,

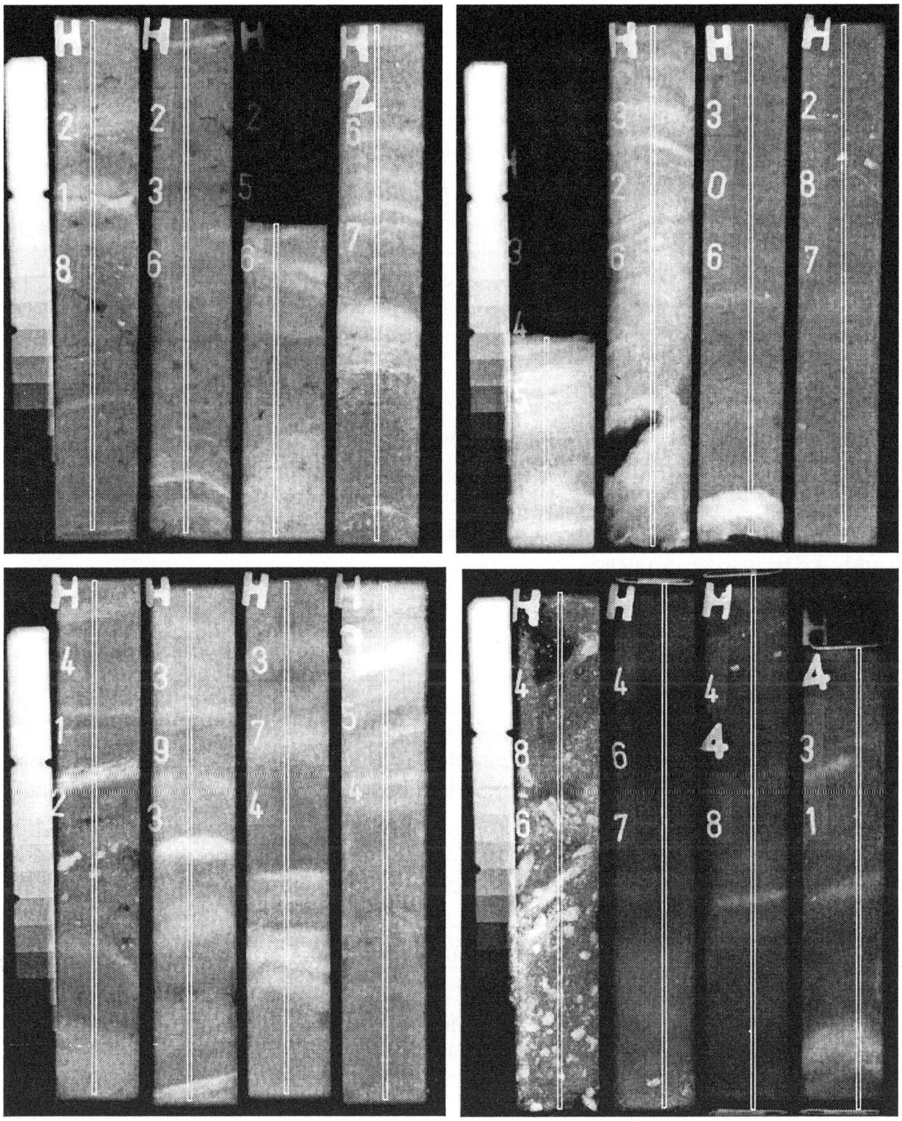

*Figure 4*. X-ray negatives of a sediment core from Hausberg Tarn, Mount Kenya. The order of the X-ray samples are from the top of the core: 218, 236, 256, 267, 287, 306, 326, 345, 354, 374, 393, 412, 431, 448, 467, 486. The bottom photograph, 486 shows the sediments adjacent to the till bottom. Light colour indicate high density, e.g. low organic content. The white rectangular boxes shows from where the profiles has been taken.

because the variability is particularly great during long-term glacier expansions, these events appear to be related to the existence of a glacier in the basin.

## 6. CONCLUSIONS

Determination of the X-ray density of lacustrine sediments yields information similar to that obtained by the organic content analyser method and the determination of weight loss at ignition. It furnishes data with a higher resolution about the sedimentary conditions at the sampling site than by standard techniques using the same laboratory time. The high resolution facilitates making distinctions between short-term events, such as slumping, from events lasting several decades.

## ACKNOWLEDGEMENTS

The Mount Kenya studies have been possible thanks to permits from the Kenyan Office of the President (No. OP/13/001/11C 192/16 and OP. 13/001/25C 48). The project has received financial support from the Swedish Natural Science Council, the National Geographical Society, Carl Mannerfelts Fond and Lagrelius Fond. We would like to thank our porters.

## REFERENCES

Axelsson, V. (1972) X-radiography of unextruded sediment cores. *Geografiska Annaler* **54A** 1, 34-36.

Axelsson, V. (1983) The use of X-ray radiographic methods in studying sedimentary properties of sediment accumulation. *Hydrobiologia* **103**, 65-69.

Baker, B.H. (1967) *Geology of the Mount Kenya Area.* Geological Survey of Kenya, report no. 79. Nairobi.

Brown, S. Bierman, P.R.. Lini, A. and Southon, J. (2000) 10 000 yr record of extreme hydrolocic events. *Geology* **4**, 335-338.

Hastenrath, S. (1984) *The Glaciers of Equatorial East Africa.* Reidel, Dordrecht, 353 p.

Karlén, W. (1976) Lacustrine sediments and thee-limit variations as indicators of Holocene climatic fluctuations in Lappland: Northern Sweden. *Geografiska Annaler* **58A**, 1-34.

Karlén, W. (1981) Lacustrine sediment studies. *Geografiska Annaler* **63A**, 273-281.

Karlén, W. Fastook, J.L. Holmgren, K. Malmström, M. Matthews, J.A. Odada, E. Risberg, J. Rosqvist, G. Sandgren, P. Shemesh' , A. and Westerberg, L.-O. (1999) Holocene glacier fluctuations on Mount Kenya, East Africa, between ~6000 cal. years BP and the present. *AMBIO* **28**, 409-418.

Leemann, A. and Niessen, F. (1994) Holocene glacial activity and climatic variations in the Swiss Alps: reconstructing a continuous record from proglacial lake sediments. *The Holocene* **4**, 259-268.

Rosqvist, G. (1995) Proglacial lacustrine sediments from El Altar, Ecuador: evidence for late-Holocene climatic Change. *The Holocene* **5**, 111-117.

von Rad, U. Schaaf, M. Michels, K.H. Schulz, H. Berger, W.H. and Sirocko, F. (1999) A 5000-yr record of climate change in varved sediments from the oxygen minimum zone off Pakistan, northeastern Arabian Sea. *Quaternary Research* **51**, 39-53.

Young, J.A.T. and Hastenrath, S. (1991) *Glaciers of the Middle East and Africa – Glaciers of Africa*. Satellite Image Atlas of Glaciers of the World. U.S. Geological Survey Professional Paper 1386-G-3, pp. G49-G70.

# Human Dimensions: Impacts and Management

# RESTORING AND PROTECTING THE AFRICAN GREAT LAKE BASIN ECOSYSTEMS-LESSONS FROM THE NORTH AMERICAN GREAT LAKES AND THE GEF

A. M. DUDA

*Global Environment Facility Secretariat, 1818 H. Street, N. W. Washington, DC, USA*

**ABSTRACT**

Continued degradation of Lakes Malawi/Nyasa, Tanganyika, and Victoria has major implications not only for their globally significant biological diversity but also for tens of millions of people depending on them for survival. Governments responsible for the lakes understood this in the early 1990s and approached the GEF for funding to begin addressing the threats. Some early lessons from implementation of GEF projects for the lakes are presented. Different approaches for each lake provided different experiences, including how to cope with war. Different institutional approaches that were chosen have broad implications for how to address complex, transboundary water problems in Africa. The role of science in addressing barriers provided by complexity are discussed, and the importance of harnessing the local scientific community is underscored. However, lack of attention to management institutions, regulatory reform, and joint management in some projects can render even the best studies politically irrelevant.

The GEF international waters focal area is introduced, with sixty some projects covering 131 participating nations. Lessons learned from some of the early projects such as the Black Sea, Danube, and Mediterranean are described. Experiences in reversing degradation in the North American Great Lakes are also presented. An institutional neutral ground is needed for jointly examining facts, for building trust among nations, for leveling the playing field, and for building the political commitment for



*E.O. Odada and D.O. Olago (eds.),*
*The East African Great Lakes: Limnology, Palaeolimnology and Biodiversity,* 537–556.

country reforms on land and water management that are needed to restore and sustain the lake basin ecosystems. The Lake Tanganyika project attempted to secure these commitments while the other two focused on single country studies and pilots. Country commitments to reduce pollution and sediment inputs and to develop joint institutions for sustainable management of the fisheries will be important considerations for additional GEF funding to reverse the growing degradation of these unique lake basin ecosystems.

## 1. INTRODUCTION

Lakes Malawi/Nyasa/Niassa, Tanganyika, and Victoria and their contributing basins serve as invaluable resources for the nine countries sharing the lake basins and the tens of millions of poor people depending on them for survival. They are globally significant freshwater ecosystems containing important areas of both terrestrial and aquatic biological diversity, and all three are becoming degraded as a result of human activities.

In the early 1990s, the riparian nations of the three lakes approached the Global Environment Facility (GEF) for assistance in reversing the degradation. GEF grant resources were provided through the United Nations Development Program (UNDP) for the Lake Tanganyika project while funding was provided through the World Bank for a Lake Malawi project and a Lake Victoria project. While the Victoria project has reached its midpoint, the other two are almost complete. Soon, governments will be seeking assistance for further interventions to restore Lake Victoria and ensure that the massive degradation being experienced there does not occur in the other two fragile lakes.

Publication of this volume provides an opportunity to draw lessons from some of the early experience of the GEF in the international waters focal area as it relates to country collaboration on transboundary water problems, and to consider lessons developed over many years in reversing the ecological decline of the North American Great Lakes. While scientific inquiry, field investigations, and simulation modeling may play important roles in diagnosis of root causes of threats to the biodiversity, a focus on studies without on-the-ground institutional reforms that lead to actions in the economic sectors that threaten the lakes will not be sufficient to reverse the existing decline. The key factor for generating the political will to undertake these necessary reforms may be similar to other GEF experiences and those of North America and Europe. That factor is the development of joint collaborative institutions that can facilitate joint fact-finding about their shared water resources, building of trust among nations, building of a shared political commitment for action over the long-term, and providing transparency in reporting national actions so that stakeholders (including NGOs and donors) may be reassured that progress is being made on these complex transboundary issues.

## 2. BACKGROUND

The complex environmental problems facing the three lakes are described in this volume and have been described before. Little by little knowledge is being compiled by researchers that contributes to a better understanding of the serious threats to each lake. In the early 1990s, Hecky and Buyenyi (1992) as well as Bootsma and Hecky (1993) described the threats to all three lakes. Site specific concerns were identified for excessive sediment affecting biodiversity in Lake Tanganyika (Cohen *et al.*, 1993); eutrophication in Lake Victoria (Hecky, 1993; Duda, 1994a; Hirji and Duda, 1995), loss of fish biodiversity in Victoria (Kaufman, 1992); and deforestation/ sedimentation as well as excessive exploitation of fisheries in Lake Malawi (Tweddle, 1992). These lakes were found to be more vulnerable to pollution and sedimentation than originally thought because of their long retention times, accelerated catchment degradation, population pressure, and the lakes' complex internal hydrodynamic processes.

As a result of the frightening pace of degradation seen by the public during the 1980s, the international attention from expatriate researchers, and domestic concerns about declining fisheries, the GEF was approached about funding work to correct the problems. The project documents identified the most serious threats to each of the lakes. The Lake Tanganyika GEF project document (UNDP,1994) listed the most immediate problems of the lake as excessive sediment from deforestation and intensive agriculture, pollution from Bujumbura, Burundi, and a long-term decline in fish catch with biodiversity of fish stocks adversely affected by the depletion, destructive fishing techniques, and the sediment. The Lake Malawi project document (World Bank,1994) indicated that the lake was not critically endangered at that point although overfishing, unsustainable fishing practices, and eroded soil from the catchment were the main problems. The Lake Victoria project faced far more complex social, economic, political, and technical barriers and was delayed in preparation. The World Bank( 1996) described the enormous growth in population around Lake Victoria, the alarming changes in the lake environment experienced the last three decades, massive blooms of algae, waterborne diseases, infestation with water hyacinth, overfishing, oxygen depletion, and introduction of an alien species, Nile Perch in the late 1950s. The environmental degradation was determined by the World Bank to be placing a present value of $270-520 million at risk to the lake communities if the large export fishery for Nile perch was lost.

These projects were prepared during the pilot phase of GEF before the creation of a GEF Council and the GEF Operational Strategy in 1995. Consequently, no guidance was available in approaches for managing the shared resources. The Lake Malawi project was submitted in 1994 as essentially a single country biodiversity project. Table 1 lists major provisions of each project as determined by the agency in cooperation with the countries. Of the GEF East Africa Lakes projects, strategies and activities vary considerably and have led to different experiences and levels of expatriate assistance in undertaking research, monitoring, planning, and capacity building functions.

*Table 1.* Major provisions of the three GEF-supported lake projects**

Lake Malawi/Nyasa- World Bank; $ 5 million GEF

Malawi, Tanzania, Mozambique

* biodiversity map and management plan for the lake
* research surveys, critical habitat, limnology/water quality sampling, fish studies
* harmonized legal framework for the 3 countries
* capacity building/training in Malawi; rehabilitate fisheries labs in other 2 nations
* strategic plan for Lake Malawi National Park

Lake Tanganyika-UNDP; $ 10 million GEF

Burundi, Tanzania, D.R. Congo, Zambia

* review and compile data on lake status for strategic plan
* establish regional program among countries:pollution control and conservation
* establish environmental education and training programs
* develop and implement mechanisms for regional coordination
* fill data gaps in sedimentation, pollution, biodiversity
* produce the strategic plan and management plans for implementation

Lake Victoria- World Bank $35 million GEF; $42 million other

Kenya, Uganda, Tanzania

* establish Lake Victoria Fisheries Organization; strengthen fisheries monitoring, extension, enforcement; conduct fisheries research
* manage and control water hyacinth
* strengthen/harmonize national regulatory frameworks for pollution control; establish lake-wide monitoring system; pilot investments in pollution control
* assess loading from catchment; improve research/info base; manage land use in improved way; afforestation pilot efforts
* improved wetlands management
* support for institutions

**From World Bank (1994); UNDP (1994); and World Bank (1996).

## 3. THE GLOBAL ENVIRONMENT FACILITY

The Global Environment Facility is an international financial entity dedicated to protecting the global environment and has initiated a diverse group of projects in over 145 countries. Two years before the 1991 Earth Summit in Rio de Janeiro, the GEF was established as a pilot program to test new approaches and innovative ways to respond to global environment challenges in its four focal areas of climate change, biodiversity conservation, ozone depletion, and international waters. In March, 1994, after 18 months of negotiations, agreement was reached in Geneva to transform the GEF from its pilot phase into a permanent financial mechanism. The restructured facility, with its several billion dollar trust fund, is open to universal participation: 166 countries have currently expressed desire to be involved. GEF builds upon a partnership among the UNDP, the UNEP, and the World Bank which are its implementing agencies and can access funding on behalf of countries for activities consistent with an Operational Strategy that the GEF Council adopted in 1995. The Council consists of 32 constituencies representing all the countries of the world, with 14 filled by donor countries and 18 filled by recipient countries.

In restructuring the GEF, governments ensured that it fully embodied the principles that were set out in the Rio conventions and in the action program from Rio known as Agenda 21. GEF has been named as the financial mechanism for the climate change convention and the biodiversity convention and is truly a mechanism for international cooperation for the purpose of providing new and additional grant and concessional funding to meet the agreed incremental costs of measures that achieve global environment benefits in the four focal areas.

The GEF Operational Strategy (GEF, 1996) was adopted by the Council to guide the preparation of country-driven projects in the GEF's four focal areas and in land degradation as it relates to the focal areas. This strategy assists the GEF Secretariat and its implementing agencies in developing work programs, business plans and budgets.

In the biodiversity area, the strategy provides a framework for development and implementation of GEF-financed activities to allow recipient countries to address the complex challenges of conservation, sustainable use, and benefit sharing related to biological diversity. It incorporates in its operational programs guidance from the Conference of the Parties of the Convention on Biological Diversity and provides a framework for monitoring and evaluation of the effectiveness of GEF-financed activities.

In the international waters focal area, GEF's objective is to contribute primarily as a catalyst to the implementation of a more comprehensive, ecosystem-based approach to managing transboundary waters and the basins draining to them. GEF agencies facilitate countries collaborating in a joint manner to focus on top priority transboundary issues, to learn how to work together, and to jointly undertake action to address the root causes of the transboundary priority water issues. The goal is to help countries to utilize the full range of technical, economic, financial, regulatory, and institutional measures needed to make changes in the way human activities are conducted in the economic sectors that create the water problems. GEF plays a catalytic role in assisting groups of countries seeking to leverage co-finance along with the expected baseline policy, institutional, and legal reforms and priority investments needed to address the transboundary problems.

As of December, 1999, GEF has allocated approximately $2.630 billion in grant finance for projects in the four focal areas and for certain multi-focal projects. Figure 1 illustrates that the bulk of funding is devoted to climate change and biodiversity projects, with international waters projects constituting 14% of the expenditures. $354 million has been allocated to 45 international waters projects and has been accompanied by another $374 million in co-finance. Twenty additional projects are under preparation with GEF finance that are projected to need an additional $185 million in grants the next two years.

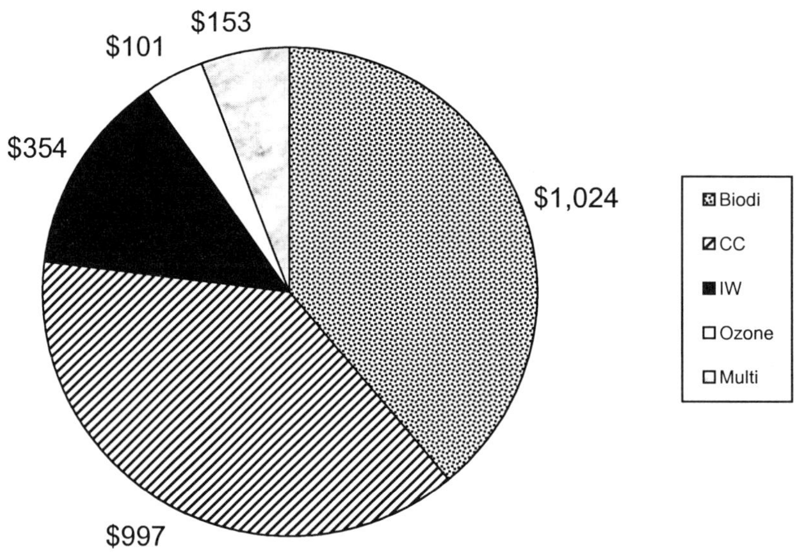


*Figure 1*. Cumulative GEF allocations by focal area in $US millions for 1992-1999(Biodi- biodiversity; CC- climate change; IW-international waters; OZ- ozone depletion; Multi- multifocal).

## 4. EARLY EXPERIENCES FROM GEF AFRICA LAKE PROJECTS

Country officials, implementing agency staff, and expatriate researchers who spurred the drive for funding for the three Africa lake projects deserve a great deal of credit for moving ahead during the early days of GEF, undertaking necessary monitoring and assessment activities to fill in gaps, and setting the stage for the determination of policy/institutional/legal reforms and investments necessary to restore and protect the lake ecosystems. Nonetheless, challenges faced each project, World Bank task managers changed 3 or 4 times in each project, and war disrupted two of the projects while post-conflict issues in Mozambique occupied attention during execution of the Lake Malawi project. Some early experiences are included in the following paragraphs.

## 4.1 Lake Malawi

The Lake Malawi project was a single country biodiversity project related to fisheries and the country of Malawi. Despite the listing of sediment as a problem from the watershed, the World Bank did not include a catchment soil erosion component in the project. A small regional component related to staff from the three countries assisting in monitoring the lake by ship cruises and refurbishing fisheries research stations in the other two countries was included through execution by the Malawi government. This did not really meet the intent of joint management of project activities with all countries on an equal footing as was later suggested in the GEF Operational Strategy. Consequently, the attention of Tanzania and Mozambique was not secured in the project. While several elements of the project did not accomplish their intended scope as described in the document, the protected area management component for producing a management plan for Lake Malawi National Park and Nankumba Peninsula, an area of world class biodiversity, was fully achieved and should be replicated in other countries and lakes.

The research conducted in limnology, water quality, and biodiversity resulted in very important new information on the lake and its basin. Draft reports on water quality by Bootsma and Hecky (1999) and on biodiversity entitled "Conservation of Biodiversity: Time to Act" by Ribbink and Hecky (1999) clearly signal a call to arms for action to address not only overfishing in the nearshore waters of the southern part of the lake but also to reduce the sediment loading coming from deforestation of the catchment and from intensive agriculture in Malawi. In particular, the sediment was found to accelerate eutrophication, to degrade benthic aquatic habitat (with less species biodiversity in the muddy areas), and to impair reproduction when the fish can not visually identify each other and spawning/egg survival is affected. Of tragic significance was that an unprecedented, massive transboundary fishkill was observed during the GEF project that likely resulted from an upwelling event. The significance of the event for the lake ecosystem, its relation to eutrophication, and implications for future kills was not determined. In addition, efforts to release more water from the lake to increase hydropower production during drought years was not addressed as part of the project nor were the implications of lower lake levels on the uniquely fragile biodiversity.

Of great significance is that the project utilized traditional African theater to spread the word about adverse ecosystem impacts to communities around the lake. This is a good example of appropriate communication techniques so that stakeholders can understand the damage being done to the ecosystem. Perhaps most importantly, the significance of land tenure reforms and sound soil conservation practices on the catchment were once and for all acknowledged for their link to the health of the lake ecosystem. No longer should there be lake projects but rather they should utilize an ecosystem approach learned many years earlier in North America and should be formulated as lake basin management projects. While needed institutional reforms were not achieved with this small project ($5 mil GEF) nor was the attention of the other two nations in the lake basin secured, the good news is that up-to-date technical information has been collected, two priority threats identified, and now the information generated by the pilot phase project is available for subsequent use in facilitating the three countries to make the policy/institutional/legal reforms and investments necessary to sustain this unique lake and the millions of people whose food and livelihoods come from it.

## 4.2 Lake Tanganyika

The multi-country Lake Tanganyika project was designed in a radically different way from the predominantly single country Lake Malawi project. UNDP took advantage of early experiences in facilitating the Danube and Black Sea basin projects to apply some of the principles in assisting Burundi, D.R. Congo, Tanzania, and Zambia in addressing their shared lake basin. With more than double the funding of the Lake Malawi project ($10 m), a great deal more could be accomplished toward building institutional commitments for joint multi-country collaboration. A project coordinating unit was established to facilitate each country participating in activities singly as well as jointly. High level officials from each nation participate in a Steering Committee that is responsible for the project. Various programs were established with the objective of helping the riparian countries produce an effective and sustainable system for managing and conserving the biodiversity of the lake. By involving local communities in its design, the programs embrace the dual needs of development and conservation so that livelihoods of the people (sustainable use of the resource of the biodiversity) can be maintained into the future. The programs varied from biodiversity to fisheries, impacts of sedimentation and catchment degradation, pollution, socio-economic issues, education, and development of a joint geographic information system (GIS).

The original planning called for production of a strategic plan for the lake. Following adoption of the GEF Operational Strategy by the GEF Council, UNDP worked with the project to modify its program of work to become more consistent with the international waters portion of the Operational Strategy. The project adopted the approach of joint fact-finding in compiling information so all countries could review it and update it through GIS technology. The results of the assessment is termed a Transboundary Diagnostic Analysis (TDA) that sets priorities for two or three top priority shared water issues. Pollution discharges in Bujumbura, Burundi and Kigoma, Tanzania were cited as hotspots for abatement activities. Excessive sediment loading from certain river basins, mostly in Burundi and D. R. Congo, and scattered elsewhere, was determined to be a priority for accelerated attention, and the overfishing issue was identified as important because of the large commercial fishery, its economic importance to certain nations, and the transboundary nature of the stock and pattern of landings and markets.

The program also adopted the formulation of a Strategic Action Program (SAP) which is not a plan but a series of activities to be implemented not only jointly but also by individual countries to address the top priority issues. Various assessments conducted under the programs built the capacity of country officials to sample and assess environmental status in the areas of biodiversity, pollution, and sedimentation. Important analyses such as the one on biodiversity by Patterson and Makin (1998) and results of investigations added to the knowledge base on accelerated degradation and helped in setting priorities. Many of the publications are available on the project web site (www.ltbp.org). The web site is essential since the project has country linkages, linkages with UNDP, GEF, and coaches from international organizations. A firewall for internal use is essential to exchange information in this remote area in preliminary form and to allow the countries to dialogue about shared issues among themselves. For those without INTERNET, CD-ROMs are produced every 3 months, and with the public portion of the website, the project promotes transparency among NGOs, government officials, different countries, and funding organizations.

Of note is the broad network in the scientific community within the countries and abroad that has been involved with the project. It would be good if this group can be harnessed into a science advisory body for subsequent joint collaborative work to ensure that sound science is utilized for improving management decisions. Similar science advisory bodies are being included in other GEF international waters projects in Africa, most recently the Lake Chad Basin GEF project and the Benguela Current Large Marine Ecosystem GEF project.

Of additional significance is that as of the end of 1999, the Lake Tanganyika governments were discussing the fourth draft of an international treaty (entitled, "The Convention on the Sustainable Management of Lake Tanganyika") to affirm their political support for the restoration and protection of the Lake Tanganyika ecosystem. The draft convention would establish a Lake Tanganyika Authority consisting of a joint Management Committee and a Secretariat to assist the nations in operationalizing sustainable management of the lake, in conserving its biological resources, and in reversing degradation of the catchment area draining to the lake. Various protocols and annexes would specify progressively more stringent country commitments as implementation proceeds. Despite unrest in D. R. Congo and Burundi, which necessitated moving the coordination office to Tanzania in the short-term, important progress has been made in understanding the technical issues of a transboundary nature, identifying hotspots for concerted action, building a joint understanding and shared ownership of their lake basin, harnessing the scientific organizations and local communities, and in setting the stage for building political commitments at the top level for joint management of the resource. The GEF project is expected to close during the latter part of 2000, and additional interventions would await the show of political commitment to reverse the degradation of the lake ecosystem.

## 4.3 Lake Victoria

Global awareness of the rapid decline of the Lake Victoria ecosystem has increased dramatically since Kaufman (1992) characterized the situation as the greatest extinction of vertebrate species that man has ever documented. The alarm over the accelerated degradation was the key driving force in the Lake Victoria GEF project being approved in the mid 1990s as the largest international waters project at $77 million in GEF's eight year history. As noted in the project document (World Bank, 1996), GEF provided $35 million in grants, the three countries received concessional loans worth $35 million from the World Bank for cofinance, and the countries contributed $7 million in-kind for the 5 year project. As described by Kaufman (1992) and by Hirji and Duda (1995) and Hecky *et al.* (1996), so many things were wrong with Lake Victoria that a massive effort on many fronts was needed in this first of several interventions over time. With its shallow nature and with the increase of population and economic activity in the lake basin (see Figure 2 from Bootsma and Hecky (1993) for a comparison of population around Lake Victoria compared to the other two lakes), it is easy to see why the decline was steeper here than the other lakes. Another driving force in spreading opinion was the interest of the global scientific community in the unique biodiversity of the lake and its shallow nature.

The complexity of the situation necessitated a different institutional structure than the other two projects. In essence, the Bank is involved with six projects in one—one grant agreement with each of the three countries and one loan agreement with each. This structure was not easily amenable to a regional, multi-country approach like Lake Tanganyika. While there is a Regional Policy and Steering Committee to resolve disputes, a "light touch" regional secretariat in Tanzania, and a Lake Victoria Fisheries Organization formed by the three countries by the Convention on Fisheries (1994), the bulk of the activities are conducted internally within each country with bilateral dialogue with the World Bank. This design has proven to be difficult to implement with a slow start in the first several years of the project. With the project's focus on building the nonexistent capacity within governments, conducting demonstration projects that address the different concerns, raising awareness, and filling in knowledge gaps, the single country focus in this umbrella project with a light touch of dialogue and sharing experiences among the three nations may be a pragmatic first step in reversing the decline of the basin ecosystem. A Panel of Scientists was envisioned to help provide scientific advice to the project. Little attention seems to have been placed on the work of such a panel, and much of the sampling/assessment is being accomplished by government ministries as part of building their capacity.

As with the Lake Malawi project, by the end of the Lake Victoria interventions in 2002, knowledge gaps should be filled, capacity to address the problems improved, awareness raised, and measures to address the problems would be tested and demonstrated for wider application. Unexpected successes such as the reduction of water hyacinth in Tanzania and Uganda as a result of biological control (weevils reared by communities) and perhaps fortuitous hydrologic conditions have added early, tangible results that stakeholders can see for themselves. These early successes coupled with many highly visible demonstrations and community-based micro-projects have built constituencies for the project that will be important for supporting political decisions on future reforms.

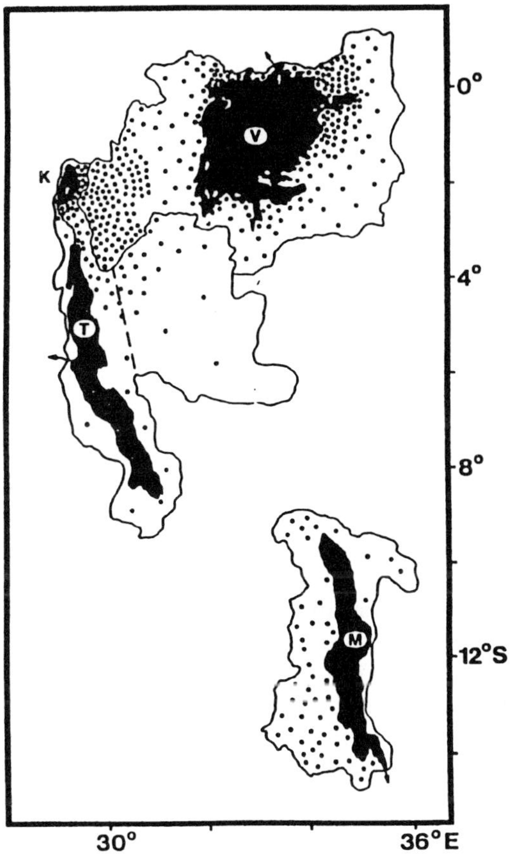


*Figure 2.* Population density in the basins of Africa's Great Lakes. One dot equals 100,000 people and are estimated for 1992. V- Lake Victoria, T- Lake Tanganyika, M- Lake Malawi (from Bootsma and Hecky, 1993).

Yet to be faced is the question of "what is next?" Will the lack of programming strategic and policy development processes of a joint, multi-country nature impair the ability to gain country commitments to undertake the policy/institutional/legal reforms and priority investments necessary on the catchment and in the lake to reverse the alarming degradation? The project focuses on building country capacity to monitor and assess the lake physical processes, water quality, and biodiversity so that a simulation model that has been developed in the project (Delft Hydraulics, 1999) might be calibrated and then provide answers on just how much investment and policy reforms are really needed to yield an acceptable lake environment.

An alternative strategy might be to enact the necessary land tenure reforms, agricultural extension and cost-sharing programs, reforestation programs, water pollution regulatory programs, fisheries extension and regulatory programs, fish levy trust programs to generate management funding from fisheries exports, water pricing reforms to support sewage collection and treatment, and support national funding for key pollution reduction investments to improve water quality. These reforms are certainly in the domestic interests of the public and would provide sustainable national benefits in addition to addressing the transboundary problems. In addition, perverse incentives such as subsidies for fertilizers on estate (industrial) plantations in certain countries and government-sponsored drainage/conversion of wetlands such as Yala Swamp in Kenya could be removed as a start towards restoration. Adaptive management approaches facilitating early implementation of these reforms and investments may yield more rapid results than a "study-model-'implement a little' approach". Without validation on a dataset independent from that used for calibration and verification based on actual changed conditions, modeling scenarios should be very carefully interpreted. A "how much is enough" strategy may delay restoration beyond the point where irreversibility may become a factor.

In the early 1990s, the World Bank provided modest grant funding to the Government of Tanzania to undertake water quality sampling and flow measurements to estimate pollution releases to Lake Victoria from Tanzania as part of a national water resources assessment. The largest contributors of nutrients and oxygen depleting substances were not the small industries or the sediment in rural rivers but rather human sewage, especially from the unsewered area of the largest city, Mwanza, during rainfall events and from areas with greater livestock (Mpendazoe *et al.*, 1994). Duda (1994a) and Hirji and Duda (1995) discussed the implications of these findings for restoration of Lake Victoria. If oxygen depletion was a major problem as was eutrophication, first steps toward restoration would be to reduce the larger contributors of oxygen depleting substances and nutrients. This strategy has worked in other lakes and at that point seemed to be a logical first step. Now that the data gaps are being filled by the GEF project, large sources of oxygen depleting substances and phosphorus would still seem to be in line for reduction. With all three countries having brewery wastewater discharges to Lake Victoria, with the sewage releases from about 30 million people eventually contributing to poor lake quality - including during storm events from unsewered periurban areas, with sewer line bypasses at poorly functioning pump stations, with a pulp and paper mill discharge, with discharges from perhaps a half dozen sugar mills, and with subsidized fertilizer used on massive industrial estates in Kenya, a political commitment to reduce this pollution loading would still seem to be the logical next step.

Perhaps the most important experience from the Lake Victoria project is the precedent set by the World Bank and the countries in matching GEF grant funding with concessional loans (funding under control of the countries, in essence) in order to demonstrate the country commitment and the country's stake in the outcome of the project. This precedent for implementation of on-the-ground activities represents packaging of funding for interventions with domestic benefits as well as global (transboundary) benefits. The sewage treatment demonstrations (including the innovative, small bore, community-based sewage collection system for periurban areas of Mwanza being funded by the project based on feasibility determined earlier by Mpendazoe *et al.* (1994) and further described by Duda (1994a)) have domestic benefits in reducing disease, and accelerating implementation provides domestic benefits for Tanzanian residents as well as for the shared waterbody. Such packages of finance that include investments in the expected baseline that are in the best development interests of a nation along with the incremental cost finance represented by the GEF truly illustrate the intent of the GEF Operational Strategy in the international waters focal area.

## 5. EARLY EXPERIENCES OF OTHER GEF WATER PROJECTS

With over 60 GEF projects under preparation or implementation in the international waters focal area, there will be a large body of experience generated within 4-5 years regarding different approaches for addressing transboundary water-related problems. Table 2 lists many of those projects, and among them are many internationally known river basins and shared marine ecosystems such as the Senegal, Niger, Nile, and Okavango basins; the Mekong, Danube, Dneiper, and Plata basins; as well as the Baltic, Mediterranean, Black Sea, Red Sea, Yellow Sea, Caribbean, Benguela, Gulf of Guinea, and Canary Current Large Marine Ecosystems. Some of these initiatives began in the early and mid 1990s, and early experiences might be of interest as subsequent interventions are planned for the African Great Lakes.

The Mediterranean program existed for 20 years before it received GEF funding to produce a TDA and a SAP as catalytic tools for accelerating decision-making on reforms and investments to restore and protect the Sea. For those 20 years, the program focus was on monitoring, research, and assessments related to the Sea only since the Barcelona Convention describing the multi-country cooperation was for marine waters. The countries were clearly ready for the transition to action as the TDA led quickly to the SAP and the SAP was institutionalized as part of a Protocol to the Convention with pollution reduction commitments for certain contaminants to be reduced certain amounts by certain deadlines. The SAP also extended the program into the river basins draining to the Sea as part of reducing land-based sources of marine pollution.

These experiences suggest that many years can be spent on research and studies that often delay implementation of common-sense solutions. The entire drainage basin rather than the shared waterbody must be subject to action in order to reverse degradation in the waterbody. In addition, a political commitment in terms of a treaty or a succession of more stringent treaties and/or protocols to the treaties are often necessary to ensure that country-based commitments to reforms and action will be implemented. All three lake projects were started with an inordinate amount of emphasis on the lake rather than its catchment. Experiences in each of the projects clearly show that the lake basin ecosystem must be addressed rather than just the lake. To this end, Rwanda and Burundi were not included in the initial Lake Victoria project due to the war and unrest of the early 1990s. The Kagera basin draining these countries is the largest tributary by far to Lake Victoria and population, sewage releases, and deforestation are high in the basin. Finally, outside of the Lake Victoria Fisheries Organization, an international legal commitment in terms of a treaty does not exist to drive restoration or joint management activities for each of the lakes.

*Table 2.* Transboundary water projects under implementation or preparation with GEF funding. *LME: Large Marine Ecosystem; ** Salt water lake.

| River Basin Projects | Lake Basin Projects | Marine Ecosystem Projects |
|---|---|---|
| Tumen | Lake Chad Basin | Benguela Current LME* |
| Danube | Lake Tanganyika | Canary Current LME* |
| Okavango | Lake Malawi/Nyasa | Plata Maritime Front |
| Mekong | Lake Victoria | Yellow Sea LME* |
| Niger | Lake Titicaca | Black Sea LME* |
| Senegal | Lake Ohrid | Bay of Bengal LME* |
| Volta | Lake Manzala | Mediterranean Sea |
| Nile | Aral Sea** | Gulf of Guinea LME* |
| San Juan | Caspian Sea** | Baltic Sea LME* |
| Bermejo | | Red Sea |
| Sao Francisco | | S. China Sea LME* |
| Upper Paraguay | | E. Asian Seas |
| Guarani Aquifer | | W. Indian Ocean |

The Baltic Sea program has also existed for 20 years. Because a sound analysis of transboundary issues exists and the equivalent of a SAP was produced in 1991 in the form of a Joint Comprehensive Program, GEF funding is directly channeled to on-the-ground implementation of nutrient reduction measures, which is the key transboundary priority. Political commitment exists among collaborating countries in terms of a convention on environment with the Helsinki Convention creating HELCOM and the fisheries convention creating the International Baltic Sea Fisheries Commission. Experience in the Baltic as well as the North American Great Lakes shows that both types of joint bodies are needed, not just one, to ensure that the correct ministries and constituencies are mobilized to restore and protect the shared resources.

Joint commissions with joint processes of dialogue and fact finding are present in many projects from the Bermejo Binational Basin to the Plata Maritime Front

Commission and the Lake Titicaca Commission. The recent Benguela Current GEF project involving South Africa, Namibia, and Angola proposed formation of a joint commission and adoption of a convention as part of its GEF international waters project to operationalize real-time management of its unique fishery resource. The Benguela project and the Lake Chad Basin GEF Project both have science advisory panels included to ensure that the scientific community in the countries can be mobilized on behalf of the shared resources and that sound science can be applied for management purposes. All three lake basin projects might benefit from a formalized, active panel from the scientific community of the collaborating nations. While the intent for establishing a science advisory body was there for Lake Victoria, its potential may not have been reached.

The most mature of GEF interventions in this focal area involve a series of projects for the Danube basin and for the six Black Sea countries. GEF assistance followed the adoption of an environmental convention for each that expressed political commitment to collaborate in restoring the shared basins and marine ecosystem. These were framework conventions much as the Lake Tanganyika draft convention is a framework convention. Each project area utilized the joint fact-finding of a TDA process and then expressed commitments for policy/institutional/legal reforms and priority investments in a SAP that was adopted at high levels of each government. Each of the projects illustrate that international waters interventions must proceed on three different levels: the joint multi-country level, the national level through inter-ministerial committees, and then implementation should proceed sub-nationally with national support. None of the three African lake projects seem to have embraced comprehensive actions at all 3 levels.

The Steering Committees for the Danube basin and the Black Sea projects included major donors, NGOs active in the areas, and the other GEF Implementing Agencies. Of significance was the involvement in the Danube basin project of NGOs at an NGO forum associated with each Steering Committee meeting and through small grants to NGOs to begin the sub-national implementation needed to address the many hotspots identified by each nation for action. In addition, the investment phase of GEF involvement is now following completion of the strategic phase that was aimed at securing political commitments to action. The World Bank has developed a Strategic Partnership on investments for the Black Sea-Danube basin aimed at the transboundary priorities. While the 1:1 ratio of concessional loans to GEF grants in the Lake Victoria project constitutes best practice for that continent, the investments for the Danube Black Sea are to designed to leverage 3 fold co-finance, including loans for investments. This should be more the norm for eastern Europe, East Asia, and Latin America where more national resources would be expected to be devoted to addressing the shared water issues.

# 6. LESSONS FROM THE NORTH AMERICAN GREAT LAKES

The first treaty addressing the shared Great Lakes of North America was enacted in 1909 and evolved over time with more stringent commitments to action through a 1972 Agreement between Canada and the U.S., a 1978 Revised Agreement, and an ecosystem-based Protocol of commitments in 1987. This parallels the institutional development in Europe's most mature case of joint management with a Rhine Commission on pollution being formed in 1950 and followed subsequently by the signing of more specific Rhine conventions in 1963, 1976, and 1999. While advocates for informal action may argue that such conventions are not for developing countries, the point remains that a political commitment to joint, equitable action is necessary to address transboundary water-related environmental problems and is also needed to provide confidence to funding organizations that commitments will be undertaken.

The experiences of the International Joint Commission (IJC - Canada and U.S.) since the 1909 Boundary Waters Treaty was signed have been detailed elsewhere (Lemarquand,1993; Duda, 1994b; Duda and LaRoche, 1997) and will only be briefly included here. The treaty established the IJC to assist the countries in resolving existing and preventing future cross-border disputes, ranging from issues of water pollution, flow diversions, structures, and habitat conservation. The commission was given more responsibility under the Great Lakes Water Quality Agreement of 1972 and country commitments to enact the necessary policy/institutional/legal reforms and investments to restore and protect the 5 great lakes were made progressively more stringent with revisions in 1978 and the Protocol in 1987. Lakes Erie, Michigan, and Ontario were brought back from serious degradation through the joint processes facilitated by the Commission Secretariat through various joint steering committees (known as Boards) and Subcommittees of the two Parties and their sub-national levels of government.

The lessons begin with political commitment at the top with an international agreement for the countries to jointly work together on their shared resources and to implement needed actions. The IJC accomplishes this through various boards of high level country officials that undertake processes of joint fact finding to review progress in cleanup over time. A board of ministry officials and a separate board representing the science community of both countries work to evaluate progress undertaken by each country in implementing its commitments to cleanup/prevention and then report this with recommendations to the Commission. The Commission has no power to execute but only to convene the Parties, report the facts to governments and the public, and when progress is slow to tender recommendations on next priorities. Through advice of its boards, including the science community and the transparency and dialogue that is conducted with the public and NGOs, Lake Erie was brought back from the "dead" as it was termed in the 1960s and significant ecosystem improvements have occurred in the other great lakes and in the basins draining to them.

For possible considerations in the restoration and protection of the African Great Lakes, several lessons stand out. Political commitment in a joint agreement is a necessary step. Activities should be conducted not only on the multi-country, international level but also at the national level through inter-ministerial committees that provide input to the work at the international level and then operationalize commitments nationally down through implementation at the sub-national level. A science advisory body is useful in facilitating management institutions to utilize sound science in their joint fact finding and technical work. The IJC found that actions at the lake-wide level were too complex and consequently actions were broken down into smaller sub-national areas i.e. basins draining to portions of the lakes that were declared hotspots for involvement with communities, industries, and NGOs in cleanup and restoration. This is consistent with hotspot approaches in the Danube, Mediterranean, and Baltic. All three African lake projects have identified certain areas of pollution inputs or rural sediment and nutrient inputs for targeting actions to specific catchments.

The IJC early on recognized the need for an ecosystem approach and included the lake basins in interventions. This approach will also be essential for Africa based on the results of the first GEF projects. Harmonization of regulatory programs as well as monitoring programs in terms of verification that progress is being made can be facilitated through these transparent, joint commission activities. Agreement on indicators to be monitored, collection techniques, quality control and quality assurance was important for North America and will be important for monitoring and evaluation purposes for the Africa lakes. Rather than just utilizing indicators of water chemistry, ecosystem indicators (certain biological indicators of flagship species and communities) were debated with the scientific community and eventually adopted to follow restoration progress. A word of caution about the use of simulation models compared to common sense in decision-making was also the subject of debate among the science community in the North American Great Lakes. Uses and abuses of modeling, over-reliance for decision-making, and limitations of application due to unverified assumptions were addressed by the Great Lakes Science Advisory Board (1986).

One final comment on lessons relates to the joint multi-country processes necessary for addressing fisheries issues versus the larger environmental issues that are so crucial for the African lakes. Just as the Baltic utilizes two commissions to address these issues, the North American Great Lakes also utilizes a fishery commission that is separate from the IJC. Established by the Convention on Great Lakes Fisheries in 1953, the Great Lakes Fishery Commission utilizes similar joint processes to facilitate political agreement on joint fisheries management issues. Each commission has its own ministries and sub-national governmental organizations responsible for implementation. When commercial fisheries are so politically and economically important as in Tanganyika and Victoria, consideration should be given to utilization of two separate joint commissions to ensure that proper focus is placed on political support for action. Further information is available on the fishery commission from its internet web site at www.glfc.org and for the IJC at its web site, www.ijc.org.

# 7. DISCUSSION

With about 50 percent of the Earth's land surface consisting of transboundary basins and 70 percent of our planet's surface in oceans, the majority of the world's water resources must be managed internationally as transboundary resources. These waters are included as priorities for funding under the Global Environment Facility, and all three of the African Great Lakes have received grant funding during the beginning of the GEF in the early 1990s. All three lakes face threats to the sustainability of their biodiversity. Lake Victoria showed the effects first due to its vulnerability over a decade ago, and the other two are now showing signs of degradation. The sheer numbers of people, the extent of economic activity, the pollution sources, and the introduction of the Nile Perch make the restoration of Lake Victoria enormously complex and in need of a different level of attention from the other two.

Transboundary water resource management is necessarily complex in nature with a wide variety of social, political, economic, physiographic, and ecological conditions to be taken into consideration. As noted in the GEF Operational Strategy and in early GEF international waters projects, an initial strategic project has been a useful tool to help countries sharing a basin to understand their shared problems, to focus on a few priorities, and then to break down complex situations into more manageable pieces for speeding understanding and implementation. The GEF-recommended processes of joint fact-finding and sharing of information that jointly produce a Transboundary Diagnostic Analysis (TDA) help to (1) break down barriers among nations so that they focus on a few priorities and (2) set the stage for countries to formulate a Strategic Action Program (SAP) of country-specific and joint actions needed to address the transboundary priorities. Only the Lake Tanganyika GEF project undertook these processes.

Experience bears out the observation that work on multiple levels of institutions is essential to address these transboundary issues. Some action is necessary on the multi-country, international level with countries that share the basin. Action is necessary at the national level through inter-ministerial committees involving sectors that create the stress on the ecosystems in producing the input to the multi-country processes as well as to translate the international political commitment down to the third institutional level, the sub-national level, for implementation on the ground by local governments, private sector, or communities. The country-specific inter-ministerial committees are proving to be key elements in GEF waters projects. In addition, involvement of stakeholders and harnessing the information that the scientific community can provide in each collaborating nation are also proving to be important factors.

Experience also shows that political commitments at the highest levels are necessary to ensure smooth operation of multi-country institutions and implementation of corrective measures on the ground. Such commitments have been expressed in basin conventions. Where no political commitments to joint actions were present, country commitments for policy/institutional/legal reforms have been slow to emerge. While there was a convention that created the Lake Victoria Fishery Commission, all three lake projects could have benefited from a formalized multi-country political commitment. The Lake Tanganyika project did include creation of such a joint institution under a multi-country convention for improving lake management as part of their SAP, implementation of which may be the subject of a subsequently-funded GEF international waters project. At the minimum, a succession of GEF-financed interventions over time is proving to be a useful approach to take countries through step-by-step processes for expressing commitments to action and then funding incremental costs of those actions.

Country commitments to action will be necessary not only for the Lake Victoria Basin Ecosystem but also for the other two Great Lake systems. These commitments run the gamut from land tenure reform, better soil conservation measures, and reforestation initiatives to implementation of pollution control measures, better management of near-shore fisheries to protect biodiversity, and possible development of deeper fisheries so that near shore fisheries may rest. Experience shows that countries will not automatically make these reforms. Joint multi-country processes and political will are necessary to make implementation a reality. With all three lake projects, the initial GEF intervention is coming to a close. While much has been learned by the studies that have been supported, the next several years will be critical for action and leadership by participating governments to make restoration and protection of these unique ecosystems a priority.

## REFERENCES

Bootsma, H.A. and Hecky, R.E. (1993) Conservation of the African Great Lakes: A limnological perspective. *Conservation Biology* **7**, 644-656.

Bootsma, H.A. and Hecky, R.E. (1999) Water quality report, Lake Malawi/Nyasa biodiversity conservation project, SADC/GEF.

Cohen, A.S., Bills, R., Cocquyt, and Calijon, A. (1993) The impact of sediment pollution on biodiversity in Lake Tanganyika. *Conservation Biology* **7**, 667-677.

Delft Hydraulics. (1999) Preparation of a preliminary Lake Victoria physical processes and water quality model. Final Report, Netherlands.

Duda, A. M. (1994a) Lake Victoria basin management: Cross sectoral and international challenges. In Hirji, R. and F. Patorni(eds.), *Proceedings of the Seminar on Water Resources Management in Tanzania*, World Bank, Washington, pp. 70-77.

Duda, A. M. (1994b) Achieving pollution prevention goals for transboundary waters through international joint commission processes. *Water Science and Technology* **30**, 223-231.

Duda, A. M. and La Roche, D.L. (1997) Joint institutional arrangements for addressing transboundary water resources issues—lessons for the GEF. *Natural Resources Forum* **21**, 127-137.

Global Environment Facility. (1996) GEF Operational Strategy, Washington.

Great Lakes Science Advisory Board. (1986) Uses, abuses, and future of Great Lakes modeling. International Joint Commission, Windsor.

Hecky, R.E. (1993) The eutrophication of Lake Victoria. *Mitt. Int. Verein. Limnol.* **25**, 39-48.

Hecky, R.E. and Bugenyi, W. (1992) Hydrology and chemistry of the African Great Lakes and water quality issues: Problems and solutions. *Mitt. Int. Verein. Limnol.* **23**, 45-54.

Hirji, R. and Duda, A. M. (1995) A comprehensive approach for managing the Lake Victoria Basin Ecosystem. In *Comprehensive Water Resources Development of the Nile Basin: Proceedings, Third Nile 2002 Conference*. Arusha. pp K-2 to K-14.

Kaufman, L. (1992) Catastrophic change in a species rich freshwater system. *Bioscience* **42**, 846-858.

Lemarquand, D. (1993) The International Joint Commission and changing Canada-U.S. boundary relations. *Natural Resources J.* **33**, 59-98.

Mpendazoe, F., Mashauri, D., Rutagemwa, D., Jackson, M., and Kayombo, S. (1994) Evaluation of point source discharges in the Lake Victoria area, Tanzania, Phase II Report to World Bank, National Environment Management Council, Tanzania.

Patterson, G. and Makin, J. (1998) The state of biodiversity in Lake Tanganyika. Lake Tanganyika Biodiversity Project, Natural Resources Institute, Wallingford.

Ribbink, A.J. and Hecky, R.E. (1999) Conservation of biodiversity: Time to act, Lake Malawi/Nyasa Biodiversity Conservation Project, SADC/GEF.

Tweddle, D. (1992) Conservation and threats to the resources of Lake Malawi. *Mitt. Int. Verein. Limnol.* **23**, 17-24.

United Nations Development Program. (1994) Pollution control and other measures to protect biodiversity in Lake Tanganyika. GEF Project Document RAF/92/G32, New York.

World Bank. (1994) Republic of Malawi: Lake Malawi/Nyasa biodiversity conservation, GEF Project Document, Report 13044 MAL, Washington.

World Bank, (1996) Kenya, Tanzania, Uganda: Lake Victoria environmental management project, Project Document, Report 15541-AFR, Washington.

# GEOLOGICAL HAZARDS AND ANTHROPOGENIC IMPACTS ON THE ENVIRONMENT IN MALAWI: LESSON FROM A CASE STUDY OF DEBRIS FLOWS IN ZOMBA

JOHN MWENELUPEMBE[1] AND HANS-GÜNTER MYLIUS[2]

[1]*Geological Survey Department, P.O. Box 27, Zomba, Malawi*

[2]*Federal Institute for Geosciences and Natural Resources, Stilleweg 2, 30655 Hannover, Germany*

**ABSTRACT**

Natural geological hazards such as flash floods, landslides, debris flows and mudflows have caused, and will continue to cause, many problems in Malawi. Landslide risks are primarily a geological problem and it is Quaternary geology that concerns our environment most. It is a typical geological task to reconstruct the old events using geological, geomorphological and dating methods and based on these results, predict possible future disasters. Our studies of old and recent debris flow events in Southern Malawi have found out that these phenomena are mainly triggered by tropical cyclones that bring heavy rainfalls. They occur in mountain slopes with certain slope instability, soil cover and discontinuity patterns of the underlying rocks. Human impacts on the environment such as large scale earthworks projects and replacement of natural forests by non endemic tree species have also played an important role in a rising risk of flash floods and debris flows. It is recommended that the most sustainable method of mitigation against these hazards is to incorporate the geological hazard zoning maps into town and regional planning schemes. Where this may not be possible, measures that can reduce or minimise the hazard impact should be adopted.

Zomba Mountain is made of a Mesozoic syenite pluton that intruded Precambrian high-grade



*E.O. Odada and D.O. Olago (eds.),*
*The East African Great Lakes: Limnology, Palaeolimnology and Biodiversity,* 557–573.

metamorphic gneisses and granulites. Weathering erosion and uplift have left the syenite standing at an elevation of 2000 meters while the surrounding area is between 700 and 1000 meters above sea level. Zomba town is located at the foot of the mountain. Steep slopes with up to 50°, the highly jointed syenite, and the thick lateritic soil covers are the main factors likely to influence rock instability. Debris flows triggered by heavy rainfall have affected Zomba since memorial times. People of Zomba and the authorities are well aware of past landslides and the continued risk. As a result from our investigations the detailed risk zoning of Zomba contributes to the disaster preparedness plan of the town.

## 1. INTRODUCTION

The Federal Institute for Geosciences and Natural Resources of Germany (BGR) and Geological Survey Department of Malawi (GSD) have undertaken a joint assessment on geological hazard risk. The area of Zomba town and the Michesi Mountains in Southern Malawi were selected for the purpose of this investigation (Fig 1). Detailed field studies of the two areas were carried out from May to August 1998 with the following goals: reconstructing old landslide events, delineating areas that are prone to landslides and proposing mitigation measures against this hazard. Another important aspect of the project was to train GSD staff in this research field. This paper gives a brief summery of the work done.

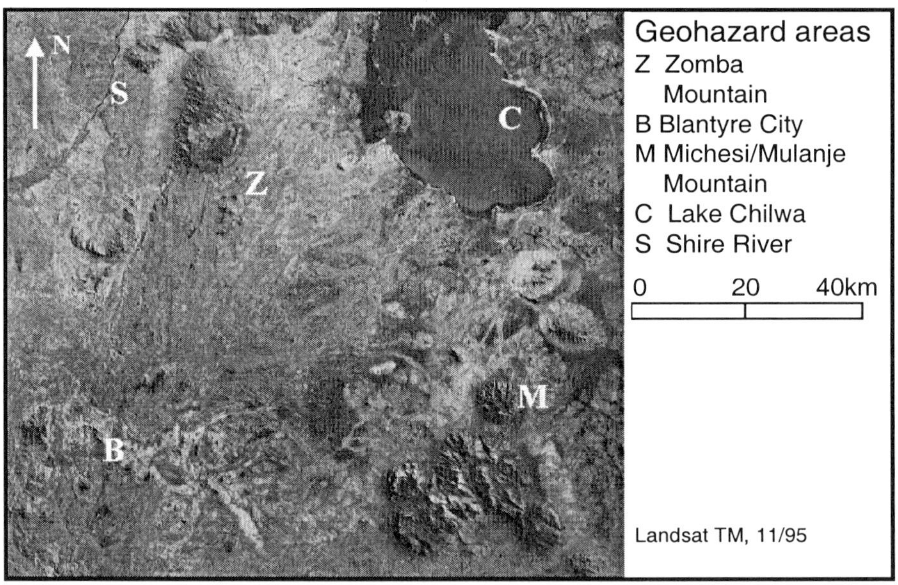


*Figure 1.* Location map of Zomba, south Malawi.

## 2. THE HISTORICAL 1946 DEBRIS FLOWS EVENT IN ZOMBA

The debris flows event most talked about occurred in 1946 in Zomba, which was at that time the Capital City of Malawi. 710 mm of rain fell in 36 hours, which caused devastating debris flows in the whole town. Twenty-two lives were lost and more than 5000 people were left homeless due to the disaster. Communication, water and power supplies were all disrupted.

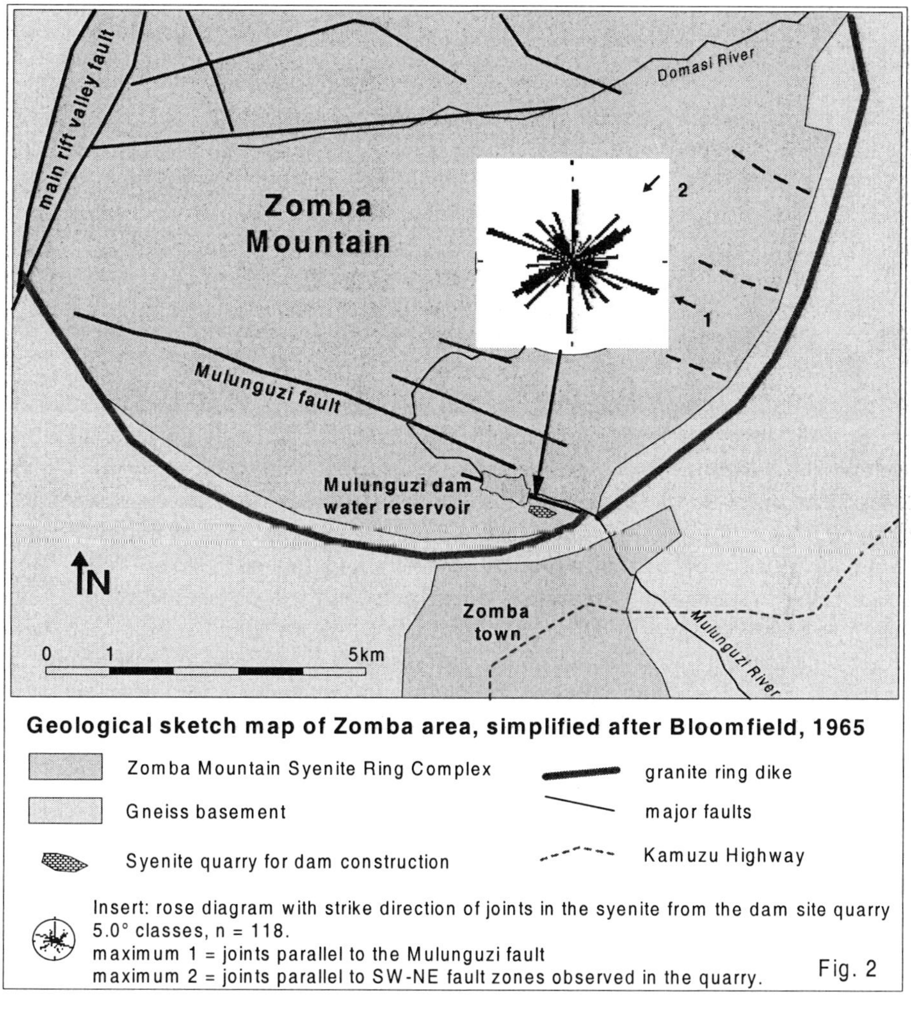


*Figure 2*. Geological sketch map of Zomba area, simplified after Bloomfield, 1965.

## 2.1 General Geology

The geology of Zomba Mountain (Figure 2) consists of a group of rocks called the Chilwa Alkaline Province, who is composed of syeno-granitic and nepheline syenite plutons of the Upper Jurassic to Lower Cretaceous age (Carter and Bennett, 1973). These form a conspicuous ring structure consisting of separate intrusions of quartz syenite, syenite, alkaline granite and micro-granite emplaced in that order (Bloomfield, 1965). The main Basement Complex rocks around the area are biotite and hornblende gneisses striking North to northeast and with high dip angles.

# 3. METHODOLOGY

A number of methods were employed to come up with sound conclusions. These included the interpretation of satellite images and aerial photographs, detailed geological-geomorphological mapping, and geotechnical investigations. Dating organic matter using carbon $^{14}C$ methods failed to be successful. Age determinations using tree rings have been more successful.

## 3.1 Remote Sensing

Interpretation of aerial photographs and satellite images yielded vast amounts of information on geology and geomorphology of the area. Not only geological features such as structural lineaments were easily mapped out. Most important for the risk assessment were the morphological features like landslides and debris flows on the slopes of the Mountain. The stereoscopic interpretation of 1:10000 aerial photographs allowed to outline scars, channels and fan areas from debris flows because the low-lines and separating ridges are clearly visible, even if they differ only by a few metres height. Different colour compositions from Landsat TM satellite image enlargements to 1:100.000 have been used. It was possible to determine the preferred tectonic features of joints and faults and secondary dyke intrusions within the syenite. No relation between the orientation of the lineaments and the landslide activity was found. However slope parallel joints might weaken the stability of the rocks.

## 3.2 Geological-Geomorphological Methods

Geological-geomorphological field mapping yielded the data for the assessment of the debris flows. Through comparative observation of the Quaternary deposits with the ancient geology, a reconstruction of the old traces of debris flows was possible. The study of faults and joint patterns in the outcrops formed an integral part of this method. The grain size, geotechnical parameters, and mineralogical composition of soils has been determined.

## 3.3 Dating

In landslide prone areas decision-makers of urban and regional planning need to know the possible rate of return of this kind of natural disasters. Providing them with comprehensive risk assessment data they can work out adequate disaster preparedness plans, guide suitable landuse, and realise mitigation measures.

Reconstruction of geomorphological features can not be complete without a well-established stratigraphical sequence. Therefore attempts were made to date some of the events wherever possible. First, documents on landslides were checked for possible age determinations of some old landslides. Old newspaper articles and reports have been sought in archives, and a number of people who witnessed some of these events were also interviewed. However, this method only allows dating events, which occurred the past 150 years.

Trees growing on old landslide deposits also give some indications of relative age of the debris flows events. Interpretation of trees annual growth rings is a method widely used in determining ages of landslides/debris flow occurrences. On trees that have been hit by objects carried by a moving debris flow scars can be observed. Trunk cross sections can be studied for age determination. The method had to be used with care as some scars were made by human activities. Trees' ages in Zomba area are restricted in time to the last 200 years. Another limitation that had to be considered was varying production of growth rings because of the climatic conditions in the humid area. Some trees may produce up to four rings per year as a result of varying intensity of the rainy season (Kayambazinthu, forestry research institute Zomba, Poschinger, 1998, pers. comm.). Information about the first trees planted in the botanical garden of Zomba in 1871, or the construction of the golf course are some of the more precise time figures.

Deep trenches that have been cut through debris flow deposits for the Zomba water supply project have been checked for remains of organic materials that could be dated with the Radiocarbon $^{14}C$ method. Due to rapid decomposition of organic matter in the subtropics no suitable material could be found.

The age determination of lichen on old boulders has had to be abandoned after getting expert advice (Niemeyer, University Hanover, pers. comm.). The investigation of endemic species requires time consuming methods like microscopic studies and results might be misleading. Some indications of ages of debris flows were also obtained from comparing geomorphological features with artificial structures of known date of installation such as aged pipelines and bridges.

A new promising approach might be the investigation of sedimentological records from Lake Malawi (Barry, pers. comm., and Johnson *et al.*, 2001). The observed thin layers of coarse grained sediments, which appear every 50 to 100 years in drill cores from the northern lake area allow the comparison, that disastrous rainfall events happened in this return period.

## 3.4 Geotechnical Approach

Some investigations were done on the properties of soil in relation to debris flows occurrence. These included studies on grain size distribution and plasticity of the soils. The former helped in understanding the soil hydrogeological/hydrological behaviour while the latter hinted on the susceptibility of soil to failure under the influence of water.

# 4. DISCUSSION OF RESULTS

## 4.1 Characteristics of Debris Flows

In Zomba Mountains almost all streams have in the past been affected by debris flows. Catchment areas are generally very small, only that from Mulunguzi River measures 18,9 km². From Mponda stream, with a catchment area of 3 km², a water rise of about 2 to 3 meters has been reported during the Zomba flood of 1946 (Ingram, 1946). Brief descriptions are given below in three sections namely source areas, transport section and deposition areas.

The reader is also referred to the map of mass movements (Fig 3.), where some of the morphological features are clearly shown. The debris flow features in the map present the two age groups of the 1946 event and older events. The old debris flows deposits probably pre date 1871, when the first trees have been planted in the botanical garden. Some of them can be still found.

### 4.1.1 Source Areas

In Zomba, recent source areas herein referred to as scars, are not easily seen. This is because of the fast colonisation of debris flows features by plants. The geomorphological features are partly inaccessible because they are well covered by thorny bushes. Most of the big scars are much older probably predating the Holocene age. However, most of the young debris flows originate from small scars located between 1750m and 1200m above mean sea level (a.m.s.l.) with Zomba town at about 900m a.m.s.l. These commonly occur in areas with thick lateritic soil cover. Some are located close to the deeply weathered rock material with intercalation of loose big boulders. Slope angles observed vary between 25° and 50°. The volume of the mostly U-shaped scars measured between 100 m³ and several hundred m³ containing material mixed up of boulder sizes between 1m and 5m in soil matrix.

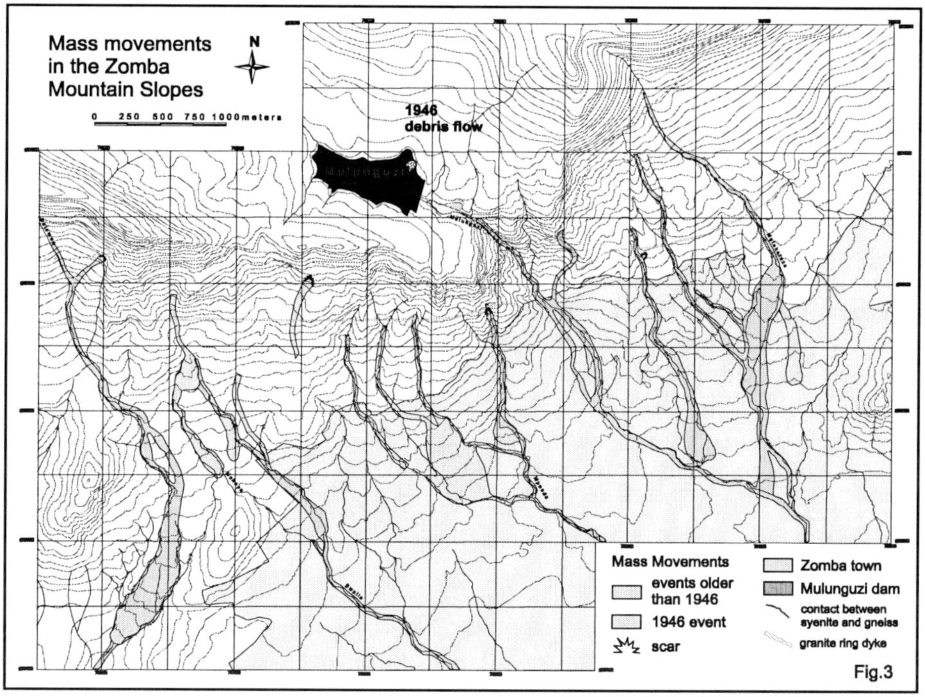


*Figure 3.* Mass movements in the Zomba mountain slopes.

### 4.1.2 Transportation Section

The transportation sections are relatively shallow in the East making deviations to the surrounding areas more likely. In the western part the valleys are quite deep, up to 10m, forming a u-shaped cross section. The channels are often bound by lateral ridges where boulders between 1m to 5 m diameter are piled up about 3m to 5 m high. Accumulations of debris flow deposits in the channel bed are generally less pronounced. Channel gradients vary between 5° to 10° with upper sections up to 25°.

### 4.1.3 Depositional Areas

This is the area where boulders, sediments and other debris flow materials are deposited. Often the maximum extension of these deposits is difficult to determine. Normally they tend to follow small morphological depressions and valleys. Muddy water spreads much further than the fan areas and may sometimes flood large areas. In Zomba deposition starts with decreasing slope angles to less than 5° around 1200m a.m.s.l. to the West while those to the East below the 1100m a.m.s.l. Deposits of the former have a high content of soil as opposed to the latter. This probably indicate that the debris flows on the Western part were relatively more viscous and that water was not available in sufficient quantities to transport the material further down slope.

## 4.2 Causes of Debris Flows

Debris flows occur due to the interplay of several factors including climate, geological-geomorphological, soil properties, vegetation cover, and human activities. There are two major main categories that control debris flows: the triggering factors and those that control or influence the occurrence of landslides.

### 4.2.1 The Triggering Factors

#### *4.2.1.1 Rainfall*

It is recognised world wide that rainfall is playing a crucial role in triggering debris flows. The same is the case for Zomba. In Malawi rains are normally associated with the passing of the Inter-Tropical Convergence Zone (ITCZ) over the country. This pattern occurs twice a year between November and March. Sometimes the ITCZ is overlain by tropical cyclones, which originate from the Indian Ocean (Figure 4). The coinciding of the two results into continuos heavy rainfall that has been correlated to most of the debris flow occurrences in Southern Malawi, and Zomba in particular.

#### *4.2.1.2 Earthquakes*

Earthquakes can trigger debris flows but no intense tremors have been observed up to now in the vicinity of the Zomba Mountain. (Chapter 4.2.2.4).

### 4.2.2 Controlling Factors

#### *4.2.2.1 Geology*

Structural pattern of rocks is the main geological factor influencing occurrence of debris flows. Two major joint sets are distinguished in the study area: tectonic and non-tectonic. Slope failure is mostly associated with the latter. This mainly results from igneous layering, exfoliation and relaxation processes of rocks on the slope. Their orientation is often random dipping parallel or sub-parallel to the slope, a condition that makes most of the slope faces unstable. The joints also act as pathways for underground water and they are favourable areas for weathering. The two factors make the rocks on the slope very susceptible to down slope failure.

It was also observed that some scars occur close to either dykes or granite ring structures. The association of debris flows to these features is not well understood. However, the most plausible explanation may be that the dykes have a greater flow of underground water and they disintegrate easily into loose blocs. These conditions may accelerate the weathering and saturation of these rocks hence inducing debris flows.

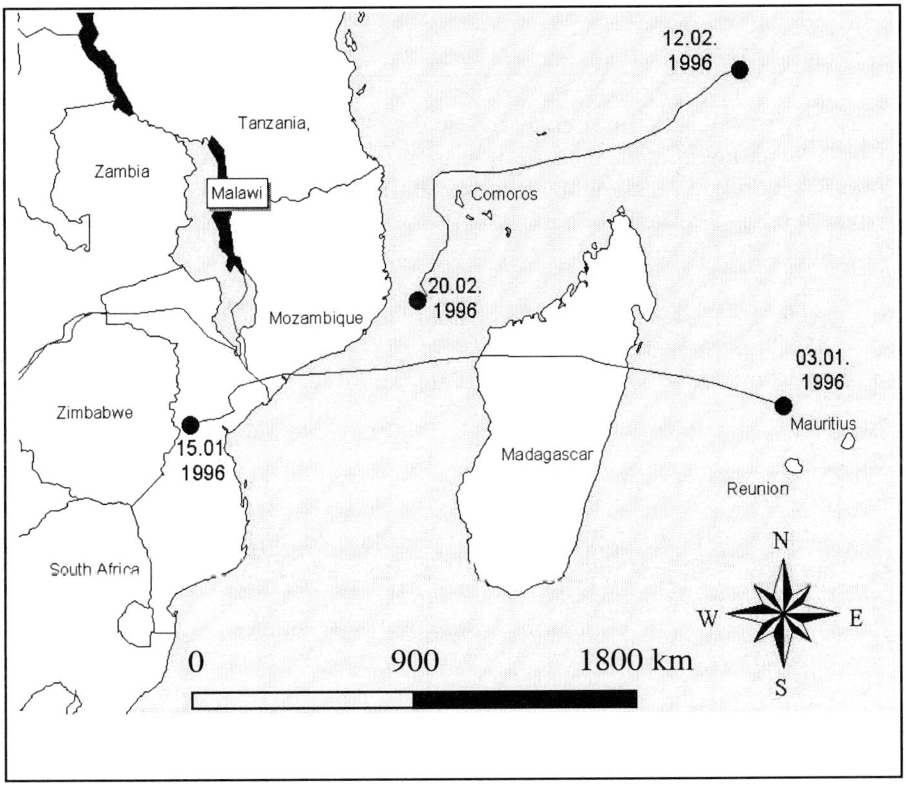


*Figure 4.* Example of tropical cyclone movement.

#### *4.2.2.2 Soil Cover*

The summarised results of grain size analysis and Atterberg limit results from Zomba Mountain soils indicate that the vast majority of samples have a fines content between 45% and 75%, with a clay component between 20% and 50%, and are therefore silts and clays, sometimes sandy or occasionally gravel. There is a significant increase in the grain size of the samples from the soil surface to the bedrock. Organic material and roots reach generally 1,2 m deep but on steep slopes, occasionally deeper. The natural moisture content of the soils varies between 19,5% and 33,9 %, with an average of 25,9%.

*Table 1.* Mechanical properties of soils from Zomba Mountain

| | max | min | average | no. of samples |
|---|---|---|---|---|
| liquid limit | 73 | 24,7 | 48,8 | 65 |
| plastic limit | 42 | 18 | 29 | 65 |
| plasticity index | 47 | 3,4 | 19,8 | 65 |
| shear strength | 35 | 21 | 29 | 59 |

Plastic and liquid limits define the boundary at which the soil changes from plastic to semi-liquid state. The relation of the two is given by a numerical value called plastic index, which is obtained by subtracting plastic limit from liquid limit. Shear strength is the internal resistance of soil or body to exerted force. It also gives information on the cohesion of particles in a mass.

The figures above show relatively low values of plasticity index and shear strength. This means there is a high probability for the binding bonds in the soil to be broken by adding some water. In the event of heavy rainfall the soil liquefies and the debris mass flows down slope. The soil is, at least locally, rather susceptible to flow.

#### *4.2.2.3 Slope Morphology*

The slope angle also contributes to the failure of slope faces. Generally, the higher the slope angle the greater the likelihood to failure. High thickness of soil or weathered rocks is increasing the risk of slope failures. In Zomba it was found out that debris flows start at slope angles of above 20°. Very steep slopes exceeded 50° to sub-vertical. The deposition areas start generally when the slope angle in the channel section is reduced to less than 5°.

#### *4.2.2.4 Seismicity*

In Malawi, earthquakes are restricted to the Rift Valley system, and related to seismically active zones located in the neighbouring countries like Mozambique, Zambia and Tanzania. The likelihood of occurrence of strong tremors decreases with increasing distance from these areas. Influence of earthquakes on slope stability decreases remarkably with distance from the epicentre. The reason for this is the absorption of short, high-energetic wavelengths by the rocks. This causes a reduction in the horizontal accelerations, which are most crucial for slope stability. The biggest earthquake in the recorded history occurred in Salima in Central Malawi in 1989 with a magnitude of 6.1.

So far, no earthquake events can be correlated to the occurrence of the studied debris flows in Southern Malawi. However this does not rule out the possibilities of experiencing landslides due to seismic activity, because the main Rift Valley Fault cuts the western limits of Zomba Mountain (Figure 2).

#### *4.2.2.5 Human Influence*

Although landslides are a natural phenomenon and normal features of landscapes experiencing dissection, their magnitude, frequency and geographical distribution have considerably been aggravated in recent times by human intervention. A few examples from Zomba Mountain slopes of such activities are briefly discussed below.

#### *4.2.2.6 Vegetation Cover*

Since 1907 in Zomba Mountain the indigenous forest species have been replaced by exotic ones. Until 1955 plantations of Mexican pine trees have replaced already 145 hectares of natural forest. These figures increased from 1118 ha to 2093 ha between 1970 to 1992 (Sitaubi *et al.*, 1996). The changes in ecological conditions in one way or the other contribute to the slope instability.

Pine trees and natural forest differ quite a lot in root density hence vary in their effects of holding unconsolidated soil. Natural forest has a higher root density than the artificial ones, hence a stronger interlocking network of holding loose soil particles. In addition vegetation takes up moisture from the upper layers of the soil and can thus reduce the overall moisture content of the soil, increasing its shear strength. This means that the natural forest has a better capacity of preventing debris slides and flows. Plants on the other hand add weight to the slope. If the added weight is large and the root network is of limited extent, the vegetation may have a de-stabilising effect instead. Dense pine forests for industrial use add much more weight than endemic forests with more scattered canopy trees.

Fires, which destroyed large areas of forest in the Zomba Mountain in 1996 and 1998 and routine logging have the same effect of reducing the forest cover and making land more susceptible to geomorphological processes such as landslides and soil erosion.

The extension of Zomba town towards the steep slopes of Zomba Mountain has led to indiscriminate cutting of the forest cover for mansion like houses. Adequate town planning should be implemented to prevent such developments. Finally the uncontrolled and further increasing cutting of trees for firewood will have unpredictable results if there is no change in public comprehension.

*4.2.2.7 Earthworks Development*

Engineering and construction works involving large land surfaces in slopes of mountain areas can have a big impact on the stability of the slopes. In Zomba Mountain slopes, several landslides have been reported due to the undercutting of the slopes for road alignment. In 1997 a small landslide blocked the lower road to Kuchawe on the Zomba mountain plateau. In March 1999, a debris slide deposit blocked the ascent road to Kuchawe. Dangerous cracks have since developed on the road due to the impact of the mass in an already unstable channel section with a slope angle of 30 to 40°.

In 1946, a landslide mass of 100.000m³ from the northern slopes of the old Mulunguzi Water Reservoir entered into the dam and caused a high wave of water from the reservoir into the river below. A new bigger reservoir is now under construction to secure water supply for Zomba town and surroundings. A number of geotechnical aspects have been addressed to avoid a repeat of the 1946 catastrophe. However attention needs to be paid to slope stability due to the thickness of the soil in the deeply weathered rocks which can easily slump. The reservoir slopes have been checked for stability, and where the safety was not high enough, berms have been excavated to unload slopes. The old 1946 sliding area has been stabilised by a supporting embankment.

# 5. MITIGATION MEASURES

Effective and successful geohazard mitigation measures aim at achieving reduction of risk from mass movements. The implementation of the measures should be based on the detailed knowledge of the nature, scale, distribution and causes of the landslide hazard. Broadly, these are grouped into active and passive. Active measures deal with engineering aspects such as design of a structure while passive measures focus on development planning and warning systems. Only mitigation measures achievable for a developing country have been considered. Realisation of appropriate engineered structures is a question of available funds.

## 5.1 Active Measures

### 5.1.1 Wider Underpassage of Bridges

Debris flow danger increases when boulders and tree trunks accumulate in the riverbed, damming the natural flow of water. This happens often in Zomba where some bridges and culverts are small sized and easily blocked by debris. Hence construction of bridges, which allow free passage of debris such as, logs and big boulders can greatly minimise loss of lives and damage to property by debris flows.

### 5.1.2 Dam Erection and Increasing Water Course Capacity

Erection of dams can also practically reduce damage to an area. The design of such a structure need to take care of the possible different sizes and directions of the potential debris flows. Although these structures do guarantee security to a certain extent, they can be more dangerous in case of breakage.

Deepening of riverbeds that are shallow is almost an impossible task without the necessary machinery. In appropriate areas the water capacity of river channels can be increased by removing logs that otherwise are accumulating potential debris flow masses. This can reduce the likelihood of having debris flows initiated or deviated to other areas.

### 5.1.3 Slope Stabilisation

There are a number of engineered slope stabilisation methods that are mostly expensive. The mountain roads however need proper slope drainage, to avoid ponding of water and accumulation of erosion materials. Properly designed tree plantations can stabilise endangered areas as well, and they are achievable.

## 5.2 Passive Measures

### 5.2.1 Landslide Zoning

Integrating the recommendations from landslide hazard zoning into land use planning of an area is considered one of the most achievable means of reducing risks. This approach is considered to be relatively cheap if applied in time. The method is based on detailed geohazard risk assessment (Figure 5). In this case all areas which may be affected by landslides or debris flows are zoned out. It is strongly recommended that the high-risk areas be kept free of human settlements. Applying hazard zoning to an already existing populated area may however, be problematic, as it is not easy to convince landowners to abandon their homes. Poschinger *et al.* (1998) refers to this situation as requiring a political settlement and not a scientific solution. Vulnerable installations such as water pipes should be erected where it is possible far away from the debris flow path. Where this is not possible, the designer should take into account the potential hazard from debris flows and use appropriate structures. This point is being emphasised because outbreaks of waterborne diseases commonly occur as a result of lack of clean water during disasters such as those caused by mass movements.

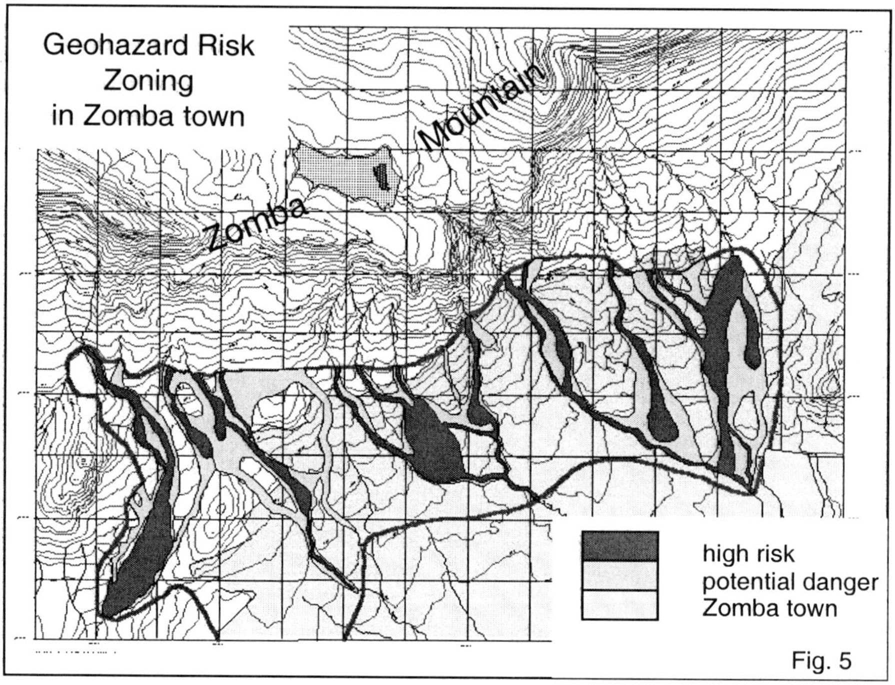


*Figure 5.* Geohazard risk zoning in Zomba town.

### 5.2.2 Warning System

A simple, effective and appropriate warning system can be devised based on landslide zoning by the Geological Survey and rainfall forecasting from the Meteorological Department. This warning can be issued when cyclones or storms have been detected and projected that they will bring rains beyond the thresholds values for debris flow initiation. However, this can only be meaningful if detailed research on the relationship between climate, landslides, soil, geology, land forms and vegetation is undertaken and the needed data bases are readily available for a disaster preparedness plan.

A National Warning System for floods already exists. The National Meteorological Department, in collaboration with the Regional Tropical Cyclone Centre in La Reunion, monitors and tracks the movement of tropical cyclones and the development of the El Niño Southern Oscillation (ENSO) effects. These observation allow to issue timely warning of the onset of strong winds and heavy rains (Meteorological Department, 1973). If geological-geomorphological parameters related to landslides can be integrated correctly into the system, then efficient debris flows warning system for Southern Malawi can be achieved. It should, however, be stressed that such systems can not replace landslide zoning, it can only be a supplementary measure.

## 6. RETURN PERIOD PREDICTIONS

Landslide prediction establishes the magnitude and frequency characteristics of recorded landslides to determine the probability of slope instability of an area. A reliable mechanism of predicting debris flows in Zomba can only be achieved if the return periods of tropical cyclones are calculated. The available data indicate that there are several cyclones per year, which approach the Mozambique coast (Table 2). Predictions for their recurrence for long periods of time have proved difficult and unreliable.

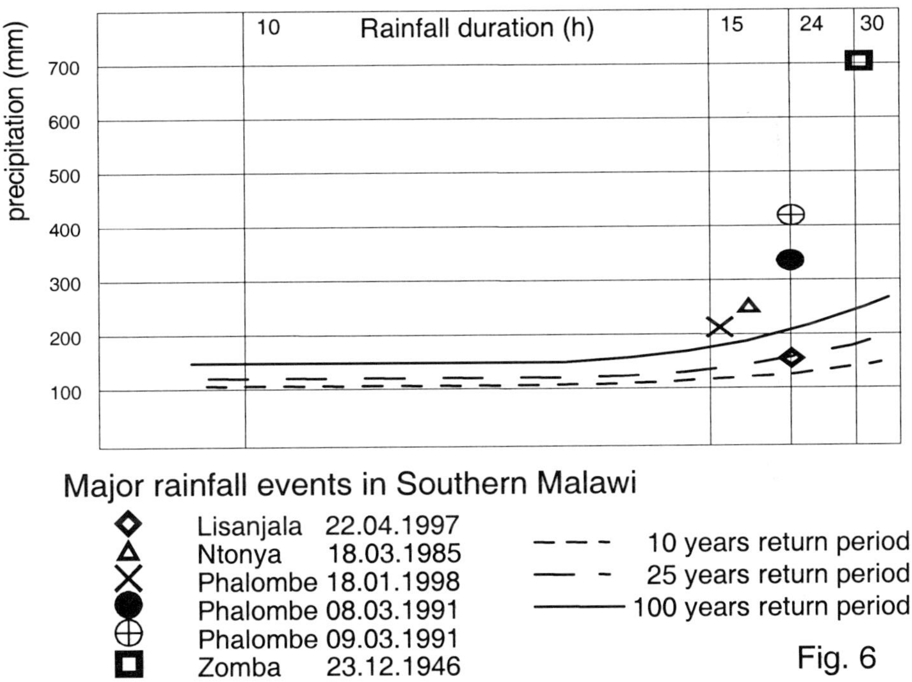


*Figure 6.* Major rainfall events in southern Malawi.

It is evident from the plot (Figure 6) that the Zomba (1946) rainfall event is far above the one hundred years return period. It is worth clarifying here that the plot was formulated using data from a number of environments, which may differ from that of Southern Malawi. Developing such methods particularly for Malawi would require a lot of data on both climate and geological-geomorphological aspects, which is scanty or not available at the moment.

*Table 2.* Number of cyclones approaching the Mozambique Coast.

| Year | Approaching Africa | Away from Africa | Total |
|---|---|---|---|
| 1991-1992 | 1 | 13 | 14 |
| 1992-1993 | 1 | 13 | 14 |
| 1993-1994 | 1 | 14 | 15 |
| 1994-1995 | 1 | 12 | 13 |
| 1995-1996 | 2 | 9 | 11 |
| 1996-1997 | 3 | 11 | 14 |

It should also be pointed out that predicting occurrences of landslides based on tropical cyclone return periods can only allow forecasting of rainfall events which trigger slope failure in the most general terms and in no way give information on the location, magnitude and timing of specific landslide events. These predictions are further supported by the probable global changes in climate. Leading climatologists argue that global changes in rainfall patterns might occur in the near future.

## 7. CONCLUSION

Heavy rainfall associated with tropical cyclones is the main trigger of all debris flows in Southern Malawi. Other controlling factors include disturbances of natural slope by human activities like destruction of forest cover and inappropriate land use.

To reduce the impact of debris flow hazard it is necessary to design structures such that they can withstand geological hazards or use required construction standards. The forest cover on Zomba mountain slopes should be protected and a limit for urban development towards the Mountain has to be established to prevent dangerous undercutting of slopes for construction purposes. Bridges should have a wider under passage and where necessary, protective dams should be erected. It is also important that steps should be taken to put in place an acceptable and appropriate early warning system in close co-operation with the Meteorological Department. The responsible local and regional authorities should work out a disaster preparedness plan.

Of great importance for a better understanding of the relationship between nature and human activities is the distribution of knowledge to the general public and in particular to decision makers in regional and town planning. An attempt is shown in Figure 7.

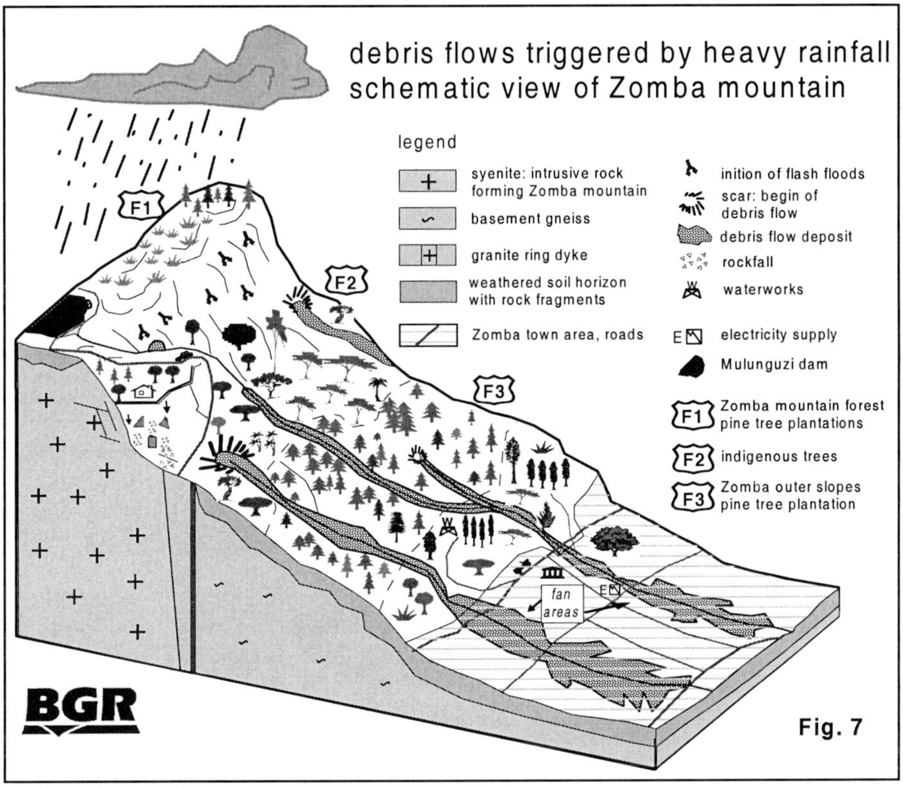


*Figure 7.* Debris flows triggered by heavy rainfall. Schematic view of Zomba mountain.

# REFERENCES


Bloomfield, K. (1965) *The Geology of the Zomba area.* - Geological Survey Department, Malawi, Bulletin No. 16, Government Printer, Zomba.

Carter , G.S. and Bennett, J.D. (1971) *The Geology and Mineral Resources of Malawi.* - Geological Survey Department, Malawi, Bulletin 6, Government Printer, Zomba.

Ingram, J.H. (1946) Procedures of an inquest held before 2nd class magistrate, Zomba.

Johnson, T.C., Barry, S.L., Chan, Y. and Wilkinson P. (2001) Decadal record of climate variability spanning the past 700 yr in the southern tropics of East Africa. *Geology,* **29**(1), 83-86.

Meteorological Department (1973) *National Tropical Cyclone Warning Procedures of Malawi*, Blantyre.

Poschinger, A., Cheyo, D.R.C. and Mwenelupembe, J.J. (1998) *Geohazards in Zomba and Michesi Mountain areas.* commissioned project report, Hanover.

Sitaubi, L.A., Mtambo, P. and Makungwa, S.D. (1996) *Forestry development and stream flows on Zomba Plateau.* Forestry Research Institute of Malawi Newsletter No. 78, Zomba.

# THE HUMAN DIMENSIONS STUDIES ON THE EAST AFRICAN LAKE REGIONS: A REVIEW

MARY MAGDALENE OPONDO
*Department of Geography, University of Nairobi, P. O. Box 30197*
*Nairobi, Kenya*

**ABSTRACT**

Several studies have been carried out under the auspices of the IDEAL (International Decade for East African Lakes) programme on the physical and biological factors affecting the East African lakes. These include aspects such as geological and sedimentary processes, climate dynamics and variability, limnology, food webs and fisheries, and paleoenvironment. However, the human dimensions research has not been part of the overall framework of the IDEAL programme. This paper makes a comparison of the human dimensions and natural science research on the East African Lakes. Further it emphasises the significance of linking up the research findings of the natural scientists, (such as those documented in the IDEAL programme) and those of the social scientists in finding in combating the impact of regional environmental change of the East African lakes.

## 1. INTRODUCTION

The IDEAL scientists have made significant contributions to research on the East Africa lakes since the late 1980s. While the number of natural scientists studying these lakes has been increasing [1], the human dimensions aspects of these lakes are often lacking. Human dimensions studies cover the intricate ways in which



*E.O. Odada and D.O. Olago (eds.),*
*The East African Great Lakes: Limnology, Palaeolimnology and Biodiversity,* 575–586.

individuals and societies not only contribute to, and are influenced by global environmental change, but also mitigate and adapt to it. The challenge for human dimensions research is to discover whether, and how there could be a de-linking of the improvement of human well being from the negative effects of the production and consumption systems that sustain life in human societies (IHDP, 1999).

The East African Lake basins and corresponding hinterlands are found over a wide geographical spread with an equally large inter-lacustrine human population. The increasing population pressure and socio-economic activities in the lake basins has led to changes in land use, water quality, biodiversity, wetlands and fisheries. The lake systems have thus been adversely affected by problems of the water hyacinth, land-based agricultural and industrial activities. This has been manifested in decreasing fish stocks, declining biodiversity, water pollution and encroachment of algae blooms. The rapid urbanisation of the lake regions has also increased their vulnerability to industrial pollution and in some areas this has already induced devastating environmental degradation. Environmental regulations regarding the discharge of effluents and sewage in the East African lakes region are either non-existent or not enforced.

The East African Lakes thus play a significant role in the complex human-environment interactions of these production and consumption systems that form the livelihood in these regions. The activities of man that are intricately linked to the great biogeochemical cycles continue to be negatively impacted by the requirements to satisfy the needs of a growing inter-lacustrine human population. These lakes should therefore be seen as being embedded into the natural cycle, such as the carbon or hydrological cycles and not merely as political and administrative boundaries. For instance, the interaction between the hydrological cycle and the satisfaction of mankind's needs account for a substantial portion of the impact of lakes upon the environment. These include the need for mobility of people, raw materials, manufactured goods, food, water, and wastes.

Globally the transportation system depends upon the burning of fossil fuels, which are directly related to changes in the carbon cycle. The burning of fossil fuels for transportation accounts for approximately 25-30 per cent of the anthropogenic emissions of carbon dioxide into the earth's atmosphere. The effective reduction of outputs like carbon dioxide can only be met by controlling the inputs of materials and energy into these lake economies. This then brings in the idea of "ecological footprint". The ecological footprint is the amount of land and water resources that are necessary in sustaining these lacustrine regions at a given level of living. Since most lake regions are dependent upon global transportation systems, their "carbon footprint" extends to the entire planet. Other complex human-environment interactions include those of agriculture with the nitrogen, sulphur, and soil nutrients cycles; construction and industrial production with the raw materials cycle; and changes in land use and land cover due to human activities in the lake basins (Moreno et. al., 1995; Leitmann, 1996; Rees and Wackernagel, 1996; Pauli, 1998; IHDP, 1999).

Therefore, these lakes influence and are simultaneously influenced by large-scale biological, physical and chemical cycles that sustain all life on earth. What modifies these cycles also modifies these lake regions and how these lakes alter the cycles affect the local, regional and global environmental change. In order to trace the ecological footprints of the East African lakes, the collaboration and integration of the findings of the natural and social scientists is a necessity. It is within this perspective that the research on the human dimensions is assessed.

## 2. SURVEY OF PUBLISHED RESEARCH STUDIES

The general trends of human dimensions research on the East African lakes are derived from a survey of studies covering a time span of ten years (1989 – 1999). A total of 89 studies were covered in the survey. The time span used is significant because it gives the inception of IDEAL as the mid-point against which the literature on human dimensions is assessed. Thus making it possible to evaluate whether there has been a significant change in the direction of these research studies.

Most of the studies in the survey were those published in academic journals (such as African Affairs, Journal of Sustainable Agriculture) and books, but some donor funded studies [2], M.A and PhD dissertations and theses were also included in the survey. Frequencies were then run on the sample to determine the lake of focus, the most common object of analysis and the disciplinary bias of the studies. Determining the most common object of analysis was based on the thematic content of the studies, which revealed the following broad categories:

- fishery resources;
- watershed management;
- water hyacinth problematic;
- environmental conflicts; and
- water pollution

However, it was not possible to classify about one-tenth of the studies into these four categories. This is because these studies dealt with a number of sub-themes making it difficult to classify them, therefore, they were grouped together as "miscellaneous". The disciplinary orientation aimed at analysing the multidisciplinary character of the studies. This involved examining whether the studies were social or natural science-based or encompassed aspects of both.

The survey reveals that almost half of these studies have been done on Lake Victoria [3], about one-third on Lake Malawi, with Lakes Kivu and Kyoga having the least (Table 1). The IDEAL studies also reflect a similar pattern.

The most common object of analysis in the human dimensions' studies is fisheries, watershed management and the water hyacinth in that order (Table 2). In contrast the IDEAL studies have tended to concentrate on sedimentary processes, paleoclimate and paleoenvironment; hydrology and physical limnology; and food webs and fisheries. However, there is a review article on the anthropogenic impacts of man in the first IDEAL [4] volume, which emphasises the need for integrated Lake Basin management plans.

*Table 1*. Lake of focus, 1989-1999.

| *Lake of Focus* | Frequency |
|---|---|
| East African Lakes | 7 |
| Lake Victoria | 42 |
| Lake Malawi | 12 |
| Lake Turkana | 11 |
| Lake Tanganyika | 7 |
| Lake Naivasha | 4 |
| Lake Albert | 2 |
| Lake Kivu | 2 |
| Lake Kyoga | 2 |
| Total | 89 |

*Table 2*. Subject of analysis, 1989-1999.

| *Subject* | Proportion (%) |
|---|---|
| Fishery Resources | 43 |
| Watershed Management | 22 |
| Water Hyacinth | 16 |
| Water Pollution | 9 |
| Miscellaneous | 10 |
| Total | 100 |

Some of the research studies assessed are multidisciplinary in nature, but most of them have a bias towards the social sciences. A few of these studies particularly those dealing with fisheries and water hyacinth have been done by natural scientists (or at least their research findings are inclined towards the natural sciences). The importance of the fishing industry for livelihoods in the Lake Regions cannot be overstated. Fisheries are an important resource, not only for nutritional purposes but also in economic terms. This industry employs millions of people in the lake basins and is a key source of export revenue (Bokea and Ikiara, 2000). Fishing, both for local consumption and commercial trade has had significant consequences on the East African lakes. Exensive and indiscriminate fishing has led to over-exploitation of the fishery resources. Increased pollution and the water hyacinth further threaten this resource, and yet it is a source of cheap protein for the majority of the poor inter-lacustrine populations. Given the importance of the fishing industry, research is bound to attract both the natural and social scientists who have began to realise that theirs would be an exercise in futility unless a multidisciplinary approach is adopted. The other socio-economic aspects covered include market networks, incomes, labour and gender implications in the fisheries sector.

Lake Victoria has probably experienced the worst effects of the water hyacinth, which at one time threatened to cover the whole lake, crippling transportation and the fishing industry. The water hyacinth is slowly being eradicated from the lake but it keeps propagating, making its removal an expensive and demanding exercise. The problem of water hyacinth appears to have encouraged greater co-operation between the natural and social scientists. One sees a clear reference in the human dimensions studies to findings from the natural sciences. The reverse however is not true, the natural scientists rarely integrate the human dimensions' research findings into their studies. Studies on the water hyacinth have sought to document the problems (mainly economic, health and transportation) caused by its rapid spread. The analysis of conditions (e.g. presence of high nutrients) and causes (such as discharges of heavy fertiliser run-off from the farmlands) which encourage the propagation of the water hyacinth demand an integration of research efforts.

The human dimensions studies on human and industrial pollution have mainly been carried out on the Lake Victoria basin, which contains the largest urban population in the East African Lake regions. Urban centres in this basin discharge toxic industrial affluent and raw sewage into the lake. Therefore, Lake Victoria has been slowly dying over the last two decades from the oversupply of nutrients and untreated sewage that have led to massive fish deaths, toxic algae blooms and the water hyacinth problem. Studies conducted under the umbrella of the IDEAL programme have examined some of the factors leading to pollution of the Lake regions and could greatly enrich the human dimensions efforts in curbing this environmental degradation. For instance according to Cohen and Reinthal, 1998 and Bootsman, 1999, deforestation of hill slopes for agricultural land, pastures, roads and urban areas leads to the destruction of watersheds. This induces increased rates of siltation that can have negative effects on lake organisms. For example, parts of the Ruzizi River watershed, north of Tanzania that have been completely burned and deforested are easily eroded during the rainy seasons.

Some of the human dimensions studies show a clear departure from the top-down approach and instead encourage participatory and integrated development approaches with the Lake communities. This involves integrating the target communities into the overall research design to encourage a people oriented approach. For instance, localising Agenda 21 [5] in Nakuru municipality in Kenya seeks to involve the local communities in plan formulation and implementation in the on-going efforts to have an ecologically sound lake-park. The vulnerability of Lake Nakuru National Park is seen in its co-existence with a booming medium-sized town leading to population encroachment on the lake's ecosystem. Community participation is seen as part of the solution towards the successful implementation of policy.

Another recurrent theme is regional environmental conflicts in the management of the East African lakes. These mainly concentrate on the real and potential conflicts generated by the sharing of the fisheries resources and the control of the water hyacinth. Such studies advocate for greater interdependence among institutional frameworks such as the Lake Victoria Environmental Management Project and Lake Victoria Fisheries Organisation [6]. The legal and policy constraints in regard to regional co-operation among the East African Lakes appear to be a popular theme, particularly for those studies conducted after mid-1990s [7]. Therefore the ecological interdependence of the East Africa Lakes not only calls for co-operation at the political level, but also among the social and natural scientists.

Generally the human dimensions' research reviewed appear to be unaware of the research activities of IDEAL. None of them have utilised the IDEAL research findings as baseline data. Yet some of the research findings have direct relevance for human dimensions research. For instance, Semazzi (1999) details how the regional climate prediction model for Lake Victoria can be applied in resource management in areas such as:

- prediction of fish environments and population dynamics in Lake Victoria;
- prediction of lake transport of potentially highly toxic chemical affluents from industries.

The human dimensions researchers could use such valuable information to make policy recommendations and sensitise the target communities.

The East Africa Lake regions, through their political, socio-economical and environmental linkages, underscore the need for an integrated management approach in containing global change. These complex lake systems must be set in the context of:

- globalization of trade – neo-liberal policies that encourage non-traditional exports (such as cut-flowers and fish) and the resultant effects on the environment and livelihood of the lake regions [8].
- aspirations of developing countries for improvements in the quality of life;
- changing political relationships among states;

It is increasingly being recognised that neo-liberal policies have encouraged the production of non-traditional exports such frozen fish fillets as a means of generating foreign exchange to reduce Africa's debt burden. During the past two decades trade in fish between the developed and developing countries has increased significantly, with the Lake Victoria fisheries emerging as the leading source for the fish export market. This export trade has had profound effects on the traditional fisheries and the different groups of people who rely on it (Jansen, 2000). However, since 1994 fish production has been on the decline. Some of the factors responsible for this are the predatory nature of the Nile Perch, the spread of the water hyacinth, the destructive fishing methods and the reduced exports to the European markets.

The contribution of the fish export trade to the local communities is a controversial issue – there have been both winners and losers with the latter appearing to be in the majority. Prior to the introduction of the fish export industry the fishing communities around the East African Lakes were assured of a cheap source of protein. However, the fish export industry has diverted this cheap source of protein to the developed countries, hindering the inter-lacustrine population's aspirations in achieving quality nutritional status. The changing political relationships among states necessitate greater co-operation and establishment of regional laws and regulations regarding the utilisation of the lake resources. Nevertheless, this has not been amicably achieved due to the membership of the various nation states in different regional blocs, often with conflicting and competing interests. A case in point is the Lomé Convention that requires the renegotiations of duties for products (previously enjoying trade preferential treatment) on a country-to-country basis.

Each of these aspects (globalization of trade, aspirations of developing countries for improvements in the quality of life and changing political relationships among states) needs to be viewed in a systems perspective, that is, as a set of complex, dynamic relationships in which change in one key component is likely to affect change in another. The globalization of the world has posed one of the greatest challenges ever to the scientific community. For scientific research to be effective in providing practical solutions, it must increasingly be involved in inter- and trans-disciplinary research in collaboration with all the stakeholders (at the grassroots, policy /micro, meso and macro-levels) (IHDP, 1999).

Global environmental change necessitates by its very nature that whatever efforts are made towards the management and conservation (sustainable development) of resources need a co-ordinated and integrated effort at all levels – local, regional, and globally. However, the integration and co-ordination of research activities which should form the basis of any policies to be formulated between the natural and social sciences has not been adequately prioritised. Lack of this is evident at the disjointed nature and, apparently, lack of control on the conservation and management of water resources in the East African Lake regions.

The dearth of dissemination of environmental research findings to the relevant stakeholders is also another obstacle. The policy makers rarely have access to these research findings, which would enable them to make sound environmental decisions on the management and conservation of the East African Lake regions. Moreover, most of the institutional mechanisms for implementation tend to use the top-down approach, thus discouraging the participation of the lake communities in projects affecting them. Thus it is necessary to consider the benefits of involving local communities in these processes. A participatory (or bottom-up) approach not only improves the quality of plans but also the acceptability of such plans by all stakeholders (Akatch, 1996; Madete, 1998; Mumma, 1999; Municipal Council of Nakuru 1999; Guijt and Shah, 1999).

## 3. CONCLUSION

There is therefore an urgent need to institute the necessary mechanisms to:

- bring about multi- and trans- disciplinary research projects among the social and natural scientists;
- incorporate the human dimensions research within the IDEAL programme;
- disseminate these findings to policy makers and relevant stakeholders;
- ensure that the research is utilised to empower the communities where research has been conducted by encouraging more participatory approaches.

This is necessary if the determination and evaluation of regional policies, regulations and compliance; and monitoring and enforcement on environmental change in the East African Lake regions is to take place.

# FOOTNOTES

1. See Johnson and Odada (1996) and Lehman (1998).
2. The United Nations Development Programme (UNDP), the World Bank, the European Union and Swedish International Development Agency are some of the donor agencies that have funded a number of studies on the East African Lakes, particularly Lake Victoria.
3. This is not surprising since Lake Victoria is not only the second largest (69,000 sq. km.) fresh water body in the world, but also has great socio-economic potential. The lake straddles Kenya, Uganda and Tanzania symbolising unity and conflict simultaneously. Moreover given the increasing population pressure and socio-economic activities in the lake basin, the impact of human dimensions are easily manifested.
4. Refer to the article by Kudhongania, A. W., Ocenodongo, D. L. and Okaronon, J. O. (1996).
5. Agenda 21 underscores the primary importance of the participation and co-operation of local authorities in fulfilling the Agenda's activities at the local level.
6. The three East African governments initiated the Lake Victoria Environmental Management Project with financial support from the Global Environmental Facility and the World Bank. The Lake Victoria Fisheries Organisation was established in 1994 and is one of the institutions of the East African Community. It aims at promoting regional co-operation in the management and sustainability of fisheries and other living resources of Lake Victoria.
7. This can be attributed to the re-establishment of the East African Community.
8. The rapid intensification of floriculture in the last decade has led to the depletion and pollution of Lake Naivasha in Kenya. The reliance of the Lake Victoria fish exports on the European Union markets was put to test when the European Union banned these exports in 1999. The impact that this has had on the economy of the Lake Victoria region (before the ban was recently lifted) has been massive, ranging from the laying off of workers to reduced revenues.

# REFERENCES

Abdulai, A. and Hazell, P. (1995) The role of agriculture in sustainable development in Africa, *Journal of Sustainable Agriculture* **7**, No. 2/3, 101-115.

Achieng', A. P. (1990) The impact of the Nile Perch, Lates niloticus (L.) on the fisheries of Lake Victoria. *Journal for Fish Biology* **37a**, 17-32.

Akatch, S. O. (ed.) (1996) *Dying Lake Victoria: A Community-based Prevention Programme* – Osienala, Initiatives Publishers, Nairobi.

Aloo, P. A. (1996) Anthropogenic impacts on fisheries resources of Lake Naivasha, in T. C. Johnson and E. O. Odada (eds.), *The Limnology, Climatology and Paleoclimatology of the East African Lakes*, Gordon and Breach Publishers, Amsterdam, pp. 325-336.

Alot, M. (2000) Focus on lake an impetus for growth, *Daily Nation*, 14 March, p.8.

Aquatic Conservation Network (1992) Resolutions of the Workshop on people, fisheries, biodiversity and the future of Lake Victoria, 17-20th August, 1992, Jinja, Uganda. *Aquatic Survival*, **1** No.3, (13), 2-5.

Bokea, C. and Ikiara, M. (2000), The macro-economy of the fishing industry in Lake Victoria (Kenya), IUCN Eastern Africa Programme Report No. 7.

Bootsma, H. (1999) Water quality research on Lake Malawi/Nyasa, *IDEAL Bulletin*, Summer 1999, pp. 4-5.

Bugenyi, F. W. B. and Magumba, K. M. (1996) The present physiochemical ecology of Lake Victoria, Uganda, in T. C. Johnson and E. O. Odada (eds.), *The Limnology, Climatology and Paleoclimatology of the East African Lakes*, Gordon and Breach Publishers, Amsterdam, pp. 141-154.

Cohen, A. S., and Reinthal, P. (1998) Biodiversity : The Great Lakes of Africa. Instructional CD ROM, W. Blundell, producer. University of Arizona, Tucson.

Cohen, A. S., Kaufman, L. and Ogutu-Ohwayo, R. (1996) Anthropogenic threats, impacts and conservation strategies in the African Great Lakes: A review, in T. C. Johnson and E. O. Odada (eds.), *The Limnology, Climatology and Paleoclimatology of the East African Lakes*, Gordon and Breach Publishers, Amsterdam, pp. 575-624.

Dribidu, E. *et. al.* (1997) Legal aspects of Lake Victoria Environmental Management that require further elaboration, National Environment Authority, UNEP/UNDP Joint Environmental Law and Institutional Project, East African Sub-Region, Uganda Component.

Geheb, K. (1995) Exploring people-environment relationships: the changing nature of small-scale fishery in the Kenyan sector of Lake Victoria, in T. Binns (ed.), *People and Environment in Africa*, John Wiley, New York, pp. 91-101.

Geheb, K. and Binns, T. (1997) Fishing farmers or farming fishermen? The quest for household income and nutritional security on the Kenyan shores of Lake Victoria. *African Affairs*, **96** ( 382), 73-93.

Grove, A. T. (1996) African river discharges and lake levels in the twentieth century, in T. C. Johnson and E. O. Odada (eds.) (1996), *The Limnology, Climatology and Paleoclimatology of the East African Lakes*. Gordon and Breach Publishers, Amsterdam, pp. 95-102.

Guijt, I. and Shah, M. K. (1999) *The Myth of Community – Gender issues in participatory development*, Intermediate Technology Publications, London.

Haaldsdottir, G. (1996) Communities' participation in inland fisheries in Sub-Saharan Africa : Fact or fiction? Paper presented at the American Anthropological Association Annual Meetings, San Francisco, November, 1996.

Harris, C. K., Wiley, D. S. and Wilson, D. C. (1996) *Tensions in the Management of Lake Victoria Fisheries*, Chapman and Hall, London.

Harris, C. K., Wiley, D. S. and Wilson, D. C. (1995) Socio-economic impacts of the introduction of foreign species into Lake Victoria Fisheries, in T. J. Pitcher and P. J. B. Hart (eds.), *The Impact of Species Change in African Lakes*, Chapman and Hall, London.

International Human Dimensions Programme on Global Environmental Change (IHDP) (1999) Industrial Transformation Science Report, IHDP Report No. 12, Amsterdam.

Jansen, E. G. (1997) Rich fisheries – Poor Fisherfolk, IUCN Eastern Africa Programme Report No. 1.

Johnson, T. C. and Odada, E. O. (eds.) (1996) *The Limnology, Climatology and Paleoclimatology of the East African Lakes*. Gordon and Breach Publishers, Amsterdam.

Katunzi, E. F. B. (1996) A review of Lake Victoria fisheries with recommendations for management and conservation, in T. C. Johnson and E. O. Odada (eds.) (1996), *The Limnology, Climatology and Paleoclimatology of the East African Lakes*. Gordon and Breach Publishers, Amsterdam, pp. 357-366.

Kudhongania, A. W., Ocenodongo, D. L. and Okaronon, J. O. (1996) Anthropogenic perturbations on the Lake Victoria ecosystem, in T. C. Johnson and E. O. Odada (eds.) (1996), *The Limnology,*

*Climatology and Paleoclimatology of the East African Lakes*. Gordon and Breach Publishers, Amsterdam, pp. 625-632.

Lamb, H. F. (1999) Crater Lake research in Ethiopia, *IDEAL Bulletin*, Summer 1999, pp. 8-9.

Lehman, J. T. (1996) Pelagic food webs in the East African Great Lakes, in T. C. Johnson and E. O. Odada (eds.) (1996), *The Limnology, Climatology and Paleoclimatology of the East African Lakes*. Gordon and Breach Publishers, Amsterdam, pp. 281-302.

Lehman, J. T. (ed.) (1998) *Environmental Change and Response in East African Lakes*. Kluwer Academic Publishers, Amsterdam.

Lehman, J. T., Halat, K. and Betz, B. (1998) Secondary production by lake fly Chaoborus in Lake Victoria, East Africa, in Lehman, J. T. (ed.) *Environmental Change and Response in East African Lakes*. Kluwer Academic Publishers, Amsterdam, pp. 221-236.

Leitmann, J. (1999) *Sustaining Cities: Environmental Planning and Management in Urban Design*. McGraw-Hill, New York.

Lukyamuzi J.K., (1997) Fighting the water hyacinth without disaster: Legal and environmental issues, Paper presented at the 16th Biennial Conference of the Weed Society for East Africa, 15-18th September, Kampala, Uganda.

Madete, A. (1998) Prospect for sustainable management of Lake Victoria, Paper presented at the Workshop on Pollution of Lake Victoria Environment, 10-12th June.

Medard, M. and Wilson, D. C. (1996) Changing economic problems for women in the Nile perch fishing communities on Lake Victoria, Paper presented at the annual meeting of the Society for Applied Anthropology, Baltimore MD 27-31.

Moreno, A. et. al., (1996) Industry, energy, and transportation: Impacts and adaptation, in R. T. Watson et. al., (eds.), *Climate Change 1995: Impacts, Adaptations, and Mitigation of Climate Change: Scientific-Technical Analyses*. Cambridge University Press, Cambridge, pp. 365-398.

Mumma A. (1999) The Lake Victoria water hyacinth: Its implications for International Environmental Conflicts (IECS) management and regional relations in East Africa, M. A. Dissertation, University of Nairobi, Kenya.

Municipal Council of Nakuru (1999) Nakuru Strategic Plan – Action Plan for Sustainable Urban Development of Nakuru Town and its Environs, Nakuru, Kenya.

Nicholson, S. E. (1996) A review of climate dynamics and climate variability in eastern Africa, in T. C. Johnson and E. O. Odada (eds.) (1996), *The Limnology, Climatology and Paleoclimatology of the East African Lakes*. Gordon and Breach Publishers, Amsterdam, pp. 25-56.

Ochieng'-Okatch, J. I. (1992) The development of Lake Victoria fisheries – Up to date and priority areas for future research, in G. E. M. Ogutu (ed.), *Proceedings of the Artisanal Fisheries of Lake Victoria, Kenya Options for Management, Production and Marketing Workshop*, 24-26th November, 1988, Kisumu, Kenya, Shrikon Publishers, Nairobi, pp. 107-111.

Ochola, W.O., Muhia, R. N. and Mwarasomba, L. I. (1999) Culture, traditions and society: The challenge to natural resource management and development – a report of Lake Victoria region socio-cultural study. National Soil and Water Conservation Programme, SIDA.

Oduor and Ligunya A. G. (1995) Promotion of stakeholders consultations and community participation, Lake Victoria Environmental Management Programme (LVEMP) Draft Report, August (1995), Kenya.

Ogutu, G. E. M. (ed.), *Proceedings of the Artisanal Fisheries of Lake Victoria, Kenya Options for Management, Production and Marketing Workshop*, 24-26th November, 1988, Kisumu, Kenya, Shrikon Publishers, Nairobi.

Oguya, F. (1998) The environmental issues affecting the Lake Victoria ecosystem and the interventions implemented by LVEMP, National Secretariat, Nairobi, Kenya.

Olindo, P. (2000) Lake Victoria: World Bank millions a wasted effort, *Daily Nation*, 30 March, p.25.

Pauli, G. (1998) *Up sizing – The Road to Zero Emissions, More Jobs, More Income, No Pollution*. Greefleaf Publishing, Sheffield.

Rees, W. E. and Wackernagel, M. (1996) Urban ecological footprints: Why cities cannot be sustainable – And why they are a key to sustainability. *Environmental Impact Assessment Review*, **16** (4-6), pp. 223-248, July-November.

Rogers, P. (1994) Hydrology and water quality, in W.B Meyer and B. L. Turner II, (eds.), *Changes in Land Use and Land Cover: A Global Perspective*, Cambridge University Press, Cambridge, pp. 231-258.

Schoeman, H. S. (1986) Rethinking of the human factor in development planning. *Journal of Contemporary African Studies*, **5**, No. 1-2, pp. 67-98.

Semazzi, F. H. M. et. al. (1999) Development of a regional climate prediction model for the Lake Victoria basin, *IDEAL Bulletin*, Summer 1999, pp. 10-11.

Spiegel, R. H. and Coulter, G. W. (1996) in T. C. Johnson and E. O. Odada (eds.) *The Limnology, Climatology and Paleoclimatology of the East African Lakes*. Gordon and Breach Publishers, Amsterdam, pp. 103-140.

Stark, R. C. (1995) Adopting multidisciplinary approaches to sustainable agricultural research, *American Journal of Alternative Agriculture*, 10, No. 4, pp. 9-17.

Theaodo, D. P. (1996) The economic sociology of Kenya's Lake Victoria beach communities : The intersection of the economy, gender and communal relations, Dissertation, Michigan University.

Tvedten, I. (1990) The difficult transition from subsistence to commercial fishing, *Maritime Anthropological Studies*, **3**, No. 1, 23-36.

Tvedten, I. and Hersough, B. (1992) Small scale fisheries in Africa, Scandinavian Institute of Development Studies, Uppsala.

Twongo, T. (1996) Growing impact of water hyacinth on nearshore environments on Lakes Victoria and Kyoga, in T. C. Johnson and E. O. Odada (eds.) (1996), *The Limnology, Climatology and Paleoclimatology of the East African Lakes*. Gordon and Breach Publishers, Amsterdam, pp. 633-642.

Wilson, D. C. (1998) Markets, networks and risks: An analysis of labour remuneration in the Lake Victoria fishing industry. *Sociological Forum* **7**, pp. 23-45

Wilson, D. C. (1996) The critical human ecology of the Lake Victoria fishing industry, Dissertation, Michigan State University.

Wilson, D. C. and Medard, M. (1996) The implications for fisheries management of the changing situation in Lake Victoria fishing communities: Preliminary findings. *Journal of African Rural and Urban Studies*, **6**, 6-14.

Wilson, D. C. et. al. (1996) Potentials for co-management of the Nile Perch fishery – Lake Victoria, Tanzania, Paper presented at the "Voices from the Commons" Conference, International Association for the Study of Common Property, Berkeley, 5-8th June.

World Bank, UNDP, CEC, FAO (1991) Fisheries and aquaculture research capabilities and needs in Africa. Studies of Kenya, Malawi, Mozambique, Zimbabwe, Mauritania, Morocco and Senegal, World Bank Technical Paper No. 149, Fisheries Series, Washington D. C.

Worthington, E. B. (1996) Early research on East African Lakes: A historical sketch, in T. C. Johnson and E. O. Odada (eds.) (1996), *The Limnology, Climatology and Paleoclimatology of the East African Lakes*. Gordon and Breach Publishers, Amsterdam, pp. 659-664.

## Advances in Global Change Research

1. P. Martens and J. Rotmans (eds.): *Climate Change: An Integrated Perspective.* 1999 ISBN 0-7923-5996-8
2. A. Gillespie and W.C.G. Burns (eds.): *Climate Change in the South Pacific: Impacts and Responses in Australia, New Zealand, and Small Island States.* 2000 ISBN 0-7923-6077-X
3. J.L. Innes, M. Beniston and M.M. Verstraete (eds.): *Biomass Burning and Its Inter-Relationships with the Climate Systems.* 2000 ISBN 0-7923-6107-5
4. M.M. Verstraete, M. Menenti and J. Peltoniemi (eds.): *Observing Land from Space: Science, Customers and Technology.* 2000 ISBN 0-7923-6503-8
5. T. Skodvin: *Structure and Agent in the Scientific Diplomacy of Climate Change.* An Empirical Case Study of Science-Policy Interaction in the Intergovernmental Panel on Climate Change. 2000 ISBN 0-7923-6637-9
6. S. McLaren and D. Kniveton: *Linking Climate Change to Land Surface Change.* 2000 ISBN 0-7923-6638-7
7. M. Beniston and M.M. Verstraete (eds.): *Remote Sensing and Climate Modeling: Synergies and Limitations.* 2001 ISBN 0-7923-6801-0
8. E. Jochem, J. Sathaye and D. Bouille (eds.): *Society, Behaviour, and Climate Change Mitigation.* 2000 ISBN 0-7923-6802-9
9. G. Visconti, M. Beniston, E.D. Iannorelli and D. Barba (eds.): *Global Change and Protected Areas.* 2001 ISBN 0-7923-6818-1
10. M. Beniston (ed.): *Climatic Change: Implications for the Hydrological Cycle and for Water Management.* 2002 ISBN 1-4020-0444-3
11. N.H. Ravindranath and J.A. Sathaye: *Climatic Change and Developing Countries.* 2002 ISBN 1-4020-0104-5; Pb 1-4020-0771-X
12. E.O. Odada and D.O. Olaga: *The East African Great Lakes: Limnology, Palaeolimnology and Biodiversity.* 2002 ISBN 1-4020-0772-8

KLUWER ACADEMIC PUBLISHERS – DORDRECHT / BOSTON / LONDON